Anatomy & Physiology | REVEALED® 3.0

An Interactive Cadaver Dissection Experience

This unique multimedia tool is designed to help you master human anatomy and physiology with:

- Content customized to your course
- Stunning cadaver specimens
- Vivid animations
- Lab practical quizzing

my Course Content
- Maximize efficiency by studying exactly what's required.
- Your instructor selects the content that's relevant to your course.

Dissection
- Peel layers of the body to reveal structures beneath the surface.

Animation
- Over 150 animations make anatomy and physiology easier to visualize and understand.

Histology
- Study interactive slides that simulate what you see in lab.

Imaging
- Correlate dissected anatomy with X-ray, MRI, and CT scans.

Quiz
- Gauge proficiency with customized quizzes and lab practicals that cover only what you need for your course.

WWW.APREVEALED.COM

Full Textbook Integration! AP|R

Icons throughout the book indicate specific McGraw-Hill Anatomy & Physiology | REVEALED 3.0 ® content that corresponds to the text and figures.

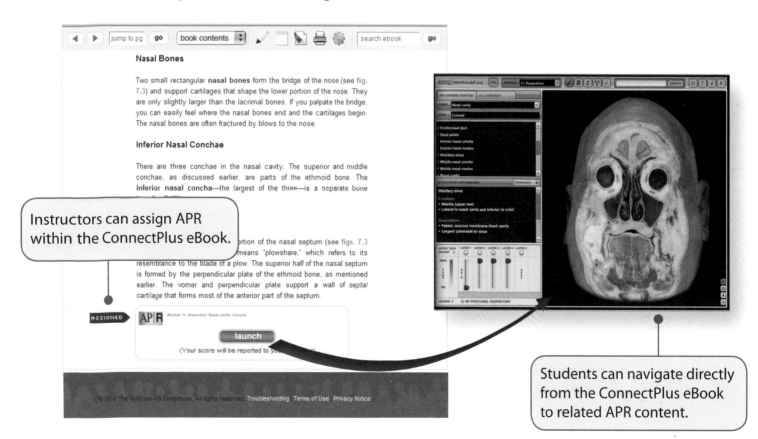

Instructors can assign APR within the ConnectPlus eBook.

Students can navigate directly from the ConnectPlus eBook to related APR content.

Hole's
ESSENTIALS OF
HUMAN ANATOMY & PHYSIOLOGY

eleventh edition

DAVID SHIER
Washtenaw Community College

JACKIE BUTLER
Grayson County College

RICKI LEWIS
Alden March Bioethics Institute

Mc Graw Hill

Connect Learn Succeed™

HOLE'S ESSENTIALS OF HUMAN ANATOMY & PHYSIOLOGY, ELEVENTH EDITION

This book is printed on acid-free paper.

4 5 6 7 8 9 0 DOW/DOW 1 0 9 8 7 6 5 4 3

ISBN 978–0–07–337815–2
MHID 0–07–337815–1

Vice President, Editor-in-Chief: *Marty Lange*
Vice President, EDP: *Kimberly Meriwether David*
Senior Director of Development: *Kristine Tibbetts*
Executive Editor: *James F. Connely*
Senior Developmental Editor: *Fran Schreiber*
Marketing Manager: *Denise M. Massar*
Senior Project Manager: *Jayne L. Klein*
Senior Buyer: *Sandy Ludovissy*
Lead Media Project Manager: *Stacy A. Vath*
Designer: *Tara McDermott*
Cover Designer: *Ellen Pettengell*
Cover Image: © *Getty Images*
Senior Photo Research Coordinator: *John C. Leland*
Photo Research: *Toni Michaels/PhotoFind, LLC*
Compositor: *Precision Graphics*
Typeface: *10.5/12 ITC Garamond STD Light*
Printer: *R. R. Donnelley*

Library of Congress Cataloging-in-Publication Data

Shier, David.
   Hole's essentials of human anatomy & physiology / David Shier, Jackie Butler, Ricki Lewis. — 11th ed.
      p. cm.
   Includes index.
      ISBN 978–0–07–337815–2 — ISBN 0–07–337815–1 (hard copy : alk. paper) 1. Human physiology. 2. Human anatomy. I. Butler, Jackie. II. Lewis, Ricki. III. Title.

[DNLM: 1. Anatomy. 2. Physiology. QS 4]
QP34.5.S49 2012
612--dc22
                                                              2010034043

www.mhhe.com

Brief Contents

Unit 1
Levels of Organization

1 Introduction to Human Anatomy and Physiology   1
2 Chemical Basis of Life   30
3 Cells   50
4 Cellular Metabolism   76
5 Tissues   94

Unit 2
Support and Movement

6 Integumentary System   116
7 Skeletal System   132
8 Muscular System   178

Unit 3
Integration and Coordination

9 Nervous System   213
10 The Senses   262
11 Endocrine System   291

Unit 4
Transport

12 Blood   318
13 Cardiovascular System   340
14 Lymphatic System and Immunity   377

Unit 5
Absorption and Excretion

15 Digestive System and Nutrition   400
16 Respiratory System   442
17 Urinary System   467
18 Water, Electrolyte, and Acid-Base Balance   489

Unit 6
The Human Life Cycle

19 Reproductive Systems   505
20 Pregnancy, Growth, Development, and Genetics   536

David Shier

David Shier has more than thirty years of experience teaching anatomy and physiology, primarily to premedical, nursing, dental, and allied health students. He has effectively incorporated his extensive teaching experience into another student-friendly revision of *Hole's Essentials of Human Anatomy and Physiology* and *Hole's Human Anatomy and Physiology*. David has published in the areas of renal and cardiovascular physiology, the endocrinology of fluid and electrolyte balance, and hypertension. A faculty member in the Life Science Department at Washtenaw Community College, he is actively involved in a number of projects dealing with assessment, articulation, and the incorporation of technology into instructional design. David holds a Ph.D. in physiology from the University of Michigan.

Jackie Butler

Jackie Butler's professional background includes work at the University of Texas Health Science Center conducting research about the genetics of bilateral retinoblastoma. She later worked at Houston's M. D. Anderson Hospital investigating remission in leukemia patients. A popular educator for more than twenty-five years at Grayson County College, Jackie teaches microbiology and human anatomy and physiology for health science majors.

Her experience and work with students of various educational backgrounds have contributed significantly to another revision of *Hole's Essentials of Human Anatomy and Physiology* and *Hole's Human Anatomy and Physiology*. Jackie Butler received her B.S. and M.S. degrees from Texas A&M University, focusing on microbiology, including courses in immunology and epidemiology.

Ricki Lewis

Ricki Lewis's career communicating science began with earning a Ph.D. in genetics from Indiana University in 1980. It quickly blossomed into writing for newspapers and magazines, and writing the introductory textbook *Life*. Since then she has taught a variety of life science courses and published the textbook *Human Genetics: Concepts and Applications*, an essay collection, and a novel about stem cells.

Since 1984 Ricki has been a genetic counselor for a large ob/gyn practice. She is active with the American Society of Human Genetics, and teaches an online course in "Genethics" at Albany Medical College.

A Note from the Authors

To the Student

Welcome! As you read this (with your eyes) and understand it (with your brain), perhaps turning to the next page (with muscle actions of your fingers, hand, forearam, and arm), you are using the human body to do so. In this eleventh edition of *Hole's Essentials of Human Anatomy and Physiology,* our goal is to provide you with an interesting and readable introduction to how all of this works! It is not simple, and there are times when it may not seem easy, but it is always fascinating, and understanding how your body works can be fun!

Many of you are on a path toward a career in health care, athletics, science, or education. We understand that many of you face the challenges of balancing family, work, and academics. Always remember that your course is not so much a hurdle along your way as it is a stepping stone. We have written this book to help you succeed in your coursework and to help prepare you to make that journey.

To the Teacher

We are authors, but first and foremost we are teachers, active in the classroom. What we and our reviewers do in class is reflected in this new edition. Students have always come first in our approach to teaching and textbook authoring, but we now feel more excited than ever about the student-oriented, teacher-friendly quality of this text.

Along with updated versions of the extra resources that students and teachers alike have found so helpful over the years (Anatomy and Physiology Revealed®, text websites, and so on), we are especially pleased to present the new Learn, Practice, Assess approach. Each chapter opens with Learning Outcomes, contains many opportunites to Practice throughout, and closes with Assessments that are closely tied to the learning outcomes. Students can use this new feature not only to focus their study efforts, but also to take an active role in monitoring their own progress toward mastering the material. All of these resources are described in more detail in the Chapter Preview beginning on page xviii.

David Shier, Jackie Butler, Ricki Lewis

New to this Edition

Global Changes

- End-of-chapter Integrative Assessments/Critical Thinking questions include reference to previous chapters.
- Practice Questions are added to the legends of selected figures.
- Clinical Terms are on the book website.
- Complex figures include the legend content in the artwork, paralleling the text.
- Many new vignettes and small boxes.
- All boxed material updated, with a more clinical focus.

Specific Changes At-a-Glance

Chapter	Topic	Change	Rationale
1	Head cavities (fig. 1.9)	Improved depth	Accuracy
1	Directional terms (fig. 1.13)	Rewritten	Clarity
1	Anatomical terms (fig. 1.14)	Rewritten	Clarity, consistency
1	Anatomical terms	Rewritten	Clarity, consistency
2	Matter and mass	Rewritten	Clarity
2	Ionically-bonded substances	Dissociate, not dissolve	Accuracy
3	Reprogrammed cells	New vignette	Update
3	Gene expression	New material	Update
3	Cell membrane (fig. 3.3)	Lipid bilayer inset added	Clarity
3	Osmosis	Rewritten	Clarity
3	Organelles	Functions added	Update, balance
4	Enzyme-substrate complex	New fig. 4.5	Clarity
4	Fate of pyruvic acid	Redrawn	Clarity
4	Catabolism of macronutrients (fig. 4.9)	Redrawn	Update
4	DNA replication (fig 4.11)	Redrawn	Accuracy, detail
5	Tissues (figs. 5.1c, d; 5.2; 5.3; 5.4; 5.5; 5.6; 5.7; 5.13; 5.14; 5.15; 5.16; 5.17; 5.18; 5.19; 5.20; 5.21; 5.22; 5.23; 5.24)	Many new micrographs and corresponding line art	Clarity, an attempt to more closely resemble the microscope slides the students will be observing in lab
5	Extracellular matrix Clinical Application	Rewritten, new figure	Update, more clinical approach
6	Itching	New vignette	New information
6	Skin (figs. 6.1; 6.2; 6.5; 6.7)	Many new micrographs and corresponding line art	Clarity
6	Skin cancer	Rewritten	Update, more clinical approach
6	Fingerprints	Rewritten	Clarity, update
6	Burns	Rule of nines added to Clinical Application	More clinical approach

Continued next page—

Specific Changes At-a-Glance —*Continued*

Chapter	Topic	Change	Rationale
6	Botox	New small box	More clinical approach
7	Skeletal system (figs. 7.1; 7.9; 7.38; 7.39; 7.40)	Many figures improved	Update, clarity
7	Joint movements	Photos of people added	More clinical approach
7	Arthritis	Box expanded into Clinical Application	Update, more clinical approach
8	Thick and thin muscle filaments	Figs. 8.1 and 8.2 redone	Accuracy, clarity
8	Motor end plate, motor units, and recruitment	Reorganized and rewritten	Clarity
9	Vegetative brain	Vignette rewritten	Update
9	Nerve impulse conduction and synaptic transmission	Distinguished better	Clarity, consistency
9	Relationship of CNS/PNS, sensory/motor	Fig. 9.2 redone	Clarity
9	Membrane and action potentials	Figs. 9.12 and 9.13	Clarity
9	Meninges	Figs. 9.21 and 9.22 redone	Clarity
9	Nerve impulse, nerve tract, axons, fibers, nerve fibers	Redundancy eliminated	Clarity, consistency
9	Lateral horn	New micrograph and line art	Clarity
9	Sensory and motor speech areas	Rewritten	Update
10	Sensation and perception	Rewritten	Clarity
10	Sound volume perception in terms of action potentials	Rewritten	Clarity
10	Clinical Applications on synesthesia and migraines	Rewritten	Update
11	Hormone secretion regulation	Rewritten	Accuracy
11	Clinical Application on diabetes	A1c testing, new glucose monitoring methods	Update, more clinical approach
12	Collection and centrifugation of blood sample	Photos added to fig. 12.1	Update, clarity
12	Blood components	Fig. 12.12 moved up	Clarity
12	Genetics Connection	Factor V Leiden replaces ITP, which is not genetic; also includes coagulation disorders	Accuracy, update
12	Blood cell formation (fig. 12.4), rbc life cycle (fig. 12.6), platelet plug (fig. 12.13)	Reworked	Update
12	Artery cross section (fig. 12.15)	New micrographs	Clarity
13	Human heart and major vessels	New photo for fig. 13.3	Clarity
13	Tachycardia/bradycardia	New small box	More clinical information
13	SA node and depolarization pathway	Fig. 13.11 redrawn	Clarity
13	Blood color	Fig. 13.21 lightened	Clarity
13	Venous valves	Fig. 13.23 colors lightened	Arrows more visible
13	Varicose veins	Rewritten and moved to veins section	Clarity
13	Major vein figures show paired veins	Figs. 13.33 and 13.35 redone	Accuracy, clarity
14	Lymphatic vessel valve	Micrograph in fig 14.3 replaced	Clarity
14	Lymphatic pathway	Detail added to fig. 14.5	Clarity, update
14	Thymus and spleen	New micrographs for figs. 14.9b and 14.10b	Clarity

Chapter	Topic	Change	Rationale
14	T cell/B cell activation	Fig. 14.13 redone and corresponding text rewritten	Clarity
14	Complement	Agglutination and neutralization added	More information
14	Primary and secondary immune response	Graphs in fig. 14.16 separated	Clarity
15	Gut microbiome	Vignette expanded	Update
15	Gastric gland cells and hepatic lobules	New micrographs for figs. 15.12b and 15.17c	Clarity
15	Movements in alimentary canal (fig. 15.4), mouth (fig. 15.6), skull (fig. 15.7), salivary glands (fig. 15.10) and stomach (fig. 15.11)	Redrawn	Clarity
15	Inflammatory bowel disease	Clinical Application rewritten	Update
16	Mechanics of inspiration	Rewritten	Clarity
16	Spirometry	Cannot measure residual volume	Clarity
16	Basic breathing rhythm	Figs. 16.16 and 16.17 redone and corresponding text rewritten	Update
16	Cystic fibrosis	Clinical Application rewritten	Update
17	Hemolytic uremic syndrome	Vignette rewritten	Update
17	Macula densa	Location, new fig. 17.7	Accuracy
17	Afferent and efferent arterioles	Anatomical differences moved to part on glomerular filtration	Accuracy, clarity
17	Net filtration pressure	Fig. 17.10 matches fig. 13.21 on capillary filtration	Consistency
18	Heatstroke	New vignette	More clinical approach
18	Water intoxication	New information in Clinical Application	Update
19	Seminiferous tubules	New micrograph in fig. 19.2c	Clarity
19	Spermatogonia and sperm	New micrograph in fig. 19.4	Clarity
19	Prostate cancer	Clinical Application rewritten	Update
19	Uterus	Fundus added	More information
19	Breast cancer	Clinical Application rewritten	Update
19	Contraceptives	Fig. 19.15 redone	Update
19	Sexually transmitted diseases	Changed to sexually transmitted infections	Update, accuracy
20	Postmortem sperm retrieval	New vignette	Update
20	Critical period	Added to discussion, new orange box	More information
20	Teratogens	Clinical Application 20.2 rewritten	Update
20	Aging	Added	More information
Appendix B	Metrics	New	Students need help making conversations to/from metric measurements.
Appendix E	Figure Questions Answers	New	Provides answers to the new figure questions

Learn, Practice, Assess!

Learn ▶

Learning Outcomes open chapters, and are closely linked to Chapter Assessments and Integrative Assessments/Critical Thinking questions found at the end of each chapter.

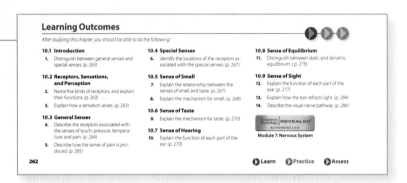

Learning Outcomes

After studying this chapter, you should be able to do the following:

10.1 Introduction
1. Distinguish between general senses and special senses. (p. 263)

10.2 Receptors, Sensations, and Perception
2. Name five kinds of receptors, and explain their functions. (p. 263)
3. Explain how a sensation arises. (p. 263)

10.3 General Senses
4. Describe the receptors associated with the senses of touch, pressure, temperature, and pain. (p. 264)
5. Describe how the sense of pain is produced. (p. 265)

10.4 Special Senses
6. Identify the locations of the receptors associated with the special senses. (p. 267)

10.5 Sense of Smell
7. Explain the relationship between the senses of smell and taste. (p. 267)
8. Explain the mechanism for smell. (p. 268)

10.6 Sense of Taste
9. Explain the mechanism for taste. (p. 270)

10.7 Sense of Hearing
10. Explain the function of each part of the ear. (p. 270)

10.8 Sense of Equilibrium
11. Distinguish between static and dynamic equilibrium. (p. 275)

10.9 Sense of Sight
12. Explain the function of each part of the eye. (p. 277)
13. Explain how the eye refracts light. (p. 284)
14. Describe the visual nerve pathway. (p. 286)

REVEALED
apcrevealed.com
Module 7: Nervous System

262

▶ Learn ▶ Practice ▶ Assess

Learning tools to help you succeed...

Check out the Chapter Preview, Foundations for Success, on page xviii. The Chapter Preview was specifically designed to help you **LEARN** how to study. It provides helpful study tips.
NEW! for this edition is a section on learning styles!

Vignettes lead into chapter content. They connect you to many areas of health care including technology, physiology, medical conditions, historical perspectives, and careers.

NEW! Anatomy and Physiology Revealed (APR) icon at the beginning of each chapter tells you which system in APR applies to this chapter.

Aids to Understanding Words help you remember scientific word meanings. Examine root words, stems, prefixes, suffices, pronunciations, and build a solid anatomy and physiology vocabulary.

Reference Plates offer vibrant detail of body structures.

Practice

Practice with a question or series of questions after major sections. They will test your understanding of the material.

Practice

1. What factors probably stimulated an early interest in the human body?
2. What kinds of activities helped promote the development of modern medical science?

Interesting applications help you practice and apply their knowledge...

AP|R NEW! Anatomy and Physiology Revealed icons found in figure legends. These icons indicate that there is a direct link to APR available in the eBook provided with Connect Plus for this title!

Q: NEW! Figure Questions allow an additional assessment. Found on key figures throughout the chapter.

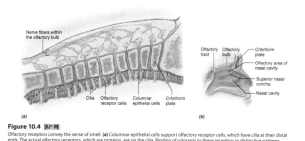

(a)   (b)

Figure 10.4 AP|R
Olfactory receptors convey the sense of smell. **(a)** Columnar epithelial cells support olfactory receptor cells, which have cilia at their distal ends. The actual olfactory receptors, which are proteins, are on the cilia. Binding of odorants to these receptors in distinctive patterns conveys the information that the brain interprets as an odor. **(b)** The olfactory area is associated with the superior nasal concha.

Figure 10.6 AP|R
Major parts of the ear. The outer ear includes the auricle, external acoustic meatus, and eardrum. The middle ear includes the auditory ossicles (malleus, incus, and stapes) and the oval window. The inner ear includes the semicircular canals and the cochlea.

Q: *How do the action potentials generated along auditory pathways compare with those on taste and smell pathways?*
Answer can be found in Appendix E on page 568.

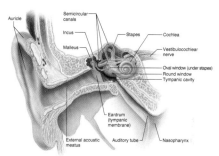
The fovea centralis of the human eye has 150,000 cones per square millimeter. In contrast, a bird of prey's eye has about a million cones per square millimeter.

Facts of Life provides interesting bits of anatomy and physiology information, adding a touch of wonder to chapter topics.

Boxed information applies ideas and facts in the narrative to clinical situations.

As a person ages, tiny, dense clumps of gel or deposits of crystal-like substances form in the vitreous humor. When these clumps cast shadows on the retina, the person sees small, moving specks in the field of vision, called *floaters*.

NEW! Clinical Applications present disorders, physiological responses to environmental factors, and other topics of general interest.

Genetics Connections explore the molecular underpinnings of familiar as well as not so familiar illnesses. Read about such topics as ion channel disorders, muscular dystrophy, and cystic fibrosis.

Clinical Application 10.1

Synesthesia: Connected Senses

"The song was full of glittering orange diamonds."
"The paint smelled blue."
"The sunset was salty."
"The pickle tasted like a rectangle."

About 1 in 2,000 people have a condition called synesthesia ("joined sensation"), in which sensation and perception mix, so that the brain perceives from another. Most comm of time evoke specific colo tary, are very specific, and a person might report th Thursday a very dark, shiny Synesthesia runs in fa ated the condition with m ferent genes. Female "syn one. Creative individuals with the condition. They i Mayer, Tori Amos, and Fran

and physicist Richard Feynman, who used to include the hues with which he visualized chemical equations on the chalkboard, to the amusement of his students. One of the co-authors of this book has it—to her, days are colors. The earliest recorded mention of synesthesia is an essay from John Locke in 1690. More and more people with synesthesia are recog-

---

Genetics Connection 8.1

Inherited Diseases of Muscle

A variety of inherited conditions affect muscle tissue. These disorders differ in the nature of the genetic defect, the type of protein that is abnormal in form or function, and the particular muscles in the body that are impaired.

The Muscular Dystrophies—Missing Proteins
A muscle cell is packed with filaments of actin and myosin. Much less abundant, but no less important, is a protein called *dystrophin*. It holds skeletal muscle cells together by linking

actin in the cell to glycoproteins in the cell membrane, which helps attach the cell to the extracellular matrix. Missing or abnormal dystrophin or the glycoproteins cause muscular dystrophies. These illnesses vary in severity and age of onset, but in all cases, muscles weaken and degenerate. Eventually, fat and connective tissue replace muscle.

Duchenne muscular dystrophy (DMD) is the most severe type of the illness. Symptoms begin by age five and affect only boys. By age thirteen, the person cannot walk, and by early adulthood he usually dies from failure of the respiratory muscles. In DMD, dystrophin is often missing. In Becker muscular dystrophy, symptoms begin in early adulthood, are less severe, and result from underproduction of dystrophin.

Charcot-Marie-Tooth Disease— A Duplicate Gene
Charcot-Marie-Tooth disease causes a slowly progress-ing weakness in the muscles of the hands and feet and a decrease in tendon reflexes in these parts. In this illness, an extra gene impairs the insulating sheath around affected nerve cells, so that nerve cells cannot adequately stimulate muscles. Physicians perform two tests—electromyography and nerve conduction velocity—to diagnose Charcot-Marie-Tooth disease. It is also possible to test for the gene mutation to confirm a diagnosis based on symptoms.

Hereditary Idiopathic Dilated Cardiomyopathy—A Tiny Glitch
This very rare inherited form of heart failure usually begins in a person's forties and is lethal in 50% of cases within five years of diagnosis, unless a heart transplant can be per-formed. The condition is caused by a tiny genetic error in a form of actin found only in cardiac muscle, where it is the predominant component of the thin filaments. The mutation disturbs actin's ability to anchor to the Z lines in heart muscle cells, preventing actin from effectively transmitting the force of contraction. As a result, the heart chambers enlarge and eventually fail.

Figure 8B
This young man has Duchenne muscular dystrophy. The condition has not yet severely limited his activities, but he shows the hypertrophied (overdeveloped) calf muscles that result from his inability to rise from a sitting position the usual way—an early sign of the illness.

Learn, Practice, Assess!

Assess ▶

Tools to help you make the connection and master anatomy & physiology!

Chapter Assessments check your understanding of the chapter's learning outcomes.

Integrative Assessments/Critical Thinking questions allow you to connect and apply information from previous chapters as well as information within the current chapter.

Chapter Summary Outlines help you review the chapter's main ideas.

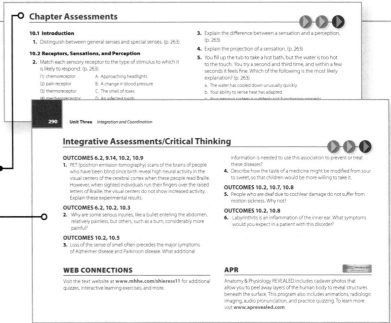

Chapter Assessments

10.1 Introduction
1. Distinguish between general senses and special senses. (p. 263)

10.2 Receptors, Sensations, and Perception
2. Match each sensory receptor to the type of stimulus to which it is likely to respond. (p. 263)

(1) chemoreceptor   A. Approaching headlights
(2) pain receptor   B. A change in blood pressure
(3) thermoreceptor   C. The smell of roses
(4) mechanoreceptor   D. An infected tooth

3. Explain the difference between a sensation and a perception. (p. 263)
4. Explain the projection of a sensation. (p. 263)
5. You fill up the tub to take a hot bath, but the water is too hot to the touch. You try a second and third time, and within a few seconds it feels fine. Which of the following is the most likely explanation? (p. 263)
 a. The water has cooled down unusually quickly.
 b. Your ability to sense heat has adapted.
 c. Your nervous system is suddenly not functioning properly.

290   **Unit Three**   *Integration and Coordination*

Integrative Assessments/Critical Thinking

OUTCOMES 6.2, 9.14, 10.2, 10.9
1. PET (positron emission tomography) scans of the brains of people who have been blind since birth reveal high neural activity in the visual centers of the cerebral cortex when these people read Braille. However, when sighted individuals run their fingers over the raised letters of Braille, the visual centers do not show increased activity. Explain these experimental results.

OUTCOMES 6.2, 10.2, 10.3
2. Why are some serious injuries, like a bullet entering the abdomen, relatively painless, but others, such as a burn, considerably more painful?

OUTCOMES 10.2, 10.5
3. Loss of the sense of smell often precedes the major symptoms of Alzheimer disease and Parkinson disease. What additional

information is needed to use this association to prevent or treat these diseases?

4. Describe how the taste of a medicine might be modified from sour to sweet, so that children would be more willing to take it.

OUTCOMES 10.2, 10.7, 10.8
5. People who are deaf due to cochlear damage do not suffer from motion sickness. Why not?

OUTCOMES 10.2, 10.8
6. Labyrinthitis is an inflammation of the inner ear. What symptoms would you expect in a patient with this disorder?

WEB CONNECTIONS

Visit the text website at www.mhhe.com/shieress11 for additional quizzes, interactive learning exercises, and more.

APR

Anatomy & Physiology REVEALED includes cadaver photos that allow you to peel away layers of the human body to reveal structures beneath the surface. This program also includes animations, radiologic imaging, audio pronunciation, and practice quizzing. To learn more visit www.aprevealed.com.

Summary Outline

10.1 Introduction (p. 263)
Sensory receptors sense changes in their surroundings.

10.2 Receptors, Sensations, and Perception (p. 263)
1. Types of receptors
 a. Each type of receptor is most sensitive to a distinct type of stimulus.
 b. The major types of receptors are chemoreceptors, pain receptors, thermoreceptors, mechanoreceptors, and photoreceptors.
2. Sensations
 a. A sensation is the awareness that sensory stimulation has occurred.
 b. A particular part of the cerebral cortex interprets every impulse reaching it in a specific way.
 c. The cerebral cortex projects a sensation back to the region of stimulation.
3. Sensory adaptation may involve receptors becoming unresponsive or inhibition along the CNS pathways leading to the sensory regions of the cerebral cortex.

10.3 General Senses (p. 264)
General senses are associated with receptors in the skin, muscles, joints, and viscera.
1. Touch and pressure senses
 a. Free ends of sensory nerve fibers are receptors for the sensation of itching.
 b. Tactile corpuscles are receptors for the sensation of light touch.
 c. Lamellated corpuscles are receptors for the sensation of heavy pressure.
2. Temperature senses
 Temperature receptors include two sets of free nerve endings that are warm and cold receptors.
3. Sense of pain
 a. Pain receptors are free nerve endings that tissue damage stimulates.
 b. Visceral pain
 (1) Pain receptors are the only receptors in viscera that provide sensations.
 (2) Pain sensations produced from visceral receptors may feel as if they are coming from some other body part, called referred pain.
 (3) Visceral pain may be referred because sensory impulses from the skin and viscera travel on common nerve pathways.
 c. Pain nerve fibers
 (1) The two main types of pain fibers are acute pain fibers and chronic pain fibers.
 (2) Acute pain fibers conduct nerve impulses rapidly. Chronic pain fibers conduct impulses more slowly.
 (3) Pain impulses are processed in the gray matter of the spinal cord and ascend to the brain.

(4) Within the brain, pain impulses pass through the reticular formation before being conducted to the cerebral cortex.
 d. Regulation of pain impulses
 (1) Awareness of pain occurs when pain impulses reach the thalamus.

ORGANIZATION

Skeletal System

Integumentary System
Vitamin D, activated in the skin, plays a role in calcium absorption and availability for bone matrix.

Lymphatic System
Cells of the immune system originate in the bone marrow.

Muscular System
Muscles pull on bones to cause movement.

Digestive System
Absorption of dietary calcium provides material for bone matrix.

Nervous System
Proprioceptors sense the position of body parts. Pain receptors warn of trauma to bone. Bones protect the brain and spinal cord.

Respiratory System
Ribs and muscles work together in breathing.

Endocrine System
Some hormones act on bone to help regulate blood calcium levels.

Urinary System
The kidneys and bones work together to help regulate blood calcium levels.

Cardiovascular System
Blood transports nutrients to bone cells. Bone helps regulate plasma calcium levels, important to heart function.

Reproductive System
The pelvis helps support the uterus during pregnancy. Bones provide a source of calcium during lactation.

Bones provide support, protection, and movement and also play a role in calcium balance.

ORGANIZATION
Illustrations
found at the end of selected chapters conceptually link the highlighted body system to every other system and reinforce the dynamic interplay among systems. These illustrations help you review chapter concepts and reinforce the big picture in learning and applying the principles of anatomy and physiology.

Teaching and Learning Supplements

Text Website – www.mhhe.com/shieress11

NEW! for the eleventh edition Ancillary Correlation Guide—instructors will find this guide invaluable. McGraw-Hill offers a variety of ancillary products to accompany our texts. The authors have gone through the ancillaries and correlated them to the specific Learning Outcome found at the beginning of each chapter! Here are the ancillaries that are correlated to the specific Learning Outcomes for *Hole's Essentials of Human Anatomy & Physiology*, Eleventh Edition:

- Textbook
- Website—**www.mhhe.com/shieress11**
- EZ Test Online
- Ph.I.L.S. 3.0
- MediaPhys 3.0
- Anatomy & Physiology Revealed
- Virtual Anatomy Dissection Review
- Student Study Guide—offers chapter overviews, chapter outcomes, focus questions, mastery tests, study activities, and mastery test answers.

McGraw-Hill's Presentation Tools

Presentation Materials for Lecture and Lab—incorporate customized lectures, visually enhanced test and quizzes, compelling course websites, or attractive printed support materials.

- **NEW!** A complete set of pre-made PowerPoints® linking **Anatomy & Physiology Revealed** to text material are now available for your use!
- **NEW!** A complete set of animation embedded PowerPoint slides are now available.
- **NEW!** Along with our online digital library containing photos, artwork, and animations, we now also offer **FlexArt.** FlexArt allows the instructor to customize artwork.
- Computerized test bank edited by the **Author Team** is powered by McGraw-Hill's flexible electronic testing program EZ Test Online.

Laboratory Manual for Hole's Essentials of Human Anatomy & Physiology, Eleventh Edition, by Terry R. Martin, Kishwaukee College, is designed to accompany the eleventh edition of *Hole's Essentials of Human Anatomy & Physiology.*

 Craft your teaching resources to match the way you teach. With McGraw-Hill Create™, **www.mcgrawhillcreate.com**, you can easily rearrange chapters, combine material from other content sources, and quickly upload content you have written such as your course syllabus or teaching notes. Find the content you need in Create by searching through thousands of leading McGraw-Hill textbooks. Arrange your book to fit your teaching style. Create even allows you to personalize your book's appearance by selecting the cover and adding your name, school, and course information. Order a Create book and you'll receive a complimentary print review copy in 3–5 business days or a complimentary electronic review copy (eComp) via email in minutes. Go to **www.mcgrawhillcreate.com** today and register to experience how McGraw-Hill Create™ empowers you to teach *your* students *your* way.

 Ph.I.L.S. 3.0 is the perfect way to reinforce key physiology concepts with powerful lab experiments. Created by Dr. Phil Stephens at Villanova University, this program offers 37 *laboratory simulations* that may be used to supplement or substitute for wet labs. All 37 labs are self-contained experiments— no lengthy instruction manual required. Users can adjust variables, view outcomes, make predictions, draw conclusions, and print lab reports. This easy-to-use software offers the flexibility to change the parameters of the lab experiment—there is no limit!

Teaching and Learning Supplements

|ANATOMY & PHYSIOLOGY

McGraw-Hill's ConnectPlus™ Anatomy & Physiology interactive learning platform provides a customizable, assignable eBook, auto-graded assessments, an adaptive diagnostic tool, lecture capture, access to instructor resources, and powerful reporting—all in an easy-to-use interface.

Learn more at
www.mcgrawhillconnect.com

Anatomy & Physiology | REVEALED® 3.0

An Interactive Cadaver Dissection Experience

This unique multimedia tool is designed to help you master human anatomy and physiology with:

- Content customized to your course
- Stunning cadaver specimens
- Vivid animations
- Lab practical quizzing

my Course Content
- Maximize efficiency by studying exactly what's required.
- Your instructor selects the content that's relevant to your course.

Dissection
- Peel layers of the body to reveal structures beneath the surface.

Animation
- Over 150 animations make anatomy and processes easier to visualize and understand.

Histology
- Study interactive slides that simulate what you see in lab.

Imaging
- Correlate dissected anatomy with X-ray, MRI, and CT scans.

Quiz
- Gauge proficiency with customized quizzes and lab practicals that cover only what you need for your course.

W W W . A P R E V E A L E D . C O M

McGraw-Hill Higher Education and Blackboard® have teamed up. What does this mean for you?

1. **Your life, simplified.** Now you and your students can access McGraw-Hill's Connect™ and Create™ right from within your Blackboard course – all with one single sign-on. Say goodbye to the days of logging in to multiple applications.

2. **Deep integration of content and tools.** Not only do you get single sign-on with Connect™ and Create™, you also get deep integration of McGraw-Hill content and content engines right in Blackboard. Whether you're choosing a book for your course or building Connect™ assignments, all the tools you need are right where you want them – inside of Blackboard.

3. **Seamless Gradebooks.** Are you tired of keeping multiple gradebooks and manually synchronizing grades into Blackboard? We thought so. When a student completes an integrated Connect™ assignment, the grade for that assignment automatically (and instantly) feeds your Blackboard grade center.

4. **A solution for everyone.** Whether your institution is already using Blackboard or you just want to try Blackboard on your own, we have a solution for you. McGraw-Hill and Blackboard can now offer you easy access to industry leading technology and content, whether your campus hosts it, or we do. Be sure to ask your local McGraw-Hill representative for details.

Acknowledgments

We would like to acknowledge the valuable contributions of the reviewers for the eleventh edition who read either portions or all of the text, and who provided detailed criticisms and ideas for improving the narrative and the illustrations. They include the following:

Diana Alagna, *Branford Hall Career Institute*

Barbara Banning, *Augusta Technical College*

Dan Bickerton, *Ogeechee Technical College*

Russell L. Blalock, *Central Georgia Technical College*

Jocelyn Cash, *Central Piedmont Community College*

Barbara A. Coles, *Wake Technical Community College*

Cynthia Conaway Mavroidis, *Northwest State Community College*

George Ealy, *Keiser University*

Christine M. Enz, *Everest Institute – Rochester*

Bridget A. Falkenstein, *Sierra College*

Maria Florez, *Lone Star College-CyFair*

Deborah Furbish, *Wake Technical Community College*

Linda Gerlock, *Middle Georgia Technical College*

Brent M. Graves, *Northern Michigan University*

Mark E. Grigsby, *Northeastern Oklahoma A&M College*

Martie Heath-Sinclair, *Hawkeye Community College*

Jean Helgeson, *Collin County Community College*

Shannon E. Henninger, *Augusta Technical College*

Eugenia Ilisei, *IIA College*

Mary Jane Keith, *Wichita State University*

Sara Lanier, *Augusta Technical College*

Dean V. Lauritzen, *City College of San Francisco*

M. Reading-Martin, *Western Nebraska Community College*

Kevin E. Morkel, *Bohecker College - Columbus*

Lisa Nagle, *Augusta Technical College*

Brent Odson, *Bohecker College-Columbus*

Susan Rohde, *Triton College*

Jimmy Rozell, *Tyler Junior College*

Amy Fenech Sandy, *Columbus Technical College*

Leba Sarkis, *Aims Community College*

Jennifer Sieszputowski, *Kirtland Community College*

Daniel A Slutsky, *American River College*

Marty Stickle, *Asheville Buncombe Technical Community College*

John F. Tarpey, *City College of San Francisco*

Rita F Waller, *Augusta Technical College*

Kara Young, *Baltimore School of Massage-York Campus*

Yu (Joy) Zhao, *American River College*

And thanks to those who prepared ancillary materials:

Simone Brito, *College of Southern Nevada*

LuAnne Clark, *Lansing Community College*

Jason LaPres, *Lone Star College*

Sheri Martin, *Central Georgia Technical College*

Patrice Parsons, *Grayson County College*

Greg Reeder, *Broward College*

Melanie Waite-Altringer, *Anoka–Ramsey Community College*

Any textbook is the result of hard work by a large team. Although we directed the revision, many "behind-the-scenes" people at McGraw-Hill were indispensable to the project. We would like to thank our editorial team of Jim Connely, Marty Lange, and Fran Schreiber; our production team, which included Jayne Klein, Sandy Ludovissy, Tara McDermott, John Leland, and Stacy Patch; copyeditor Wendy Nelson and freelance photo researcher Toni Michaels; and most of all, John Hole, for giving us the opportunity and freedom to continue his classic work. We also thank our wonderfully patient families for their support.

David Shier
Jackie Butler
Ricki Lewis

Contents

Chapter Preview: Foundations for Success   *xviii*

Unit 1
Levels of Organization

Chapter 1
Introduction to Human Anatomy and Physiology 1

1.1 Introduction 2
1.2 Anatomy and Physiology 3
1.3 Levels of Organization 3
1.4 Characteristics of Life 4
1.5 Maintenance of Life 5
1.6 Organization of the Human Body 8
1.7 Anatomical Terminology 14

Reference Plates
The Human Organism 22

Chapter 2
Chemical Basis of Life 30

2.1 Introduction 31
2.2 Structure of Matter 31
2.3 Chemical Constituents of Cells 39

Chapter 3
Cells 50

3.1 Introduction 51
3.2 Composite Cell 52
3.3 Movements Through Cell Membranes 60
3.4 The Cell Cycle 67

Chapter 4
Cellular Metabolism 76

4.1 Introduction 77
4.2 Metabolic Reactions 77
4.3 Control of Metabolic Reactions 79
4.4 Energy for Metabolic Reactions 80
4.5 Metabolic Pathways 82
4.6 DNA (Deoxyribonucleic Acid) 83
4.7 Protein Synthesis 85

Chapter 5
Tissues 94

5.1 Introduction 95
5.2 Epithelial Tissues 95
5.3 Connective Tissues 102
5.4 Types of Membranes 110
5.5 Muscle Tissues 110
5.6 Nervous Tissues 111

Unit 2
Support and Movement

Chapter 6
Integumentary System 116

6.1 Introduction 117
6.2 Skin and Its Tissues 117
6.3 Accessory Structures of the Skin 122
6.4 Regulation of Body Temperature 125
6.5 Healing of Wounds 125

Chapter 7
Skeletal System 132

7.1 Introduction 133
7.2 Bone Structure 133
7.3 Bone Development and Growth 135
7.4 Bone Function 137
7.5 Skeletal Organization 142
7.6 Skull 144
7.7 Vertebral Column 149
7.8 Thoracic Cage 153
7.9 Pectoral Girdle 155
7.10 Upper Limb 155
7.11 Pelvic Girdle 158
7.12 Lower Limb 161
7.13 Joints 164

Reference Plates
Human Skull 175

Chapter 8
Muscular System 178

8.1 Introduction 179
8.2 Structure of a Skeletal Muscle 179
8.3 Skeletal Muscle Contraction 182
8.4 Muscular Responses 187
8.5 Smooth Muscle 191
8.6 Cardiac Muscle 191
8.7 Skeletal Muscle Actions 192
8.8 Major Skeletal Muscles 194

Unit 3
Integration and Coordination

Chapter 9

Nervous System 213

9.1   Introduction 214
9.2   General Functions of the Nervous System 215
9.3   Neuroglia 216
9.4   Neurons 216
9.5   The Synapse 221
9.6   Cell Membrane Potential 222
9.7   Nerve Impulses 227
9.8   Synaptic Transmission 228
9.9   Impulse Processing 228
9.10 Types of Nerves 230
9.11 Nerve Pathways 231
9.12 Meninges 232
9.13 Spinal Cord 234
9.14 Brain 236
9.15 Peripheral Nervous System 246
9.16 Autonomic Nervous System 250

Chapter 10

The Senses 262

10.1 Introduction 263
10.2 Receptors, Sensations, and Perception 263
10.3 General Senses 264
10.4 Special Senses 267
10.5 Sense of Smell 267
10.6 Sense of Taste 269
10.7 Sense of Hearing 270
10.8 Sense of Equilibrium 275
10.9 Sense of Sight 276

Chapter 11

Endocrine System 291

11.1 Introduction 292
11.2 General Characteristics of the Endocrine System 292
11.3 Hormone Action 293
11.4 Control of Hormonal Secretions 296
11.5 Pituitary Gland 297
11.6 Thyroid Gland 301
11.7 Parathyroid Glands 303
11.8 Adrenal Glands 304
11.9 Pancreas 306
11.10 Other Endocrine Glands 309
11.11 Stress and Health 311

Unit 4
Transport

Chapter 12

Blood 318

12.1 Introduction 319
12.2 Blood Cells 319
12.3 Blood Plasma 327
12.4 Hemostasis 330
12.5 Blood Groups and Transfusions 333

Chapter 13

Cardiovascular System 340

13.1 Introduction 341
13.2 Structure of the Heart 342
13.3 Heart Actions 347
13.4 Blood Vessels 353
13.5 Blood Pressure 359
13.6 Paths of Circulation 363
13.7 Arterial System 363
13.8 Venous System 369

Chapter 14

Lymphatic System and Immunity 377

14.1 Introduction 378
14.2 Lymphatic Pathways 378
14.3 Tissue Fluid and Lymph 380
14.4 Lymph Movement 381
14.5 Lymph Nodes 381
14.6 Thymus and Spleen 382
14.7 Body Defenses Against Infection 384
14.8 Innate (Nonspecific) Defenses 384
14.9 Adaptive (Specific) Defenses, or Immunity 386

Unit 5
Absorption and Excretion

Chapter 15

Digestive System and
Nutrition 400

15.1 Introduction 401
15.2 General Characteristics of the
       Alimentary Canal 401
15.3 Mouth 403
15.4 Salivary Glands 408
15.5 Pharynx and Esophagus 408
15.6 Stomach 410
15.7 Pancreas 413
15.8 Liver 415
15.9 Small Intestine 420
15.10 Large Intestine 424
15.11 Nutrition and Nutrients 428

Chapter 16

Respiratory System 442

16.1 Introduction 443
16.2 Organs of the Respiratory System 443
16.3 Breathing Mechanism 450
16.4 Control of Breathing 456
16.5 Alveolar Gas Exchanges 459
16.6 Gas Transport 460

Chapter 17

Urinary System 467

17.1 Introduction 468
17.2 Kidneys 468
17.3 Urine Formation 472
17.4 Urine Elimination 481

Chapter 18

Water, Electrolyte, and
Acid-Base Balance 489

18.1 Introduction 490
18.2 Distribution of Body Fluids 490
18.3 Water Balance 492
18.4 Electrolyte Balance 493
18.5 Acid-Base Balance 497
18.6 Acid-Base Imbalances 500

Unit 6
The Human Life Cycle

Chapter 19

Reproductive Systems 505

19.1 Introduction 506
19.2 Organs of the Male Reproductive
       System 506
19.3 Hormonal Control of Male
       Reproductive Functions 513
19.4 Organs of the Female Reproductive
       System 516
19.5 Hormonal Control of Female
       Reproductive Functions 522
19.6 Mammary Glands 525
19.7 Birth Control 526
19.8 Sexually Transmitted Infections 530

Chapter 20

Pregnancy, Growth, Development,
and Genetics 536

20.1 Introduction 537
20.2 Pregnancy 537
20.3 Prenatal Period 541
20.4 Postnatal Period 553
20.5 Aging 555
20.6 Genetics 556

Appendix A
AIDS TO UNDERSTANDING WORDS 564

Appendix B
METRIC MEASUREMENT SYSTEM AND
CONVERSIONS 565

Appendix C
PERIODIC TABLE OF ELEMENTS 566

Appendix D
CHANGES OCCURRING IN THE HEART DURING
A CARDIAC CYCLE 567

Appendix E
FIGURE QUESTION ANSWERS 568

Glossary 569
Credits 585
Application Index 587
Subject Index 589

Chapter Preview

Foundations for Success

The Chapter Preview not only provides great study tips to offer a foundation for success, but it also offers tips on how to utilize this particular text.

A photo on the opening page for each chapter generates interest.

OPENING VIGNETTE

Beginning each chapter is a vignette that discusses current events or research news relating to the subject matter in the chapter. These vignettes demonstrate applications of the concepts learned in the study of anatomy and physiology.

It is a beautiful day. You can't help but stare wistfully out the window, the scent of spring blooms and sound of birds making it impossible to concentrate on what the instructor is saying. Gradually, the lecture fades as you become aware of your own breathing, the beating of your heart, and the sweat that breaks out on your forehead in response to the radiant heat from the glorious day. Suddenly your reverie is cut short—the instructor has dropped a human anatomy and physiology textbook on your desk. You jump. Your heart hammers and a flash of fear grips your chest—but you soon realize what has happened and recover.

The message is clear: pay attention. So you do, tuning out the great outdoors and focusing on the lecture. In this course, you will learn all about the events that you have just experienced, including your response to the sudden stimulation of the instructor's wake-up call. This is a good reason to learn about how to stay focused in the course.

Learning Outcomes

After studying this chapter, you should be able to do the following:

Each chapter begins with a list of outcomes indicating the knowledge you should gain as you work through the chapter. (Note the blue learn arrow.) These outcomes are intended to help you master the similar outcomes set by your instructor. The outcomes will be tied directly to assessments of knowledge gained.

P.1 Introduction

1. Explain the importance of an individualized approach to learning.

P.2 Strategies for Your Success

2. Summarize what you should do before attending class.

3. Identify student activities that enhance classroom experience.

4. List and describe several study techniques that can facilitate learning new material.

 Learn   **Practice**   **Assess**

Aids to Understanding Words
(Appendix A on page 564 has a complete list of Aids to Understanding Words.)

This section introduces building blocks of words that your instructor may assign. Learning them is a good investment of your time, because they can be used over and over and apply to many of the terms you will use in your career. Appendix A (p. 564) has a comprehensive list of these prefixes, suffixes, and root words.

ana- [up] *ana*tomy: the study of breaking up the body into its parts.

multi- [many] *multi*tasking: performing several tasks simultaneously.

physio- [relationship to nature] *physio*logy: the study of how body parts function.

P.1 INTRODUCTION

Each chapter begins with an overview that tells you what to expect and why the subject matter is important.

Studying the human body can be overwhelming at times. The new terminology, used to describe body parts and how they work, can make it seem as if you are studying a foreign language. Learning all the parts of the body, along with the composition of each part, and how each part fits with the other parts to make the whole requires memorization. Understanding the way each body part works individually, as well as body parts working together, requires a higher level of knowledge, comprehension, and application. Identifying underlying structural similarities, from the macroscopic to the microscopic levels of body organization, taps more subtle critical thinking skills. This chapter will catalyze success in this active process of learning. (Remember that while the skills and tips discussed in this chapter relate to learning anatomy and physiology, they can be applied to other subjects.)

Learning occurs in different ways or modes. Most students use several modes (multimodal), but are more comfortable and use more effectively one or two learning styles. Some students prefer to read the written word to remember it and the concept it describes or to actually write the words; others learn best by looking at visual representations, such as photographs and drawings. Still others learn most effectively by hearing the information or explaining it to someone else. For some learners, true understanding remains elusive until a principle is revealed in a laboratory or clinical setting that provides a memorable context and engages all of the senses.

This text is balanced among the learning styles; read-write learners will appreciate the lists, definitions (glossary), and tables; visual learners will discover in the pages of text many diagrams, flow charts, and figures, all with consistent and purposeful use of color (in figures where bones are color-coded, for example, a particular bone is always the same color); auditory learners will find pronunciations whenever new scientific terms are introduced, so that they may "sound out" the new vocabulary;

and kinesthetic learners will appreciate real-life examples and applications to relate to their own activities.

After each major section, a question or series of questions tests your understanding of the material and enables you to practice using the information. (Note the green practice arrow.) If you cannot answer the question(s), you should reread that section, being particularly on the lookout for the answer(s).

Check Your Recall
1. List some difficulties a student may experience when studying the human body.
2. List the ways that people learn.

P.2 STRATEGIES FOR YOUR SUCCESS

Major divisions within a chapter are called "A-heads." They are numbered sequentially in very large, purple type and identify major content areas.

Many strategies for academic success are common sense, but it might help to review them. You may encounter new and helpful methods of learning.

Before Class

The major divisions are subdivided into "B-heads," which are identified by large, black type. These will help you organize the concepts upon which the major divisions are built.

Before attending class, prepare by reading and outlining or taking notes on the assigned pages of the text. If outlining, leave adequate space between entries to allow room for note-taking during lectures. Or, fold each page of notes taken before class in half so that class notes can be written on the blank side of the paper across from the reading notes on the same topic. This introduces the topics of the next class lecture, as well as new terms. Some students team a vocabulary list with each chapter's notes. The outline or notes from the reading can be taken to class and expanded during the lecture. At a minimum, the student should at least skim through the text, reading A-heads, B-heads, and the summary outline to become acquainted with the topics and vocabulary in advance of class attendance.

As you read, you may feel the need for a "study break" or to "chill out." Other times, you may just need to shift gears. Try the following. Throughout the book are shaded boxes that present sidelights to the main text. Indeed, some of these may cover topics that your instructor chooses to highlight. Read them! They are interesting, informative, and a change of pace.

Health-care workers repeatedly monitor patients' *vital signs*—observable body functions that reflect essential metabolic activities. Vital signs indicate that a person is alive. Assessment of vital signs includes measuring body temperature and blood pressure and monitoring rates and types of pulse and breathing movements. Absence of vital signs signifies death. A person who has died displays no spontaneous muscular movements, including those of the breathing muscles and beating heart. A dead body does not respond to stimuli and has no reflexes, such as the knee-jerk reflex and the pupillary reflexes of the eye. Brain waves cease with death, as demonstrated by a flat electroencephalogram (EEG), which signifies a lack of electrical activity in the brain.

The skeleton of an average 160-pound body weighs about 29 pounds.

Genetics Connection 16.1

Cystic Fibrosis

"Woe to that child which when kissed on the forehead tastes salty. He is bewitched and soon must die." So went a seventeenth-century British saying about a child with cystic fibrosis (CF). Until recently, salty skin, foul stools, and poor weight gain ("failure to thrive") were typically the first symptoms of CF. Today most new cases are detected before birth, using genetic tests. The disease, inherited from two carrier parents, affects about 30,000 people in the United States and 70,000 worldwide. It isn't known how many people have mild forms of the disease, merely with symptoms of frequent respiratory infection. More than 1,000 mutations can cause CF, so severity varies widely.

In 1938, physicians first described CF as a defect in channels leading from certain glands. This causes formation of extremely thick, sticky mucus, which encourages infections by microorganisms not otherwise common in the lungs. A clogged pancreas prevents digestive juices from reaching the intestines and thus impairs absorption of nutrients.

In the 1930s, life expectancy for a child with CF was five years, but by 1960 it became possible to treat the symptoms. Antibiotics control the respiratory infections, and daily "bronchial drainage" exercises shake the stifling mucus free from the lungs of infants. Older children and adults wear a vibrating vest for half-hour stretches two to four times a day to shake the mucus free. Some people multitask, taking daily antibiotics in a nebulizer as they wear the vest. Digestive enzymes mixed into soft foods enhance nutrient absorption.

The gene that is mutant in CF normally encodes a protein called the "cystic fibrosis transmembrane regulator," or CFTR for short. It is an ion channel that controls chloride transport out of cells. In severe CF, the chloride channel is missing one crucial amino acid, and is so deformed that it fails to function. The abnormal handling of chloride ions thickens the mucus. Organs become clogged.

Discovery of the most common CFTR mutation in 1989 enabled development of more targeted treatments. Some drugs allow more chloride to leave the cells lining the lungs. Two new drugs, still experimental, are small molecules that escort abnormal CFTR protein to the cell surface, where it apparently functions. The drugs act as "correctors," saving the errant CFTR proteins from being dismantled before they can reach the cell surface.

Life with severe CF is difficult. One little girl did not mind the twice-daily vibrating vest, or even the feeding tube she needed at night to pack in nutrients. But she hated the measures to avoid respiratory infections, especially in summertime. She had to stay away from hoses, which harbor lung-loving *Pseudomonas* bacteria. Bonfires or cookouts could expose her to lung-clogging particulates in the air. She couldn't even go into a pool—too little chlorine would invite bacterial infections, and too much would irritate her lungs. But unlike children of a generation ago, her disease is controlled enough that she will likely live well into adulthood.

Clinical Application 15.1

Dental Caries

Sticky foods, such as caramel, lodge between the teeth and in the crevices of molars, feeding bacteria such as *Actinomyces, Streptococcus mutans,* and *Lactobacillus*. These microorganisms metabolize carbohydrates in the food, producing acid by-products that destroy tooth enamel and dentin. The bacteria also produce sticky substances that hold them in place.

If a person eats a candy bar but does not brush the teeth soon afterward, the acid-forming bacteria may decay tooth enamel, creating a condition called *dental caries*. Unless a dentist cleans and fills the resulting cavity that forms where enamel is destroyed, the damage will spread to the underlying dentin.

Dental caries can be prevented in several ways:

1. Brush and floss teeth regularly.
2. Have regular dental exams and cleanings.
3. Talk with your dentist about receiving a fluoride treatment. Fluoride is added to the water supply in many communities. Fluoride is incorporated into the enamel's chemical structure, strengthening it.
4. The dentist may apply a sealant to children's and adolescents' teeth where crevices might hold onto decay-causing bacteria. The sealant is a coating that keeps acids from eating away at tooth enamel.

Remember when you were very young and presented with a substantial book for the first time? You were likely intimidated by its length, but were reassured that there were "a lot of pictures." This book has many illustrations too, all designed to help you master the material and become that person who you would want treating you.

Photographs and Line Art

The heading above this box is a "C-head." Sometimes subdivisions have so many parts that the book goes to this third level of organization. This heading is presented in a slightly smaller, italicized font.

Photographs provide a realistic view of anatomy.

Because line art can present different positions, layers, or perspectives, it can provide a unique view.

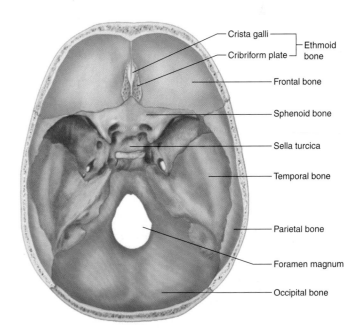

Macroscopic to Microscopic

Many figures show anatomical structures in a manner that is macroscopic to microscopic (or vice versa).

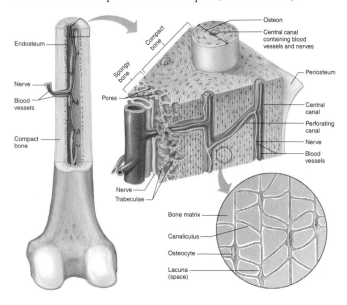

Anatomical Structures

Some figures illustrate the locations of anatomical structures.

Flow Charts

Flow charts depict sequences of related events, steps of pathways, and complex concepts, easing comprehension. Other figures may show physiological processes.

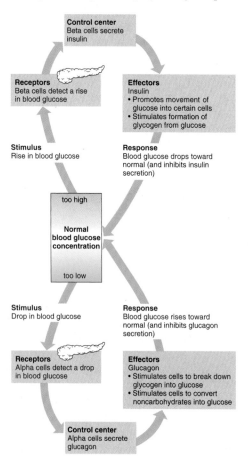

Other figures illustrate the functional relationships of anatomical structures.

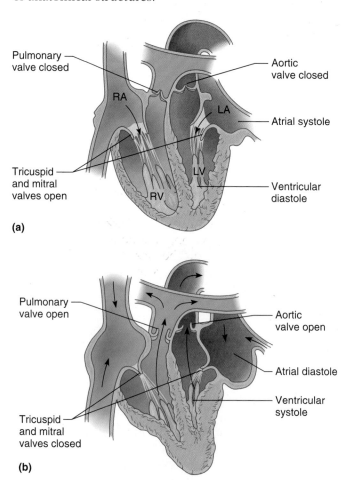

Organizational Tables

Organizational tables can help "put it all together," but are not a substitute for reading the text or having good lecture notes.

Table 5.6	Muscle and Nervous Tissues	
Type	**Function**	**Location**
Skeletal muscle tissue (striated)	Voluntary movements of skeletal parts	Muscles usually attached to bones
Smooth muscle tissue (lacks striations)	Involuntary movements of internal organs	Walls of hollow internal organs
Cardiac muscle tissue (striated)	Heart movements	Heart muscle
Nervous tissue	Sensory reception and conduction of electrical impulses	Brain, spinal cord, and peripheral nerves

It is critical that you attend class regularly, and be on time—even if the instructor's notes are posted on the Web, and the information is in the textbook. For many learners, hearing and writing new information is a better way to retain facts than just scanning notes on a computer screen. Attending lectures and discussion sections also provides more detailed and applied analysis of the subject matter, as well as a chance to ask questions.

During Class

Be alert and attentive in class. Take notes by adding to either the outline or notes taken while reading. Auditory learners benefit from recording the lectures and listening to them while driving or doing chores. This is called **multitasking**—doing more than one activity at a time.

Participate in class discussions, asking questions of the instructor and answering questions he or she poses. All of the students are in the class to learn, and many will be glad someone asked a question others would not be comfortable asking. Such student response can alert the instructor to topics that are misunderstood or not understood at all. However, respect class policy. Due to time constraints and class size, asking questions may be more appropriate after a large lecture class or during tutorial (small group) sessions.

After Class

In learning complex material, expediency is critical. Organize, edit, and review notes as soon after class as possible, fleshing out sections where the lecturer got ahead of the listener. Highlighting or underlining (in color, for visual learners) the key terms, lists, important points and major topics make them stand out, which eases both daily reviews and studying for exams.

Lists

Organizing information into lists or categories can minimize information overload, breaking it into manageable chunks. For example, when studying the muscles of the thigh it is easier to learn the insertion, origin, action, and nerve supply of the four muscles making up the quadriceps femoris as a group, because they all have the same insertion, action, and nerve supply . . . they differ only in their origins.

Mnemonic Devices

Another method for remembering information is the **mnemonic device.** One type of mnemonic device is a list of words, forming a phrase, in which the first letter of each word corresponds to the first letter of each word that must be remembered. For example, ***Frequent parade often tests soldiers' endurance*** stands for the skull bones **f**rontal, **p**arietal, **o**ccipital, **t**emporal, **s**phenoid, and **e**thmoid. Another type of mnemonic device is a word formed by the first letters of the items to be remembered. For example, ***ipmat*** represents the stages in the cell cycle: **i**nterphase, **p**rophase, **m**etaphase, **a**naphase, and **t**elophase.

Study Groups

Forming small study groups helps some students. Together the students review course material and compare notes. Working as a team and alternating leaders allows students to verbalize the information. Individual students can study and master one part of the assigned material, and then explain it to the others in the group, which incorporates the information into the memory of the speaker. Hearing the material spoken aloud also helps the auditory learner. Be sure to use anatomical and physiological terms, in explanations and everyday conversation, until they become part of your working vocabulary, rather than intimidating jargon. Most important of all—the group must stay on task, and not become a vehicle for social interaction. Your instructor may have suggestions or guidelines for setting up study groups.

Flash Cards

Flash cards may seem archaic in this computer age, but they are still a great way to organize and master complex and abundant information. The act of writing or drawing on a note card helps the tactile learner. Master a few new cards each day, and review cards from previous days, and use them all again at the end of the semester to prepare for the comprehensive final exam. They may even come in handy later, such as in studying for exams for admission to medical school or graduate school. Divide your deck in half and flip half of the cards so that the answer rather than the question is showing. Mix and shuffle them. Get used to identifying a structure or process from a description as well as giving a description when provided with a process or structure. This is more like what will be expected of you in the real world of the health-care professional.

Manage Your Time

For each hour in the classroom, most students will spend at least three hours outside of class studying. Many of you have important obligations outside of class, such as jobs and family responsibilities. As important as these are, you still need to master this material on your path to becoming a health-care professional. Good time management skills are therefore essential in your study of human anatomy and physiology. In addition to class, lab, and study time, multitask. Spend time waiting for a ride, in a doctor's office, or on line reviewing notes or reading the text.

Daily repetition is helpful, so scheduling several short study periods each day can replace an end-of-semester crunch to cram for an exam. This does not take the place of time to prepare for the next class. Thinking about these suggestions for learning now can maximize study time throughout the semester, and, hopefully, lead to academic success. A working knowledge of the structure and function of the human body provides the foundation for all careers in the health sciences.

Check Your Recall

3. Why is it important to prepare before attending class?
4. Name two ways to participate in class discussions.
5. List several aids for remembering information.

Summary Outline

> A summary of the chapter provides an outline to review major ideas and is a tool for organizing thoughts.

P.1 Introduction (page xix)

Try a variety of methods to study the human body.

P.2 Strategies for Your Success (page xix)

While strategies for academic success seem to be common sense, you might benefit from reminders of study methods.

1. Before class
   Read the assigned text material prior to the corresponding class meeting.
   a. Photographs give a realistic view and line art shows different perspectives.

   b. Macroscopic to microscopic show increase in detail.
   c. Flow charts depict sequences and steps.
   d. Figures of anatomical structures show locations.
   e. Organizational charts/tables summarize text.
2. During class
   Take notes and participate in class discussions.
3. After class
   a. Organize, edit, and review class notes.
   b. Mnemonic devices aid learning.
      (1) The first letters of the words to remember begin words of an easily recalled phrase.
      (2) The first letters of the items to be remembered form a word.
   c. Small study groups reviewing and vocalizing material can divide and conquer the learning task.
   d. Making flash cards helps the tactile learner.
   e. Time management skills encourage scheduled studying, including daily repitition instead of cramming for exams.

Chapter Assessments

> Chapter assessments that are tied directly to the learning outcomes allow you to assess your mastery of the material. (Note the purple assess arrow.)

P.1 Introduction

1. Explain why the study of the human body can be overwhelming. (p. xix)

P.2 Strategies for Success

2. Methods to prepare for class include: (p. xix)
   a. reading the chapter.
   b. outlining the chapter.
   c. taking notes on the assigned reading.
   d. making a vocabulary list.
   e. all of the above.
3. Describe how you can participate in class discussions. (p. xxiii)
4. Forming the phrase *"I passed my anatomy test"* to remember the cell cycle (interphase, prophase, metaphase, anaphase, telophase) is a _____ device. (p. xxiii)
5. Name a benefit and a drawback of small study groups. (p. xxiii)
6. Explain the value of repetition in learning and preparation for exams. (p. xxiv)

Integrative Assessments/Critical Thinking

A textbook is inherently linear. This text begins with Chapter 1 and ends with Chapter 20. Understanding physiology and the significance of anatomy, however, requires you to be able to recall previous concepts. Toward this end, we have included in the Integrative Assessments/Critical Thinking section references to sections from earlier chapters. Making connections is what it is all about!

OUTCOME P.1

1. Which study methods are most successful for you?

OUTCOMES P.1, P.2

2. Design a personalized study schedule.

Check out the text website at **www.mhhe.com/shieress11** for additional study tools. There is also information about the applicable Anatomy & Physiology Revealed® CD-ROM.

WEB CONNECTIONS

Visit the text website at **www.mhhe.com/shieress11** for additional quizzes, interactive learning exercises, and more.

APR

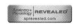

Anatomy & Physiology REVEALED® includes cadaver photos that allow you to peel away layers of the human body to reveal structures beneath the surface. This program also includes animations, radiologic imaging, audio pronunciations, and practice quizzing. To learn more visit **www.aprevealed.com**.

1

Introduction to Human Anatomy and Physiology

The mummy's toe. She lived between 1069 and 664 B.C. in Thebes, a city in ancient Egypt. Only pieces of her skeleton remain, held in place with plaster, glue, and linen. Yet, the telltale bones reveal a little of what her life was like.

The shape of the pelvic bones indicates that the person was female. She was 50 to 60 years old when she died, according to the way the bony plates of her skull fit together and the lines of mineral deposition in a well-preserved tooth. Among the preserved bones from the skull, pelvis, upper limbs, and right lower limbs, the right big toe stands out, for it ends in a prosthesis, a manufactured replacement for a skeletal part. Was it purely cosmetic, or did it work?

The mummy's toe tip is wooden and painted a dark brown, perhaps to blend in with her skin color. A long part and two smaller parts anchor the structure to the stump. Seven leather strings once attached it to the foot, and it even bears a fake nail. Connective tissue and skin grew over the prosthesis, revealing that her body had accepted the replacement part, and the shape of the prosthesis was remarkably like that of a real toe. Signs of wear indicate that it was indeed used. Modern-day scientists made replicas of the toe and volunteers who were missing the same toe tried them out, demonstrating that the mummy's toe must have been crucial for balance and locomotion.

The replacement toe is evidence of sophisticated medical technology. Modern-day medical sleuths obtained computerized tomog-

A wooden toe on an ancient Egyptian mummy reveals sophisticated knowledge of human anatomy and physiology from long ago.

raphy (CT) scans of the remnants of the mummy. They detected poor mineral content in the toe, plus calcium deposits in the largest blood vessel, the aorta, suggesting impaired circulation to the feet. Perhaps the mummy in life suffered from type 2 diabetes mellitus, which can impede circulation to the toes. If gangrene had set in, healers might have amputated the affected portion of the toe, replacing it with a very reasonable facsimile.

The ancient Egyptians made other replacement parts, including ears, noses, feet, and lower limbs. Today prosthetic toes are made of silicones, which are plastic-like materials. People use them who have lost digits to injury, cancer, or, perhaps like the ancient Egyptian woman, diabetes.

Learning Outcomes

After studying this chapter, you should be able to do the following:

1.1 Introduction
1. Identify some of the early discoveries that led to our understanding of the body. (p. 2)

1.2 Anatomy and Physiology
2. Explain how anatomy and physiology are related. (p. 3)

1.3 Levels of Organization
3. List the levels of organization in the human body and the characteristics of each. (p. 3)

1.4 Characteristics of Life
4. List and describe the major characteristics of life. (p. 4)
5. Give examples of metabolism. (p. 4)

1.5 Maintenance of Life
6. List and describe the major requirements of organisms. (p. 5)
7. Explain the importance of homeostasis to survival. (p. 5)

8. Describe the parts of a homeostatic mechanism and explain how they function together. (p. 6)

1.6 Organization of the Human Body
9. Identify the locations of the major body cavities. (p. 8)
10. List the organs located in each major body cavity. (p. 8)
11. Name and identify the locations of the membranes associated with the thoracic and abdominopelvic cavities. (p. 10)

 Learn   **Practice**   **Assess**

12. Name the major organ systems, and list the organs associated with each. (p. 12)

13. Describe the general functions of each organ system. (p. 12)

1.7 Anatomical Terminology

14. Properly use the terms that describe relative positions, body sections, and body regions. (p. 14)

Anatomy & Physiology **REVEALED®**
aprevealed.com

Module 1: Body Orientation

Aids to Understanding Words   *(Appendix A on page 564 has a complete list of Aids to Understanding Words.)*

append- [to hang something] *append*icular: Pertaining to the limbs.

cardi- [heart] peri*cardi*um: Membrane that surrounds the heart.

cran- [helmet] *cran*ial: Pertaining to the portion of the skull that surrounds the brain.

dors- [back] *dors*al: Position toward the back.

homeo- [same] *homeo*stasis: Maintenance of a stable internal environment.

-logy [study of] physio*logy*: Study of body functions.

meta- [change] *meta*bolism: Chemical changes that occur within the body.

pariet- [wall] *pariet*al membrane: Membrane that lines the wall of a cavity.

pelv- [basin] *pelv*ic cavity: Basin-shaped cavity enclosed by the pelvic bones.

peri- [around] *peri*cardial membrane: Membrane that surrounds the heart.

pleur- [rib] *pleur*al membrane: Membrane that encloses the lungs and lines the thoracic cavity.

-stasis [standing still] homeo*stasis*: Maintenance of a stable internal environment.

-tomy [cutting] ana*tomy*: Study of structure, which often involves cutting or removing body parts.

1.1 INTRODUCTION

Modern medicine began with long-ago observations on the function, and malfunction, of the human body. The study of the human body probably began with our earliest ancestors, who must have been curious about how their bodies worked, as we are today. At first their interests most likely concerned injuries and illnesses, because healthy bodies demand little attention from their owners. Early healers relied heavily on superstitions and notions about magic. However, as healers tried to help the sick, they began to discover useful ways of examining and treating the human body. They observed the effects of injuries, noticed how wounds healed, and examined cadavers to determine causes of death. They also found that certain herbs and potions could sometimes be used to treat coughs, headaches, fevers, and other common signs of illness.

Over time, people began to believe that humans could understand forces that caused natural events. They began observing the world around them more closely, asking questions and seeking answers. This set the stage for the development of modern medical science.

As techniques for making accurate observations and performing careful experiments evolved, knowledge of the human body expanded rapidly (fig. 1.1). At the same time, early medical providers coined many new terms to name body parts, describe their locations, and explain their functions and interactions. These terms, most of which originated from Greek and Latin words, formed the basis for the language of anatomy and physiology that persists today. (The names of some modern medical and applied sciences are listed on pages 17–19.)

Practice

1. What factors probably stimulated an early interest in the human body?

2. What kinds of activities helped promote the development of modern medical science?

Figure 1.1

The study of the human body has a long history, as evidenced by this illustration from the second book of *De Humani Corporis Fabrica* by Andreas Vesalius, issued in 1543. (Note the similarity to the anatomical position, described later in this chapter on page 14.)

1.2 ANATOMY AND PHYSIOLOGY

Anatomy (ah-nat′o-me) is the branch of science that deals with the structure (morphology) of body parts—their forms and how they are organized. **Physiology** (fiz″e-ol′o-je), on the other hand, concerns the functions of body parts—what they do and how they do it.

The topics of anatomy and physiology are difficult to separate because the structures of body parts are so closely associated with their functions. Body parts form a well-organized unit—the human organism—and each part functions in the unit's operation. A particular body part's function depends on the way the part is constructed—that is, how its subparts are organized. For example, the organization of the parts in the human hand with its long, jointed fingers makes it easy to grasp objects; the hollow chambers of the heart are adapted to pump blood through tubular blood vessels; the shape of the mouth enables it to receive food; and teeth are shaped to break solid foods into small pieces (fig. 1.2).

Anatomy and physiology are ongoing as well as ancient fields. Researchers frequently discover new information about physiology, particularly at the molecular level since the human genome was sequenced in 2001. Less frequently they discover new parts of human anatomy, such as a small piece of connective tissue between the upper part of the spinal cord and a muscle at the back of the head. This connective tissue bridge may trigger pain impulses in certain types of tension headaches.

> By discovering which of our 20,500 or so genes are active in particular diseases, researchers are finding commonalities among illnesses that are not apparent on the whole-body level, suggesting new treatments. These connections form what researchers call a "diseasome."

(a)   **(b)**

Figure 1.2
The structures of body parts make possible their functions: **(a)** The hand is adapted for grasping, **(b)** the mouth for receiving food. (Arrows indicate movements associated with these functions.)

Practice
3. Why is it difficult to separate the topics of anatomy and physiology?
4. List several examples that illustrate how the structure of a body part makes possible its function.

1.3 LEVELS OF ORGANIZATION

Until the invention of magnifying lenses and microscopes about 400 years ago, anatomists were limited in their studies to what they could see with the unaided eye—large parts. But with these new tools, investigators discovered that larger body structures are made up of smaller parts, which in turn are composed of even smaller ones.

Figure 1.3 shows the levels of organization that modern-day scientists recognize. All materials, including those that make up the human body, are composed of chemicals. Chemicals consist of microscopic particles called **atoms,** which join to form **molecules.** Small molecules can combine in complex ways to form larger **macromolecules.**

In the human and other organisms, the basic unit of structure and function is a **cell,** which is microscopic. Although cells vary in size, shape, and specialized functions, all share certain characteristics. For instance, all cells of humans and other complex organisms contain structures called **organelles** (or″gah-nelz′) that carry out specific activities. Organelles are composed of aggregates of macromolecules, such as proteins, carbohydrates, lipids, and nucleic acids.

Cells may be organized into layers or other structures that have common functions. Such a group of cells forms a **tissue.** Groups of different tissues that interact form **organs**—complex structures with specialized functions—and groups of organs that function closely together compose **organ systems.** Organ systems make up an **organism** (or′gah-nizm), which is a living thing.

Body parts can be described in terms of different levels of organization, such as the *atomic level,* the *molecular level,* or the *cellular level.* Furthermore, body parts differ in complexity from one level to the next. That is, atoms are less complex than molecules, molecules are less complex than organelles, tissues are less complex than organs, and so forth.

Chapters 2–6 discuss these levels of organization in more detail. Chapter 2 describes the atomic and molecular levels. Chapter 3 deals with organelles and cellular structures and functions, and chapter 4 explores cellular metabolism. Chapter 5 describes tissues and presents membranes (linings) as examples of organs, and chapter 6 considers the skin and its accessory organs as an

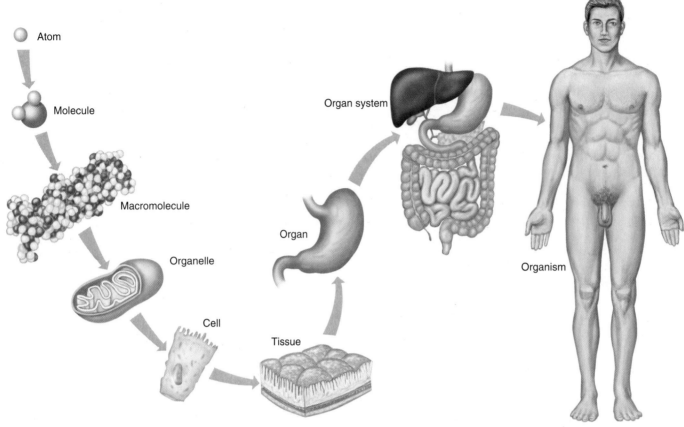

Atom

Molecule

Macromolecule

Organelle

Cell

Tissue

Organ

Organ system

Organism

Figure 1.3

A human body is composed of parts within parts, with increasing complexity.

example of an organ system. In the remaining chapters, the structures and functions of each of the other organ systems are described in detail.

Practice

5. How does the human body illustrate levels of organization?

6. What is an organism?

7. How do body parts at different levels of organization vary in complexity?

1.4 CHARACTERISTICS OF LIFE

Before beginning a more detailed study of anatomy and physiology, it is helpful to consider some of the traits humans share with other organisms, particularly with other animals. As living organisms, we can move and respond to our surroundings. We start out as small indi-viduals and then grow, eventually becoming able to reproduce. We gain energy by taking in or ingesting food, by breaking it down or digesting it, and by absorb-ing and assimilating it. The absorbed substances circu-late throughout the internal environment of our bodies. We can then, by the process of respiration, release the energy in these nutrients for use in such vital functions as growth and repair of body parts. Finally, we excrete wastes from the body. All of these processes involve **metabolism** (mĕ-tab'o-lizm), the sum total of all of the chemical reactions in the body that break substances down and build them up. The reactions of metabolism enable us to acquire and use energy to fuel life pro-cesses. Table 1.1 summarizes the characteristics of life.

Practice

8. What are the characteristics of life?

9. How are the characteristics of life dependent on metabolism?

Table 1.1	Characteristics of Life
Process	**Examples**
Movement	Change in position of the body or of a body part; motion of an internal organ
Responsiveness	Reaction to a change inside or outside the body
Growth	Increase in body size without change in shape
Reproduction	Production of new organisms and new cells
Respiration	Obtaining oxygen, removing carbon dioxide, and releasing energy from foods (Some forms of life do not use oxygen in respiration.)
Digestion	Breakdown of food substances into simpler forms that can be absorbed and used
Absorption	Passage of substances through membranes and into body fluids
Circulation	Movement of substances in body fluids
Assimilation	Changing absorbed substances into chemically different forms
Excretion	Removal of wastes produced by metabolic reactions

1.5 MAINTENANCE OF LIFE

The structures and functions of almost all body parts help maintain life. Even an organism's reproductive structures, whose primary function is to ensure that its species will continue into the future, may contribute to survival. For example, sex hormones help to strengthen bones.

Requirements of Organisms

Being alive requires certain environmental factors, including the following:

1. **Water** is the most abundant chemical in the body. It is required for many metabolic processes and provides the environment in which most of them take place. Water also transports substances within the organism and is important in regulating body temperature.
2. **Foods** are substances that provide the body with necessary chemicals (nutrients) in addition to water. Some of these chemicals are used as energy sources, others supply raw materials for building new living matter, and still others help regulate vital chemical reactions.
3. **Oxygen** is a gas that makes up about one-fifth of ordinary air. It is used to release energy from

food substances. This energy, in turn, drives metabolic processes.

4. **Heat** is a form of energy. It is a product of metabolic reactions, and the degree of heat present partly determines the rate at which these reactions occur. Generally, the more heat, the more rapidly chemical reactions take place. (*Temperature* is a measure of the degree of heat.)
5. **Pressure** is an application of force to something. For example, the force on the outside of the body due to the weight of air above it is called *atmospheric pressure*. In humans, this pressure is important in breathing. Similarly, organisms living under water are subjected to *hydrostatic pressure*—a pressure a liquid exerts—due to the weight of water above them. In humans, heart action produces blood pressure (another form of hydrostatic pressure), which forces blood through blood vessels.

Health-care workers repeatedly monitor patients' *vital signs*—observable body functions that reflect essential metabolic activities. Vital signs indicate that a person is alive. Assessment of vital signs includes measuring body temperature and blood pressure and monitoring rates and types of pulse and breathing movements. Absence of vital signs signifies death. A person who has died displays no spontaneous muscular movements, including those of the breathing muscles and beating heart. A dead body does not respond to stimuli and has no reflexes, such as the knee-jerk reflex and the pupillary reflexes of the eye. Brain waves cease with death, as demonstrated by a flat electroencephalogram (EEG), which signifies a lack of electical activity in the brain.

Organisms require water, food, oxygen, heat, and pressure, but these alone are not enough to ensure survival. Both the quantities and the qualities of such factors are also important. For example, the volume of water entering and leaving an organism must be regulated, as must the concentration of oxygen in body fluids. Similarly, survival depends on the quality as well as the quantity of food available—that is, food must supply the correct nutrients in adequate amounts.

Homeostasis

Factors in the external environment may change. If an organism is to survive, however, conditions within the fluid surrounding its body cells, which compose its **internal environment,** must remain relatively stable (fig. 1.4). In other words, body parts function only when the concentrations of water, nutrients, and oxygen and the conditions of heat and pressure remain within certain narrow limits. This condition of a stable internal environment is called **homeostasis** (ho″me-ō-sta′sis).

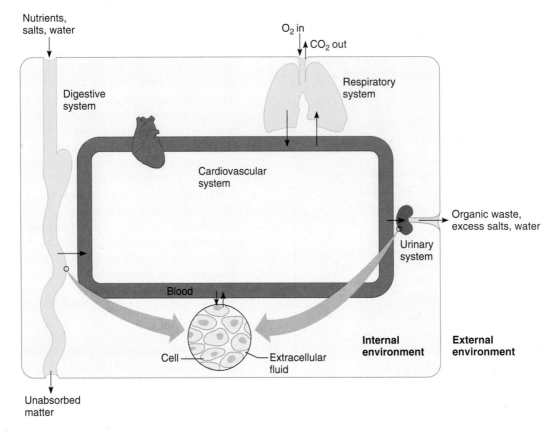

Nutrients, salts, water

O₂ in
CO₂ out

Digestive system

Respiratory system

Cardiovascular system

Organic waste, excess salts, water

Urinary system

Blood

Internal environment

External environment

Cell

Extracellular fluid

Unabsorbed matter

Figure 1.4

Our cells lie within an internal fluid environment (extracellular fluid). Concentrations of water, nutrients, and oxygen in the internal environment must be maintained within certain ranges to sustain life.

The body maintains homeostasis through a number of self-regulating control systems, or **homeostatic mechanisms,** that share the following three components (fig. 1.5):

- **Receptors** provide information about specific conditions (stimuli) in the internal environment.
- A **set point** tells what a particular value should be, such as body temperature at 37°C (Celsius) or 98.6°F (Fahrenheit). More about metric equivalents can be found in Appendix B (p. 565), since metric units are used throughout this text.
- **Effectors** cause responses that alter conditions in the internal environment.

A homeostatic mechanism works as follows. If the receptors measure deviations from the set point, effectors are activated that can return conditions toward normal. As conditions return toward normal, the deviation from the set point progressively lessens and the effectors are gradually shut down. Such a response is called a **negative feedback** (neg′ah-tiv fēd′bak) mechanism, both because the deviation from the set point is corrected (moves in the opposite or negative direction) and because the correction reduces the action of the effectors. This latter aspect is important because it prevents a correction from going too far.

To better understand this idea of negative feedback, imagine a room equipped with a furnace and an air conditioner (fig. 1.6). Suppose the room temperature is to remain near 20°C (68°F), so the thermostat is adjusted to an operating level, or set point, of 20°C. Because a thermostat senses temperature changes, it will signal the furnace to start and the air conditioner to stop whenever the room temperature drops below the set point. If the temperature rises above the set point,

Control center (set point)

(Change is compared to the set point.)

Receptors

Effectors (muscles or glands)

Stimulus (Change occurs in internal environment.)

Response (Change is corrected.)

Figure 1.5

A homeostatic mechanism monitors a particular aspect of the internal environment and corrects any changes back to the value indicated by the set point.

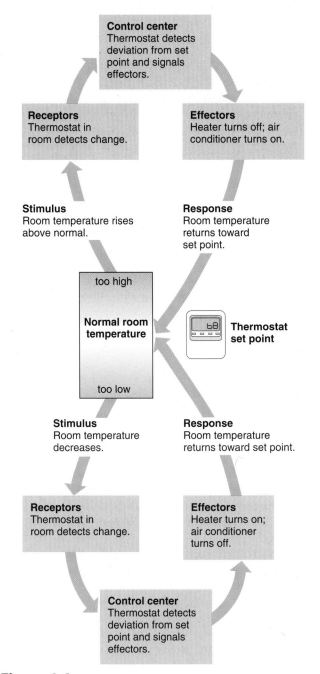

Figure 1.6

A thermostat signals an air conditioner and a furnace to turn on or off to maintain a relatively stable room temperature. This system is an example of a homeostatic mechanism.

Q: *What would happen to room temperature if the set point were turned up?*

Answer can be found in Appendix E on page 568.

the thermostat will stop the furnace and start the air conditioner. As a result, the room will maintain a relatively constant temperature.

A similar homeostatic mechanism regulates body temperature. Temperature receptors are scattered throughout the body. The "thermostat" is a temperature-sensitive region in a temperature control center of the brain. In healthy persons, the set point of the brain's thermostat is at or near 37°C (98.6°F).

If a person is exposed to cold and body temperature begins to drop, the temperature receptors sense this change and the temperature control center triggers heat-generating and heat-conserving activities. For example, small groups of muscles are stimulated to contract involuntarily, an action called *shivering.* Such muscular contractions produce heat, which helps warm the body. At the same time, blood vessels in the skin are signaled to constrict so that less warm blood flows through them. In this way, deeper tissues retain heat that might otherwise be lost.

If a person is becoming overheated, the brain's temperature control center triggers a series of changes that promote loss of body heat. Sweat glands in the skin secrete perspiration, and as this fluid evaporates from the surface, heat is carried away and the skin is cooled. At the same time, the brain center dilates blood vessels in the skin. This action allows the blood carrying heat from deeper tissues to reach the surface, where heat is lost to the outside (fig. 1.7). The brain stimulates an increase in heart rate, which sends a greater volume of blood into surface vessels, and an increase in breathing rate, which allows the lungs to expel more heat-carrying air. Body temperature regulation is discussed further in chapter 6 (p. 125).

Another homeostatic mechanism regulates the blood pressure in the blood vessels (arteries) leading away from the heart. In this instance, pressure-sensitive receptors in the walls of these vessels sense changes in blood pressure and signal a pressure control center in the brain. If blood pressure is above the set point, the brain signals the heart chambers to contract more slowly and with less force. This decreased heart action sends less blood into the blood vessels, decreasing the pressure inside them. If blood pressure falls below the set point, the brain center signals the heart to contract more rapidly and with greater force. As a result, the pressure in the vessels increases. Chapter 13 (pp. 361–362) discusses regulation of blood pressure in more detail.

Human physiology offers many other examples of homeostatic mechanisms. All work by the same general process as the two preceding examples. Just as anatomical terms are used repeatedly throughout this book, so can the basic principles of a homeostatic mechanism be applied to the different organ systems. Homeostatic mechanisms maintain a relatively constant internal environment, yet physiological values may vary slightly in a person from time to time or from one individual to the next. Therefore, both normal values for an individual and the *normal range* for the general population are clinically important.

Most feedback mechanisms in the body are negative. However, sometimes change stimulates further change. A process that moves conditions away from the normal state is called a *positive feedback mechanism.* In blood clotting, for example, the chemicals that carry out clotting stimulate more clotting, minimizing bleeding (see chapter 12, p. 331). Another positive feedback

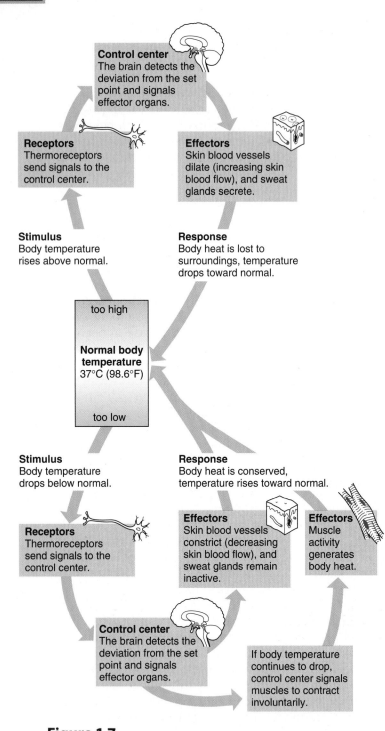

Figure 1.7

A homeostatic mechanism regulates body temperature.

mechanism increases the strength of uterine contractions during childbirth, helping to bring the new individual into the world.

Positive feedback mechanisms usually produce unstable conditions, which might seem incompatible with homeostasis. However, the examples of positive feedback associated with normal health have very specific functions and are short-lived.

1.6 ORGANIZATION OF THE HUMAN BODY

The human organism is a complex structure composed of many parts. Its major features include several body cavities, layers of membranes within these cavities, and a variety of organ systems.

Body Cavities

The human organism can be divided into an **axial** (ak′se-al) portion, which includes the head, neck, and trunk, and an **appendicular** (ap″en-dik′u-lar) portion, which includes the upper and lower limbs. Within the axial portion are the **cranial cavity,** which houses the brain; the **vertebral canal,** which contains the spinal cord within the sections of the backbone (vertebrae); the **thoracic** (tho-ras′ik) **cavity;** and the **abdominopelvic** (ab-dom″ĭ-no-pel′vik) **cavity.** The organs within these last two cavities are called **viscera** (vis′er-ah) (fig. 1.8*a*).

A broad, thin skeletal (voluntary) muscle called the **diaphragm** separates the thoracic cavity from the abdominopelvic cavity. The thoracic cavity wall is composed of skin, skeletal muscles, and various bones.

A region called the **mediastinum** (me″de-as-ti′num) separates the thoracic cavity into two compartments, which contain the right and left lungs. The remaining thoracic viscera—heart, esophagus, trachea, and thymus—are located within the mediastinum (fig. 1.8*b*).

The abdominopelvic cavity, which includes an upper abdominal portion and a lower pelvic portion, extends from the diaphragm to the floor of the pelvis. Its wall consists primarily of skin, skeletal muscles, and bones. The viscera within the **abdominal cavity** include the stomach, liver, spleen, gallbladder, kidneys, and most of the small and large intestines.

The **pelvic cavity** is the portion of the abdominopelvic cavity enclosed by the hip bones (see chapter 7, p. 158). It contains the terminal portion of the large intestine, the urinary bladder, and the internal reproductive organs.

Smaller cavities within the head include (fig. 1.9):

1. **Oral cavity,** containing the teeth and tongue.
2. **Nasal cavity,** located within the nose and divided into right and left portions by a nasal septum.

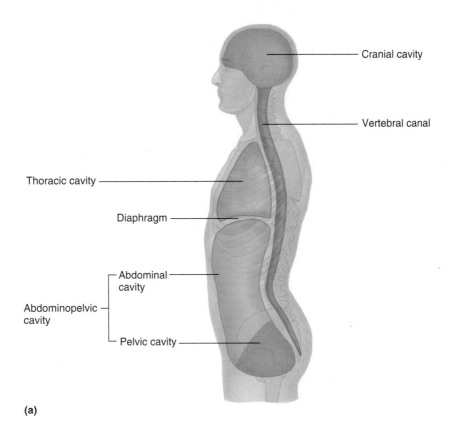

Cranial cavity

Vertebral canal

Thoracic cavity

Diaphragm

Abdominal cavity

Abdominopelvic cavity

Pelvic cavity

(a)

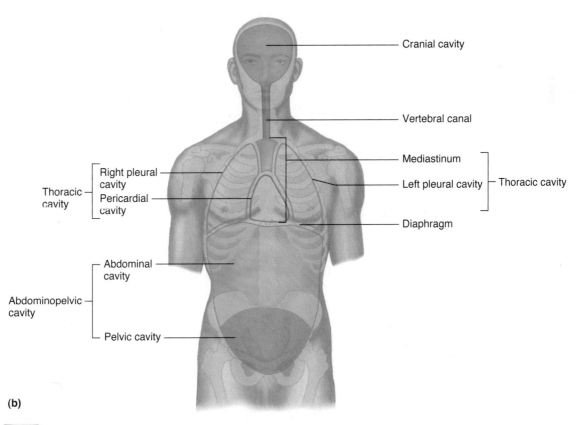

Cranial cavity

Vertebral canal

Mediastinum

Right pleural cavity

Left pleural cavity

Thoracic cavity

Thoracic cavity

Pericardial cavity

Diaphragm

Abdominal cavity

Abdominopelvic cavity

Pelvic cavity

(b)

Figure 1.8 AP|R

Major body cavities. **(a)** Lateral view. **(b)** Anterior view.

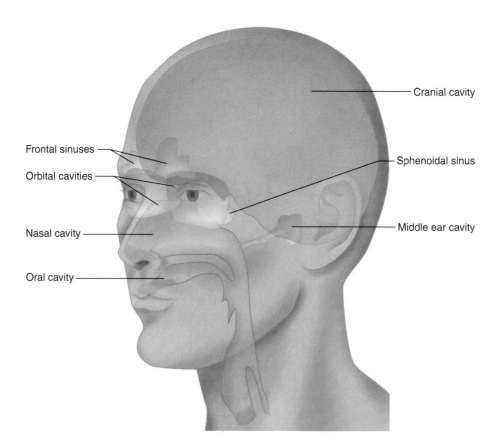

Cranial cavity

Frontal sinuses

Orbital cavities

Sphenoidal sinus

Nasal cavity

Middle ear cavity

Oral cavity

Figure 1.9

The cavities within the head include the cranial, oral, nasal, orbital, and middle ear cavities, as well as several sinuses. (Not all of the sinuses are shown.)

Several air-filled *sinuses* connect to the nasal cavity (see chapter 7, pp. 144–148). These include the frontal and sphenoidal sinuses shown in figure 1.9.

3. **Orbital cavities,** containing the eyes and associated skeletal muscles and nerves.
4. **Middle ear cavities,** containing the middle ear bones.

Thoracic and Abdominopelvic Membranes

The walls of the right and left thoracic compartments, which contain the lungs, are lined with a membrane called the *parietal pleura* (fig. 1.10). A similar membrane, called the *visceral pleura,* covers each lung. (Note: **Parietal** [pah-ri′ĕ-tal] refers to the membrane attached to the wall of a cavity; **visceral** [vis′er-al] refers to the membrane that is deeper—toward the interior—and covers an internal organ, such as a lung.)

The parietal and visceral **pleural membranes** (ploo′ral mem′brānz) are separated by a thin film of watery fluid (serous fluid), which they secrete. While no actual space normally exists between these membranes, the potential space between them is called the *pleural cavity* (see figs. 1.8*b* and 1.10).

The heart, which is located in the broadest portion of the mediastinum, is surrounded by **pericardial** (per″ĭ-kar′de-al) **membranes.** A thin *visceral pericardium* covers the heart's surface and is separated from a thicker *parietal pericardium* by a small volume of fluid. The *pericardial cavity* (see figs. 1.8*b* and 1.10) is the potential space between these membranes.

In the abdominopelvic cavity, the lining membranes are called **peritoneal** (per″ĭ-to-ne′al) **membranes.** A *parietal peritoneum* lines the wall, and a *visceral peritoneum* covers each organ in the abdominal cavity (fig. 1.11). The *peritoneal cavity* is the potential space between these membranes.

Practice

13. What does *viscera* mean?
14. Which organ occupies the cranial cavity? the vertebral canal?
15. Name the cavities of the head.
16. Describe the membranes associated with the thoracic and abdominopelvic cavities.

Vertebra

Spinal cord

Plane of section

Mediastinum

Aorta

Left lung

Esophagus

Right lung

Rib

Heart

Visceral pleura

Pleural cavity

Visceral pericardium

Parietal pleura

Anterior

Pericardial cavity

Sternum

Parietal pericardium

Figure 1.10 AP|R

A transverse section through the thorax reveals the serous membranes associated with the heart and lungs (superior view).

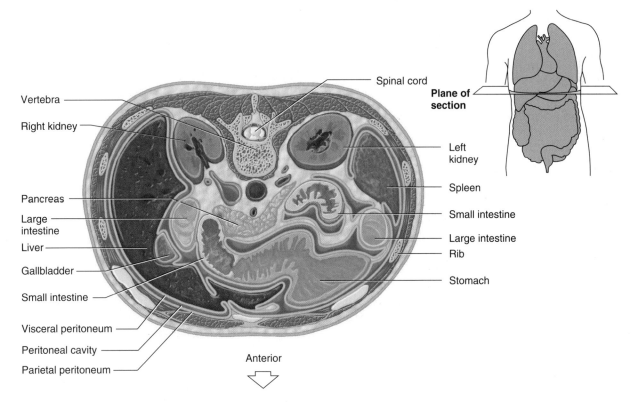

Spinal cord

Plane of section

Vertebra

Right kidney

Left kidney

Spleen

Pancreas

Small intestine

Large intestine

Large intestine

Liver

Rib

Gallbladder

Stomach

Small intestine

Visceral peritoneum

Peritoneal cavity

Anterior

Parietal peritoneum

Figure 1.11 AP|R

Transverse section through the abdomen (superior view).

Organ Systems

The human organism consists of several organ systems. Each system includes a set of interrelated organs that work together, allowing each system to provide specialized functions that contribute to homeostasis (fig. 1.12). As you read about each system, you may want to consult the illustrations of the human torso in the Reference Plates (see pp. 23–29) and locate some of the organs described.

Body Covering

Organs of the **integumentary** (in-teg-u-men'tar-e) **system** (see chapter 6) include the skin and various accessory organs, such as the hair, nails, sweat glands, and sebaceous glands. These parts protect underlying tissues, help regulate body temperature, house a variety of sensory receptors, and synthesize certain products.

Support and Movement

The organs of the skeletal and muscular systems (see chapters 7 and 8) support and move body parts. The **skeletal** (skel'ĕ-tal) **system** consists of bones, as well as ligaments and cartilages that bind bones together. These parts provide frameworks and protective shields for softer tissues, are attachments for muscles, and act with muscles when body parts move. Tissues within bones also produce blood cells and store inorganic salts.

Muscles are the organs of the **muscular** (mus'ku-lar) **system.** By contracting and pulling their ends closer together, muscles provide forces that move body parts. They also maintain posture and are the main source of body heat.

Integration and Coordination

For the body to act as a unit, its parts must be integrated and coordinated. The nervous and endocrine systems control and adjust various organ functions, thus helping to maintain homeostasis.

The **nervous** (ner'vus) **system** (see chapter 9) consists of the brain, the spinal cord, nerves, and sense organs (see chapter 10). The cells of the nervous system communicate with each other and with muscles and glands using chemical signals called *neurotransmitters.* Each neurotransmitter exerts a relatively short-term effect on its target. Some nerve cells are specialized sensory receptors that detect changes inside and outside the body. Other nerve cells receive information from these sensory receptors and interpret and respond to that information. Still other nerve cells extend from the brain or spinal cord to muscles or glands, stimulating them to contract or to secrete products.

The **endocrine** (en'do-krin) **system** (see chapter 11) includes all the glands that secrete chemical messengers called *hormones.* The hormones, in turn, move away from the glands in body fluids, such as blood or tissue fluid (fluid from the spaces within tissues). A particular hormone affects only a particular group of cells, called its *target cells.* A hormone alters the metabolism of its target cells. Compared to nerve impulses, hormonal effects occur over a relatively longer time period. Organs of the endocrine system include the hypothalamus of the brain; the pituitary, thyroid, parathyroid, and adrenal glands; and the pancreas, ovaries, testes, pineal gland, and thymus.

Transport

Two organ systems transport substances throughout the internal environment. The **cardiovascular** (kahr''de-o-vas'ku-lur) **system** (see chapters 12 and 13) includes the heart, arteries, veins, capillaries, and blood. The heart is a muscular pump that helps force blood through the blood vessels. Blood transports gases, nutrients, hormones, and wastes. It carries oxygen from the lungs and nutrients from the digestive organs to all body cells, where these biochemicals are used in metabolic processes. Blood also transports hormones and carries wastes from body cells to the excretory organs, where the wastes are removed from the blood and released to the outside.

The **lymphatic** (lim-fat'ik) **system** (see chapter 14) is closely related to the cardiovascular system. It is composed of the lymphatic vessels, lymph nodes, thymus, spleen, and a fluid called *lymph.* This system transports some of the tissue fluid back to the bloodstream and carries certain fatty substances away from the digestive organs and into the bloodstream. Cells of the lymphatic system are called lymphocytes, and they defend the body against infections by removing disease-causing microorganisms and viruses from tissue fluid.

Absorption and Excretion

Organs in several systems absorb nutrients and oxygen and excrete various wastes. For example, the organs of the **digestive** (di-jest'iv) **system** (see chapter 15) receive foods from the outside. Then they break down food molecules into simpler forms that can pass through cell membranes and be absorbed. Materials that are not absorbed are transported back to the outside and eliminated. Certain digestive organs also produce hormones and thus function as parts of the endocrine system. The digestive system includes the mouth, tongue, teeth, salivary glands, pharynx, esophagus, stomach, liver, gallbladder, pancreas, small intestine, and large intestine. Chapter 15 also discusses nutrition.

The organs of the **respiratory** (re-spi'rah-to''re) **system** (see chapter 16) move air in and out and exchange gases between the blood and the air. More specifically, oxygen passes from the air within the lungs into the blood, and carbon dioxide leaves the blood and enters the air. The nasal cavity, pharynx, larynx, trachea, bronchi, and lungs are parts of this system.

Figure 1.12 AP|R

The organ systems in humans interact, maintaining homeostasis.

The **urinary** (u'rǐ-ner''e) **system** (see chapter 17) consists of the kidneys, ureters, urinary bladder, and urethra. The kidneys remove wastes from blood and help maintain the body's water and salt (electrolyte) concentrations. The product of these activities is urine. Other portions of the urinary system store urine and transport it outside the body. Chapter 18 discusses the urinary system's role in maintaining water and electrolyte concentrations and the acidity of the internal environment.

Reproduction

Reproduction is the process of producing offspring (progeny). Cells reproduce when they divide and give rise to new cells. However, the **reproductive** (re''produk'tiv) **system** of an organism produces whole new organisms like itself (see chapter 19).

The male reproductive system includes the scrotum, testes, epididymides, ductus deferentia, seminal vesicles, prostate gland, bulbourethral glands, penis, and urethra. These parts produce and maintain sperm cells (spermatozoa). Components of the male reproductive system also transfer sperm cells into the female reproductive tract.

The female reproductive system consists of the ovaries, uterine tubes, uterus, vagina, clitoris, and vulva. These organs produce and maintain the female sex cells (egg cells, or oocytes), transport the female sex cells within the female reproductive system, and can receive the male sex cells (sperm cells) for the possibility of fertilizing an egg. The female reproductive system also supports development of embryos, carries fetuses to term, and functions in the birth process.

> ### Practice
> 17. Name and list the organs of the major organ systems.
> 18. Describe the general functions of each organ system.

1.7 ANATOMICAL TERMINOLOGY

To communicate effectively with one another, researchers and clinicians have developed a set of precise terms to describe anatomy. These terms concern the relative positions of body parts, relate to imaginary planes along which cuts may be made, and describe body regions.

Use of such terms assumes that the body is in the **anatomical position.** This means that the body is standing erect, face forward, with the upper limbs at the sides and the palms forward. Note that the terms "right" and "left" refer to the right and left of the body in anatomical position.

Relative Positions

Terms of relative position describe the location of one body part with respect to another. They include the following (many of these terms are illustrated in fig. 1.13):

1. **Superior** means that a body part is above another part. (The thoracic cavity is superior to the abdominopelvic cavity.)
2. **Inferior** means that a body part is below another body part. (The neck is inferior to the head.)
3. **Anterior** (or *ventral*) means toward the front. (The eyes are anterior to the brain.)
4. **Posterior** (or *dorsal*) means toward the back. (The pharynx is posterior to the oral cavity.)
5. **Medial** refers to an imaginary midline dividing the body into equal right and left halves. A body part is medial if it is closer to the midline than another part. (The nose is medial to the eyes.)
6. **Lateral** means toward the side, away from the imaginary midline. (The ears are lateral to the eyes.)

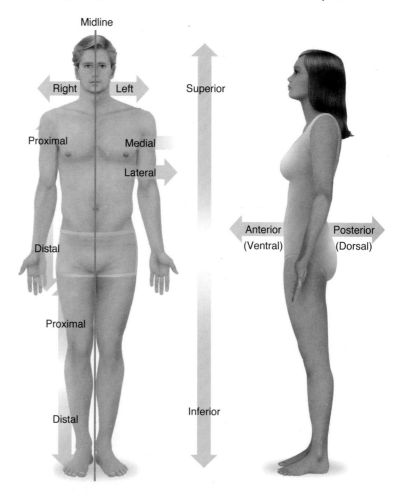

Figure 1.13 AP|R

Relative positional terms describe a body part's location with respect to other body parts.

Q: *Which is more lateral, the hand or the hip?*

Answer can be found in Appendix E on page 568.

7. **Bilateral** refers to paired structures, one of which is on each side. (The lungs are bilateral.)
8. **Ipsilateral** refers to structures on the same side. (The right lung and the right kidney are ipsilateral.)
9. **Contralateral** refers to structures on the opposite side. (A patient with a fractured bone in the right leg would have to bear weight on the contralateral—in this case, left—lower limb.)
10. **Proximal** describes a body part that is closer to a point of attachment to the trunk than another body part. (The elbow is proximal to the wrist.) *Proximal* may also refer to another reference point, such as the proximal tubules, which are closer to the filtering structures in the kidney.
11. **Distal** is the opposite of proximal. It means that a particular body part is farther from a point of attachment to the trunk than another body part is. (The fingers are distal to the wrist.) Distal may also refer to another reference point, such as decreased blood flow distal to occlusion of a coronary artery.
12. **Superficial** means situated near the surface. (The epidermis is the superficial layer of the skin.) *Peripheral* also means outward or near the surface. It describes the location of certain blood vessels and nerves. (The nerves that branch from the brain and spinal cord are peripheral nerves.)
13. **Deep** describes parts that are more internal than superficial parts. (The dermis is the deep layer of the skin.)

Body Sections

Observing the relative locations and organization of internal body parts requires cutting or sectioning the body along various planes (fig. 1.14). The following terms describe such planes and the sections that result:

1. **Sagittal** refers to a lengthwise plane that divides the body into right and left portions. If a sagittal plane passes along the midline and thus divides the body into equal parts, it is called *median* (midsagittal). A sagittal section lateral to midline is called *parasagittal.*
2. **Transverse** (or *horizontal*) refers to a plane that divides the body into superior and inferior portions.
3. **Frontal** (or *coronal*) refers to a plane that divides the body into anterior and posterior portions.

Sometimes, a cylindrical organ such as a long bone is sectioned. In this case, a cut across the structure is called a *cross section,* an angular cut is an *oblique section,* and a lengthwise cut is a *longitudinal section* (fig. 1.15). Clinical Application 1.1 discusses using computerized tomography to view body sections.

Figure 1.14 AP|R

Observation of internal parts requires sectioning the body along various planes.

A section along the median plane

Median (midsagittal) plane

Parasagittal plane

Transverse (horizontal) plane

A section along a transverse plane

A section along a frontal plane

Frontal (coronal) plane

Computerized Tomography

Radiologists use a procedure called *computerized tomography,* or CT scanning, to visualize internal organ sections (fig. 1A). In this procedure, an X-ray-emitting device moves around the body region being examined. At the same time, an X-ray detector moves in the opposite direction on the other side. As the devices move, an X-ray beam passes through the body from hundreds of different angles. Since tissues and organs of varying composition within the body absorb X rays differently, the amount of X ray reaching the detector varies from position to position. A computer records the measurements from the X-ray detector, and combines them mathematically to create a sectional image of the internal body parts that can be viewed on a monitor.

(a)

(b)

Figure 1A

Falsely colored CT (computerized tomography) scans of **(a)** the head and **(b)** the abdomen. *Note:* These are not shown in correct relative size.

(a) (b) (c)

Figure 1.15

Cylindrical parts may be cut in **(a)** cross section, **(b)** oblique section, or **(c)** longitudinal section.

Body Regions

A number of terms designate body regions. The abdominal area, for example, is subdivided into the following nine regions, as figure 1.16*a* shows:

1. The **epigastric region** is the upper middle portion.
2. The **right** and **left hypochondriac regions** lie on each side of the epigastric region.
3. The **umbilical region** is the middle portion.
4. The **right** and **left lumbar regions** lie on each side of the umbilical region.
5. The **hypogastric region** is the lower middle portion.
6. The **right** and **left iliac regions** (right and left inguinal regions) lie on each side of the hypogastric region.

The abdominal area is also often subdivided into four quadrants, as figure 1.16*b* shows.

The following adjectives are commonly used to refer to various body regions, some of which are illustrated in figure 1.17:

abdominal (ab-dom'ĭ-nal) The region between the thorax and pelvis.
acromial (ah-kro'me-al) The point of the shoulder.
antebrachial (an"te-bra'ke-al) The forearm.
antecubital (an"te-ku'bĭ-tal) The space in front of the elbow.
axillary (ak'sĭ-ler"e) The armpit.
brachial (bra'ke-al) The arm.
buccal (buk'al) The cheek.
carpal (kar'pal) The wrist.
celiac (se'le-ak) The abdomen.

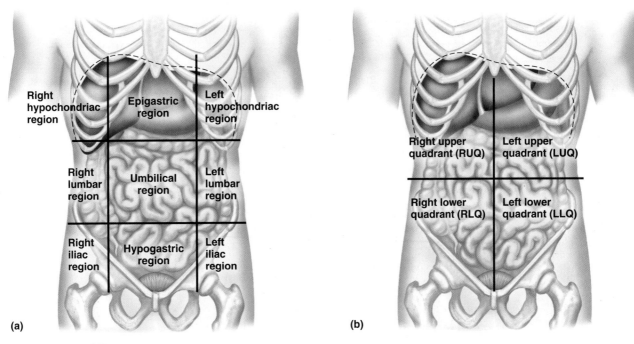

(a)   **(b)**

Figure 1.16

The abdominal area is commonly subdivided in two ways: **(a)** into nine regions and **(b)** into four quadrants.

cephalic (sĕ-fal'ik) The head.
cervical (ser'vĭ-kal) The neck.
costal (kos'tal) The ribs.
coxal (kok'sal) The hip.
crural (krōōr'al) The leg.
cubital (ku'bĭ-tal) The elbow.
digital (dij'ĭ-tal) The finger or toe.
dorsal (dor'sal) The back.
femoral (fem'or-al) The thigh.
frontal (frun'tal) The forehead.
genital (jen'ĭ-tal) The reproductive organs.
gluteal (gloo'te-al) The buttocks.
inguinal (ing'gwĭ-nal) The groin—the depressed area
   of the abdominal wall near the thigh.
lumbar (lum'bar) The loin—the region of the lower
   back between the ribs and the pelvis.
mammary (mam'er-e) The breast.
mental (men'tal) The chin.
nasal (na'zal) The nose.
occipital (ok-sip'ĭ-tal) The lower posterior region of
   the head.
oral (o'ral) The mouth.
orbital (or'bi-tal) The bony socket of the eye.
palmar (pahl'mar) The palm of the hand.
patellar (pah-tel'ar) The front of the knee.
pectoral (pek'tor-al) The chest.
pedal (ped'al) The foot.
pelvic (pel'vik) The pelvis.
perineal (per''ĭ-ne'al) The perineum—the region between
   the anus and the external reproductive organs.
plantar (plan'tar) The sole of the foot.
popliteal (pop''lĭ-te'al) The area behind the knee.

sacral (sa'kral) The posterior region between the hip
   bones.
sternal (ster'nal) The middle of the thorax, anteriorly.
sural (su'ral) The calf of the leg.
tarsal (tahr'sal) The instep of the foot.
umbilical (um-bil'ĭ-kal) The navel.
vertebral (ver'te-bral) The spinal column.

Practice

19. Describe the anatomical position.
20. Using the appropriate terms, describe the relative
    positions of several body parts.
21. Describe the three types of body sections.
22. Name the nine regions of the abdomen.

Some Medical and Applied Sciences

cardiology (kar''de-ol'o-je) Branch of medical science
   dealing with the heart and heart diseases.
cytology (si-tol'o-je) Study of the structure, function,
   and abnormalities of cells. Cytology and histology
   are subdivisions of microscopic anatomy.
dermatology (der''mah-tol'o-je) Study of the skin and its
   diseases.
endocrinology (en''do-krĭ-nol'o-je) Study of
   hormones, hormone-secreting glands, and their
   diseases.
epidemiology (ep''ĭ-de''me-ol'o-je) Study of the
   factors determining the distribution and frequency

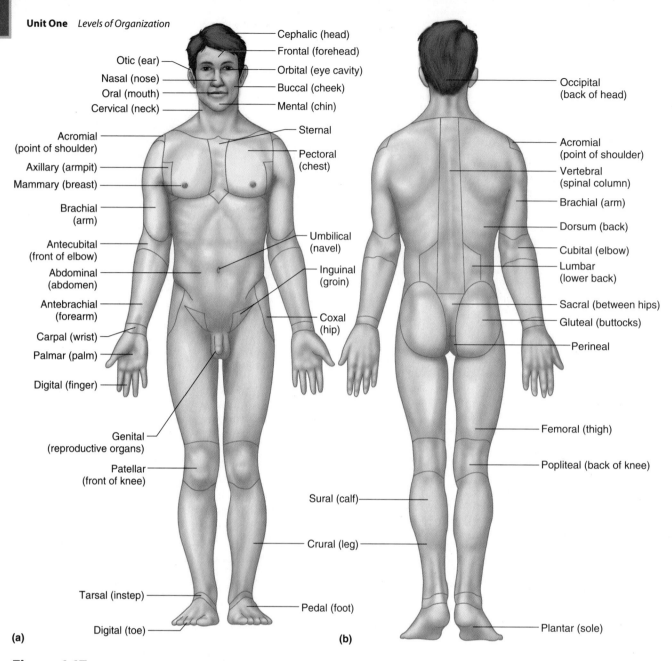

Figure 1.17

Some terms used to describe body regions. **(a)** Anterior regions. **(b)** Posterior regions.

of health-related conditions in a defined human population.

gastroenterology (gas″tro-en″ter-ol′o-je) Study of the stomach and intestines and their diseases.

geriatrics (jer″e-at′riks) Branch of medicine dealing with older individuals and their medical problems.

gerontology (jer″on-tol′o-je) Study of the aging process.

gynecology (gi″ně-kol′o-je) Study of the female reproductive system and its diseases.

hematology (hēm″ah-tol′o-je) Study of the blood and blood diseases.

histology (his-tol′o-je) Study of the structure and function of tissues. Histology and cytology are subdivisions of microscopic anatomy.

immunology (im″u-nol′o-je) Study of the body's resistance to infectious disease.

neonatology (ne″o-na-tol′o-je) Study of newborns and the treatment of their disorders.

nephrology (ně-frol′o-je) Study of the structure, function, and diseases of the kidneys.

neurology (nu-rol′o-je) Study of the nervous system and its disorders.

obstetrics (ob-stet′riks) Branch of medicine dealing with pregnancy and childbirth.

oncology (ong-kol′o-je) Study of cancers.

ophthalmology (of″thal-mol′o-je) Study of the eye and eye diseases.

orthopedics (or″tho-pe′diks) Branch of medicine dealing with the muscular and skeletal systems and their problems.

otolaryngology (o″to-lar″in-gol′o-je) Study of the ear, throat, and larynx, and their diseases.

pathology (pah-thol′o-je) Study of structural and functional changes that disease causes.

pediatrics (pe″de-at′riks) Branch of medicine dealing with children and their diseases.

pharmacology (fahr″mah-kol′o-je) Study of drugs and their uses in the treatment of disease.

podiatry (po-di′ah-tre) Study of the care and treatment of feet.

psychiatry (si-ki′ah-tre) Branch of medicine dealing with the mind and its disorders.

radiology (ra″de-ol′o-je) Study of X rays and radioactive substances and their uses in the diagnosis and treatment of diseases.

toxicology (tok″sĭ-kol′o-je) Study of poisonous substances and their effects upon body parts.

urology (u-rol′o-je) Branch of medicine dealing with the urinary system, apart from the kidneys (nephrology) and the male reproductive system, and their diseases.

Summary Outline

1.1 Introduction (p. 2)

1. Early interest in the human body probably developed as people became concerned about injuries and illnesses.
2. Primitive doctors began to learn how certain herbs and potions affected body functions.
3. The belief that humans could understand forces that caused natural events led to the development of modern science.
4. A set of terms originating from Greek and Latin words is the basis for the language of anatomy and physiology.

1.2 Anatomy and Physiology (p. 3)

1. Anatomy describes the form and organization of body parts.
2. Physiology considers the functions of anatomical parts.
3. The function of a body part depends on the way it is constructed.

1.3 Levels of Organization (p. 3)

The body is composed of parts with different levels of complexity.
1. Matter is composed of atoms.
2. Atoms join to form molecules.
3. Organelles are built of groups of large molecules (macromolecules).
4. Cells, which contain organelles, are the basic units of structure and function that form the body.
5. Cells are organized into tissues.
6. Tissues are organized into organs.
7. Organs that function closely together compose organ systems.
8. Organ systems constitute the organism.
9. Beginning at the atomic level, these levels of organization differ in complexity from one level to the next.

1.4 Characteristics of Life (p. 4)

Characteristics of life are traits all organisms share.
1. These characteristics include:
   a. Movement—changing body position or moving internal parts.
   b. Responsiveness—sensing and reacting to internal or external changes.
   c. Growth—increasing size without changing shape.
   d. Reproduction—producing offspring.
   e. Respiration—obtaining oxygen, using oxygen to release energy from foods, and removing gaseous wastes.
   f. Digestion—breaking down food substances into component nutrients that the intestine can absorb.
   g. Absorption—moving substances through membranes and into body fluids.
   h. Circulation—moving substances through the body in body fluids.
   i. Assimilation—changing substances into chemically different forms.
   j. Excretion—removing body wastes.
2. Acquisition and use of energy constitute metabolism.

1.5 Maintenance of Life (p. 5)

The structures and functions of body parts maintain the life of the organism.
1. Requirements of organisms
   a. Water is used in many metabolic processes, provides the environment for metabolic reactions, and transports substances.
   b. Food supplies energy, raw materials for building new living matter, and chemicals necessary in vital reactions.
   c. Oxygen releases energy from food materials. This energy drives metabolic reactions.
   d. Heat is a product of metabolic reactions and helps govern the rates of these reactions.
   e. Pressure is an application of force to something. In humans, atmospheric and hydrostatic pressures help breathing and blood movements, respectively.
2. Homeostasis
   a. If an organism is to survive, the conditions within its body fluids must remain relatively stable.
   b. Maintenance of a stable internal environment is called *homeostasis*.
   c. Homeostatic mechanisms help regulate body temperature and blood pressure.
   d. Homeostatic mechanisms act through negative feedback.

1.6 Organization of the Human Body (p. 8)

1. Body cavities
   a. The axial portion of the body includes the cranial cavity, the vertebral canal, the thoracic cavity, and the abdominopelvic cavity.
   b. The diaphragm separates the thoracic and abdominopelvic cavities.
   c. The organs in a body cavity are called *viscera*.
   d. The mediastinum separates the thoracic cavity into right and left compartments.
   e. Body cavities in the head include the oral, nasal, orbital, and middle ear cavities.
2. Thoracic and abdominopelvic membranes
   a. Thoracic membranes
      (1) Pleural membranes line the thoracic cavity (parietal pleura) and cover each lung (visceral pleura).

(2) Pericardial membranes surround the heart (parietal pericardium) and cover its surface (visceral pericardium).

(3) The pleural and pericardial cavities are the potential spaces between the respective parietal and visceral membranes.

b. Abdominopelvic membranes

(1) Peritoneal membranes line the abdominopelvic cavity (parietal peritoneum) and cover the organs inside (visceral peritoneum).

(2) The peritoneal cavity is the potential space between the parietal and visceral membranes.

3. Organ systems

The human organism consists of several organ systems. Each system includes a set of interrelated organs.

a. Body covering

(1) The integumentary system includes the skin, hair, nails, sweat glands, and sebaceous glands.

(2) It protects underlying tissues, regulates body temperature, houses sensory receptors, and synthesizes various substances.

b. Support and movement

(1) Skeletal system

(a) The skeletal system is composed of bones, as well as cartilages and ligaments that bind bones together.

(b) It provides a framework, protective shields, and attachments for muscles. It also produces blood cells and stores inorganic salts.

(2) Muscular system

(a) The muscular system includes the muscles of the body.

(b) It moves body parts, maintains posture, and produces body heat.

c. Integration and coordination

(1) Nervous system

(a) The nervous system consists of the brain, spinal cord, nerves, and sense organs.

(b) It receives impulses from sensory parts, interprets these impulses, and acts on them by stimulating muscles or glands to respond.

(2) Endocrine system

(a) The endocrine system consists of glands that secrete hormones.

(b) Hormones help regulate metabolism.

(c) This system includes the hypothalamus of the brain and the pituitary, thyroid, parathyroid, and adrenal glands, as well as the pancreas, ovaries, testes, pineal gland, and thymus.

d. Transport

(1) Cardiovascular system

(a) The cardiovascular system includes the heart, which pumps blood, and the blood vessels, which carry blood to and from body parts.

(b) Blood transports oxygen, nutrients, hormones, and wastes.

(2) Lymphatic system

(a) The lymphatic system is composed of lymphatic vessels, lymph fluid, lymph nodes, thymus, and spleen.

(b) It transports lymph fluid from tissues to the bloodstream, carries certain fatty substances away from the digestive organs, and aids in defending the body against disease-causing agents.

e. Absorption and excretion

(1) Digestive system

(a) The digestive system receives foods, breaks down food molecules into nutrients that can pass through cell membranes, and eliminates materials that are not absorbed.

(b) It includes the mouth, tongue, teeth, salivary glands, pharynx, esophagus, stomach, liver, gallbladder, pancreas, small intestine, and large intestine.

(c) Some digestive organs produce hormones.

(2) Respiratory system

(a) The respiratory system takes in and sends out air and exchanges gases between the air and blood.

(b) It includes the nasal cavity, pharynx, larynx, trachea, bronchi, and lungs.

(3) Urinary system

(a) The urinary system includes the kidneys, ureters, urinary bladder, and urethra.

(b) It filters wastes from the blood and helps maintain water and electrolyte concentrations and the acidity of the internal environment.

f. Reproduction

(1) The reproductive systems produce new organisms.

(2) The male reproductive system includes the scrotum, testes, epididymides, ductus deferentia, seminal vesicles, prostate gland, bulbourethral glands, urethra, and penis, which produce, maintain, and transport male sex cells (sperm cells).

(3) The female reproductive system includes the ovaries, uterine tubes, uterus, vagina, clitoris, and vulva, which produce, maintain, and transport female sex cells (oocytes).

1.7 Anatomical Terminology (p. 14)

Terms with precise meanings help investigators communicate effectively.

1. Relative positions

These terms describe the location of one part with respect to another part.

2. Body sections

Body sections are planes along which the body may be cut to observe the relative locations and organization of internal parts.

3. Body regions

Special terms designate various body regions.

Chapter Assessments

1.1 Introduction

1. Briefly describe the early discoveries that led to our understanding of the human body. (p. 2)

1.2 Anatomy and Physiology

2. Explain the difference between anatomy and physiology. (p. 3)

3. Identify relationships between the form and the function of body parts. (p. 3)

1.3 Levels of Organization

4. List the levels of organization within the human body and describe the characteristics of each. (p. 3)

1.4 Characteristics of Life

5. List and describe ten characteristics of life. (p. 4)

6. Define *metabolism* and give examples. (p. 4)

1.5 Maintenance of Life

7. List and describe five requirements of organisms. (p. 5)

8. Define *homeostasis,* and explain its importance. (p. 5)

9. Identify the parts of a general physiological control system and explain how they work together. (p. 6)

10. Explain the control of body temperature. (p. 7)

11. Describe a homeostatic mechanism that helps regulate blood pressure. (p. 7)

1.6 Organization of the Human Body

12. Explain the difference between the axial and appendicular portions of the body. (p. 8)

13. Identify the cavities within the axial portion of the body. (p. 8)

14. Define *viscera*. (p. 8)

15. Describe the mediastinum and its contents. (p. 8)

16. List the cavities of the head and the contents of each cavity. (p. 8)

17. Name the body cavity that houses each of the following organs: (p. 8)

    a. Stomach   e. Trachea   h. Esophagus
    b. Heart   f. Rectum   i. Spleen
    c. Brain   g. Spinal cord   j. Urinary bladder
    d. Liver

18. Distinguish between a parietal and a visceral membrane. (p. 10)

19. Name the major organ systems, and describe the general functions of each. (p. 12)

20. List the major organs that compose each organ system. (p. 12)

1.7 Anatomical Terminology

21. Write complete sentences using each of the following terms to correctly describe the relative locations of specific body parts: (p. 14)

    a. Superior   e. Medial   i. Superficial
    b. Inferior   f. Lateral   j. Peripheral
    c. Anterior   g. Proximal   k. Deep
    d. Posterior   h. Distal

22. Sketch the outline of a human body, and use lines to indicate an example of each of the following sections: (p. 15)

    a. Sagittal   b. Transverse   c. Frontal

23. Sketch the abdominal area, and indicate the locations of the following regions: (p. 16)

    a. Epigastric   c. Hypogastric   e. Lumbar
    b. Umbilical   d. Hypochondriac f. Iliac

24. Provide the common name for the region to which each of the following terms refers: (p. 16)

    a. Acromial   g. Crural   m. Occipital   s. Plantar
    b. Antebrachial h. Femoral   n. Orbital   t. Popliteal
    c. Axillary   i. Genital   o. Otic   u. Sacral
    d. Buccal   j. Gluteal   p. Palmar   v. Tarsal
    e. Celiac   k. Inguinal   q. Pectoral   w. Umbilical
    f. Coxal   l. Mental   r. Pedal   x. Vertebral

Integrative Assessments/Critical Thinking

OUTCOMES 1.2, 1.3, 1.4, 1.5

1. Which characteristics of life does a computer have? Why is a computer not alive?

OUTCOMES 1.2, 1.3, 1.4, 1.5, 1.6

2. Put the following in order, from smallest and simplest to largest and most complex, and describe their individual roles in homeostasis: organ, molecule, organelle, atom, organ system, tissue, organism, cell, macromolecule.

OUTCOMES 1.4, 1.5

3. What environmental characteristics would be necessary for a human to survive on another planet?

OUTCOMES 1.5, 1.6

4. In health, body parts interact to maintain homeostasis. Illness can threaten the maintenance of homeostasis, requiring treatment. What treatments might be used to help control a patient's (*a*) body temperature, (*b*) blood oxygen concentration, and (*c*) water content?

OUTCOMES 1.5, 1.6, 1.7

5. How might health-care professionals provide the basic requirements of life to an unconscious patient? Describe the body parts involved in the treatment, using correct directional and regional terms.

OUTCOME 1.6

6. Suppose two individuals develop benign (noncancerous) tumors that produce symptoms because they occupy space and crowd adjacent organs. If one of these persons has the tumor in the thoracic cavity and the other has the tumor in the abdominopelvic cavity, which person would be likely to develop symptoms first? Why? Which might be more immediately serious? Why?

OUTCOMES 1.6, 1.7

7. If a patient complained of a "stomachache" and pointed to the umbilical region as the site of discomfort, which organs located in this region might be the source of the pain?

WEB CONNECTIONS

Visit the text website at **www.mhhe.com/shieress11** for additional quizzes, interactive learning exercises, and more.

APR

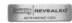

Anatomy & Physiology REVEALED includes cadaver photos that allow you to peel away layers of the human body to reveal structures beneath the surface. This program also includes animations, radiologic imaging, audio pronunciations, and practice quizzing. To learn more visit **www.aprevealed.com**.

The Human Organism

The series of illustrations that follows shows the major parts of the human torso. The first plate illustrates the anterior surface and reveals the superficial muscles on one side. Each subsequent plate exposes deeper organs, including those in the thoracic, abdominal, and pelvic cavities.

Chapters 6–19 of this textbook describe the organs and organ systems of the human organism in detail. As you read them, refer to these plates to visualize the locations of various organs and the three-dimensional relationships among them.

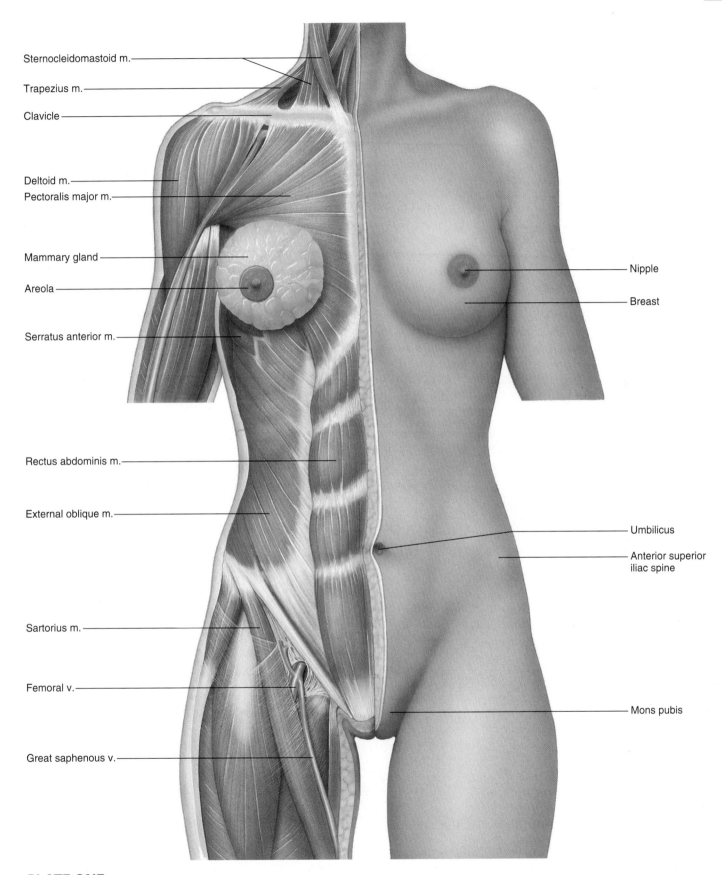

Sternocleidomastoid m.

Trapezius m.

Clavicle

Deltoid m.

Pectoralis major m.

Mammary gland

Areola

Serratus anterior m.

Rectus abdominis m.

External oblique m.

Sartorius m.

Femoral v.

Great saphenous v.

Nipple

Breast

Umbilicus

Anterior superior iliac spine

Mons pubis

PLATE ONE

Human female torso showing the anterior surface on one side and the superficial muscles exposed on the other side. (*m.* stands for *muscle,* and *v.* stands for *vein.*)

Larynx

Sternocleidomastoid m.

Clavicle

Deltoid m.

Pectoralis major m.

Latissimus dorsi m.

Rectus abdominis m.

External oblique m.

Internal oblique m.

Transversus abdominis m.

Femoral n.

Femoral a.

Femoral v.

Common carotid a.

Internal jugular v.

Thyroid gland

External intercostal m.

Coracobrachialis m.

Pectoralis minor m.

Long head biceps brachii m.

Short head biceps brachii m.

Serratus anterior m.

Transversus abdominis m.

Linea alba

Rectus abdominis m. (cut)

Tensor fasciae latae m.

Sartorius m.

Rectus femoris m.

Great saphenous v.

PLATE TWO

Human male torso with the deeper muscle layers exposed.

(*n.* stands for *nerve, a.* stands for *artery, m.* stands for *muscle,* and *v.* stands for *vein.*)

Common carotid a.

Internal jugular v.

External jugular v.

Subscapularis m.

Teres major m.

Latissimus dorsi m.

External intercostal mm.

Internal intercostal mm.

Liver

Falciform ligament

Gallbladder

Greater omentum

Urinary bladder

Inguinal canal

Spermatic cord

Penis

Sartorius m.

Thyroid cartilage

Thyroid gland

Trachea

Subclavian v.

Coracobrachialis m.

Sternum

Left lung

Pericardial sac

Diaphragm

Stomach

External oblique m.

Internal oblique m.

Transversus abdominis m.

Anterior superior iliac spine

Small intestine

Femoral n.

Femoral a.

Femoral v.

PLATE THREE

Human male torso with the deep muscles removed and the abdominal viscera exposed.

(*n.* stands for *nerve, a.* stands for *artery, m.* stands for *muscle, mm.* stands for *muscles,* and *v.* stands for *vein.*)

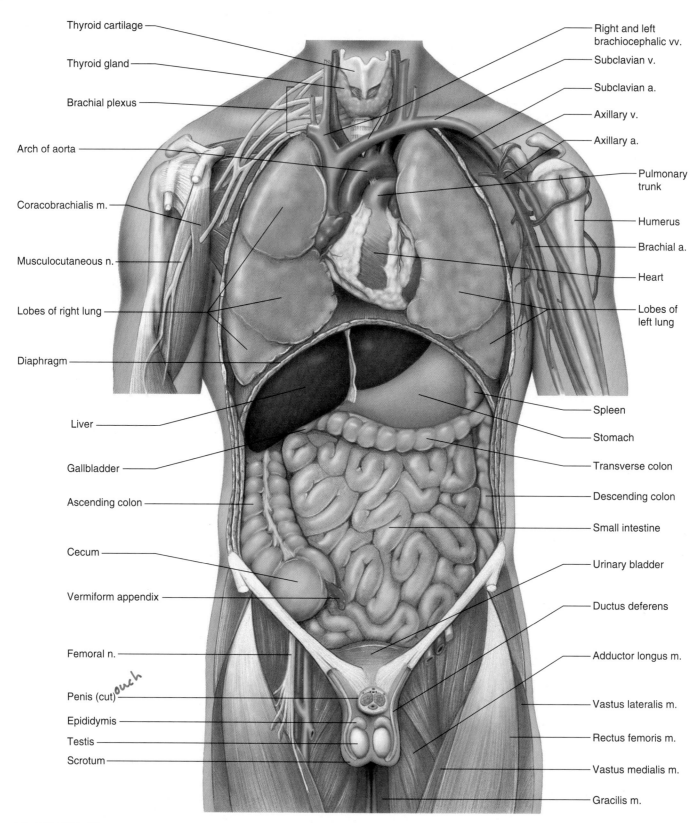

Thyroid cartilage

Thyroid gland

Brachial plexus

Arch of aorta

Coracobrachialis m.

Musculocutaneous n.

Lobes of right lung

Diaphragm

Liver

Gallbladder

Ascending colon

Cecum

Vermiform appendix

Femoral n.

Penis (cut) *ouch*

Epididymis

Testis

Scrotum

Right and left
brachiocephalic vv.

Subclavian v.

Subclavian a.

Axillary v.

Axillary a.

Pulmonary
trunk

Humerus

Brachial a.

Heart

Lobes of
left lung

Spleen

Stomach

Transverse colon

Descending colon

Small intestine

Urinary bladder

Ductus deferens

Adductor longus m.

Vastus lateralis m.

Rectus femoris m.

Vastus medialis m.

Gracilis m.

PLATE FOUR

Human male torso with the thoracic and abdominal viscera exposed.

(*n.* stands for *nerve, a.* stands for *artery, m.* stands for *muscle, v.* stands for *vein,* and *vv.* stands for *veins.*)

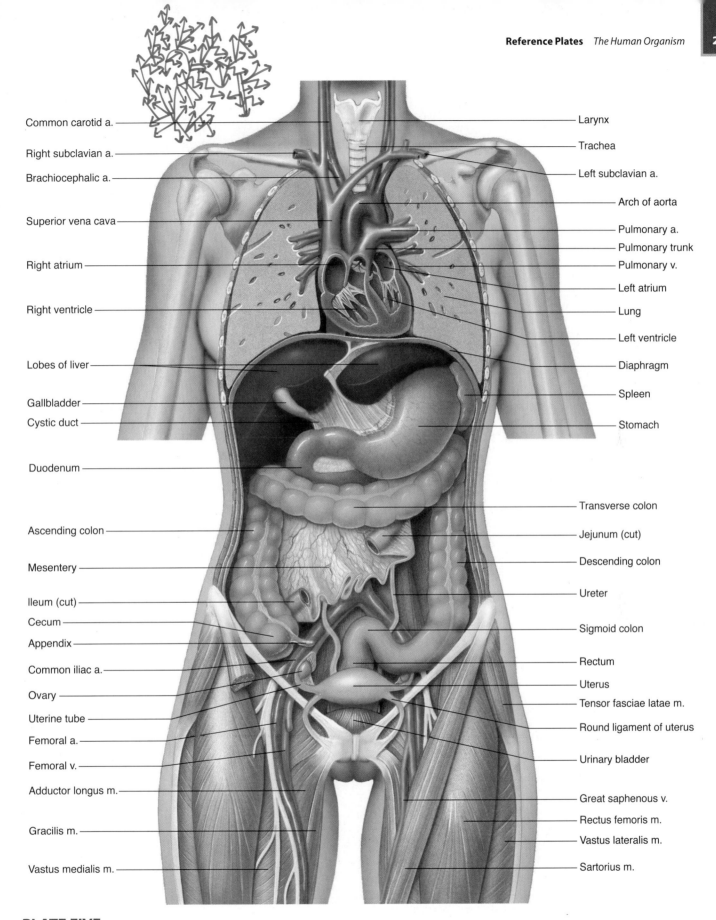

Common carotid a.

Right subclavian a.

Brachiocephalic a.

Superior vena cava

Right atrium

Right ventricle

Lobes of liver

Gallbladder

Cystic duct

Duodenum

Ascending colon

Mesentery

Ileum (cut)

Cecum

Appendix

Common iliac a.

Ovary

Uterine tube

Femoral a.

Femoral v.

Adductor longus m.

Gracilis m.

Vastus medialis m.

Larynx

Trachea

Left subclavian a.

Arch of aorta

Pulmonary a.

Pulmonary trunk

Pulmonary v.

Left atrium

Lung

Left ventricle

Diaphragm

Spleen

Stomach

Transverse colon

Jejunum (cut)

Descending colon

Ureter

Sigmoid colon

Rectum

Uterus

Tensor fasciae latae m.

Round ligament of uterus

Urinary bladder

Great saphenous v.

Rectus femoris m.

Vastus lateralis m.

Sartorius m.

PLATE FIVE

Human female torso with the lungs, heart, and small intestine sectioned and the liver reflected (lifted back).
(*a.* stands for *artery*, *m.* stands for *muscle*, and *v.* stands for *vein*.)

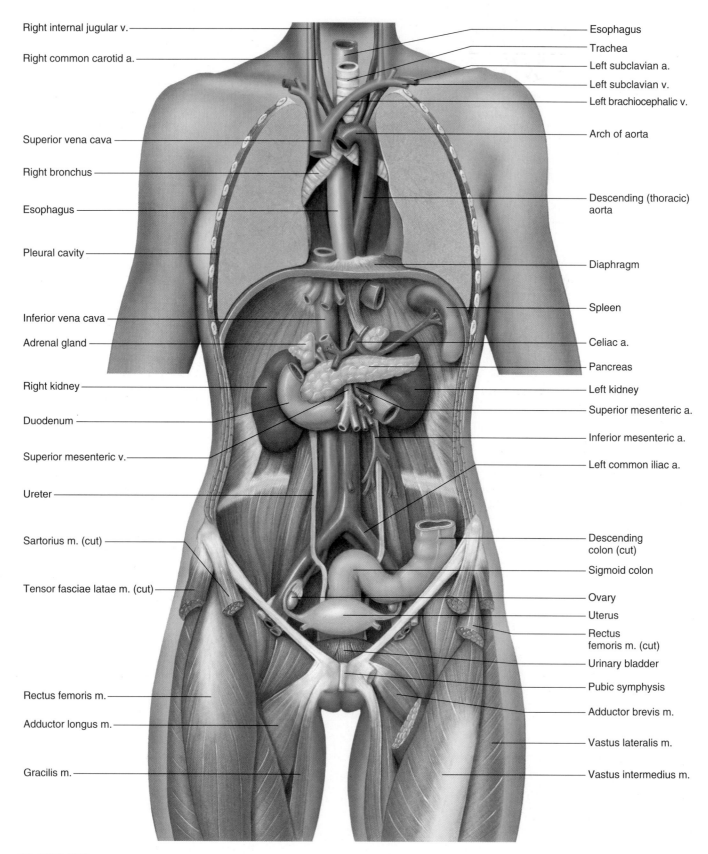

Right internal jugular v.

Right common carotid a.

Superior vena cava

Right bronchus

Esophagus

Pleural cavity

Inferior vena cava

Adrenal gland

Right kidney

Duodenum

Superior mesenteric v.

Ureter

Sartorius m. (cut)

Tensor fasciae latae m. (cut)

Rectus femoris m.

Adductor longus m.

Gracilis m.

Esophagus

Trachea

Left subclavian a.

Left subclavian v.

Left brachiocephalic v.

Arch of aorta

Descending (thoracic) aorta

Diaphragm

Spleen

Celiac a.

Pancreas

Left kidney

Superior mesenteric a.

Inferior mesenteric a.

Left common iliac a.

Descending colon (cut)

Sigmoid colon

Ovary

Uterus

Rectus femoris m. (cut)

Urinary bladder

Pubic symphysis

Adductor brevis m.

Vastus lateralis m.

Vastus intermedius m.

PLATE SIX

Human female torso with the heart, stomach, liver, and parts of the intestine and lungs removed.

(*a.* stands for *artery, m.* stands for *muscle,* and *v.* stands for *vein.*)

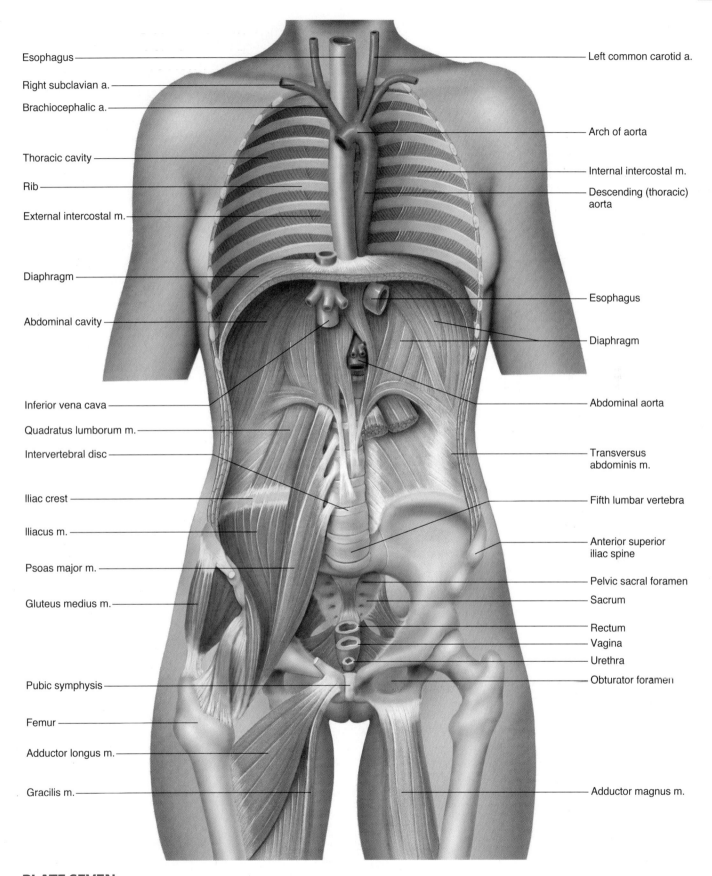

Esophagus

Right subclavian a.

Brachiocephalic a.

Thoracic cavity

Rib

External intercostal m.

Diaphragm

Abdominal cavity

Inferior vena cava

Quadratus lumborum m.

Intervertebral disc

Iliac crest

Iliacus m.

Psoas major m.

Gluteus medius m.

Pubic symphysis

Femur

Adductor longus m.

Gracilis m.

Left common carotid a.

Arch of aorta

Internal intercostal m.

Descending (thoracic) aorta

Esophagus

Diaphragm

Abdominal aorta

Transversus abdominis m.

Fifth lumbar vertebra

Anterior superior iliac spine

Pelvic sacral foramen

Sacrum

Rectum

Vagina

Urethra

Obturator foramen

Adductor magnus m.

PLATE SEVEN

Human female torso with the thoracic, abdominal, and pelvic viscera removed.

(*a.* stands for *artery,* and *m.* stands for *muscle.*)

2

Chemical Basis of Life

The chloracne rash associated with dioxin poisoning includes blackheads and pus-filled cysts.

A presidential poisoning. Chemicals such as nutrients help a human body, but chemicals that are toxins (poisons) harm it. The strange case of Victor Yushchenko, president of Ukraine, provided an unplanned experiment in how the body handles one particular toxin—a type of molecule called a dioxin.

A heated political campaign raged in Ukraine in the fall of 2004. Presidential candidate Yushchenko supported increased contact with Europeans, while his opponents supported an alliance with Russia. On September 5, Yushchenko fell acutely ill at a dinner in Kiev with senior Ukrainian officials. He was diagnosed with pancreatitis and edema (swelling), which was initially attributed to bad sushi and/or a viral infection.

Three weeks later, a new symptom revealed the nature of the poison—Yushchenko developed an extremely disfiguring rash, called chloracne. The rash results from the body's attempt to rid itself of the poison, sending it from the bloodstream to adipose tissue, and then to sebaceous (oil) glands in the skin, where it is eliminated from the body.

Dioxins came to public attention during the Vietnam conflict in the 1960s, when, as components of herbicides, they caused chlor-

acne in exposed individuals. Seventeen types of dioxins are known. Yushchenko's blood had the normal trace amounts of sixteen of the dioxins, but 50,000 times the normal background amount of the other. The purified nature of the dioxin and its abundance were sure signs that someone had poisoned his food.

Yushchenko agreed to become part of an experiment. He allowed researchers to periodically analyze his blood, skin, urine, sweat, feces, and adipose tissue for the telltale dioxin, to learn about dioxin poisoning. After three years, the researchers found that about 60% of the poison passed through Yushchenko's body unaltered, and most of it, plus two chemicals into which it was broken down, was eliminated in feces. Medical authorities concluded that the poisoning had been intended to kill Yushchenko, but he survived—and won the election.

Learning Outcomes

After studying this chapter, you should be able to do the following:

2.1 Introduction
1. Give examples of how the study of living materials requires an understanding of chemistry. (p. 31)

2.2 Structure of Matter
2. Describe the relationships among matter, atoms, and molecules. (p. 31)
3. Describe how atomic structure determines how atoms interact. (p. 32)

4. Explain how molecular and structural formulas symbolize the composition of compounds. (p. 37)
5. Describe three types of chemical reactions. (p. 37)
6. Explain what acids, bases, and buffers are. (p. 38)
7. Define pH and be able to use the pH scale. (p. 39)

2.3 Chemical Constituents of Cells
8. List the major inorganic chemicals common in cells and identify the functions of each. (p. 40)
9. Describe the general functions of the four main groups of organic chemicals in cells. (p. 41)

Module 2: Cells & Chemistry

Aids to Understanding Words   *(Appendix A on page 564 has a complete list of Aids to Understanding Words.)*

di- [two] *di*saccharide: Compound whose molecules are composed of two sugar units bound together.

glyc- [sweet] *glyc*ogen: Complex carbohydrate composed of many glucose molecules bound together in a particular way.

lip- [fat] *lip*ids: Group of organic compounds that includes fats.

-lyt [dissolvable] electro*lyt*e: Substance that releases ions in water.

mono- [one] *mono*saccharide: Compound whose molecule consists of a single sugar unit.

poly- [many] *poly*unsaturated: Molecule that has many double bonds between its carbon atoms.

sacchar- [sugar] mono*sacchar*ide: Molecule consisting of a single sugar unit.

syn- [together] *syn*thesis: Process by which substances are united to form a new type of substance.

2.1 INTRODUCTION

At the cellular level of organization, chemistry, in a sense, becomes biology. A cell's working parts—its organelles—are intricate assemblies of molecules. Because the molecules that build the cells that build tissues and organs are themselves composed of atoms, the study of anatomy and physiology begins with chemistry.

Chemistry is the branch of science that considers the composition of matter and how this composition changes. Understanding chemistry is essential for understanding anatomy and physiology because body structures and functions result from chemical changes within cells. Indeed, the human body is composed of chemicals, including salts, water, proteins, carbohydrates, lipids, and nucleic acids. All of the food that we eat, liquids that we drink, and medications that we may take when we are sick are chemicals.

2.2 STRUCTURE OF MATTER

Matter is anything that has weight and takes up space. This includes all the solids, liquids, and gases in our surroundings, as well as inside our bodies. Matter consists of chemicals.

> Strictly speaking, matter has *mass* and takes up space. **Mass** is related to the amount of a substance, whereas **weight** refers to how heavy it is. If your weight on earth is 150 pounds, on the moon it would be only 25 pounds, but your mass would be the same in both places. That is, your composition is the same on the moon as it is on earth, but you weigh less on the moon, because the force of gravity is lower there. Since we are dealing with life on earth, and a constant gravity, we can consider mass and weight as roughly equivalent. Many of our students find it easier to think in terms of weight rather than mass.

Elements and Atoms

All matter is composed of fundamental substances called **elements** (el'ĕ-mentz). Examples include such common materials as iron, copper, silver, gold, aluminum, carbon, hydrogen, and oxygen. Although some elements exist in a pure form, they are found more frequently in combination with other elements.

Living organisms require about twenty elements. Of these, oxygen, carbon, hydrogen, and nitrogen make up more than 95% (by weight) of the human body (table 2.1). As the table shows, a one- or two-letter symbol represents each element.

Elements are composed of tiny particles called **atoms** (at'omz), which are the smallest complete units

Table 2.1	Elements in the Human Body	
Major Elements	**Symbol**	**Approximate Percentage of the Human Body (by weight)**
Oxygen	O	65.0
Carbon	C	18.5
Hydrogen	H	9.5
Nitrogen	N	3.2
Calcium	Ca	1.5
Phosphorus	P	1.0
Potassium	K	0.4
Sulfur	S	0.3
Chlorine	Cl	0.2
Sodium	Na	0.2
Magnesium	Mg	0.1
		Total 99.9%
Trace Elements		
Chromium	Cr	Together less than 0.1%
Cobalt	Co	
Copper	Cu	
Fluorine	F	
Iodine	I	
Iron	Fe	
Manganese	Mn	
Zinc	Zn	

of elements. Atoms of an element are similar to each other, but they differ from the atoms that make up other elements. Atoms vary in size, weight, and the ways they interact with other atoms. Some atoms can combine with atoms like themselves or with other atoms by forming attractions called **chemical bonds,** whereas other atoms cannot form such bonds.

Atomic Structure

An atom consists of a central portion, called the **nucleus,** and one or more **electrons** (e-lek′tronz) that constantly move around it. The nucleus contains one or more relatively large particles called **protons** (pro′tonz). The nucleus also usually contains one or more **neutrons** (nu′tronz), which are similar in size to protons.

Electrons, which are extremely small, each carry a single, negative electrical charge (e^-), whereas protons each carry a single, positive electrical charge (p^+). Neutrons are uncharged and thus are electrically neutral (n^0) (fig. 2.1).

Because the nucleus contains the protons, it is always positively charged. However, the number of electrons outside the nucleus equals the number of protons. Therefore, a complete atom is electrically uncharged, or neutral.

The atoms of different elements have different numbers of protons. The number of protons in the atoms of a particular element is called the element's **atomic number.** Hydrogen, for example, whose atoms each have one proton, has the atomic number 1; carbon, whose atoms each have six protons, has the atomic number 6.

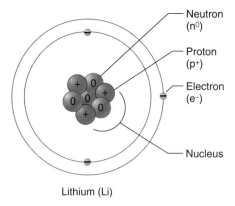

Lithium (Li)

Figure 2.1 AP|R

An atom consists of subatomic particles. In an atom of the element lithium, three electrons move around a nucleus that consists of three protons and four neutrons.

The **atomic weight** of an atom of an element approximately equals the number of protons and neutrons in its nucleus; electrons have very little weight. Thus, the atomic weight of hydrogen, with one proton and no neutrons, is 1, whereas the atomic weight of carbon, with six protons and six neutrons, is 12 (table 2.2). In other words, an atom of carbon weighs about twelve times more than an atom of hydrogen.

All the atoms of a particular element have the same atomic number because they have the same number of protons and electrons. However, the atoms of an element vary in the number of neutrons in their nuclei; thus, they vary in atomic weight. For example, all oxygen atoms have eight protons in their nuclei, but these atoms may have eight, nine, or ten neutrons, corresponding to,

Table 2.2	Atomic Structure of Elements 1 Through 12							
Element	Symbol	Atomic Number	Atomic Weight	Protons	Neutrons	Electrons in Shells		
						First	Second	Third
Hydrogen	H	1	1	1	0	1		
Helium	He	2	4	2	2	2 (inert)		
Lithium	Li	3	7	3	4	2	1	
Beryllium	Be	4	9	4	5	2	2	
Boron	B	5	11	5	6	2	3	
Carbon	C	6	12	6	6	2	4	
Nitrogen	N	7	14	7	7	2	5	
Oxygen	O	8	16	8	8	2	6	
Fluorine	F	9	19	9	10	2	7	
Neon	Ne	10	20	10	10	2	8 (inert)	
Sodium	Na	11	23	11	12	2	8	1
Magnesium	Mg	12	24	12	12	2	8	2

respectively, atomic weights of 16, 17, and 18. Atoms of an element with different atomic weights are called **isotopes** (iʹso-tōps) of that element. Because a sample of an element is likely to include more than one isotope, the atomic weight of the element is often considered to be the average weight of the isotopes present. (See Appendix C, Periodic Table of the Elements, p. 566.)

How atoms interact reflects their number of electrons. Because the number of electrons in an atom is equal to its number of protons, all the isotopes of a particular element have the same number of electrons and react chemically in the same manner. Therefore, any of the isotopes of oxygen can play the same role in an organism's metabolic reactions.

Isotopes may be stable, or they may have unstable atomic nuclei that decompose, releasing energy or pieces of themselves. Unstable isotopes are called *radioactive* because they emit energetic particles, and the energy or atomic fragments they give off are called *radiation*.

Three common forms of radiation are alpha (α), beta (β), and gamma (γ). Alpha radiation consists of particles from atomic nuclei, each of which includes two protons and two neutrons, that travel slowly and can weakly penetrate matter. Beta radiation consists of much smaller particles (electrons) that travel more rapidly and penetrate matter more deeply. Gamma radiation is similar to X-ray radiation and is the most penetrating of these forms.

Each kind of radioactive isotope produces one or more forms of radiation, and each becomes less radioactive at a particular rate. The time required for an isotope to lose one-half of its radioactivity is called its *half-life*. Thus, the isotope of iodine called iodine-131, which emits one-half of its radiation in 8.1 days, has a half-life of 8.1 days. Half-lives vary greatly. The half-life of phosphorus-32 is 14.3 days; that of cobalt-60 is 5.26 years; and that of radium-226 is 1,620 years. Clinical Application 2.1 discusses some practical applications of radioactive isotopes.

Practice

1. What are elements?
2. Which elements are most common in the human body?
3. Where are electrons, protons, and neutrons located in an atom?
4. What is the difference between atomic number and atomic weight?

Bonding of Atoms

Atoms can attach to other atoms by forming chemical bonds. The chemical behavior of atoms results from interactions among their electrons. When atoms form chemical bonds, they gain, lose, or share electrons.

The electrons of an atom occupy one or more areas of space, called *shells,* around the nucleus (see table 2.2). For the elements up to atomic number 18, the maximum number of electrons that each of the first three inner shells can hold is as follows:

First shell (closest to the nucleus)	2 electrons
Second shell	8 electrons
Third shell	8 electrons

More complex atoms may have as many as eighteen electrons in the third shell. Simplified diagrams, such as those in figure 2.2, depict electron locations within the shells of atoms.

The electrons in the outermost shell of an atom determine its chemical behavior. Atoms such as helium, whose outermost electron shells are filled, have stable structures and are chemically inactive, or **inert** (see table 2.2). Atoms such as hydrogen or lithium, whose outermost electron shells are incompletely filled, tend to gain, lose, or share electrons in ways that empty or fill their outer shells. This enables them to achieve stable structures.

Atoms that gain or lose electrons become electrically charged and are called **ions** (iʹonz). An atom of sodium, for example, has eleven electrons: two in the first shell, eight in the second shell, and one in the third shell (fig. 2.3). This atom tends to lose the electron from its outer shell, which leaves the second (now the outermost) shell filled and the new form stable (fig. 2.4a). In the process, sodium is left with eleven protons (11^+) in its nucleus and only ten electrons (10^-). As a result, the atom develops a net electrical charge of 1^+ and is called a sodium ion, symbolized Na^+.

A chlorine atom has seventeen electrons, with two in the first shell, eight in the second shell, and seven in the third shell. An atom of this type tends to accept a single electron, filling its outer shell and achieving stability (fig. 2.4a). In the process, the chlorine atom is left with seventeen protons (17^+) in its nucleus and eighteen electrons (18^-). The atom develops a net electrical charge of 1^- and is called a chloride ion, symbolized Cl^-.

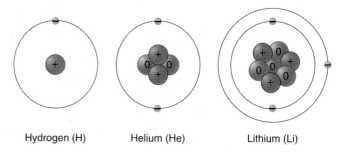

Hydrogen (H)   Helium (He)   Lithium (Li)

Figure 2.2

Electrons orbit the atomic nucleus. The single electron of a hydrogen atom is located in its first shell. The two electrons of a helium atom fill its first shell. Two of the three electrons of a lithium atom are in the first shell, and one is in the second shell.

Clinical Application 2.1

Radioactive Isotopes: Helpful and Harmful

Radioactive chemicals are useful in studying life processes and in diagnosing and treating some diseases. Atomic radiation is detected with special equipment, such as a scintillation counter (fig. 2A). A radioactive isotope can be introduced into an organism and then traced as it enters into metabolic activities. For example, the human thyroid gland is unique in using the element iodine in its metabolism. Therefore, radioactive iodine-131 is used to study thyroid functions and to evaluate thyroid disease (fig. 2B). Doctors use thallium-201, which has a half-life of 73.5 hours, to assess heart conditions, and gallium-67, with a half-life of 78 hours, to detect and monitor the progress of certain cancers and inflammatory diseases.

Atomic radiation also can change chemical structures and in this way alter vital cellular processes. For this reason, doctors sometimes use radioactive isotopes, such as cobalt-60, to treat cancers. The radiation from the cobalt preferentially kills the rapidly dividing cancer cells.

Exposure to radiation can cause disease, such as certain cancers. The transfer of energy as radiation is emitted damages DNA in ways that kill cells or make them cancerous. Exposure to ultraviolet radiation in sunlight, for example, causes skin cancer, and excess medical X rays or gamma rays increase the risk of developing cancer in a variety of body parts, including salivary glands, bone, kidneys, and blood.

Figure 2A
Scintillation counters detect radioactive isotopes.

(a)

Larynx

Thyroid gland

Trachea

(b)

Figure 2B
(a) Scan of the thyroid gland 24 hours after the patient received radioactive iodine. Note how closely the scan in **(a)** resembles the shape of the thyroid gland, shown in **(b)**.

Sodium atom contains
11 electrons (e⁻)
11 protons (p⁺)
12 neutrons (n⁰)

Atomic number = 11
Atomic weight = 23

Figure 2.3

A sodium atom.

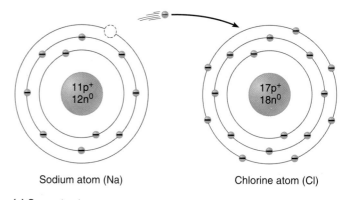

Sodium atom (Na)   Chlorine atom (Cl)

(a) Separate atoms
If a sodium atom loses an electron to a chlorine atom, the sodium atom becomes a sodium ion (Na^+), and the chlorine atom becomes a chloride ion (Cl^-).

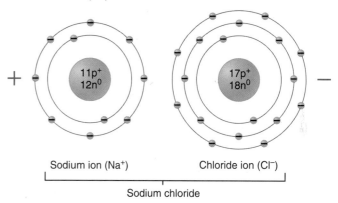

Sodium ion (Na^+)   Chloride ion (Cl^-)

Sodium chloride

(b) Bonded ions
These oppositely charged particles attract electrically and join by an ionic bond.

(c) Salt crystal
Ionically bonded substances form arrays such as a crystal of NaCl.

Figure 2.4

An ionic bond forms when one atom gains and another atom loses electrons **(a)** and then oppositely charged ions attract **(b)**. Ionically bonded substances may form crystals **(c)**.

Because oppositely charged ions attract, sodium and chloride ions react to form a type of chemical bond called an **ionic bond** (electrovalent bond). Sodium ions (Na^+) and chloride ions (Cl^-) uniting in this manner form the compound sodium chloride (NaCl), or table salt (fig. 2.4*b*). Some ions have an electrical charge greater than 1—for example, Ca^{+2} (or Ca^{++}).

Ionically bound substances do not form discrete molecules—instead, they form arrays, such as crystals of sodium chloride (fig. 2.4*c*). The molecular formulas for compounds like sodium chloride (NaCl) give the relative amounts of each element present. Atoms may also bond by sharing electrons rather than by exchanging them. A hydrogen atom, for example, has one electron in its first shell but requires two electrons to achieve a stable structure (fig. 2.5). It may fill this shell by combining with another hydrogen atom in such a way that the two atoms share a pair of electrons. The two electrons then encircle the nuclei of both atoms, and each atom achieves a stable form. The chemical bond between the atoms that share electrons is called a **covalent bond.**

Carbon atoms, which have two electrons in their first shells and four electrons in their second shells, form covalent bonds when they unite with other atoms. In fact, carbon atoms (and certain other atoms) may bond in such a way that two atoms share one or more pairs of electrons. If one pair of electrons is shared, the resulting bond is called a *single covalent bond;* if two pairs of electrons are shared, the bond is called a *double covalent bond. Triple covalent bonds* are also possible between some atoms.

Different types of chemical bonds share electrons to different degrees. At one extreme is the ionic bond in which atoms gain or lose electrons. At the other extreme is the covalent bond in which the electrons are shared

equally. In between lies the covalent bond in which electrons are not shared equally, resulting in a molecule whose shape gives an uneven distribution of charges. Such a molecule is called **polar.** Unlike an ion, a polar molecule has an equal number of protons and electrons, but more of the electrons are found at one end of the molecule, making that end slightly negative, while the other end of the molecule is left slightly positive. Typically, polar covalent bonds form where hydrogen

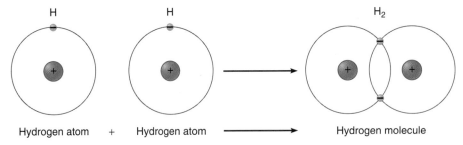

Figure 2.5

A hydrogen molecule forms when two hydrogen atoms share a pair of electrons. A covalent bond forms between the atoms.

atoms bond to oxygen or nitrogen atoms. Water is an important polar molecule (fig. 2.6a).

The attraction of the positive hydrogen end of a polar molecule to the negative nitrogen or oxygen end of another polar molecule is called a **hydrogen bond.** Hydrogen bonds are relatively weak. For exam-

(a)

Slightly negative end

Slightly positive ends

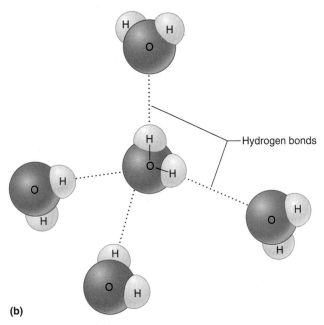

Hydrogen bonds

(b)

Figure 2.6

Water is a polar molecule. **(a)** Water molecules have equal numbers of electrons and protons but are polar because the electrons are shared unequally, creating slightly negative ends and slightly positive ends. **(b)** Hydrogen bonding connects water molecules.

ple, below 0°C the hydrogen bonds between the water molecules shown in figure 2.6b are strong enough to form ice. Above 0°C, however, increased molecular movement breaks the hydrogen bonds, and water becomes a liquid. At body temperature (37°C), hydrogen bonds are important in protein and nucleic acid structure. In these cases, many hydrogen bonds form between polar regions of different parts of a single, very large molecule (see figs. 2.18 and 2.21). Together, these individually weak bonds provide strength. The contribution of hydrogen bonds to protein and nucleic acid structure is described in section 2.3, pages 44 and 46.

Molecules and Compounds

When two or more atoms bond, they form a new kind of particle called a **molecule** (mol'ĕ-kūl). If atoms of the same element bond, they produce molecules of that element. Gases of hydrogen, oxygen, and nitrogen consist of such molecules (see fig. 2.5).

When atoms of different elements bond, they form molecules called **compounds.** Two atoms of hydrogen, for example, can bond with one atom of oxygen to produce a molecule of the compound water (H_2O) (fig. 2.7). Table sugar (*sucrose*), baking soda, natural gas, beverage alcohol, and most drugs are compounds.

A molecule of a compound always consists of definite kinds and numbers of atoms. A molecule of water, for instance, always has two hydrogen atoms and one oxygen atom. If two hydrogen atoms bond with two oxygen atoms, the compound formed is not water, but hydrogen peroxide (H_2O_2). Table 2.3 summarizes the characteristics of the particles of matter discussed so far.

Practice

5. What is an ion?

6. Describe two ways that atoms bond with other atoms.

7. Distinguish between an ion and a polar molecule.

8. Distinguish between a molecule and a compound.

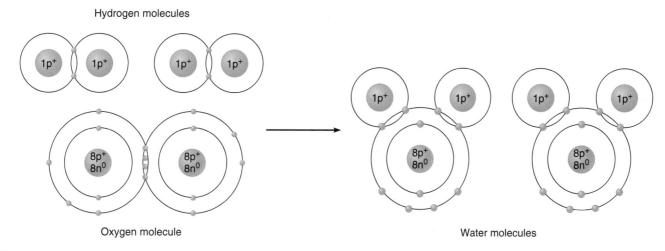

Hydrogen molecules

Oxygen molecule

Water molecules

Figure 2.7

Hydrogen molecules can combine with oxygen molecules, forming water molecules. The shared electrons represent covalent bonds.

Formulas

A **molecular formula** (mo-lek'u-lar for'mu-lah) represents the numbers and types of atoms in a molecule. Such a formula displays the symbols for the elements in the molecule and the number of atoms of each element. For example, the molecular formula for water is H_2O, which means that each water molecule consists of two atoms of hydrogen and one atom of oxygen (fig. 2.8). The molecular formula for the sugar *glucose* is $C_6H_{12}O_6$, indicating that each glucose molecule consists of six atoms of carbon, twelve atoms of hydrogen, and six atoms of oxygen.

Usually, the atoms of each element form a specific number of covalent bonds. Hydrogen atoms form single

$$H-H \qquad O=O \qquad \begin{matrix} H \quad H \\ \diagdown \diagup \\ O \end{matrix} \qquad O=C=O$$

$$H_2 \qquad O_2 \qquad H_2O \qquad CO_2$$

Figure 2.8

Structural and molecular formulas for molecules of hydrogen, oxygen, water, and carbon dioxide. Note the double covalent bonds. (Triple covalent bonds are also possible between some atoms.)

bonds, oxygen atoms form two bonds, nitrogen atoms form three bonds, and carbon atoms form four bonds. Symbols and lines can depict bonds as follows:

$$-H \qquad -O- \qquad \diagup N \diagdown \qquad -\overset{|}{\underset{|}{C}}-$$

These representations show how atoms are joined and arranged in various molecules. Single lines represent single bonds, and double lines represent double bonds. Illustrations of this type are called **structural formulas** (struk'cher-al for'mu-lahz) (fig. 2.8). Three-dimensional models of structural formulas use different colors for the different kinds of atoms (fig. 2.9).

Chemical Reactions

Chemical reactions form or break bonds between atoms, ions, or molecules, generating new chemical combinations. For example, when two or more atoms (reactants) bond to form a more complex structure (product), the reaction is called **synthesis** (sin'thĕ-sis). Such a reaction is symbolized in this way:

$$A + B \longrightarrow AB$$

Table 2.3	Some Particles of Matter
Particle	**Characteristics**
Atom	Smallest particle of an element that has the properties of that element
Electron (e⁻)	Extremely small particle; carries a negative electrical charge and is in constant motion around the nucleus of an atom
Proton (p⁺)	Relatively large particle; carries a positive electrical charge and is found within the nucleus of an atom
Neutron (n⁰)	Relatively large particle; uncharged and thus electrically neutral; found within the nucleus of an atom
Molecule	Particle formed by the chemical union of two or more atoms
Ion	Atom or molecule that is electrically charged because it has gained or lost one or more electrons

(a) A water molecule (H₂O), with the white parts depicting
hydrogen atoms and the red part representing oxygen.

(b) A glucose molecule ($C_6H_{12}O_6$), in which the black
parts represent carbon atoms.

Figure 2.9
Three-dimensional molecular models depict spatial relationships
of the constituent atoms.

If the bonds within a reactant molecule break so
that simpler molecules, atoms, or ions form, the reaction
is called **decomposition** (de″kom-po-zish′un). Decomposition is symbolized as follows:

$$AB \longrightarrow A + B$$

Synthesis, which requires energy, is particularly
important in the growth of body parts and the repair
of worn or damaged tissues, which require buildup of
larger molecules from smaller ones. In contrast, decomposition occurs when food molecules are digested into
smaller ones that can be absorbed.

A third type of chemical reaction is an **exchange
reaction.** In this reaction, parts of two different types of
molecules trade positions as bonds are broken and new
bonds are formed. The reaction is symbolized as follows:

$$AB + CD \longrightarrow AD + CB$$

An example of an exchange reaction is when an acid
reacts with a base, producing water and a salt. Acids
and bases are described in the next section.

Many chemical reactions are reversible. This means
that the product (or products) of the reaction can
change back to the reactant (or reactants) that originally
underwent the reaction. A **reversible reaction** is symbolized with a double arrow:

$$A + B \rightleftharpoons AB$$

Whether a reversible reaction proceeds in one direction or the other depends on such factors as the relative

proportions of the reactant (or reactants) and product
(or products), as well as the amount of available energy.
Particular atoms or molecules that can change the rate
(not the direction) of a reaction without being consumed in the process, called **catalysts,** speed many
chemical reactions in the body so that they proceed fast
enough to sustain the activities of life.

Acids and Bases

When ionically bound substances are placed in water,
the slightly negative and positive ends of the water molecules cause the ions to leave each other and interact
with the water molecules instead. For example, sodium
chloride (NaCl) releases sodium ions (Na^+) and chloride
ions (Cl^-) when it is placed in water:

$$NaCl \longrightarrow Na^+ + Cl^-$$

In this way, the polarity of water dissociates salts in
the internal environment (fig. 2.10). Since the resulting
solution contains electrically charged particles (ions),
it will conduct an electric current. Substances that
release ions in water are, therefore, called **electrolytes**
(e-lek′tro-lītz).

Acids are electrolytes that release hydrogen ions
(H^+) in water. For example, in water, the compound
hydrochloric acid (HCl) releases hydrogen ions (H^+) and
chloride ions (Cl^-):

$$HCl \longrightarrow H^+ + Cl^-$$

Electrolytes that release ions that bond with hydrogen
ions are called **bases.** For example, the compound

Salt crystal

Na^+

Cl^-

Ions in
solution

Figure 2.10 AP|R
The polar nature of water molecules dissociates sodium chloride
(NaCl) in water, releasing sodium ions (Na^+) and chloride ions (Cl^-).

sodium hydroxide (NaOH) releases hydroxide ions (OH⁻) when placed in water:

$$NaOH \longrightarrow Na^+ + OH^-$$

The hydroxide ions, in turn, can bond with hydrogen ions to form water; thus, sodium hydroxide is a base. Many bases are present in the body fluids, but because of the way they react in water, the concentration of hydroxide ions is a good estimate of the total base concentration. (Note: Some ions, such as OH⁻, consist of two or more atoms. However, such a group behaves as a unit and usually remains unchanged during a chemical reaction.)

The concentrations of hydrogen ions (H⁺) and hydroxide ions (OH⁻) in body fluids greatly affect the chemical reactions that control certain physiological functions, such as blood pressure and breathing rate. Since their concentrations are inversely related (if one goes up, the other goes down), we need to keep track of only one of them. A value called **pH** measures hydrogen ion concentration.

The pH scale ranges from 0 to 14. A solution with a pH of 7.0, the midpoint of the scale, contains equal numbers of hydrogen and hydroxide ions and is said to be *neutral*. A solution that contains more hydrogen ions than hydroxide ions has a pH less than 7.0 and is *acidic*. A solution with fewer hydrogen ions than hydroxide ions has a pH greater than 7.0 and is *basic* (alkaline).

Figure 2.11 indicates the pH values of some common substances. Each whole number on the pH scale represents a tenfold difference in hydrogen ion concentration, and as the hydrogen ion concentration increases, the pH number gets smaller. Thus, a solution with a pH of 6 has ten times the hydrogen ion concentration of a solution with a pH of 7. This means that relatively small changes in pH can reflect large changes in hydrogen ion concentration.

Buffers are chemicals that resist pH change. They combine with hydrogen ions when these ions are in excess, or they donate hydrogen ions when these ions are depleted. Buffers and the regulation of the hydrogen ion concentration in body fluids are discussed further in chapter 18 (pp. 498–499).

> The pH of human blood is about 7.4, and ranges from 7.35 to 7.45 (see fig. 2.11). If the pH drops below 7.35, the person has *acidosis*; if it rises above 7.45, the condition is *alkalosis*. Without medical intervention, a person usually cannot survive if blood pH drops to 6.9 or rises to 7.8 for more than a few hours. Homeostatic mechanisms like those described in chapter 1 (p. 6) regulate pH of the internal environment.

Practice

9. What is a molecular formula? A structural formula?
10. Describe three kinds of chemical reactions.
11. Compare the characteristics of an acid with those of a base.
12. What does pH measure?
13. What is a buffer?

2.3 CHEMICAL CONSTITUENTS OF CELLS

Chemicals, including those that enter into metabolic reactions or are produced by them, can be divided into two large groups. Chemicals that include both carbon and hydrogen atoms are called **organic** (or-gan′ik). The rest are **inorganic** (in″or-gan′ik).

Inorganic substances usually dissolve in water and dissociate to release ions; thus, they are *electrolytes*. Many organic compounds also dissolve in water, but they are more likely to dissolve in organic liquids, such as ether or alcohol. Organic substances that dissolve in water usually do not release ions and are therefore called *nonelectrolytes*.

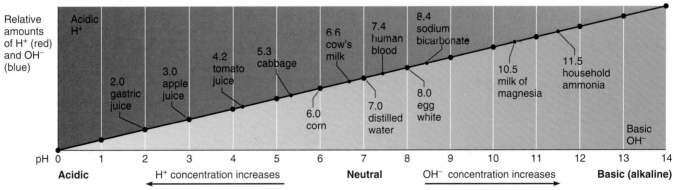

Figure 2.11

The pH scale measures hydrogen ion (H⁺) concentration. As the concentration of H⁺ increases, a solution becomes more acidic, and the pH value decreases. As the concentration of ions that bond with H⁺ (such as hydroxide ions) increases, a solution becomes more basic (alkaline), and the pH value increases. The pH values of some common substances are shown.

Q: *What is the pH of the internal environment?*

Answer can be found in Appendix E on page 568.

Inorganic Substances

Among the inorganic substances common in cells are water, oxygen, carbon dioxide, and a group of compounds called salts.

Water

Water is the most abundant compound in living material and accounts for about two-thirds of the weight of an adult human. It is the major component of blood and other body fluids, including those in cells.

Water is an important *solvent* because many substances readily dissolve in it. A substance dissolved in a liquid, such as water, is called a *solute*. When it dissolves it is broken down into smaller and smaller pieces, eventually to molecular-sized particles, which may be ions. If two or more types of solutes are dissolved, they are much more likely to react with one another than were the original large pieces. Consequently, most metabolic reactions occur in water.

Water also plays an important role in moving chemicals in the body. For example, the aqueous (watery) portion of blood carries many vital substances, such as oxygen, sugars, salts, and vitamins, from the organs of digestion and respiration to the body cells.

Water can absorb and transport heat. Blood carries heat released from muscle cells during exercise from deeper parts of the body to the surface, where it may be lost to the outside.

Oxygen

Molecules of oxygen (O_2) enter the body through the respiratory organs and are transported throughout the body by the blood. The red blood cells bind and carry most of the oxygen. Cellular organelles use oxygen to release energy from the sugar glucose and other nutrients. The released energy drives the cell's metabolic activities.

Carbon Dioxide

Carbon dioxide (CO_2) is a simple, carbon-containing compound of the inorganic group. It is produced as a waste product when certain metabolic processes release energy, and it is exhaled from the lungs.

Salts

A salt is a compound composed of oppositely charged ions, such as sodium (Na^+) and chloride (Cl^-), which is the familiar table salt NaCl. Salts are abundant in tissues and fluids. They provide many necessary ions, including sodium (Na^+), chloride (Cl^-), potassium (K^+), calcium (Ca^{+2}), magnesium (Mg^{+2}), phosphate (PO_4^{-3}), carbonate (CO_3^{-2}), bicarbonate (HCO_3^-), and sulfate (SO_4^{-2}). These ions are important in metabolic processes, including transport of substances into and out of cells, muscle contraction, and nerve impulse conduction. Table 2.4 summarizes the functions of some of the inorganic substances in the body.

Table 2.4	Inorganic Substances Common in the Body	
Substance	**Symbol or Formula**	**Functions**
I. Inorganic molecules		
Water	H_2O	Major component of body fluids (chapter 12, p. 327); medium in which most biochemical reactions occur; transports chemicals (chapter 12, p. 329); helps regulate body temperature (chapter 6, p. 125)
Oxygen	O_2	Used in energy release from glucose molecules (chapter 4, p. 80)
Carbon dioxide	CO_2	Waste product that results from metabolism (chapter 4, p. 80); reacts with water to form carbonic acid (chapter 16, p. 460)
II. Inorganic ions		
Bicarbonate ions	HCO_3^-	Helps maintain acid-base balance (chapter 18, p. 498)
Calcium ions	Ca^{+2}	Necessary for bone development (chapter 7, p. 137), muscle contraction (chapter 8, p. 183), and blood clotting (chapter 12, p. 333)
Carbonate ions	CO_3^{-2}	Component of bone tissue (chapter 7, p. 141)
Chloride ions	Cl^-	Helps maintain water balance (chapter 18, p. 496)
Magnesium ions	Mg^{+2}	Component of bone tissue (chapter 7, p. 141); required for certain metabolic processes (chapter 15, p. 434)
Phosphate ions	PO_4^{-3}	Required for synthesis of ATP, nucleic acids, and other vital substances (chapter 4, pp. 80, 84); component of bone tissue (chapter 7, p. 141); helps maintain polarization of cell membranes (chapter 9, p. 222)
Potassium ions	K^+	Required for polarization of cell membranes (chapter 9, p. 222)
Sodium ions	Na^+	Required for polarization of cell membranes (chapter 9, p. 222); helps maintain water balance (chapter 18, p. 492)
Sulfate ions	SO_4^{-2}	Helps maintain polarization of cell membranes (chapter 9, p. 222)

Practice

14. How do inorganic and organic molecules differ?
15. How do electrolytes and nonelectrolytes differ?
16. Name the inorganic substances common in body fluids.

Organic Substances

Important groups of organic chemicals in cells include carbohydrates, lipids, proteins, and nucleic acids.

Carbohydrates

Carbohydrates (kar″bo-hi′drātz) provide much of the energy that cells require. They supply materials to build certain cell structures and often are stored as reserve energy supplies.

Carbohydrate molecules consist of atoms of carbon, hydrogen, and oxygen. These molecules usually have twice as many hydrogen as oxygen atoms—the same ratio of hydrogen to oxygen as in water molecules (H_2O). This ratio is easy to see in the molecular formulas of the carbohydrates glucose ($C_6H_{12}O_6$) and sucrose ($C_{12}H_{22}O_{11}$).

The carbon atoms of carbohydrate molecules join in chains whose lengths vary with the type of carbohydrate. For example, carbohydrates with shorter chains are called **sugars.**

Sugars with 6 carbon atoms (hexoses) are examples of *simple sugars,* or **monosaccharides** (mon″o-sak′ah-rīdz). The simple sugars include glucose, fructose, and galactose, as well as the 5-carbon sugars ribose and deoxyribose. Figure 2.12 illustrates the structural formulas of glucose.

In *complex carbohydrates,* a number of simple sugar molecules link to form molecules of varying sizes (fig. 2.13). Some complex carbohydrates, such as sucrose (table sugar) and lactose (milk sugar), are *double sugars,* or **disaccharides** (di-sak′ah-rīdz), whose molecules each consist of two simple sugar building blocks. Other complex carbohydrates are made up of many simple sugar units joined to form **polysaccharides** (pol″e-sak′ah-rīdz), such as plant starch. Animals,

(a) Some glucose molecules ($C_6H_{12}O_6$) have a straight chain of carbon atoms.

(b) More commonly, glucose molecules form a ring structure.

(c) This shape symbolizes the ring structure of a glucose molecule.

Figure 2.12

Structural formulas depict a molecule of glucose.

(a) Monosaccharide   (b) Disaccharide

(c) Polysaccharide

Figure 2.13

Carbohydrate molecules vary in size. **(a)** A monosaccharide molecule consists of one building block with 6 carbon atoms. **(b)** A disaccharide molecule consists of two of these building blocks. **(c)** A polysaccharide molecule consists of many such building blocks.

including humans, synthesize a polysaccharide similar to starch called *glycogen.*

Lipids

Lipids (lip′idz) are organic substances that are insoluble in water but soluble in certain organic solvents, such as ether and chloroform. Lipids include a variety of compounds—fats, phospholipids, and steroids—that have vital functions in cells. The most common lipids are fats.

Fats are used primarily to store energy for cellular activities. Fat molecules can supply more energy, gram for gram, than carbohydrate molecules.

Like carbohydrates, fat molecules are composed of carbon, hydrogen, and oxygen atoms. However, fats have a much smaller proportion of oxygen atoms than do carbohydrates. The formula for the fat tristearin, $C_{57}H_{110}O_6$, illustrates these characteristic proportions.

The building blocks of fat molecules are **fatty acids** and **glycerol** (glis′er-ol). Each glycerol molecule bonds with three fatty acid molecules to produce a single fat, or *triglyceride,* molecule (fig. 2.14).

The glycerol portions of all fat molecules are identical, but fats are diverse because there are many kinds of fatty acids. Fatty acid molecules differ in the lengths of their carbon atom chains, which usually have an even number of carbon atoms. The chains also vary in the way the carbon atoms bond. In some cases, the carbon atoms all join by single carbon-carbon bonds. This type of fatty acid is *saturated;* that is, each carbon atom is bound to as many hydrogen atoms as possible and

is thus saturated with hydrogen atoms. Other fatty acid chains have not bound the maximum number of hydrogen atoms. These fatty acids, therefore, have one or more double bonds between carbon atoms, because a carbon atom must form four bonds to be stable. Fatty acid molecules with double bonds are *unsaturated,* and those with many double-bonded carbon atoms are *poly-unsaturated.* Similarly, fat molecules that contain only saturated fatty acids are called *saturated fats,* and those that include unsaturated fatty acids are called *unsaturated fats.* The triglyceride molecule in figure 2.14 is an example of an unsaturated fat.

A **phospholipid** (fos″fo-lip′id) molecule is similar to a fat molecule in that it consists of a glycerol portion and fatty acid chains (fig. 2.15a,b). A phospholipid, however, has only two fatty acid chains; in place of the third is a portion that includes a phosphate group. The phosphate portion is soluble in water (hydrophilic) and forms the "head" of the molecule, while the fatty acid portion is insoluble in water (hydrophobic) and forms a "tail" (fig. 2.15c). Phospholipids are important in cellular structures.

Steroid (ste′roid) molecules are complex structures that include four connected rings of carbon atoms (fig. 2.16). Among the more important steroids are cholesterol, which is in all body cells and is used to synthesize other steroids: sex hormones, such as estrogen, progesterone, and testosterone; and several hormones from the adrenal glands. Chapters 11 and 19 discuss these steroids.

Table 2.5 lists the three important groups of lipids and their characteristics.

Glycerol portion

Fatty acid portions

Figure 2.14

A triglyceride molecule (fat) consists of a glycerol portion and three fatty acid portions. This is an example of an unsaturated fat. The double bond between carbon atoms in the unsaturated fatty acid is shown in red.

Q: *Why is it incorrect to say that fat is another word for lipid?*

Answer can be found in Appendix E on page 568.

Figure 2.15

Fats and phospholipids. **(a)** A fat molecule (triglyceride) consists of a glycerol and three fatty acids. **(b)** In a phospholipid molecule, a phosphate-containing group replaces one fatty acid. **(c)** Schematic representation of a phospholipid.

(a) A fat molecule

Glycerol portion

(b) A phospholipid molecule (the unshaded portion may vary)

Glycerol portion   Phosphate portion

Water-insoluble "tail"

Water-soluble "head"

(c) Schematic representation of a phospholipid molecule

Figure 2.16

Steroid structure. **(a)** The general structure of a steroid. **(b)** The structural formula for cholesterol, a steroid widely distributed in the body.

(a) General structure of a steroid

(b) Cholesterol

Table 2.5	Important Groups of Lipids	
Group	**Basic Molecular Structure**	**Characteristics**
Triglycerides	Three fatty acid molecules bound to a glycerol molecule	Most common lipids in body; stored in fat tissue as an energy supply; fat tissue also provides insulation beneath the skin
Phospholipids	Two fatty acid molecules and a phosphate group bound to a glycerol molecule	Used as structural components in cell membranes; abundant in liver and parts of the nervous system
Steroids	Four connected rings of carbon atoms	Widely distributed in the body and have a variety of functions; include cholesterol, hormones of adrenal cortex, sex hormones, bile acids, and vitamin D

Saturated fats are more abundant in fatty foods that are solids at room temperature, such as butter, lard, and most animal fats. Unsaturated fats, on the other hand, are in foods that are liquid at room temperature, such as soft margarine and seed oils (corn, sesame, soybean, sunflower, and peanut). Exceptions are coconut and palm kernel oils, which are high in saturated fats. The most heart-healthy fats are olive and canola oils, which are monounsaturated—that is, they have one carbon-carbon double bond.

Manufacturers of prepared foods sometimes harden vegetable oils by adding hydrogen atoms in a way that produces "trans" fat. A diet high in saturated fats, trans fats, and cholesterol increases the risk of developing atherosclerosis, in which fatty deposits obstruct the inner linings of arteries. Some cities have banned the use of trans fats in restaurant food.

Proteins

Proteins (pro'tēnz) have a wide variety of functions in the body. Many serve as structural materials, energy sources, or hormones. Others combine with carbohydrates (to form glycoproteins) and function as receptors on cell surfaces, allowing cells to respond to particular kinds of molecules that bind to them. Proteins called *antibodies* detect and destroy foreign substances in the body. Metabolism could not occur fast enough to support life were it not for *enzymes,* which catalyze specific chemical reactions. (Enzymes are discussed in more detail in chapter 4, p. 79.)

Like carbohydrates and lipids, proteins are composed of atoms of carbon, hydrogen, and oxygen. In addition, proteins always contain nitrogen atoms, and some proteins contain sulfur atoms. The building blocks

Figure 2.17

Amino acid structure.
(a) An amino acid has an amino group, a carboxyl group, and a hydrogen atom that are common to all amino acid molecules, and a specific R group.
(b) Some representative amino acids and their structural formulas. Each type of amino acid molecule has a particular shape due to its R group.

(a) General structure of an amino acid. The portion common to all amino acids is within the oval. It includes the amino group (—NH₂) and the carboxyl group (—COOH). The "R" group, or the "rest of the molecule," is what makes each amino acid unique.

(b) Cysteine. Cysteine has an R group that contains sulfur.

Phenylalanine. Phenylalanine has a complex R group.

of proteins are **amino acids,** each of which has an *amino group* (—NH₂) at one end and a *carboxyl group* (—COOH) at the other (fig. 2.17*a*). Amino acids also have a *side chain,* or *R group* ("R" may be thought of as the "rest of the molecule"). The composition of the R group distinguishes one type of amino acid from another (fig. 2.17*b*).

Twenty different amino acids make up the proteins of most living organisms. The amino acids join in polypeptide chains that vary in length from less than 100 to more than 5,000 amino acids. A protein molecule consists of one or more polypeptide chains.

> A human body has more than 200,000 types of proteins, but only about 20,500 genes, which are the instructions for production of particular polypeptides. The numbers are not the same because parts of some genes encode sequences of amino acids found in more than one type of protein. It is a little like assembling a large and diverse wardrobe by combining a few basic pieces of clothing in different ways.

Proteins have several levels of structure: Primary, secondary, and tertiary levels are shown in figure 2.18*a–c*. Hydrogen bonding and even covalent bonding between atoms in different parts of the polypeptide give the final protein a distinctive three-dimensional shape, or **conformation** (fig. 2.19). The conformation of a protein determines its function. Some proteins are long and fibrous, such as the keratin proteins that form hair, or fibrin, the protein whose threads form a blood clot. Many proteins are globular and function as enzymes, ion channels, carrier proteins, or receptors. Myoglobin and hemoglobin, which transport oxygen in muscle and blood, respectively, are globular.

In many cases, slight, reversible changes in conformation may occur as part of the protein's normal function. For example, some of the proteins involved

in muscle contraction exert a pulling force as a result of such a shape change, leading to movement. Such changes in shape are reversible, so the protein can perform its function again and again.

When hydrogen bonds in a protein break as a result of exposure to excessive heat, radiation, electricity, pH changes, or various chemicals, a protein's unique shape may be changed dramatically, or *denatured*. Such proteins lose their special properties. For example, heat denatures the protein in egg white (albumin), changing it from a liquid to a solid. This is an irreversible change—a hard-boiled egg cannot return to its uncooked, runny state. Similarly, cellular proteins that are denatured may be permanently altered and lose their functions.

Proteins with more than one polypeptide chain have a fourth level of conformation, the *quaternary structure*. The constituent polypeptides are connected, often forming a very large protein (see fig. 2.18*d*). Hemoglobin is a quaternary protein made up of four separate polypeptide chains.

For most proteins, the conformation, which determines its function, is always the same for a given amino acid sequence or primary structure. Thus, it is the amino acid sequence that ultimately determines the role of a protein in the body. Genes, made of nucleic acid, contain the information for the amino acid sequences of all the body's proteins in a form that the cell can decode.

Nucleic Acids

Nucleic acids (nu-kle′ik as′idz) form genes and take part in protein synthesis. These molecules are generally very large. They include atoms of carbon, hydrogen, oxygen, nitrogen, and phosphorus, which form building blocks called **nucleotides.** Each nucleotide consists of a 5-carbon *sugar* (ribose or deoxyribose), a *phosphate group,* and one of several *nitrogenous* (nitrogen-containing) *bases* (fig. 2.20). A nucleic acid molecule consists of a chain of many nucleotides (polynucleotide chain).

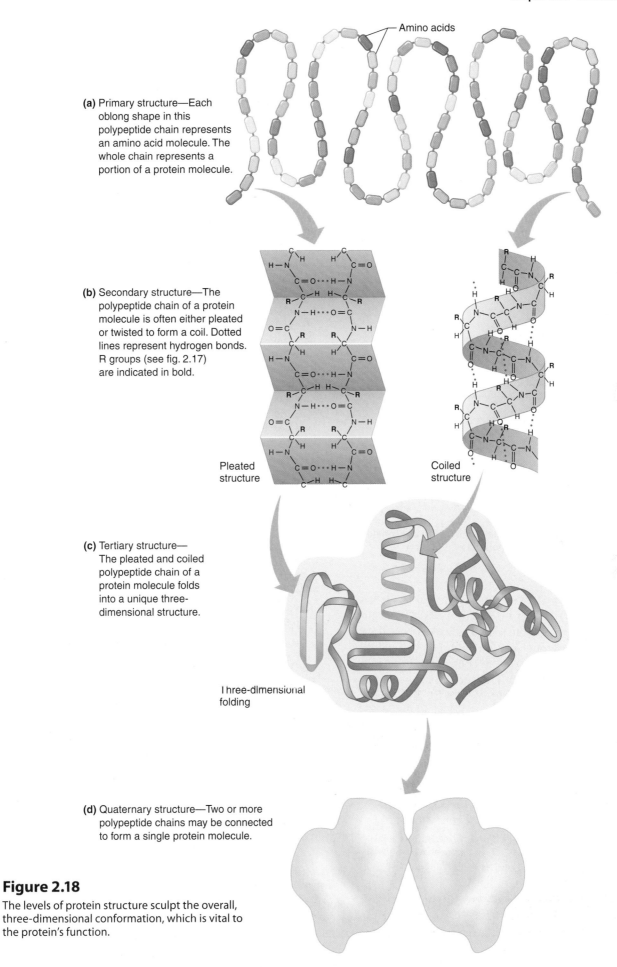

Amino acids

(a) Primary structure—Each oblong shape in this polypeptide chain represents an amino acid molecule. The whole chain represents a portion of a protein molecule.

(b) Secondary structure—The polypeptide chain of a protein molecule is often either pleated or twisted to form a coil. Dotted lines represent hydrogen bonds. R groups (see fig. 2.17) are indicated in bold.

Pleated structure

Coiled structure

(c) Tertiary structure—The pleated and coiled polypeptide chain of a protein molecule folds into a unique three-dimensional structure.

Three-dimensional folding

(d) Quaternary structure—Two or more polypeptide chains may be connected to form a single protein molecule.

Figure 2.18

The levels of protein structure sculpt the overall, three-dimensional conformation, which is vital to the protein's function.

Figure 2.19

A model of a portion of the protein collagen. The complex shape of a protein is characteristic of that protein and determines its functional properties.

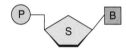

Figure 2.20

A nucleotide consists of a 5-carbon sugar (S = sugar), a phosphate group (P = phosphate), and a nitrogenous base (B = base).

Nucleic acids are of two types. One type—**RNA (ribonucleic acid)** (ri″bo-nu-kle′ik as′id)—is composed of molecules whose nucleotides have ribose. RNA usually is a single polynucleotide chain, but it can fold into various shapes that enable it to control when certain genes are accessed (fig. 2.21*a*). The second type—**DNA (deoxyribonucleic acid)** (de-ok′si-ri″bo-nu-kle″ik

as′id)—has deoxyribose and forms a double polynucleotide chain. The two chains are held together by hydrogen bonds (fig. 2.21*b*).

DNA molecules store information in a type of molecular code created by the sequences of the four types of nitrogenous bases. Cells use this information to synthesize protein molecules. RNA molecules carry out protein synthesis. (Nucleic acids are discussed in more detail in chapter 4, pp. 83–89.) Certain nucleotides, such as adenosine triphosphate (ATP), have another role providing energy to chemical reactions (fig. 2.22). ATP is discussed further in chapter 4 (p. 80). Table 2.6 summarizes the four groups of organic compounds. Clinical Application 2.2 discusses the use of biomarkers (both organic and inorganic compounds) in disease diagnosis, indicators of toxin exposure, and forensics.

> Recall that water molecules are polar. Many larger molecules, including carbohydrates, proteins, and nucleic acids, have polar regions as well and dissolve easily in water as a result. Unlike electrolytes, however, they do not dissociate when they dissolve in water—they remain intact. Such water-soluble molecules are said to be hydrophilic (they "like" water).
>
> Molecules that lack polar regions, such as triglycerides and steroids, do not dissolve in water ("oil and water don't mix"). Such molecules do dissolve in lipid and are said to be lipophilic (they "like" lipid).
>
> Water solubility and lipid solubility are important factors in drug delivery and in the movement of substances throughout the body. Much of this is discussed further in chapter 3 (pp. 53–54).

Practice

17. Compare the chemical composition of carbohydrates, lipids, proteins, and nucleic acids.
18. How does an enzyme affect a chemical reaction?
19. What is the chemical basis of the great diversity of proteins?
20. What are the functions of nucleic acids?

Table 2.6	Organic Compounds in Cells			
Compound	**Elements Present**	**Building Blocks**	**Functions**	**Examples**
Carbohydrates	C,H,O	Simple sugars	Provide energy, cell structure	Glucose, starch
Lipids	C,H,O (often P)	Glycerol, fatty acids, phosphate groups	Provide energy, cell structure	Triglycerides, phospholipids, steroids
Proteins	C,H,O,N (often S)	Amino acids	Provide cell structure, enzymes, energy	Albumins, hemoglobin
Nucleic acids	C,H,O,N,P	Nucleotides	Store information for protein synthesis; control cell activities	RNA, DNA

Clinical Application 2.2

Biomarkers

A biomarker is a chemical in the body that indicates a disease process or exposure to a toxin. Tests that measure levels of specific biomarkers may be used to indicate increased risk of developing a particular disease, to help diagnose a disease, or to select a treatment. Familiar biomarkers are the types of blood serum cholesterol. If the level of low-density lipoproteins (LDL) is elevated and the level of high-density lipoproteins (HDL) is low compared to healthy population values, then the individual may be at increased risk of developing heart and blood vessel disease. Considered along with other factors such as family history, the biomarker test results may lead a physician to prescribe an LDL-lowering statin drug.

A biomarker must meet certain criteria. Practical considerations include how easy it is to obtain and the cost, compared to the cost for treating disease detected much later. A blood or urine test, for example, is a simple way to sample a biomarker. A biomarker must also demonstrate sensitivity and specificity. Sensitivity is the ability to detect disease only when it is really present. Specificity is the test's ability to exclude the disease in a patient who does not actually have it.

A biomarker must also have the same predictive power in different individuals. For example, C-reactive protein is a biomarker of inflammation measured in a blood sample. It is being investigated as a marker for coronary heart disease,

because inflammation is part of this condition. However, in some families, C-reactive protein level does not correlate with increased risk of coronary heart disease. It is not considered to be as closely associated with heart disease as is elevated LDL cholesterol.

The four major types of chemicals in a human body—carbohydrates, lipids, proteins, and nucleic acids—form the basis of many biomarker tests. Examples include carbohydrates in nipple fluid that may indicate breast cancer; LDL in blood serum; prostate-specific antigen protein in a blood sample indicating an enlarged prostate gland; and DNA mutations or levels of specific mRNA molecules indicating a disease.

New biomarker tests evaluate several chemicals in a body fluid sample. To assess exposure to tobacco smoke, for example, a biomarker panel measures carbon monoxide and biochemicals that the body produces as it breaks down carcinogens in cigarette smoke. Biomarker panels are also valuable in describing a cancer so that an appropriate drug can be prescribed.

A different use of biomarkers is in forensics. Researchers identify the types of proteins in samples of biological material to determine the source. Peripheral blood, menstrual blood, saliva, mucus, tears, vaginal secretions, and semen all have distinctive protein signatures. In some cases knowing the nature of a biological substance at a crime scene can provide a powerful clue.

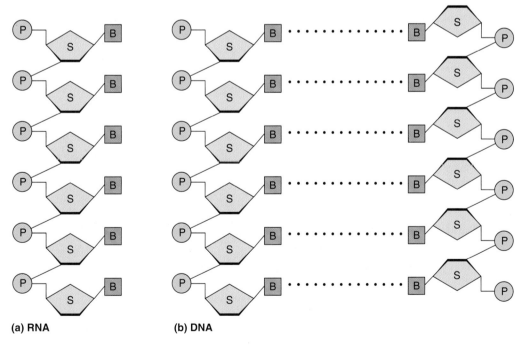

(a) RNA (b) DNA

Figure 2.21

A schematic representation of nucleic acid structure. A nucleic acid molecule consists of (a) one (RNA) or (b) two (DNA) polynucleotide chains. DNA chains are held together by hydrogen bonds (dotted lines), and they entwine, forming a double helix.

Figure 2.22

An ATP (adenosine triphosphate) molecule consists of an adenine, a ribose, and three phosphates. The wavy lines connecting the last two phosphates represent high-energy chemical bonds. When broken, these bonds release energy, which the cell uses for metabolic processes.

Summary Outline

2.1 Introduction (p. 31)

Chemistry describes the composition of substances and how chemicals react with each other. The human body is composed of chemicals.

2.2 Structure of Matter (p. 31)

1. Elements and atoms
   a. Matter is composed of elements.
   b. Some elements occur in pure form, but many are found combined with other elements.
   c. Elements are composed of atoms, which are the smallest complete units of elements.
   d. Atoms of different elements have characteristic size, weight, and ways of interacting.
2. Atomic structure
   a. An atom consists of one or more electrons surrounding a nucleus, which contains one or more protons and usually one or more neutrons.
   b. Electrons are negatively charged, protons are positively charged, and neutrons are uncharged.
   c. A complete atom is electrically neutral.
   d. An element's atomic number is equal to the number of protons in each atom. The atomic weight is equal to the number of protons plus the number of neutrons in each atom.
   e. Isotopes are atoms with the same atomic number but different atomic weights.
   f. Some isotopes are radioactive.
3. Bonding of atoms
   a. When atoms combine, they gain, lose, or share electrons.
   b. Electrons occupy shells around a nucleus.
   c. Atoms with completely filled outer shells are inert, but atoms with incompletely filled outer shells tend to gain, lose, or share electrons and thus achieve stable structures.
   d. Atoms that lose electrons become positively charged ions. Atoms that gain electrons become negatively charged ions.
   e. Ions with opposite electrical charges attract and form ionic bonds. Atoms that share electrons form covalent bonds.
   f. A polar covalently bonded molecule has an uneven distribution of charges.
   g. The attraction of positive to negative portions of polar covalent molecules is called a hydrogen bond.
4. Molecules and compounds
   a. Two or more atoms of the same element may bond to form a molecule of that element. Atoms of different elements may bond to form a molecule of a compound.
   b. Molecules consist of definite kinds and numbers of atoms.
5. Formulas
   a. A molecular formula represents the numbers and types of atoms in a molecule.
   b. A structural formula depicts the arrangement of atoms within a molecule.
6. Chemical reactions
   a. A chemical reaction breaks or forms bonds between atoms, ions, or molecules.
   b. Three types of chemical reactions are: synthesis, in which larger molecules form from smaller particles; decomposition, in which larger molecules are broken down into smaller particles; and exchange reactions, in which the parts of two different molecules trade positions.
   c. Many reactions are reversible. The direction of a reaction depends on the proportions of reactants and end products, the energy available, and the presence of catalysts.
7. Acids and bases
   a. Compounds that release ions in water are electrolytes.
   b. Electrolytes that release hydrogen ions are acids, and those that release hydroxide or other ions that react with hydrogen ions are bases.
   c. A value called pH represents a solution's concentration of hydrogen ions (H^+) and hydroxide ions (OH^-).
   d. A solution with equal numbers of H^+ and OH^- is neutral and has a pH of 7.0. A solution with more H^+ than OH^- is acidic and has a pH less than 7.0. A solution with fewer H^+ than OH^- is basic and has a pH greater than 7.0.
   e. Each whole number on the pH scale represents a tenfold difference in the hydrogen ion concentration.
   f. Buffers are chemicals that resist pH change.
   g. The pH in the internal environment is regulated.

2.3 Chemical Constituents of Cells (p. 39)

Molecules that have carbon and hydrogen atoms are organic and are usually nonelectrolytes. Other molecules are inorganic and are usually electrolytes.

1. Inorganic substances
   a. Water is the most abundant compound in the body and is a solvent in which chemical reactions occur. Water transports chemicals and heat.
   b. Oxygen releases energy from glucose and other nutrients. This energy drives metabolism.
   c. Carbon dioxide is produced when metabolism releases energy.
   d. Salts provide a variety of ions that metabolic processes require.
2. Organic substances
   a. Carbohydrates provide much of the energy that cells require and also contribute to cell structure. Their basic building blocks are simple sugar molecules.
   b. Lipids, such as fats, phospholipids, and steroids, supply energy and build cell parts. The basic building blocks of fats are molecules of glycerol and fatty acids.
   c. Proteins serve as structural materials, energy sources, hormones, cell surface receptors, and enzymes.
      (1) The building blocks of proteins are amino acids.
      (2) Proteins vary in the numbers, types, and sequences of their amino acids.
      (3) The amino acid chain of a protein molecule folds into a complex shape (conformation) that is maintained largely by hydrogen bonds.
      (4) Excessive heat, radiation, electricity, altered pH, or various chemicals can denature proteins.
   d. Nucleic acids are the genetic material and control cellular activities.
      (1) Nucleic acid molecules are composed of nucleotides.
      (2) The two types of nucleic acids are RNA and DNA.
      (3) DNA molecules store information that cell parts use to construct specific protein molecules. RNA molecules help synthesize proteins.

Chapter Assessments

2.1 Introduction

1. Define *chemistry*. (p. 31)

2.2 Structure of Matter

2. Define *matter*. (p. 31)

3. Explain the relationship between elements and atoms. (p. 31)

4. List the four most abundant elements in the human body. (p. 31)

5. Describe the parts of an atom and where they are found within the atom. (p. 32)

6. Explain why a complete atom is electrically neutral. (p. 32)

7. Define *atomic number, atomic weight,* and *isotope.* (p. 32)

8. Explain how electrons are distributed within the electron shells of an atom. (p. 33)

9. An ionic bond forms when (p. 33)
   a. atoms share electrons.
   b. positively charged and negatively charged parts of polar covalent molecules attract.
   c. ions with opposite electrical charges attract.
   d. two atoms exchange protons.
   e. an element has two types of isotopes.

10. Explain the relationship between molecules and compounds. (p. 36)

11. Show the difference between a molecular formula and a structural formula. (p. 37)

12. The formula $C_6H_{12}O_6$ means _____. (p. 37)

13. Three major types of chemical reactions are _____ , _____ , and _____. (p. 37)

14. Explain what a reversible reaction is. (p. 38)

15. Define *catalyst.* (p. 38)

16. Define *acid* and *base.* (p. 38)

17. Explain what pH measures, and describe the pH scale. (p. 39)

18. Define *buffer.* (p. 39)

2.3 Chemical Constituents of Cells

19. Distinguish between inorganic and organic substances. (p. 39)

20. Distinguish between electrolytes and nonelectrolytes. (p. 39)

21. Describe the roles water and oxygen play in the human body. (p. 40)

22. List several ions in body fluids. (p. 40)

23. Describe the general characteristics of carbohydrates. (p. 41)

24. Distinguish between simple sugars and complex carbohydrates. (p. 41)

25. Describe the general characteristics of lipids, and list the three main kinds of lipids. (p. 42)

26. A triglyceride molecule consists of (p. 42)
   a. cholesterol and 3 fatty acids.   d. 3 glycerols and 1 fatty acid.
   b. 3 monosaccharides.   e. 3 fatty acids and 1 glycerol.
   c. 3 amino acids.

27. Explain the difference between saturated and unsaturated fats. (p. 42)

28. A hydrophilic molecule dissolves in (p. 42)
   a. lipid but not water.   d. both lipid and water.
   b. water but not lipid.   e. alcohol and protein.
   c. neither lipid nor water.

29. List at least three functions of proteins. (p. 43)

30. Describe four levels of protein structure. (p. 44)

31. Explain how protein molecules may denature. (p. 44)

32. Describe the structure of nucleic acids. (p. 44)

33. Explain the major functions of nucleic acids. (p. 46)

Integrative Assessments/Critical Thinking

OUTCOME 2.2

1. An advertisement for a cosmetic powder claims that the product is "chemical-free." Explain why this is impossible.

2. The thyroid gland metabolizes iodine, the most common form of which has a molecular weight of 127 (^{127}I). A physician wants to use a radioactive isotope of iodine (^{123}I) to test whether a patient's thyroid gland is metabolizing normally. Based on what you know about how atoms react, do you think this physician's plan makes sense or not?

OUTCOMES 2.2, 2.3

3. What acidic and basic substances do you encounter in your everyday activities? What acidic foods do you eat regularly? What basic foods do you eat?

OUTCOME 2.3

4. A topping for ice cream contains fructose, hydrogenated soybean oil, salt, and cellulose. What types of chemicals are in it?

5. At a restaurant, a waiter recommends a sparkling carbonated beverage, claiming that it contains no carbohydrates. The product label lists water and fructose as ingredients. Is the waiter correct?

6. How would you explain the dietary importance of amino acids and proteins to a person who is following a diet composed primarily of carbohydrates?

7. A friend, while frying some eggs, points to the change in the egg white (which contains a protein called albumin) and explains that if the conformation of a protein changes, it will no longer have the same properties and will lose its ability to function. Do you agree or disagree with this statement?

WEB CONNECTIONS

Visit the text website at **www.mhhe.com/shieress11** for additional quizzes, interactive learning exercises, and more.

APR

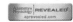

Anatomy & Physiology REVEALED includes cadaver photos that allow you to peel away layers of the human body to reveal structures beneath the surface. This program also includes animations, radiologic imaging, audio pronunciations, and practice quizzing. To learn more visit **www.aprevealed.com**.

3

Cells

Reprogramming a cell. The first signs of amyotrophic lateral sclerosis (ALS), also known as Lou Gehrig's disease, are subtle—a foot may drag, clothing may feel heavy on the body, or an exercise usually done with ease may become difficult. An actor was fired from a starring television role because his slurred speech was attributed to drunkenness; a teacher retired when he could no longer hold chalk or pens. Usually within five years of noticing these first signs, failure of the motor neurons that stimulate muscles becomes so widespread that breathing becomes impossible.

ALS currently has no treatment. Part of the reason is that because neurons do not divide, they cannot be grown long enough in laboratory culture to observe what goes wrong in ALS. A new technology called cellular reprogramming, however, can take a specialized cell type back to a stage at which it can specialize in any of several ways. Then, by adding certain chemical factors, researchers can guide the specialization toward the cell type that is affected in a certain disease. Cells can be reinvented in this way because they all contain the same complete set of genes. Such a reprogrammed cell is like a stem cell, but it does not require derivation from an embryo—and it grows in a lab dish.

For ALS, cells taken from arm skin of two women in their eighties who have mild cases of the disease were reprogrammed to specialize as motor neurons. Researchers can now observe the very first inklings of the disease. Such knowledge can be used to identify new drug targets and develop new drugs.

Neurons cannot be cultured in a laboratory dish for very long, and therefore are difficult to study (400x). Researchers can study them by using a new technique called reprogramming, which stimulates one cell type to become another.

ALS was the first of dozens of diseases now represented by reprogrammed cells, including inherited immune deficiencies, diabetes, blood disorders, and Parkinson's disease. In the future, reprogrammed cells might be used therapeutically to replace abnormal cells. First, though, researchers must learn how to control the integration of reprogrammed cells into tissues and organs in the body.

Learning Outcomes

After studying this chapter, you should be able to do the following:

3.1 Introduction

1. Explain how cells differ from one another. (p. 51)

3.2 Composite Cell

2. Explain how the structure of a cell membrane makes possible its functions. (p. 53)

3. Describe each type of organelle, and explain its function. (p. 55)

4. Describe the parts of the cell nucleus and its parts. (p. 60)

3.3 Movements Through Cell Membranes

5. Explain how substances move into and out of cells. (p. 60)

3.4 The Cell Cycle

6. Explain why regulation of the cell cycle is important to health. (p. 67)

7. Describe the cell cycle. (p. 69)

8. Explain how stem cells and progenitor cells make possible growth and repair of tissues. (p. 71)

9. Explain how two differentiated cell types can have the same genetic information, but different appearances and functions. (p. 71)

Module 2: Cells & Chemistry

 Learn　 **Practice**　 **Assess**

Aids to Understanding Words

(Appendix A on page 564 has a complete list of Aids to Understanding Words.)

cyt- [cell] *cyt*oplasm: Fluid (cytosol) and organelles that occupy the space between the cell membrane and the nuclear envelope.

endo- [within] *endo*plasmic reticulum: Complex of membranous structures within the cytoplasm.

hyper- [above] *hyper*tonic: Solution that has a greater osmotic pressure than body fluids.

hypo- [below] *hypo*tonic: Solution that has a lesser osmotic pressure than body fluids.

inter- [between] *inter*phase: Stage between the end of one cell division and the beginning of the next.

iso- [equal] *iso*tonic: Solution that has the same osmotic pressure as body fluids.

mit- [thread] *mit*osis: Process of cell division when threadlike chromosomes become visible within a cell.

phag- [to eat] *phag*ocytosis: Process by which a cell takes in solid particles.

pino- [to drink] *pino*cytosis: Process by which a cell takes in tiny droplets of liquid.

-som [body] ribo*som*e: Tiny, spherical structure that consists of protein and RNA and functions in protein synthesis.

3.1 INTRODUCTION

Recipe for a human being: cells, their products, and fluids. A cell, as the unit of life, is a world unto itself. To build a human, about 75 trillion cells connect and interact, forming dynamic tissues, organs, and organ systems.

The cells that make up an adult human body have similarities and distinctions. They consist of the same basic structures, yet vary considerably in the number and distribution of their component structures, and in size and shape. The three-dimensional forms of cells make possible their functions, as figure 3.1 illustrates. For instance, some nerve cells have long, threadlike extensions that transmit electrical impulses from one part of the body to another. Epithelial cells that line the inside of the mouth are thin, flattened, and tightly packed into a tile-like layer that protects cells beneath them. Muscle cells, which pull structures closer together, are slender and rodlike. The precise alignment of the protein fibers in muscle cells provides the strength to withstand the contraction that moves the structures to which they attach.

A cell continually carries out activities essential for life, as well as more specialized functions, and adapts to changing conditions. The genes control a cell's actions and responses. Nearly all cells have a full set of genetic instructions (the genome), yet they use only some of this information. Like a person accessing only a small part of the Internet to learn something, a cell accesses only some of the vast store of information in the genome to survive and specialize.

(a) A nerve cell's long extensions enable it to transmit electrical impulses from one body part to another.

(b) The sheetlike organization of epithelial cells enables them to protect underlying cells.

(c) The alignment of contractile proteins within muscle cells enables them to contract, pulling closer together the structures to which they attach.

Figure 3.1
Cells vary in structure and function.

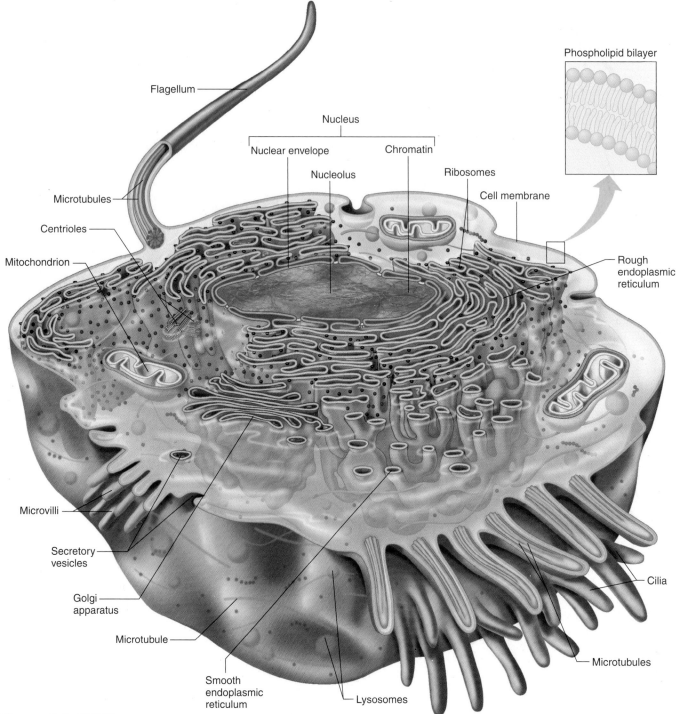

Figure 3.2 AP|R

A composite cell illustrates the organelles and other structures found in cells. Specialized cells differ in the numbers and types of organelles, reflecting their functions. Organelles are not drawn to scale.

3.2 COMPOSITE CELL

Cells vary greatly in size, shape, content, and function, and therefore describing a "typical" cell is challenging. The cell shown in figure 3.2 and described in this chapter is a composite cell that includes many known cell structures. In reality, any given cell has most, but perhaps not all, of these structures, and cells have differing numbers of some of them.

Under the light microscope, a properly applied stain reveals three basic cell parts: the **cell membrane** (sel mem′-brān) that encloses the cell, the **nucleus** (nu′kle-us) that houses the genetic material and controls cellular activities, and the **cytoplasm** (si′to-plazm) that fills out the cell.

Within the cytoplasm are specialized structures called **organelles** (or-gan-elz′), which can be seen clearly only under the higher magnification of electron microscopes. Organelles are suspended in a liquid called *cytosol*. They are not static and still, as figure 3.2 might suggest. Some organelles move within the cell, and even those that appear not to move are the sites of ongoing biochemical activity. Organelles perform specific functions, such as partitioning off biochemicals that might harm other cell parts; dismantling debris; processing secretions; and extracting energy from nutrients.

Practice

1. Give three examples of how a cell's shape makes possible the cell's function.
2. Name the three major parts of a cell and their functions.
3. Define organelles and explain their general functions in a cell.

Cell Membrane

The cell membrane (also called the *plasma membrane*) is more than a simple boundary surrounding the cellular contents. It is an actively functioning part of the living material. The cell membrane regulates movement of substances in and out of the cell and is the site of much biological activity. Many of a cell's actions that enable it to survive and to interact with other cells use a molecular communication process called signal transduction. A series of molecules that are part of the cell membrane form pathways that detect signals from outside the cell and transmit them inward, where yet other molecules orchestrate the cell's response. The cell membrane also helps cells attach to certain other cells, which is important in forming tissues.

General Characteristics

The cell membrane is extremely thin, flexible, and somewhat elastic. It typically has complex surface features with many outpouchings and infoldings that increase surface area (fig. 3.2). In addition to maintaining cell integrity, the cell membrane is **selectively permeable** (se-lek′tiv-le per′me-ah-bl) (also known as *semipermeable* or *differentially permeable*), which means that only certain substances can enter or leave the cell.

Cell Membrane Structure

A cell membrane is composed mainly of lipids and proteins, with fewer carbohydrates. Its basic framework is a double layer, or *bilayer*, of phospholipid molecules. Each phospholipid molecule includes a phosphate group and two fatty acids bound to a glycerol molecule (see chapter 2, p. 42). The water-soluble phosphate "heads" form the surfaces of the membrane, and the water-insoluble fatty acid "tails" make up the interior of the membrane. The lipid molecules can move sideways within the plane of the membrane. The two membrane layers form a soft and flexible, but stable, fluid film.

The cell membrane's interior is oily because it consists largely of the fatty acid tails of the phospholipid molecules (fig. 3.3). Molecules such as oxygen and carbon dioxide,

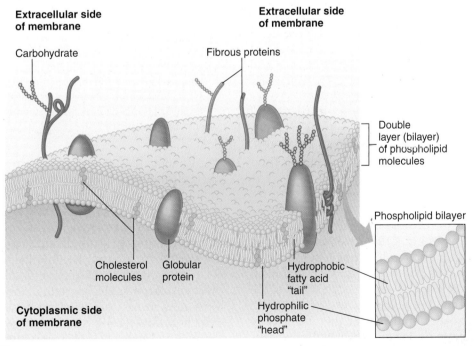

Figure 3.3 AP|R

The cell membrane is composed primarily of phospholipids (and some cholesterol), with proteins embedded throughout the lipid bilayer. Parts of the membrane-associated proteins that extend from the outer surface help to establish the identity of the cell as part of a particular tissue, organ, and person.

Clinical Application 3.1

Too Little or Too Much Pain

The ten-year-old boy amazed the people on the streets of the small northern Pakistani town. He was completely unable to feel pain and had become a performer, stabbing knives through his arms and walking on hot coals to entertain crowds. Several other people in this community, where relatives often married relatives, were also unable to feel pain. Researchers studied the connected families and discovered a mutation that alters sodium channels on certain nerve cells. The mutation blocks the channels so that the message to feel pain cannot be sent. The boy died at age thirteen from jumping off a roof.

His genes could protect him from pain, but pain protects against injury by providing a warning.

A different mutation affecting the same sodium channels causes very different symptoms. In "burning man syndrome," the channels become hypersensitive, opening and flooding the body with pain easily, in response to exercise, an increase in room temperature, or just putting on socks. In another condition, "paroxysmal extreme pain disorder," the sodium channels stay open too long, causing excruciating pain in the rectum, jaw, and eyes. Researchers are using the information from these genetic studies to develop new painkillers.

which are soluble in lipids, can easily pass through this bilayer. However, the bilayer is impermeable to water-soluble molecules, which include amino acids, sugars, proteins, nucleic acids, and various ions. Cholesterol molecules embedded in the cell membrane's interior help make the membrane less permeable to water-soluble substances, while their rigid structure stabilizes the membrane.

A cell membrane includes a few types of lipid molecules, but many kinds of proteins, which provide special functions. Membrane proteins are classified according to their positions. Membrane-spanning (transmembrane) proteins extend through the lipid bilayer and may protrude from one or both faces. Peripheral membrane proteins associate mostly with one side of the bilayer. Membrane proteins also vary in shape—they may be globular, rodlike, or fibrous. The cell membrane has been described as a "fluid mosaic" because its proteins are embedded in an oily background and therefore can move, like ships on a sea. Some lipids outside the bilayer join, forming "rafts" to which proteins that function together may cluster, easing their interactions.

Membrane proteins have a variety of functions. Some form receptors on the cell surface that bind incoming hormones or growth factors, starting signal transduction. Receptors are structures that have specific shapes that fit and hold certain molecules. Many receptors are partially embedded in the cell membrane. Other proteins transport ions or molecules across the cell membrane. Some membrane proteins form ion channels in the phospholipid bilayer that allow only particular ions to enter or leave. Ion channels are specific for calcium (Ca^{+2}), sodium (Na^+), potassium (K^+), or

chloride (Cl^-). A cell membrane may have a few thousand ion channels specific for each of these ions.

Many ion channels open or close like a gate under specific conditions, such as a change in electrical forces across the membrane of a nerve cell, or receiving biochemical messages from inside or outside the cell. Clinical Application 3.1 discusses how ion channels are involved in feeling—or not feeling—pain.

> Ten million or more ions can pass through an ion channel in one second!

> Drugs may act by affecting ion channels, and abnormal ion channels cause certain disorders. In cystic fibrosis, for example, abnormal chloride channels in cells lining the lung passageways and ducts of the pancreas cause the symptoms. Sodium channels also malfunction. The overall result: Salt trapped inside cells draws moisture in and thickens surrounding mucus.

Proteins that extend inward from the inner face of the cell membrane anchor it to the protein rods and tubules that support the cell from within. Proteins that extend from the outer surface of the cell membrane mark the cell as part of a particular tissue or organ belonging to a particular person. This identification as self is important for the functioning of the immune system (see chapter 14, p. 386). Many of these proteins are attached to carbohydrates, forming glycoproteins. Another type of protein on a cell's surface is a cellular adhesion molecule (CAM), which guides a cell's interactions with other cells. For example, a series of CAMs helps a white blood cell move to the site of an injury, such as a splinter in the skin.

Cytoplasm

The cytoplasm is the gel-like material in which organelles are suspended—it makes up most of a cell's volume. When viewed through a light microscope, cytoplasm usually appears as a clear jelly with specks scattered throughout. However, an electron microscope, which provides much greater magnification and the ability to distinguish fine detail (resolution), reveals that the cytoplasm contains networks of membranes and organelles suspended in the clear liquid *cytosol*. Cytoplasm also includes abundant protein rods and tubules that form a framework, or **cytoskeleton** (si″to-skel′e-ten), meaning "cell skeleton."

Most cell activities occur in the cytoplasm, where nutrients are received, processed, and used. The following organelles have specific functions in carrying out these activities:

1. **Endoplasmic reticulum** (en′do-plaz′mik rĕ-tik′u-lum) The endoplasmic reticulum (ER) is a complex organelle composed of membrane-bounded, flattened sacs, elongated canals, and fluid-filled, bubblelike sacs called *vesicles*. These membranous parts are interconnected and communicate with the cell membrane, the nuclear envelope, and other organelles. The ER provides a vast tubular network that transports molecules from one cell part to another. It winds from the nucleus out toward the cell membrane.

    The endoplasmic reticulum participates in the synthesis of protein and lipid molecules. These molecules may leave the cell as secretions or be used within the cell for such functions as producing new ER or cell membrane as the cell grows. The ER acts as a quality control center for the cell. Its chemical environment enables a forming protein to start to fold into the shape necessary for its function. The ER can identify and dismantle a misfolded protein, much as a defective toy might be pulled from an assembly line at a factory and discarded.

    In many places, the ER's outer membrane is studded with many tiny, spherical structures called *ribosomes*, which give the ER a textured appearance when viewed with an electron microscope (fig. 3.4a,b). These parts of the ER are called *rough ER*. The ribosomes are sites of protein synthesis and are found in the cytoplasm as well as associated with ER. Proteins being synthesized move through ER tubules to another organelle, the Golgi apparatus, for further processing.

    As the ER nears the cell membrane, it widens and ribosomes become sparse and then are no longer associated with the ER. This section of the ER is called *smooth ER* (fig. 3.4c). Along the smooth ER are enzymes that are important in lipid synthesis, absorption of fats from the digestive tract, and the metabolism of drugs. Cells that break down drugs and alcohol, such as liver cells, have extensive networks of smooth ER.

2. **Ribosomes** (ri′bo-sōmz) Ribosomes, where protein synthesis occurs, are attached to ER membranes or are scattered throughout the cytoplasm. Clusters of ribosomes in the cytoplasm, called *polysomes,* enable a cell to quickly manufacture proteins required in large amounts. All ribosomes are composed of protein and RNA molecules. Ribosomes provide enzymatic activity as well as a structural support for the RNA molecules that come together as the cell links amino acids to form proteins, discussed in chapter 4 (pp. 85–89).

3. **Golgi apparatus** (gol′je ap″ah-ra′tus) The Golgi apparatus is a stack of about six flattened, membranous sacs. This organelle refines, packages, and transports proteins synthesized on ribosomes associated with the ER. Proteins arrive at the Golgi apparatus enclosed in vesicles (sacs) composed of the ER membrane. These vesicles fuse with the membrane at the innermost end of the Golgi apparatus, which is specialized to receive glycoproteins.

    As glycoproteins pass from layer to layer through the stacks of Golgi membrane, they are modified chemically. Sugar molecules may be added or removed. When the altered glycoproteins reach the outermost layer, they are packaged in bits of Golgi membrane, which bud off and form bubblelike transport vesicles. Such a vesicle may then move to and fuse with the cell membrane, releasing its contents to the outside as a secretion (figs. 3.2 and 3.5). This process is called *exocytosis* (see page 65).

Researchers are creating "artificial organelles" to use in industrial processes and to better understand how cells work. So far, artificial ribosomes and Golgi apparatuses have been constructed, with ER coming soon. These parts of the secretory network can produce protein if given genetic information. The first protein manufactured on artificial ribosomes was firefly luciferase, responsible for the insect's famous "glow."

Practice

4. What is a *selectively permeable membrane?*
5. Describe the chemical structure of a cell membrane.
6. What are the functions of the endoplasmic reticulum?
7. What are the functions of the Golgi apparatus?
8. Explain how organelles and other structures interact to secrete substances from the cell.

ER membrane
Ribosomes

(a)

Membranes

Membranes

Ribosomes

(b)   (c)

Figure 3.4

The endoplasmic reticulum is the site of protein and lipid synthesis, and serves as a transport system. **(a)** A transmission electron micrograph of rough endoplasmic reticulum (ER) (28,500×). **(b)** Rough ER is dotted with ribosomes, whereas **(c)** smooth ER lacks ribosomes.

4. **Mitochondria** (mi″to-kon′dre-ah; *sing.* mi″to-kon′dre-on) Mitochondria are elongated, fluid-filled sacs that vary in size and shape. They can move slowly through the cytoplasm and reproduce by dividing. A mitochondrion has an outer and an inner layer (figs. 3.2 and 3.6). The inner layer is folded extensively into partitions called *cristae.* Connected to the cristae are enzymes that control some of the chemical reactions that release energy from certain nutrient molecules in a process called cellular respiration. Mitochondria are the major sites of chemical reactions that capture and store this energy in the chemical bonds of adenosine triphosphate (ATP). A cell can easily use energy stored as ATP. This is why very active cells, such as muscle cells, have many thousands of mitochondria. (Chapter 4, p. 82, describes this energy-releasing function in more detail.) Mitochondria resemble bacterial cells and contain a small amount of their own DNA.

5. **Lysosomes** (li′so-sōmz) Lysosomes, the "garbage disposals of the cell," are tiny membranous sacs (see fig. 3.2). They bud off of sections of Golgi membranes and have an acidic pH that enables certain enzymes to function. The powerful lysosomal enzymes break down nutrient molecules or foreign particles. Certain white blood cells, for example, can engulf bacteria, which are then digested by the lysosomal enzymes. In liver cells, lysosomes break down cholesterol, toxins, and drugs. Lysosomes also destroy worn cellular parts. Genetics Connection 3.1 describes disorders that result from deficiencies of lysosomal enzymes. AP|R

6. **Peroxisomes** (pĕ-roks′ĭ-sōmz) These membranous sacs are abundant in liver and kidney cells. They house enzymes that catalyze (speed) a variety of biochemical reactions, including breakdown of hydrogen peroxide (a by-product of metabolism) and fatty acids; and detoxification of alcohol.

Figure 3.5

The Golgi apparatus processes secretions. **(a)** A transmission electron micrograph of a Golgi apparatus (48,500×). **(b)** The Golgi apparatus consists of membranous sacs that continually receive vesicles from the endoplasmic reticulum and produce vesicles that enclose secretions.

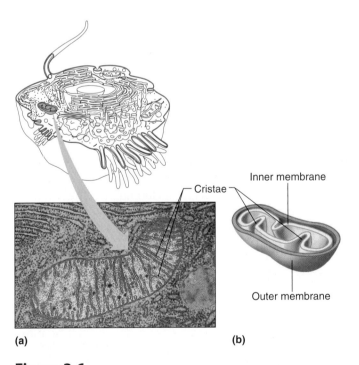

(a)   (b)

Figure 3.6

A mitochondrion is a major site of energy reactions. **(a)** A transmission electron micrograph of a mitochondrion (28,000×). **(b)** Cristae partition this saclike organelle.

7. **Microfilaments and microtubules** Microfilaments and microtubules are two types of thin, threadlike strands in the cytoplasm. They form the cytoskeleton and are also part of certain structures that have specialized activities.

   Microfilaments are tiny rods of a protein called actin. They form meshworks or bundles, and provide cell motility (movement). In muscle cells, for example, microfilaments aggregate to form *myofibrils*, which help these cells contract (see chapter 8, p. 179).

   Microtubules are long, slender tubes with diameters two or three times those of microfilaments (fig. 3.7). Microtubules are composed of molecules of a globular protein called tubulin, attached in a spiral to form a long tube. They are important in cell division. Intermediate filaments lie between microfilaments and microtubules in diameter, and are made of different proteins in different cell types. They are abundant in skin cells and neurons, but scarce in other cell types.

8. **Centrosome** (sen′tro-sōm) The centrosome is a structure near the Golgi apparatus and nucleus. It is nonmembranous and consists of two hollow cylinders, called *centrioles,* which are composed of microtubules organized in nine groups of

Genetics Connection 3.1

Lysosomal Storage Diseases

Hunter Kelly was born in 1997, the son of NFL quarterback Jim Kelly and his wife Jill. At first, Hunter cried frequently and had difficulty feeding, and his limbs seemed stiff. He became less alert, and motor skill development slowed and then stopped. At nine months old, Hunter was diagnosed with Krabbe disease, which he had inherited from his parents, who are carriers. The lysosomes in Hunter's cells could not make an enzyme that is necessary to produce myelin, a lipid that insulates neurons. As a result, there was a buildup of the biochemical that the enzyme normally acts on to form myelin, and his neurons did not have enough myelin.

Unfortunately, by the time of diagnosis, damage to Hunter's nervous system was already advanced. He ceased moving and responding, lost hearing and vision, and had to be fed by tube. Hunter Kelly lived for eight years. Had he been born today, he would have been tested for Krabbe disease along with dozens of other such "inborn errors of metabolism" with a few drops of blood taken from his heel shortly after birth. A stem cell transplant from a donor's umbilical cord blood may have prevented his symptoms.

Lysosomes house 43 different types of enzymes, and so 43 different types of disorders fall under the heading of "lysosomal storage diseases," a subtype of inborn errors. Each enzyme must be present within a certain concentration range in order for the cell to function properly. Although each of the disorders is rare, together they affect about 10,000 people worldwide. Some lysosomal storage diseases can be treated in any of three ways, depending upon the nature of the abnormality: replacing the enzyme, using a drug to reduce the biochemical buildup, or using a drug that can unfold and correctly refold a misfolded enzyme, enabling it to function.

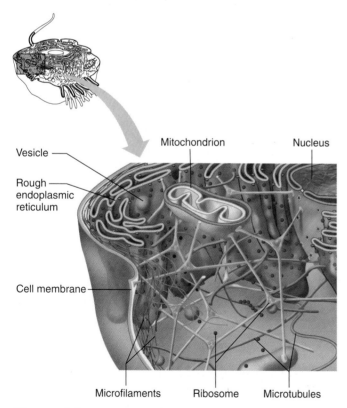

Figure 3.7

The cytoskeleton provides an inner framework for a cell. Microtubules built of tubulin and microfilaments built of actin help maintain the shape of a cell by forming a scaffolding beneath the cell membrane and in the cytoplasm. A cell's shape is critical to its function.

three (figs. 3.2 and 3.8). The centrioles lie at right angles to each other. During mitosis, the centrioles distribute chromosomes to newly forming cells.

9. **Cilia and flagella** Cilia and flagella are motile structures that extend from the surfaces of certain cells. They are composed of microtubules in a "9 + 2" array, similar to centrioles but with two additional microtubules in the center. Cilia and flagella are similar structures that differ mainly in length and abundance.

Cilia fringe the free surfaces of some epithelial (lining) cells. Each cilium is tiny and hairlike, and is attached beneath the cell membrane (see fig. 3.2). Cilia form in precise patterns. They move in a coordinated "to-and-fro" manner, so that rows of them beat in succession, producing a wave of motion. This wave moves fluids, such as mucus, over the surface of certain tissues, including those that form the inner linings of the respiratory tubes (fig. 3.9*a*).

Flagella are much longer than cilia, and usually a cell has only one. A flagellum moves in an undulating wave, which begins at its base. The tail of a sperm cell is a flagellum that enables the cell to "swim." The sperm tail is the only flagellum in humans (fig. 3.9*b*).

Cilia do more than wave. Certain cilia lining the respiratory tubes have receptors that detect bitter chemicals. These cilia initiate signaling that alters the wave pattern so that the bitter substance—possibly a poison—is sent out of the body. The receptors are identical to those that function in the sense of taste.

(a)

Centriole
(cross section)

Centriole
(longitudinal section)

(b)

Figure 3.8

Centrioles. **(a)** Transmission electron micrograph of the two centrioles in a centrosome (120,000×). **(b)** The centrioles lie at right angles to one another. These structures help to apportion the chromosomes of a dividing cell into two cells.

Cilia

Flagellum

(a)

(b)

Figure 3.9

Cilia and flagella provide movement. **(a)** Cilia are motile, hairlike extensions that fringe the surfaces of certain cells, including those that form the inner lining of the respiratory tubes (5,800×). Cilia remove debris from the respiratory tract with their sweeping, to-and-fro movement. **(b)** Flagella form the tails of these human sperm cells, enabling them to "swim" (840×).

10. **Vesicles** (ves′ĭ-klz) Vesicles are membranous sacs that store or transport substances within a cell. Larger vesicles that contain mostly water form from part of the cell membrane folding inward and pinching off, carrying liquid or solid material formerly outside the cell into the cytoplasm. Smaller vesicles shuttle material from the rough ER to the Golgi apparatus as part of secretion (see fig. 3.2).

Practice

9. Describe a mitochondrion.

10. What is the function of a lysosome?

11. How do microfilaments and microtubules differ?

12. Identify a structure that consists of microtubules.

13. What is a centrosome and what does it do?

14. Locate cilia and flagella and explain what they are composed of and what they do.

Cell Nucleus

The nucleus houses the genetic material (DNA), which directs all cell activities (figs. 3.2 and 3.10). It is a large, roughly spherical structure enclosed in a double-layered **nuclear envelope,** which consists of inner and outer lipid bilayer membranes. The nuclear envelope has protein-lined channels called *nuclear pores* that allow certain molecules to exit the nucleus. A nuclear pore is not just a hole, but a complex opening formed from 100 or so types of proteins. A nuclear pore is small enough to let out the RNA molecules that carry genes' messages, but not large enough to let out the DNA itself, which must remain in the nucleus to maintain the genetic instruction set.

The nucleus contains a fluid, called *nucleoplasm,* in which the following structures are suspended:

1. **Nucleolus** (nu-kle′o-lus) A nucleolus ("little nucleus") is a small, dense body composed largely of RNA and protein. It has no surrounding membrane and forms in specialized regions of certain chromosomes. Ribosomes form in the nucleolus and then migrate through nuclear pores to the cytoplasm.
2. **Chromatin** Chromatin consists of loosely coiled fibers of DNA and protein that condense to form structures called **chromosomes** (kro′mo-sōmz).

The DNA contains the information for protein synthesis. When the cell begins to divide, chromatin fibers coil tightly, and individual chromosomes become visible when stained and viewed under a light microscope. At other times, chromatin unwinds locally to permit the information in certain genes (DNA sequences) to be accessed.

Table 3.1 summarizes the structures and functions of organelles.

Practice

15. Identify the structure that separates the nuclear contents from the cytoplasm.
16. What is produced in the nucleolus?
17. Describe chromatin and how it changes.

3.3 MOVEMENTS THROUGH CELL MEMBRANES

The cell membrane is a selective barrier that controls which substances enter and leave the cell. Movements of substances into and out of cells include passive mechanisms that do not require cellular energy (diffusion, facilitated diffusion, osmosis, and filtration) and active mechanisms that use cellular energy (active transport, endocytosis, and exocytosis).

Passive Mechanisms

Diffusion

Diffusion (dĭ-fu′zhun) (also called *simple diffusion*) is the tendency of molecules or ions in a liquid or air

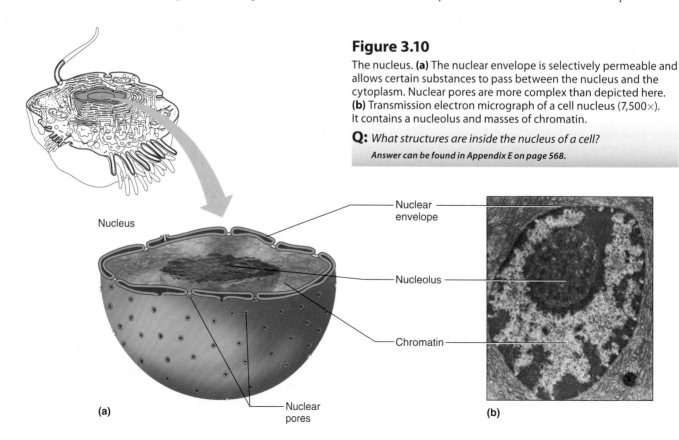

Figure 3.10

The nucleus. **(a)** The nuclear envelope is selectively permeable and allows certain substances to pass between the nucleus and the cytoplasm. Nuclear pores are more complex than depicted here. **(b)** Transmission electron micrograph of a cell nucleus (7,500×). It contains a nucleolus and masses of chromatin.

Q: *What structures are inside the nucleus of a cell?*

Answer can be found in Appendix E on page 568.

Nucleus
Nuclear envelope
Nucleolus
Chromatin
Nuclear pores
(a)
(b)

Table 3.1	Structures and Functions of Cell Parts	
Cell Part(s)	**Structure**	**Function**
Cell membrane	Membrane composed of protein and lipid molecules	Maintains integrity of cell and controls passage of materials into and out of cell
Endoplasmic reticulum	Complex of interconnected membrane-bounded sacs and canals	Transports materials within cell, provides attachment for ribosomes, and synthesizes lipids
Ribosomes	Particles composed of protein and RNA molecules	Synthesize proteins
Golgi apparatus	Group of flattened, membranous sacs	Packages protein molecules for transport and secretion
Mitochondria	Membranous sacs with inner partitions	Release energy from nutrient molecules and change energy into a usable form
Lysosomes	Membranous sacs	Digest worn cellular parts or substances that enter cells
Peroxisomes	Membranous sacs	House enzymes that catalyze diverse reactions, including breakdown of hydrogen peroxide and fatty acids, and alcohol detoxification
Microfilaments and microtubules	Thin rods and tubules	Support the cytoplasm and help move substances and organelles within the cytoplasm
Centrosome	Nonmembranous structure composed of two rodlike centrioles	Helps distribute chromosomes to new cells during cell division
Cilia and flagella	Motile projections attached beneath the cell membrane	Cilia propel fluid over cellular surfaces, and a flagellum enables a sperm cell to move
Vesicles	Membranous sacs	Contain and transport various substances
Nuclear envelope	Double membrane that separates the nuclear contents from the cytoplasm	Maintains integrity of nucleus and controls passage of materials between nucleus and cytoplasm
Nucleolus	Dense, nonmembranous body composed of protein and RNA	Site of ribosome synthesis
Chromatin	Fibers composed of protein and DNA	Contains information for synthesizing proteins

solution to move from regions of higher concentration to regions of lower concentration, thus becoming more evenly distributed, or more *diffuse*. Diffusion occurs because molecules and ions are in constant motion. Each particle travels in a separate path along a straight line until it collides and bounces off another particle, changing direction, colliding again, and changing direction once more. At body temperature, small molecules such as water move more than a thousand miles per hour. A single molecule may collide with other molecules a million times each second.

Collisions are less likely if there are fewer particles, so there is a net movement of particles from a region of higher concentration to a region of lower concentration. The difference in concentration is called a *concentration gradient,* and diffusion occurs down a concentration gradient. With time, the concentration of a given substance becomes uniform throughout a solution, *diffusional equilibrium* (dĭ-fu′zhunl e″kwĭ-lib′re-um). Although random movements continue, there is no further net movement, and the concentration of a substance is uniform throughout the solution.

Sugar (a solute) in a sugar cube put in a glass of water (a solvent), can be used to illustrate diffusion (fig. 3.11). At first the sugar remains highly concentrated

at the bottom of the glass. Diffusion moves the sugar molecules from the area of high concentration and disperses them into solution among the moving water molecules. Eventually, the sugar molecules become uniformly distributed in the water.

Diffusion of a substance across a membrane can happen only if (1) the membrane is permeable to that substance, and (2) a concentration gradient exists such that the substance is at a higher concentration on one side of the membrane or the other (fig. 3.12). Consider oxygen and carbon dioxide, to which cell membranes are permeable. In the body, oxygen diffuses into cells and carbon dioxide diffuses out of cells, but equilibrium is never reached. Intracellular oxygen is always low because oxygen is constantly used up in metabolic reactions. Extracellular oxygen is maintained at a high level by homeostatic mechanisms in the respiratory and cardiovascular systems. Thus, a concentration gradient always allows oxygen to diffuse into cells.

The level of carbon dioxide, which is a metabolic waste product, is always high inside cells. Homeostasis maintains a lower extracellular carbon dioxide level, so a concentration gradient always favors carbon dioxide diffusing out of cells (fig. 3.13).

(a) **(b)** **(c)** **(d)**

Time ⟶

Figure 3.11

A dissolving sugar cube illustrates diffusion. **(a–c)** A sugar cube placed in water slowly disappears as the sugar molecules dissolve and then diffuse from regions where they are more concentrated toward regions where they are less concentrated. **(d)** Eventually the sugar molecules are distributed evenly throughout the water.

Permeable
membrane

● Solute molecule
● Water molecule

A B   A B   A B

(1) **(2)** **(3)**

Figure 3.12 AP|R

Time ⟶

Diffusion is a passive movement of molecules. **(1)** A membrane permeable to water and solute molecules separates a container into two compartments. Compartment *A* contains both types of molecules, while compartment *B* contains only water molecules. **(2)** As a result of molecular motions, solute molecules tend to diffuse from compartment *A* into compartment *B*. Water molecules tend to diffuse from compartment *B* into compartment *A*. **(3)** Eventually equilibrium is reached.

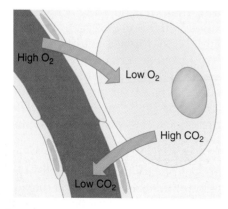

High O_2

Low O_2

High CO_2

Low CO_2

Figure 3.13

Diffusion enables oxygen to enter cells and carbon dioxide to leave.

Dialysis is a technique that uses diffusion to separate small molecules from larger ones in a liquid. The artificial kidney uses a variant of this process—*hemodialysis*—to treat patients suffering from kidney damage or failure. An artificial kidney (dialyzer) passes blood from a patient through a long, coiled tubing composed of porous cellophane. The size of the pores allows smaller molecules carried in the blood, such as the waste material urea, to exit through the tubing, while larger molecules, such as those of blood proteins, remain inside the tubing. The tubing is submerged in a tank of dialyzing fluid (wash solution), which contains varying concentrations of different chemicals. Altering the concentrations of molecules in the dialyzing fluid can control which molecules diffuse out of blood and which remain in it.

Facilitated Diffusion

Substances that are not able to pass through the lipid bilayer need the help of membrane proteins to get across, a process known as **facilitated diffusion** (fah-sil″ĭ-ta‾t′ed dĭ-fu′zhun) (fig. 3.14). One form of facilitated diffusion uses the ion channels and pores described earlier. Molecules such as glucose and amino acids are not lipid-soluble, but are too large to pass through membrane channels. They enter cells by another form of facilitated diffusion that uses a carrier molecule. For example, a glucose molecule outside a cell combines with a special protein carrier molecule at the surface of the cell membrane. The union of the glucose and the carrier molecule changes the shape of the carrier, enabling it to move glucose to the other side of the membrane. The carrier releases the glucose and then returns to its original shape and picks up another glucose molecule. The hormone *insulin,* discussed in chapter 11 (p. 308), promotes facilitated diffusion of glucose through the membranes of certain cells.

Facilitated diffusion is similar to simple diffusion in that it only moves molecules from regions of higher concentration toward regions of lower concentration. The number of carrier molecules in the cell membrane limits the rate of facilitated diffusion, which occurs in most cells.

Osmosis

Osmosis (oz-mo′sis) is the movement of water across a selectively permeable membrane into a compartment containing solute that cannot cross the same membrane. The mechanism of osmosis is complex, but in part involves the diffusion of water down its concentration gradient. In the following example, assume that the selectively permeable membrane is permeable to water molecules (the solvent) but impermeable to protein molecules (the solute).

In solutions, a higher concentration of solute (protein in this case) means a lower concentration of water; a lower concentration of solute means a higher concentration of water. This is because solute molecules take up space that water molecules would otherwise occupy.

Like molecules of other substances, molecules of water diffuse from areas of higher concentration to areas of lower concentration. In figure 3.15, the greater concentration of protein in compartment *A* means that the water concentration there is less than the concentration of pure water in compartment *B*. Therefore, water diffuses from compartment *B* across the selectively permeable membrane and into compartment *A*. In other

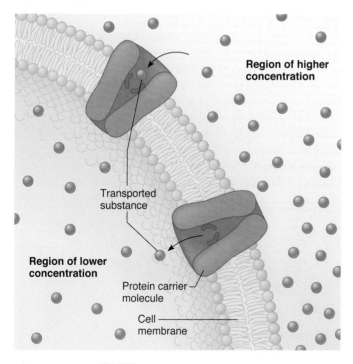

Figure 3.14 AP|R

Facilitated diffusion uses carrier molecules to transport some substances into or out of cells, from a region of higher concentration to one of lower concentration.

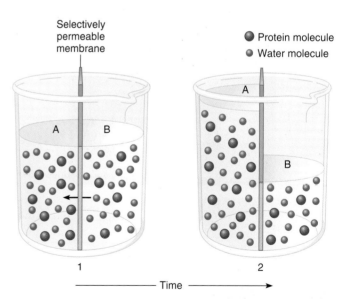

Figure 3.15

Osmosis. **(1)** A selectively permeable membrane separates the container into two compartments. At first, compartment *A* contains a higher concentration of protein (and a lower concentration of water) than compartment *B*. Water moves by osmosis from compartment *B* into compartment *A*. **(2)** The membrane is impermeable to proteins, so equilibrium can be reached only by movement of water. As water accumulates in compartment *A*, the water level on that side of the membrane rises.

words, water moves from compartment *B* into compartment *A* by osmosis. Protein, on the other hand, cannot move out of compartment *A* because the selectively permeable membrane is impermeable to it. Note in figure 3.15 that as osmosis occurs, the water level on side *A* rises. This ability of osmosis to generate enough pressure to lift a volume of water is called *osmotic pressure.*

The greater the concentration of impermeant solute particles (protein in this case) in a solution, the *greater* the osmotic pressure. Water always tends to move toward solutions of greater osmotic pressure. That is, water moves by osmosis toward regions of trapped solute—whether in a laboratory exercise or in the body.

Cell membranes are generally permeable to water, so water equilibrates by osmosis throughout the body, and the concentration of water and solutes everywhere in the intracellular and extracellular fluids is essentially the same. Therefore, the osmotic pressure of the intracellular and extracellular fluids is the same. Any solution that has the same osmotic pressure as body fluids is called **isotonic** (fig. 3.16*a*).

Solutions that have a higher osmotic pressure than body fluids are called **hypertonic.** If cells are put into a hypertonic solution, water moves by osmosis out of the cells into the surrounding solution, and the cells shrink (fig. 3.16*b*). Conversely, cells put into a **hypotonic** solution, which has a lower osmotic pressure than body fluids, gain water by osmosis, and therefore they swell (fig. 3.16*c*).

If the concentration of solute in solutions that are infused into body tissues or blood is not controlled, osmosis may swell or shrink cells, impairing their function. For instance, if red blood cells are placed in distilled water (which is hypotonic to them), the cells gain water by osmosis, and they may burst (hemolyze). Yet red blood cells exposed to 0.9% NaCl solution (normal saline) do not change shape because this solution is isotonic to human cells. A red blood cell in a hypertonic solution shrinks.

Filtration

Molecules move through membranes by diffusion because of random movements. In other instances, the process of **filtration** (fil-tra′shun) forces molecules through membranes.

Filtration is commonly used to separate solids from water. One method is to pour a mixture of solids and water onto filter paper in a funnel. The paper is a porous membrane through which the small water molecules can pass, leaving the larger solid particles behind. *Hydrostatic pressure,* created by the weight of water due to gravity, forces the water molecules through to the other side. A familiar example of filtration is making coffee by the drip method.

In the body, tissue fluid forms when water and small dissolved substances are forced out through the thin, porous walls of blood capillaries, but larger particles, such as blood protein molecules, are left inside (fig. 3.17). The force for this movement comes from blood pressure, generated largely by heart action, which is greater inside the vessel than outside it. However, the impermeant proteins tend to hold water in blood vessels by osmosis, thus preventing formation of excess tissue fluid, a condition called **edema.** Filtration also helps the kidneys cleanse blood.

(a)

(b)

(c)

Figure 3.16 AP|R

When red blood cells are placed **(a)** in an isotonic solution, the cells maintain their characteristic shapes. **(b)** In a hypertonic solution, cells shrink. **(c)** In a hypotonic solution, cells swell and may burst (5,000×).

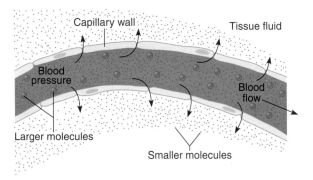

Figure 3.17 AP|R

Blood pressure forces smaller molecules through tiny openings in the capillary wall. The larger molecules remain inside.

Active Mechanisms

When molecules or ions pass through cell membranes by diffusion or facilitated diffusion, their net movements are from regions of higher concentration to regions of lower concentration. Sometimes, however, particles move from a region of lower concentration to one of higher concentration. This requires energy, which comes from cellular metabolism and, specifically, from a molecule called adenosine triphosphate (ATP). The situation is a little like requiring a push to enter a crowded room.

Active Transport

Active transport (ak′tiv trans′port) is a process that moves particles through membranes from a region of lower concentration to a region of higher concentration. Sodium ions, for example, can diffuse slowly through cell membranes, but their concentration typically remains much greater outside cells than inside cells. This is because active transport continually moves sodium ions through cell membranes from regions of lower concentration (inside) to regions of higher concentration (outside).

Active transport is similar to facilitated diffusion in that it uses specific carrier molecules in cell membranes (fig. 3.18). It differs from facilitated diffusion in that particles move from regions of low concentration to regions of high concentration, and energy from ATP is required. Up to 40% of a cell's energy supply may be used to actively transport particles through its membranes.

The carrier molecules in active transport are proteins with binding sites that combine with the particles being transported. Such a union triggers the release of energy, and this alters the shape of the carrier protein. As a result, the "passenger" particles move through the membrane. Once on the other side, the transported particles are released, and the carriers can accept other passenger molecules at their binding sites. Because these carrier proteins transport substances from regions of low concentration to regions of high concentration, they are sometimes called "pumps."

Particles that are actively transported across cell membranes include sugars and amino acids as well as sodium, potassium, calcium, and hydrogen ions. Some of these substances are actively transported into cells, and others are actively transported out.

(a)

(b)

Figure 3.18 AP|R

Active transport moves molecules against their concentration gradient. **(a)** During active transport, a molecule or an ion combines with a carrier protein, whose shape changes as a result. **(b)** This process, which requires cellular energy, transports the particle across the cell membrane.

Q: *What are two requirements for active transport to occur?*
Answer can be found in Appendix E on page 568.

Endocytosis and Exocytosis

Two processes use cellular energy to move substances into or out of a cell without actually crossing the cell membrane. In **endocytosis** (en″do-si-to′sis), molecules or other particles that are too large to enter a cell by diffusion or active transport are conveyed in a vesicle that forms from a portion of the cell membrane. In **exocytosis** (ek″so-si-to′sis), the reverse process secretes a substance stored in a vesicle from the cell. Nerve cells use exocytosis to release the neurotransmitter chemicals that signal other nerve cells, muscle cells, or glands.

Endocytosis and exocytosis can act together, bringing a particle into a cell, and escorting it, in a vesicle, out of the cell at another place in the cell membrane. This process is called transcytosis. HIV enters the body this way.

Endocytosis happens in three ways: pinocytosis, phagocytosis, and receptor-mediated endocytosis. In **pinocytosis** (pi″no-si-to′sis), meaning "cell drinking," cells take in tiny droplets of liquid from their surroundings, as a small portion of the cell membrane indents. The open end of the tubelike part that forms seals off and produces a small vesicle, which detaches from the surface and moves into the cytoplasm. Eventually the vesicular membrane breaks down, and the liquid inside becomes part of the cytoplasm. In this way, a cell can take in water and the particles dissolved in it, such as proteins, that otherwise might be too large to enter.

Phagocytosis (fag″o-si-to′sis), meaning "cell eating," is similar to pinocytosis, but the cell takes in solids rather than liquids. Certain types of white blood cells are called *phagocytes* because they can take in solid particles such as bacteria and cellular debris. When a phagocyte first encounters a particle, the particle attaches to the phagocyte's cell membrane. This stimulates a portion of the membrane to project outward, surround the particle, and slowly draw it inside the cell. The part of the membrane surrounding the particle detaches from the cell's surface, forming a vesicle containing the particle (fig. 3.19).

Once a particle has been phagocytized inside a cell, a lysosome combines with the newly formed vesicle and lysosomal digestive enzymes decompose the contents. The products of this decomposition may then diffuse out of the lysosome and into the cytoplasm. Exocytosis may expel any remaining residue.

Pinocytosis and phagocytosis engulf any molecules in the vicinity of the cell membrane. In contrast, **receptor-mediated endocytosis** moves very specific kinds of particles into the cell. In this process, protein molecules extend through the cell membrane to the outer surface, where they form receptors to which only specific molecules from outside the cell, called their *ligands,* can bind (fig. 3.20). Cholesterol molecules enter cells by receptor-mediated endocytosis. Table 3.2 summarizes the types of movements into and out of cells.

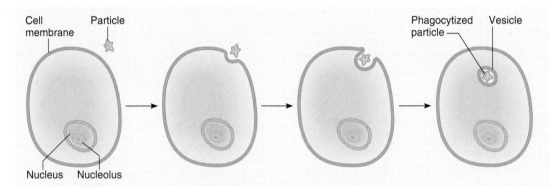

Figure 3.19

A cell may use phagocytosis to take in a solid particle from its surroundings.

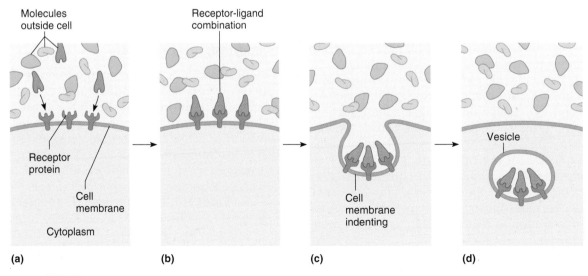

Figure 3.20 AP|R

Receptor-mediated endocytosis brings specific molecules into a cell. **(a,b)** A specific molecule binds to a receptor protein, forming a receptor-ligand combination. **(c)** The binding of the ligand to the receptor protein stimulates the cell membrane to indent. **(d)** Continued indentation forms a vesicle, which encloses and then transports the molecule into the cytoplasm.

Table 3.2 Movements Through Cell Membranes

Process	Characteristics	Source of Energy	Example
Passive mechanisms			
Diffusion	Molecules move through the phospholipid bilayer from regions of higher concentration to regions of lower concentration.	Molecular motion	Exchange of oxygen and carbon dioxide in the lungs
Facilitated diffusion	Molecules move across the membrane through channels or by carrier molecules from a region of higher concentration to one of lower concentration.	Molecular motion	Movement of glucose through a cell membrane
Osmosis	Water molecules move through a selectively permeable membrane toward the solution with more impermeant solute (greater osmotic pressure).	Molecular motion	Distilled water entering a cell
Filtration	Smaller molecules are forced through porous membranes from regions of higher pressure to regions of lower pressure.	Hydrostatic pressure	Water molecules leaving blood capillaries
Active mechanisms			
Active transport	Carrier molecules transport molecules or ions through membranes from regions of lower concentration toward regions of higher concentration.	Cellular energy (ATP)	Movement of various ions, sugars, and amino acids through membranes
Endocytosis			
Pinocytosis	Membrane engulfs droplets of liquid from surroundings.	Cellular energy	Membrane-forming vesicles containing large particles dissolved in water
Phagocytosis	Membrane engulfs particles from surroundings.	Cellular energy	White blood cell engulfing bacterial cell
Receptor-mediated endocytosis	Membrane engulfs selected molecules combined with receptor proteins.	Cellular energy	Cell removing cholesterol molecules from its surroundings
Exocytosis	Vesicle fuses with membrane and releases contents outside of the cell.	Cellular energy	Neurotransmitter release

Practice

22. Explain the mechanism that maintains unequal concentrations of ions on opposite sides of a cell membrane.

23. How are facilitated diffusion and active transport similar and different?

24. What is the difference between endocytosis and exocytosis?

25. Explain how receptor-mediated endocytosis is more specific than pinocytosis or phagocytosis.

3.4 THE CELL CYCLE

The series of changes that a cell undergoes from the time it forms until it divides is called the *cell cycle* (fig. 3.21). This cycle may seem simple: A newly formed cell grows for a time and then divides to form two new cells, which in turn may grow and divide. Yet the phases and timing of the cycle are quite complex. The major phases are interphase, mitosis, and cytoplasmic division (cytokinesis). Then the resulting "daughter" cells may undergo further changes that make them specialize. Groups of special proteins interact at certain times in the cell cycle, called *checkpoints,* in ways that control the cell cycle. Of particular importance is the restriction checkpoint that determines a cell's fate—whether it will continue in the cell cycle and divide, move into a non-dividing stage as a specialized cell, or die.

The cell cycle is very precisely regulated. Stimulation from a hormone or growth factor may trigger cell division. This occurs, for example, when the breasts develop into milk-producing glands during pregnancy. Disruption of the cell cycle can affect health: If cell division is too infrequent, a wound cannot heal; if too frequent, a cancer grows. Clinical Application 3.2 discusses cancer.

Most cells do not normally divide continually. If grown in the laboratory, most types of human cells divide only forty to sixty times. Presumably, such controls operate in the body too. Some cells, such as those that line the small intestine, may divide the maximum

Clinical Application 3.2

Cancer

Cancer is a group of closely related diseases that can affect many different organs. One in three of us will develop cancer. These conditions result from changes in genes (mutations) that alter the cell cycle. Cancers share the following characteristics:

1. **Hyperplasia** is uncontrolled cell division. Normal cells divide a set number of times, signaled by the shortening of chromosome tips. Cancer cells make *telomerase*. As a result, cells are not signaled to stop dividing.
2. **Dedifferentiation** is loss of the specialized structures and functions of the normal type of cell from which the cancer cells descend (fig. 3A).
3. **Invasiveness** is the ability of cancer cells to break through boundaries, called *basement membranes,* that separate cell layers.
4. **Angiogenesis** is the ability of cancer cells to induce extension of nearby blood vessels, which nourish the cells and remove wastes, enabling the cancer to grow.
5. **Metastasis** is the spread of cancer cells to other tissues. Normal cells usually aggregate in groups of similar kinds. Some cancer cells can move into the bloodstream or lymphatic system to establish tumors elsewhere in the body.

Mutations in certain genes cause cancers. Such a mutation may activate a cancer-causing oncogene (a gene that normally controls mitotic rate) or inactivate a protective gene called a tumor suppressor. A person may inherit one abnormal cancer-causing gene variant, present in all cells, and develop cancer when the second copy of that gene mutates in a cell of the organ that will be affected. This second muta-

tion may occur in response to an environmental trigger. That is, cancer usually entails mutations in somatic (non-sex) cells.

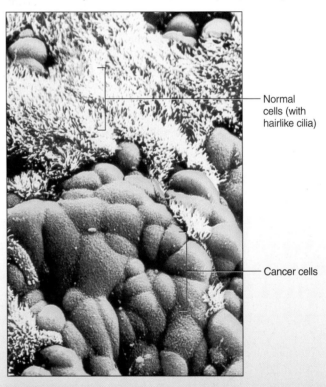

Normal cells (with hairlike cilia)

Cancer cells

Figure 3A

The absence of cilia on these cancer cells, compared to the nearby cilia-fringed cells from which they arose, is one sign of their dedifferentiation (2,250×).

Figure 3.21

The cell cycle is divided into interphase, when cellular components duplicate, and cell division (mitosis and cytokinesis), when the cell splits in two, distributing its contents into two daughter cells. Interphase is divided into two gap phases (G_1 and G_2) when specific molecules and structures duplicate, and a synthesis phase (S), when the genetic material replicates. Mitosis can be considered in stages—prophase, metaphase, anaphase, and telophase.

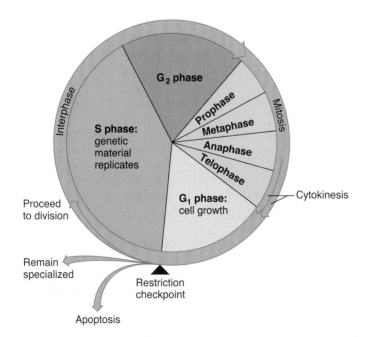

number of times. Others, such as nerve cells, normally do not divide. A cell "knows" when to stop dividing because of a built-in "clock" in the form of the chromosome tips. These chromosome regions, called *telomeres,* shorten with each cell division. When the telomeres shorten to a certain length, the cell no longer divides. An enzyme called telomerase keeps telomeres long in cell types that must continually divide, such as certain cells in bone marrow.

Studies show that chronic psychological or emotional stress, obesity, and persistent elevated blood glucose levels can hasten telomere shortening.

Interphase

Before a cell actively divides, it must grow and duplicate much of its contents, so that two "daughter" cells can form from one. This period of preparedness is called **interphase.**

Once thought to be a time of rest, interphase is actually a time of great synthetic activity, when the cell obtains and utilizes nutrients to manufacture new living material and maintain its routine functions. The cell duplicates membranes, ribosomes, lysosomes, and mitochondria. Perhaps most importantly, the cell in interphase takes on the tremendous task of replicating its genetic material. This is important so that each of the two new cells will have a complete set of genetic instructions.

Interphase is considered in phases. DNA is replicated during the S (or synthesis) phase, which is bracketed by two gap (or growth) periods, called G_1 and G_2, when other structures are duplicated.

Cell Division

A cell can divide in two ways. One type of cell division is *meiosis,* which is part of *gametogenesis,* the formation of egg cells (in the female) and sperm cells (in the male). Because an egg fertilized by a sperm must have the normal complement of 46 chromosomes, both the egg and the sperm must first halve their normal chromosome number to 23 chromosomes. Meiosis, through a process called reduction division, accomplishes this. Only a few cells undergo meiosis, which is discussed in chapter 19 (pp. 507–509 and 517).

The other, much more common form of cell division increases cell number, which is necessary for growth and development and for wound healing. It consists of two separate processes: (1) division of the nucleus, called **mitosis** (mi-to′sis), and (2) division of the cytoplasm, called **cytokinesis** (si″to-ki-ne′sis).

Division of the nucleus must be very precise because it contains the DNA. Each new cell resulting from mitosis must have a complete and accurate copy of this information to survive. DNA replicates during interphase, but it is equally distributed into two cells in mitosis.

Mitosis is described in stages, but the process is actually continuous (fig. 3.22). Stages, however, indicate the sequence of major events. The stages are:

1. **Prophase** One of the first indications that a cell is going to divide is that the chromosomes become visible in the nucleus when stained. This happens because the genetic material coils very tightly. Because the cell has gone through S phase, each prophase chromosome is composed of two identical structures (chromatids), which are temporarily attached at a region on each called the *centromere.*

   The centrioles of the centrosome replicate just before mitosis begins. During prophase, the two newly formed centriole pairs move to opposite ends of the cell. Soon the nuclear envelope and the nucleolus break up, disperse, and are no longer visible. Microtubules are assembled from tubulin proteins in the cytoplasm and associate with the centrioles and chromosomes. A spindle-shaped array of microtubules (spindle fibers) forms between the centrioles as they move apart.

2. **Metaphase** The chromosomes line up about midway between the centrioles, as a result of microtubule activity. Spindle fibers attach to the centromeres of each chromosome so that a fiber from one pair of centrioles contacts one centromere, and a fiber from the other pair of centrioles attaches to the other centromere.

3. **Anaphase** The centromeres are pulled apart. As the chromatids separate, they become individual chromosomes that move in opposite directions, once again guided by microtubule activity. The spindle fibers shorten and pull their attached chromosomes toward the centrioles at opposite ends of the cell.

4. **Telophase** The final stage of mitosis begins when the chromosomes complete their migration toward the centrioles. It is much like the reverse of prophase. As the chromosomes approach the centrioles, they elongate and unwind from rods into threadlike chromatin. A nuclear envelope forms around each chromosome set, and nucleoli appear within the new nuclei. Finally, the microtubules disassemble into free tubulin molecules.

Cytoplasmic Division

Cytoplasmic division (cytokinesis) begins during anaphase, when the cell membrane starts to constrict around the middle of the cell. This constriction, called a *cleavage furrow,* continues through telophase. Contraction of a ring of microfilaments, which assemble in the cytoplasm and attach to the inner surface of the cell membrane, divides the cytoplasm. The contractile ring forms at right angles to the microtubules that pulled the chromosomes to opposite sides of the cell during mitosis. The ring pinches inward, separating the two newly

Late Interphase
Cell has passed the restriction checkpoint and completed DNA replication, as well as replication of centrioles and mitochondria, and synthesis of extra membrane.

Early Interphase
of daughter cells— a time of normal cell growth and function.

(a)

Restriction checkpoint

Nuclear envelope
Chromatin fibers
Centrioles

Microtubules

Prophase
Chromosomes condense and become visible. Nuclear envelope and nucleolus disperse. Spindle apparatus forms.

(b)

Centromere
Spindle fiber

Late Prophase

Sister chromatids

Cleavage furrow

(e)

Chromosomes

Nuclear envelopes

Telophase and Cytokinesis
Nuclear envelopes begin to reassemble around two daughter nuclei. Chromosomes decondense. Spindle disappears. Division of the cytoplasm into two cells.

(d)

(c)

Anaphase
Sister chromatids separate to opposite poles of cell. Events begin which lead to cytokinesis.

Metaphase
Chromosomes align along equator, or metaphase plate of cell.

	Mitosis
	Cytokinesis
	G₁ phase
	S phase
	G₂ phase

G₁ phase, S phase, G₂ phase — Interphase

Figure 3.22 AP|R

Mitosis and cytokinesis produce two cells from one. **(a)** During interphase, before mitosis, chromosomes are visible only as chromatin fibers. A single pair of centrioles is present, but not visible at this magnification. **(b)** In prophase, as mitosis begins, chromosomes have condensed and are easily visible when stained. The centrioles have replicated, and each pair moves to an opposite end of the cell. The nuclear envelope and nucleolus disappear, and spindle fibers associate with the centrioles and the chromosomes. **(c)** In metaphase, the chromosomes line up midway between the centrioles. **(d)** In anaphase, the centromeres are pulled apart by the spindle fibers, and the chromatids, now individual chromosomes, move in opposite directions. **(e)** In telophase, chromosomes complete their migration and unwind to become chromatin fibers, the nuclear envelope re-forms, and microtubules disassemble. Cytokinesis, which actually began during anaphase, continues during telophase. Not all chromosomes are shown in these drawings. (Micrographs approximately 360×)

formed nuclei and distributing about half of the organelles into each new cell. The newly formed cells may differ slightly in size and number of organelles, but they contain identical DNA.

Practice

26. Outline the cell cycle.
27. Explain regulation of the cell cycle.
28. Describe the events that occur during mitosis.

Cell Differentiation

Cells come from preexisting cells, by the processes of mitosis and cytokinesis. Cell division explains how a fertilized egg develops into an individual consisting of trillions of cells, of at least 260 specialized types. The process of specialization is called **differentiation.**

The ability to generate new cells is essential to the growth and repair of tissues. Cells that retain the ability to divide repeatedly without specializing, called **stem cells,** allow for this continual growth and renewal. A stem cell divides mitotically to yield either two daughter cells like itself (stem cells), or one daughter cell that is a stem cell and one that becomes partially specialized, termed a **progenitor cell.** The ability of a stem cell to divide and give rise to at least one other stem cell is called self-renewal. A progenitor cell's daughter cells can become any of a few cell types. For example, a neural stem cell divides to give rise to another stem cell and a neural progenitor cell. The progenitor cell then can divide, and its daughter cells differentiate, becoming nervous tissue. All of the differentiated cell types in a human body arise through such lineages of stem and progenitor cells. Figure 3.23 depicts a few of them.

Many organs have stem or progenitor cells that are stimulated to divide when injury or illness occurs. This action replaces cells, promoting healing. For example, one in 10,000 to 15,000 bone marrow cells is a stem cell, which can give rise to blood as well as several other cell types. Stem cells in organs may have been set aside in the embryo or fetus as repositories of future healing. Certain stem cells can travel to replace injured or dead cells in response to signals sent from the site of damage.

Throughout development, cells progressively specialize by utilizing different parts of the complete genetic instructions, or genome, that are present in each cell. That is, some genes are "turned on" in certain cells, and other genes are turned on in other cell types. In this way, for example, an immature bone cell forms from a progenitor cell by manufacturing proteins that bind bone mineral and an enzyme required for bone formation. An immature muscle cell, in contrast, forms from a muscle progenitor cell by accumulating contractile proteins. The bone progenitor does not produce contractile proteins, nor does the muscle progenitor produce mineral-binding proteins. The final differentiated cell is like a database from which only some information is accessed.

Cell Death

A cell that does not divide or specialize has another option—it may die. **Apoptosis** (ap″o-to′sis) is a form of cell death that is a normal part of development, rather than the result of injury or disease. Apoptosis sculpts organs from naturally overgrown tissues. In the fetus, for example, apoptosis carves away webbing between developing fingers and toes, removes extra brain cells, and preserves only those immune system cells that recognize the body's cell surfaces. After birth, apoptosis occurs after a sunburn—it peels away damaged skin cells that might otherwise turn cancerous.

A cell in the throes of apoptosis goes through characteristic steps. It rounds up and bulges, the nuclear membrane breaks down, chromatin condenses, and enzymes cut the chromosomes into many equal-size pieces of DNA. Finally, the cell shatters into membrane-enclosed fragments, and a scavenger cell engulfs and destroys them. Apoptosis is a little like cleaning up a very messy room by placing garbage in many garbage bags, which are then removed.

Practice

29. Why must cells divide and specialize?
30. Distinguish between a stem cell and a progenitor cell.
31. How are new cells generated and how do they specialize?
32. How is cell death a normal part of development?

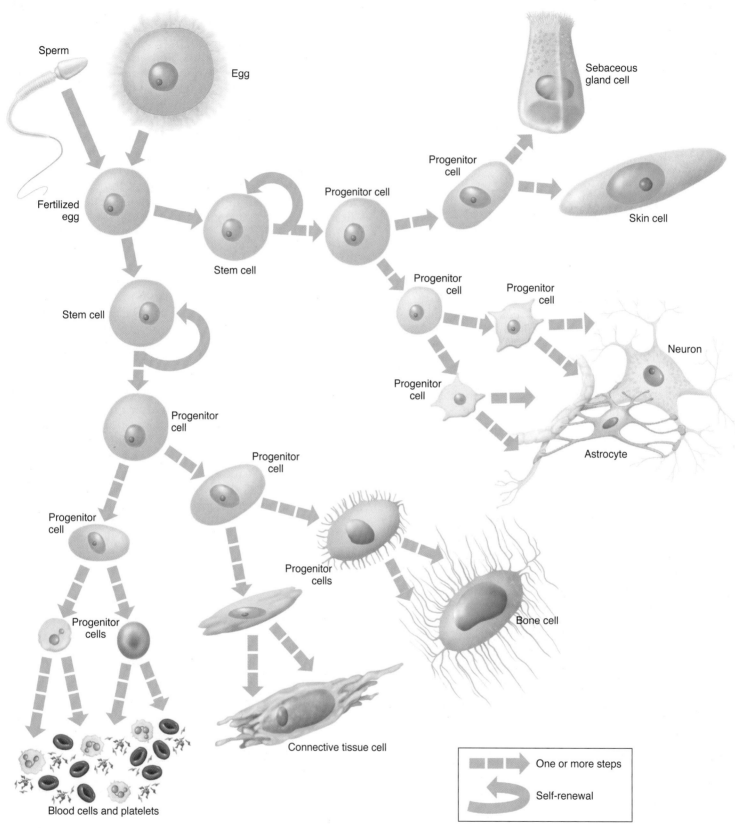

Figure 3.23

Cells specialize along cell lineage pathways. All cells in the human body ultimately descend from stem cells, through the processes of mitosis and differentiation. This simplified view depicts a few of the pathways that cells follow, grouping the cell types by the closeness of their lineages. Imagine the complexity of the lineages of the more than 260 human cell types!

Summary Outline

3.1 Introduction (p. 51)

Cells vary considerably in size, shape, and function. The shapes of cells make possible their functions.

3.2 Composite Cell (p. 52)

A cell includes a cell membrane, cytoplasm, and a nucleus. Organelles perform specific functions. The nucleus controls overall cell activities because it contains DNA, the genetic material.

1. Cell membrane
   a. The cell membrane forms the outermost limit of the living material.
   b. It is a selectively permeable passageway that controls the entrance and exit of substances. Its molecules transmit signals.
   c. The cell membrane includes protein, lipid, and carbohydrate molecules.
   d. The cell membrane's framework is mainly a bilayer of phospholipid molecules.
   e. Molecules that are soluble in lipids pass through the cell membrane easily, but water-soluble molecules do not.
   f. Proteins function as receptors on membrane surfaces and form channels for the passage of ions and molecules.
   g. Patterns of surface carbohydrates associated with membrane proteins enable certain cells to recognize one another.
2. Cytoplasm
   a. Cytoplasm contains networks of membranes, organelles, and the rods and tubules of the cytoskeleton, suspended in cytosol.
   b. The endoplasmic reticulum is a tubular communication system in the cytoplasm that transports lipids and proteins.
   c. Ribosomes function in protein synthesis.
   d. The Golgi apparatus packages glycoproteins for secretion.
   e. Mitochondria contain enzymes that catalyze reactions that release energy from nutrient molecules.
   f. Lysosomes contain digestive enzymes that decompose substances.
   g. Peroxisomes house enzymes that catalyze breakdown of hydrogen peroxide and fatty acids, and detoxification of alcohol.
   h. Microfilaments (actin) and microtubules (tubulin) aid cellular movements and support and stabilize the cytoplasm and organelles. Together they form the cytoskeleton. Microtubules also form centrioles, cilia, and flagella.
   i. The centrosome contains centrioles that aid in distributing chromosomes during cell division.
   j. Cilia and flagella are motile extensions from cell surfaces.
   k. Vesicles contain substances that recently entered the cell or that are to be secreted from the cell.
3. Cell nucleus
   a. The nucleus is enclosed in a double-layered nuclear envelope.
   b. It contains a nucleolus, where ribosomes are produced.
   c. It contains chromatin, which is composed of loosely coiled fibers of DNA and protein. As chromatin fibers condense during cell division, chromosomes become visible.

3.3 Movements Through Cell Membranes (p. 60)

The cell membrane is a barrier through which substances enter and leave a cell.

1. Passive mechanisms do not require cellular energy.
   a. Diffusion
      (1) Diffusion is the movement of molecules or ions from regions of higher concentration to regions of lower concentration.
      (2) In the body, diffusion exchanges oxygen and carbon dioxide.
   b. Facilitated diffusion
      (1) In facilitated diffusion, ion channels and pores or special carrier molecules move substances through the cell membrane.
      (2) This process moves substances only from regions of higher concentration to regions of lower concentration.
   c. Osmosis
      (1) Osmosis is the movement of water across a selectively permeable membrane into a compartment containing solute that cannot cross the same membrane.
      (2) Osmotic pressure increases as the number of impermeant particles dissolved in a solution increases.
      (3) A solution is isotonic to a cell when it has the same osmotic pressure as the cell.
      (4) Cells lose water when placed in hypertonic solutions and gain water when placed in hypotonic solutions.
   d. Filtration
      (1) Filtration is the movement of molecules through membranes from regions of higher hydrostatic pressure to regions of lower hydrostatic pressure.
      (2) Blood pressure causes filtration through porous capillary walls, forming tissue fluid.
2. Active mechanisms require cellular energy.
   a. Active transport
      (1) Active transport moves particles through membranes from a region of lower concentration to a region of higher concentration.
      (2) It requires cellular energy from ATP and carrier molecules in the cell membrane.
   b. Endocytosis and exocytosis
      (1) Endocytosis may convey large particles into a cell. Exocytosis is the reverse of endocytosis.
      (2) In pinocytosis, a cell membrane engulfs tiny droplets of liquid.
      (3) In phagocytosis, a cell membrane engulfs solid particles.
      (4) Receptor-mediated endocytosis moves specific types of particles into cells.

3.4 The Cell Cycle (p. 67)

The cell cycle includes interphase, mitosis, and cytoplasmic division. It is highly regulated.

1. Interphase
   a. During interphase, a cell duplicates membranes, ribosomes, organelles, and DNA.
   b. Interphase terminates when mitosis begins.
2. Cell Division
   a. Meiosis is a form of cell division that forms sex cells.
   b. Mitosis is the division and distribution of genetic material to new cells, increasing cell number.
   c. The stages of mitosis are prophase, metaphase, anaphase, and telophase.
3. Cytoplasmic division distributes cytoplasm into two portions, completing about the same time as mitosis.
4. Differentiation is cell specialization. Stem cells provide new cells for growth and repair. They give rise to other stem cells and to progenitor cells that begin to specialize. Differentiation reflects the activation of specific sets of genes.
5. A cell that does not divide or differentiate may undergo apoptosis, a form of cell death that is a normal part of development.

Chapter Assessments

3.1 Introduction

1. An adult human body consists of about _____ cells. (p. 51)
   a. 2 billion
   b. 50 billion
   c. 75 trillion
   d. 8 quadrillion
   e. 50 million

2. Define *cell*. (p. 51)

3. Discuss how cells differ from one another. (p. 51)

3.2 Composite Cell

4. The three major parts of a cell are _____. (p. 52)
   a. the nucleus, the nucleolus, and the nuclear envelope
   b. the nucleus, cytoplasm, and the cell membrane
   c. a nerve cell, an epithelial cell, and a muscle cell
   d. endoplasmic reticulum, Golgi apparatus, and ribosomes
   e. cytoplasm, organelles, and chromatin

5. Explain the general function of organelles. (p. 53)

6. Define *selectively permeable*. (p. 53)

7. Describe the structure of a cell membrane and explain how this structural organization provides the membrane's function. (p. 53)

8. List three functions of membrane proteins. (p. 54)

9. Match the following structures with their definitions: (pp. 55–59)

   (1) Golgi apparatus
   (2) mitochondria
   (3) peroxisomes
   (4) cilia
   (5) endoplasmic reticulum
   (6) cytoskeleton
   (7) vesicles
   (8) ribosomes

   A. Sacs that contain enzymes that catalyze a variety of specific biochemical reactions
   B. Structures on which protein synthesis occurs
   C. Structures that house the reactions that release energy from nutrients
   D. A network of microfilaments and microtubules that supports and shapes a cell
   E. A structure that modifies, packages, and exports glycoproteins
   F. Membrane-bounded sacs
   G. A network of membranous channels and sacs where lipids and proteins are synthesized
   H. Hairlike structures that extend from certain cell surfaces and wave about

10. List the parts of the nucleus and explain why each is important. (p. 60)

3.3 Movements Through Cell Membranes

11. Distinguish between active and passive mechanisms of movement across cell membranes. (p. 60)

12. Match the transport mechanisms with their descriptions. (pp. 60–66)

    (1) diffusion
    (2) facilitated diffusion
    (3) filtration
    (4) active transport
    (5) endocytosis
    (6) exocytosis

    A. The cell membrane engulfs a particle or substance, drawing it into the cell in a vesicle
    B. Movement down the concentration gradient with a carrier protein, without energy input
    C. Movement down the concentration gradient without a carrier protein or energy input
    D. A particle or substance leaves a cell in a vesicle that merges with the cell membrane
    E. Movement against the concentration gradient with energy input
    F. Hydrostatic pressure forces substances through membranes

13. Define *osmosis*. (p. 63)

14. Distinguish between hypertonic, hypotonic, and isotonic solutions. (p. 64)

15. Explain how phagocytosis differs from receptor-mediated endocytosis. (p. 66)

3.4 The Cell Cycle

16. Explain why it is important for the cell cycle to be highly regulated. (p. 67)

17. Distinguish between interphase and mitosis. (p. 69)

18. The period of the cell cycle when DNA replicates is ____. (p. 69)
    a. G_1 phase
    b. G_2 phase
    c. S phase
    d. prophase
    e. telophase

19. Explain how meiosis differs from mitosis. (p. 69)

20. _____ occur simultaneously. (p. 69)
    a. G_1 phase and G_2 phase
    b. Interphase and mitosis
    c. Cytokinesis and telophase
    d. Prophase and metaphase
    e. Meiosis and mitotic metaphase

21. Describe the events of mitosis in sequence. (p. 69)

22. Define *differentiation*. (p. 71)

23. A stem cell _____. (p. 71)
    a. undergoes apoptosis
    b. self-renews
    c. is differentiated
    d. gives rise to only fully differentiated daughter cells
    e. forms from a progenitor cell

24. Describe the steps of apoptosis. (p. 71)

Integrative Assessments/Critical Thinking

OUTCOME 3.2

1. Why does a muscle cell contain many mitochondria, and why does a white blood cell contain many lysosomes?
2. Organelles compartmentalize a cell, much as a department store displays related items together. What advantage does such compartmentalization offer a large cell? Cite two examples of organelles and the activities they compartmentalize.
3. Exposure to tobacco smoke immobilizes cilia, and they eventually disappear. How might this effect explain why smokers have an increased incidence of coughing and respiratory infections?

OUTCOME 3.3

4. Which process—diffusion, osmosis, or filtration—is utilized in the following situations?
   a. Injection of a drug that is hypertonic to the tissues stimulates pain.
   b. The urea concentration in the dialyzing fluid of an artificial kidney is decreased.
5. What characteristic of cell membranes may explain why fat-soluble substances such as chloroform and ether rapidly affect cells?

OUTCOME 3.4

6. New treatments for several conditions are being developed using stem cells in medical waste, such as biopsy material, teeth, menstrual blood, umbilical cords, and fatty tissue removed in liposuction. For example, fat samples from injured horses are used to grow stem cells to treat tendon injuries. Explain how the two defining characteristics of stem cells enable them to be used to replace damaged or diseased tissue, so that the new tissue functions as opposed to forming a scar.
7. Explain how the cell cycle of a cancer cell is abnormal.

WEB CONNECTIONS

Visit the text website at **www.mhhe.com/shieress11** for additional quizzes, interactive learning exercises, and more.

APR

Anatomy & Physiology REVEALED includes cadaver photos that allow you to peel away layers of the human body to reveal structures beneath the surface. This program also includes animations, radiologic imaging, audio pronunciation, and practice quizzing. To learn more visit **www.aprevealed.com**.

4

Cellular Metabolism

Arsenic poisoning. Disrupting the body's ability to extract energy from nutrients can have serious effects on health. Arsenic is a chemical element that, if present in the body in excess, shuts down metabolism. It can do so suddenly or gradually.

Given in one large dose, arsenic causes chest pain, vomiting, diarrhea, shock, coma, and death. In contrast, a series of many small doses causes dark skin lesions that feel as if they are burning, numb hands and feet, and skin cancer. It may progress to paralysis and organ failure. Such gradual poisoning, called arsenicosis, may occur from contact with pesticides or environmental pollutants. The world's largest outbreak of arsenicosis, however, is due to a natural exposure.

When the World Bank and UNICEF began tapping into aquifers in India and Bangladesh in the late 1960s, they were trying to supply clean water to areas ravaged by sewage and industrial waste released from rivers subject to cycles of floods and droughts. Millions of people had already perished from diarrheal diseases due to the poor sanitation. But digging wells to provide clean water backfired when workers unwittingly penetrated a layer of sediment naturally rich in arsenic. The chemical has been leaching into the water in at least 2 million wells in Bangladesh alone ever since, reaching levels 30 times the safety limit set by the World Health Organization. Effects on health took several years to show up. When they did, the people thought arsenicosis was contagious. In addition to their physical pain, affected individuals bore the psychological pain of being shunned.

Chronic exposure to arsenic in drinking water causes the burning, colored lesions of arsenicosis.

Arsenic damages the body by binding to bonds between sulfur atoms in proteins. The effects on metabolism largely stem from impairment of an enzyme that helps the breakdown products of glucose enter the mitochondria, where energy is extracted. Cells run out of energy.

Today UNICEF is helping the people of India and Bangladesh avoid arsenic poisoning. They are diagnosing and treating arsenicosis, and providing tanks to store rainwater. A vast education campaign has done much to quell the stigma of arsenicosis. One program in West Bengal teaches women and teens how to recognize arsenic-tainted sediments and gives the people kits to test the water in new wells being drilled.

Learning Outcomes

After studying this chapter, you should be able to do the following:

4.1 Introduction
1. Briefly explain the function of metabolism. (p. 77)

4.2 Metabolic Reactions
2. Compare and contrast anabolism and catabolism. (p. 77)

4.3 Control of Metabolic Reactions
3. Describe how enzymes control metabolic reactions. (p. 79)

4.4 Energy for Metabolic Reactions
4. Explain how cellular respiration releases chemical energy. (p. 80)
5. Describe how energy in the form of ATP becomes available for cellular activities. (p. 80)

4.5 Metabolic Pathways
6. Describe the general metabolic pathways of carbohydrates, lipids, and proteins. (p. 82)

4.6 DNA (Deoxyribonucleic Acid)
7. Describe how DNA molecules store genetic information. (p. 83)

8. Describe how DNA molecules are replicated. (p. 84)

4.7 Protein Synthesis
9. Describe the steps of protein synthesis. (p. 85)

Module 2: Cells & Chemistry

Aids to Understanding Words *(Appendix A on page 564 has a complete list of Aids to Understanding Words.)*

an- [without] *an*aerobic respiration: Respiratory process that does not require oxygen.

ana- [up] *ana*bolism: Cellular processes that use smaller molecules to build larger ones.

cata- [down] *cata*bolism: Cellular processes that break larger molecules into smaller ones.

mut- [change] *mut*ation: Change in genetic information.

-zym [causing to ferment] en*zym*e: Protein that speeds a chemical reaction without itself being consumed.

4.1 INTRODUCTION

A cell is a very busy place. Thousands of chemical reactions occur inside cells to carry on the activities of life, and also to provide the specialized characteristics of different cell types. A constant energy supply is required to fuel all of this activity. In addition to energy, a cell requires information—the instruction manual of DNA, the genetic material.

Cellular metabolism is the set of chemical reactions that acquire, store, and release energy in cells. The energy comes from the chemical bonds of nutrient molecules from the diet. A cell uses some of that energy to copy DNA when the cell divides and uses energy to access genetic information to construct proteins from amino acid building blocks. Special proteins called **enzymes** (en'zīmz) are vital to all of these activities. They control each of the interrelated reactions of metabolism.

4.2 METABOLIC REACTIONS

The metabolic reactions that control a cell's use of energy are of two major types. The reactions of **anabolism** (an"ah-bol'lizm) build up (synthesize) larger molecules from smaller ones, requiring input of energy. The reactions of **catabolism** (kă-tab'o-lizm) break down (decompose) larger molecules into smaller ones, releasing energy. The reactions of anabolism and catabolism together, constitute **metabolism** (mĕ-tab'o-lizm).

Anabolism

Anabolism provides the biochemicals required for cell growth and repair. An anabolic reaction called **dehydration synthesis** (de"hi-dra'shun sin'the-sis), for example, joins many simple sugar molecules (monosaccharides) to form larger molecules of glycogen. When monosaccharides join, an —OH (hydroxyl group) from one monosaccharide molecule and an —H (hydrogen atom) from an —OH group of another are removed. As the —H and —OH react to produce a water molecule (H_2O), the monosaccharides are joined by a shared oxygen atom (fig. 4.1). This process repeats and the molecular chain extends.

Dehydration synthesis also links glycerol and fatty acid molecules in fat (adipose) cells to form fat molecules (triglycerides). In this case, three hydrogen atoms are removed from a glycerol molecule, and an —OH group is removed from each of three fatty acid molecules (fig. 4.2). The result is three water molecules and a single fat molecule. Shared oxygen atoms bind the glycerol and fatty acid portions.

In cells, dehydration synthesis joins amino acid molecules, forming protein molecules. When two amino

Monosaccharide + Monosaccharide   Disaccharide + Water

Figure 4.1

Building up and breaking down molecules. A disaccharide is formed from two monosaccharides in a dehydration synthesis reaction (arrows to the right). In the reverse reaction, hydrolysis, a disaccharide is broken down into two monosaccharides (arrows to the left).

Figure 4.2

Forming a fat. A glycerol molecule and three fatty acid molecules react, yielding a fat molecule (triglyceride) in a dehydration synthesis reaction (arrows to the right). In the reverse reaction, hydrolysis, a triglyceride is broken down into three fatty acids and a glycerol (arrows to the left).

acid molecules unite, an —OH from one and an —H from the —NH₂ group of another are removed. A water molecule forms, and the amino acid molecules are joined by a bond between a carbon atom and a nitrogen atom, called a *peptide bond* (fig. 4.3). Two bound amino acids form a *dipeptide,* and many linked in a chain form a *polypeptide.* Generally, a polypeptide that has a specific function and consists of 100 or more amino acids is considered a *protein.* Some protein molecules consist of more than one polypeptide.

Nucleic acids are also formed by dehydration synthesis joining nucleotides. This process is described later in the chapter.

Catabolism

Catabolic reactions break down larger molecules into smaller ones. An example of a catabolic reaction is **hydrolysis** (hi-drol′ĭ-sis), which breaks down carbohydrates, lipids, and proteins, and splits a water molecule in the process. For instance, hydrolysis of a disaccharide such as sucrose yields two monosaccharides (glucose and fructose) as a molecule of water splits:

$$C_{12}H_{22}O_{11} + H_2O \longrightarrow C_6H_{12}O_6 + C_6H_{12}O_6$$
(Sucrose) (Water)   (Glucose) (Fructose)

In this case, the bond between the simple sugars that form sucrose breaks. Then, the water supplies a hydrogen atom to one sugar and a hydroxide group to the other. Hydrolysis is the opposite of dehydration synthesis (see figs. 4.1, 4.2, and 4.3). Each of these reactions is reversible and can be summarized as follows:

Hydrolysis ⟶

Disaccharide + Water ⇄ Monosaccharide + Monosaccharide

⟵ Dehydration synthesis

Hydrolysis is responsible for digestion. Specifically, it breaks down carbohydrates into monosaccharides, fats into glycerol and fatty acids, proteins into amino acids, and nucleic acids into nucleotides. (Chapter 15, pages 428–431, discusses digestion in more detail.)

Practice

1. What are the general functions of anabolism and catabolism?

2. What are the products of the anabolism of monosaccharides, glycerol and fatty acids, and amino acids?

3. Distinguish between dehydration synthesis and hydrolysis.

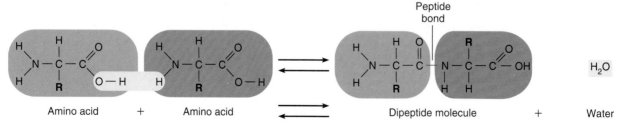

Figure 4.3

Peptide bonds link amino acids. When dehydration synthesis unites two amino acid molecules, a peptide bond forms between a carbon atom and a nitrogen atom, resulting in a dipeptide molecule (arrows to the right). In the reverse reaction, hydrolysis, a dipeptide molecule is broken down into two amino acids (arrows to the left).

4.3 CONTROL OF METABOLIC REACTIONS

Specialized cells, such as nerve, muscle, or blood cells, carry out distinctive chemical reactions. However, all cells perform certain basic chemical reactions, such as buildup and breakdown of carbohydrates, lipids, proteins, and nucleic acids. These reactions include hundreds of specific chemical changes that occur rapidly—yet in a coordinated fashion—thanks to enzymes.

Enzyme Action

Like other chemical reactions, metabolic reactions require energy to proceed. The temperature conditions in cells, however, are usually such that chemical reactions proceed too slowly to support life. Enzymes make these reactions possible by lowering the amount of energy, called the *activation energy,* required to start these reactions. In this way, enzymes speed metabolic reactions. This acceleration is called *catalysis,* and an enzyme is a catalyst. Enzyme molecules are not consumed in the reactions they catalyze and can function repeatedly. Therefore, a few enzyme molecules can have a powerful effect.

Each enzyme is specific, acting only on a particular type of molecule, called its **substrate** (sub′strāt). Many enzymes are named after their substrates, with *-ase* as a suffix. A lipase, for example, catalyzes a reaction that breaks down a lipid. Another example of an enzyme is *catalase.* Its substrate is hydrogen peroxide, which is a toxic by-product of certain metabolic reactions. Catalase speeds breakdown of hydrogen peroxide into water and oxygen, preventing accumulation of hydrogen peroxide, which can damage cells.

Specific enzymes catalyze each of the hundreds of different chemical reactions that constitute cellular metabolism. Every cell contains hundreds of different enzymes, and each enzyme must recognize its specific substrate. This ability of an enzyme to identify its substrate arises from the three-dimensional shape, or conformation, of the enzyme molecule. Each enzyme's polypeptide chain twists and coils into a unique conformation that fits the particular shape of its substrate molecule.

During an enzyme-catalyzed reaction, part of the enzyme molecule called the **active site** temporarily combines with parts of the substrate molecules, forming an enzyme-substrate complex (fig. 4.4). This interaction strains certain chemical bonds in the substrates, altering their orientation so that they require less energy to react. The reaction proceeds, product forms, and the enzyme is released in its original conformation.

An enzyme-catalyzed reaction can be summarized as follows:

$$\begin{array}{c}\text{Substrate}\\\text{molecules}\end{array} + \begin{array}{c}\text{Enzyme}\\\text{molecule}\end{array} \rightarrow \begin{array}{c}\text{Enzyme-}\\\text{substrate}\\\text{complex}\end{array} \rightarrow \begin{array}{c}\text{Product}\\\text{(changed}\\\text{substrates)}\end{array} + \begin{array}{c}\text{Enzyme}\\\text{molecule}\end{array}$$

Many enzyme-catalyzed reactions are reversible, and in some cases the same enzyme catalyzes the reaction in both directions. The speed of the reaction depends partly on the number of enzyme and substrate molecules in the cell. A reaction is faster if the concentrations of the enzyme or the substrate increase. Enzyme efficiency varies greatly. Some enzymes can catalyze only a few reactions per second, whereas others can catalyze many thousands.

Figure 4.4 AP|R

An enzyme-catalyzed reaction. (Many enzyme-catalyzed reactions, as depicted here, are reversible.) In the forward reaction (dark-shaded arrows), **(a)** the shapes of the substrate molecules fit the shape of the enzyme's active site. **(b)** When the substrate molecules temporarily combine with the enzyme, a chemical reaction proceeds. **(c)** The result is a product molecule and an unaltered enzyme. The active site changes shape as the substrate binds, such that formation of the enzyme-substrate complex is more like a hand fitting into a glove, which has some flexibility, than a key fitting into a lock.

Factors That Alter Enzymes

Almost all enzymes are proteins, and like other proteins, they can be denatured by exposure to heat, radiation, electricity, certain chemicals, or fluids with extreme pH values. Some enzymes become inactive at 41°C (105.8°F). Some poisons denature enzymes. Cyanide, for instance, can interfere with respiratory enzymes and damage cells by halting their energy-obtaining reactions.

Cofactors and Coenzymes

Some enzymes become active only when they combine with a nonprotein component called a **cofactor.** It may be an ion of an element, such as copper, iron, or zinc, or a small organic molecule, called a **coenzyme** (ko-en′zīm). Many coenzymes are vitamin molecules. An example of a coenzyme is coenzyme A, which takes part in cellular respiration, discussed in section 4.4 (p. 82).

Practice

4. What is an *enzyme?*

5. How does an enzyme recognize its substrate?

6. List factors that affect the speed of an enzyme-controlled reaction.

7. List factors that can denature enzymes.

4.4 ENERGY FOR METABOLIC REACTIONS

Energy is the capacity to change something; it is the ability to do work. We recognize energy by what it can do. Common forms of energy are heat, light, sound, electrical energy, mechanical energy, and chemical energy. Most metabolic processes use chemical energy.

Release of Chemical Energy

Chemical energy is held in the bonds between the atoms of molecules and is released when these bonds are broken. Burning is an example of an intervention that can break the bonds of chemicals in the environment (outside the body). Burning begins by applying heat. As the chemical burns, bonds break, and energy escapes as heat and light.

Cells "burn" glucose molecules in a process called **oxidation** (ok″sĭ-da′shun). The energy released from breaking the bonds of glucose powers the anabolic reactions of cells that build molecules. However, oxidation inside cells and burning outside of cells differ. Burning requires a relatively large input of energy,

most of which escapes as heat or light. In cells, enzymes reduce the activation energy required for the oxidation that occurs in the reactions of *cellular respiration.* These reactions release the energy in the bonds of nutrient molecules. Cells can capture about 40% of the energy released from breaking chemical bonds in cellular respiration and transfer it to special energy-carrying molecules. The rest of the liberated energy escapes as heat, which helps maintain body temperature.

Practice

8. Define *energy.*

9. Explain how oxidation inside cells differs from burning in the outside environment.

Cellular Respiration

Cellular respiration consists of three distinct, yet interconnected, series of reactions: **glycolysis, the citric acid cycle,** and the **electron transport chain,** shown schematically in figure 4.5. Glucose and oxygen are required for cellular respiration; the products of these reactions include CO_2, water, and energy:

$$C_6H_{12}O_6 + O_2 \longrightarrow 6CO_2 + 6H_2O + \text{energy}$$

Although most of the energy released in cellular respiration is lost as heat, almost half is captured in the form of high-energy electrons that the cell can use to synthesize **ATP (adenosine triphosphate).**

ATP

Each ATP molecule includes a chain of three chemical groups called phosphates (fig. 4.6). As energy is released during cellular respiration, some of it is captured in the bond of the end phosphate. When energy is required for a metabolic reaction, this terminal phosphate bond breaks, releasing the stored energy. (The second phosphate bond is high-energy too.) The cell uses ATP for a variety of functions, including active transport and synthesis of various compounds (anabolism).

An ATP molecule that has lost its terminal phosphate becomes an ADP (adenosine diphosphate) molecule. The ADP can be converted back into ATP by the addition of energy and a third phosphate. Thus, as figure 4.7 shows, ATP and ADP molecules shuttle back and forth between the energy-releasing reactions of cellular respiration and the energy-utilizing reactions of the cell.

Glycolysis

Cellular respiration begins with glycolysis, literally "the breaking of glucose" (see fig. 4.5). Glycolysis occurs in

Glycolysis AP|R

1 The 6-carbon sugar glucose is broken down in the cytosol into two 3-carbon pyruvic acid molecules with a net gain of 2 ATP and the release of high-energy electrons.

Citric Acid Cycle AP|R

2 The 3-carbon pyruvic acids generated by glycolysis enter the mitochondria separately. Each loses a carbon (generating CO_2) and is combined with a coenzyme to form a 2-carbon acetyl coenzyme A (acetyl CoA). More high-energy electrons are released.

3 Each acetyl CoA combines with a 4-carbon oxaloacetic acid to form the 6-carbon citric acid, for which the cycle is named. For each citric acid, a series of reactions removes 2 carbons (generating 2 CO_2's), synthesizes 1 ATP, and releases more high-energy electrons. The figure shows 2 ATP, resulting directly from 2 turns of the cycle per glucose molecule that enters glycolysis.

Electron Transport Chain AP|R

4 The high-energy electrons still contain most of the chemical energy of the original glucose molecule. Special carrier molecules bring the high-energy electrons to a series of enzymes that store much of the remaining energy in more ATP molecules. The other products are heat and water. The function of oxygen as the final electron acceptor in this last step is why the overall process is called aerobic respiration.

Figure 4.5

Glycolysis takes place in the cytosol and does not require oxygen. Aerobic respiration takes place in the mitochondria, and only in the presence of oxygen. The products of glycolysis and aerobic respiration include ATP, heat, carbon dioxide, and water. Glycolysis generates 2 ATP, the citric acid cycle generates 2 ATP, and the electron transport chain releases 32 to 34 ATP molecules. Thus, the total yield of ATP molecules per glucose molecule is 36 to 38, depending on the cell type.

Q: *Where in a cell does glycolysis occur?*
Answer can be found in Appendix E on page 568.

Figure 4.6

Phosphate bonds contain the energy stored in ATP.

Figure 4.7

ATP provides energy for metabolic reactions. Cellular respiration generates ATP.

Figure 4.8

A metabolic pathway consists of a series of enzyme-controlled reactions leading to formation of a product.

the cytosol (the liquid portion of the cytoplasm), and because it does not directly require oxygen, it is sometimes referred to as the **anaerobic** (an″a-er-o′bik) phase of cellular respiration.

Aerobic Respiration

The fate of pyruvic acid depends on oxygen availability. If oxygen is not present, then pyruvic acid enters an anaerobic pathway that yields lactic acid and limited energy (see fig. 8.10). If oxygen is present in sufficient quantity, the pyruvic acid generated by glycolysis can enter the more energy-efficient pathways of **aerobic respiration** (a″er-o′bik res″pǐ-ra′shun) in the mitochondria. After doing so, each molecule of pyruvic acid loses a carbon atom and binds a coenzyme to form a molecule of acetyl CoA, which can then combine with a 4-carbon compound to enter the citric acid cycle and then the electron transport chain. The final acceptor of electrons passed along the electron transport chain is oxygen. This is why the pathway is called "aerobic." (Because the reactions of the electron transport chain add phosphates to form ATP, they are also known as oxidative phosphorylation.) The aerobic reactions yield up to 36 ATP molecules per glucose (see fig. 4.5).

For each glucose molecule that is completely broken down, up to 38 molecules of ATP are produced. Two of these ATP molecules come from glycolysis and the rest form during the aerobic phase. About half the energy released goes to ATP synthesis, while the rest ends up as heat. Complete oxidation of glucose also produces carbon dioxide and water. The carbon dioxide is eventually exhaled, and the water becomes part of the internal environment.

In humans, metabolism does not generate enough water to meet daily needs, so we must drink water to survive. In contrast, a small desert rodent, the kangaroo rat, can survive entirely on the water produced by aerobic respiration.

Practice

10. What is the general function of ATP in metabolism?
11. Describe what happens during glycolysis.
12. What is the function of oxygen in cellular respiration?
13. What are the final products of cellular respiration?

4.5 METABOLIC PATHWAYS

Like cellular respiration, anabolic and catabolic reactions occur in a specific sequence of enzyme-catalyzed steps. Such a sequence of reactions is called a **metabolic pathway** (fig. 4.8).

For some pathways, the enzymes are positioned in the exact sequence as that of the reactions they control. For example, the enzymes responsible for aerobic respiration reside in tiny, stalked sections of the inner membranes (cristae) of mitochondria, in the sequence in which they function.

Recall that the rate of an enzyme-controlled reaction usually increases if either the number of substrate molecules or the number of enzyme molecules increases. However, the rate of a metabolic pathway is often determined by a *regulatory enzyme* responsible for one of its steps. The number of molecules of a regulatory enzyme is limited. Consequently, the enzyme supply can become saturated with substrate molecules whenever the substrate concentration exceeds a certain level. Once the enzyme is saturated, increasing the number of substrate molecules will no longer affect the reaction rate.

A regulatory enzyme that controls the whole pathway is called a *rate-limiting enzyme,* and is generally the first enzyme in a series. This position is important because if an enzyme at some other point in the sequence were rate-limiting, an intermediate chemical in the pathway might accumulate.

This section has dealt with the metabolism of glucose, a carbohydrate. Carbohydrates, fats, and proteins are the macronutrients, so named because they are required in large amounts. Fats and proteins can also be broken down to release energy for ATP synthesis. For all three types of macronutrients, the final process is aerobic respiration, and the most common entry point is into the citric acid cycle as acetyl coenzyme A (acetyl CoA) (fig. 4.9). These metabolic pathways and their regulation are described further in chapter 15 (pp. 428–431).

Practice

14. What is a *metabolic pathway?*
15. What is a rate-limiting enzyme?

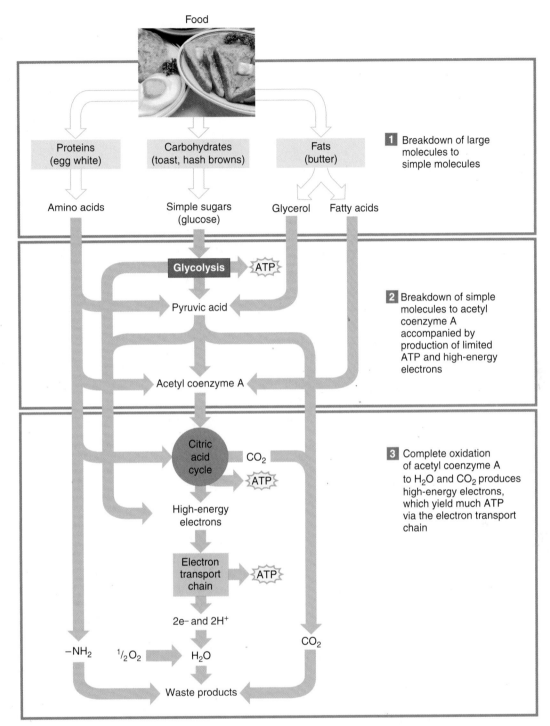

Figure 4.9

A summary of the breakdown (catabolism) of proteins, carbohydrates, and fats.

4.6 DNA (DEOXYRIBONUCLEIC ACID)

Enzymes control essential metabolic reactions. Therefore, cells must have instructions for producing enzymes as well as other types of proteins. The sequences of building blocks of **DNA (deoxyribonucleic acid)** molecules hold the information to manufacture proteins in the form of a *genetic code*.

Genetic Information

DNA molecules pass from parents to offspring when a sperm fertilizes an egg. As an offspring develops, mitosis passes the information in the DNA sequences of the chromosomes to new cells. A complete set of genetic instructions constitutes the **genome** (je′nōm). All cells except the sex cells contain two copies of the genome, one from each parent. Segments of the genome that encode proteins are called **genes** (jēnz). Only about 2% of the human genome encodes protein; much of the rest controls when and where genes become active to guide protein synthesis.

DNA Molecules

Recall from chapter 2 (p. 44) that the building blocks of nucleic acids are nucleotides (see fig. 2.20, p. 46). They are joined so that the sugars and phosphates alternate, forming a long "backbone" to the polynucleotide chain (see fig. 2.21*b*, p. 47). In DNA there are two such polynucleotide chains.

In a DNA molecule, the nitrogenous bases project from the backbone and bind weakly to the bases of the other strand (fig. 4.10). The resulting structure is like a ladder in which the uprights represent the alternating sugar and phosphate backbones of the two strands, and the rungs represent the nitrogenous bases. In a nucleotide, the DNA base may be one of four types: *adenine* (A), *thymine* (T), *cytosine* (C), or *guanine* (G). A gene is

a sequence of nucleotide bases along one DNA strand that specifies a particular protein's amino acid sequence.

The nitrogenous bases of DNA pair in specific ways: adenine only to thymine, and cytosine only to guanine. For example, a DNA strand with the base sequence A, C, G, C lies opposite and then joins a strand with the sequence T, G, C, G (see the upper region of DNA in fig. 4.10). These pairs—A with T, and G with C—are called *complementary base pairs*. The long DNA molecule forms a double helix. It is so long that it has to fold to fit inside a cell's nucleus. The two strands of a DNA double helix run in opposite orientations ("head-to-tail").

A molecule of DNA is typically millions of base pairs long. The great length of DNA molecules may seem quite a challenge to copy, or replicate, when a cell divides, but a set of special enzymes accurately and rapidly carries out this process.

> If unwound, the DNA in a single cell would stretch to about the height of an average man. It obviously winds very tightly!

DNA Replication

When a cell divides, each newly formed cell must have a copy of the original cell's genetic information (DNA) so it will be able to synthesize the proteins to build cellular parts and metabolize. This copying is called **replication** (rep″lĭ-ka′shun). It takes place during interphase of the cell cycle.

The double-stranded structure of DNA makes replication possible. As replication begins, hydrogen bonds between complementary base pairs in each DNA molecule break (fig. 4.11). The double helix unwinds and pulls apart, exposing the nitrogenous bases. Then an enzyme called DNA polymerase brings in new DNA nucleotides, and they form complementary pairs with the exposed bases. Other enzymes knit together the sugar-phosphate backbone. In this way, a new strand of complementary nucleotides forms along each of the old strands. This replication produces two complete DNA molecules, each with one old strand of the original molecule and one new strand. During mitosis, the two DNA molecules that form the two chromatids of each of the chromosomes separate so that one of these two DNA molecules passes to each of the new cells. Clinical Application 4.1 describes errors in DNA replication that cause sequence changes, a type of mutation.

Figure 4.10 AP|R

DNA structure. The molecular "ladder" of a double-stranded DNA molecule twists into a double helix. The ladder's "rungs" consist of complementary base pairs held together by hydrogen bonds— A with T (or T with A) and G with C (or C with G).

Practice

16. Distinguish between *gene* and *genome*.
17. Why must DNA molecules replicate?
18. List the steps of DNA replication.

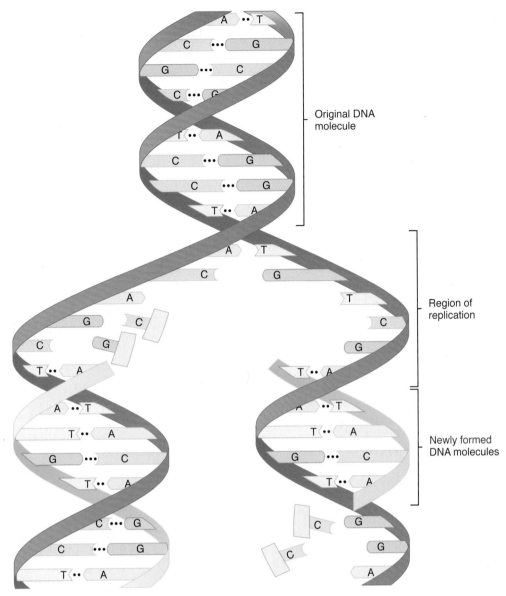

Figure 4.11 AP|R

DNA replication forms two double helices from one. When a DNA molecule replicates, its original strands separate locally. A new strand of complementary nucleotides forms along each original strand.

4.7 PROTEIN SYNTHESIS

DNA provides the genetic instructions that a cell requires to synthesize proteins. Manufacturing proteins is a multi-step, enzyme-catalyzed process.

The Genetic Code—Instructions for Making Proteins

Cells can synthesize specific proteins because the sequence of nucleotide bases in the DNA of genes specifies a particular sequence of amino acid building blocks of a protein molecule. This correspondence of gene and protein building block sequence is called the **genetic code.**

Each of the twenty types of amino acids in a biological protein is represented in a DNA molecule by a particular sequence of three nucleotides. The DNA sequence G, G, T represents one type of amino acid; G, C, A represents another; and T, T, A another. Other nucleotide sequences encode the instructions for beginning or ending the synthesis of a protein molecule. Thus, the sequence of nucleotides in a DNA molecule denotes the order of amino acids of a protein molecule, as well as where to start or stop that protein's synthesis.

Transcription

DNA molecules are confined to a cell's nucleus to maintain the genetic information, but protein synthesis occurs in the cytoplasm. The genetic information reaches the

Clinical Application 4.1

Mutations

It is easy to make an error when typing a paragraph consisting of several hundred letters. DNA replication, which is similar to copying such a paragraph, is also error-prone. A newly replicated gene may have too many or too few bases, or an "A" where the complementary base should be a "C." Fortunately, cells have several mechanisms that scan newly replicated DNA, detect mutations, and correct them. When this DNA repair fails, health may suffer. Mutations may occur spontaneously or may be induced by agents called mutagens, such as certain toxic chemicals or ionizing radiation.

Inherited illnesses result from mutations. Certain genetic tests can detect a particular mutation before symptoms of the associated illness begin. This is possible because the mutated gene is present from the time of conception. For example, a genetic test may identify the mutation that causes the neurological disorder Huntington disease in an eighteen-year-old, even though the symptoms—personality and cognitive changes and uncontrollable, constant movements—probably will not appear for another two decades or more. Predictive testing is controversial, particularly when the illness is not treatable.

Not all mutations are harmful. About 1% of the individuals in European populations have a mutation that makes them immune to HIV infection. The gene that is mutant normally encodes a protein to which the virus must bind in order to enter immune system cells. Without this protein, the virus cannot bind to and enter human cells. In Asian and African populations, this mutation is very rare.

---

cytoplasm by being copied into molecules of RNA (ribonucleic acid), which can exit the nucleus because they are much shorter than the DNA that composes chromosomes and they are single-stranded. The process of synthesizing RNA is called **transcription. Messenger RNA (mRNA)** is the type of RNA that carries a gene's message out of the nucleus. Other types of RNA also help to build proteins.

RNA (ribonucleic acid) molecules differ from DNA molecules in several ways (table 4.1). RNA molecules are single-stranded, and their nucleotides include the sugar ribose rather than deoxyribose. Like DNA, each RNA nucleotide includes one of four nitrogenous bases. However, whereas adenine, cytosine, and guanine nucleotides are in both DNA and RNA, thymine nucleotides are only in DNA. In place of thymine nucleotides, RNA molecules have *uracil* (U) nucleotides.

The enzyme RNA polymerase synthesizes mRNA following the rules of complementary base pairing. For example, the DNA sequence A, T, G, C, G specifies the complementary mRNA bases U, A, C, G, C (fig. 4.12). Specific DNA sequences outside the actual genes signal which of the two DNA strands contains the information to build a protein. RNA polymerase also recognizes sequences in the DNA that indicate where the gene begins, where it stops, and the correct direction to read the DNA, just like a sentence. When the RNA polymerase reaches the end of the gene, it releases the newly formed mRNA. Transcription is complete.

> Many genes are actually transcribed in "pieces" called exons. Sections called introns are spliced out of the mRNA before a protein is synthesized. Cells can transcribe and assemble different combinations of exons from a particular gene, creating slightly different forms of a protein that can function in different cell types or under specific conditions.

Table 4.1	**A Comparison of DNA and RNA Molecules**	
	DNA	**RNA**
Main location	Part of chromosomes, in nucleus	Cytoplasm
5-carbon sugar	Deoxyribose	Ribose
Basic molecular structure	Double-stranded	Single-stranded
Nitrogenous bases included	Adenine, thymine, cytosine, guanine	Adenine, uracil, cytosine, guanine
Major functions	Contains genetic code for protein synthesis; replicates prior to cell division	mRNA carries transcribed DNA information to cytoplasm and acts as template for synthesis of protein molecules; tRNA carries amino acids to mRNA

Translation

Each amino acid in a protein is specified by three contiguous bases in the DNA sequence. Those amino acids, in the proper order, are represented by a series of three-base sequences, called **codons** (ko′donz), in mRNA (table 4.2). In addition to the mRNA codons that specify amino acids, the sequence AUG represents the "start" of a gene, and three other mRNA base sequences indicate "stop." To guide protein synthesis, an mRNA molecule must leave the nucleus and associate with a ribosome in the cytoplasm. There the series of codons on mRNA are translated from the "language" of nucleic acids to the "language" of amino acids. This process of protein synthesis is appropriately called **translation.**

Building a protein molecule requires ample supplies of the correct amino acids in the cytoplasm and positioning them in the order specified along a strand of mRNA. A second kind of RNA molecule called **transfer RNA (tRNA)** correctly aligns amino acids, which are then linked by an enzyme to form proteins (fig. 4.13). Like mRNA, tRNA is synthesized in the nucleus and sent into the cytoplasm, where it assists in constructing a protein molecule.

Because twenty different types of amino acids form biological proteins, at least twenty different types of tRNA molecules must be available, one for each type of amino acid. Each type of tRNA has a region at one end that consists of three nucleotides that form complementary base pairs with a specific mRNA codon. The three nucleotides in the tRNA are called an **anticodon**

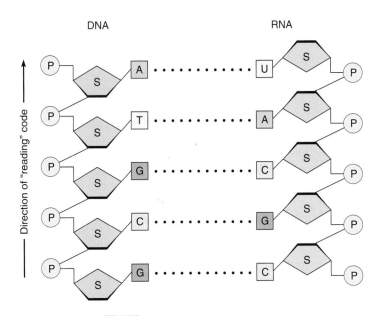

DNA   RNA

Direction of "reading" code

Figure 4.12   AP|R

Transcription of RNA from DNA. When an RNA molecule is synthesized beside a strand of DNA, complementary nucleotides bond as in a double-stranded DNA molecule, with one exception: RNA contains uracil nucleotides (U) in place of thymine nucleotides (T).

(an″tĭ-ko′don). In this way, tRNA carries its amino acid to a correct position on an mRNA strand. This action occurs on a ribosome (see fig. 4.13).

There are 64 possible types of tRNA anticodons, because there are this many possible triplets. Some amino acids may bind to more than one type of tRNA.

Table 4.2	Codons (mRNA Three-Base Sequences)											
		SECOND LETTER										
		U		**C**		**A**		**G**				
	U	UUU	phenylalanine (phe)	UCU	serine (ser)	UAU	tyrosine (tyr)	UGU	cysteine (cys)	U		
		UUC		UCC		UAC		UGC		C		
		UUA	leucine (leu)	UCA		UAA	STOP	UGA	STOP	A		
		UUG		UCG		UAG	STOP	UGG	tryptophan (trp)	G		
FIRST LETTER	**C**	CUU	leucine (leu)	CCU	proline (pro)	CAU	histidine (his)	CGU	arginine (arg)	U	**THIRD LETTER**	
		CUC		CCC		CAC		CGC		C		
		CUA		CCA		CAA	glutamine (gln)	CGA		A		
		CUG		CCG		CAG		CGG		G		
	A	AUU	isoleucine (ile)	ACU	threonine (thr)	AAU	asparagine (asn)	AGU	serine (ser)	U		
		AUC		ACC		AAC		AGC		C		
		AUA		ACA		AAA	lysine (lys)	AGA	arginine (arg)	A		
		AUG	START methionine (met)	ACG		AAG		AGG		G		
	G	GUU	valine (val)	GCU	alanine (ala)	GAU	aspartic acid (asp)	GGU	glycine (gly)	U		
		GUC		GCC		GAC		GGC		C		
		GUA		GCA		GAA	glutamic acid (glu)	GGA		A		
		GUG		GCG		GAG		GGG		G		

As protein synthesis begins, a ribosome binds an mRNA molecule. A tRNA molecule with the complementary anticodon holding its amino acid forms hydrogen bonds with the first mRNA codon. A second tRNA then binds the next codon, bringing its amino acid to an adjacent site on the ribosome. Then a peptide bond forms between the two amino acids, beginning a chain. The first tRNA molecule is released from its amino acid and is recycled to the cytoplasm (fig. 4.14). This process repeats as the ribosome moves along the mRNA molecule. The amino acids delivered by the tRNA molecules are added one at a time to the developing protein. Enzymes control protein synthesis.

As the protein molecule forms, it folds into its unique conformation and is then released to become a separate functional molecule. Correct protein folding is essential to health. In cells, misfolded proteins are threaded through spool-shaped structures called *proteasomes.* Here, they are either refolded into the

functional conformation, or destroyed if they are too abnormal.

A gene that is transcribed and translated into a protein is said to be *expressed.* The types and amounts of proteins in a cell, which can change with changing conditions, largely determine the function a cell performs in the body. Gene expression is the basis for cell differentiation, described in chapter 3 (p. 71). Genetics Connection 4.1 considers the roles that the analysis of gene expression will play in health care.

Practice

19. Define *genetic code.*
20. What is the function of DNA?
21. How is genetic information carried from the nucleus to the cytoplasm?
22. List the steps of protein synthesis.

Figure 4.13

Protein synthesis. DNA information is transcribed into mRNA, which in turn is translated into a sequence of amino acids. The inset shows some examples of the correspondence between mRNA codons and the specific amino acids that they encode.

Q: *What is the name of the molecule that carries DNA information so that it can be translated into protein?*

Answer can be found in Appendix E on page 568.

1 The transfer RNA molecule for the last amino acid added holds the growing polypeptide chain and is attached to its complementary codon on mRNA.

2 A second tRNA binds complementarily to the next codon, and in doing so brings the next amino acid into position on the ribosome. A peptide bond forms, linking the new amino acid to the growing polypeptide chain.

3 The tRNA molecule that brought the last amino acid to the ribosome is released to the cytoplasm, and will be used again. The ribosome moves to a new position at the next codon on mRNA.

4 A new tRNA complementary to the next codon on mRNA brings the next amino acid to be added to the growing polypeptide chain.

Figure 4.14 AP|R

A closer look at protein synthesis. Molecules of transfer RNA (tRNA) attach to and carry specific amino acids, aligning them in the sequence determined by the codons of mRNA. These amino acids, connected by peptide bonds, form a polypeptide chain of a protein molecule. Protein synthesis occurs on ribosomes.

Genetics Connection 4.1

Beyond the Human Genome Project: Personalizing Medicine

Researchers sequenced "the" human genome in 2001, but now that the genomes of several hundred people have been sequenced, it is becoming clear that although we are all very much alike, genetically speaking, we also vary in many ways. Sequences of A, T, C, and G are not the only types of information in genomes.

Profiling Gene Expression

To analyze gene expression (where and when genes are accessed to produce proteins), researchers use small squares of glass or nylon called DNA microarrays, or "chips," to immobilize many genes of interest. For example, a DNA chip for cardiovascular disease includes thousands of genes whose

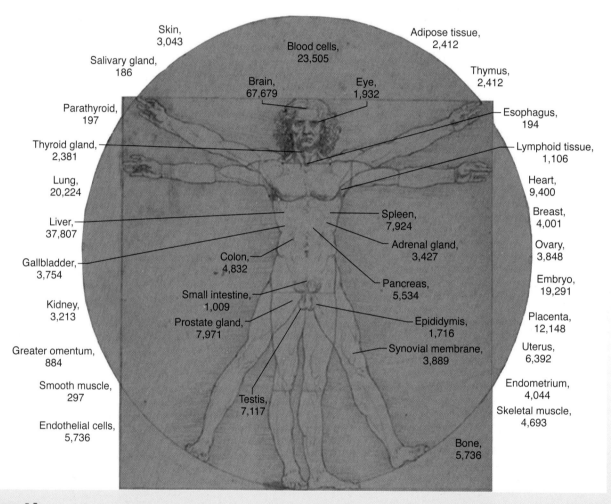

Figure 4A

Different types of cells express different sets of genes. DNA microarrays (chips) are used to identify the genes that are expressed in particular tissues and organs. This illustration depicts the numbers of genes expressed in the designated body parts. Note that the liver, which carries out thousands of biochemical reactions, uses the information in 37,807 genes, compared to the 2,412 genes expressed in less active adipose (fat) tissue. (The numbers can exceed the total number of genes in the genome because some DNA sequences can encode parts of different proteins.)

protein products control blood pressure, blood clotting, and the synthesis, transport, and metabolism of cholesterol and other lipids. The cell type affected in the condition is sampled, and its messenger RNA molecules are collected and copied, using an enzyme, into DNA molecules. The mRNAs from a specialized cell reflect the types of proteins manufactured there. The DNA copies are labeled and then added to the DNA chip. Where the sample sequence is complementary to the DNA on the chip, a label molecule lights up. The resulting pattern of fluorescent spots reveals, to a computer, which genes are expressed. This information enables researchers to view disease at a molecular level, which can suggest new types of treatments. Figure 4A summarizes information from many gene expression analyses.

DNA chips allow researchers to make compelling comparisons. A muscle cell from a person with diabetes mellitus expresses different genes than a muscle cell from a person who does not have diabetes. For certain cancers, DNA chips can predict which drugs will most likely be effective for an individual, and how likely the cancer is to spread.

Identifying Our Genetic Differences

The human genome sequence indicates that we are 99.9% similar to each other, and researchers are focusing on the tiny percentage of difference. One way to do this is to identify "single nucleotide polymorphisms," or SNPs (pronounced "snips"). A SNP is a single base site in the genome that differs in more than 1% of a population (fig. 4B). ("Polymorphism" means "many forms.") A mutation may also affect a single DNA base, but it is much rarer than a SNP. The human

genome is riddled with millions of SNPs, about one for every 1,000 DNA bases.

A SNP by itself can be helpful or harmful—or, in most instances, have no effect at all. But the patterns generated by many SNPs are valuable for their predictive power. "Genome-wide association studies" compare the patterns of a million SNPs across the genome among people with a particular illness and among similar people without the illness. The differences in SNPs can lead researchers to genes whose encoded proteins may be involved in the illness, and the SNP patterns can also be used to assess risk.

Copy Number Variants

Genomes include much more than protein-encoding DNA sequences—they are also riddled with many short repeated sequences. The number of repeats of particular sequences, called a *copy number variant* (fig. 4C), is another form of genetic information, although we do not yet completely understand it. A repeated sequence may range from a few bases to millions, and copies may be anywhere in the genome. Copy number variants may hold clues to health. For example, the number of copies of a particular gene influences susceptibility to HIV infection by controlling the structure of the receptor protein that HIV uses to enter a human cell. We still have much to learn about the significance of SNPs, gene expression differences, and copy number variants. What is becoming clear, though, is that there is a range of variations in human genomes that we can call "normal."

Moe   A C C T C T A T C T C A A C G C

Larry   A C C T C T A T C A T C A T C T C A A C G C

Curly   A C C T C T A T C A T C A T C A T C A T C T C A A C G C

Figure 4C
Copy number differences are very common among human genomes. In this example, the 3-base sequence A T C varies in copy number on this hypothetical section of chromosome for these three individuals. In actuality, such repeated sequences may be millions of bases long. Their significance is not well understood, but they are one way to distinguish individuals.

A C C T C T A T C T C A A C G C G   96% of population

A C C T C T A T A T C A A C G C G   4% of population = SNP

Figure 4B
SNPs distinguish individuals. A SNP is a site in the genome that differs in more than 1% of a population.

Summary Outline

4.1 Introduction (p. 77)

A cell continuously carries on thousands of metabolic reactions that maintain life and enable specialization. Cellular metabolism acquires, stores, and releases energy.

4.2 Metabolic Reactions (p. 77)

1. Anabolism
   a. Anabolism builds large molecules from smaller molecules.
   b. In dehydration synthesis, water forms, and smaller molecules join by sharing atoms.
   c. Carbohydrates are synthesized from monosaccharides, fats from glycerol and fatty acids, proteins from amino acids, and nucleotides from nucleic acids.
2. Catabolism
   a. Catabolism breaks down larger molecules into smaller ones.
   b. In hydrolysis, a water molecule is split as an enzyme breaks the bond between two parts of a molecule.
   c. Hydrolysis breaks down carbohydrates into monosaccharides, fats into glycerol and fatty acids, proteins into amino acids, and nucleic acids into nucleotides.

4.3 Control of Metabolic Reactions (p. 79)

Enzymes control metabolic reactions, which include many specific chemical changes.

1. Enzyme action
   a. Enzymes lower the amount of energy required to start chemical reactions.
   b. Enzymes are molecules that promote metabolic reactions without being consumed (catalysis).
   c. An enzyme acts upon a specific substrate molecule.
   d. The shape of an enzyme molecule fits the shape of its substrate.
   e. When an enzyme combines with its substrate, the substrate changes, a product forms, and the enzyme is released in its original form.
   f. The speed of an enzyme-controlled reaction depends partly upon the number of enzyme and substrate molecules and the enzyme's efficiency.
2. Factors that alter enzymes
   a. Almost all enzymes are proteins. Harsh conditions cause them to lose their shape, or denature.
   b. Heat, radiation, electricity, certain chemicals, and extreme pH values denature enzymes.

4.4 Energy for Metabolic Reactions (p. 80)

Energy is the capacity to do work. Common forms of energy include heat, light, sound, electrical energy, mechanical energy, and chemical energy.

1. Release of chemical energy
   a. Most metabolic processes use chemical energy released when chemical bonds break.
   b. The energy released from glucose breakdown during cellular respiration drives the reactions of cellular metabolism.
2. Cellular respiration
   a. ATP
      (1) Energy is captured in the bond of the terminal phosphate of each ATP.
      (2) When a cell requires energy, the terminal phosphate bond of an ATP molecule breaks, releasing stored energy.
      (3) An ATP molecule that loses its terminal phosphate becomes ADP.
      (4) An ADP molecule that captures energy and a phosphate becomes ATP.

   b. Glycolysis
      (1) The first phase of glucose decomposition does not directly require oxygen (anaerobic).
      (2) Some of the energy released is transferred to ATP.
   c. Aerobic respiration
      (1) The second phase of glucose decomposition requires oxygen.
      (2) Many more ATP molecules form during this phase than during the anaerobic phase.
      (3) The final products of glucose breakdown are carbon dioxide, water, and energy (ATP and heat).

4.5 Metabolic Pathways (p. 82)

A sequence of enzyme-controlled reactions constitutes a metabolic pathway.

1. Regulatory enzymes in limited numbers set rates of metabolic pathways.
2. Regulatory enzymes become saturated when substrate concentrations exceed a certain level.

4.6 DNA (Deoxyribonucleic Acid) (p. 83)

DNA molecules contain information that instructs a cell how to synthesize enzymes and other proteins.

1. Genetic information
   a. Inherited traits result from DNA information passed from parents to offspring.
   b. A complete set of genetic instructions is a genome.
   c. A gene is a DNA sequence that contains the information for making a particular protein.
2. DNA molecules
   a. A DNA molecule consists of two strands of nucleotides wound into a double helix.
   b. The nucleotides of a DNA strand are in a particular sequence.
   c. The nucleotides of each strand pair with those of the other strand in a complementary fashion (A with T and G with C).
3. DNA replication
   a. When a cell divides, each new cell requires a copy of the older cell's genetic information.
   b. DNA molecules replicate during interphase of the cell cycle.
   c. Each new DNA molecule has one old strand and one new strand.

4.7 Protein Synthesis (p. 85)

Genes provide instructions for making proteins, which take part in many aspects of cell function.

1. The genetic code—instructions for making proteins
   a. A sequence of DNA nucleotides encodes a sequence of amino acids.
   b. RNA molecules transfer genetic information from the nucleus to the cytoplasm.
   c. Transcription
      (1) RNA molecules are usually single-stranded; they contain ribose instead of deoxyribose and uracil nucleotides in place of thymine nucleotides.
      (2) Messenger RNA (mRNA) molecules consist of nucleotide sequences that are complementary to those of exposed strands of DNA.
      (3) Messenger RNA molecules associate with ribosomes and provide patterns for the synthesis of protein molecules.
   d. Translation
      (1) A ribosome binds to an mRNA molecule.
      (2) Molecules of tRNA position amino acids along a strand of mRNA.
      (3) Amino acids released from the tRNA molecules join and form a protein molecule that folds into a unique shape.

Chapter Assessments

4.1 Introduction
1. Explain the relationship between genes and cellular metabolism. (p. 77)
2. Explain why enzymes are important in the body. (p. 77)

4.2 Metabolic Reactions
3. Distinguish between anabolism and catabolism. (p. 77)
4. Distinguish between dehydration synthesis and hydrolysis. (p. 77)

4.3 Control of Metabolic Reactions
5. Describe how an enzyme interacts with its substrate. (p. 79)
6. Define *active site*. (p. 79)
7. The process of changing the shape of an enzyme to the point where it loses function is called _____. (p. 80)

   a. active site   d. denaturation
   b. substrate   e. conformation
   c. product
8. Define *cofactor*. (p. 80)

4.4 Energy for Metabolic Reactions
9. Explain how oxidation of molecules inside cells differs from burning materials outside of cells. (p. 80)
10. Explain the importance of ATP, and the relationship of ATP to ADP. (p. 80)
11. Distinguish between anaerobic and aerobic phases of cellular respiration. (p. 82)
12. Match the parts of cellular respiration to their associated activities. (p. 82)

    (1) electron transport chain   A. Glucose molecules are broken down into pyruvic acid.
    (2) glycolysis   B. Carrier molecules and enzymes extract energy and store it as ATP, releasing water and heat.
    (3) citric acid cycle   C. Acetyl CoA molecules are broken down to release CO_2 and high-energy electrons.

13. Identify the final acceptor of the electrons released in the reactions of cellular respiration. (p. 82)

4.5 Metabolic Pathways
14. Define *metabolic pathway*. (p. 82)
15. Explain how one enzyme can control the rate of a metabolic pathway. (p. 82)
16. Identify the cellular respiration pathway where glucose, fats, and proteins commonly enter. (p. 82)

4.6 DNA (Deoxyribonucleic Acid)
17. Distinguish between a gene and a genome. (p. 84)
18. DNA information provides instructions for the cell to ____. (p. 84)

    a. manufacture carbohydrate molecules   c. manufacture RNA from amino acids
    b. extract energy   d. synthesize protein molecules
                                                 e. synthesize lipids

19. Explain why DNA replication is essential. (p. 84)
20. Describe the events of DNA replication. (p. 84)

4.7 Protein Synthesis
21. If a strand of DNA has the sequence A T G C G A T C C G C, then the sequence of an mRNA molecule transcribed from it is _____. (p. 86)
22. Distinguish between transcription and translation. (p. 86)
23. Describe the function of a ribosome in protein synthesis. (p. 87)
24. Calculate the number of amino acids encoded by a DNA sequence of 27 nucleotides. (p. 87)
25. Define *gene expression*. (p. 88)

Integrative Assessments/Critical Thinking

OUTCOMES 2.2, 4.4, 4.6, 4.7
1. The chapter discusses several specific types of chemical bonds. Describe each of the following, and explain why each is important.
   a. High-energy phosphate bond
   b. Peptide bond
   c. The bond between an mRNA codon and a tRNA anticodon

OUTCOMES 4.2, 4.5
2. How can the same biochemical be both a reactant (a starting material) and a product?

OUTCOMES 4.3, 4.4
3. What effect might changes in the pH of body fluids or body temperature that accompany illness have on enzymes?

OUTCOME 4.4
4. After finishing a grueling marathon, a runner exclaims, "Whew, I think I've used up all my ATP!" Could this be possible?

OUTCOME 4.5
5. Explain how proteins assist in DNA replication.

WEB CONNECTIONS

Visit the text website at **www.mhhe.com/shieress11** for additional quizzes, interactive learning exercises, and more.

APR

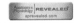

Anatomy & Physiology REVEALED includes cadaver photos that allow you to peel away layers of the human body to reveal structures beneath the surface. This program also includes animations, radiologic imaging, audio pronunciations, and practice quizzing. To learn more visit **www.aprevealed.com**.

5

Tissues

(a)

(b)

*Henrietta Lacks (**a**) checked into Johns Hopkins Hospital in 1951, where she was treated for cervical cancer. A sample of her cells went on to become a standard cell line used in laboratories all over the world (**b**).*

Donating tissue for research. Who owns your cells? Thanks to several cases, facilities that store human cells and tissues for research purposes today carefully inform donors how parts of themselves might be used.

The most famous cell donor was Henrietta Lacks, who died at age 31 in 1951 at Johns Hopkins Hospital from a fast-spreading cancer of her cervix. Before she died, a researcher interested in growing cells took a sample of Henrietta's tumor—without her knowledge. He quickly learned that the cells divided remarkably fast, something he had been seeking for years to use as a research tool. Dubbed "HeLa" in standard shorthand for a cell culture's source, the cells went on to become laboratory standards, far outliving Henrietta.

HeLa cells were shot into space, bombarded with radiation, and sent to the depths of the oceans, to investigate what a human body might tolerate. They were used in thousands of experiments, including those that led to development of the polio vaccine, *in vitro* fertilization, chemotherapy, animal cloning, innumerable drugs, gene maps, and they replaced animals in cosmetic testing. Henrietta's family was oblivious to their use. When her husband and children finally learned that her cells were being used all over the world, they imagined parts of her living on in the dark basement of the hospital where she died, the victim of mad scientists.

Other cases followed Henrietta's. In 1990, John Moore sued the University of California for profiting from his spleen cells, removed to treat leukemia years earlier. He lost. In 2006, so did William Catalona, who developed a biomarker test for prostate cancer using tumor cells from thousands of patients. When Dr. Catalona switched universities, he asked

his patients to request that their tissues be transferred too so he could continue working. But a court ruled in favor of the first institution, concluding that allowing patients to move their samples would adversely affect cell and tissue banks.

Thanks to these and other cases, cell and tissue banks today ensure that tissue donors know their rights. The informed consent form for the National Institutes of Health's Coriell Cell Repository, for example, states: *"Submission of my sample to the Repository may give scientists valuable research material that can help them to develop new diagnostic tests, new treatments, and new ways to prevent disease."*

Learning Outcomes

After studying this chapter, you should be able to do the following:

5.1 Introduction

1. List the four major tissue types, and tell where each is located in the body. (p. 95)

5.2 Epithelial Tissues

2. Describe the general characteristics and functions of epithelial tissues. (p. 95)

3. Name the types of epithelium, and for each type, identify an organ in which that type is found. (p. 96)

4. Explain how glands are classified. (p. 101)

5.3 Connective Tissues

5. Compare and contrast the general components, cells, fibers, and extracellular matrix (where applicable) in each type of connective tissue. (p. 104)

6. Describe the major functions of each type of connective tissue. (p. 104)

5.4 Types of Membranes

7. Distinguish among the four major types of membranes. (p. 110)

5.5 Muscle Tissues

8. Distinguish among the three types of muscle tissues. (p. 110)

5.6 Nervous Tissues

9. Describe the general characteristics and functions of nervous tissues. (p. 111)

Module 3: Tissues

 Learn **Practice** **Assess**

Aids to Understanding Words

(Appendix A on page 564 has a complete list of Aids to Understanding Words.)

adip- [fat] *adip*ose tissue: Tissue that stores fat.

chondr- [cartilage] *chondr*ocyte: Cartilage cell.

-cyt [cell] osteo*cyt*e: Bone cell.

epi- [upon] *epi*thelial tissue: Tissue that covers all free body surfaces.

-glia [glue] neuro*glia*: Cells that support neurons; part of nervous tissue.

inter- [between] *inter*calated disc: Band between adjacent cardiac muscle cells.

macr- [large] *macr*ophage: Large phagocytic cell.

os- [bone] *os*seous tissue: Bone tissue.

pseud- [false] *pseud*ostratified epithelium: Tissue with cells that appear to be in layers, but are not.

squam- [scale] *squam*ous epithelium: Tissue with flattened or scalelike cells.

strat- [layer] *strat*ified epithelium: Tissue with cells in layers.

5.1 INTRODUCTION

Cells, the basic units of structure and function in the human organism, are organized into groups called **tissues** (tish'uz). Each type of tissue is composed of similar cells specialized to carry on a particular function.

The tissues of the human body are of four major types: *epithelial, connective, muscle,* and *nervous.* Epithelial tissues form protective coverings and function in secretion and absorption. Connective tissues support soft body parts and bind structures together. Muscle tissues produce body movements, and nervous tissues conduct impulses that help control and coordinate body activities.

Table 5.1 compares the four major tissue types. Throughout this chapter, simplified line drawings (for example, fig. 5.1*a*) are included with each micrograph (for example, fig. 5.1*b*) to emphasize the distinguishing characteristics of the specific tissue, as well as a locator icon (an example of where in the body that particular tissue may be found).

Practice

1. What is a tissue?
2. List the four major types of tissues.

5.2 EPITHELIAL TISSUES AP|R

General Characteristics

Epithelial (ep"ĭ-the'le-al) **tissues** are found throughout the body. Epithelium covers organs, forms the inner lining of body cavities, and lines hollow organs. It always has a *free (apical) surface* exposed to the outside or internally to an open space. The underside of this tissue is anchored to connective tissue by a thin, nonliving layer called the **basement membrane.**

As a rule, epithelial tissues lack blood vessels. However, nutrients diffuse to epithelium from underlying connective tissues, which have abundant blood vessels.

Epithelial cells readily divide. As a result, injuries heal rapidly as new cells replace lost or damaged ones. For example, skin cells and cells that line the stomach and intestines are continually damaged and replaced.

Epithelial cells are tightly packed. Consequently, these cells form effective protective barriers in such structures as the outer layer of the skin and the lining of the mouth. Other epithelial functions include secretion, absorption, and excretion.

Epithelial tissues are classified according to shapes and numbers of layers of cells. Epithelial tissues that are composed of thin, flattened cells are *squamous;* those with cube-shaped cells are *cuboidal;* and those with

Table 5.1	Types of Tissue		
Type	**Function**	**Location**	**Distinguishing Characteristics**
Epithelial	Protection, secretion, absorption, excretion	Cover body surface, cover and line internal organs, compose glands	Lack blood vessels, readily divide; cells are tightly packed
Connective	Bind, support, protect, fill spaces, store fat, produce blood cells	Widely distributed throughout body	Mostly have good blood supply; cells are farther apart than epithelial cells
Muscle	Movement	Attached to bones, in the walls of hollow internal organs, heart	Able to contract in response to specific stimuli
Nervous	Transmit impulses for coordination, regulation, integration, and sensory reception	Brain, spinal cord, nerves	Cells communicate with each other and other body parts

Free surface
of tissue

Simple
squamous
epithelium

Basement
membrane

Nucleus

Connective
tissue

(a)

(b)

Free surface
of simple
squamous
epithelium

Nucleus

(c)

(d)

Figure 5.1

Simple squamous epithelium consists of a single layer of tightly packed, flattened cells. **(a)** and **(b)** side view (400×), **(c)** and **(d)** surface view (250×). In one example, it forms the air sacs of the lungs.

tall, elongated cells are *columnar;* those with single layers of cells are *simple;* those with two or more layers of cells are *stratified.* In the following descriptions, note that the free surfaces of epithelial cells are modified in ways that reflect their specialized functions.

Practice

3. List the general characteristics of epithelial tissues.
4. Describe the classification of epithelium in terms of shape and number of layers of cells.

Simple Squamous Epithelium

Simple squamous (skwa′mus) **epithelium** consists of a single layer of thin, flattened cells. These cells fit tightly together, somewhat like floor tiles, and their nuclei are usually broad and thin (fig. 5.1).

Substances pass rather easily through simple squamous epithelium, which is common at sites of diffusion and filtration. For instance, simple squamous epithelium lines the air sacs (alveoli) of the lungs where oxygen and carbon dioxide are exchanged. It also forms the walls of capillaries, lines the insides of blood and lymph vessels, and covers the membranes that line body cavities. However, because it is so thin and delicate, simple squamous epithelium is easily damaged.

Simple Cuboidal Epithelium

Simple cuboidal epithelium consists of a single layer of cube-shaped cells. These cells usually have centrally located, spherical nuclei (fig. 5.2).

Simple cuboidal epithelium covers the ovaries and lines most of the kidney tubules and the ducts of certain glands, where the free surface faces the hollow channel or *lumen.* In the kidneys, this tissue functions in secretion and absorption; in glands, it secretes glandular products.

(a)　　　　　　　　　　　　　　　　　(b)

Figure 5.2 AP|R

Simple cuboidal epithelium consists of a single layer of tightly packed, cube-shaped cells (630×). In one example, it lines the kidney tubules.

Q: *Is this section through the kidney tubules a cross section or a longitudinal section?*

Answer can be found in Appendix E on page 568.

(a)　　　　　　　　　　　　　　　　　(b)

Figure 5.3 AP|R

Simple columnar epithelium consists of a single layer of elongated cells (400×). In one example, it lines the small intestine.

Simple Columnar Epithelium

The cells of **simple columnar epithelium** are elongated; that is, they are longer than they are wide. This tissue is composed of a single layer of cells with elongated nuclei usually located at about the same level, near the basement membrane (fig. 5.3). The cells of this tissue can be ciliated or nonciliated. *Cilia* extend from the free surfaces of the cells and move constantly (see chapter 3, p. 58). In the female reproductive tract, cilia aid in moving the egg cell through the uterine tube to the uterus.

Nonciliated simple columnar epithelium lines the uterus and portions of the digestive tract, including the stomach and the small and large intestines. Because its cells are elongated, this tissue is thick, which enables it to protect underlying tissues. Simple columnar epithelium also secretes digestive fluids and absorbs nutrients from digested food.

Simple columnar cells, specialized for absorption, often have many tiny, cylindrical processes extending from their surfaces. These processes, called *microvilli,* increase the surface area of the cell membrane where it is exposed to substances being absorbed.

Typically, specialized, flask-shaped glandular cells are scattered among the columnar cells of simple columnar epithelium. These cells, called *goblet cells,* secrete a protective fluid, called *mucus,* onto the free surface of the tissue (see fig. 5.3).

Pseudostratified Columnar Epithelium

The cells of **pseudostratified** (soo"do-strat'ĭ-fīd) **columnar epithelium** appear to be stratified or layered, but they are not. Instead, the nuclei lie at two or more levels in the row of aligned cells. However, the cells, which vary in shape, all reach the basement membrane, even though some of them may not contact the free surface.

Pseudostratified columnar epithelial cells commonly have cilia, which extend from the free surfaces of the cells. Goblet cells scattered throughout this tissue secrete mucus, which the cilia sweep away (fig. 5.4).

Pseudostratified columnar epithelium lines the passages of the respiratory system. Here, the mucus-covered linings are sticky and trap dust and microorganisms that enter with the air. The cilia move the mucus and its captured particles upward and out of the airways.

Stratified Squamous Epithelium

The many cell layers of **stratified squamous epithelium** make this tissue relatively thick. Cells divide in the deeper layers, and newer cells push older ones farther outward, where they flatten (fig. 5.5). In naming stratified epithelial tissues based on shape of cells, the appearance of the top layer of cells is used.

Stratified squamous epithelium forms the outer layer of the skin (*epidermis*). As skin cells age, they accumulate a type of protein called *keratin,* and then harden and die. This "keratinization" produces a covering of dry, tough, protective material that prevents water and other substances from escaping underlying tissues and blocks various chemicals and microorganisms from entering (see fig. 6.2, p. 119).

Stratified squamous epithelium also lines the oral cavity, esophagus, vagina, and anal canal. In these parts, the tissue is not keratinized; it stays soft and moist, and the cells on its free surfaces remain alive.

Stratified Cuboidal Epithelium

Stratified cuboidal epithelium consists of two or three layers of cuboidal cells that form the lining of a lumen (fig. 5.6). The layering of the cells provides more protection than the single layer affords.

Stratified cuboidal epithelium lines the larger ducts of the mammary glands, sweat glands, salivary glands, and pancreas. It also forms the lining of developing ovarian follicles and seminiferous tubules, which are parts of the female and male reproductive systems, respectively.

Stratified Columnar Epithelium

Stratified columnar epithelium consists of several layers of cells (fig. 5.7). The superficial cells are columnar, whereas the basal layers consist of cuboidal cells. Small amounts of stratified columnar epithelium are found in the male urethra and ductus deferens and in parts of the pharynx.

Transitional Epithelium

Transitional epithelium is specialized to change in response to increased tension. It forms the inner lining of the urinary bladder and lines the ureters and the superior urethra. When the wall of one of these organs contracts, the tissue consists of several layers of cuboidal cells; however, when the organ is distended, the tissue stretches, and the physical relationships among the cells change (fig. 5.8). In addition to providing an expandable lining, transitional epithelium forms a barrier that helps prevent the contents of the urinary tract from diffusing back into the internal environment.

(a)   (b)

Cilia (free surface of tissue)

Cytoplasm

Goblet cell

Nucleus

Basement membrane

Connective tissue

Figure 5.4

Pseudostratified columnar epithelium appears stratified because the cell nuclei are located at different levels (1,000×). In one example, it lines the passages of the respiratory system.

(a)   (b)

Figure 5.5 AP|R

Stratified squamous epithelium consists of many layers of cells (65×).
In one example, it lines the oral cavity and esophagus.

(a)   (b)

Figure 5.6 AP|R

Stratified cuboidal epithelium consists of two to three layers of cube-shaped cells surrounding
a lumen (600×), such as in the salivary glands.

(a)   (b)

Figure 5.7

Stratified columnar epithelium consists of a superficial layer of columnar cells overlying several layers of cuboidal cells (230×). In one example, it is found in the male urethra.

(a)   (b)

(c)   (d)

Figure 5.8   AP|R

Transitional epithelium. (**a** and **b**) When the organ wall contracts, transitional epithelium is unstretched and consists of many layers (675×). (**c** and **d**) When the organ is distended, the tissue stretches and appears thinner (675×). Transitional epithelium lines the urinary bladder and the ureters and part of the urethra.

Up to 90% of all human cancers are *carcinomas,* growths that originate in epithelium. Most carcinomas begin on surfaces that contact the external environment, such as skin, linings of the airways, or linings of the stomach or intestine. This observation suggests that the more common cancer-causing agents may not deeply penetrate tissues. Carcinomas may also arise internally, such as in a duct in a breast or in the prostate gland.

Glandular Epithelium

Glandular epithelium is composed of cells specialized to produce and secrete substances into ducts or into body fluids. Such cells are usually found within columnar or cuboidal epithelium, and one or more of these cells constitute a *gland.* Glands that secrete their products into ducts that open onto surfaces, such as the skin or the lining of the digestive tract, are called **exocrine glands.** Glands that secrete their products into tissue fluid or blood are called **endocrine glands.** (Endocrine glands are discussed in chapter 11.)

Exocrine glands are classified according to the ways these glands secrete their products (fig. 5.9). Glands that release fluid by exocytosis are called **merocrine** (mer′o-krin) **glands.** Glands that lose small portions of their glandular cell bodies during secretion are called **apocrine** (ap′o-krin) **glands.** Glands that release entire cells that disintegrate to release cell secretions are called **holocrine** (ho′lo-krin) **glands.** Table 5.2 summarizes these glands and their secretions.

Most exocrine secretory cells are merocrine, and they can be further subclassified based on their secretion of serous fluid or mucus. *Serous fluid* is typically watery, and has a high concentration of enzymes. Serous cells secreting this fluid, which lubricates, are commonly associated with the visceral and parietal membranes of the thoracic and abdominopelvic cavities. The thicker fluid, *mucus,* is rich in the glycoprotein *mucin* and abundantly secreted by cells, for protection, in the inner linings of the digestive, respiratory, and reproductive systems. Mucous cells and goblet cells secrete mucus, but in different parts of the body. Table 5.3

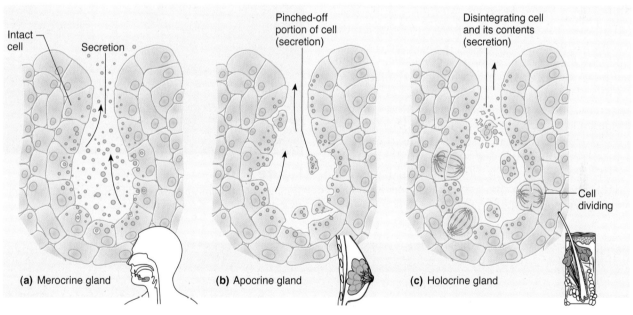

Figure 5.9

Types of exocrine glands. **(a)** AP|R Merocrine glands release secretions without losing cytoplasm. **(b)** AP|R Apocrine glands lose small portions of their cell bodies during secretion. **(c)** AP|R Holocrine glands release entire cells filled with secretory products.

Table 5.2	Types of Exocrine Glandular Secretions	
Type of Gland	**Description of Secretion**	**Example**
Merocrine glands	A fluid product released through the cell membrane by exocytosis	Salivary glands, pancreatic glands, sweat glands of the skin
Apocrine glands	Cellular product and portions of the free ends of glandular cells pinch off during secretion	Mammary glands, ceruminous glands lining the external ear canal
Holocrine glands	Disintegrated entire cells filled with secretory products	Sebaceous glands of the skin

summarizes the characteristics of the different types of epithelial tissues.

Practice

5. Describe the special functions of each type of epithelium.
6. Distinguish between exocrine glands and endocrine glands.
7. Explain how exocrine glands are classified.
8. Distinguish between serous fluid and mucus.

5.3 CONNECTIVE TISSUES AP|R

General Characteristics

Connective (kŏ-nek'tiv) **tissues** bind structures, provide support and protection, serve as frameworks, fill spaces, store fat, produce blood cells, protect against infections, and help repair tissue damage. Connective tissue cells are farther apart than epithelial cells, and they have an abundance of **extracellular matrix** (eks''trah-sel'u-lar ma'triks) between them. This extracellular matrix is composed of *protein fibers,* and a *ground substance* consisting of nonfibrous protein and other molecules, and fluid. The consistency of the extracellular matrix varies from fluid to semisolid to solid. Clinical Application 5.1 discusses the extracellular matrix and its relationship to disease. Most connective tissue cells can divide. These tissues have varying degrees of vascularity, but in most cases they have good blood supplies and are well nourished. Some connective tissues, such as bone and cartilage, are quite rigid. Loose connective tissue and dense connective tissue are more flexible.

Major Cell Types

Connective tissues contain a variety of cell types. Some cells are called *fixed cells* because they reside in the tissue for an extended period of time. These include fibroblasts and mast cells. Other cells, such as macrophages, are *wandering cells.* They move through and appear in tissues temporarily, usually in response to an injury or infection.

Fibroblasts (fi'bro-blastz) are the most common type of fixed cell in connective tissue. These large, star-shaped cells produce fibers by secreting proteins into the extracellular matrix of connective tissues (fig. 5.10).

Macrophages (mak'ro-fājez), or histiocytes, originate as white blood cells (see chapter 12, p. 325) and are almost as numerous as fibroblasts in some connective

Figure 5.10

Scanning electron micrograph of a fibroblast (4,000×), the most abundant cell type of connective tissue.

Table 5.3	Epithelial Tissues	
Type	**Function**	**Location**
Simple squamous epithelium	Filtration, diffusion, osmosis; covers surface	Air sacs of the lungs, walls of capillaries, linings of blood and lymph vessels
Simple cuboidal epithelium	Secretion, absorption	Surface of ovaries, linings of kidney tubules, and linings of ducts of certain glands
Simple columnar epithelium	Absorption, secretion, protection	Linings of uterus, stomach, and intestines
Pseudostratified columnar epithelium	Protection, secretion, movement of mucus	Linings of respiratory passages
Stratified squamous epithelium	Protection	Outer layer of skin, linings of oral cavity, throat, vagina, and anal canal
Stratified cuboidal epithelium	Protection	Linings of larger ducts of mammary glands, sweat glands, salivary glands, and pancreas
Stratified columnar epithelium	Protection, secretion	Part of the male urethra and parts of the pharynx
Transitional epithelium	Distensibility, protection	Inner lining of urinary bladder and linings of ureters and part of urethra
Glandular epithelium	Secretion	Salivary glands, sweat glands, endocrine glands

tissues. They are specialized to carry on phagocytosis. Macrophages can move about and function as scavenger and defensive cells that clear foreign particles from tissues (fig. 5.11).

Mast cells are large and widely distributed in connective tissues. They are usually near blood vessels (fig. 5.12). Mast cells release *heparin,* which prevents blood clotting, and *histamine,* which promotes some of the reactions associated with inflammation and allergies (see chapter 14, p. 585).

Figure 5.11

Macrophages are scavenger cells common in connective tissues. This scanning electron micrograph shows a number of macrophages engulfing parts of a larger cell (3,300×).

Cell being engulfed

Macrophage

Figure 5.12

Scanning electron micrograph of a mast cell (6,600×), which releases heparin and histamine.

Connective Tissue Fibers

Fibroblasts produce three types of connective tissue fibers: collagenous fibers, elastic fibers, and reticular fibers. Of these, collagenous and elastic fibers are the most abundant.

Collagenous (kol-laj′ĕ-nus) **fibers** are thick threads of the protein *collagen.* They are grouped in long, parallel bundles, and are flexible but only slightly elastic. More importantly, they have great tensile strength—that is, they resist considerable pulling force. Thus, collagenous fibers are important components of body parts that hold structures together, such as **ligaments** (which connect bones to bones) and **tendons** (which connect muscles to bones).

Tissue containing abundant collagenous fibers is called *dense connective tissue.* It appears white, and for this reason collagenous fibers are sometimes called *white fibers.* In contrast, *loose connective tissue* has fewer collagenous fibers.

Elastic fibers are composed of a springlike protein called *elastin.* These thin fibers branch, forming complex networks. Elastic fibers are weaker than collagenous fibers, but they are easily stretched or deformed and will resume their original lengths and shapes when the force acting on them is removed. Elastic fibers are common in body parts normally subjected to stretching, such as the vocal cords. They are sometimes called *yellow fibers* because tissues well supplied with them appear yellowish.

Reticular fibers are thin collagenous fibers. They are highly branched and form delicate supporting networks in a variety of tissues, including those of the spleen. Table 5.4 summarizes the components of connective tissue, and their functions.

When skin is exposed to prolonged and intense sunlight, connective tissue fibers lose elasticity, and the skin stiffens and becomes leathery. In time, the skin may sag and wrinkle. Collagen injections may temporarily smooth out wrinkles. However, collagen applied as a cream to the skin does not combat wrinkles because collagen molecules are far too large to penetrate the skin.

Table 5.4	Components of Connective Tissue
Cell Type	**Function**
Fibroblasts	Produce fibers
Macrophages	Carry on phagocytosis
Mast cells	Secrete heparin and histamine
Tissue Fibers	**Function**
Collagenous	Hold structures together with great tensile strength
Elastic	Stretch easily
Reticular	Lend delicate support

Clinical Application 5.1

The Body's Glue: The Extracellular Matrix

Rather than being just "filler" between cells, the extracellular matrix (ECM) is a complex and changing mix of molecules that modifies the tissue to suit different organs and conditions. Not only does the ECM serve as a scaffolding to organize cells into tissues, but it relays the biochemical signals that control cell division, differentiation, repair, and migration.

The ECM has two basic components: the basement membrane that covers cell surfaces, and the rest of the material between cells, called the interstitial matrix. The basement membrane is mostly composed of tightly packed collagenous fibers from which large, cross-shaped glycoproteins called laminins extend. The laminins (and other glycoproteins such as fibronectin, the proteoglycans, and tenascin) traverse the interstitial matrix and contact receptors, called integrins, on other cells (fig. 5A). In this way, the ECM connects cells into tissues. At least twenty types of collagen and precursors of hormones, enzymes, growth factors, and immune system biochemicals (cytokines) compose the various versions of the ECM. The precursor molecules are activated under certain conditions.

The components of the ECM are always changing, as its cells synthesize proteins while enzymes called proteases break down specific proteins. The balance of components is important to maintaining and repairing organ structure. Disrupt the balance, and disease can result. Following are three common examples.

Cancer

The spread of a cancerous growth takes advantage of the normal ability of fibroblasts to contract as they close a wound, where they are replaced with normal epithelium.

Chemical signals from cancer cells make fibroblasts more contractile, and they take on the characteristics of cancer cells. At the same time, alterations in laminins loosen the connections of the fibroblasts to surrounding cells. This abnormal flexibility enables the changed fibroblasts to migrate, helping the cancer spread. Normally, fibroblasts secrete abundant collagen.

Liver Fibrosis

In fibrosis, a part of all chronic liver diseases, collagen deposition increases so that the ECM exceeds its normal 3% of the organ. Healthy liver ECM sculpts a framework that supports the epithelial and vascular tissues of the organ. In response to a damaging agent such as a virus, alcohol, or a toxic drug, hepatic stellate cells secrete collagenous fibers in the areas where the epithelium and blood vessels meet. Such limited fibrosis seals off the affected area, preventing its spread. But if the process continues—if an infection is not treated or the noxious stimulus not removed—the ECM grows and eventually blocks the interaction between liver cells and the bloodstream. The liver tissue can harden, a dangerous condition called *cirrhosis*.

Heart Failure and Atherosclerosis

The heart's ECM organizes cells into a three-dimensional network that coordinates their contractions into the rhythmic heartbeat necessary to pump blood. It consists of collagen, fibronectin, laminin, and elastin surrounding cardiac muscle cells and myofibroblasts, and is also in the walls of arteries. Heart failure and atherosclerosis reflect imbalances of collagen production and degradation. As in the liver,

Categories of Connective Tissue

Connective tissue is divided into two major categories. *Connective tissue proper* includes loose connective tissue and dense connective tissue. The *specialized connective tissues* include cartilage, bone, and blood.

Loose Connective Tissue

Loose connective tissue includes areolar tissue, adipose tissue, and reticular connective tissue. **Areolar** (ah-re′o-lar) **tissue** forms delicate, thin membranes throughout the body. The cells of this tissue, mainly fibroblasts, are located some distance apart and are separated by a gel-like ground substance that contains many collagenous and elastic fibers that fibroblasts

secrete (fig. 5.13). Areolar tissue binds the skin to the underlying organs and fills spaces between muscles. It lies beneath most layers of epithelium, where its many blood vessels nourish nearby epithelial cells.

Adipose (ad′ĭ-pōs) **tissue,** or fat, develops when certain cells (adipocytes) store fat as droplets in their cytoplasm and enlarge (fig. 5.14). When such cells become so abundant that they crowd other cell types, they form adipose tissue. Adipose tissue lies beneath the skin, in spaces between muscles, around the kidneys, behind the eyeballs, in certain abdominal membranes, on the surface of the heart, and around certain joints. Adipose tissue cushions joints and some organs, such as the kidneys. It also insulates beneath the skin, and it stores energy in fat molecules.

ECM buildup walls off an area where circulation is blocked, but if it continues, the extra scaffolding stiffens the heart, which can lead to heart failure. In atherosclerosis, excess ECM accumulates on the interior linings of arteries, blocking blood flow. During a myocardial infarction (heart attack), collagen synthesis and deposition increase in affected and nonaffected heart parts, which is why damage can continue even after pain stops.

The ECM was once considered a barrier to be overcome in developing drugs, but today several new drugs actually target components of the ECM. For example, drugs that affect the ECM may prevent a cancer's spread.

Figure 5A

The extracellular matrix (ECM) is a complex and dynamic meshwork of various proteins and glycoproteins. Collagen is abundant. Other common components include integrins that anchor the ECM to cells, proteoglycans, and fibronectin. The ECM may also include precursors of growth factors, hormones, enzymes, and cytokines. It is vital to maintaining the specialized characteristics of tissues and organs.

(a)　　　　(b)

Figure 5.13 AP|R

Areolar tissue contains numerous fibroblasts that produce collagenous and elastic fibers (800×). It is widespread in the body.

(a)   (b)

Fat droplet
Cell membrane
Nucleus

Figure 5.14

Adipose tissue cells contain large fat droplets that push the nuclei close to the cell membranes (180×). Adipose tissue beneath the skin provides insulation.

The average adult has 40 to 50 billion fat cells.

Overeating and lack of exercise can increase the size of adipose cells, leading to being overweight or obese. During periods of fasting, however, fat supplies energy, and adipocytes lose fat, shrink, and become more like fibroblasts.

Reticular connective tissue is composed of thin, collagenous fibers in a three-dimensional network. It helps provide the framework of certain internal organs, such as the liver and spleen.

Dense Connective Tissue

Dense connective tissue consists of many closely packed, thick, collagenous fibers and a fine network of elastic fibers. It has few cells, most of which are fibroblasts (fig. 5.15).

Collagenous fibers of dense connective tissue are very strong, enabling the tissue to withstand pulling forces. It often binds body parts in the form of tendons and ligaments. This type of tissue is also in the protective white layer of the eyeball and in the deeper skin layers. The blood supply to dense connective tissue is poor, slowing tissue repair. **AP|R**

(a)   (b)

Fibroblasts
Collagenous fibers

Figure 5.15

Dense connective tissue consists largely of tightly packed collagenous fibers (500×). Examples of this tissue include parts of tendons and ligaments.

Cartilage

Cartilage (kar′ti-lij) is a rigid connective tissue. It provides support, frameworks, and attachments, protects underlying tissues, and forms structural models for many developing bones.

Cartilage extracellular matrix is abundant and is largely composed of collagenous fibers embedded in a gel-like ground substance. Cartilage cells, or **chondrocytes** (kon′dro-sītz), occupy small chambers called *lacunae* and lie completely within the extracellular matrix (fig. 5.16).

A cartilaginous structure is enclosed in a covering of connective tissue called the *perichondrium.* Nutrients diffuse to cartilage cells from blood vessels in the perichondrium. Because this tissue does not have a direct blood supply, cartilage heals slowly and chondrocytes do not divide frequently.

Different types of extracellular matrix distinguish three types of cartilage. **Hyaline cartilage,** the most common type, has very fine collagenous fibers in its extracellular matrix and looks somewhat like white glass (fig. 5.16). It is found on the ends of bones in many joints, in the soft part of the nose, and in the supporting rings of the respiratory passages. Hyaline cartilage is also important in the development and growth of most bones (see chapter 7, p. 136).

Elastic cartilage has a dense network of elastic fibers and thus is more flexible than hyaline cartilage (fig. 5.17). It provides the framework for the external ears and for parts of the larynx.

(a)　(b)

Figure 5.16 AP|R

Cartilage cells (chondrocytes) are located in lacunae, which are in turn surrounded by extracellular matrix containing very fine collagenous fibers (400×). This is hyaline cartilage, the most common type.

(a)　(b)

Figure 5.17

Elastic cartilage contains many elastic fibers in its extracellular matrix (1,200×). In one example, it is part of the external ear.

Fibrocartilage, a very tough tissue, has many collagenous fibers (fig. 5.18). It is a shock absorber for structures that are subjected to pressure. For example, fibrocartilage forms pads (intervertebral discs) between the individual bones (vertebrae) of the spinal column. It also cushions bones in the knees and in the pelvic girdle.

> Between ages thirty and seventy, a person's nose may lengthen and widen by as much as half an inch, and the ears may lengthen by a quarter inch, because the cartilage in these areas continues to grow as we age.

Bone

Bone is the most rigid connective tissue. Its hardness is largely due to mineral salts, such as calcium phosphate and calcium carbonate, between cells. This extracellular matrix also has many collagenous fibers, which are flexible and reinforce the mineral components of bone.

Bone internally supports body structures. It protects vital parts in the cranial and thoracic cavities, and is an attachment for muscles. Bone also contains red marrow, which forms blood cells. It stores and releases inorganic chemicals such as calcium and phosphorus.

Bone matrix is deposited in thin layers called *lamellae,* which form concentric patterns around tiny longitudinal tubes called *central canals,* or Haversian canals (fig. 5.19). Bone cells, or **osteocytes** (os'te-o-sītz), are located in lacunae, which are rather evenly spaced within the lamellae.

In a bone, the osteocytes and layers of extracellular matrix, which are concentrically clustered around a central canal, form a cylinder-shaped unit called an **osteon** (os'te-on), or Haversian system. Many osteons cemented together form the substance of bone.

Each central canal contains a blood vessel, which places every bone cell near a nutrient supply. In addition, bone cells have many cytoplasmic processes that extend outward and pass through very small tubes in the extracellular matrix called *canaliculi.* These cellular processes connect with the processes of nearby cells. As a result, materials can move rapidly between blood vessels and bone cells. Thus, in spite of its inert appearance, bone is a very active tissue that heals much more rapidly than does injured cartilage. (The microscopic structure of bone is described in more detail in chapter 7, p. 134.)

Blood

Blood transports a variety of materials between interior body cells and those that exchange substances with the external environment. In this way, blood helps maintain stable internal environmental conditions. Blood is composed of *formed elements* suspended in a fluid extracellular matrix called *blood plasma.* The formed elements include *red blood cells, white blood cells,* and cell fragments called *platelets* (fig. 5.20). Most blood cells form in red marrow within the hollow parts of certain long bones. Chapter 12 describes blood in detail. Table 5.5 lists the characteristics of the connective tissues.

Practice

13. Describe the general characteristics of cartilage.
14. Explain why injured bone heals more rapidly than injured cartilage.
15. What are the major components of blood?

(a)   Chondrocyte — Nucleus — Collagenous fiber — Extracellular matrix   (b)

Figure 5.18 AP|R

Fibrocartilage contains many large collagenous fibers in its extracellular matrix (400×). In one example, it forms the pads between vertebrae.

Osteon

Lamella

Central canal

Lacuna

Canaliculi

(a)

(b)

(c)

Figure 5.19

Bone tissue. **(a)** Bone matrix is deposited in concentric layers around central canals. **(b)** Micrograph of bone tissue (200×). **(c)** Artificially colored scanning electron micrograph of an osteocyte within a lacuna (6,000×).

Table 5.5	Connective Tissues	
Type	**Function**	**Location**
Loose connective tissue		
Areolar tissue	Binds organs	Beneath skin, between muscles, beneath epithelial tissues
Adipose tissue	Protects, insulates, stores fat	Beneath skin, around kidneys, behind eyeballs, on surface of heart
Reticular connective tissue	Supports	Walls of liver and spleen
Dense connective tissue	Binds organs	Tendons, ligaments, deeper layers of skin
Hyaline cartilage	Supports, protects, provides framework	Ends of bones, nose, rings in the walls of respiratory passages
Elastic cartilage	Supports, protects, provides flexible framework	Framework of external ear and parts of larynx
Fibrocartilage	Supports, protects, absorbs shock	Between bony parts of spinal column, parts of pelvic girdle and knee
Bone	Supports, protects, provides framework	Bones of skeleton
Blood	Transports substances, helps maintain stable internal environment	Throughout body within a closed system of blood vessels and heart chambers

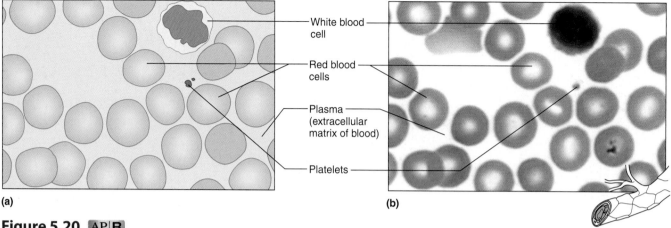

White blood cell

Red blood cells

Plasma (extracellular matrix of blood)

Platelets

(a)　　　　　　　　　　　　　　　(b)

Figure 5.20 AP|R

Blood tissue consists of red blood cells, white blood cells, and platelets suspended in a fluid extracellular matrix (1,000×).

Q: *What is the consistency of the extracellular matrix of this tissue?*

Answer can be found in Appendix E on page 568.

5.4 TYPES OF MEMBRANES

After discussing epithelial and connective tissues, membranes are better understood. **Epithelial membranes** are thin, sheetlike structures composed of epithelium and underlying connective tissue, covering body surfaces and lining body cavities. The three major types of epithelial membranes are *serous, mucous,* and *cutaneous.*

Serous (se′rus) **membranes** line body cavities that lack openings to the outside. These membranes form the inner linings of the thorax (parietal pleura) and abdomen (parietal peritoneum), and they cover the organs in these cavities (visceral pleura and visceral peritoneum, respectively, as shown in figs. 1.10 and 1.11, p. 11). A serous membrane consists of a layer of simple squamous epithelium and a thin layer of loose connective tissue. The cells of a serous membrane secrete watery *serous fluid,* which lubricates membrane surfaces.

Mucous (mu′cus) **membranes** line cavities and tubes that open to the outside of the body, including the oral and nasal cavities and the tubes of the digestive, respiratory, urinary, and reproductive systems. A mucous membrane consists of epithelium overlying a layer of loose connective tissue. Goblet cells within a mucous membrane secrete *mucus.*

Another epithelial membrane is the **cutaneous** (ku-ta′ne-us) **membrane,** more commonly called skin. It is described in detail in chapter 6.

A different type of membrane, composed entirely of connective tissues, is a **synovial** (si-nove-al) **membrane.** It lines joints and is discussed further in chapter 7 (p. 164).

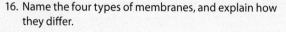

Practice

16. Name the four types of membranes, and explain how they differ.

5.5 MUSCLE TISSUES AP|R

General Characteristics

Muscle (mus′el) **tissues** are able to contract. Their elongated cells, sometimes called *muscle fibers,* can shorten and thicken. As they contract, muscle fibers pull at their attached ends, which moves body parts. The three types of muscle tissue—skeletal, smooth, and cardiac—are introduced here and discussed in more detail in chapter 8.

Approximately 40% of the body is skeletal muscle, and almost another 10% is smooth muscle or cardiac muscle.

Skeletal Muscle Tissue

Skeletal muscle tissue is found in muscles that usually attach to bones and can be controlled by conscious effort. For this reason, it is often called *voluntary* muscle tissue. The long, threadlike cells of skeletal muscle have alternating light and dark cross-markings called *striations.* Each cell has many nuclei (fig. 5.21). A nerve cell must stimulate a muscle cell to contract, and then the muscle cell relaxes when stimulation stops.

The muscles built of skeletal muscle tissue move the head, trunk, and limbs. They enable us to make facial expressions, write, talk, sing, chew, swallow, and breathe.

Smooth Muscle Tissue

Smooth muscle tissue is so called because its cells do not have striations. Smooth muscle cells are shorter than skeletal muscle cells and are spindle-shaped, each with a single, centrally located nucleus (fig. 5.22). This tissue composes the walls of hollow internal organs, such as the stomach, intestine, urinary bladder, uterus, and blood vessels. Unlike skeletal muscle, smooth muscle cannot

Striations

Nuclei

Portion of a muscle fiber

(a)

(b)

Figure 5.21

Skeletal muscle tissue is composed of striated muscle fibers with many nuclei (700×). It provides voluntary movement.

Cytoplasm

Nucleus

(a)

(b)

Figure 5.22

Smooth muscle tissue consists of spindle-shaped cells, each with a large nucleus (1,000×). It provides involuntary movement, such as the propulsion of food along the digestive tract.

be stimulated to contract by conscious efforts. Thus, its actions are *involuntary*. For example, smooth muscle tissue moves food through the digestive tract, constricts blood vessels, and empties the urinary bladder.

Cardiac Muscle Tissue

Cardiac muscle tissue is only in the heart. Its cells, which are striated and branched, are joined end to end, forming complex networks. Each cardiac muscle cell has a single nucleus (fig. 5.23). Where one cell touches another cell is a specialized intercellular junction called an *intercalated disc,* discussed further in chapter 8, p. 192.

Cardiac muscle, like smooth muscle, is controlled involuntarily. This tissue makes up the bulk of the heart and pumps blood through the heart chambers and into blood vessels.

Practice

17. List the general characteristics of muscle tissues.

18. Distinguish among skeletal, smooth, and cardiac muscle tissues.

5.6 NERVOUS TISSUES AP|R

Nervous (ner′vus) **tissues** are found in the brain, spinal cord, and peripheral nerves. The basic cells are called **neurons** (nu′ronz), or nerve cells (fig. 5.24). Neurons sense certain types of changes in their surroundings. They respond by transmitting electrical impulses along cellular processes called *axons.* As a result of the patterns by which neurons connect and communicate with each other and with muscle and gland cells,

Striations

Nucleus

Intercalated disc

(a)

(b)

Figure 5.23

Cardiac muscle cells are branched and interconnected, with a single nucleus each (400×). Their contraction is involuntary.

Nuclei of neuroglia

Cellular process

Cytoplasm

Cell membrane

Nucleus

(a)

(b)

Figure 5.24 AP|R

Several processes are part of this neuron, enabling it to communicate with other neurons (500×).

they can coordinate, regulate, and integrate many body functions.

In addition to neurons, nervous tissue includes **neuroglia** (nu-ro′gle-ah), shown in figure 5.24. Neuroglia divide and are crucial to the functioning of neurons. These cells support and bind the components of nervous tissue, carry on phagocytosis, and help supply nutrients to neurons by connecting them to blood vessels. They also play a role in cell-to-cell communication. Nervous

tissue is discussed in more detail in chapter 9. Table 5.6 summarizes the general characteristics of muscle and nervous tissues.

Practice

19. Describe the general characteristics of nervous tissues.
20. Distinguish between neurons and neuroglia.

The cells of different tissues vary greatly in their abilities to divide. Cells that divide continuously include the epithelial cells of the skin and the inner lining of the digestive tract, and the connective tissue cells that form blood cells in red bone marrow. However, skeletal and cardiac muscle cells and nerve cells do not usually divide at all after differentiating.

Fibroblasts respond rapidly to injuries by increasing in number and fiber production. They are often the principal agents of repair in tissues that have limited abilities to regenerate. For instance, fibroblasts form scar tissue after a heart attack occurs. Many organs include pockets of stem or progenitor cells that can divide and replace damaged, differentiated cells, under certain conditions.

Table 5.6	Muscle and Nervous Tissues	
Type	**Function**	**Location**
Skeletal muscle tissue (striated)	Voluntary movements of skeletal parts	Muscles usually attached to bones
Smooth muscle tissue (lacks striations)	Involuntary movements of internal organs	Walls of hollow internal organs
Cardiac muscle tissue (striated)	Heart movements	Heart muscle
Nervous tissue	Sensory reception and conduction of electrical impulses	Brain, spinal cord, and peripheral nerves

Summary Outline

5.1 Introduction (p. 95)

Tissues are groups of cells with specialized structural and functional roles. The four major types of human tissue are epithelial, connective, muscle, and nervous.

5.2 Epithelial Tissues (p. 95)

1. General characteristics
   a. Epithelial tissue covers organs, lines cavities and hollow organs, and is the major tissue of glands.
   b. Epithelium is anchored to connective tissue by a basement membrane, lacks blood vessels, consists of tightly packed cells, and is replaced continuously.
   c. It functions in protection, secretion, absorption, and excretion.
   d. Epithelial tissues are classified according to shape and number of layers of cells.
2. Simple squamous epithelium
   a. This tissue consists of a single layer of thin, flattened cells.
   b. It functions in gas exchange in the lungs and lines blood and lymph vessels and various body cavities.
3. Simple cuboidal epithelium
   a. This tissue consists of a single layer of cube-shaped cells.
   b. It carries on secretion and absorption in the kidneys and various glands.
4. Simple columnar epithelium
   a. This tissue is composed of elongated cells whose nuclei are near the basement membrane.
   b. It lines the uterus and digestive tract.
   c. Many absorbing cells have microvilli.
   d. This tissue has goblet cells that secrete mucus.
5. Pseudostratified columnar epithelium
   a. Nuclei located at two or more levels give this tissue a stratified appearance.
   b. Cilia that are part of this tissue move mucus over the surface of the tissue.
   c. It lines passages of the respiratory system.
6. Stratified squamous epithelium
   a. This tissue is composed of many layers of cells.
   b. It protects underlying cells.
   c. It forms the outer layer of the skin and lines the oral cavity, esophagus, vagina, and anal canal.
7. Stratified cuboidal epithelium
   a. This tissue is composed of two or three layers of cube-shaped cells.
   b. It lines the larger ducts of the mammary glands, sweat glands, salivary glands, and pancreas.
   c. It protects.
8. Stratified columnar epithelium
   a. The top layer of cells in this tissue are column-shaped. Cube-shaped cells make up the bottom layers.
   b. It is in the male urethra and ductus deferens and parts of the pharynx.
   c. This tissue protects and secretes.
9. Transitional epithelium
   a. This tissue is specialized to become distended.
   b. It lines the urinary bladder, ureters, and superior urethra.
10. Glandular epithelium
    a. Glandular epithelium is composed of cells that are specialized to secrete substances.
    b. A gland consists of one or more cells.
       (1) Exocrine glands secrete into ducts.
       (2) Endocrine glands secrete into tissue fluid or blood.
    c. Exocrine glands are classified according to the composition of their secretions.
       (1) Merocrine glands secrete fluid without loss of cytoplasm.
          (a) Serous cells secrete watery fluid with a high enzyme content.
          (b) Mucous cells secrete mucus.
       (2) Apocrine glands lose portions of their cells during secretion.
       (3) Holocrine glands release cells filled with secretory products.

5.3 Connective Tissues (p. 102)

1. General characteristics
   a. Connective tissue connects, supports, protects, provides frameworks, fills spaces, stores fat, produces blood cells, protects against infection, and helps repair damaged tissues.
   b. Connective tissue cells usually have considerable extracellular matrix between them.
   c. This extracellular matrix consists of fibers, a ground substance, and fluid.

d. Major cell types
   (1) Fibroblasts produce collagenous and elastic fibers.
   (2) Macrophages are phagocytes.
   (3) Mast cells may release heparin and histamine, and usually are near blood vessels.
e. Connective tissue fibers
   (1) Collagenous fibers are composed of collagen and have great tensile strength.
   (2) Elastic fibers are composed of elastin and are very elastic.
   (3) Reticular fibers are thin collagenous fibers.
2. Categories of connective tissue
   Connective tissue proper includes loose connective tissue and dense connective tissue. Specialized connective tissue includes cartilage, bone, and blood.
   a. Loose connective tissue
      (1) Areolar tissue forms thin membranes between organs and binds them. It is beneath the skin and between muscles.
      (2) Adipose tissue stores fat, cushions, and insulates. It is found beneath the skin, in certain abdominal membranes, behind the eyeballs, and around the kidneys, heart, and various joints.
      (3) Reticular connective tissue is composed of thin, collagenous fibers. It helps provide the framework of the liver and spleen.
   b. Dense connective tissue
      (1) This tissue is largely composed of strong, collagenous fibers.
      (2) It is found in the tendons, ligaments, white portions of the eyes, and the deeper skin layer.
   c. Cartilage
      (1) Cartilage provides a supportive framework for various structures.
      (2) Its extracellular matrix is composed of fibers and a gel-like ground substance.
      (3) Cartilaginous structures are enclosed in a perichondrium, which contains blood vessels.
      (4) Cartilage lacks a direct blood supply and is slow to heal.
      (5) Major types are hyaline cartilage, elastic cartilage, and fibrocartilage.
   d. Bone
      (1) The extracellular matrix of bone contains mineral salts and collagen.
      (2) Its cells are usually organized in concentric circles around central canals. Canaliculi connect the cells.
      (3) Bone is an active tissue that heals rapidly.
   e. Blood
      (1) Blood transports substances and helps maintain a stable internal environment.
      (2) Blood is composed of red blood cells, white blood cells, and platelets suspended in plasma.
      (3) Blood cells develop in red marrow in the hollow parts of long bones.

5.4 Types of Membranes (p. 110)

1. Epithelial membranes are composed of epithelium and underlying connective tissue. Serous, mucous, and cutaneous membranes are epithelial membranes.
2. Serous membranes, composed of epithelium and loose connective tissue, are membranes that line body cavities lacking openings to the outside. The cells of a serous membrane secrete serous fluid to lubricate membrane surfaces.
3. Mucous membranes, composed of epithelium and loose connective tissue, are membranes that line body cavities opening to the outside. Goblet cells within these membranes secrete mucus.
4. Another epithelial membrane, the cutaneous membrane, is the external body covering commonly called skin.
5. Synovial membranes, composed entirely of connective tissues, line joints.

5.5 Muscle Tissues (p. 110)

1. General characteristics
   a. Muscle tissues contract, moving structures that are attached to them.
   b. The three types are skeletal, smooth, and cardiac muscle tissues.
2. Skeletal muscle tissue
   a. Muscles containing this tissue usually are attached to bones and controlled by conscious effort.
   b. Muscle cells, also called muscle fibers, are long and threadlike.
   c. Muscle cells contract when stimulated by nerve cells, and then relax when stimulation stops.
3. Smooth muscle tissue
   a. This tissue is in the walls of hollow internal organs.
   b. It is involuntarily controlled.
4. Cardiac muscle tissue
   a. This tissue is found only in the heart.
   b. Cells are joined by intercalated discs and form branched networks.

5.6 Nervous Tissues (p. 111)

1. Nervous tissues are in the brain, spinal cord, and peripheral nerves.
2. Neurons (nerve cells)
   a. Neurons sense changes and respond by transmitting electrical impulses to other neurons or to muscles or glands.
   b. They coordinate, regulate, and integrate body activities.
3. Neuroglia
   a. Some of these cells bind and support nervous tissue.
   b. Others carry on phagocytosis.
   c. Still others connect neurons to blood vessels.
   d. Some are involved in cell-to-cell communication.

Chapter Assessments

5.1 Introduction

1. Which of the following is a major tissue type in the body? (p. 95)
   a. epithelial   d. connective
   b. nervous   e. all of the above
   c. muscle

2. Indicate where each major type of tissue can be found in the body. (p. 95)

5.2 Epithelial Tissues

3. A general characteristic of epithelial tissues is that _____. (p. 95)
   a. numerous blood vessels are present
   b. cells are spaced apart
   c. cells readily divide
   d. there is much extracellular matrix between cells
   e. they contain collagenous fibers

4. Explain how the structure of epithelial tissues provides for the functions of epithelial tissues. (p. 95)

5. Match the epithelial tissue to an organ in which the tissue is found. (pp. 96–101)

(1) simple squamous epithelium

(2) simple cuboidal epithelium

(3) simple columnar epithelium

(4) pseudostratified columnar epithelium

(5) stratified squamous epithelium

(6) stratified cuboidal epithelium

(7) stratified columnar epithelium

(8) transitional epithelium

(9) glandular epithelium

A. lining of intestines

B. lining of ducts of mammary glands

C. lining of urinary bladder

D. salivary glands

E. air sacs of lungs

F. respiratory passages

G. ductus deferens

H. lining of kidney tubules

I. outer layer of skin

6. Distinguish between exocrine and endocrine glands. (p. 101)

7. A gland that secretes substances out of cells by exocytosis is a(n) _____. (p. 101)

a. merocrine gland

b. apocrine gland

c. holocrine gland

5.3 Connective Tissues

8. Define *extracellular matrix*. (p. 102)

9. Describe three major types of connective tissue cells. (p. 102)

10. Distinguish between collagen and elastin. (p. 103)

11. Compare and contrast the different types of loose connective tissue. (p. 104)

12. Define *dense connective tissue*. (p. 106)

13. Explain why injured dense connective tissue and cartilage are usually slow to heal. (p. 106)

14. Name the types of cartilages and describe their differences and similarities. (p. 107)

15. Describe how bone cells are organized in bone tissue. (p. 108)

16. The fluid extracellular matrix of blood is called _____. (p. 108)

a. white blood cells

b. red blood cells

c. platelets

d. plasma

e. bone marrow

5.4 Types of Membranes

17. Identify the locations of four types of membranes in the body and indicate the types of tissues making up each membrane. (p. 110)

5.5 Muscle Tissues

18. Compare and contrast skeletal, smooth, and cardiac muscle tissues in terms of location, cell appearance, and control. (p. 110)

5.6 Nervous Tissues

19. Distinguish between the functions of neurons and neuroglia. (p. 111)

Integrative Assessments/Critical Thinking

OUTCOMES 3.4, 5.2, 5.3, 5.5, 5.6

1. Cancer-causing agents (carcinogens) usually act on dividing cells. Which of the four major types of tissues would carcinogens most influence? Least influence?

OUTCOME 5.3

2. Collagen and elastin are added to many beauty products. What type of tissues are they normally part of?

3. Joints such as the elbow, shoulder, and knee contain considerable amounts of cartilage and dense connective tissue. How does this explain why joint injuries are often slow to heal?

4. Disorders of collagen are characterized by deterioration of connective tissues. Why would you expect such diseases to produce widely varying symptoms?

WEB CONNECTIONS

Visit the text website at **www.mhhe.com/shieress11** for additional quizzes, interactive learning exercises, and more.

APR

Anatomy & Physiology REVEALED includes cadaver photos that allow you to peel away layers of the human body to reveal structures beneath the surface. This program also includes animations, radiologic imaging, audio pronunciations, and practice quizzing. To learn more, visit **www.aprevealed.com**.

6

Integumentary System

Itching. Few sensations are as intense as an itch. For millions of people, itching is much more than an annoyance. Each year in the United States, nearly 33 million people consult dermatologists about itchy skin rashes.

Itching occurs in more than fifty diseases, from psychiatric conditions to cancer. It may also happen in response to an insect bite or exposure to a poisonous plant, or arise as a side effect of taking a particular drug. Itching may interfere significantly with quality of life, and even hasten death by chronically disturbing sleep. A case report in a medical journal describes a woman who, unable to feel pain due to shingles, actually scratched through her skull in response to intense itching. Nearly half of all people undergoing hemodialysis to treat kidney failure experience severe itching.

Why do we itch? It may be a holdover from long-ago times when scratching an itch was a way of ridding the outside of the body of disease-carrying insects. Whatever the reason, itching is more than just a less intense form of pain, as has been thought. It involves fewer nerve fibers over a larger area than a typical pain response.

Itching has been difficult to study for several reasons. Our reporting of the intensity of itching is highly subjective. Another problem is that testing treatments may cause pain. Most small animal models, such as rodents, itch from different stimuli than do humans. For this reason, some researchers study itching and scratching in monkeys. In one set of experiments, researchers induced itching in the animals'

Scratching is a common response to an itch.

feet, then traced the responding neurons to a specific site at the base of the spinal cord where certain cells sense pain and itch. Scratching the itch quieted these nerve cells—but only for 5 to 10 seconds. Learning how scratching affects nerve cells might provide a new target for drug developers.

While researchers are still unsure of the origins of itching, sufferers can seek topical relief with oatmeal baths and lubricating creams. Antihistamine drugs work only on hives, which are a rare cause of itching. Drugs used to treat seizures and depression may help some severe cases of itching. The International Forum for the Study of Itch, founded in 2005, brings together dermatologists and neurologists to tackle the problem.

Learning Outcomes

After studying this chapter, you should be able to do the following:

6.1 Introduction

1. Describe what constitutes an organ, and name the large organ of the integumentary system. (p. 117)

6.2 Skin and Its Tissues

2. List the general functions of the skin. (p. 117)

3. Describe the structure of the layers of the skin. (p. 117)

4. Summarize the factors that determine skin color. (p. 120)

6.3 Accessory Structures of the Skin

5. Describe the anatomy and physiology of each accessory structure of the skin. (p. 122)

6.4 Regulation of Body Temperature

6. Explain how the skin helps regulate body temperature. (p. 125)

6.5 Healing of Wounds

7. Describe wound healing. (p. 125)

Module 4: Integumentary System

 Learn   Practice   Assess

Aids to Understanding Words   *(Appendix A on page 564 has a complete list of Aids to Understanding Words.)*

cut- [skin] sub*cut*aneous: Beneath the skin.

derm- [skin] *derm*is: Inner layer of the skin.

epi- [upon] *epi*dermis: Outer layer of the skin.

follic- [small bag] hair *follic*le: Tubelike depression in which a hair develops.

kerat- [horn] *kerat*in: Protein produced as epidermal cells die and harden.

melan- [black] *melan*in: Dark pigment produced by certain cells.

seb- [grease] *seb*aceous gland: Gland that secretes an oily substance.

6.1 INTRODUCTION

Two or more types of tissues grouped together and performing specialized functions constitute an **organ** (see fig. 1.3, p. 4). The skin is the largest organ in the body by weight. Skin and its various accessory structures (hair, fingernails, sensory receptors, and glands) make up the **integumentary** (in-teg-u-men'tar-e) **system.** The skin forms a barrier between ourselves and the outside, and is a strong yet flexible covering for our bodies.

If the skin of a 150-pound person were spread out flat, it would cover approximately 20 square feet.

6.2 SKIN AND ITS TISSUES

The skin is one of the most versatile organs of the body. It is vital in maintaining homeostasis. In addition to providing a protective covering, the skin helps regulate body temperature, retards water loss from deeper tissues, houses sensory receptors, synthesizes various biochemicals, and excretes small amounts of wastes.

The skin plays a role in the production of vitamin D, which is necessary for normal bone and tooth development. This vitamin can be obtained in the diet or can form from a substance (dehydrocholesterol) that is synthesized by cells in the digestive system. When dehydrocholesterol reaches the skin by means of the blood and is exposed to ultraviolet light from the sun, it is converted to vitamin D.

The skin includes two distinct layers (fig. 6.1). The outer layer, called the **epidermis** (ep"i-der'mis), is composed of stratified squamous epithelium. The inner layer, or **dermis** (der'mis), is thicker than the epidermis. It includes connective tissue consisting of collagen and elastic fibers, epithelial tissue, smooth muscle tissue, nervous tissue, and blood. A *basement membrane* anchors the epidermis to the dermis and separates these two skin layers.

Various treatments temporarily smooth facial wrinkles. "Botox" is an injection of a very dilute solution of botulinum toxin. Produced by the bacterium *Clostridium botulinum,* the toxin causes food poisoning. It also blocks nerve activation of the facial muscles that control smiling, frowning, and squinting. After three months, though, the facial nerves contact the muscles at different points, and the wrinkles return. (Botox used at higher doses to treat neuromuscular conditions can cause adverse effects.) Other anti-wrinkle treatments include chemical peels and dermabrasion to reveal new skin surface; collagen injections; and transplants of subcutaneous fat from the buttocks to the face.

Beneath the dermis is loose connective tissue, predominantly adipose tissue, that binds the skin to the underlying organs, forming the **subcutaneous** (sub"ku-ta'ne-us) **layer** (hypodermis). As its name indicates, this layer is beneath the skin and not a true layer of the skin. The collagen and elastic fibers of this layer are continuous with those of the dermis. Most of these fibers run parallel to the surface of the skin, extending in all directions. As a result, no sharp boundary separates the dermis and the subcutaneous layer. The adipose tissue of the subcutaneous layer insulates, helping to conserve body heat and impeding the entrance of heat from the outside. The subcutaneous layer also contains the major blood vessels that supply the skin and underlying adipose tissue.

Practice

1. List the general functions of the skin.
2. Name the tissue in the outer layer of the skin.
3. Name the tissues in the inner layer of the skin.
4. Name the tissues in the subcutaneous layer beneath the skin.
5. What are the functions of the subcutaneous layer?

Epidermis

Dermis

Subcutaneous
layer

Hair shaft
Sweat
Sweat gland pore
Stratum corneum
Stratum basale
Capillary
Dermal papilla
Basement membrane
Sebaceous gland
Arrector pili muscle
Sweat gland duct
Hair follicle
Sweat gland
Nerve cell process
Adipose tissue
Blood vessels
Muscle

(a)

Epidermis

Dermis

Hair follicle

Arrector pili
muscle

Sebaceous
gland

(b)

Figure 6.1 AP|R

Skin. **(a)** The skin is an organ that includes two layers, the epidermis and dermis, that lie atop a subcutaneous ("beneath the skin") layer. A section of skin. **(b)** This light micrograph depicts the layered structure of the skin (100×).

Intradermal injections are injected into the skin. *Subcutaneous injections* are administered through a hollow needle into the subcutaneous layer beneath the skin. Subcutaneous injections and *intramuscular injections,* administered into muscles, are sometimes called hypodermic injections.

Some substances are introduced through the skin by means of an adhesive transdermal patch that includes a small reservoir containing a drug. The drug passes from the reservoir through a permeable membrane at a known rate. It then diffuses into the epidermis and enters the blood vessels of the dermis. Transdermal patches deliver drugs that protect against motion sickness, alleviate chest pain associated with heart disease, and lower blood pressure. A transdermal patch that delivers nicotine is used to help people stop smoking.

Epidermis

The epidermis lacks blood vessels because it is composed entirely of stratified squamous epithelium. However, the deepest layer of epidermal cells, called the *stratum basale* (stra′tum ba′sal), or stratum germinativum, is close to the dermis and is nourished by dermal blood vessels (fig. 6.1*a*). As the cells of this layer divide and grow, the older epidermal cells are pushed away from the dermis toward the skin surface. The farther the cells move, the poorer their nutrient supply becomes, and in time they die.

The older cells (keratinocytes) harden in a process called **keratinization** (ker″ah-tin″ĭ-za′shun). The cytoplasm fills with strands of a tough, fibrous, waterproof *keratin* protein. As a result, many layers of tough, tightly packed dead cells accumulate in the outermost epidermis, forming a layer called the *stratum corneum* (kor′ne-um). These dead cells are eventually shed.

The thickness of the epidermis varies from region to region. In most areas, only four layers can be distinguished: the *stratum basale, stratum spinosum* (spi-no′sum), *stratum granulosum* (gran″u-lo′sum), and *stratum corneum.* An additional layer, the *stratum lucidum* (loo′sid-um), is in the thickened and hairless (glabrous) skin of the palms and soles (fig. 6.2). The stratum lucidum may be missing where the epidermis is thin over the rest of the body.

In healthy skin, production of epidermal cells is generally closely balanced with loss of dead cells from the stratum corneum, so that the skin does not wear away completely. In fact, the rate of cell division increases where the skin is rubbed or pressed regularly, causing growth of thickened areas called *calluses* on the palms and soles, and keratinized conical masses on the toes called *corns.*

The epidermis has important protective functions. It shields the moist underlying tissues against excess water loss, mechanical injury, and the effects of harmful chemicals. When intact, the epidermis also keeps out disease-causing microorganisms (pathogens).

Specialized cells in the epidermis called *melanocytes* produce **melanin** (mel′ah-nin), a dark pigment that provides skin color (fig. 6.3*a*). Melanin also absorbs ultraviolet radiation in sunlight, preventing it from causing mutations in the DNA of skin cells and other damaging effects. Melanocytes lie in the deepest portion of the epidermis. They are the only cells that can produce melanin, but the pigment may also appear in nearby epidermal cells. Melanocytes have long, pigment-containing cellular extensions that pass upward between neighboring epidermal cells. These extensions transfer melanin granules into neighboring cells by a process called *cytocrine secretion.*

(a)　　　　　　　　　　　　　　　　　　　　(b)

Stratum corneum
Stratum lucidum
Stratum granulosum
Stratum spinosum
Stratum basale
Basement membrane
Dermal papilla
Dermis

Figure 6.2 AP|R

Epidermis of thick skin. **(a)** The layers of the epidermis are distinguished by changes in cells as they are pushed toward the surface of the skin. **(b)** Light micrograph of skin (500×).

Q: *Where is thick skin found on the body?*

Answer can be found in Appendix E on page 568.

(a)

(b)

Figure 6.3

Melanocytes produce melanin. **(a)** This transmission electron micrograph shows a melanocyte with pigment-containing granules (10,600×). **(b)** A melanocyte may have pigment-containing extensions that pass between epidermal cells and transfer pigment into them. Note that much of the melanin is deposited above the nucleus, where the pigment can absorb UV radiation from outside before the DNA is damaged.

The neighboring epidermal cells may actually contain more melanin than the melanocytes (fig. 6.3b). When melanin reaches the keratinized cells, it breaks into pieces, which provide the pigmentation. Clinical Application 6.1 discusses skin cancer arising from melanocytes and other epidermal cells.

Skin Color

Skin color is due largely to melanin. All people have about the same number of melanocytes in their skin. Differences in skin color result from differences in the amount of melanin that melanocytes produce and in the distribution and size of the pigment granules. Skin color

Clinical Application 6.1

Skin Cancer

Skin cancer usually arises in nonpigmented epithelial cells in the deep layer of the epidermis, or from melanocytes. Skin cancers originating from epithelial cells are called *cutaneous carcinomas* (squamous cell carcinoma or basal cell carcinoma); those arising from melanocytes are *cutaneous melanomas* (melanocarcinomas or malignant melanomas) (fig. 6A).

Cutaneous carcinomas are the most common type of skin cancer, occurring most frequently in light-skinned people over forty years of age. Those who are regularly exposed to sunlight, such as farmers, sailors, athletes, and sun worshippers, are at increased risk. Cutaneous carcinomas may result from failure of apoptosis, which normally peels away sun-damaged cells.

Cutaneous carcinomas often develop from hard, dry, scaly growths (lesions) that have reddish bases. They may be flat or raised and usually firmly adhere to the skin. They are most common on the neck, face, or scalp. Cutaneous carcinomas grow slowly and are usually cured with surgical removal or radiation treatment.

Cutaneous melanomas are pigmented with melanin, often with a variety of colored areas, such as variegated brown, black, gray, or blue. Melanomas usually have irregular rather than smooth outlines, and may feel bumpy. The "ABCDE" rule provides a checklist for melanoma: A for asymmetry; B for border (irregular); C for color (more than one); D for diameter (more than 6 millimeters); and E for elevation.

People of any age may develop cutaneous melanomas. These cancers are caused by short, intermittent exposure to high-intensity sunlight, such as a severe sunburn in a person who usually stays indoors. Melanoma is not associated with sustained sun exposure, as are the other types of skin cancers.

A cutaneous melanoma, usually appearing on the back or limbs, may arise from normal-appearing skin or from a mole (nevus). The lesion spreads horizontally through the skin, but may thicken and grow downward, invading deeper tissues. A melanoma surgically removed while in its horizontal growth phase may be arrested, but once it thickens and spreads into deeper tissues, it becomes difficult to treat.

To reduce the risk of developing skin cancer, avoid exposing the skin to high-intensity sunlight, use sunscreens and sunblocks, and examine the skin regularly. Report any "ABCDE" lesions to a physician at once.

(a)

(b)

(c)

Figure 6A

Skin cancer. **(a)** Squamous cell carcinoma. **(b)** Basal cell carcinoma. **(c)** Malignant melanoma.

is mostly genetically determined. If genes instruct melanocytes to produce abundant melanin, the skin is dark.

More than a hundred genes affect pigmentation of the skin, hair, and irises.

Environmental and physiological factors also influence skin color. Sunlight, ultraviolet light from sunlamps, and X rays stimulate production of additional pigment. Blood in the dermal vessels may affect skin color as physiological changes occur. When blood is well oxygenated, the blood pigment (hemoglobin) is bright red, making the skin of light-complexioned people appear pinkish.

On the other hand, when blood oxygen concentration is low, hemoglobin is dark red, and the skin appears bluish—a condition called *cyanosis*. Other physiological factors affect skin color. For example, a diet high in yellow vegetables may turn skin orange-yellow, because these foods are rich in a pigment called beta-carotene. Biochemical imbalances may also affect skin color. In newborns who have jaundice, for example, buildup of a substance called bilirubin turns the skin yellowish.

Practice

6. Explain how the epidermis is formed.
7. Distinguish between the stratum basale and the stratum corneum.
8. What is the function of melanin?
9. What factors influence skin color?

Dermis

The boundary between the epidermis and dermis is uneven because epidermal ridges project inward and conical projections of dermis, called dermal papillae, extend into the spaces between the ridges (see fig. 6.1*a*). Dermal papillae can be found in skin all over the body, but they are most abundant in the hands and feet. The friction ridges formed by the dermal papillae leave a patterned impression when a finger is pressed against a surface—a fingerprint. Genes determine fingerprint patterns, but the environment can alter them in a few situations. Fingerprint patterns can change slightly as a fetus presses the forming ridges against the uterine wall, which is why the fingerprints of identical twins are not exactly alike. Certain drugs used to treat cancer can erase fingerprints.

The dermis binds the epidermis to underlying tissues (see fig. 6.1*a*). It is largely composed of dense connective tissue that includes tough collagenous fibers and elastic fibers within a gel-like ground substance. Networks of these fibers give the skin toughness and elasticity.

Dermal blood vessels supply nutrients to all skin cells. These vessels also help regulate body temperature, as explained later in this chapter on page 125.

Epidermal cells can die if their blood supply from the dermis, which brings nutrients, is blocked. For example, when a person lies in one position for a prolonged period, the weight of the body pressing against the bed blocks the skin's blood supply. If cells die, the tissues begin to break down (necrosis), and a *pressure ulcer* (also called a decubitus ulcer or bedsore) may appear.

Pressure ulcers usually form in the skin overlying bony projections, such as on the hip, heel, elbow, or shoulder. Frequently changing body position or massaging the skin to stimulate blood flow in regions associated with bony prominences can prevent pressure ulcers.

Nerve cell processes are scattered throughout the dermis. Motor cell processes carry impulses out from the brain or spinal cord to dermal muscles and glands. Sensory cell processes carry impulses away from specialized sensory receptors, such as touch receptors in the dermis, and into the brain or spinal cord. Specialized sensory receptors are discussed in chapter 10 (p. 264). The dermis also contains accessory structures including hair follicles, sebaceous (oil-producing) glands, and sweat glands (see fig. 6.1*a*).

To create a tattoo, very fine needles inject inks into the dermis. The color is permanent, because dermis cells are not shed, unlike cells of the epidermis. To remove a tattoo, a laser shatters the ink molecules, and the immune system removes the resulting debris. Before laser removal became available in the late 1980s, unwanted tattoos were scraped, frozen, or cut away—all painful procedures.

Practice

10. What types of tissues make up the dermis?
11. What are the functions of these tissues?

6.3 ACCESSORY STRUCTURES OF THE SKIN

Nails

Nails are protective coverings on the ends of the fingers and toes. Each nail consists of a *nail plate* that overlies a surface of skin called the *nail bed*. Specialized epithelial cells that are continuous with the epithelium of the skin produce the nail bed. The whitish, thickened, half-moon–shaped region (lunula) at the base of a nail plate covers the most actively growing region. The epithelial cells here divide, and the newly formed cells become keratinized. This gives rise to tiny, keratinized scales that become part of the nail plate, pushing it forward over the nail bed. The keratin of nails is harder than that produced by the epidermal stratum corneum. In time, the nail plate extends beyond the end of the nail bed and with normal use gradually wears away (fig. 6.4).

The thumbnail grows the slowest; the middle nail grows the fastest.

Hair Follicles

Hair is present on all skin surfaces except the palms, soles, lips, nipples, and parts of the external reproductive organs. Each hair develops from a group of epidermal cells at the base of a tubelike depression called a **hair follicle** (hār fol′i-kl) (figs. 6.1 and 6.5). This follicle

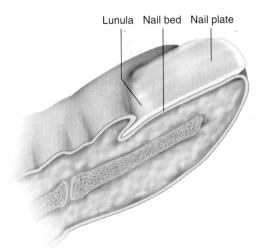

Lunula   Nail bed   Nail plate

Figure 6.4   AP|R

Nails grow from epithelial cells that divide and become keratinized, forming the rest of the nail.

Q: *What is the most actively growing region of the nail?*

Answer can be found in Appendix E on page 568.

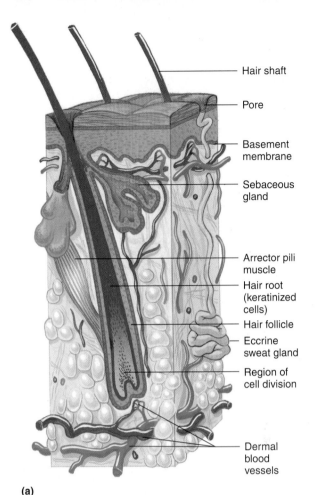

- Hair shaft
- Pore
- Basement membrane
- Sebaceous gland
- Arrector pili muscle
- Hair root (keratinized cells)
- Hair follicle
- Eccrine sweat gland
- Region of cell division
- Dermal blood vessels

(a)

- Hair follicle
- Hair root
- Adipose tissue
- Region of cell division

(b)

Figure 6.5

Hair follicle. **(a)** A hair grows from the base of a hair follicle when epidermal cells divide and older cells move outward and become keratinized. **(b)** Light micrograph of a hair follicle (175×).

extends from the surface into the dermis and contains the *hair root.* The epidermal cells at its base are nourished from dermal blood vessels in a projection of connective tissue at the deep end of the follicle. As these epidermal cells divide and grow, older cells are pushed toward the surface. The cells that move upward and away from their nutrient supply become keratinized and die. Their remains constitute the structure of a developing *hair shaft* that extends away from the skin surface (fig. 6.6). In other words, a hair is composed of dead epidermal cells.

Genes determine hair color by directing the type and amount of pigment that epidermal melanocytes produce. Dark hair has more of the brownish-black **eumelanin** (u-mel'ah-nin), while blonde hair and red hair have more of the reddish-yellow **pheomelanin** (fe"o-mel'ah-nin). The white hair of a person with the inherited condition *albinism* lacks melanin altogether. A mixture of pigmented and unpigmented hair usually appears gray.

A bundle of smooth muscle cells, forming the *arrector pili muscle,* attaches to each hair follicle (see figs. 6.1*a* and 6.5*a*). This muscle is positioned so that a short hair within the follicle stands on end when the muscle contracts. If a person is emotionally upset or very cold, nerve impulses may stimulate the arrector pili muscles to contract, causing gooseflesh, or goose bumps.

Just above the "bulge" region at the base of a hair follicle are stem cells that can give rise to epidermal cells of hair or skin. The first clue to the existence of these cells was that new skin in burn patients arises from hair follicles. Manipulating these stem cells could someday treat baldness (alopecia) or extreme hairiness (hirsutism).

Figure 6.6

This scanning electron micrograph shows a hair emerging from the epidermis (875×).

Sebaceous Glands

Sebaceous glands (se-ba′shus glandz) contain groups of specialized epithelial cells and are usually associated with hair follicles (figs. 6.5*a* and 6.7). They are holocrine glands (see chapter 5, p. 101) that secrete an oily mixture of fatty material and cellular debris called *sebum* through small ducts into the hair follicles. Sebum helps keep the hair and skin soft, pliable, and waterproof.

> Many teens are all too familiar with a disorder of the sebaceous glands called *acne* (acne vulgaris). Overactive and inflamed glands in some body regions become plugged, producing blackheads (comedones), or surrounded by small, red elevations producing pimples (pustules).

Sweat Glands

Sweat (swet) **glands,** or sudoriferous glands, are exocrine glands that are widespread in the skin. Each gland consists of a tiny tube that originates as a ball-shaped coil in the deeper dermis or superficial subcutaneous layer. The coiled portion of the gland is closed at its deep end and is lined with sweat-secreting epithelial cells.

The most numerous sweat glands, the **eccrine** (ek′rin) **glands,** respond throughout life to body temperature elevated by environmental heat or physical exercise (see fig. 6.5*a*). These glands are common on the forehead, neck, and back, where they produce profuse sweat on hot days or during intense physical activity. They also release moisture that appears on the palms and soles when a person is emotionally stressed.

The fluid (sweat) that eccrine glands secrete is carried away by a tube (duct) that opens at the surface as

Figure 6.7

A sebaceous gland secretes sebum into a hair follicle, shown here in oblique section (200×).

a *pore.* Sweat is mostly water, but it also contains small amounts of salt and wastes, such as urea and uric acid. Thus, sweating is also an excretory function.

Other sweat glands, known as **apocrine glands,** become active at puberty. Although they are currently called apocrine, these glands secrete by the same mechanism as eccrine glands, usually when a person is emotionally upset, frightened, in pain, or during sexual arousal. In adults, the apocrine glands are most numerous in the axillary regions and groin. Ducts of these glands open into hair follicles. The secretions of these glands develop a scent as they are metabolized by skin bacteria.

Other sweat glands are structurally and functionally modified to secrete specific fluids, such as the *ceruminous glands* of the external ear canal that secrete earwax. The female *mammary glands* that secrete milk are another example of modified sweat glands (see chapter 20, p. 552).

> The average square inch (6.45 square centimeters) of skin holds 650 sweat glands, 20 blood vessels, 60,000 melanocytes, and more than a thousand nerve endings.

Practice

12. Describe the structure of the nail bed.
13. Explain how a hair forms.
14. What is the function of the sebaceous glands?
15. Distinguish between the eccrine and apocrine sweat glands.

6.4 REGULATION OF BODY TEMPERATURE

Regulation of body temperature is vitally important because even slight shifts can disrupt the rates of metabolic reactions. Normally the temperature of deeper body parts remains close to a set point of 37°C (98.6°F). Maintenance of a stable temperature requires that the amount of heat the body loses be balanced by the amount it produces. The skin plays a key role in the homeostatic mechanism that regulates body temperature.

Heat is a product of cellular metabolism; thus, the more active cells of the body are the major heat producers. Examples include skeletal and cardiac muscle cells, and cells of the liver.

When body temperature rises above the set point, the nervous system stimulates structures in the skin and other organs to release heat. For example, during physical exercise, active muscles release heat, which the blood carries away. The warmed blood reaches the part of the brain (the hypothalamus) that controls the body's temperature set point, which signals muscles in the walls of dermal blood vessels to relax. As these vessels dilate (vasodilation), more blood enters them, and some of the heat in the blood escapes to the outside.

At the same time as the skin loses heat, the nervous system stimulates the eccrine sweat glands to become active and to release sweat onto the skin surface. As this fluid evaporates (changes from a liquid to a gas), it carries heat away from the surface, cooling the skin further.

When body temperature drops below the set point, as may occur in a very cold environment, the brain triggers different responses in the skin structures. Muscles in the walls of dermal blood vessels are stimulated to contract; this decreases the flow of heat-carrying blood through the skin and helps reduce heat loss. Also, the sweat glands remain inactive, decreasing heat loss by evaporation. If body temperature continues to drop, the nervous system may stimulate muscle cells in the skeletal muscles throughout the body to contract slightly. This action requires an increase in the rate of cellular respiration and releases heat as a by-product. If this response does not raise body temperature to normal, small groups of muscles may rhythmically contract with greater force, causing the person to shiver, generating more heat. Chapter 1 introduced this type of homeostatic mechanism (fig. 1.7, p. 8).

Deviation from the normal range for body temperature impairs health and may be lethal. People with severe spinal cord injuries can no longer control body temperature, which fluctuates depending upon the environment.

In hypothermia, core body temperature falls below 95°F. The body becomes so cold that it cannot maintain function. Symptoms of worsening hypothermia include a gradual loss of coordination, stiffening muscles, confusion, fatigue, and slow, shallow breathing. When core temperature falls to 87.8°F, the skin turns a bluish-gray, weakness intensifies, and consciousness fades.

In hyperthermia, core body temperature exceeds 106°F. The skin becomes hot, dry, and flushed, and the person becomes weak, dizzy, and nauseous, with headache and a rapid, irregular pulse. The vignette that opens chapter 18 (p. 489) describes heatstroke.

Practice

16. Why is regulation of body temperature so important?
17. How does the body lose excess heat?
18. Which actions help the body conserve heat?

6.5 HEALING OF WOUNDS

A wound and the area surrounding it usually become red and painfully swollen. This is the result of *inflammation,* which is a normal response to injury or stress. Blood vessels in affected tissues dilate and become more permeable, allowing fluids to leak into the damaged tissues. Inflamed skin may become reddened, warm, swollen, and painful to touch (table 6.1). However, the dilated blood vessels provide the tissues with more nutrients and oxygen, which aids healing.

The specific events in healing depend on the nature and extent of the injury. If a break in the skin is shallow, epithelial cells along its margin are stimulated to divide more rapidly than usual, and the newly formed cells fill the gap.

If the injury extends into the dermis or subcutaneous layer, blood vessels break, and the released blood forms a

Table 6.1	Inflammation
Symptom	Cause
Redness	Vasodilation, more blood in area
Heat	Large amount of blood accumulating in area and as a by-product of increased metabolic activity in tissue
Swelling	Increased permeability of blood vessels, fluids leaving blood go into tissue spaces (edema)
Pain	Injury to neurons and increased pressure from edema

Clinical Application 6.2

Burns

A few hours outside on a sunny summer day, without use of sunscreen, may result in a minor sunburn. The slightly burned skin warms and reddens (erythema) as dermal blood vessels dilate. Mild edema may swell the exposed, tender skin, and a few days later the surface layer of skin may peel. A burn injuring only the epidermis is a *superficial partial-thickness* (first degree) *burn*. Healing usually takes a few days to two weeks, with no scarring.

More serious is a burn that destroys some epidermis as well as some underlying dermis. This is a *deep partial-thickness* (second degree) *burn*. Fluid escapes from damaged dermal capillaries, accumulating beneath the outer layer of epidermal cells, forming blisters. The injured region becomes moist and firm and may vary from dark red to waxy white. Such a burn usually happens as a result of exposure to hot objects, hot liquids, flames, or burning clothing.

The extent of healing of a deep partial-thickness burn depends upon which accessory structures of the skin survive the injury, which is possible if they are deep in the dermis. These structures include hair follicles, sweat glands, and sebaceous glands, as well as epithelial cells that divide and extend onto the surface of the injured dermis, spreading over it and forming new epidermis.

Most severe is a burn that destroys the epidermis, the dermis, and the accessory structures of the skin. This is a *full-thickness* (third degree) *burn*. The injured skin becomes dry and leathery, and may vary in color from red to black to white. A full-thickness burn usually occurs as a result of prolonged exposure to hot objects, flames, or corrosive chemicals. Most of the epithelial cells in the affected region are destroyed, and the skin heals only if epithelial cells divide and grow inward from the margin of the burn. If the injured area is extensive, it may require a transplant, using a thin layer of skin from an unburned region of the body (an autograft), cadaveric skin (a homograft), or a skin substitute such as tissue-engineered skin.

The treatment of a burn patient requires estimating the extent of the body's affected surface. Physicians use the "rule of nines," subdividing the skin's surface into regions, each accounting for 9% (or some multiple of 9%) of the total surface area (fig. 6B). This estimate is important in planning to replace body fluids and electrolytes lost from injured tissues and for covering the burned area with skin or skin substitutes.

---

clot in the wound. The blood clot and dried tissue fluids form a *scab* that covers and protects underlying tissues. Before long, fibroblasts migrate into the injured region and begin secreting new collagenous fibers that bind the edges of the wound. Suturing or otherwise closing a large break in the skin speeds this process.

As healing continues, blood vessels extend into the area beneath the scab. Phagocytic cells remove dead cells and other debris. Eventually, the damaged tissues are replaced, and the scab sloughs off. If the wound is deep, extensive production of collagenous fibers may form an elevation above the normal epidermal surface, called a *scar*.

In large, open wounds, healing may be accompanied by formation of small, rounded masses called *granulations* that develop in the exposed tissues. A granulation consists of a new branch of a blood vessel and a cluster of collagen-secreting fibroblasts that the vessel nourishes. In time, some of the blood vessels are resorbed, and the fibroblasts move away, leaving a scar largely composed of collagen fibers. Clinical Application 6.2 describes healing of burned tissue.

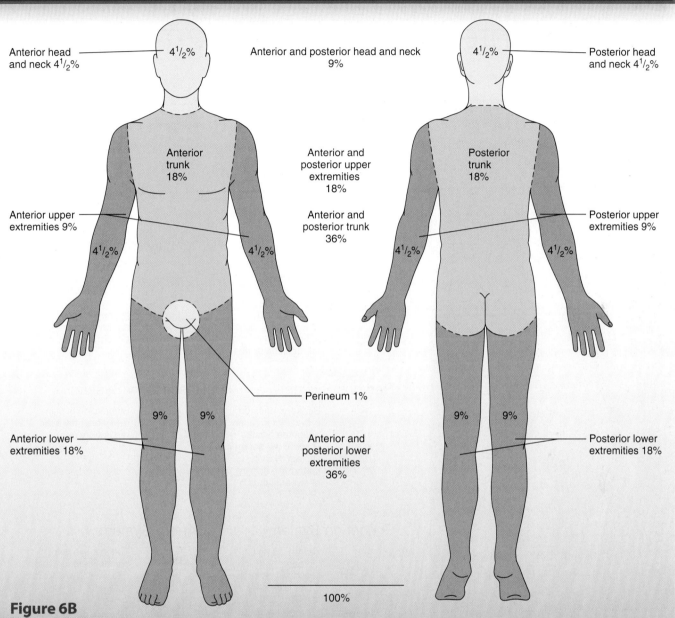

Figure 6B

As an aid for estimating the extent of damage burns cause, the body is subdivided into regions, each representing 9% (or some multiple of 9%) of the total skin surface area.

Practice

19. What is the tissue response to inflammation?

20. Distinguish between the activities necessary to heal a wound in the epidermis and those necessary to heal a wound in the dermis.

21. Explain the role of phagocytic cells in wound healing.

22. Define *granulation*.

Common Skin Disorders

acne (ak'ne) Disease of the sebaceous glands that produces blackheads and pimples.

alopecia (al″o-pe'she-ah) Hair loss, usually sudden.

athlete's foot (ath'-lētz foot) Fungus (*Tinea pedis*) infection usually in the skin of the toes and soles.

birthmark (berth' mark) Congenital blemish or spot on the skin, visible at birth or soon after.

boil (boil) Bacterial infection (furuncle) of the skin, produced when bacteria enter a hair follicle.

Integumentary System

Skeletal System

Vitamin D activated by the skin helps provide calcium needed for bone matrix.

Muscular System

Involuntary muscle contractions (shivering) work with the skin to control body temperature. Muscles act on facial skin to create expressions.

Nervous System

Sensory receptors provide information about the outside world to the nervous system. Nerves control the activity of sweat glands.

Endocrine System

Hormones help to increase skin blood flow during exercise. Other hormones stimulate either the synthesis or the decomposition of subcutaneous fat.

Cardiovascular System

Skin blood vessels play a role in regulating body temperature.

Lymphatic System

The skin, acting as a barrier, provides an important first line of defense for the immune system.

Digestive System

Excess calories may be stored as subcutaneous fat. Vitamin D activated by the skin stimulates dietary calcium absorption.

Respiratory System

Stimulation of skin receptors may alter respiratory rate.

Urinary System

The kidneys help compensate for water and electrolytes lost in sweat.

Reproductive System

Sensory receptors play an important role in sexual activity and in the suckling reflex.

The skin provides protection, contains sensory receptors, and helps control body temperature.

carbuncle (kar'bung-kl) Bacterial infection, similar to a boil, that spreads into the subcutaneous tissues.

cyst (sist) Liquid-filled sac or capsule.

dermatitis (der″mah-ti'tis) Inflammation of the skin.

eczema (ek'zĕ-mah) Noncontagious skin rash that produces itching, blistering, and scaling.

erythema (er″ĭ-the'mah) Reddening of the skin due to dilation of dermal blood vessels in response to injury or inflammation.

herpes (her'pēz) Infectious disease of the skin, caused by the herpes simplex virus and characterized by recurring formations of small clusters of vesicles.

impetigo (im″pĕ-ti'go) Contagious disease of bacterial origin, characterized by pustules that rupture and become covered with loosely held crusts.

keloid (ke'loid) Elevated, enlarging fibrous scar usually initiated by an injury.

mole (mōl) Benign skin tumor (nevus) that is usually pigmented; colors range from brown to black.

pediculosis (pĕ-dik″u-lo'sis) Disease produced by an infestation of lice.

pruritus (proo-ri'tus) Itching of the skin.

psoriasis (so-ri'ah-sis) Chronic skin disease characterized by red patches covered with silvery scales.

pustule (pus'tūl) Elevated, pus-filled area on the skin.

scabies (ska'bēz) Disease resulting from an infestation of mites.

seborrhea (seb″o-re'ah) Hyperactivity of the sebaceous glands, causing greasy skin and dandruff.

ulcer (ul'ser) Open sore.

urticaria (ur″tĭ-ka're-ah) Allergic reaction of the skin that produces reddish, elevated patches (hives).

vitiligo (vit'ĭ-li'go) Loss of melanocytes in parts of the epidermis, producing whitened areas of skin.

wart (wort) Flesh-colored, raised area caused by a viral infection.

Summary Outline

6.1 Introduction (p. 117)

An organ is formed by two or more tissue types grouped together and performing specialized functions. The skin, the largest organ in the body, is part of the integumentary system.

6.2 Skin and Its Tissues (p. 117)

Skin is a protective covering, helps regulate body temperature, retards water loss, houses sensory receptors, synthesizes various biochemicals, and excretes wastes. It is composed of an epidermis and a dermis separated by a basement membrane. Beneath the skin is the subcutaneous layer that binds the skin to underlying organs, stores fat, and contains blood vessels that supply the skin.

1. Epidermis
   a. The epidermis is stratified squamous epithelium that lacks blood vessels.
   b. The deepest layer of the epidermis, called the stratum basale, contains cells that divide.
   c. Epidermal cells undergo keratinization as they mature and are pushed toward the surface.
   d. The outermost layer, called the stratum corneum, is composed of dead epidermal cells.
   e. The epidermis protects underlying tissues against water loss, mechanical injury, and the effects of harmful chemicals.
   f. Melanin protects underlying cells from the effects of ultraviolet light.
   g. Melanocytes transfer melanin to nearby epidermal cells.
   h. Melanin provides skin color.
      (1) All people have about the same number of melanocytes.
      (2) Skin color is due largely to the amount of melanin and the distribution and size of pigment granules in the epidermis.
      (3) Environmental and physiological factors, as well as genes, influence skin color.

2. Dermis
   a. The dermis binds the epidermis to underlying tissues.
   b. Dermal blood vessels supply nutrients to all skin cells and help regulate body temperature.
   c. Nerve cell processes are scattered throughout the dermis.
      (1) Some dermal nerve cell processes carry impulses to muscles and glands of the skin.
      (2) Other dermal nerve cell processes are associated with sensory receptors in the skin, and carry impulses to the brain and spinal cord.
   d. The dermis also has hair follicles, sebaceous glands, and sweat glands.

6.3 Accessory Structures of the Skin (p. 122)

1. Nails
   a. Nails are protective covers on the ends of fingers and toes.
   b. Specialized epidermal cells that are keratinized make up nails.
   c. The keratin of nails is harder than that produced by the skin's epidermal cells.
2. Hair follicles
   a. Each hair develops from epidermal cells at the base of a tubelike hair follicle.
   b. As newly formed cells develop and grow, older cells are pushed toward the surface and undergo keratinization.
   c. Hair color is determined by genes that direct the amount of eumelanin or pheomelanin produced by melanocytes associated with hair follicles.
   d. A bundle of smooth muscle cells is attached to each hair follicle.
3. Sebaceous glands
   a. Sebaceous glands are usually associated with hair follicles.
   b. Sebaceous glands secrete sebum, which helps keep the skin and hair soft and waterproof.

4. Sweat glands
   a. Each sweat gland is a coiled tube.
   b. Sweat is primarily water but also contains salts and wastes.
   c. Eccrine sweat glands respond to elevated body temperature; apocrine glands respond to emotional upset.

6.4 Regulation of Body Temperature (p. 125)

Regulation of body temperature is vital because heat affects the rates of metabolic reactions. Normal temperature of deeper body parts is close to a set point of 37°C (98.6°F).
   1. When body temperature rises above the normal set point, dermal blood vessels dilate and sweat glands secrete sweat.

2. When body temperature drops below the normal set point, dermal blood vessels constrict and sweat glands become inactive.
3. If body temperature continues to drop, skeletal muscles involuntarily contract.

6.5 Healing of Wounds (p. 125)

Skin injuries trigger inflammation. The affected area becomes red, warm, swollen, and tender.
   1. Dividing epithelial cells fill in shallow cuts in the epidermis.
   2. Clots close deeper cuts, sometimes leaving a scar where connective tissue replaces skin.
   3. Granulations form in large, open wounds as part of the healing process.

Chapter Assessments

6.1 Introduction

1. Two or more types of tissues grouped together and performing specialized functions defines a(n) _____. (p. 117)
   a. organelle   c. organ   e. organism
   b. cell   d. organ system

2. The largest organ(s) in the body is (are) the _____. (p. 117)
   a. liver   c. lungs   e. brain
   b. intestines   d. skin

6.2 Skin and Its Tissues

3. Functions of the skin include _____. (p. 117)
   a. retarding water loss   d. excretion
   b. body temperature regulation   e. all of the above
   c. sensory reception

4. Describe how skin plays a role in the production of vitamin D. (p. 117)

5. The epidermis is composed of layers of _____ tissue. (p. 117)

6. The _____ layer of epidermal cells contains older keratinized cells and dead cells. (p. 119)
   a. stratum corneum   d. stratum spinosum
   b. stratum lucidum   e. stratum basale
   c. stratum granulosum

7. Discuss the function of melanin, other than providing color to the skin. (p. 119)

8. List and describe the influence of each factor affecting skin color. (p. 120)

9. The dermis is composed primarily of what kind of tissue? (p. 122)

6.3 Accessory Structures of the Skin

10. Describe how nails are formed, and relate the structure of nails to their function. (p. 122)

11. Distinguish between a hair and a hair follicle. (p. 122)

12. Sebaceous glands are _____ glands that secrete _____. (p. 124)

13. Compare and contrast eccrine and apocrine sweat glands. (p. 124)

6.4 Regulation of Body Temperature

14. Explain how body heat is produced. (p. 125)

15. Explain how sweat glands help regulate body temperature. (p. 125)

16. Describe the body's responses to decreasing body temperature. (p. 125)

6.5 Healing of Wounds

17. Explain how the healing of superficial breaks in the skin differs from the healing of deeper wounds. (p. 125)

Integrative Assessments/Critical Thinking

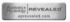

OUTCOMES 5.3, 6.2, 6.4

1. A premature infant typically lacks subcutaneous adipose tissue, and the small body has a relatively large surface area compared to its volume. How do these factors affect the ability of a premature infant to regulate its body temperature?

OUTCOME 6.2

2. Everyone's skin contains about the same number of melanocytes, even though people have many different skin colors. How is this possible?

3. Which of the following would result in the more rapid absorption of a drug: a subcutaneous injection or an intradermal injection? Why?

OUTCOMES 6.2, 6.5

4. How is it protective for skin to peel after a severe sunburn?

5. As a rule, a superficial partial-thickness burn is more painful than one involving deeper tissues. How would you explain this observation?

OUTCOMES 6.2, 6.3, 6.4, 6.5

6. What special problems would result from the loss of 50% of a person's functional skin surface? How might this person's environment be modified to partially compensate for such a loss?

WEB CONNECTIONS

Visit the text website at **www.mhhe.com/shieress11** for additional quizzes, interactive learning exercises, and more.

APR

Anatomy & Physiology REVEALED includes cadaver photos that allow you to peel away layers of the human body to reveal structures beneath the surface. This program also includes animations, radiologic imaging, audio pronunciation, and practice quizzing. To learn more, visit **www.aprevealed.com**.

7

Skeletal System

Preventing "fragility fractures." Skeletal health is a matter of balance. Before age thirty, cells that form new bone tissue counter cells that break it down, so that living bone is in a constant state of remodeling. Then the balance shifts so that bone is lost, especially in women past menopause, due to hormonal changes. This imbalance may progress to osteopenia or the more severe osteoporosis.

A "fragility fracture" is a telltale sign of dangerously low bone density. This is a fracture that happens after a fall from less than standing height, which a strong, healthy skeleton could resist. Fragility fractures occur in 1.5 million people in the United States each year, yet despite this warning sign, only one-fourth to one-third of them are followed up with bone scans and treatment to build new bone tissue. Since 1995, five new drugs have become available to treat osteoporosis. One class, the bisphosphonates, actually builds new bone.

Osteopenia and osteoporosis are common. The surgeon general estimates that half of all people over age fifty have one of these conditions, which amounts to 10 million with osteoporosis and another 35 million with osteopenia. Screening is advised for all individuals over age sixty-five, as well as for those with risk factors. The most telling predictor is a previous fragility fracture. Other risk factors include a family history of osteoporosis, recent height loss, and older age.

Spaces in bones enlarge when a person has osteoporosis. The portion of a vertebra on the left is normal; the one on the right has been weakened by osteoporosis.

These two conditions are not just concerns of people approaching retirement age. Researchers think that what puts people at risk is failure to attain maximal possible bone density by age thirty. To keep bones as strong as possible for as long as possible, it is essential to get at least 30 minutes of exercise daily (some of which should be weight-bearing), consume enough daily calcium (1,000–1,200 mg) and vitamin D (400–1,000 IU), and not smoke. There is much you can do to promote skeletal health—at any age.

Learning Outcomes

After studying this chapter, you should be able to do the following:

7.1 Introduction
1. List the active tissues in a bone. (p. 133)

7.2 Bone Structure
2. Describe the macroscopic and microscopic structure of a long bone, and list the functions of these parts. (p. 133)

7.3 Bone Development and Growth
3. Distinguish between intramembranous and endochondral bones, and explain how such bones develop and grow. (p. 135)

7.4 Bone Function
4. Discuss the major functions of bones. (p. 137)

7.5 Skeletal Organization
5. Distinguish between the axial and appendicular skeletons, and name the major parts of each. (p. 142)

7.6–7.12 Skull—Lower Limb
6. Locate and identify the bones and the major features of the bones that compose the skull, vertebral column, thoracic cage, pectoral girdle, upper limb, pelvic girdle, and lower limb. (pp. 144–163)

7.13 Joints
7. Classify joints according to the type of tissue binding the bones together, describe the different joint characteristics, and name an example of each joint type. (p. 164)
8. List six types of synovial joints, and describe the actions of each. (p. 165)
9. Explain how skeletal muscles produce movements at joints, and identify several types of joint movements. (p. 167)

Anatomy & Physiology | REVEALED®
aprevealed.com

Module 5: Skeletal System

Aids to Understanding Words   *(Appendix A on page 564 has a complete list of Aids to Understanding Words.)*

acetabul- [vinegar cup] *acetabul*um: Depression of the hip bone that articulates with the head of the femur.

ax- [axis] *ax*ial skeleton: Upright portion of the skeleton that supports the head, neck, and trunk.

-blast [bud] osteo*blast*: Cell that will form bone tissue.

carp- [wrist] *carp*als: Wrist bones.

-clast [break] osteo*clast*: Cell that breaks down bone tissue.

condyl- [knob] *condyl*e: Rounded, bony process.

corac- [a crow's beak] *corac*oid process: Beaklike process of the scapula.

cribr- [sieve] *cribr*iform plate: Portion of the ethmoid bone with many small openings.

crist- [crest] *crist*a galli: Bony ridge that projects upward into the cranial cavity.

fov- [pit] *fov*ea capitis: Pit in the head of a femur.

glen- [joint socket] *glen*oid cavity: Depression in the scapula that articulates with the head of a humerus.

inter- [among, between] *inter*vertebral disc: Structure between vertebrae.

intra- [inside] *intra*membranous bone: Bone that forms within sheetlike masses of connective tissue.

meat- [passage] auditory *meat*us: Canal of the temporal bone that leads inward to parts of the ear.

odont- [tooth] *odont*oid process: Toothlike process of the second cervical vertebra.

poie- [make, produce] hemato*poie*sis: Process that forms blood cells.

7.1 INTRODUCTION

Halloween skeletons and the skull-and-crossbones symbol for poison and pirates may make bones seem like lifeless objects. However, bone consists of a variety of very active, living tissues: bone tissue, cartilage, dense connective tissue, blood, and nervous tissue. Bones are not only very much alive but also multifunctional. Bones, the organs of the skeletal system, provide points of attachment for muscles, protect and support softer tissues, house blood-producing cells, store inorganic salts, and form passageways for blood vessels and nerves.

7.2 BONE STRUCTURE

The bones of the skeletal system differ greatly in size and shape. However, they are similar in structure, development, and function.

Bone Classification

Bones are classified according to their shapes—long, short, flat, or irregular.

- **Long bones** have long longitudinal axes and expanded ends. Examples of long bones are the forearm and thigh bones.
- **Short bones** are somewhat cubelike, with roughly equal lengths and widths. The bones of the wrists and ankles are this type.
- **Flat bones** are platelike structures with broad surfaces, such as the ribs, scapulae, and some bones of the skull.
- **Irregular bones** have a variety of shapes and are usually connected to several other bones. Irregular bones include the vertebrae that compose the backbone and many facial bones.

In addition to these four groups of bones, some authorities recognize a fifth group called **sesamoid bones** or **round bones.** These bones are usually small and nodular and are embedded in tendons adjacent to joints. The kneecap (patella) is a sesamoid bone.

Parts of a Long Bone

The femur, the long bone in the thigh, illustrates the structure of bone (fig. 7.1). At each end of such a bone is an expanded portion called an **epiphysis** (e-pif′ĭ-sis) (plural, *epiphyses*), which articulates (forms a joint) with another bone. The epiphysis that is nearest to the trunk of the body is called the proximal epiphysis. The one that is farthest from the trunk of the body is called the distal epiphysis. On its outer surface, the articulating portion of the epiphysis is coated with a layer of hyaline cartilage called **articular cartilage** (ar-tik′u-lar kar′tĭ-lij). The shaft of the bone, between the epiphyses, is called the **diaphysis** (di-af′ĭ-sis).

A tough, vascular covering of dense connective tissue called the **periosteum** (per″e-os′te-um) completely encloses the bone, except for the articular cartilage on the bone's ends. The periosteum is firmly attached to the bone, and periosteal fibers are continuous with the connecting ligaments and tendons. The periosteum also helps form and repair bone tissue.

A bone's shape makes possible the bone's functions. For example, bony projections called *processes* provide sites where ligaments and tendons attach; grooves and openings form passageways for blood vessels and nerves; and a depression of one bone may articulate with a process of another.

The wall of the diaphysis is mainly composed of tightly packed tissue called **compact bone** (kom′pakt bōn), or cortical bone. This type of bone has a continuous extracellular matrix with no spaces. The epiphyses, in contrast, are composed largely of **spongy bone** (spun′je bōn), or cancellous bone, with thin layers of

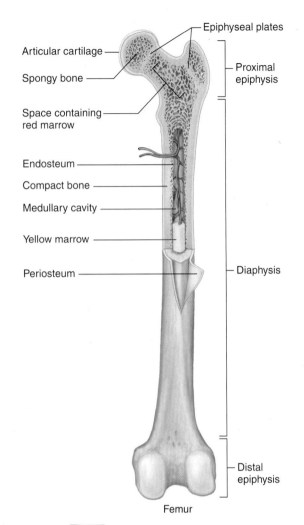

Figure 7.1 AP|R

Major parts of a long bone. This is a femur, the long bone in the thigh.

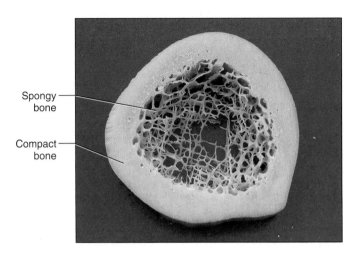

Figure 7.2 AP|R

This cross section of a long bone reveals a layer of spongy bone beneath a layer of compact bone.

compact bone on their surfaces. Spongy bone consists of numerous branching bony plates called **trabeculae** (trah-bek′u-le). Irregular connecting spaces between these plates help reduce the bone's weight (fig. 7.2). The bony plates are most highly developed in the regions of the epiphyses that are subjected to compressive forces. Both compact and spongy bone are strong and resist bending.

Compact bone in the diaphysis of a long bone forms a semirigid tube, which has a hollow chamber called the **medullary cavity** (med′u-lār″e kav′ĭ-te) that is continuous with the spaces of the spongy bone. A thin layer of cells called the **endosteum** (en-dos′te-um) lines these areas, and a specialized type of soft connective tissue called **marrow** (mar′o) fills them.

Microscopic Structure

Recall from chapter 5 (p. 108) that bone cells called *osteocytes* occupy very small, bony chambers called *lacunae,* which form concentric circles around *central*

canals (Haversian canals). Osteocytes communicate with nearby cells by means of cellular processes passing through *canaliculi* (fig. 7.3; see fig. 5.19, p. 109). The extracellular matrix of bone tissue is largely composed of collagen and inorganic salts (calcium phosphate). Collagen gives bone its strength and resilience, and inorganic salts make it hard and resistant to crushing.

In compact bone, the osteocytes and layers of extracellular matrix concentrically clustered around a central canal form a cylinder-shaped unit called an *osteon* (Haversian system). Many of these units cemented together form the substance of compact bone.

Each central canal contains blood vessels (usually capillaries) and nerve fibers surrounded by loose connective tissue. The blood in these vessels nourishes bone cells associated with the central canal.

Central canals extend longitudinally through bone tissue, and transverse *perforating canals* (Volkmann's canals) connect them. Perforating canals contain larger blood vessels and nerves by which the smaller blood vessels and nerve fibers in central canals communicate with the surface of the bone and the medullary cavity (fig. 7.3).

Spongy bone is also composed of osteocytes and extracellular matrix, but the bone cells do not aggregate around central canals. Instead, the cells lie within the *trabeculae* and get nutrients from substances diffusing into canaliculi that lead to the surface of these thin, bony plates.

Practice

1. Explain how bones are classified.
2. List five major parts of a long bone.
3. How do compact and spongy bone differ in structure?
4. Describe the microscopic structure of compact bone.

Figure 7.3 AP|R

Compact bone is composed of osteons cemented together by bone matrix. Drawing is not to scale. Extensions from osteocytes communicate through tunnel-like canaliculi.

7.3 BONE DEVELOPMENT AND GROWTH

Parts of the skeletal system begin to form during the first few weeks of prenatal development, and bony structures continue to develop and grow into adulthood. Bones form by replacing existing connective tissues in either of two ways: (1) Intramembranous bones originate between sheetlike layers of connective tissues; (2) Endochondral bones begin as masses of cartilage that are later replaced by bone tissue (fig. 7.4).

Intramembranous Bones

The broad, flat bones of the skull are **intramembranous bones** (in″trah-mem′brah-nus bōnz) . During their development, membranelike layers of unspecialized, or relatively undifferentiated, connective tissues appear at the sites of the future bones. Then, some of the partially differentiated progenitor cells enlarge and further differentiate into bone-forming cells called **osteoblasts** (os′te-o-blastz). The osteoblasts become active and

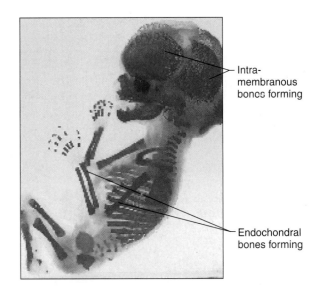

Figure 7.4

Intramembranous bones in the fetus form by replacing unspecialized connective tissue. Endochondral bones form from cartilage "models" that are gradually replaced with the harder tissue of bone. Note the stained, developing bones of this fourteen-week fetus.

deposit bony matrix around themselves, forming spongy bone tissue in all directions within the layers of connective tissues. When extracellular matrix completely surrounds osteoblasts, they are called **osteocytes.** Eventually, cells of the membranous tissues that persist outside the developing bone give rise to the periosteum. Osteoblasts on the inside of the periosteum form a layer of compact bone over the surface of the newly formed spongy bone. The formation of bone is called **ossification** (os″ĭ-fi-ka′shun).

Endochondral Bones

Most of the bones of the skeleton are **endochondral bones** (en″do-kon′dral bōnz). They develop in the fetus from masses of hyaline cartilage shaped like future bony structures. These cartilaginous models grow rapidly for a time and then begin to change extensively.

In a long bone, changes begin in the center of the diaphysis, where the cartilage slowly breaks down and disappears (fig. 7.5). At about the same time, a periosteum forms from connective tissue that encircles the developing diaphysis. Blood vessels and osteoblasts from the periosteum invade the disintegrating cartilage, and spongy bone forms in its place. This region of bone formation is called the *primary ossification center,* and bone tissue develops from it toward the ends of the cartilaginous structure. Meanwhile, osteoblasts from the periosteum deposit a thin layer of compact bone around the primary ossification center.

The epiphyses of the developing bone remain cartilaginous and continue to grow. Later, *secondary ossification centers* appear in the epiphyses, and spongy bone forms in all directions from them. As spongy bone is deposited in the diaphysis and in the epiphysis, a band of cartilage called the **epiphyseal plate** (ep″ĭ-fiz′e-al plāt), or metaphysis, remains between these two ossification centers.

The cartilaginous tissue of the epiphyseal plate includes layers of young cells that are undergoing mitosis and producing new cells. As these cells enlarge and extracellular matrix forms around them, the cartilaginous plate thickens, lengthening the bone. At the same time, calcium salts accumulate in the extracellular matrix adjacent to the oldest cartilaginous cells, and as the extracellular matrix calcifies, the cells begin to die.

In time, large, multinucleated cells called **osteoclasts** (os′te-o-klastz) break down the calcified extracellular matrix. These large cells originate in bone marrow when certain single-nucleated white blood cells (monocytes) fuse.

Osteoclasts secrete an acid that dissolves the inorganic component of the calcified matrix, and their lysosomal enzymes digest the organic components. After osteoclasts remove the extracellular matrix, bone-building osteoblasts invade the region and deposit new bone tissue in place of the calcified cartilage.

A long bone continues to lengthen while the cartilaginous cells of the epiphyseal plates are active (fig. 7.6). However, once the ossification centers of the diaphysis

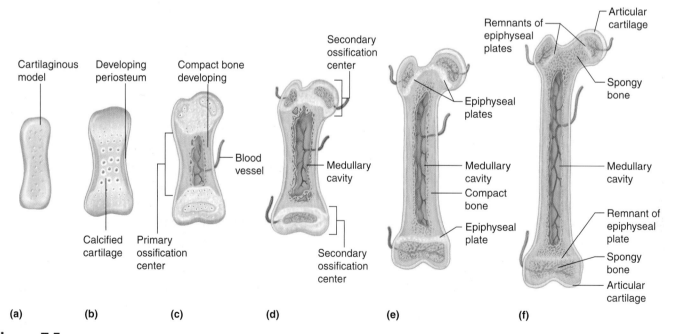

(a) Cartilaginous model   **(b)** Developing periosteum — Calcified cartilage, Primary ossification center   **(c)** Compact bone developing — Blood vessel   **(d)** Secondary ossification center, Medullary cavity, Secondary ossification center   **(e)** Epiphyseal plates, Medullary cavity, Compact bone, Epiphyseal plate   **(f)** Remnants of epiphyseal plates, Articular cartilage, Spongy bone, Medullary cavity, Remnant of epiphyseal plate, Spongy bone, Articular cartilage

Figure 7.5

Major stages (**a–d** fetal, **e** child, **f** adult) in the development of an endochondral bone. (Relative bone sizes are not to scale.)

Figure 7.6

Radiograph of the knee showing epiphyseal plates (arrows) in a child's bones indicates that the bones are still lengthening.

and epiphyses meet and the epiphyseal plates ossify, lengthening is no longer possible in that end of the bone.

A developing long bone thickens as compact bone is deposited on the outside, just beneath the periosteum. As this compact bone forms on the surface, osteoclasts erode other bone tissue on the inside. The resulting space becomes the medullary cavity of the diaphysis, which later fills with marrow. The bone in the central regions of the epiphyses and diaphysis remains spongy, and hyaline cartilage on the ends of the epiphyses persists throughout life as articular cartilage.

> If an epiphyseal plate is damaged before it ossifies, elongation of the long bone may prematurely cease, or if growth continues, it may be uneven. Therefore, injuries to the epiphyses of a young person's bones are of special concern. Surgery is used on an epiphysis to equalize growth of bones developing at very different rates.

Homeostasis of Bone Tissue

After the intramembranous and endochondral bones form, the actions of osteoclasts and osteoblasts continually remodel them. Throughout life, osteoclasts resorb bone matrix and osteoblasts replace it. Hormones that regulate blood calcium help control these opposing processes of *resorption* and *deposition* of matrix (see chapter 11, pp. 302–303). As a result, the total mass of bone tissue of an adult skeleton normally remains nearly constant, even though 3–5% of bone calcium is exchanged each year.

Factors Affecting Bone Development, Growth, and Repair AP|R

A number of factors influence bone development, growth, and repair. These include nutrition, hormonal secretions, and physical exercise. For example, vitamin D is necessary for proper absorption of calcium in the small intestine. In the absence of this vitamin, calcium (provided it is present through dietary consumption) is poorly absorbed, and the inorganic salt portion of bone matrix lacks calcium, softening and thereby deforming bones. Growth hormone secreted by the pituitary gland stimulates division of the cartilage cells in the epiphyseal plates. Sex hormones stimulate ossification of the epiphyseal plates. Physical exercise pulling on muscular attachments to bones stresses the bones, stimulating the bone tissue to thicken and strengthen. Clinical Application 7.1 describes repair of a fractured bone.

> Astronauts experience a 1% loss of bone per month in space. Under microgravity conditions, osteoblast activity decreases and osteoclast activity increases, with greater loss in spongy compared to compact bone. Researchers predict that a 50% bone loss could occur on a spaceflight that lasts several years, such as a mission to Mars.

Practice

5. Describe the development of an intramembranous bone.
6. Explain how an endochondral bone develops.
7. Explain how osteoclasts and osteoblasts remodel bone.
8. Explain how nutritional factors, hormones, and physical exercise affect bone development and growth.

7.4 BONE FUNCTION

Bones shape, support, and protect body structures. They also aid body movements, house tissues that produce blood cells, and store inorganic salts.

Support and Protection

Bones give shape to structures such as the head, face, thorax, and limbs. They also provide support and protection. For example, the bones of the lower limbs, pelvis, and backbone support the body's weight. The bones of the skull protect the eyes, ears, and brain. Bones of the rib cage and shoulder girdle protect the heart and lungs, whereas the bones of the pelvic girdle protect the lower abdominal and internal reproductive organs.

Clinical Application 7.1

Bone Fractures

A *fracture* is a break in a bone. A fracture is classified by its cause as a traumatic, spontaneous, or pathologic fracture and by the nature of the break as a greenstick, fissured, comminuted, transverse, oblique, or spiral fracture (fig. 7A). A broken bone exposed to the outside by an opening in the skin is termed a compound (open) fracture.

When a bone breaks, blood vessels in it rupture, and the periosteum is likely to tear. Blood escaping from the broken vessels spreads through the damaged area and soon forms a blood clot, or *hematoma*. Vessels in surrounding tissues dilate, swelling and inflaming the tissues.

Within days or weeks, developing blood vessels and large numbers of osteoblasts originating in the periosteum invade the hematoma. The osteoblasts rapidly divide in the regions close to the new blood vessels, building spongy bone nearby. Granulation tissue develops, and in regions farther from a blood supply, fibroblasts produce masses of fibrocartilage. Meanwhile, phagocytic cells begin to remove the blood clot, as well as any dead or damaged cells in the affected area. Osteoclasts also appear and resorb bone fragments, aiding in "cleaning up" debris.

In time, fibrocartilage fills the gap between the ends of the broken bone. This mass, termed a *cartilaginous callus,*

A *greenstick* fracture is incomplete, and the break occurs on the convex surface of the bend in the bone.

A *fissured* fracture is an incomplete longitudinal break.

A *comminuted* fracture is complete and fragments the bone.

A *transverse* fracture is complete, and the break occurs at a right angle to the axis of the bone.

An *oblique* fracture occurs at an angle other than a right angle to the axis of the bone.

A *spiral* fracture is caused by excessive twisting of a bone.

Figure 7A
Various types of fractures.

is later replaced by bone tissue in much the same way that the hyaline cartilage of a developing endochondral bone is replaced. That is, the cartilaginous callus breaks down, blood vessels and osteoblasts invade the area, and a *bony callus* fills the space.

Typically, more bone is produced at the site of a healing fracture than is necessary to replace the damaged tissues. Osteoclasts remove the excess, and the result is a bone shaped much like the original (fig. 7B).

Physicians can help the bone-healing process. The first casts to immobilize fractured bones were introduced in Philadelphia in 1876, and soon after, doctors began using screws and plates internally to align healing bone parts. Today, orthopedic surgeons also use rods, wires, and nails. These devices have become lighter and smaller; many are built of titanium. A device called a hybrid fixator treats a broken leg using metal pins internally to align bone pieces. The pins are anchored to a metal ring device worn outside the leg.

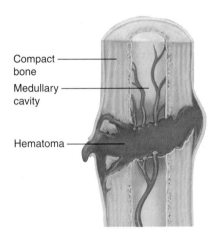

(a) Blood escapes from ruptured blood vessels and forms a hematoma.

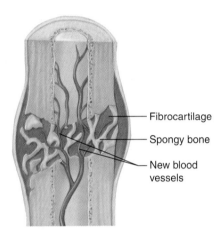

(b) Spongy bone forms in regions close to developing blood vessels, and fibrocartilage forms in more distant regions.

(c) A bony callus replaces fibrocartilage.

(d) Osteoclasts remove excess bony tissue, restoring new bone structure much like the original.

Figure 7B
Major steps **(a–d)** in repair of a fracture.

Body Movement

When limbs or other body parts move, bones and muscles interact as simple mechanical devices called **levers** (lev′erz). A lever has four basic components: (1) a rigid bar or rod, (2) a fulcrum or pivot on which the bar turns, (3) an object moved against resistance, and (4) a force that supplies energy for the movement of the bar.

The actions of bending and straightening the upper limb at the elbow illustrate bones and muscles functioning as levers. When the upper limb bends, the forearm bones represent the rigid bar, the elbow joint is the fulcrum, the hand is moved against the resistance provided by the weight, and the force is supplied by muscles on the anterior side of the arm (fig. 7.7*a*). One of these muscles, the *biceps brachii,* is attached by a tendon to a projection on a bone (radius) in the forearm, a short distance below the elbow.

When the upper limb straightens at the elbow, the forearm bones again serve as the rigid bar, the elbow joint serves as the fulcrum, and the hand moves against the resistance by pulling on the rope to raise the weight (fig. 7.7*b*). However, this time the *triceps brachii,* a muscle located on the posterior side of the arm, supplies the force. A tendon of this muscle attaches to a projection on a forearm bone (ulna) at the point of the elbow.

Blood Cell Formation

The process of blood cell formation, called **hematopoiesis** (he″mă-to-poi-e′sis), begins in the *yolk sac,* which lies outside the human embryo (see chapter 20, p. 547). Later in development, blood cells are manufactured in the liver and spleen, and still later they form in bone marrow.

Marrow is a soft, netlike mass of connective tissue within the medullary cavities of long bones, in the irregular spaces of spongy bone, and in the larger central canals of compact bone tissue. It is of two kinds: red and yellow. *Red marrow* functions in the formation of red blood cells (erythrocytes), white blood cells (leukocytes), and blood platelets. Red marrow's color comes from the oxygen-carrying pigment **hemoglobin** in the red blood cells.

In an infant, red marrow occupies the cavities of most bones. With increasing age, yellow marrow replaces much of it. *Yellow marrow* stores fat; it is not active in blood cell production. In an adult, red marrow is primarily found in the spongy bone of the skull, ribs, sternum, clavicles, vertebrae, and hip bones. However, if the body requires more blood, yellow marrow can be replaced by extensions of red bone marrow from elsewhere in the bone, which then reverts to yellow marrow when there is enough or a surplus of blood. Chapter 12 (pp. 321, 324, and 327) describes blood cell formation in more detail.

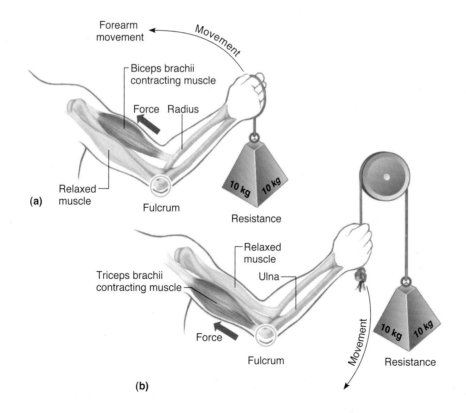

Figure 7.7

Levers and movement. **(a)** When the forearm bends at the elbow or **(b)** when the forearm straightens at the elbow, the bones and muscles function as a lever.

Bone marrow transplants have been used for more than half a century to enable people with any of several dozen types of blood diseases to tolerate high levels of chemotherapy drugs. In a bone marrow transplant, a hollow needle and syringe remove normal red marrow cells from the spongy bone of a donor, or stem cells (which can give rise to specialized blood cells) are separated out from the donor's bloodstream. Stem cells from the umbilical cord of a newborn can be used in place of bone marrow.

Cells are selected as donor cells based on their having a pattern of molecules on their surfaces that closely matches the pattern on the recipient's cells. In 30% of bone marrow transplants, the donor is a blood relative. The cells are injected into the bloodstream of the recipient, whose own marrow has been intentionally destroyed with radiation or chemotherapy. If all goes well, the donor cells travel to the spaces within bones that red marrow normally occupies, where they replenish the blood supply with healthy cells. About 15% of the time, patients die from infection because their immune systems reject the transplant, or because the transplanted tissue attacks the recipient, a condition called graft-versus-host disease.

Safer than a bone marrow transplant is an autologous stem cell transplant. Stem cells are taken from a patient's bloodstream, set aside in the laboratory, and then high doses of chemotherapy or radiation are used to destroy the rest of the bone marrow. Then the patient's own ("autologous") stem cells are infused. They migrate to the bone marrow and reconstitute a disease-free blood-forming system that the patient's immune system does not reject.

Storage of Inorganic Salts

Bones store calcium. The extracellular matrix of bone tissue is rich in calcium salts, mostly in the form of calcium phosphate. Vital metabolic processes require calcium. When the blood is low in calcium, parathyroid hormone stimulates osteoclasts to break down bone tissue, which releases calcium salts from the extracellular matrix into the blood. A high blood calcium level inhibits osteoclast activity, and calcitonin from the thyroid gland stimulates osteoblasts to form bone tissue, storing excess calcium in the extracellular matrix (fig. 7.8). Chapter 11 (pp. 302–303) describes the details of this homeostatic mechanism. Maintaining sufficient blood calcium levels is important in muscle contraction, nerve impulse conduction, blood clotting, and other physiological processes.

In addition to storing calcium and phosphorus, bone tissue contains smaller amounts of magnesium, sodium, potassium, and carbonate ions. Bones also accumulate certain harmful metallic elements such as lead, radium, or strontium, which are not normally present in the body but are sometimes accidentally ingested.

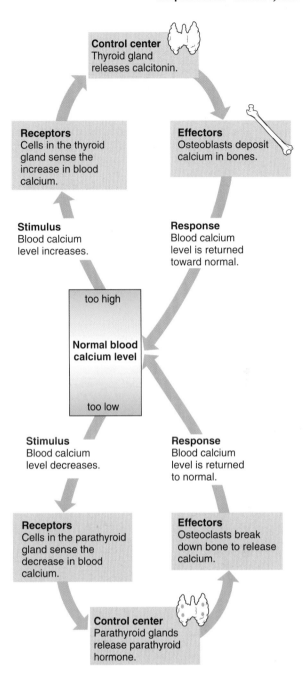

Figure 7.8 AP|R

Hormones regulate deposition and resorption of bone calcium.

Q: *What three components of a homeostatic mechanism (see fig. 1.5) are shown in this figure?*

Answer can be found in Appendix E on page 568.

Practice

9. Name the major functions of bones.
10. Distinguish between the functions of red marrow and yellow marrow.
11. List the substances normally stored in bone tissue.

7.5 SKELETAL ORGANIZATION

For purposes of study, it is convenient to divide the skeleton into two major portions—an axial skeleton and an appendicular skeleton (fig. 7.9). The **axial skeleton** consists of the bony and cartilaginous parts that support and protect the organs of the head, neck, and trunk. These parts include:

1. **Skull.** The skull is composed of the **cranium** (kra′ne-um), or brain case, and the *facial bones.*
2. **Hyoid bone.** The hyoid (hi′oid) bone is located in the neck between the lower jaw and the larynx.

It supports the tongue and is an attachment for certain muscles that help move the tongue during swallowing.

3. **Vertebral column.** The vertebral column (backbone) consists of many vertebrae separated by cartilaginous *intervertebral discs.* Near its distal end, several vertebrae fuse to form the **sacrum** (sa′krum), which is part of the pelvis. The **coccyx** (kok′siks), a small, rudimentary tailbone composed of several fused vertebrae, is attached to the end of the sacrum.
4. **Thoracic cage.** The thoracic cage protects the organs of the thoracic cavity and the upper

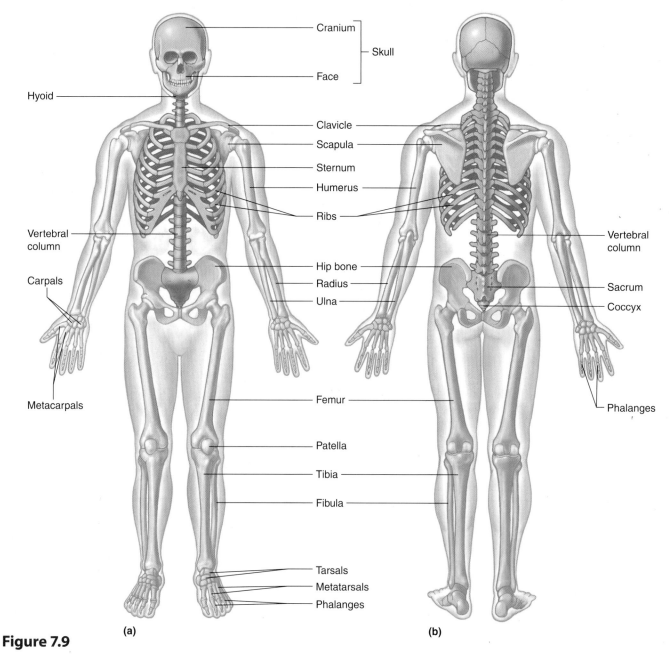

(a)

(b)

Figure 7.9

Major bones of the skeleton. **(a)** Anterior view. **(b)** Posterior view. The axial portion is shown in red, and the appendicular portions are shown in yellow.

abdominal cavity. It is composed of twelve pairs of **ribs,** which articulate posteriorly with thoracic vertebrae. The thoracic cage also includes the **sternum** (ster′num), or breastbone, to which most of the ribs attach anteriorly.

The **appendicular skeleton** consists of the bones of the upper and lower limbs and the bones that anchor the limbs to the axial skeleton. It includes:

1. **Pectoral** (pek′to-ral) **girdle.** The pectoral girdle is formed by a **scapula** (scap′u-lah), or shoulder blade, and a **clavicle** (klav′ĭ-k′l), or collarbone, on both sides of the body. The pectoral girdle connects the bones of the upper limbs to the axial skeleton and aids in upper limb movements.
2. **Upper limbs.** Each upper limb consists of a **humerus** (hu′mer-us), or arm bone, two forearm bones—a **radius** (ra′de-us) and an **ulna** (ul′nah)—and a hand. The humerus, radius, and ulna articulate with each other at the elbow joint. At the distal end of the radius and ulna is the hand. There are eight **carpals** (kar′pals), or wrist bones. The five bones of the palm are called **metacarpals** (met″ah-kar′pals), and the fourteen finger bones are called **phalanges** (fah-lan′jēz; singular, *phalanx,* fa′lanks).

3. **Pelvic girdle.** The pelvic girdle is formed by two hip bones attached to each other anteriorly and to the sacrum posteriorly. They connect the bones of the lower limbs to the axial skeleton and, with the sacrum and coccyx, form the **pelvis.**
4. **Lower limbs.** Each lower limb consists of a **femur** (fe′mur), or thigh bone, two leg bones—a large **tibia** (tib′e-ah) and a slender **fibula** (fib′u-lah)—and a foot. The femur and tibia articulate with each other at the knee joint, where the **patella** (pah-tel′ah) covers the anterior surface. At the distal ends of the tibia and fibula is the foot. There are seven **tarsals** (tahr′sals), or ankle bones. The five bones of the instep are called **metatarsals** (met″ah-tahr′sals), and the fourteen bones of the toes (like the fingers) are called **phalanges.**

Table 7.1 lists the bones of the adult skeleton, and table 7.2 lists terms that describe skeletal structures.

The skeleton of an average 160-pound body weighs about 29 pounds.

Table 7.1	Bones of the Adult Skeleton

1. Axial Skeleton
   a. Skull
      8 cranial bones
      | frontal 1 | temporal 2 |
      | parietal 2 | sphenoid 1 |
      | occipital 1 | ethmoid 1 |

      14 facial bones
      | maxilla 2 | lacrimal 2 |
      | zygomatic 2 | nasal 2 |
      | palatine 2 | vomer 1 |
      | inferior nasal concha 2 | |
      | mandible 1 | |

      22 bones

   b. Middle ear bones
      malleus 2
      incus 2
      stapes 2

      6 bones

   c. Hyoid
      hyoid bone 1

      1 bone

   d. Vertebral column
      cervical vertebrae 7
      thoracic vertebrae 12
      lumbar vertebrae 5
      sacrum 1
      coccyx 1

      26 bones

   e. Thoracic cage
      rib 24
      sternum 1

      25 bones

2. Appendicular Skeleton
   a. Pectoral girdle
      scapula 2
      clavicle 2

      4 bones

   b. Upper limbs
      humerus 2
      radius 2
      ulna 2
      carpal 16
      metacarpal 10
      phalanx 28

      60 bones

   c. Pelvic girdle
      hip bone 2

      2 bones

   d. Lower limbs
      femur 2
      tibia 2
      fibula 2
      patella 2
      tarsal 14
      metatarsal 10
      phalanx 28

      60 bones

      Total   206 bones

Table 7.2	Terms Used to Describe Skeletal Structures	
Term	**Definition**	**Examples**
Condyle (kon'dīl)	Rounded process that usually articulates with another bone	Occipital condyle of occipital bone (fig. 7.13)
Crest (krest)	Narrow, ridgelike projection	Iliac crest of ilium (fig. 7.28)
Epicondyle (ep''ĭ-kon'dīl)	Projection situated above a condyle	Medial epicondyle of humerus (fig. 7.24)
Facet (fas'et)	Small, nearly flat surface	Rib facet of thoracic vertebra (fig. 7.17)
Fontanel (fon''tah-nel')	Soft spot in the skull where membranes cover the space between bones	Anterior fontanel between frontal and parietal bones (fig. 7.16)
Foramen (fo-ra'men)	Opening through a bone that usually is a passageway for blood vessels, nerves, or ligaments	Foramen magnum of occipital bone (fig. 7.13)
Fossa (fos'ah)	Relatively deep pit or depression	Olecranon fossa of humerus (fig. 7.24)
Fovea (fo've-ah)	Tiny pit or depression	Fovea capitis of femur (fig. 7.30)
Head (hed)	Enlargement on the end of a bone	Head of humerus (fig. 7.24)
Meatus (me-a'tus)	Tubelike passageway within a bone	External acoustic meatus of ear (fig. 7.12)
Process (pros'es)	Prominent projection on a bone	Mastoid process of temporal bone (fig. 7.12)
Sinus (si'nus)	Cavity within a bone	Frontal sinus of frontal bone (fig. 7.15)
Spine (spīn)	Thornlike projection	Spine of scapula (fig. 7.23)
Suture (soo'cher)	Interlocking line of union between bones	Lambdoid suture between occipital and parietal bones (fig. 7.12)
Trochanter (tro-kan'ter)	Relatively large process	Greater trochanter of femur (fig. 7.30)
Tubercle (tu'ber-kl)	Small, knoblike process	Greater tubercle of humerus (fig. 7.24)
Tuberosity (tu''bĕ-ros'ĭ-te)	Knoblike process usually larger than a tubercle	Radial tuberosity of radius (fig. 7.25)

Practice

12. Distinguish between the axial and appendicular skeletons.
13. List the bones of the axial skeleton and of the appendicular skeleton.

7.6 SKULL

A human skull usually consists of twenty-two bones that, except for the lower jaw, are firmly interlocked along *sutures* (soo'cherz) (fig. 7.10). Eight of these interlocked bones make up the cranium, and fourteen form the facial skeleton. The **mandible** (man'dĭ-b'l), or lower jawbone, is a movable bone held to the cranium by ligaments. (Three other bones in each middle ear are discussed in chapter 10, p. 271.) Reference plates 8–11, on pages 175–177, show the human skull and its parts.

Cranium

The **cranium** encloses and protects the brain, and its surface provides attachments for muscles that make chewing and head movements possible. Some of the cranial bones contain air-filled cavities called *paranasal sinuses,* lined with mucous membranes and connected by passageways

to the nasal cavity (fig. 7.11). Sinuses reduce the skull's weight and increase the intensity of the voice by serving as resonant sound chambers.

The eight bones of the cranium, shown in figures 7.10 and 7.12, are:

1. **Frontal bone.** The frontal (frun'tal) bone forms the anterior portion of the skull above the eyes. On the upper margin of each orbit (the bony socket of the eye), the frontal bone is marked by a *supraorbital foramen* (or *supraorbital notch* in some skulls), through which blood vessels and nerves pass to the tissues of the forehead. In the frontal bone are two *frontal sinuses,* one above each eye near the midline (see fig. 7.11).
2. **Parietal bones.** One parietal (pah-ri'ĕ-tal) bone is located on each side of the skull just behind the frontal bone (fig. 7.12). Together, the parietal bones form the bulging sides and roof of the cranium. They are fused at the midline along the *sagittal suture,* and they meet the frontal bone along the *coronal suture.*
3. **Occipital bone.** The occipital (ok-sip'ĭ-tal) bone joins the parietal bones along the *lambdoid* (lam'doid) *suture* (figs. 7.12 and 7.13). It forms the back of the skull and the base of the cranium. Through a large opening on its lower surface called the *foramen magnum* pass nerve fibers from the

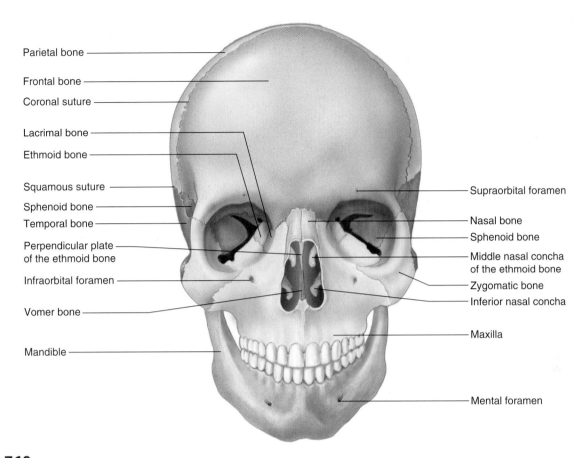

Parietal bone

Frontal bone

Coronal suture

Lacrimal bone

Ethmoid bone

Squamous suture

Sphenoid bone

Temporal bone

Perpendicular plate of the ethmoid bone

Infraorbital foramen

Vomer bone

Mandible

Supraorbital foramen

Nasal bone

Sphenoid bone

Middle nasal concha of the ethmoid bone

Zygomatic bone

Inferior nasal concha

Maxilla

Mental foramen

Figure 7.10
Anterior view of the skull.

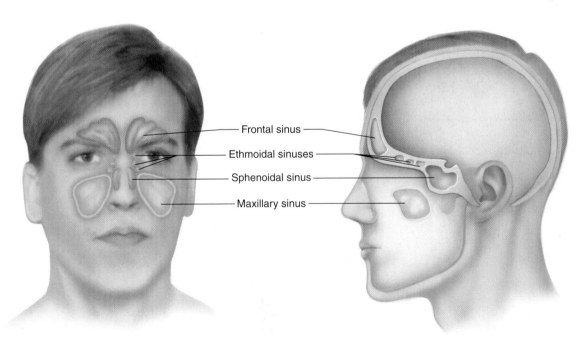

Frontal sinus

Ethmoidal sinuses

Sphenoidal sinus

Maxillary sinus

Figure 7.11
Locations of the paranasal sinuses.

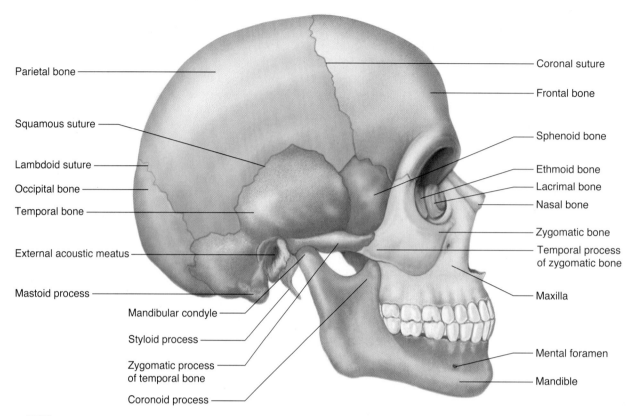

Figure 7.12
Right lateral view of the skull.

Parietal bone

Squamous suture

Lambdoid suture

Occipital bone

Temporal bone

External acoustic meatus

Mastoid process

Mandibular condyle

Styloid process

Zygomatic process
of temporal bone

Coronoid process

Coronal suture

Frontal bone

Sphenoid bone

Ethmoid bone

Lacrimal bone

Nasal bone

Zygomatic bone

Temporal process
of zygomatic bone

Maxilla

Mental foramen

Mandible

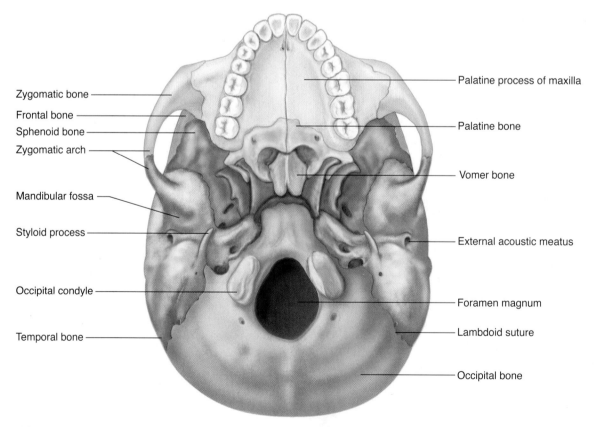

Figure 7.13
Inferior view of the skull.

Zygomatic bone

Frontal bone

Sphenoid bone

Zygomatic arch

Mandibular fossa

Styloid process

Occipital condyle

Temporal bone

Palatine process of maxilla

Palatine bone

Vomer bone

External acoustic meatus

Foramen magnum

Lambdoid suture

Occipital bone

brain, which enter the vertebral canal to become part of the spinal cord. Rounded processes called *occipital condyles,* located on each side of the foramen magnum, articulate with the first vertebra (atlas) of the vertebral column.

4. **Temporal bones.** A temporal (tem'po-ral) bone on each side of the skull joins the parietal bone along a *squamous suture* (see figs. 7.10 and 7.12). The temporal bones form parts of the sides and the base of the cranium. Located near the inferior margin is an opening, the *external acoustic meatus,* which leads inward to parts of the ear. The temporal bones have depressions called the *mandibular fossae* that articulate with condyles of the mandible. Below each external acoustic meatus are two projections—a rounded *mastoid process* and a long, pointed *styloid process.* The mastoid process provides an attachment for certain muscles of the neck, whereas the styloid process anchors muscles associated with the tongue and pharynx. A *zygomatic process* projects anteriorly from the temporal bone, joins the *zygomatic bone,* and helps form the prominence of the cheek.

5. **Sphenoid bone.** The sphenoid (sfe'noid) bone is wedged between several other bones in the anterior portion of the cranium (figs. 7.12 and 7.13). This bone helps form the base of the cranium, the sides of the skull, and the floors and sides of the orbits. Along the midline in the cranial cavity,

a portion of the sphenoid bone indents to form the saddle-shaped *sella turcica* (sel'ah tur'si-ka). The pituitary gland occupies this depression. The sphenoid bone also contains two *sphenoidal sinuses* (see fig. 7.11).

6. **Ethmoid bone.** The ethmoid (eth'moid) bone is located in front of the sphenoid bone (figs. 7.12 and 7.14). It consists of two masses, one on each side of the nasal cavity, which are joined horizontally by thin *cribriform* (krib'rĭ-form) *plates.* These plates form part of the roof of the nasal cavity (fig. 7.14).

Projecting upward into the cranial cavity between the cribriform plates is a triangular process of the ethmoid bone called the *crista galli* (kris'tă gal'li) (cock's comb). Membranes that enclose the brain attach to this process (figs. 7.14 and 7.15). Portions of the ethmoid bone also form sections of the cranial floor, the orbital walls, and the nasal cavity walls. A *perpendicular plate* projects downward in the midline from the cribriform plates and forms most of the nasal septum (fig. 7.15).

Delicate scroll-shaped plates called the *superior nasal conchae* (kong'ke) and the *middle nasal conchae* project inward from the lateral portions of the ethmoid bone toward the perpendicular plate (see fig. 7.10). The lateral portions of the ethmoid bone contain many small air spaces, the *ethmoidal sinuses* (see fig. 7.11).

Figure 7.14
Floor of the cranial cavity, viewed from above.

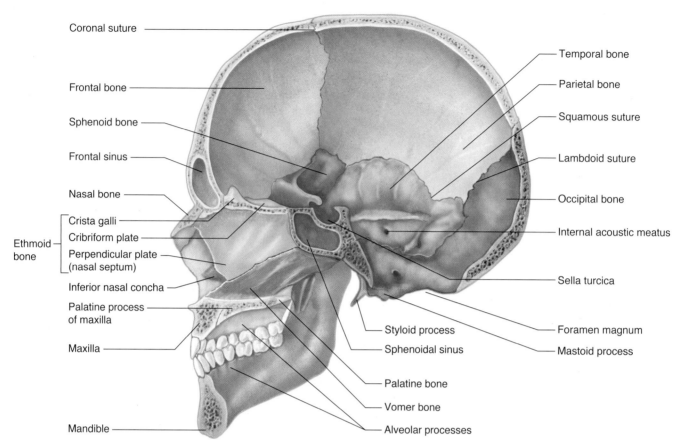

Coronal suture

Frontal bone

Sphenoid bone

Frontal sinus

Nasal bone

Crista galli
Cribriform plate
Ethmoid bone
Perpendicular plate (nasal septum)

Inferior nasal concha

Palatine process of maxilla

Maxilla

Mandible

Temporal bone

Parietal bone

Squamous suture

Lambdoid suture

Occipital bone

Internal acoustic meatus

Sella turcica

Styloid process

Sphenoidal sinus

Palatine bone

Vomer bone

Alveolar processes

Foramen magnum

Mastoid process

Figure 7.15 AP|R
Sagittal section of the skull.

Facial Skeleton

The **facial skeleton** consists of thirteen immovable bones and a movable lower jawbone. These bones form the basic shape of the face and provide attachments for muscles that move the jaw and control facial expressions. The bones of the facial skeleton are:

1. **Maxillae.** The maxillae (mak-sil′e; singular, *maxilla,* mak-sil′ah) form the upper jaw (see figs. 7.12 and 7.13). Portions of these bones compose the anterior roof of the mouth (*hard palate*), the floors of the orbits, and the sides and floor of the nasal cavity. They also contain the sockets of the upper teeth. Inside the maxillae, lateral to the nasal cavity, are *maxillary sinuses,* the largest of the sinuses (see fig. 7.11).

   During development, portions of the maxillae called *palatine processes* grow together and fuse along the midline to form the anterior section of the hard palate. The inferior border of each maxillary bone projects downward, forming an *alveolar* (al-ve′o-lar) *process* (fig. 7.15). Together, these processes form a horseshoe-shaped *alveolar arch*

(dental arch). Teeth occupy cavities in this arch (dental alveoli). Dense connective tissue binds teeth to the bony sockets.

> Sometimes fusion of the palatine processes of the maxillae is incomplete at birth; the result is a *cleft palate.* Infants with a cleft palate may have trouble suckling because of the opening between the oral and nasal cavities. A temporary prosthetic device (artificial palate) may be inserted into the mouth, or a special type of nipple can be placed on bottles, so the child can eat and drink, until surgery can be performed to correct the condition.

2. **Palatine bones.** The L-shaped palatine (pal′ah-tīn) bones are located behind the maxillae (see figs. 7.13 and 7.15). The horizontal portions form the posterior section of the hard palate and the floor of the nasal cavity. The perpendicular portions help form the lateral walls of the nasal cavity.

3. **Zygomatic bones.** The zygomatic (zi″go-mat′ik) bones form the prominences of the cheeks below and to the sides of the eyes (see figs. 7.12 and 7.13).

These bones also help form the lateral walls and the floors of the orbits. Each bone has a *temporal process,* which extends posteriorly to join the zygomatic process of a temporal bone. Together, these processes form a *zygomatic arch.*

4. **Lacrimal bones.** A lacrimal (lak′rĭ-mal) bone is a thin, scalelike structure located in the medial wall of each orbit between the ethmoid bone and the maxilla (see figs. 7.10 and 7.12).

5. **Nasal bones.** The nasal (na′zal) bones are long, thin, and nearly rectangular (see figs. 7.10 and 7.12). They lie side by side and are fused at the midline, where they form the bridge of the nose.

6. **Vomer bone.** The thin, flat vomer (vo′mer) bone is located along the midline within the nasal cavity (see figs. 7.10 and 7.15). Posteriorly, it joins the perpendicular plate of the ethmoid bone, and together they form the nasal septum.

7. **Inferior nasal conchae.** The inferior nasal conchae are fragile, scroll-shaped bones attached to the lateral walls of the nasal cavity (see figs. 7.10 and 7.15). Like the superior and middle conchae, the inferior conchae support mucous membranes in the nasal cavity.

8. **Mandible.** The mandible is a horizontal, horseshoe-shaped body with a flat portion projecting upward at each end (see figs. 7.10 and 7.12). This projection is divided into two processes—a posterior *mandibular condyle* and an anterior *coronoid process.* The mandibular condyles articulate with the mandibular fossae of the temporal bones (see fig. 7.13), whereas the coronoid processes provide attachments for muscles used in chewing. A curved bar of bone on the superior border of the mandible, the *alveolar arch,* contains the hollow sockets (dental alveoli) that bear the lower teeth.

Infantile Skull

At birth the skull is incompletely developed, with fibrous membranes connecting the cranial bones. These membranous areas of incomplete intramembranous ossification are called **fontanels** (fon″tah nelz′) or, more commonly, soft spots (fig. 7.16). They permit some movement between the bones, so that the developing skull is partially compressible and can slightly change shape. This enables an infant's skull to more easily pass through the birth canal. Eventually the fontanels close as the cranial bones grow together.

Other characteristics of an infantile skull include a relatively small face with a prominent forehead and large orbits. The jaw and nasal cavity are small, the sinuses are incompletely formed, and the frontal bone is in two parts. The skull bones are thin, but they are also somewhat flexible and thus are less easily fractured than adult skull bones.

Practice

14. Locate and name each of the bones of the cranium.
15. Locate and name each of the facial bones.
16. Explain how an adult skull differs from that of an infant.

7.7 VERTEBRAL COLUMN

The **vertebral column** extends from the skull to the pelvis and forms the vertical axis of the skeleton. It is composed of many bony parts, called **vertebrae** (ver′tĕ-brā), that are separated by masses of fibrocartilage called *intervertebral discs* and are connected to one another by ligaments (fig. 7.17). The vertebral column supports the head and trunk of the body. It also protects the spinal cord, which passes through a *vertebral canal* formed by openings in the vertebrae.

A Typical Vertebra

The vertebrae in various regions of the vertebral column have special characteristics, yet they also have features in common. A typical vertebra has a drum-shaped *body,* which forms the thick, anterior portion of the bone (fig. 7.18). A longitudinal row of these vertebral bodies supports the weight of the head and trunk. The intervertebral discs, which separate adjacent vertebral bodies, cushion and soften the forces from movements such as walking and jumping.

Projecting posteriorly from each vertebral body are two short stalks called *pedicles* (ped′ĭ-k′lz). Two plates called *laminae* (lam′i-ne) arise from the pedicles and fuse in the back to become a *spinous process.* The pedicles, laminae, and spinous process together complete a bony *vertebral arch* around the *vertebral foramen,* through which the spinal cord passes.

> If the laminae of the vertebrae fail to unite during development, the vertebral arch remains incomplete, causing a condition called *spina bifida.* The contents of the vertebral canal protrude outward. This problem occurs most frequently in the lumbosacral region. Spina bifida is associated with folic acid deficiency in certain genetically susceptible individuals.

Between the pedicles and laminae of a typical vertebra is a *transverse process,* which projects laterally and posteriorly. Ligaments and muscles are attached to the dorsal spinous process and the transverse processes. Projecting upward and downward from each vertebral arch are *superior* and *inferior articular processes.* These processes bear cartilage-covered facets by which each vertebra is joined to the one above and the one below it.

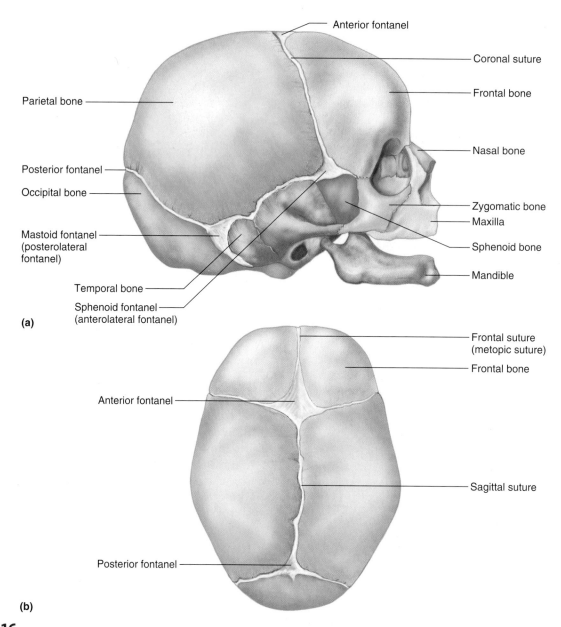

Figure 7.16

Fontanels. **(a)** Right lateral view and **(b)** superior view of the infantile skull.

On the lower surfaces of the vertebral pedicles are notches that align with adjacent vertebrae to form openings called *intervertebral foramina* (in″ter-ver′tĕ-bral fo-ram′ĭ-nah) (see fig. 7.17). These openings provide passageways for spinal nerves.

Cervical Vertebrae

Seven **cervical vertebrae** compose the bony axis of the neck (see fig. 7.17). The transverse processes of these vertebrae are distinctive because they have *transverse*

foramina, which are passageways for blood vessels to and from the brain (fig. 7.18a). Also, the spinous processes of the second through the fifth cervical vertebrae are uniquely forked (bifid). These processes provide attachments for muscles.

Two of the cervical vertebrae are of special interest: the atlas and the axis (fig. 7.19). The first vertebra, or **atlas** (at′las), supports the head. On its superior surface are two kidney-shaped *facets* that articulate with the occipital condyles.

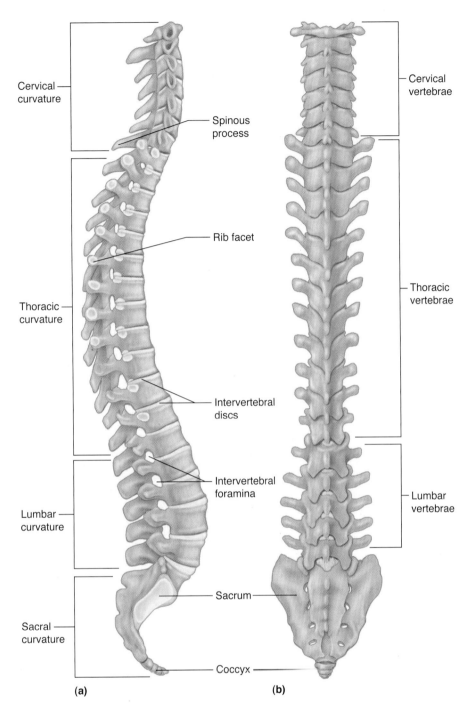

Figure 7.17
The curved vertebral column consists of many vertebrae separated by intervertebral discs. **(a)** Right lateral view. **(b)** Posterior view.

The second cervical vertebra, or **axis** (ak′sis), bears a toothlike *dens* (odontoid process) on its body. This process projects upward and lies in the ring of the atlas. As the head is turned from side to side, the atlas pivots around the dens.

Giraffes and humans have the same number of vertebrae in their necks—seven.

Thoracic Vertebrae

The twelve **thoracic vertebrae** are larger than the cervical vertebrae (see fig. 7.17). Each vertebra has a long, pointed spinous process, which slopes downward, and facets on the sides of its body, which articulate with a rib (see fig. 7.18*b*).

Beginning with the third thoracic vertebra and moving inferiorly, the bodies of these bones increase in

Figure 7.18

Superior view of **(a)** a cervical vertebra, **(b)** a thoracic vertebra, and **(c)** a lumbar vertebra.

size. Thus, they are adapted to bear increasing loads of body weight.

Lumbar Vertebrae

Five **lumbar vertebrae** are in the small of the back (loin) (see fig. 7.17). These vertebrae are adapted with larger and stronger bodies to support more weight than the vertebrae above them (see fig. 7.18*c*).

Sacrum

The **sacrum** (sa′krum) is a triangular structure, composed of five fused vertebrae, that forms the base of the vertebral column (fig. 7.20). The spinous processes of these fused bones form a ridge of *tubercles*. To the sides of the tubercles are rows of openings, the *posterior sacral foramina,* through which nerves and blood vessels pass.

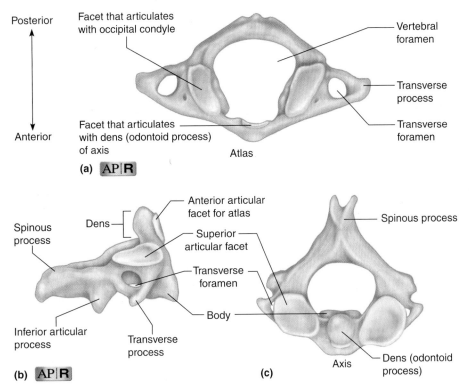

Posterior

Facet that articulates with occipital condyle

Vertebral foramen

Transverse process

Transverse foramen

Facet that articulates with dens (odontoid process) of axis

Anterior

Atlas

(a) AP|R

Spinous process

Dens

Anterior articular facet for atlas

Superior articular facet

Transverse foramen

Body

Inferior articular process

Transverse process

Spinous process

Dens (odontoid process)

Axis

(b) AP|R

(c)

Figure 7.19

Atlas and axis. **(a)** Superior view of the atlas. **(b)** Right lateral view and **(c)** superior view of the axis.

The vertebral foramina of the sacral vertebrae form the *sacral canal,* which continues through the sacrum to an opening of variable size at the tip, called the *sacral hiatus* (sa'kral hi-a'tus). On the ventral surface of the sacrum, four pairs of *anterior sacral foramina* provide passageways for nerves and blood vessels.

Coccyx

The **coccyx** (kok'siks), or tailbone, is the lowest part of the vertebral column and is usually composed of four fused vertebrae (fig. 7.20). Ligaments attach it to the margins of the sacral hiatus.

Changes in the intervertebral discs can cause back problems. Each disc is composed of a tough outer layer of fibrocartilage and an elastic central mass. With age, these discs degenerate— the central masses lose firmness, and the outer layers thin and weaken, developing cracks. Extra pressure, as when a person falls or lifts a heavy object, can break the outer layer of a disc, squeezing out the central mass. Such a rupture may press on the spinal cord or on a spinal nerve that branches from it. This condition—a ruptured or herniated disc—may cause back pain and numbness or the loss of muscular function in the parts that the affected spinal nerve innervates.

Practice

17. Describe the structure of the vertebral column.

18. Describe a typical vertebra.

19. Explain how the structures of cervical, thoracic, and lumbar vertebrae differ.

7.8 THORACIC CAGE

The **thoracic cage** includes the ribs, the thoracic vertebrae, the sternum, and the costal cartilages that attach the ribs to the sternum (fig. 7.21). These bones support the pectoral girdle and upper limbs, protect the viscera in the thoracic and upper abdominal cavities, and play a role in breathing.

Ribs

The usual number of **ribs** is twenty-four—one pair attached to each of the twelve thoracic vertebrae. The first seven rib pairs, *true ribs* (vertebrosternal ribs), join the sternum directly by their costal cartilages. The remaining five pairs are called *false ribs,* because their

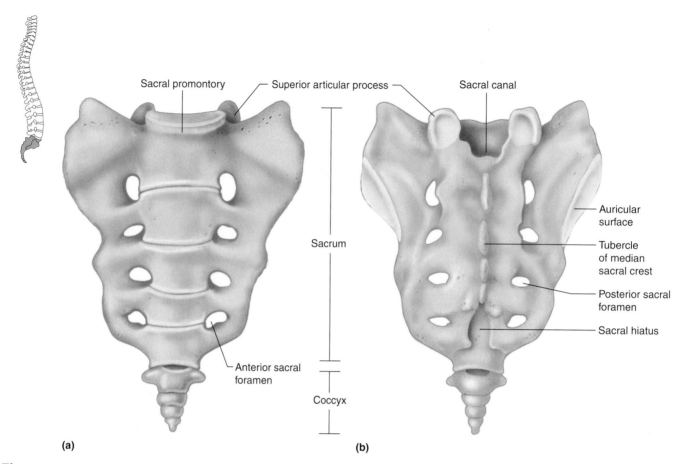

Figure 7.20

Sacrum and coccyx. **(a)** Anterior view and **(b)** posterior view.

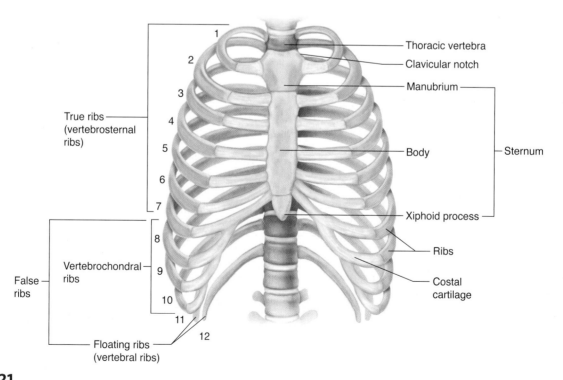

Figure 7.21

The thoracic cage includes the ribs, the thoracic vertebrae, the sternum, and the costal cartilages that attach the ribs to the sternum.

cartilages do not reach the sternum directly. Instead, the cartilages of the upper three false ribs (vertebrochondral ribs) join the cartilages of the seventh rib. The last two (or sometimes three) rib pairs are called *floating ribs* (vertebral ribs) because they have no cartilaginous attachments to the sternum.

A typical rib has a long, slender shaft, which curves around the chest and slopes downward. On the posterior end is an enlarged *head* by which the rib articulates with a *facet* on the body of its own vertebra and with the body of the next higher vertebra. A *tubercle,* close to the head of the rib, articulates with the transverse process of the vertebra.

Sternum

The **sternum,** or breastbone, is located along the midline in the anterior portion of the thoracic cage (fig. 7.21). This flat, elongated bone develops in three parts—an upper *manubrium* (mah-nu′bre-um), a middle *body,* and a lower *xiphoid* (zīf′oid) *process* that projects downward. The manubrium articulates with the clavicles by facets on its superior border.

Practice

20. Which bones compose the thoracic cage?
21. What are the differences among true, false, and floating ribs?
22. Name the three parts of the sternum.

7.9 PECTORAL GIRDLE

The **pectoral girdle,** or shoulder girdle, is composed of four parts—two clavicles and two scapulae (fig. 7.22). Although the word *girdle* suggests a ring-shaped structure, the pectoral girdle is an incomplete ring. It is open in the back between the scapulae, and the sternum separates its bones in front. The pectoral girdle supports the upper limbs and is an attachment for several muscles that move them.

Clavicles

The **clavicles,** or collarbones, are slender, rodlike bones with elongated S shapes (fig. 7.22). Located at the base of the neck, they run horizontally between the manubrium and the scapulae.

The clavicles brace the freely movable scapulae, helping to hold the shoulders in place. They also provide attachments for muscles of the upper limbs, chest, and back.

Scapulae

The **scapulae** (skap′u-le), or shoulder blades, are broad, somewhat triangular bones located on either side of the upper back (figs. 7.22 and 7.23). A *spine* divides the posterior surface of each scapula into unequal portions. This spine leads to two processes—an *acromion* (ah-kro′me-on) *process* that forms the tip of the shoulder and a *coracoid* (kor′ah-koid) *process* that curves anteriorly and inferiorly to the clavicle. The acromion process articulates with the clavicle and provides attachments for muscles of the upper limb and chest. The coracoid process also provides attachments for upper limb and chest muscles. Between the processes is a depression called the *glenoid cavity* (glenoid fossa of the scapula) that articulates with the head of the arm bone (humerus).

Practice

23. Which bones form the pectoral girdle?
24. What is the function of the pectoral girdle?

7.10 UPPER LIMB

The bones of the upper limb form the framework of the arm, forearm, and hand. They also provide attachments for muscles, and they function in levers that move limb parts. These bones include a humerus, a radius, an ulna, carpals, metacarpals, and phalanges (see fig. 7.9).

Humerus

The **humerus** is a long bone that extends from the scapula to the elbow (fig. 7.24). At its upper end is a smooth, rounded *head* that fits into the glenoid cavity of the scapula. Just below the head are two processes—a *greater tubercle* on the lateral side and a *lesser tubercle* on the anterior side. These tubercles provide attachments for muscles that move the upper limb at the shoulder. Between them is a narrow furrow, the *intertubercular groove.*

The narrow depression along the lower margin of the humerus head that separates it from the tubercles is called the *anatomical neck.* Just below the head and the tubercles is a tapering region called the *surgical neck,* so named because fractures commonly occur there. Near the middle of the bony shaft on the lateral side is a rough, V-shaped area called the *deltoid tuberosity.* It provides an attachment for the muscle (deltoid) that raises the upper limb horizontally to the side.

At the lower end of the humerus are two smooth *condyles* (a lateral *capitulum* and a medial *trochlea*) that articulate with the radius on the lateral side and the ulna on the medial side. Above the condyles on either side are *epicondyles,* which provide attachments for muscles and ligaments of the elbow. Between the epicondyles anteriorly is a depression, the *coronoid* (kor′o-noid) *fossa,* that receives a process of the ulna (coronoid process) when the elbow bends. Another depression on

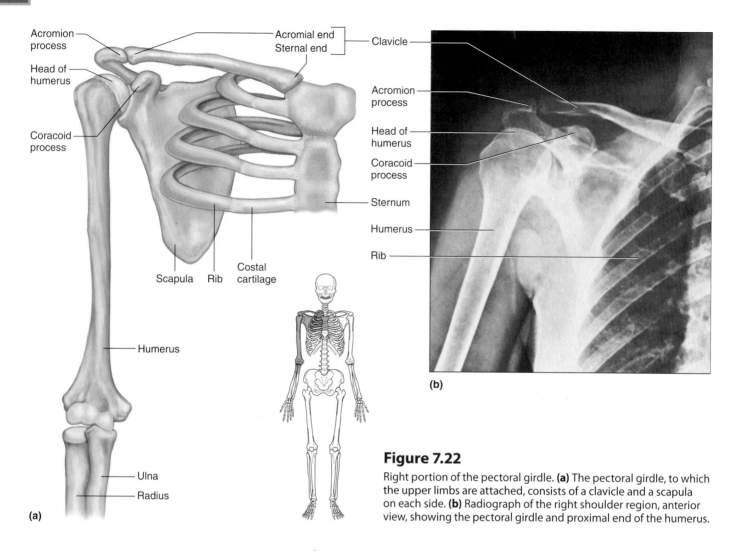

Acromion process

Head of humerus

Coracoid process

Acromial end
Sternal end

Clavicle

Acromion process

Head of humerus

Coracoid process

Sternum

Humerus

Rib

Scapula Rib Costal cartilage

Humerus

Ulna

Radius

(a)

(b)

Figure 7.22

Right portion of the pectoral girdle. **(a)** The pectoral girdle, to which the upper limbs are attached, consists of a clavicle and a scapula on each side. **(b)** Radiograph of the right shoulder region, anterior view, showing the pectoral girdle and proximal end of the humerus.

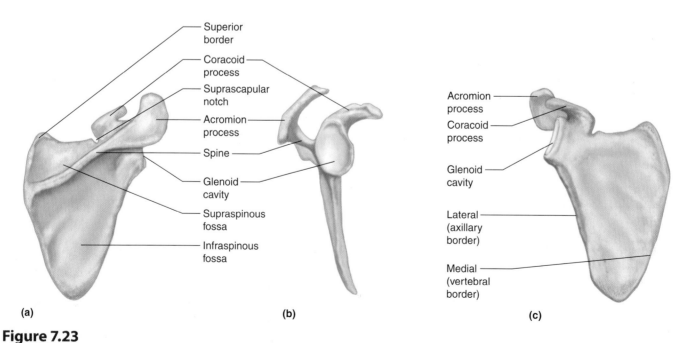

Superior border

Coracoid process

Suprascapular notch

Acromion process

Spine

Glenoid cavity

Supraspinous fossa

Infraspinous fossa

Acromion process

Coracoid process

Glenoid cavity

Lateral (axillary border)

Medial (vertebral border)

(a)   (b)   (c)

Figure 7.23

Right scapula. **(a)** Posterior surface. **(b)** Lateral view showing the glenoid cavity that articulates with the head of the humerus. **(c)** Anterior surface.

Greater tubercle

Intertubercular groove

Lesser tubercle

Head

Anatomical neck

Surgical neck

Greater tubercle

Deltoid tuberosity

Coronoid fossa

Lateral epicondyle

Capitulum

Olecranon fossa

Medial epicondyle

Trochlea

Lateral epicondyle

(a)

(b)

Figure 7.24

Right humerus. **(a)** Anterior surface. **(b)** Posterior surface.

the posterior surface, the *olecranon* (o″lek′ra-non) *fossa,* receives an ulnar process (olecranon process) when the elbow straightens.

Radius

The **radius,** located on the thumb side of the forearm, extends from the elbow to the wrist and crosses over the ulna when the hand is turned so that the palm faces backward (fig. 7.25). A thick, disclike *head* at the upper end of the radius articulates with the humerus and a notch of the ulna (radial notch). This arrangement allows the radius to rotate.

On the radial shaft just below the head is a process called the *radial tuberosity.* It is an attachment for a muscle (biceps brachii) that bends the upper limb at the elbow. At the distal end of the radius, a lateral *styloid* (sti′loid) *process* provides attachments for ligaments of the wrist.

Ulna

The **ulna** is longer than the radius and overlaps the end of the humerus posteriorly (fig. 7.25). At its proximal end, the ulna has a wrenchlike opening, the *trochlear* (trok′le-ar)

notch, that articulates with the humerus. Two processes on either side of this notch, the *olecranon process* and the *coronoid process,* provide attachments for muscles.

At the distal end of the ulna, its knoblike *head* articulates laterally with a notch of the radius (ulnar notch) and with a disc of fibrocartilage inferiorly. This disc, in turn, joins a wrist bone (triquetrum). A medial *styloid process* at the distal end of the ulna provides attachments for wrist ligaments.

Hand

The hand is made up of the wrist, palm, and fingers. The skeleton of the wrist consists of eight small **carpal bones** firmly bound in two rows of four bones each. The resulting compact mass is called a *carpus* (kar′pus). The carpus articulates with the radius and with the fibrocartilaginous disc on the ulnar side. Its distal surface articulates with the metacarpal bones. Figure 7.26 names the individual bones of the carpus.

Five **metacarpal bones,** one in line with each finger, form the framework of the palm or *metacarpus* (met″ah-kar′pus) of the hand. These bones are cylindrical, with

Figure 7.25 AP|R

Right radius and ulna. **(a)** The head of the radius articulates with the radial notch of the ulna, and the head of the ulna articulates with the ulnar notch of the radius. **(b)** Lateral view of the proximal end of the ulna.

rounded distal ends that form the knuckles of a clenched fist. They are numbered 1–5, beginning with the metacarpal of the thumb (fig. 7.26). The metacarpals articulate proximally with the carpals and distally with the phalanges.

The **phalanges** are the finger bones. Each finger has three phalanges—a proximal, a middle, and a distal phalanx—except the thumb, which has two (it lacks a middle phalanx).

Practice

25. Locate and name each of the bones of the upper limb.
26. Explain how the bones of the upper limb articulate.

7.11 PELVIC GIRDLE

The **pelvic girdle** consists of two hip bones (coxal bones, pelvic bones, or innominate bones) that articulate with each other anteriorly and with the sacrum pos-

teriorly. The sacrum, coccyx, and pelvic girdle together form the bowl-shaped **pelvis** (fig. 7.27). The pelvic girdle supports the trunk of the body, provides attachments for the lower limbs, and protects the urinary bladder, the distal end of the large intestine, and the internal reproductive organs.

Each hip bone develops from three parts—an ilium, an ischium, and a pubis (fig. 7.28). These parts fuse in the region of a cup-shaped cavity called the *acetabulum* (as″ĕ-tab′u-lum). This depression, on the lateral surface of the hip bone, receives the rounded head of the femur (thigh bone).

The **ilium** (il′e-um), which is the largest and uppermost portion of the hip bone, flares outward, forming the prominence of the hip. The margin of this prominence is called the *iliac crest*.

Posteriorly, the ilium joins the sacrum at the *sacroiliac* (sa″kro-il′e-ak) *joint*. Anteriorly, a projection of the ilium, the *anterior superior iliac spine,* can be felt lateral to the groin and provides attachments for ligaments and muscles.

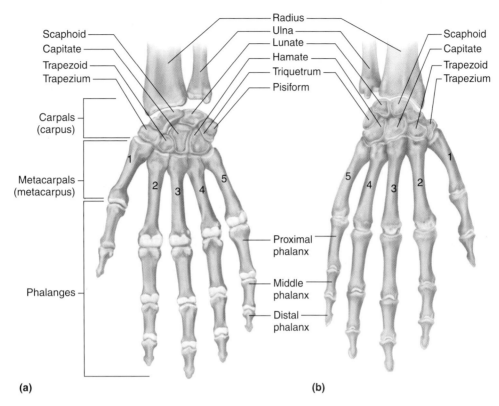

Figure 7.26

Right hand. **(a)** Anterior view. **(b)** Posterior view.

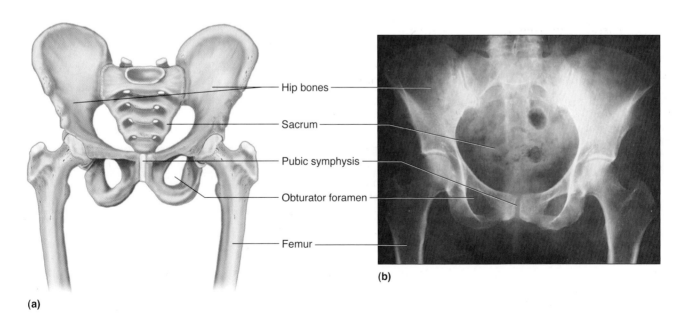

Figure 7.27

Pelvic girdle. **(a)** The pelvic girdle is formed by two hip bones. The pelvis includes the pelvic girdle as well as the sacrum and the coccyx. **(b)** Radiograph of the pelvic girdle showing the sacrum, coccyx, and proximal ends of the femurs.

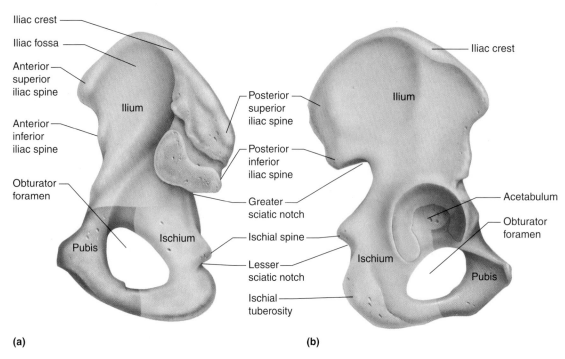

Figure 7.28

Right hip bone. **(a)** Medial surface. **(b)** Lateral view.

The **ischium** (is′ke-um), which forms the lowest portion of the hip bone, is L-shaped, with its angle, the *ischial tuberosity,* pointing posteriorly and downward. This tuberosity has a rough surface that provides attachments for ligaments and lower limb muscles. It also supports the weight of the body during sitting. Above the ischial tuberosity, near the junction of the ilium and ischium, is a sharp projection called the *ischial spine.* The distance between the ischial spines is the shortest diameter of the pelvic outlet.

The **pubis** (pu′bis) constitutes the anterior portion of the hip bone. The two pubic bones join at the midline, forming a joint called the *pubic symphysis* (pu′bik sim′fi-sis). The angle these bones form below the symphysis is the *pubic arch* (fig. 7.29).

A portion of each pubis passes posteriorly and downward to join an ischium. Between the bodies of these bones on either side is a large opening, the *obturator foramen,* which is the largest foramen in the skeleton (see figs. 7.27 and 7.28).

If a line were drawn along each side of the pelvis from the sacral promontory downward and anteriorly to the upper margin of the pubic symphysis, it would mark the *pelvic brim* (linea terminalis) (fig. 7.29). Table 7.3 summarizes some differences in the female and male pelves and other skeletal structures.

Table 7.3	Differences Between the Female and Male Skeletons
Part	**Differences**
Skull	Female skull is smaller and lighter, with less conspicuous muscular attachments. Female facial area is rounder, jaw is smaller, and mastoid process is less prominent than those of a male.
Pelvic girdle	Female hip bones are lighter, thinner, and have less obvious muscular attachments. The female obturator foramina and acetabula are smaller and farther apart than those of a male.
Pelvic cavity	Female pelvic cavity is wider in all diameters and is shorter, roomier, and less funnel-shaped. The distances between the female ischial spines and ischial tuberosities are greater than in a male.
Sacrum	Female sacrum is wider, the first sacral vertebra projects forward to a lesser degree, and the sacral curvature is bent more sharply posteriorly than in a male.
Coccyx	Female coccyx is more movable than that of a male.

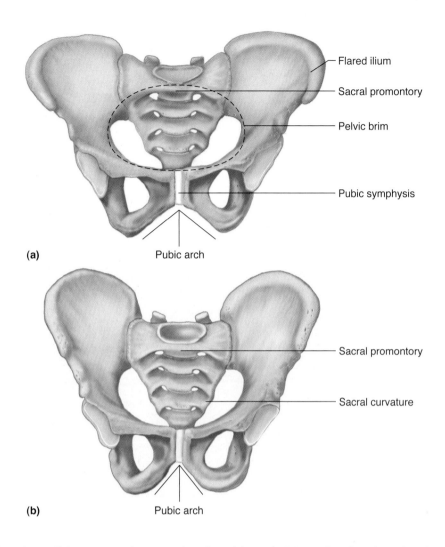

- Flared ilium
- Sacral promontory
- Pelvic brim
- Pubic symphysis

(a) Pubic arch

- Sacral promontory
- Sacral curvature

(b) Pubic arch

Figure 7.29

The female pelvis is usually wider in all diameters and roomier than that of the male. **(a)** Female pelvis. **(b)** Male pelvis.

Q: *What are some of the specific differences between the male pelvis and female pelvis?*

Answer can be found in Appendix E on page 568.

Practice

27. Locate and name each bone that forms the pelvis.
28. Name the bones that fuse to form a hip bone.

7.12 LOWER LIMB

Bones of the lower limb form the frameworks of the thigh, leg, and foot. They include a femur, a tibia, a fibula, tarsals, metatarsals, and phalanges (see fig. 7.9).

Femur

The **femur,** or thigh bone, is the longest bone in the body and extends from the hip to the knee (fig. 7.30). A large, rounded *head* at its proximal end projects medi-

ally into the acetabulum of the hip bone. On the head, a pit called the *fovea capitis* marks the attachment of a ligament (ligamentum capitis). Just below the head are a constriction, or *neck,* and two large processes—a superior, lateral *greater trochanter* and an inferior, medial *lesser trochanter.* These processes provide attachments for muscles of the lower limbs and buttocks.

The strongest bone in the body, the femur, is hollow. Ounce for ounce, it has greater pressure tolerance and bearing strength than a rod of equivalent size made of cast steel.

At the distal end of the femur, two rounded processes, the *lateral* and *medial condyles,* articulate with the tibia of the leg. A **patella,** or kneecap, also articulates with the femur on its distal anterior surface (see

Figure 7.30

Right femur. **(a)** Anterior surface. **(b)** Posterior surface.

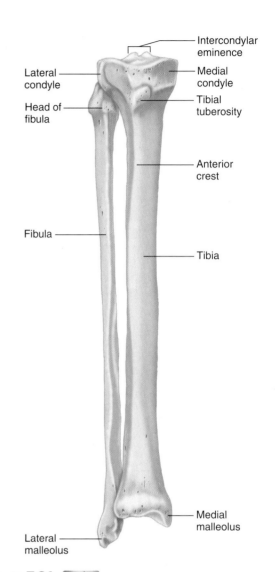

Figure 7.31 AP|R

Right tibia and fibula, anterior view.

fig. 7.9). It is located in a tendon that passes anteriorly over the knee.

Hip fracture is one of the more serious causes of hospitalization among elderly persons. The site of hip fracture is most commonly the neck of a femur or the region between the trochanters of a femur. Often a hip fracture is a cause of a fall, rather than the result of a fall.

Tibia

The **tibia,** or shin bone, is the larger of the two leg bones and is located on the medial side (fig. 7.31). Its proximal end is expanded into *medial* and *lateral condyles,* which have concave surfaces and articulate with the condyles of the femur. Below the condyles, on the anterior surface, is

a process called the *tibial tuberosity,* which provides an attachment for the *patellar ligament*—a continuation of the patella-bearing tendon.

At its distal end, the tibia expands to form a prominence on the inner ankle called the *medial malleolus* (mah-le'o-lus), which is an attachment for ligaments. On its lateral side is a depression that articulates with the fibula. The inferior surface of the tibia's distal end articulates with a large bone (the talus) in the ankle.

Fibula

The **fibula** is a long, slender bone located on the lateral side of the tibia (fig. 7.31). Its ends are slightly enlarged into a proximal *head* and a distal *lateral malleolus.* The head articulates with the tibia just below the lateral condyle; however, it does not enter into the knee joint

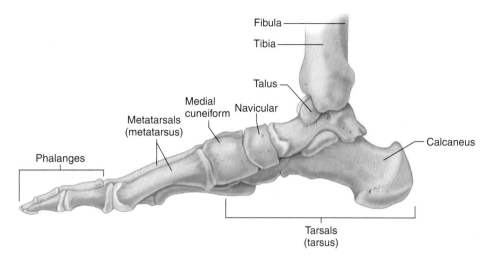

Figure 7.32

Right foot. The talus moves freely where it articulates with the tibia and fibula.

and does not bear any body weight. The lateral malleolus articulates with the ankle and protrudes on the lateral side.

Foot

The foot is made up of the ankle, the instep, and the toes. The ankle, or *tarsus* (tahr′sus), is composed of seven **tarsal bones** (figs. 7.32 and 7.33). These bones are arranged so that one of them, the **talus** (ta′lus), can move freely where it joins the tibia and fibula. The remaining tarsal bones are bound firmly together, forming a mass supporting the talus. Figure 7.33 names the individual bones of the tarsus.

The largest of the tarsals, the **calcaneus** (kal-ka′ne-us), or heel bone, is below the talus, where it projects backward to form the base of the heel. The calcaneus helps support body weight and provides an attachment for the muscles that move the foot.

The instep, or *metatarsus* (met″ah-tar′sus), consists of five elongated **metatarsal bones** that articulate with the tarsus. They are numbered 1–5, beginning on the medial side (fig. 7.33). The heads at the distal ends of these bones form the ball of the foot. The tarsals and metatarsals are arranged and bound by ligaments to form the arches of the foot. A longitudinal arch extends from the heel to the toe, and a transverse arch stretches across the foot. These arches provide a stable, springy base for the body. Sometimes, however, the tissues that bind the metatarsals weaken, producing fallen arches, or flat feet.

The **phalanges** of the toes, which are similar to those of the fingers, align and articulate with the metatarsals. Each toe has three phalanges—a proximal, a middle, and a distal phalanx—except the great toe, which lacks a middle phalanx.

Practice

29. Locate and name each of the bones of the lower limb.
30. Explain how the bones of the lower limb articulate.
31. Describe how the foot is adapted to support the body.

Figure 7.33

Right foot, viewed superiorly.

7.13 JOINTS

Joints (articulations) are functional junctions between bones. They bind parts of the skeletal system, make possible bone growth, permit parts of the skeleton to change shape during childbirth, and enable the body to move in response to skeletal muscle contractions. Joints vary considerably in structure and function. If classified according to the degree of movement they make possible, joints can be immovable (synarthrotic), slightly movable (amphiarthrotic), or freely movable (diarthrotic). Joints also can be grouped according to the type of tissue (fibrous, cartilaginous, or synovial) that binds the bones together at each junction. Currently, this structural classification by tissue type is more commonly used.

> Typically an adult human body has 230 joints.

Fibrous Joints

Fibrous (fi′brus) **joints** lie between bones that closely contact one another. A thin layer of dense connective tissue joins the bones at such joints, as in a *suture* between a pair of flat bones of the skull (fig. 7.34). Generally, no appreciable movement (synarthrotic) takes place at a fibrous joint. Some fibrous joints, such as the joint in the leg between the distal ends of the tibia and fibula, have limited movement (amphiarthrotic).

Cartilaginous Joints

Hyaline cartilage, or fibrocartilage, connects the bones of **cartilaginous** (kar″tĭ-lah′jin-us) **joints.** For example, joints of this type separate the vertebrae of the vertebral column. Each intervertebral disc is composed of a band of fibrocartilage (annulus fibrosus) surrounding a pulpy or gelatinous core (nucleus pulposus). The disc absorbs shocks and helps equalize pressure between vertebrae when the body moves (see fig. 7.17).

Due to the slight flexibility of the discs, cartilaginous joints allow limited movement (amphiarthrotic), as when the back is bent forward or to the side or is twisted. Other examples of cartilaginous joints include the pubic symphysis and the first rib with the sternum.

Synovial Joints

Most joints in the skeletal system are **synovial** (sĭ-no′ve-al) **joints,** which allow free movement (diarthrotic). They are more complex structurally than fibrous or cartilaginous joints.

(a)

Connective tissue

(b)

Figure 7.34

Fibrous joints. **(a)** The fibrous joints between the bones of the skull are immovable and are called sutures. **(b)** A thin layer of connective tissue connects the bones at the suture.

The articular ends of the bones in a synovial joint are covered with a thin layer of hyaline cartilage (articular cartilage). A tubular capsule of dense connective tissue holds the bones of a synovial joint together. This *joint capsule* is composed of an outer layer of ligaments and an inner lining of *synovial membrane,* which secretes *synovial fluid* (fig. 7.35). Synovial fluid has a consistency similar to that of uncooked egg white, enabling it to lubricate joints.

Some synovial joints have flattened, shock-absorbing pads of fibrocartilage called **menisci** (mě-nis′ke) (singular, *meniscus*) between the articulating surfaces of the bones (fig. 7.36). Such joints may also have fluid-filled sacs called **bursae** (ber′se) associated with them. Each bursa is lined with synovial membrane, which may be continuous with the synovial membrane of a nearby joint cavity. Bursae are commonly located between tendons and underlying bony prominences, as in the patella of the knee or the olecranon process of the elbow. They aid the movement of tendons that glide over these bony parts or over other tendons. Figure 7.36

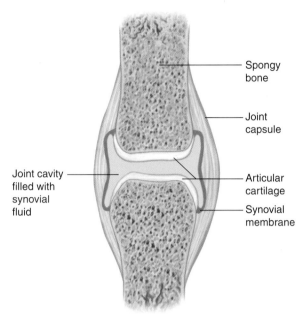

Figure 7.35
The generalized structure of a synovial joint.

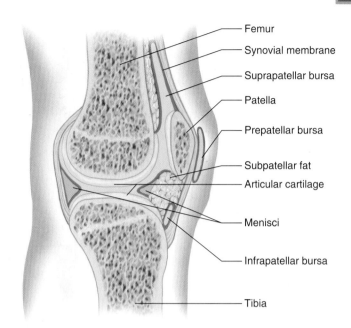

Figure 7.36
Menisci separate the articulating surfaces of the femur and tibia. Several bursae are associated with the knee joint.

shows and names some of the bursae associated with the knee.

Based on the shapes of their parts and the movements they permit, synovial joints are classified as follows:

1. A **ball-and-socket joint,** or **spheroidal joint,** consists of a bone with a globular or slightly egg-shaped head that articulates with the cup-shaped cavity of another bone. Such a joint allows the widest range of motion, permitting movements in all planes, as well as rotational movement around a central axis. The shoulder and hip have joints of this type (fig. 7.37*a*).
2. In a **condylar joint,** or **ellipsoidal joint,** the ovoid condyle of one bone fits into the elliptical cavity of another bone, as in the joints between the metacarpals and phalanges (fig. 7.37*b*). This type of joint permits a variety of movements in different planes; rotational movement, however, is not possible.
3. The articulating surfaces of **plane joints,** or **gliding joints,** are nearly flat or slightly curved. Most of the joints in the wrist and ankle, as well as those between the articular processes of adjacent vertebrae, belong to this group (fig. 7.37*c*). They allow sliding and twisting movements. The sacroiliac joints and the joints formed by ribs 2–7 connecting with the sternum are also plane joints.
4. In a **hinge joint,** the convex surface of one bone fits into the concave surface of another, as in the elbow and the joints of the phalanges (fig. 7.37*d*). Such a joint resembles the hinge of a door in that it permits movement in one plane only.
5. In a **pivot joint,** or **trochoid joint,** the cylindrical surface of one bone rotates within a ring formed of bone and ligament. Movement is limited to the rotation around a central axis. The joint between the atlas and the dens of the axis is of this type (fig. 7.37*e*).
6. A **saddle joint,** or **sellar joint,** forms between bones whose articulating surfaces have both concave and convex regions. The surface of one bone fits the complementary surface of the other. This physical relationship permits a variety of movements, as in the joint between the carpal (trapezium) and metacarpal of the thumb (fig. 7.37*f*).

Table 7.4 summarizes the types of joints. Clinical Application 7.2 discusses injuries and conditions that affect the joints.

Types of Joint Movements

Skeletal muscle action produces movements at synovial joints. Typically, one end of a muscle is attached to a relatively immovable or fixed part on one side of a joint, and the other end of the muscle is fastened to a

(a) Ball-and-socket joint

Hip bone

Head of femur in acetabulum

Femur

(b) Condylar joint

Metacarpal

Phalanx

(c) Plane joint

Carpals

(d) Hinge joint

Humerus

Radius

Ulna

(e) Pivot joint

Dens

Transverse ligament

Atlas

Axis

(f) Saddle joint

First metacarpal

Trapezium

Figure 7.37 AP|R

Types and examples of synovial (freely movable) joints.

Table 7.4	Types of Joints		
Type of Joint	**Description**	**Possible Movements**	**Examples**
Fibrous	Articulating bones are fastened together by a thin layer of dense connective tissue.	None or slight twisting	Suture between bones of skull, joint between the distal ends of tibia and fibula
Cartilaginous	Articulating bones are connected by hyaline cartilage or fibrocartilage.	Limited movement, as when back is bent or twisted	Joints between the bodies of vertebrae, pubic symphysis
Synovial	Articulating ends of bones are surrounded by a joint capsule of ligaments and synovial membranes; ends of articulating bones are covered by hyaline cartilage and separated by synovial fluid.	Allow free movement (see the following list)	
1. Ball-and-socket	Ball-shaped head of one bone articulates with cup-shaped cavity of another.	Movements in all planes and rotation	Shoulder, hip
2. Condylar	Oval-shaped condyle of one bone articulates with elliptical cavity of another.	Variety of movements in different planes, but no rotation	Joints between the metacarpals and phalanges
3. Plane	Articulating surfaces are nearly flat or slightly curved.	Sliding or twisting	Joints between various bones of wrist and ankle, sacroiliac joints, joints between ribs 2–7 and sternum
4. Hinge	Convex surface of one bone articulates with concave surface of another.	Flexion and extension	Elbow, joints of phalanges
5. Pivot	Cylindrical surface of one bone articulates with ring of bone and ligament.	Rotation around a central axis	Joint between the atlas and dens of the axis
6. Saddle	Articulating surfaces have both concave and convex regions; the surface of one bone fits the complementary surface of another.	Variety of movements, mainly in two planes	Joint between the carpal and metacarpal of thumb

movable part on the other side. When the muscle contracts, its fibers pull its movable end (*insertion*) toward its fixed end (*origin*) and a movement occurs at the joint. The following terms describe movements at joints (figs. 7.38, 7.39, and 7.40).

flexion (flek′shun) Bending parts at a joint so that the angle between them decreases and the parts come closer together (bending the knee).

extension (ek-sten′shun) Moving parts at a joint so that the angle between them increases and the parts move farther apart (straightening the knee).

dorsiflexion (dor″sĭ-flek′shun) Movement at the ankle that brings the foot closer to the shin (walking on heels).

plantar flexion (plan′tar flek′shun) Movement at the ankle that brings the foot farther from the shin (walking or standing on toes).

hyperextension (hi″per-ek-sten′shun) A term sometimes used to describe the extension of the parts at a joint beyond the anatomical position (bending the head back beyond the upright position); often used to describe an abnormal extension beyond the normal range of motion, resulting in injury.

abduction (ab-duk′shun) Moving a part away from the midline (lifting the upper limb horizontally to form a right angle with the side of the body).

adduction (ah-duk′shun) Moving a part toward the midline (returning the upper limb from the horizontal position to the side of the body).

rotation (ro-ta′shun) Moving a part around an axis (twisting the head from side to side).

circumduction (ser″kum-duk′shun) Moving a part so that its end follows a circular path (moving the finger in a circular motion without moving the hand).

pronation (pro-na′shun) Turning the hand so that the palm is downward or facing posteriorly (in anatomical position).

supination (soo″pĭ-na′shun) Turning the hand so that the palm is upward or facing anteriorly (in anatomical position).

eversion (e-ver′zhun) Turning the foot so the plantar surface faces laterally.

inversion (in-ver′zhun) Turning the foot so the plantar surface faces medially.

retraction (re-trak′shun) Moving a part backward (pulling the head backward).

protraction (pro-trak′shun) Moving a part forward (thrusting the head forward).

Joint Disorders

Joints have a tough job. They must support weight, provide a great variety of body movements, and are used frequently. In addition to this normal wear and tear, these structures are sometimes subjected to injury from trauma, overuse, infection, a misplaced immune system attack, or degeneration. The following joint problems are common.

Sprains

Sprains result from overstretching or tearing the connective tissues, including cartilage, ligaments, and tendons, that are associated with a joint, but they do not dislocate the articular bones. Sprains of the wrist and ankle usually result from forceful wrenching or twisting. For example, inverting an ankle too far can sprain it by stretching the ligaments on its lateral surface. Severe injuries may pull these tissues loose from their attachments.

A sprained joint is painful and swollen, restricting movement. Immediate treatment of a sprain is rest; more serious cases require medical attention. However, immobilization of a joint, even for a brief period, causes bone resorption and weakens ligaments. Consequently, exercise may help strengthen the joint.

Bursitis

Overuse of a joint or stress on a bursa may cause *bursitis,* an inflammation of a bursa. The bursa between the calcaneus and the Achilles tendon may become inflamed as a result of a sudden increase in physical activity using the feet. Similarly, a form of bursitis called tennis elbow affects the bursa between the olecranon process and the skin. Bursitis is treated with rest. Medical attention may be necessary.

Arthritis

The term *arthritis* covers a group of disorders that cause inflamed, swollen, and painful joints. More than a hundred different types of arthritis affect millions of people worldwide. The most common types of arthritis are rheumatoid arthritis (RA), osteoarthritis (OA), and Lyme arthritis.

Rheumatoid Arthritis (RA)

Rheumatoid arthritis, an autoimmune disorder (a condition in which the immune system attacks the body's healthy tissues), is painful and debilitating. The synovial membrane of a joint becomes inflamed and thickened. Then the articular cartilage is damaged, and fibrous tissue infiltrates, interfering with joint movements. Over time the joints ossify, fusing the articulating bones. RA may affect many joints or only a few. It is often accompanied by muscular atrophy, fatigue, and other symptoms.

Several types of drugs are used to treat RA. They include nonsteroidal anti-inflammatory drugs (NSAIDs) such as aspirin and ibuprofen, and the COX-2 inhibitors, which relieve inflammation without the gastrointestinal side effects; corticosteroids; and disease-modifying antirheumatic drugs, which are the only ones that actually slow the course of the disease. Joints severely damaged by RA may be surgically replaced with synthetic joints.

Osteoarthritis (OA)

Osteoarthritis is a degenerative disorder that may result from aging or a poorly healed injury, or it may be inherited. Articular cartilage softens and disintegrates gradually, roughening the articular surfaces. Joints become painful, and movement becomes restricted. OA usually affects the most active joints, such as those of the fingers, hips, knees, and the lower parts of the vertebral column.

NSAIDs are used to treat osteoarthritis. Some people find relief with glucosamine and hyaluronic acid supplements, but these substances have not been evaluated in clinical trials long enough to show long-term benefits. Exercise can keep osteoarthritic joints more flexible.

Lyme Arthritis

Lyme disease is a bacterial infection contracted from a tick bite. It causes intermittent arthritis of several joints, usually weeks after the initial symptoms of rash, fatigue, and flulike aches and pains. Lyme arthritis was first observed in Lyme, Connecticut, where an astute woman alerted a prominent rheumatologist to the fact that many of her young neighbors had what appeared to be the very rare juvenile form of rheumatoid arthritis. Researchers then traced the illness to a tick-borne bacterial infection. Antibiotic treatment from the onset of symptoms can prevent the arthritis. Other types of bacteria can cause arthritis, too.

Figure 7.38 AP|R

Joint movements illustrating abduction, adduction, hyperextension, extension, flexion, dorsiflexion, and plantar flexion.

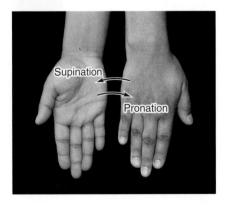

Figure 7.39 AP|R

Joint movements illustrating circumduction, rotation, supination, and pronation.

Figure 7.40 AP|R
Joint movements illustrating inversion, eversion, protraction, retraction, elevation, and depression.

elevation (el″ĕ-va′shun) Raising a part (shrugging the shoulders).
depression (de-presh′un) Lowering a part (drooping the shoulders).

Practice

32. Describe the characteristics of the three major types of joints.

33. Name six types of synovial joints.

34. What terms describe movements possible at synovial joints?

Injuries to the elbow, shoulder, and knee are commonly diagnosed and treated using a procedure called *arthroscopy.* Arthroscopy enables a surgeon to visualize the interior of a joint and perform diagnostic or therapeutic procedures, guided by the image on a video screen. An arthroscope is a thin, tubular instrument about 25 centimeters long containing optical fibers that transmit an image. The surgeon inserts the device through a small incision in the joint capsule. Arthroscopy is far less invasive than conventional surgery. Many runners have undergone uncomplicated arthroscopy and raced just weeks later.

Skeletal System

Integumentary System

Vitamin D, activated in the skin, plays a role in calcium absorption and availability for bone matrix.

Lymphatic System

Cells of the immune system originate in the bone marrow.

Muscular System

Muscles pull on bones to cause movement.

Digestive System

Absorption of dietary calcium provides material for bone matrix.

Nervous System

Proprioceptors sense the position of body parts. Pain receptors warn of trauma to bone. Bones protect the brain and spinal cord.

Respiratory System

Ribs and muscles work together in breathing.

Endocrine System

Some hormones act on bone to help regulate blood calcium levels.

Urinary System

The kidneys and bones work together to help regulate blood calcium levels.

Cardiovascular System

Blood transports nutrients to bone cells. Bone helps regulate plasma calcium levels, important to heart function.

Reproductive System

The pelvis helps support the uterus during pregnancy. Bones provide a source of calcium during lactation.

Bones provide support, protection, and movement and also play a role in calcium balance.

Summary Outline

7.1 Introduction (p. 133)

Individual bones are the organs of the skeletal system. A bone contains active tissues.

7.2 Bone Structure (p. 133)

Bone structure reflects its function.

1. Bones are classified according to their shapes, including long, short, flat, and irregular.
2. Parts of a long bone
   a. Epiphyses at each end are covered with articular cartilage and articulate with other bones.
   b. The shaft of a bone is called the diaphysis.
   c. Except for the articular cartilage, a bone is covered by a periosteum.
   d. Compact bone has a continuous extracellular matrix with no gaps.
   e. Spongy bone has irregular interconnecting spaces between bony plates, trabeculae, that reduce the weight of bone.
   f. Both compact and spongy bone are strong and resist bending.
   g. The diaphysis contains a medullary cavity filled with marrow.
3. Microscopic structure
   a. Compact bone contains osteons cemented together.
   b. Central canals contain blood vessels that nourish the cells of osteons.
   c. Diffusion from the surface of the thin, bony plates nourishes the cells of spongy bone.

7.3 Bone Development and Growth (p. 135)

1. Intramembranous bones
   a. Intramembranous bones develop from layers of unspecialized connective tissues.
   b. Osteoblasts within the membranous layers form bone tissue.
   c. Mature bone cells are called osteocytes.
2. Endochondral bones
   a. Endochondral bones develop as hyaline cartilage that is later replaced by bone tissue.
   b. The primary ossification center appears in the diaphysis, whereas secondary ossification centers appear in the epiphyses.
   c. An epiphyseal plate remains between the primary and secondary ossification centers.
   d. The epiphyseal plates are responsible for lengthening.
   e. Long bones continue to lengthen until the epiphyseal plates are ossified.
   f. Growth in thickness is due to ossification beneath the periosteum.
3. Homeostasis of bone tissue
   a. Osteoclasts break down bone matrix, and osteoblasts deposit bone matrix to continually remodel bone.
   b. The total mass of bone remains nearly constant.
4. Factors affecting bone development, growth, and repair include nutrition, hormonal secretions, and physical exercise.

7.4 Bone Function (p. 137)

1. Support and protection
   a. Bones shape and form body structures.
   b. Bones support and protect softer underlying tissues.
2. Body movement
   a. Bones and muscles function together as levers.
   b. A lever consists of a bar, a pivot (fulcrum), a resistance, and a force that supplies energy.
3. Blood cell formation
   a. At different ages, hematopoiesis occurs in the yolk sac, liver and spleen, and red bone marrow.
   b. Red marrow houses developing red blood cells, white blood cells, and blood platelets. Yellow marrow stores fat.

4. Storage of inorganic salts
   a. Bones store calcium in the extracellular matrix of bone tissue, which contains large quantities of calcium phosphate.
   b. When blood calcium is low, osteoclasts break down bone, releasing calcium salts. When blood calcium is high, osteoblasts form bone tissue and store calcium salts.
   c. Bone stores small amounts of magnesium, sodium, potassium, and carbonate ions.

7.5 Skeletal Organization (p. 142)

1. The skeleton can be divided into axial and appendicular portions.
2. The axial skeleton consists of the skull, hyoid bone, vertebral column, and thoracic cage.
3. The appendicular skeleton consists of the pectoral girdle, upper limbs, pelvic girdle, and lower limbs.

7.6 Skull (p. 144)

The skull consists of twenty-two bones: eight cranial bones and fourteen facial bones.

1. Cranium
   a. The cranium encloses and protects the brain.
   b. Some cranial bones contain air-filled paranasal sinuses.
   c. Cranial bones include the frontal bone, parietal bones, occipital bone, temporal bones, sphenoid bone, and ethmoid bone.
2. Facial skeleton
   a. Facial bones form the basic shape of the face and provide attachments for muscles.
   b. Facial bones include the maxillae, palatine bones, zygomatic bones, lacrimal bones, nasal bones, vomer bone, inferior nasal conchae, and mandible.
3. Infantile skull
   a. Fontanels connect incompletely developed bones.
   b. The proportions of the infantile skull are different from those of an adult skull.

7.7 Vertebral Column (p. 149)

The vertebral column extends from the skull to the pelvis and protects the spinal cord. It is composed of vertebrae separated by intervertebral discs.

1. A typical vertebra
   a. A typical vertebra consists of a body and a bony vertebral arch, which surrounds the spinal cord.
   b. Notches on the upper and lower surfaces provide intervertebral foramina through which spinal nerves pass.
2. Cervical vertebrae
   a. Transverse processes bear transverse foramina.
   b. The atlas (first vertebra) supports and balances the head.
   c. The dens of the axis (second vertebra) provides a pivot for the atlas when the head is turned from side to side.
3. Thoracic vertebrae
   a. Thoracic vertebrae are larger than cervical vertebrae.
   b. Facets on the sides articulate with the ribs.
4. Lumbar vertebrae
   a. The vertebral bodies are large and strong.
   b. They support more body weight than other vertebrae.
5. Sacrum
   a. The sacrum is a triangular structure formed of five fused vertebrae.
   b. Vertebral foramina form the sacral canal.
6. Coccyx
   a. The coccyx, composed of four fused vertebrae, forms the lowest part of the vertebral column.
   b. It acts as a shock absorber when a person sits.

7.8 Thoracic Cage (p. 153)

The thoracic cage includes the ribs, thoracic vertebrae, sternum, and costal cartilages. It supports the pectoral girdle and upper limbs, protects viscera, and functions in breathing.

1. Ribs
   a. Twelve pairs of ribs attach to the twelve thoracic vertebrae.
   b. Costal cartilages of the true ribs join the sternum directly. Those of the false ribs join it indirectly or not at all.
   c. A typical rib has a shaft, a head, and tubercles that articulate with the vertebrae.
2. Sternum
   a. The sternum consists of a manubrium, body, and xiphoid process.
   b. It articulates with the clavicles.

7.9 Pectoral Girdle (p. 155)

The pectoral girdle is composed of two clavicles and two scapulae. It forms an incomplete ring that supports the upper limbs and provides attachments for muscles.

1. Clavicles
   a. Clavicles are rodlike bones located between the sternum and the scapulae.
   b. They hold the shoulders in place and provide attachments for muscles.
2. Scapulae
   a. The scapulae are broad, triangular bones.
   b. They articulate with the humerus of each upper limb and provide attachments for muscles.

7.10 Upper Limb (p. 155)

Bones of the upper limb form the framework, provide the attachments for muscles, and function in levers that move the limb and its parts.

1. Humerus
   a. The humerus extends from the scapula to the elbow.
   b. It articulates with the radius and ulna at the elbow.
2. Radius
   a. The radius is located on the thumb side of the forearm between the elbow and the wrist.
   b. It articulates with the humerus, ulna, and wrist.
3. Ulna
   a. The ulna is longer than the radius and overlaps the humerus posteriorly.
   b. It articulates with the radius laterally and with a disc of fibrocartilage inferiorly.
4. Hand
   a. The wrist is composed of eight carpal bones that form a carpus.
   b. The palm or metacarpus includes five metacarpal bones and fourteen phalanges compose the fingers.

7.11 Pelvic Girdle (p. 158)

The pelvic girdle consists of two hip bones that articulate with each other anteriorly and with the sacrum posteriorly.

1. The sacrum, coccyx, and pelvic girdle form the bowl-shaped pelvis.
2. Each hip bone consists of an ilium, ischium, and pubis, which are fused in the region of the acetabulum.
   a. The ilium
      (1) The ilium is the largest portion of the hip bone.
      (2) It joins the sacrum at the sacroiliac joint.
   b. The ischium
      (1) The ischium is the lowest portion of the hip bone.
      (2) It supports the body weight when sitting.
   c. The pubis
      (1) The pubis is the anterior portion of the hip bone.
      (2) The pubic bones are fused anteriorly at the pubic symphysis.

7.12 Lower Limb (p. 161)

Bones of the lower limb provide frameworks of the thigh, leg, and foot.

1. Femur
   a. The femur extends from the hip to the knee.
   b. The patella articulates with the femur's anterior surface.
2. Tibia
   a. The tibia is located on the medial side of the leg.
   b. It articulates proximally with the femur and distally with the talus of the ankle.
3. Fibula
   a. The fibula is located on the lateral side of the tibia.
   b. It articulates with the ankle but does not bear body weight.
4. Foot
   a. The ankle consists of the tarsus formed by the talus and six other tarsals.
   b. The instep or metatarsus includes five metatarsals, and fourteen phalanges compose the toes.

7.13 Joints (p. 164)

Joints can be classified according to degree of movement as well as according to the type of tissue that binds the bones together.

1. Fibrous joints
   a. Bones at fibrous joints are tightly joined by a layer of dense connective tissue.
   b. Little (amphiarthrotic) or no movement (synarthrotic) occurs at a fibrous joint.
2. Cartilaginous joints
   a. A layer of cartilage joins the bones of cartilaginous joints.
   b. Such joints allow limited movement (amphiarthrotic).
3. Synovial joints
   a. The bones of a synovial joint are covered with hyaline cartilage and held together by a fibrous joint capsule.
   b. The joint capsule consists of an outer layer of ligaments and an inner lining of synovial membrane.
   c. Pads of fibrocartilage, menisci, act as shock absorbers in some synovial joints.
   d. Bursae are located between tendons and underlying bony prominences.
   e. Synovial joints that allow free movement (diarthrotic) include ball-and-socket, condylar, plane, hinge, pivot, and saddle.
4. Types of joint movements
   a. Muscles fastened on either side of a joint produce the movements of synovial joints.
   b. Joint movements include flexion, extension, dorsiflexion, plantar flexion, hyperextension, abduction, adduction, rotation, circumduction, pronation, supination, eversion, inversion, retraction, protraction, elevation, and depression.

Chapter Assessments

7.1 Introduction

1. Active, living tissues found in bone include _____. (p. 133)
   a. blood
   b. nervous tissue
   c. dense connective tissue
   d. bone tissue
   e. all of the above

7.2 Bone Structure

2. Sketch a typical long bone, and label its epiphyses, diaphysis, medullary cavity, periosteum, and articular cartilages. On the sketch, designate the locations of compact and spongy bone. (p. 133)

3. Discuss the functions of the parts labeled in the sketch you made for question 2. (p. 133)

4. Differentiate between the microscopic structure of compact bone and spongy bone. (p. 134)

7.3 Bone Development and Growth

5. Explain how the development of intramembranous bone differs from that of endochondral bone. (p. 135)

6. _____ are mature bone cells, whereas _____ are bone-forming cells and _____ are bone-resorbing cells. (p. 135)

7. Explain the function of an epiphyseal plate. (p. 136)

8. Physical exercise pulling on muscular attachments to bones stimulates _____. (p. 137)

7.4 Bone Function

9. Give several examples of how bones support and protect body parts. (p. 137)

10. List and describe other functions of bones. (p. 137)

7.5 Skeletal Organization

11. Bones of the head, neck, and trunk compose the _____ skeleton; bones of the limbs and their attachments compose the _____ skeleton. (p. 142)

7.6–7.12 (Skull–Lower Limb)

12. Name the bones of the cranium and the facial skeleton. (pp. 144–149)

13. Describe a typical vertebra, and distinguish among the cervical, thoracic, and lumbar vertebrae. (pp. 149–152)

14. Name the bones that compose the thoracic cage. (p. 153)

15. The clavicle and scapula form the _____ girdle, whereas the hip bones and sacrum form the _____ girdle. (pp. 155 and 158)

16. Name the bones of the upper and lower limbs. (pp. 155–163)

17. Match the parts listed on the left with the bones listed on the right. (pp. 144–163)

(1) Foramen magnum	A. Maxilla
(2) Mastoid process	B. Occipital bone
(3) Palatine process	C. Temporal bone
(4) Sella turcica	D. Femur
(5) Deltoid tuberosity	E. Fibula
(6) Greater trochanter	F. Humerus
(7) Lateral malleolus	G. Radius
(8) Medial malleolus	H. Sternum
(9) Radial tuberosity	I. Tibia
(10) Xiphoid process	J. Sphenoid bone

7.13 Joints

18. Describe and give an example of a fibrous joint, a cartilaginous joint, and a synovial joint. (p. 164)

19. Name an example of each type of synovial joint, and describe the parts of the joint as they relate to the movement(s) allowed by that particular joint. (p. 165)

20. Joint movements occur when a muscle contracts and the muscle fibers pull the muscle's movable end of attachment to the bone, the _____, toward its fixed end, the _____. (p. 167)

21. Match the movement on the left with the appropriate description on the right. (pp. 167–170)

(1) Rotation	A. turning palm upward
(2) Supination	B. decreasing angle between parts
(3) Extension	C. moving part forward
(4) Eversion	D. moving part around axis
(5) Protraction	E. moving part toward midline
(6) Flexion	F. turning foot so plantar surface faces laterally
(7) Pronation	G. increasing angle between parts
(8) Abduction	H. lowering a part
(9) Depression	I. turning palm downward
(10) Adduction	J. moving part away from midline

Integrative Assessments/Critical Thinking

OUTCOMES 5.3, 7.2, 7.6

1. How does the structure of a bone make it strong yet lightweight?

OUTCOMES 5.3, 7.13

2. How would you explain to an athlete why damaged joint ligaments and cartilages are so slow to heal following an injury?

OUTCOMES 7.3, 7.4, 7.11

3. Suppose archaeologists discover human skeletal remains in Ethiopia. Examination of the bones suggests that the remains represent four types of individuals. Two of the skeletons have bone densities that are 30% less than those of the other two skeletons. The skeletons with the lower bone mass also have broader front pelvic bones. Within the two groups defined by bone mass, smaller skeletons have bones with evidence of epiphyseal plates, but larger bones have only a thin line where the epiphyseal plates should be. Give the age group and gender of the individuals in this find.

OUTCOMES 7.3, 7.10, 7.12

4. When a child's bone is fractured, growth may be stimulated at the epiphyseal plate of that bone. What problems might this extra growth cause in an upper or lower limb before the growth of the other limb compensates for the difference in length?

WEB CONNECTIONS

Visit the text website at **www.mhhe.com/shieress11** for additional quizzes, interactive learning exercises, and more.

APR

Anatomy & Physiology REVEALED includes cadaver photos that allow you to peel away layers of the human body to reveal structures beneath the surface. This program also includes animations, radiologic imaging, audio pronunciation, and practice quizzing. To learn more visit **www.aprevealed.com**.

Human Skull

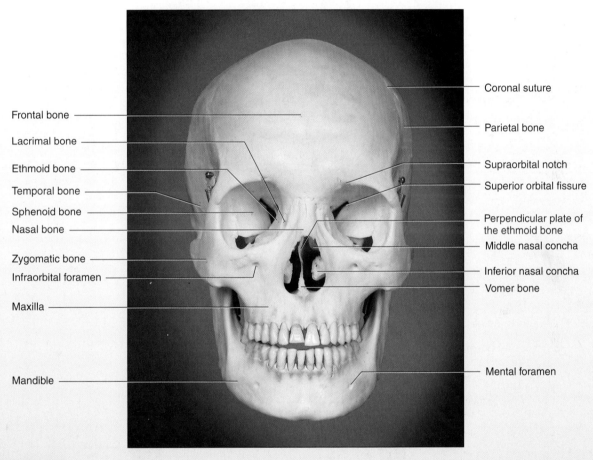

Frontal bone

Lacrimal bone

Ethmoid bone

Temporal bone

Sphenoid bone

Nasal bone

Zygomatic bone

Infraorbital foramen

Maxilla

Mandible

Coronal suture

Parietal bone

Supraorbital notch

Superior orbital fissure

Perpendicular plate of
the ethmoid bone

Middle nasal concha

Inferior nasal concha

Vomer bone

Mental foramen

PLATE EIGHT

The skull, anterior view.

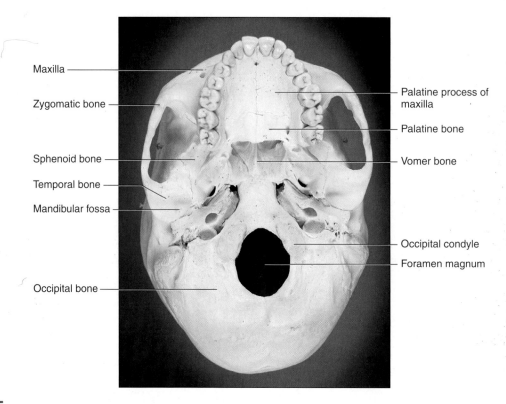

Maxilla

Zygomatic bone

Sphenoid bone

Temporal bone

Mandibular fossa

Occipital bone

Palatine process of maxilla

Palatine bone

Vomer bone

Occipital condyle

Foramen magnum

PLATE NINE

The skull, inferior view.

Frontal bone

Ethmoid bone
- Crista galli
- Cribriform plate

Temporal bone

Parietal bone

Occipital bone

Sphenoid bone

Sella turcica

Foramen magnum

PLATE TEN

The skull, floor of the cranial cavity.

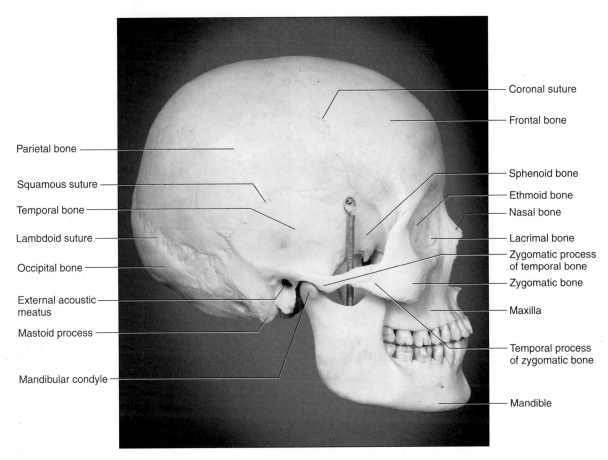

Coronal suture

Frontal bone

Parietal bone

Sphenoid bone

Squamous suture

Ethmoid bone

Temporal bone

Nasal bone

Lambdoid suture

Lacrimal bone

Occipital bone

Zygomatic process
of temporal bone

External acoustic
meatus

Zygomatic bone

Mastoid process

Maxilla

Temporal process
of zygomatic bone

Mandibular condyle

Mandible

PLATE ELEVEN
The skull, lateral view.

8

Muscular System

Double the muscle. The newborn had an astonishing appearance—his prominent arm and thigh muscles looked as if he'd been weight lifting in the womb. When the child reached five years of age, his muscles were twice normal size, and he could lift weights heavier than many adults could lift. He also had half the normal amount of body fat.

The boy's muscle cells cannot produce a protein called myostatin, which normally stops stem cells from developing into muscle cells. In this boy a mutation turned off this genetic brake, and as a result his muscles bulge, their cells both larger and more numerous than those in the muscles of a normal child. The boy is healthy so far, but because myostatin is also normally made in cardiac muscle, he may develop heart problems.

Other species with myostatin mutations are well known. Naturally "double-muscled" cattle and sheep are valued for their high weights early in life. Chicken breeders lower myostatin production to yield meatier birds, and "mighty mice" with silenced myostatin genes are used in basic research to study muscle overgrowth. In clinical applications, researchers are investigating ways to block myostatin activity to stimulate muscle growth to reverse muscle-wasting from AIDS, cancer, and muscular dystrophy. Myostatin is also of interest in athletics. Hypothetically, infants could be tested to identify those with myostatin gene variants that predict athletic prowess, given the right training. Myostatin could also be abused to enhance athletic performance.

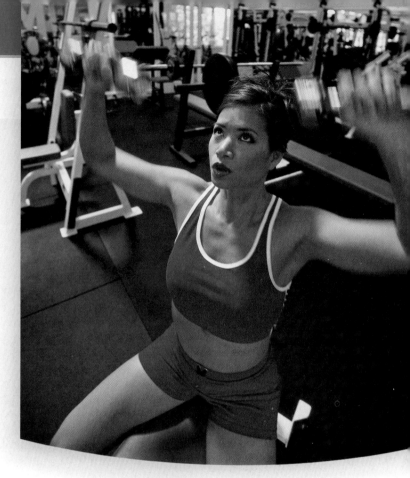

For those of us not endowed with genetically doubled muscles, regular resistance training (weight training) can strengthen muscles.

For those of us not endowed with double-muscle mutations, resistance (weight) training can increase the ratio of muscle to fat in our bodies, which offers several benefits. Because muscle cells burn calories at three times the rate of fat cells, a lean body is more energetically efficient. Weight training increases muscle strength and bone density; lowers blood pressure; decreases the risks of developing arthritis, osteoporosis, and diabetes mellitus; and is even associated with improved self-esteem and fewer sick days.

Learning Outcomes

After studying this chapter, you should be able to do the following:

8.1 Introduction
1. List various outcomes of muscle actions. (p. 179)

8.2 Structure of a Skeletal Muscle
2. Identify the structures that make up a skeletal muscle. (p. 179)
3. Identify the major parts of a skeletal muscle fiber, and the function of each. (p. 179)
4. Discuss nervous stimulation of a skeletal muscle. (p. 182)

8.3 Skeletal Muscle Contraction
5. Identify the major events of skeletal muscle fiber contraction. (p. 183)
6. List the energy sources for muscle fiber contraction. (p. 184)
7. Describe how oxygen debt develops. (p. 186)

8. Describe how a muscle may become fatigued. (p. 187)

8.4 Muscular Responses
9. Distinguish between a twitch, recruitment, and a sustained contraction. (p. 188)
10. Explain how muscular contractions move body parts and help maintain posture. (p. 190)

8.5 Smooth Muscle
11. Distinguish between the structures and functions of multiunit smooth muscle and visceral smooth muscle. (p. 191)
12. Compare the contraction mechanisms of skeletal and smooth muscle fibers. (p. 191)

8.6 Cardiac Muscle
13. Compare the contraction mechanisms of cardiac and skeletal muscle fibers. (p. 192)

8.7 Skeletal Muscle Actions
14. Explain how the attachments, locations, and interactions of skeletal muscles make different movements possible. (p. 192)

8.8 Major Skeletal Muscles
15. Identify and locate the major skeletal muscles of each body region. (pp. 194–207)
16. Identify the actions of the major skeletal muscles of each body region. (pp. 194–207)

Module 6: Muscular System

 Learn **Practice** **Assess**

Aids to Understanding Words   *(Appendix A on page 564 has a complete list of Aids to Understanding Words.)*

calat- [something inserted] inter*calat*ed disc: Membranous band that connects cardiac muscle cells.

erg- [work] syn*erg*ist: Muscle that works with a prime mover to produce a movement.

hyper- [over, more] muscular *hyper*trophy: Enlargement of muscle fibers.

inter- [between] *inter*calated disc: Membranous band that connects cardiac muscle cells.

laten- [hidden] *laten*t period: Time between application of a stimulus and the beginning of a muscle contraction.

myo- [muscle] *myo*fibril: Contractile structure within a muscle cell.

sarco- [flesh] *sarco*plasm: Material (cytoplasm) within a muscle fiber.

syn- [together] *syn*ergist: Muscle that works with a prime mover to produce a movement.

tetan- [stiff] *tetan*ic contraction: Sustained muscular contraction.

-troph [well fed] muscular hyper*troph*y: Enlargement of muscle fibers.

8.1 INTRODUCTION

Talking and walking, breathing and sneezing—in fact, all movements—require muscles. Muscles are organs composed of specialized cells that use the chemical energy stored in nutrients to pull on structures to which they are attached. Muscular actions also provide muscle tone, propel body fluids and food, generate the heartbeat, and distribute heat.

Muscles are of three types—skeletal muscle, smooth muscle, and cardiac muscle, as described in chapter 5 (pp. 110–111). This chapter focuses mostly on skeletal muscle, which attaches to bones and is under conscious control. Smooth muscle and cardiac muscle are discussed briefly.

8.2 STRUCTURE OF A SKELETAL MUSCLE

A skeletal muscle is an organ of the muscular system. It is composed of skeletal muscle tissue, nervous tissue, blood, and other connective tissues.

Connective Tissue Coverings

Layers of dense connective tissue called **fascia** (fash′e-ah) separate an individual skeletal muscle from adjacent muscles and hold it in position (fig. 8.1). This connective tissue surrounds each muscle and may project beyond its end to form a cordlike tendon. Fibers in a tendon may intertwine with those in a bone's periosteum, attaching the muscle to the bone. In other cases, the connective tissue forms broad fibrous sheets called **aponeuroses** (ap″o-nu-ro′sez), which may attach to bone or to the coverings of adjacent muscles (see figs. 8.17*a* and 8.19).

The layer of connective tissue that closely surrounds a skeletal muscle is called *epimysium* (fig. 8.1). Other layers of connective tissue, called *perimysium,* extend inward from the epimysium and separate the muscle tissue into small compartments. These compartments contain bundles of skeletal muscle fibers called *fascicles* (fasciculi). Each muscle fiber within a fascicle (fasciculus) lies within a layer of connective tissue in the form of a thin covering called *endomysium.* Layers of connective tissue, therefore, enclose and separate all parts of a skeletal muscle. This organization allows the parts to move somewhat independently. Many blood vessels and nerves pass through these layers.

In *tendinitis,* a tendon (the attachment of a muscle to a bone) becomes painfully inflamed and swollen following injury or the repeated stress of athletic activity. If rest, physical therapy, and anti-inflammatory drugs do not alleviate tendinitis, then ultrasound can be applied to break up scar tissue. In *tenosynovitis,* the connective tissue sheath of the tendon (the tenosynovium) is inflamed. The tendons most commonly affected are those associated with the joint capsules of the shoulder, elbow, and hip and those that move the hand, thigh, and foot.

Skeletal Muscle Fibers

A skeletal muscle fiber is a single cell that contracts (shortens) in response to stimulation and then relaxes when the stimulation ends. Each skeletal muscle fiber is a thin, elongated cylinder with rounded ends, and it may extend the full length of the muscle. Just beneath its cell membrane (or *sarcolemma*), the cytoplasm (or *sarcoplasm*) of the fiber has many small, oval nuclei and mitochondria (fig. 8.1). The sarcoplasm also contains many threadlike **myofibrils** (mi″o-fi′brilz) that lie parallel to one another.

Myofibrils play a fundamental role in muscle contraction. They consist of two kinds of protein filaments—thick ones composed of the protein **myosin** (mi′o-sin) and thin ones mainly composed of the protein **actin** (ak′tin) (figs. 8.2 and 8.3). (Two other thin filament proteins, troponin and tropomyosin, are discussed later on page 183.) The organization of these filaments produces the characteristic alternating light and dark *striations,* or bands, of a skeletal muscle fiber.

Figure 8.1

A skeletal muscle is composed of a variety of tissues, including layers of connective tissue. Fascia covers the surface of the muscle, epimysium lies beneath the fascia, and perimysium extends into the structure of the muscle where it separates muscle cells into fascicles. Endomysium separates individual muscle fibers.

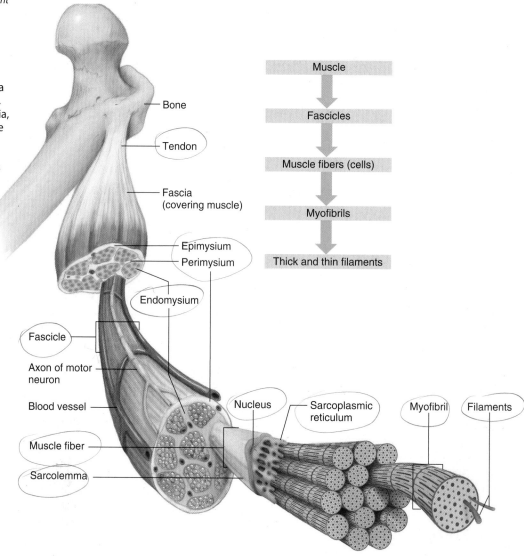

Bone
Tendon
Fascia (covering muscle)
Epimysium
Perimysium
Endomysium
Fascicle
Axon of motor neuron
Blood vessel
Nucleus
Sarcoplasmic reticulum
Myofibril
Filaments
Muscle fiber
Sarcolemma

Muscle → Fascicles → Muscle fibers (cells) → Myofibrils → Thick and thin filaments

Figure 8.2 AP|R

Skeletal muscle fiber. **(a)** A skeletal muscle fiber contains many myofibrils, each consisting of **(b)** repeating units called sarcomeres. The characteristic striations of a sarcomere reflect the organization of actin and myosin filaments.

Skeletal muscle fiber
Sarcoplasmic reticulum
Myosin (thick) filaments
Actin (thin) filaments
Myofibril
Sarcomere
Z line
H zone
Z line
M line
I band
A band
I band
A band
(a)
(b)

(a)

Sarcomere

A band

I band I band

Z line Z line

Thin (actin) filaments Thick (myosin) filaments

(b)

Figure 8.3

A sarcomere is a functional unit of muscle contraction.
(a) Micrograph (16,000×). **(b)** The spatial relationship of thin and thick filaments in a sarcomere makes contraction possible.

The striations of skeletal muscle form a repeating pattern of units called **sarcomeres** (sar′ko-měrz) along each muscle fiber. The myofibrils may be thought of as sarcomeres joined end-to-end (fig. 8.2). Muscle fibers, and in a way muscles themselves, may be considered a collection of sarcomeres. Sarcomeres are discussed later as the functional units of muscle contraction (p. 183).

The striation pattern of skeletal muscle fibers has two main parts. The first, the *I bands* (the light bands), are composed of thin actin filaments directly attached to structures called *Z lines*. The second part of the striation pattern consists of the *A bands* (the dark bands), which are composed of thick myosin filaments overlapping thin actin filaments. The A band consists of a region where the thick and thin filaments overlap, and a central region (*H zone*) consisting only of thick filaments, plus a thickening known as the *M line* (fig. 8.2). The M line consists of proteins that help hold the thick filaments in place. A sarcomere extends from one Z line to the next (figs. 8.2 and 8.3).

Within the sarcoplasm of a muscle fiber is a network of membranous channels that surrounds each myofibril and runs parallel to it (fig. 8.4). These membranes form the **sarcoplasmic reticulum,** which corresponds to the endoplasmic reticulum of other cells. Another set of membranous channels, called **transverse tubules** (T tubules), extends inward as invaginations from the fiber's membrane and passes all the way through the fiber. Thus, each tubule opens to the

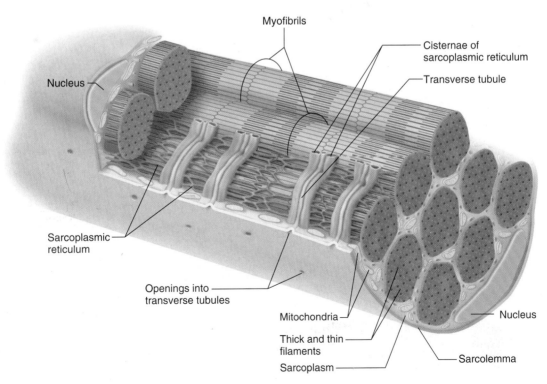

Myofibrils

Cisternae of sarcoplasmic reticulum

Transverse tubule

Nucleus

Sarcoplasmic reticulum

Openings into transverse tubules

Mitochondria

Nucleus

Thick and thin filaments

Sarcolemma

Sarcoplasm

Figure 8.4

Within the sarcoplasm of a skeletal muscle fiber are a network of sarcoplasmic reticulum and a system of transverse tubules.

outside of the muscle fiber and contains extracellular fluid. Furthermore, each transverse tubule lies between two enlarged portions of the sarcoplasmic reticulum called *cisternae,* near the region where the actin and myosin filaments overlap. The sarcoplasmic reticulum and transverse tubules activate the muscle contraction mechanism when the muscle fiber is stimulated.

> Muscle fibers and their associated connective tissues are flexible but can tear if overstretched. This type of injury, common in athletes, is called *muscle strain.* The seriousness of the injury depends on the degree of damage the tissues sustain. If the strain is mild, only a few muscle fibers are injured, the fascia remains intact, and loss of function is minimal. In a severe strain, however, many muscle fibers as well as the fascia tear, and muscle function may be completely lost. Such a severe strain is painful and produces discoloration and swelling.

Practice

1. Describe how connective tissue is part of a skeletal muscle.
2. Describe the general structure of a skeletal muscle fiber.
3. Explain why skeletal muscle fibers appear striated.
4. Explain the relationship between the sarcoplasmic reticulum and the transverse tubules.

Neuromuscular Junction

Recall from chapter 5 (p. 111) that neurons (nerve cells) play a role in body communication by conducting electrical impulses. Neurons that control effectors are called **motor neurons.** The opening vignette to chapter 3 (p. 50) describes amyotrophic lateral sclerosis (ALS, or Lou Gehrig's disease), which affects motor neurons that control skeletal muscles.

Each skeletal muscle fiber is functionally (but not physically) connected to the axon of a motor neuron that passes outward from the brain or the spinal cord, in much the same way that you can talk into a cell phone although your mouth is not in direct physical contact with it. This functional connection is called a **synapse** (sin′aps). Neurons communicate with the cells that they control by releasing chemicals, called **neurotransmitters** (nu″ro-trans′mit-erz), at synapses. Normally, a skeletal muscle fiber contracts only upon stimulation by a motor neuron.

The synapse between the motor neuron and the muscle fiber that it controls is called a **neuromuscular junction.** Here, the muscle fiber membrane is specialized to form a **motor end plate.** In this region of the muscle fiber, nuclei and mitochondria are abundant, and the cell membrane (sarcolemma) is extensively folded (fig. 8.5).

The end of the motor neuron forms fine projections into recesses of the muscle fiber membrane. The cyto-

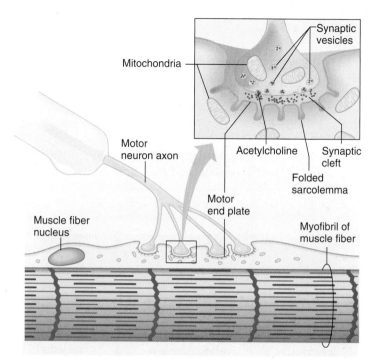

Figure 8.5 AP|R

A neuromuscular junction includes the end of a motor neuron and the motor end plate of a muscle fiber.

Q: *How does neurotransmitter released into the synaptic cleft reach the muscle fiber membrane?*

Answer can be found in Appendix E on page 568.

plasm at the distal ends of these motor neuron axons is rich in mitochondria and contains many tiny vesicles (synaptic vesicles) that store neurotransmitters.

When an electrical impulse traveling from the brain or spinal cord reaches the end of a motor neuron axon, some of the vesicles release neurotransmitter molecules into the gap (synaptic cleft) between the neuron and the motor end plate of the muscle fiber. Diffusion of neurotransmitter to the muscle fiber membrane stimulates the muscle fiber to contract.

Practice

5. Which two structures approach each other at a neuromuscular junction?
6. Describe a motor end plate.
7. What is the function of a neurotransmitter?

8.3 SKELETAL MUSCLE CONTRACTION

A muscle fiber contraction is a complex interaction of organelles and molecules in which myosin binds to actin and exerts a pulling force. The result is a movement within the myofibrils in which the filaments of actin and myosin slide past one another, increasing the amount of overlap. This action shortens the muscle fiber, which then pulls on its attachments.

Role of Myosin and Actin

A myosin molecule is composed of two twisted protein strands with globular parts called cross-bridges projecting outward along their lengths. Many of these molecules together compose a myosin (thick) filament (fig. 8.6). An actin molecule is a globular structure with a binding site to which the myosin cross-bridges can attach. Many actin molecules twist into a double strand (helix), forming an actin (thin) filament. The proteins **troponin** and **tropomyosin** are also part of the actin filament (fig. 8.6).

The sarcomere is considered the functional unit of skeletal muscles because we can describe the contraction of an entire skeletal muscle in terms of the shortening of the sarcomeres within its muscle fibers. The force that shortens the sarcomeres comes from the cross-bridges pulling on the thin filaments. A myosin cross-bridge can attach to an actin binding site and bend slightly, pulling on the actin filament. Then the head can release, straighten, combine with another binding site further down the actin filament, and pull again (fig. 8.7).

The **sliding filament model** of muscle contraction includes all of these actin-myosin interactions and is named for how the sarcomeres shorten. Thick and thin filaments do not change length. Rather, they slide past one another, with the thin filaments moving toward the center of the sarcomere from both ends (fig. 8.8).

The globular parts of the myosin filaments contain an enzyme, **ATPase,** which catalyzes the breakdown of ATP to ADP and phosphate (see chapter 4, p. 80). This reaction provides energy that puts the myosin cross-bridge in a "cocked" position. When a cocked cross-bridge binds to actin, it pulls on the thin filament. After the cross-bridge pulls, another ATP binding to the cross-bridge causes it to be released from actin even before the ATP splits. The ATPase then catalyzes the breakdown of

ATP to ADP and phosphate, putting the myosin cross-bridge in a "cocked" position again. This cycle repeats as long as ATP is available as an energy source and as long as the muscle fiber is stimulated to contract.

Stimulus for Contraction AP|R

A skeletal muscle fiber normally does not contract until a neurotransmitter stimulates it. The neurotransmitter in muscle contraction is **acetylcholine** (as″ĕ-til-ko′lēn). This neurotransmitter is synthesized in the cytoplasm of the motor neuron and stored in vesicles at the distal end of the motor neuron axons. When an impulse (described in chapter 9, p. 222) reaches the end of a motor neuron axon, some of the vesicles release their acetylcholine into the space (synaptic cleft) between the motor neuron axon and the motor end plate (see fig. 8.5).

Acetylcholine diffuses rapidly across the synaptic cleft and binds to specific protein molecules (receptors) in the muscle fiber membrane, increasing membrane permeability to sodium ions. Entry of these charged particles into the muscle cell stimulates an electrical impulse much like the impulse on the motor neuron. The impulse passes in all directions over the surface of the muscle fiber membrane and travels through the transverse tubules, deep into the fiber, until it reaches the sarcoplasmic reticulum (see fig. 8.4).

The sarcoplasmic reticulum contains a high concentration of calcium ions. In response to a muscle impulse, the membranes of the cisternae become more permeable to these ions, and the calcium ions diffuse into the sarcoplasm of the muscle fiber.

When a high concentration of calcium ions is in the sarcoplasm, troponin and tropomyosin interact in a way that exposes binding sites on actin. As a result, linkages form between the actin and myosin filaments, and the muscle fiber contracts (see figs. 8.7 and 8.8). The contraction, which also requires ATP, continues as long as nerve impulses release acetylcholine.

When nervous stimulation ceases, two events lead to muscle relaxation. First, the acetylcholine that stimulated the muscle fiber is rapidly decomposed by the enzyme **acetylcholinesterase** (as′ĕ-til-ko″lin-es′ter-ās). This enzyme is present at the neuromuscular junction on the membranes of the motor end plate. Acetylcholinesterase prevents a single impulse on a motor neuron from continuously stimulating the muscle fiber.

The second event in muscle relaxation takes place once acetylcholine is broken down and the stimulus to the muscle fiber ceases. Calcium ions are actively transported back into the sarcoplasmic reticulum, which decreases the calcium ion concentration of the sarcoplasm. The linkages between actin and myosin filaments break, and consequently, the muscle fiber relaxes. Table 8.1 summarizes the major events leading to muscle contraction and relaxation.

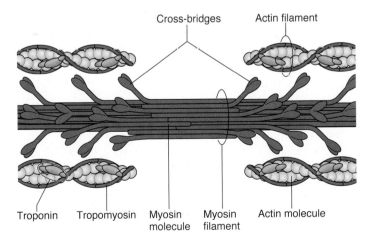

Figure 8.6

Thick filaments are composed of the protein myosin, and thin filaments are composed primarily of the protein actin. Myosin molecules have cross-bridges that extend toward nearby actin filaments.

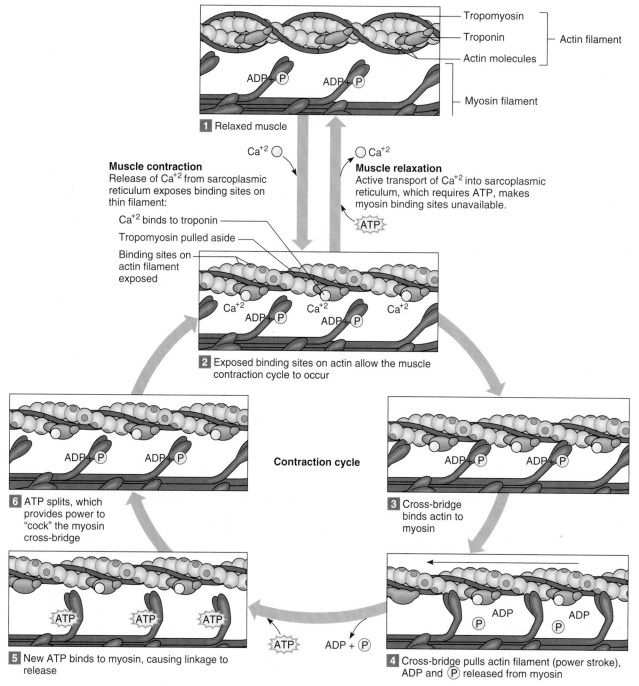

Figure 8.7 AP|R

The sliding filament model. **(1)** and **(2)** When calcium ion concentration rises, binding sites on actin filaments open, and cross-bridges attach. **(3)** and **(4)** Upon binding to actin, cross-bridges spring from the cocked position and pull on actin filaments. **(5)** ATP binds to the cross-bridge (but is not yet broken down), causing the cross-bridge to release from the actin filament. **(6)** ATP breakdown provides energy to "cock" the unattached myosin cross-bridge. As long as ATP and calcium ions are present, the cycle continues. When calcium ion concentration is low, the muscle remains relaxed.

Practice

8. Explain how an impulse on a motor neuron can trigger a muscle contraction.

9. Explain how the filaments of a myofibril interact during muscle contraction.

Energy Sources for Contraction

ATP molecules supply the energy for muscle fiber contraction. However, a muscle fiber has only enough ATP to enable it to contract for a very short time. Therefore, when a fiber is active, ATP must be regenerated.

Sarcomere

A band

Z line   Z line

Thin filaments   Thick filaments

1 Relaxed

2 Contracting

3 Fully contracted

(a)

Sarcomere

A band

Z line   Z line

(b)

Figure 8.8 AP|R

When a skeletal muscle contracts, **(a)** individual sarcomeres shorten as thin filaments slide past thick filaments toward the center of the sarcomere. **(b)** This transmission electron micrograph shows a sarcomere shortening during muscle contraction (23,000×).

Q: *What happens to the length of the thick and thin filaments during contraction?*

Answer can be found in Appendix E on page 568.

Table 8.1	Major Events of Muscle Contraction and Relaxation

Muscle Fiber Contraction

1. An impulse travels down a motor neuron axon.
2. The motor neuron terminal releases the neurotransmitter acetylcholine (ACh).
3. ACh binds to ACh receptors.
4. The sarcolemma is stimulated, and an impulse travels over the surface of the muscle fiber and deep into the fiber through the transverse tubules.
5. The muscle impulse reaches the sarcoplasmic reticulum, and calcium channels open.
6. Calcium ions diffuse from the sarcoplasmic reticulum into the sarcoplasm and bind to troponin molecules.
7. Tropomyosin molecules move and expose specific sites on actin.
8. Actin and myosin form linkages.
9. Thin (actin) filaments are pulled toward the center of the sarcomere by myosin cross-bridges.
10. The muscle fiber shortens as a contraction occurs.

Muscle Fiber Relaxation

1. Acetylcholinesterase decomposes acetylcholine, and the muscle fiber membrane is no longer stimulated.
2. Calcium ions are actively transported into the sarcoplasmic reticulum.
3. ATP breaks linkages between actin and myosin filaments without breakdown of the ATP itself.
4. Breakdown of ATP "cocks" the cross-bridges
5. Troponin and tropomyosin molecules inhibit the interaction between myosin and actin filaments.
6. The muscle fiber remains relaxed, yet ready, until stimulated again.

The initial source of energy available to a contracting muscle comes from existing ATP molecules in the cell. Almost immediately, however, cells must regenerate ATP from ADP and phosphate. The molecule that makes this possible is **creatine phosphate** (kre′ah-tin fos′fāt). Like ATP, creatine phosphate contains high-energy phosphate bonds, and it is four to six times more abundant in muscle fibers than ATP. Creatine phosphate, however, cannot directly supply energy to a cell's energy-utilizing reactions. Instead, it stores excess energy released from the mitochondria. When ATP supply is sufficient, an enzyme in the mitochondria (creatine phosphokinase) catalyzes the synthesis of creatine phosphate, which stores excess energy in its phosphate bonds (fig. 8.9).

As ATP decomposes, the energy from creatine phosphate can be transferred to ADP molecules, converting them back into ATP. Active muscle, however, rapidly

Figure 8.9

Creatine phosphate may be used to replenish ATP stores when ATP levels in a muscle cell are low.

exhausts the supply of creatine phosphate. When this happens, the muscle fibers use cellular respiration of glucose as an energy source for synthesizing ATP.

Oxygen Supply and Cellular Respiration

Glycolysis can take place in the absence of oxygen, as discussed in chapter 4 (p. 82). However, the more complete breakdown of glucose occurs in the mitochondria and requires oxygen. The blood carries the oxygen from the lungs to body cells to support this aerobic respira-

tion. Red blood cells carry the oxygen, loosely bound to molecules of **hemoglobin,** the pigment responsible for the red color of blood.

Another pigment, **myoglobin,** is synthesized in muscle cells and imparts the reddish-brown color of skeletal muscle tissue. Like hemoglobin, myoglobin can combine loosely with oxygen. This ability to temporarily store oxygen reduces a muscle's requirement for a continuous blood supply during muscular contraction (fig. 8.10).

Oxygen Debt

When a person is resting or is moderately active, the respiratory and cardiovascular systems can usually supply sufficient oxygen to skeletal muscles to support aerobic respiration. However, this is not the case when skeletal muscles are used strenuously for even a minute or two. In this situation, muscle fibers must increasingly use anaerobic respiration to obtain energy.

In one form of anaerobic respiration, glucose molecules are broken down by glycolysis to yield *pyruvic acid* (see chapter 4, p. 81). Because the oxygen supply is low, however, the pyruvic acid reacts to produce *lactic acid,* which may accumulate in the muscles (fig. 8.10). Lactic acid diffuses into the bloodstream and eventually

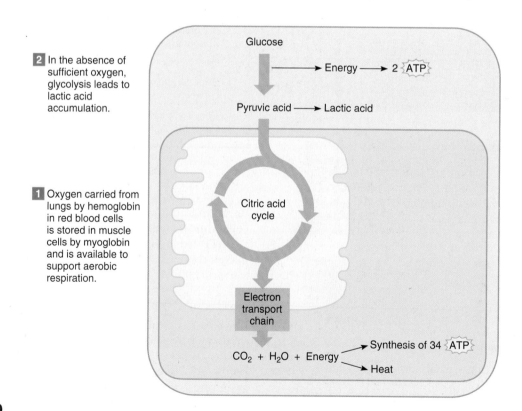

Figure 8.10

The oxygen required to support aerobic respiration is carried in the blood and stored in myoglobin. In the absence of sufficient oxygen, pyruvic acid is converted to lactic acid by anaerobic respiration. The maximum number of ATPs generated per glucose molecule varies with cell type; in skeletal muscle, it is 36 (2 + 34).

reaches the liver. In liver cells, reactions requiring ATP synthesize glucose from lactic acid.

During strenuous exercise, available oxygen is used primarily to synthesize the ATP the muscle fiber requires to contract, rather than to make ATP for synthesizing glucose from lactic acid. Consequently, as lactic acid accumulates, a person develops an **oxygen debt** (ok'sĭ-jen det) that must be repaid. Oxygen debt equals the amount of oxygen liver cells require to convert the accumulated lactic acid into glucose, plus the amount muscle cells require to restore ATP and creatine phosphate to their original concentrations. The conversion of lactic acid back into glucose is slow. Repaying an oxygen debt following vigorous exercise may take several hours.

The metabolic capacity of a muscle may change with physical training. With high-intensity exercise that depends more on glycolysis for ATP, a muscle synthesizes more glycolytic enzymes, and its capacity for glycolysis increases. With aerobic exercise, more capillaries and mitochondria form, and the muscle's capacity for aerobic respiration is greater. Table 8.2 summarizes muscle metabolism, and Clinical Application 8.1 discusses abuse of steroid drugs to enhance muscle performance.

Muscle Fatigue

A muscle exercised strenuously for a prolonged period may lose its ability to contract, a condition called *fatigue.* Interruption in the muscle's blood supply or, rarely, lack of acetylcholine in motor neuron axons may cause fatigue. However, fatigue is most likely to arise at least in part from accumulation of lactic acid in the muscle following anaerobic respiration. The lactic acid buildup lowers pH, and as a result, muscle fibers no longer respond to stimulation.

Occasionally a muscle becomes fatigued and cramps at the same time. A cramp is a painful condition in which a muscle undergoes a sustained involuntary contraction. Cramps are thought to occur when changes in the extracellular fluid surrounding the muscle fibers and their motor neurons somehow trigger uncontrolled stimulation of the muscle.

Several hours after death, skeletal muscles partially contract and become rigid, fixing the joints in place. This condition, *rigor mortis,* may continue for 72 hours or more. It results from an increase in membrane permeability to calcium ions and a decrease in ATP in muscle fibers, which prevents relaxation. The actin and myosin filaments of the muscle fibers remain linked until the muscles begin to decompose.

Heat Production

Less than half of the energy released in cellular respiration is available for use in metabolic processes; the rest becomes heat. Although all active cells generate heat, muscle tissue is a major heat source because muscle is such a large proportion of the total body mass. Blood transports heat generated in muscle to other tissues, which helps maintain body temperature.

Practice

10. Which biochemicals provide the energy to regenerate ATP?
11. What are the sources of oxygen for aerobic respiration?
12. How are lactic acid, oxygen debt, and muscle fatigue related?
13. What is the relationship between cellular respiration and heat production?

8.4 MUSCULAR RESPONSES

One way to observe muscle contraction is to remove a single muscle fiber from a skeletal muscle and connect it to a device that records changes in the fiber's length. Such experiments usually require an electrical device that can produce stimuli of varying strengths and frequencies.

Threshold Stimulus

When an isolated muscle fiber is exposed to a series of stimuli of increasing strength, the fiber remains unresponsive until a certain strength of stimulation called

Table 8.2	Muscle Metabolism	
Type of Exercise	Low to moderate intensity: Blood flow provides sufficient oxygen for cellular requirements	High intensity: Oxygen supply is not sufficient for cellular requirements
Pathway Used	Glycolysis, leading to pyruvic acid formation and aerobic respiration	Glycolysis, leading to lactic acid formation
ATP Production	36 ATP per glucose for skeletal muscle	2 ATP per glucose
Waste Product	Carbon dioxide is exhaled	Lactic acid accumulates

Clinical Application 8.1

Steroids and Athletes—An Unhealthy Combination

It seems that not a year goes by without a few famous athletes confessing to, or being caught using, steroid hormones to bulk up their muscles to improve performance. High school and college athletes abuse steroids too. Athletes who abuse steroids seek the hormone's ability to increase muscular strength. They are caught when the steroids or their breakdown products are detected in urine or when natural testosterone levels plummet in a negative feedback response to the outside hormone supply (fig. 8A). But improved performance today may have consequences tomorrow. Steroids hasten adulthood, stunting height and causing early hair loss. In males, excess steroid hormones lead to breast development, and in females to a deepened voice, hairiness, and a male physique. The kidneys, liver, and heart may be damaged, and atherosclerosis may develop because steroids raise LDL and lower HDL—the opposite of a healthy cholesterol profile. In males, the body mistakes the synthetic steroids for the natural hormone and lowers its own production of testosterone. Infertility may result. Steroids can also cause psychiatric symptoms, including delusions, depression, and violence.

Anabolic steroids have been used for medical purposes since the 1930s, to treat underdevelopment of the testes and the resulting testosterone deficiency, anemia, and muscle-wasting disorders. Today, they are used to treat wasting associated with AIDS.

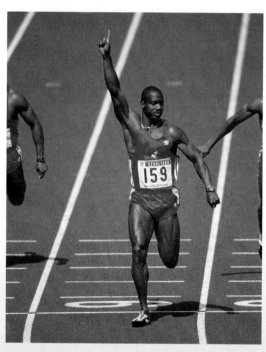

Figure 8A

Sprinter Ben Johnson ran away with the gold medal in the 100-meter race at the 1988 Summer Olympics—but then had to return the award when traces of a steroid drug showed up in his urine. Drug abuse continues to be a problem among amateur as well as professional athletes.

---

the **threshold stimulus** (thresh′old stim′u-lus) is applied. Once threshold is reached, an electrical impulse is generated, that spreads throughout the muscle fiber, releasing enough calcium ions from the sarcoplasmic reticulum to activate cross-bridge binding and contract that fiber. A single impulse in a motor neuron normally releases enough ACh to bring the muscle fibers in its motor unit to threshold, generating a muscle impulse in each muscle fiber.

Recording of a Muscle Contraction

The contractile response of a single muscle fiber to a muscle impulse is called a **twitch.** A twitch consists of a period of contraction, during which the fiber pulls at its attachments, followed by a period of relaxation, during which the pulling force declines. These events can be recorded in a pattern called a myogram (fig. 8.11). Note that a twitch has a brief delay between the time

Figure 8.11

A myogram of a single muscle twitch.

Use and Disuse of Skeletal Muscles

Skeletal muscles are very responsive to use and disuse. Forcefully exercised muscles enlarge, which is called *muscular hypertrophy.* Conversely, an unused muscle undergoes *atrophy,* decreasing in size and strength.

The way a muscle responds to use also depends on the type of exercise. A muscle contracting with lower intensity, during swimming or running, activates *slow-twitch fibers,* which are oxidative and thus fatigue-resistant. With use, these specialized muscle fibers develop more mitochondria, and more extensive capillary networks envelop them. Such changes increase the slow-twitch fibers' ability to resist fatigue during prolonged exercise, although their sizes and strengths may remain unchanged.

Forceful exercise, such as weight lifting, in which a muscle exerts more than 75% of its maximum tension, utilizes *fast-twitch fibers,* which may be glycolytic and thus fatigable. In response to strenuous exercise, these fibers produce new filaments of actin and myosin, the diameters of the muscle fibers increase, and the entire muscle enlarges. However, the muscular hypertrophy does not produce new muscle fibers.

The strength of a muscular contraction is directly proportional to the diameter of the activated muscle fibers.

Consequently, an enlarged muscle can produce stronger contractions than before. Such a change, however, does not increase the muscle's ability to resist fatigue during activities like swimming or running.

If regular exercise stops, the capillary networks shrink, and the number of mitochondria within the muscle fibers drops. The number of actin and myosin filaments decreases, and the entire muscle atrophies. Such atrophy commonly occurs when accidents or diseases block motor nerve impulses from reaching muscle fibers. An unused muscle may shrink to less than half its usual size within a few weeks .

The fibers of muscles whose motor neurons are severed not only shrink, but also may fragment and, in time, be replaced by fat or fibrous connective tissue. However, reinnervation within the first few months following an injury may restore muscle function.

Astronauts experience muscle atrophy and impaired performance with long-term exposure to the microgravity environment of space. Customized workouts using special resistance equipment can minimize the changes in muscle structure and function. Otherwise, loss of muscle mass can make a thirty-year-old's muscles work like those of an eighty-year-old.

of stimulation and the beginning of contraction. This is the **latent period,** which in human muscle may be less than 2 milliseconds.

When a muscle fiber is brought to threshold under a given set of conditions, it tends to contract completely, such that each twitch generates the same force. This has been referred to as an *all-or-none* response. This is misleading, however, because in normal use of muscles, the force generated by muscle fibers and by whole muscles must vary.

Understanding the contraction of individual muscle fibers is important for understanding how muscles work, but such contractions by themselves are of little significance in day-to-day activities. Rather, the actions we need to perform usually require the contraction of multiple muscle fibers simultaneously. To record how a whole muscle responds to stimulation, a skeletal muscle can be removed from a frog or other small animal and mounted on a special device. The muscle is then stimulated electrically, and when it contracts, it pulls on a lever. The lever's movement is recorded as a myogram. Because the myogram results from the combined

twitches of muscle fibers taking part in the contraction, it looks essentially the same as the twitch contraction depicted in figure 8.11. Clinical Application 8.2 describes two types of twitches—the fatigue-resistant slow twitch and the fatigable fast twitch. Muscle fibers are either slow twitch or fast twitch.

The skeletal muscles of an average person have about half fast-twitch and half slow-twitch muscle fibers. In contrast, the muscles of an Olympic sprinter typically have more than 80% fast-twitch muscle fibers, and those of an Olympic marathoner, more than 90% slow-twitch muscle fibers.

Contractions of whole muscles enable us to perform everyday activities, but the force generated by those contractions must be controlled. For example, holding a styrofoam cup of coffee firmly enough that it does not slip through our fingers, but not so forcefully as to crush it, requires precise control of contractile force. In the whole muscle, the degree of tension developed reflects (1) the frequency at which individual muscle fibers are stimulated and (2) how many fibers take part in the overall contraction of the muscle.

Summation

The force that a muscle fiber can generate is not limited to the maximum force of a single twitch. A muscle fiber exposed to a series of stimuli of increasing frequency reaches a point when it is unable to completely relax before the next stimulus in the series arrives. When this happens, the force of individual twitches combines by the process of **summation.** When the resulting forceful, sustained contraction lacks even partial relaxation, it is called a **tetanic** (tĕ-tan′ik) **contraction,** or tetanus (fig. 8.12).

Recruitment of Motor Units

While summation increases the force of contraction of a single muscle fiber, a whole muscle can generate more force if more muscle fibers are involved in the contraction. A muscle fiber usually has a single motor end plate. The axons of motor neurons, however, are densely branched. By means of these branches, one motor neuron may connect to many muscle fibers. Together, a motor neuron and the muscle fibers that it controls constitute a **motor unit** (mo′tor u′nit) (fig. 8.13).

Each motor unit is a functional unit because an impulse in its motor neuron will cause all of the muscle fibers in that motor unit to contract at the same time. An increase in the number of motor units being activated during a contraction is called **recruitment** (re-krōōt′ment).

A whole muscle is composed of many motor units controlled by different motor neurons, which respond to different intensities of stimulation. That is, some motor neurons are more easily brought to threshold than others. If only the more sensitive motor neurons reach threshold, few motor units contract. At higher intensities of stimulation, other motor neurons are brought to threshold, and more motor units are activated. As the intensity of stimulation increases, recruitment of motor units continues until, finally, all motor units in that muscle are activated and the muscle contracts with maximal tension.

Sustained Contractions

Summation and recruitment together can produce a *sustained contraction* of increasing strength. Sustained contractions of whole muscles enable us to perform everyday activities. Such contractions are responses to a rapid series of impulses transmitted from the brain and spinal cord on motor neuron axons.

Even when a muscle appears to be at rest, its fibers undergo some sustained contraction. This is called **muscle tone** (tonus). Muscle tone is a response to nerve impulses that originate repeatedly from the spinal cord and stimulate a few muscle fibers. Muscle tone is particularly important in maintaining posture. If muscle tone is suddenly lost, as happens when a person loses consciousness, the body collapses.

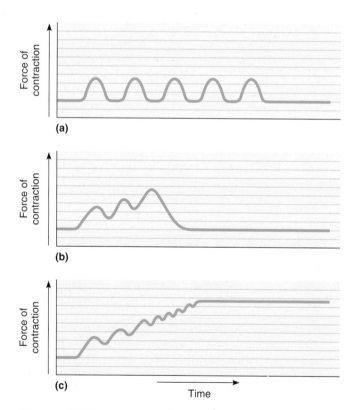

(a)

(b)

(c)

Time

Figure 8.12

Myograms of **(a)** a series of twitches, **(b)** summation, and **(c)** a tetanic contraction. Note that stimulation frequency increases from one myogram to the next.

Figure 8.13

Portions of two motor units. Muscle fibers within a motor unit are innervated by a single neuron and may be distributed throughout the muscle.

When skeletal muscles contract very forcefully, they may generate up to fifty pounds of pull for each square inch of muscle cross section. Large muscles, such as those in the thigh, can pull with several hundred pounds of force. This force may be so great that the tendons of muscles tear away from their attachments to the bones (*muscle pull*).

Practice

14. Define *threshold stimulus*.
15. What is a motor unit?
16. Distinguish between a twitch and a sustained contraction.
17. What is recruitment?
18. How is muscle tone maintained?

8.5 SMOOTH MUSCLE

The contractile mechanism of smooth muscles is essentially the same as for skeletal muscles. The cells of smooth muscle, however, have some important structural and functional differences from the other types of muscle.

Smooth Muscle Fibers

Recall from chapter 5 (p. 110) that smooth muscle cells are elongated, with tapering ends. Smooth muscle cells contain filaments of actin and myosin in myofibrils that extend the lengths of the cells, but these filaments are organized differently and more randomly than those in skeletal muscle. Therefore, smooth muscle cells lack striations (and appear "smooth" under the microscope). The sarcoplasmic reticulum in these cells is not well developed.

The two major types of smooth muscles are multiunit and visceral. In **multiunit smooth muscle,** the muscle fibers are separate rather than organized into sheets. Smooth muscle of this type is found in the irises of the eyes and in the walls of blood vessels. Typically, multiunit smooth muscle tissue contracts only in response to stimulation by neurons or certain hormones.

Visceral smooth muscle is composed of sheets of spindle-shaped cells in close contact with one another (see fig. 5.22, p. 111). This more common type of smooth muscle is found in the walls of hollow organs, such as the stomach, intestines, urinary bladder, and uterus.

Fibers of visceral smooth muscles can stimulate each other. When one fiber is stimulated, the impulse moving over its surface may excite adjacent fibers, which in turn stimulate still others. Visceral smooth muscles also display *rhythmicity,* a pattern of repeated contractions.

Rhythmicity is due to self-exciting fibers that deliver spontaneous impulses periodically into surrounding muscle tissue. These two features—transmission of impulses from cell to cell and rhythmicity—are largely responsible for the wavelike motion, called **peristalsis,** that occurs in certain tubular organs, such as the intestines, and helps force the contents of these organs along their lengths.

Smooth Muscle Contraction

Smooth muscle contraction resembles skeletal muscle contraction in a number of ways. Both mechanisms include reactions of actin and myosin, both are triggered by membrane impulses and an increase in intracellular calcium ions, and both use energy from ATP. However, these two types of muscle tissue also have significant differences.

Recall that acetylcholine is the neurotransmitter in skeletal muscle. Two neurotransmitters affect smooth muscle—acetylcholine and norepinephrine. Each of these neurotransmitters stimulates contractions in some smooth muscles and inhibits contractions in others (see chapter 9, p. 228). Also, a number of hormones affect smooth muscle, stimulating contractions in some cases and altering the degree of response to neurotransmitters in others.

Smooth muscle is slower to contract and to relax than skeletal muscle. On the other hand, smooth muscle can maintain a forceful contraction longer with a given amount of ATP. Also, unlike skeletal muscle, smooth muscle fibers can change length without changing tautness. As a result, smooth muscles in the stomach and intestinal walls can stretch as these organs fill, yet maintain the pressure inside these organs.

Practice

19. Describe two major types of smooth muscle.
20. What special characteristics of visceral smooth muscle make peristalsis possible?
21. How does smooth muscle contraction differ from skeletal muscle contraction?

8.6 CARDIAC MUSCLE

Cardiac muscle is found only in the heart. Its mechanism of contraction is essentially the same as that of skeletal and smooth muscle, but with some important differences.

Cardiac muscle is composed of branching, striated cells interconnected in three-dimensional networks (see fig. 5.23, p. 112). Each cell has many filaments of actin and myosin, organized similarly to those in skeletal muscle. A cardiac muscle cell also has a sarcoplasmic reticulum, many mitochondria, and a system of transverse tubules. However, the cisternae of cardiac muscle fibers are less well developed and store less calcium than those of skeletal muscle. On the other hand, the transverse tubules of cardiac muscle are larger, and they release many calcium ions into the sarcoplasm in response to muscle impulses. This extra calcium from the transverse tubules comes from the extracellular fluid and causes cardiac muscle twitches to be longer than skeletal muscle twitches.

The opposing ends of cardiac muscle cells are connected by structures called *intercalated discs*. These are elaborate junctions between cardiac muscle cell membranes. Intercalated discs allow muscle impulses to pass freely so that they travel rapidly from cell to cell, triggering contraction. The discs help to join cells and to transmit the force of contraction from cell to cell. Thus, when one portion of the cardiac muscle network is stimulated, the resulting impulse passes to the other parts of the network, and the whole structure contracts as a functional unit.

Cardiac muscle is also self-exciting and rhythmic. Consequently, a pattern of contraction and relaxation repeats, causing the rhythmic contractions of the heart.

Table 8.3 summarizes the characteristics of the three types of muscle tissue. Genetics Connection 8.1 considers several inherited diseases that affect the muscular system.

Practice

22. How is cardiac muscle similar to smooth muscle?
23. How is cardiac muscle similar to skeletal muscle?
24. What is the function of intercalated discs?
25. What characteristic of cardiac muscle contracts the heart as a unit?

8.7 SKELETAL MUSCLE ACTIONS

Skeletal muscles provide a variety of body movements, as described in chapter 7 (pp. 167–170). Each muscle's movement depends largely on the kind of joint it is associated with and the way the muscle attaches on either side of that joint.

Origin and Insertion

Recall that bones forming movable joints function as levers (see chapter 7, p. 140). One end of a skeletal muscle usually fastens to a relatively immovable or fixed part at a movable joint, and the other end connects to a movable part on the other side of that joint. The immovable end of the muscle is called its **origin** (or'ĭ-jin), and the movable end is its **insertion** (in-ser'shun). When a muscle contracts, its insertion is pulled toward its origin.

Some muscles have more than one origin or insertion. The *biceps brachii* in the arm, for example, has two origins. This is reflected in the name *biceps,* which means "two heads." (Note: The head of a muscle is the part nearest its origin.) One head of the muscle attaches to the coracoid process of the scapula, and the other head arises from a tubercle above the glenoid cavity of the scapula. The muscle extends along the anterior surface of the humerus and is inserted by means of a tendon on the radial tuberosity of the radius. When the biceps brachii contracts, its insertion is pulled toward its origin, and the forearm flexes at the elbow (fig. 8.14).

The movements termed *flexion* and *extension* describe changes in the angle between bones that meet at a joint. For example, flexion of the elbow refers to a movement of the forearm that bends the elbow, or decreases the angle. Alternatively, one could say that flexion of the elbow results from the action of the biceps brachii on the radius of the forearm.

Table 8.3	Types of Muscle Tissue		
	Skeletal	**Smooth**	**Cardiac**
Major Location	Skeletal muscles	Walls of hollow viscera, blood vessels	Wall of the heart
Major Function	Movement of bones at joints, maintenance of posture	Movement of viscera, peristalsis, vasoconstriction	Pumping action of the heart
Cellular Characteristics			
Striations	Present	Absent	Present
Nucleus	Many nuclei	Single nucleus	Single nucleus
Special features	Well-developed transverse tubule system	Lacks transverse tubules	Well-developed transverse tubule system; intercalated discs separating adjacent cells
Mode of Control	Voluntary	Involuntary	Involuntary
Contraction Characteristics	Contracts and relaxes rapidly	Contracts and relaxes slowly; self-exciting; rhythmic	Network of cells contracts as a unit; self-exciting; rhythmic

Inherited Diseases of Muscle

A variety of inherited conditions affect muscle tissue. These disorders differ in the nature of the genetic defect, the type of protein that is abnormal in form or function, and the particular muscles in the body that are impaired.

The Muscular Dystrophies—Missing Proteins

A muscle cell is packed with filaments of actin and myosin. Much less abundant, but no less important, is a protein called *dystrophin*. It holds skeletal muscle cells together by linking actin in the cell to glycoproteins in the cell membrane, which helps attach the cell to the extracellular matrix. Missing or abnormal dystrophin or the glycoproteins cause muscular dystrophies. These illnesses vary in severity and age of onset, but in all cases, muscles weaken and degenerate. Eventually, fat and connective tissue replace muscle.

Duchenne muscular dystrophy (DMD) is the most severe type of the illness (fig. 8B). Symptoms begin by age five and affect only boys. By age thirteen, the person cannot walk, and by early adulthood he usually dies from failure of the respiratory muscles. In DMD, dystrophin is often missing. In Becker muscular dystrophy, symptoms begin in early adulthood, are less severe, and result from underproduction of dystrophin.

Figure 8B
This young man has Duchenne muscular dystrophy. The condition has not yet severely limited his activities, but he shows the hypertrophied (overdeveloped) calf muscles that result from his inability to rise from a sitting position the usual way—an early sign of the illness.

Charcot-Marie-Tooth Disease— A Duplicate Gene

Charcot-Marie-Tooth disease causes a slowly progressing weakness in the muscles of the hands and feet and a decrease in tendon reflexes in these parts. In this illness, an extra gene impairs the insulating sheath around affected nerve cells, so that nerve cells cannot adequately stimulate muscles. Physicians perform two tests—electromyography and nerve conduction velocity—to diagnose Charcot-Marie-Tooth disease. It is also possible to test for the gene mutation to confirm a diagnosis based on symptoms.

Hereditary Idiopathic Dilated Cardiomyopathy—A Tiny Glitch

This very rare inherited form of heart failure usually begins in a person's forties and is lethal in 50% of cases within five years of diagnosis, unless a heart transplant can be performed. The condition is caused by a tiny genetic error in a form of actin found only in cardiac muscle, where it is the predominant component of the thin filaments. The mutation disturbs actin's ability to anchor to the Z lines in heart muscle cells, preventing actin from effectively transmitting the force of contraction. As a result, the heart chambers enlarge and eventually fail.

Since students often find it helpful to think of movements in terms of the specific actions of the muscles involved, we may also describe flexion and extension in these terms. Thus, the action of the biceps brachii may be described as "flexion of the forearm at the elbow," and the action of the quadriceps group as "extension of the leg at the knee." We believe this occasional departure from strict anatomical terminology eases understanding and learning.

Interaction of Skeletal Muscles

Skeletal muscles almost always function in groups. Consequently, a particular body movement requires more than contracting a single muscle; instead, after learning to make a particular movement, the person initiates the movement consciously, and the nervous system stimulates the appropriate group of muscles.

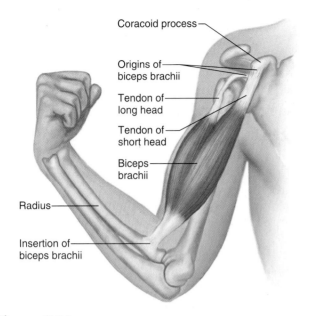

Coracoid process

Origins of biceps brachii

Tendon of long head

Tendon of short head

Biceps brachii

Radius

Insertion of biceps brachii

Figure 8.14
The biceps brachii has two heads that originate on the scapula. A tendon inserts this muscle on the radius.

Careful observation of body movements indicates the special roles of muscles. For instance, when the upper limb is lifted horizontally away from the side, a contracting *deltoid* muscle provides most of the movement and is said to be the **prime mover** (prīm mōōv′er), also referred to as an **agonist** (ag′o-nist). However, while a prime mover is acting, certain nearby muscles are also contracting. In the case of the contracting deltoid muscle, nearby muscles help hold the shoulder steady and in this way make the prime mover's action more effective. Muscles that contract and assist the prime mover are called **synergists** (sin′er-jistz).

Certain muscles act as **antagonists** (an-tag′o-nistz) to prime movers. These muscles can resist a prime mover's action and cause movement in the opposite direction. For example, the antagonist of the prime mover that raises the upper limb can lower the upper limb, or the antagonist of the prime mover that bends the upper limb can straighten it (see fig. 7.7, p. 140). If both a prime mover and its antagonist contract simultaneously with equal force, the part they act upon remains rigid. Consequently, smooth body movements depend on antagonists relaxing and, thus, giving way to the prime movers whenever the prime movers contract. Once again, the nervous system controls these complex actions.

Sometimes the relationship between two muscles changes. For example, the pectoralis major and latissimus dorsi are antagonistic for flexion and extension of the shoulder. However, they are synergistic for medial rotation of the shoulder. Thus, any role of a muscle must be learned in the context of a particular movement.

Practice

26. Distinguish between the origin and the insertion of a muscle.
27. Define *prime mover*.
28. What is the function of a synergist? an antagonist?

8.8 MAJOR SKELETAL MUSCLES

This section discusses the locations, actions, and attachments of some of the major skeletal muscles. (Figures 8.15 and 8.16 and reference plates 1 and 2, pp. 23–24, show the locations of the superficial skeletal muscles—those near the surface.)

Note that the names of these muscles often describe them. A name may indicate a muscle's relative size, shape, location, action, number of attachments, or the direction of its fibers, as in the following examples:

pectoralis major Of large size (major) and located in the pectoral region (chest).

deltoid Shaped like a delta or triangle.

extensor digitorum Extends the digits (fingers or toes).

biceps brachii Having two heads (biceps) or points of origin and located in the brachium (arm).

sternocleidomastoid Attached to the sternum, clavicle, and mastoid process.

external oblique Located near the outside, with fibers that run obliquely (in a slanting direction).

Muscles of Facial Expression

A number of small muscles that lie beneath the skin of the face and scalp enable us to communicate feelings through facial expression (fig. 8.17*a*). Many of these muscles, located around the eyes and mouth, are responsible for such expressions as surprise, sadness, anger, fear, disgust, and pain. As a group, the muscles of facial expression join the bones of the skull to connective tissue in various regions of the overlying skin. They include:

epicranius (ep″ĭ-kra′ne-us) Composed of two parts, the *frontalis* (frun-ta′lis) and the *occipitalis* (ok-sip″ĭ-ta′lis)

orbicularis oculi (or-bik′u-la-rus ok′u-li)

orbicularis oris (or-bik′u-la-rus o′ris)

buccinator (buk′sĭ-na″tor)

zygomaticus (zi″go-mat′ik-us)

platysma (plah-tiz′mah)

Table 8.4 lists the origins, insertions, and actions of the muscles of facial expression. (The muscles that move the eyes are listed in chapter 10, pp. 278–279.)

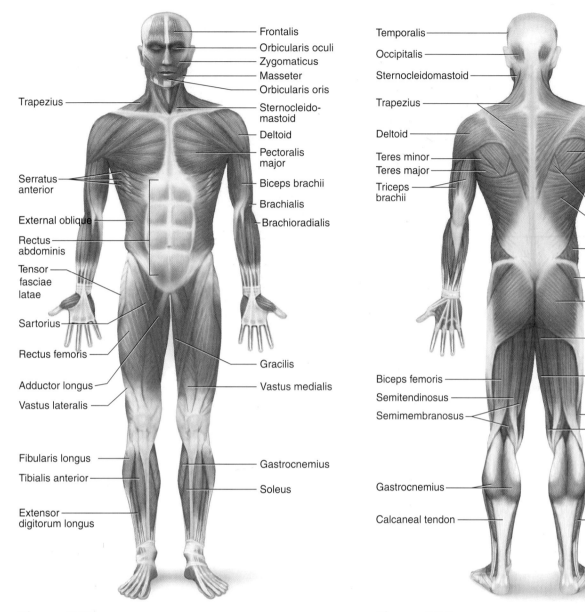

Figure 8.15
Anterior view of superficial skeletal muscles.

Figure 8.16
Posterior view of superficial skeletal muscles.

Table 8.4	Muscles of Facial Expression AP|R		
Muscle	**Origin**	**Insertion**	**Action**
Epicranius	Occipital bone	Skin and muscles around eye	Raises eyebrow
Orbicularis oculi	Maxillary and frontal bones	Skin around eye	Closes eye
Orbicularis oris	Muscles near the mouth	Skin of lips	Closes and protrudes lips
Buccinator	Outer surfaces of maxilla and mandible	Orbicularis oris	Compresses cheeks inward
Zygomaticus	Zygomatic bone	Orbicularis oris	Raises corner of mouth
Platysma	Fascia in upper chest	Lower border of mandible	Draws angle of mouth downward

Epicranius— {
  Frontalis
  Occipitalis
}

Epicranial aponeurosis

Temporalis

Orbicularis oculi

Zygomaticus

Masseter

Sternocleidomastoid

Buccinator

Orbicularis oris

Platysma

(a) AP|R

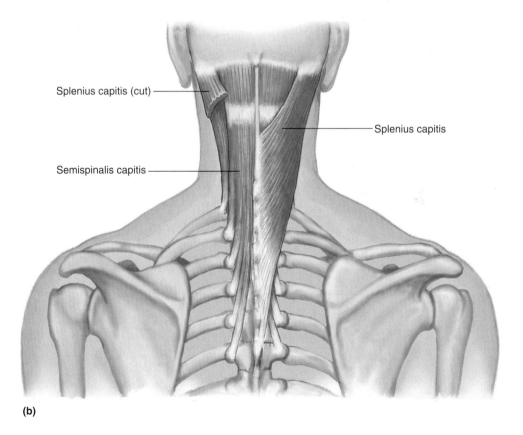

Splenius capitis (cut)

Semispinalis capitis

Splenius capitis

(b)

Figure 8.17

Muscles of the face and neck. **(a)** Muscles of facial expression and mastication. **(b)** Posterior view of muscles that move the head.

The human body has more than 600 distinct skeletal muscles. The face alone includes 60 muscles, more than 40 of which are used to frown, and 20 to smile. Thinner than a thread and barely visible, the stapedius in the middle ear is the body's smallest muscle. In contrast is the gluteus maximus, the largest muscle, located in the buttock. The sartorius, which pulls on the thigh, is the longest muscle in the body.

Muscles of Mastication

Muscles attached to the mandible produce chewing movements. Two pairs of these muscles elevate the mandible, a motion used in biting. These muscles are the *masseter* (mas-se′ter) and the *temporalis* (tem-po-ra′lis) (fig. 8.17*a*). Table 8.5 lists the origins, insertions, and actions of the muscles of mastication.

Grinding the teeth, a common response to stress, may strain the temporomandibular joint—the articulation between the mandibular condyle of the mandible and the mandibular fossa of the temporal bone. This condition, called temporomandibular joint syndrome (TMJ syndrome), may produce headache, earache, and pain in the jaw, neck, or shoulder.

Muscles That Move the Head

Head movements result from the actions of paired muscles in the neck and upper back. These muscles flex, extend, and rotate the head. They include (fig. 8.17):

sternocleidomastoid (ster″no-kli″do-mas′toid)
splenius capitis (sple′ne-us kap′ĭ-tis)
semispinalis capitis (sem″e-spi-na′lis kap′ĭ-tis)

Table 8.6 lists the origins, insertions, and actions of muscles that move the head.

Muscles That Move the Pectoral Girdle

The muscles that move the pectoral girdle are closely associated with those that move the arm. A number of these chest and shoulder muscles connect the scapula to nearby bones and move the scapula upward, downward, forward, and backward. They include (figs. 8.18 and 8.19):

trapezius (trah-pe′ze-us)
rhomboid major (rom-boid′)
levator scapulae (le-va′tor scap′u-lē)
serratus anterior (ser-ra′tus an-te′re-or)
pectoralis minor (pek″to-ra′lis)

Table 8.7 lists the origins, insertions, and actions of the muscles that move the pectoral girdle.

Muscles That Move the Arm

The arm is one of the more freely movable parts of the body. Muscles that connect the humerus to various regions of the pectoral girdle, ribs, and vertebral column make these movements possible (figs. 8.18, 8.19, 8.20, and 8.21). These muscles can be grouped according to their primary actions—flexion, extension, abduction, and rotation—as follows:

Flexors
coracobrachialis (kor″ah-ko-bra′ke-al-is)
pectoralis major (pek″to-ra′lis)

Extensors
teres major (te′rēz)
latissimus dorsi (lah-tis′ĭ-mus dor′si)

Table 8.5	Muscles of Mastication AP|R		
Muscle	**Origin**	**Insertion**	**Action**
Masseter	Lower border of zygomatic arch	Lateral surface of mandible	Elevates mandible
Temporalis	Temporal bone	Coronoid process and lateral surface of mandible	Elevates mandible

Table 8.6	Muscles That Move the Head AP|R		
Muscle	**Origin**	**Insertion**	**Action**
Sternocleidomastoid	Anterior surface of sternum and upper surface of clavicle	Mastoid process of temporal bone	Pulls head to one side, pulls head toward chest, or raises sternum
Splenius capitis	Spinous processes of lower cervical and upper thoracic vertebrae	Mastoid process of temporal bone	Rotates head, bends head to one side, or brings head into an upright position
Semispinalis capitis	Processes of lower cervical and upper thoracic vertebrae	Occipital bone	Extends head, bends head to one side, or rotates head

Table 8.7	Muscles That Move the Pectoral Girdle AP|R		
Muscle	**Origin**	**Insertion**	**Action**
Trapezius	Occipital bone and spines of cervical and thoracic vertebrae	Clavicle; spine and acromion process of scapula	Rotates scapula and raises arm; raises scapula; pulls scapula medially or pulls scapula and shoulder downward
Rhomboid major	Spines of upper thoracic vertebrae	Medial border of scapula	Raises and adducts scapula
Levator scapulae	Transverse processes of cervical vertebrae	Medial margin of scapula	Elevates scapula
Serratus anterior	Outer surfaces of upper ribs	Ventral surface of scapula	Pulls scapula anteriorly and downward
Pectoralis minor	Sternal ends of upper ribs	Coracoid process of scapula	Pulls scapula anteriorly and downward or raises ribs

Abductors

supraspinatus (su″prah-spi′na-tus)
deltoid (del′toid)

Rotators

subscapularis (sub-scap′u-lar-is)
infraspinatus (in″frah-spi′na-tus)
teres minor (te′rēz)

Table 8.8 lists the origins, insertions, and actions of muscles that move the arm.

Muscles That Move the Forearm

Muscles that connect the radius or ulna to the humerus or pectoral girdle produce most of the forearm movements. A group of muscles located along the anterior surface of the humerus flexes the elbow, and a single posterior muscle extends this joint. Other muscles move the radioulnar joint and rotate the forearm.

Muscles that move the forearm include (figs. 8.20, 8.21, and 8.22):

Figure 8.18 AP\|R

Muscles of the posterior shoulder. The right trapezius is removed to show underlying muscles.

Table 8.8	Muscles That Move the Arm AP R		
Muscle	**Origin**	**Insertion**	**Action**
Coracobrachialis	Coracoid process of scapula	Shaft of humerus	Flexes and adducts arm
Pectoralis major	Clavicle, sternum, and costal cartilages of upper ribs	Intertubercular groove of humerus	Pulls arm anteriorly and across chest, rotates humerus, or adducts arm
Teres major	Lateral border of scapula	Intertubercular groove of humerus	Extends humerus or adducts and rotates arm medially
Latissimus dorsi	Spines of sacral, lumbar, and lower thoracic vertebrae, iliac crest, and lower ribs	Intertubercular groove of humerus	Extends and adducts arm and rotates humerus inwardly, or pulls shoulder downward and posteriorly
Supraspinatus	Posterior surface of scapula	Greater tubercle of humerus	Abducts arm
Deltoid	Acromion process, spine of scapula, and clavicle	Deltoid tuberosity of humerus	Abducts arm, extends or flexes humerus
Subscapularis	Anterior surface of scapula	Lesser tubercle of humerus	Rotates arm medially
Infraspinatus	Posterior surface of scapula	Greater tubercle of humerus	Rotates arm laterally
Teres minor	Lateral border of scapula	Greater tubercle of humerus	Rotates arm laterally

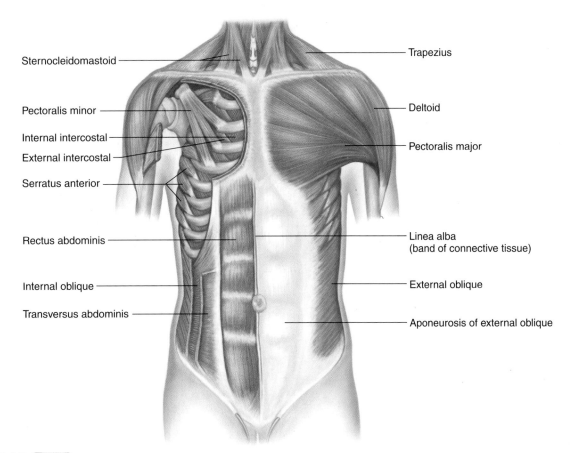

Figure 8.19 AP R

Muscles of the anterior chest and abdominal wall. The right pectoralis major is removed to show the pectoralis minor.

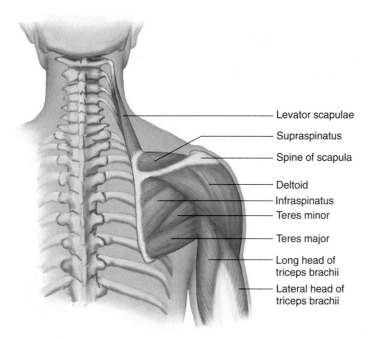

Levator scapulae
Supraspinatus
Spine of scapula
Deltoid
Infraspinatus
Teres minor
Teres major
Long head of triceps brachii
Lateral head of triceps brachii

Figure 8.20 AP|R
Muscles of the posterior surface of the scapula and arm.

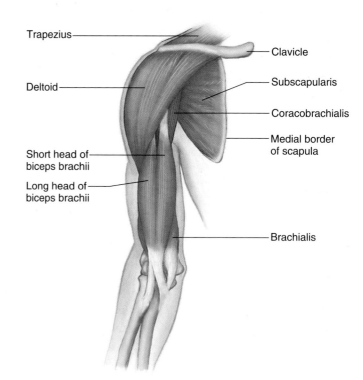

Trapezius
Clavicle
Deltoid
Subscapularis
Coracobrachialis
Medial border of scapula
Short head of biceps brachii
Long head of biceps brachii
Brachialis

Figure 8.21 AP|R
Muscles of the anterior shoulder and arm, with the rib cage removed.

Flexors
biceps brachii (bi′seps bra′ke-i)
brachialis (bra′ke-al-is)
brachioradialis (bra″ke-o-ra″de-a′lis)

Extensor
triceps brachii (tri′seps bra′ke-i)

Rotators
supinator (su′pĭ-na-tor)
pronator teres (pro-na′tor te′rēz)
pronator quadratus (pro-na′tor kwod-ra′tus)

Table 8.9 lists the origins, insertions, and actions of muscles that move the forearm.

Muscles That Move the Hand

Many muscles move the hand. They originate from the distal end of the humerus and from the radius and ulna. The two major groups of these muscles are flexors on the anterior side of the forearm and extensors on the posterior side. These muscles include (figs. 8.22 and 8.23):

Flexors
flexor carpi radialis (flex′sor kar-pi′ ra″de-a′lis)
flexor carpi ulnaris (flex′sor kar-pi′ ul-na′ris)
palmaris longus (pal-ma′ris long′gus)
flexor digitorum profundus (flex′sor dij″ĭ-to′rum pro-fun′dus)

Extensors
extensor carpi radialis longus (eks-ten′sor kar-pi′ ra″de-a′lis long′gus)
extensor carpi radialis brevis (eks-ten′sor kar-pi′ ra″de-a′lis brev′ĭs)

Biceps brachii
Brachialis
Supinator
Pronator teres
Brachioradialis
Extensor carpi radialis longus
Flexor carpi radialis
Palmaris longus
Flexor carpi ulnaris
Pronator quadratus

Figure 8.22 AP|R
Muscles of the anterior forearm.

Table 8.9	Muscles That Move the Forearm AP|R		
Muscle	**Origin**	**Insertion**	**Action**
Biceps brachii	Coracoid process and tubercle above glenoid cavity of scapula	Radial tuberosity of radius	Flexes forearm at elbow and rotates hand laterally
Brachialis	Anterior shaft of humerus	Coronoid process of ulna	Flexes forearm at elbow
Brachioradialis	Distal lateral end of humerus	Lateral surface of radius above styloid process	Flexes forearm at elbow
Triceps brachii	Tubercle below glenoid cavity and lateral and medial surfaces of humerus	Olecranon process of ulna	Extends forearm at elbow
Supinator	Lateral epicondyle of humerus and crest of ulna	Lateral surface of radius	Rotates forearm laterally
Pronator teres	Medial epicondyle of humerus and coronoid process of ulna	Lateral surface of radius	Rotates forearm medially
Pronator quadratus	Anterior distal end of ulna	Anterior distal end of radius	Rotates forearm medially

extensor carpi ulnaris (eks-ten′sor kar-pi′ ul-na′ris)
extensor digitorum (eks-ten′sor dij″ĭ-to′rum)

Table 8.10 lists the origins, insertions, and actions of muscles that move the hand.

Figure 8.23 AP\|R
Muscles of the posterior forearm.

Labels: Triceps brachii; Brachioradialis; Extensor carpi radialis longus; Flexor carpi ulnaris; Extensor carpi radialis brevis; Extensor carpi ulnaris; Extensor digitorum

Muscles of the Abdominal Wall

Bone supports the walls of the chest and pelvic regions, but not those of the abdomen. Instead, the anterior and lateral walls of the abdomen are composed of layers of broad, flattened muscles. These muscles connect the rib cage and vertebral column to the pelvic girdle. A band of tough connective tissue called the **linea alba** extends from the xiphoid process of the sternum to the pubic symphysis (see fig. 8.19). It is an attachment for some of the abdominal wall muscles.

Contraction of these muscles decreases the size of the abdominal cavity and increases the pressure inside. These actions help press air out of the lungs during forceful exhalation and aid in the movements of defecation, urination, vomiting, and childbirth.

The abdominal wall muscles include (see fig. 8.19):

external oblique (eks-ter′nal o-blēk′)
internal oblique (in-ter′nal o-blēk′)
transversus abdominis (trans-ver′sus ab-dom′ĭ-nis)
rectus abdominis (rek′tus ab-dom′ĭ-nis)

Table 8.11 lists the origins, insertions, and actions of muscles of the abdominal wall.

Muscles of the Pelvic Outlet

Two muscular sheets—a deeper **pelvic diaphragm** and a more superficial **urogenital diaphragm**—span the outlet of the pelvis. The pelvic diaphragm forms the floor of the pelvic cavity, and the urogenital diaphragm fills the space within the pubic arch (see fig. 7.29, p. 161). The muscles of the male and female pelvic outlets include (fig. 8.24):

Pelvic diaphragm
levator ani (le-va′tor ah-ni′)

Table 8.10	Muscles That Move the Hand		
Muscle	**Origin**	**Insertion**	**Action**
Flexor carpi radialis	Medial epicondyle of humerus	Base of second and third metacarpals	Flexes and abducts wrist
Flexor carpi ulnaris	Medial epicondyle of humerus and olecranon process	Carpal and metacarpal bones	Flexes and adducts wrist
Palmaris longus	Medial epicondyle of humerus	Fascia of palm	Flexes wrist
Flexor digitorum profundus	Anterior surface of ulna	Bases of distal phalanges in fingers 2–5	Flexes distal joints of fingers
Extensor carpi radialis longus	Distal end of humerus	Base of second metacarpal	Extends wrist and abducts hand
Extensor carpi radialis brevis	Lateral epicondyle of humerus	Base of second and third metacarpals	Extends wrist and abducts hand
Extensor carpi ulnaris	Lateral epicondyle of humerus	Base of fifth metacarpal	Extends and adducts wrist
Extensor digitorum	Lateral epicondyle of humerus	Posterior surface of phalanges in fingers 2–5	Extends fingers

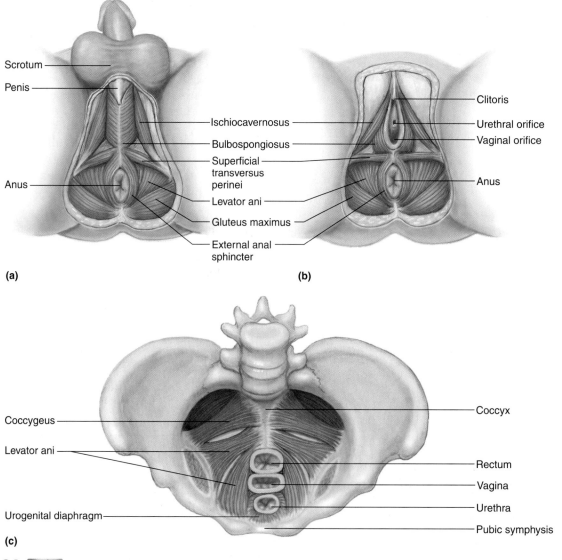

(a)

(b)

(c)

Figure 8.24 AP|R

External view of muscles of **(a)** the male pelvic outlet and **(b)** the female pelvic outlet. **(c)** Internal view of the female pelvic and urogenital diaphragms.

Table 8.11	Muscles of the Abdominal Wall		
Muscle	**Origin**	**Insertion**	**Action**
External oblique	Outer surfaces of lower ribs	Outer lip of iliac crest and linea alba	Tenses abdominal wall and compresses abdominal contents
Internal oblique	Crest of ilium and inguinal ligament	Cartilages of lower ribs, linea alba, and crest of pubis	Tenses abdominal wall and compresses abdominal contents
Transversus abdominis	Costal cartilages of lower ribs, processes of lumbar vertebrae, lip of iliac crest, and inguinal ligament	Linea alba and crest of pubis	Tenses abdominal wall and compresses abdominal contents
Rectus abdominis	Crest of pubis and pubic symphysis	Xiphoid process of sternum and costal cartilages	Tenses abdominal wall and compresses abdominal contents; also flexes vertebral column

Urogenital diaphragm
superficial transversus perinei (su″per-fish′al trans-ver′sus per″ĭ-ne′i)
bulbospongiosus (bul″bo-spon″je-o′sus)
ischiocavernosus (is″ke-o-kav″er-no′sus)

Table 8.12 lists the origins, insertions, and actions of pelvic outlet muscles.

Muscles That Move the Thigh

Muscles that move the thigh are attached to the femur and to some part of the pelvic girdle. These muscles are in anterior and posterior groups. Muscles of the anterior group primarily flex the thigh; those of the posterior group extend, abduct, or rotate the thigh. The muscles in these groups include (figs. 8.25, 8.26, and 8.27):

Anterior group
psoas major (so′as)
iliacus (il′e-ak-us)

Posterior group
gluteus maximus (gloo′te-us mak′si-mus)
gluteus medius (gloo′te-us me′de-us)
gluteus minimus (gloo′te-us min′ĭ-mus)
tensor fasciae latae (ten′sor fash′e-e lah-tē)

Still another group of muscles attached to the femur and pelvic girdle adduct the thigh. They include (figs. 8.25 and 8.27):

adductor longus (ah-duk′tor long′gus)
adductor magnus (ah-duk′tor mag′nus)
gracilis (gras′il-is)

Table 8.13 lists the origins, insertions, and actions of muscles that move the thigh.

Muscles That Move the Leg

Muscles that move the leg connect the tibia or fibula to the femur or to the pelvic girdle. They can be separated into two major groups—those that flex the knee and those that extend the knee. Muscles of these groups

Table 8.12	Muscles of the Pelvic Outlet		
Muscle	**Origin**	**Insertion**	**Action**
Levator ani	Pubic bone and ischial spine	Coccyx	Supports pelvic viscera and provides sphincter-like action in anal canal and vagina
Superficial transversus perinei	Ischial tuberosity	Central tendon	Supports pelvic viscera
Bulbospongiosus	Central tendon	Males: Urogenital diaphragm and fascia of the penis	Males: Assists emptying of urethra
		Females: Pubic arch and root of clitoris	Females: Constricts vagina
Ischiocavernosus	Ischial tuberosity	Pubic arch	Assists function of bulbospongiosus

Figure 8.25 AP|R

Muscles of the anterior right thigh. (Note that the vastus intermedius is a deep muscle not visible in this view.)

Figure 8.26

Muscles of the lateral right thigh.

Table 8.13	Muscles That Move the Thigh AP|R		
Muscle	**Origin**	**Insertion**	**Action**
Psoas major	Lumbar intervertebral discs, bodies and transverse processes of lumbar vertebrae	Lesser trochanter of femur	Flexes thigh
Iliacus	Iliac fossa of ilium	Lesser trochanter of femur	Flexes thigh
Gluteus maximus	Sacrum, coccyx, and posterior surface of ilium	Posterior surface of femur and fascia of thigh	Extends thigh
Gluteus medius	Lateral surface of ilium	Greater trochanter of femur	Abducts and rotates thigh medially
Gluteus minimus	Lateral surface of ilium	Greater trochanter of femur	Abducts and rotates thigh medially
Tensor fasciae latae	Anterior iliac crest	Fascia of thigh	Abducts, flexes, and rotates thigh medially
Adductor longus	Pubic bone near pubic symphysis	Posterior surface of femur	Adducts, flexes, and rotates thigh laterally
Adductor magnus	Ischial tuberosity	Posterior surface of femur	Adducts, extends, and rotates thigh laterally
Gracilis	Lower edge of pubic symphysis	Medial surface of tibia	Adducts thigh, flexes and rotates lower limb medially

Figure 8.27 AP|R

Muscles of the posterior right thigh.

Labels (top to bottom, left and right):
- Gluteus medius
- Gluteus maximus
- Vastus lateralis covered by fascia
- Biceps femoris
- Adductor magnus
- Gracilis
- Semitendinosus
- Semimembranosus
- Sartorius
- Gastrocnemius

include the hamstring group and the quadriceps femoris group (figs. 8.25, 8.26, and 8.27):

Flexors

biceps femoris (bi′seps fem′or-is)
semitendinosus (sem″e-ten′dĭ-no-sus)
semimembranosus (sem″e-mem′brah-no-sus)
sartorius (sar-to′re-us)

Extensor

quadriceps femoris group (kwod′rĭ-seps fem′or-is)
    Composed of four parts—the rectus femoris, vastus lateralis, vastus medialis, and vastus intermedius.

Table 8.14 lists the origins, insertions, and actions of muscles that move the leg.

Muscles That Move the Foot

A number of muscles that move the foot are in the leg. They attach the femur, tibia, and fibula to bones of the foot, move the foot upward (dorsiflexion) or downward (plantar flexion), and turn the sole of the foot medial (inversion) or lateral (eversion). These muscles include (figs. 8.28, 8.29, and 8.30):

Dorsal flexors

tibialis anterior (tib″e-a′lis an-te′re-or)
fibularis (peroneus) tertius (fib″u-la′ris ter′shus)
extensor digitorum longus (eks-ten′sor dij″ĭ-to′rum long′gus)

Table 8.14	Muscles That Move the Leg AP|R		
Muscle	**Origin**	**Insertion**	**Action**
Sartorius	Anterior superior iliac spine	Medial surface of tibia	Flexes leg and thigh, abducts thigh, rotates thigh laterally, and rotates leg medially
Hamstring group			
Biceps femoris	Ischial tuberosity and posterior surface of femur	Head of fibula and lateral condyle of tibia	Flexes leg, extends thigh
Semitendinosus	Ischial tuberosity	Medial surface of tibia	Flexes leg, extends thigh
Semimembranosus	Ischial tuberosity	Medial condyle of tibia	Flexes leg, extends thigh
Quadriceps femoris group			
Rectus femoris	Spine of ilium and margin of acetabulum	Patella by the tendon, which continues as patellar ligament to tibial tuberosity	Extends leg at knee
Vastus lateralis	Greater trochanter and posterior surface of femur	Patella by the tendon, which continues as patellar ligament to tibial tuberosity	Extends leg at knee
Vastus medialis	Medial surface of femur	Patella by the tendon, which continues as patellar ligament to tibial tuberosity	Extends leg at knee
Vastus intermedius	Anterior and lateral surfaces of femur	Patella by the tendon, which continues as patellar ligament to tibial tuberosity	Extends leg at knee

Figure 8.28
Muscles of the anterior right leg.

Figure 8.29
Muscles of the lateral right leg. (Note that the tibialis posterior is a deep muscle not visible in this view.)

Plantar flexors
gastrocnemius (gas″trok-ne′me-us)
soleus (so′le-us)
flexor digitorum longus (flek′sor dij″ĭ-to′rum long′gus)

Invertor
tibialis posterior (tib″e-a′lis pos-tēr′e-or)

Evertor
fibularis (peroneus) longus (fib″u-la′ris long′gus)

Table 8.15 lists the origins, insertions, and actions of muscles that move the foot.

Practice

29. What information is imparted in a muscle's name?

30. Which muscles provide facial expressions? ability to chew? head movements?

31. Which muscles move the pectoral girdle? abdominal wall? pelvic outlet? the arm, forearm, and hand? the thigh, leg, and foot?

Semitendinosus
Semimembranosus
Gracilis
Sartorius
Biceps femoris

Gastrocnemius:
Medial head
Lateral head
Fibularis longus

Soleus

Calcaneal tendon

Flexor digitorum longus

Calcaneus

Figure 8.30 AP|R
Muscles of the posterior right leg.

Table 8.15	Muscles That Move the Foot AP|R		
Muscle	**Origin**	**Insertion**	**Action**
Tibialis anterior	Lateral condyle and lateral surface of tibia	Tarsal bone (cuneiform) and first metatarsal	Dorsiflexion and inversion of foot
Fibularis tertius	Anterior surface of fibula	Dorsal surface of fifth metatarsal	Dorsiflexion and eversion of foot
Extensor digitorum longus	Lateral condyle of tibia and anterior surface of fibula	Dorsal surfaces of second and third phalanges of the four lateral toes	Dorsiflexion and eversion of foot and extension of toes
Gastrocnemius	Lateral and medial condyles of femur	Posterior surface of calcaneus	Plantar flexion of foot and flexion of leg at knee
Soleus	Head and shaft of fibula and posterior surface of tibia	Posterior surface of calcaneus	Plantar flexion of foot
Flexor digitorum longus	Posterior surface of tibia	Distal phalanges of the four lateral toes	Plantar flexion and inversion of foot, and flexion of the four lateral toes
Tibialis posterior	Lateral condyle and posterior surface of tibia, and posterior surface of fibula	Tarsal and metatarsal bones	Plantar flexion and inversion of foot
Fibularis longus	Lateral condyle of tibia and head and shaft of fibula	Tarsal and metatarsal bones	Plantar flexion and eversion of foot; also supports arch

Muscular System

Muscles provide the force for moving body parts.

Integumentary System

The skin increases heat loss during skeletal muscle activity.

Lymphatic System

Muscle action pumps lymph through lymphatic vessels.

Skeletal System

Bones provide attachments that allow skeletal muscles to cause movement.

Digestive System

Skeletal muscles are important in swallowing. The digestive system absorbs nutrients needed for muscle contraction.

Nervous System

Neurons control muscle contractions.

Respiratory System

Breathing depends on skeletal muscles. The lungs provide oxygen for body cells and excrete carbon dioxide.

Endocrine System

Hormones help increase blood flow to exercising skeletal muscles.

Urinary System

Skeletal muscles help control expulsion of urine from the urinary bladder.

Cardiovascular System

The heart pumps as a result of cardiac muscle contraction. Blood flow delivers oxygen and nutrients and removes wastes.

Reproductive System

Skeletal muscles are important in sexual activity.

Summary Outline

8.1 Introduction (p. 179)

The three types of muscle tissue are skeletal, smooth, and cardiac.

8.2 Structure of a Skeletal Muscle (p. 179)

Individual muscles are the organs of the muscular system. They include skeletal muscle tissue, nervous tissue, blood, and connective tissues.

1. Connective tissue coverings
   a. Fascia covers skeletal muscles.
   b. Other connective tissues attach muscles to bones or to other muscles.
   c. A network of connective tissue extends throughout the muscular system.
2. Skeletal muscle fibers
   a. Each skeletal muscle fiber is a single muscle cell, which is the unit of contraction.
   b. The cytoplasm contains mitochondria, sarcoplasmic reticulum, and myofibrils of actin and myosin.
   c. The organization of actin and myosin filaments produces striations.
   d. Transverse tubules extend inward from the cell membrane and associate with the sarcoplasmic reticulum.
3. Neuromuscular junction
   a. Motor neurons stimulate muscle fibers to contract.
   b. In response to a nerve impulse, the end of a motor neuron axon secretes a neurotransmitter, which stimulates the muscle fiber to contract.

8.3 Skeletal Muscle Contraction (p. 182)

Muscle fiber contraction results from a sliding movement of actin and myosin filaments.

1. Role of myosin and actin
   a. Cross-bridges of myosin filaments form linkages with actin filaments.
   b. The reaction between actin and myosin filaments generates the force of contraction.
2. Stimulus for contraction
   a. Acetylcholine released from the distal end of a motor neuron axon stimulates a skeletal muscle fiber.
   b. Acetylcholine causes the muscle fiber to conduct an impulse over the surface of the fiber that reaches deep within the fiber through the transverse tubules.
   c. A muscle impulse signals the sarcoplasmic reticulum to release calcium ions.
   d. Linkages form between actin and myosin, and the myosin cross-bridges pull on actin filaments, shortening the fiber.
   e. The muscle fiber relaxes when cross-bridges release from actin (ATP is needed, but is not broken down) and when calcium ions are actively transported (requiring ATP breakdown) back into the sarcoplasmic reticulum.
   f. Acetylcholinesterase breaks down acetylcholine.
3. Energy sources for contraction
   a. ATP supplies the energy for muscle fiber contraction.
   b. Creatine phosphate stores energy that can be used to synthesize ATP.
   c. ATP is needed for muscle relaxation.
4. Oxygen supply and cellular respiration
   a. Aerobic respiration requires oxygen.
   b. Red blood cells carry oxygen to body cells.
   c. Myoglobin in muscle cells temporarily stores oxygen.

5. Oxygen debt
   a. During rest or moderate exercise, muscles receive enough oxygen to respire aerobically.
   b. During strenuous exercise, oxygen deficiency may cause lactic acid to accumulate.
   c. Oxygen debt is the amount of oxygen required to convert accumulated lactic acid to glucose and to restore supplies of ATP and creatine phosphate.
6. Muscle fatigue
   a. A fatigued muscle loses its ability to contract.
   b. Muscle fatigue is usually due to accumulation of lactic acid.
7. Heat production
   a. More than half of the energy released in cellular respiration is lost as heat.
   b. Muscle action is an important source of body heat.

8.4 Muscular Responses (p. 187)

1. Threshold stimulus is the minimal stimulus required to elicit a muscular contraction.
2. Recording a muscle contraction
   a. A twitch is a single, short contraction reflecting stimulation of a muscle fiber.
   b. A myogram is a recording of an electrically stimulated isolated muscle.
   c. The latent period, the time between stimulus and responding muscle contraction, is followed by a period of contraction and a period of relaxation.
3. Summation
   a. A rapid series of stimuli may produce summation of twitches.
   b. Forceful, sustained contraction without relaxation is a tetanic contraction.
4. Recruitment of motor units
   a. One motor neuron and the muscle fibers associated with it constitute a motor unit.
   b. All the muscle fibers of a motor unit contract together.
   c. Recruitment increases the number of motor units being activated in a whole muscle.
   d. The many motor units in a whole muscle are controlled by different motor neurons which respond to different thresholds of stimulation.
   e. At a low intensity of stimulation, small numbers of motor units contract.
   f. At increasing intensities of stimulation, other motor units are recruited until the muscle contracts with maximal force.
5. Sustained contractions
   a. Summation and recruitment together can produce a sustained contraction of increasing strength.
   b. Even when a muscle is at rest, its fibers usually remain partially contracted.

8.5 Smooth Muscle (p. 191)

The contractile mechanism of smooth muscle is similar to that of skeletal muscle.

1. Smooth muscle fibers
   a. Smooth muscle cells contain filaments of actin and myosin, less organized than those in skeletal muscle.
   b. Types include multiunit smooth muscle and visceral smooth muscle.
   c. Visceral smooth muscle displays rhythmicity and is self-exciting.

2. Smooth muscle contraction
   a. Two neurotransmitters—acetylcholine and norepinephrine— and hormones affect smooth muscle function.
   b. Smooth muscle can maintain a contraction longer with a given amount of energy than can skeletal muscle.
   c. Smooth muscles can change length without changing tension.

8.6 Cardiac Muscle (p. 191)
1. Like skeletal muscle cells, cardiac muscle cells have actin and myosin filaments that are well-organized and striated.
2. Cardiac muscle twitches last longer than skeletal muscle twitches.
3. Intercalated discs connect cardiac muscle cells.
4. A network of fibers contracts as a unit.
5. Cardiac muscle is self-exciting and rhythmic.

8.7 Skeletal Muscle Actions (p. 192)
The type of movement a skeletal muscle produces depends on the way the muscle attaches on either side of a joint.
1. Origin and insertion
   a. The immovable end of a skeletal muscle is its origin, and the movable end is its insertion.
   b. Some muscles have more than one origin.
2. Interaction of skeletal muscles
   a. Skeletal muscles function in groups.
   b. A prime mover is responsible for most of a movement. Synergists aid prime movers. Antagonists can resist the action of a prime mover.
   c. Smooth movements depend on antagonists giving way to the actions of prime movers.

8.8 Major Skeletal Muscles (p. 194)
1. Muscles of facial expression
   a. These muscles lie beneath the skin of the face and scalp and are used to communicate feelings through facial expression.
   b. They include the epicranius, orbicularis oculi, orbicularis oris, buccinator, zygomaticus, and platysma.
2. Muscles of mastication
   a. These muscles attach to the mandible and are used in chewing.
   b. They include the masseter and temporalis.
3. Muscles that move the head
   a. Muscles in the neck and upper back move the head.
   b. They include the sternocleidomastoid, splenius capitis, and semispinalis capitis.
4. Muscles that move the pectoral girdle
   a. Most of these muscles connect the scapula to nearby bones and closely associate with muscles that move the arm.
   b. They include the trapezius, rhomboid major, levator scapulae, serratus anterior, and pectoralis minor.

5. Muscles that move the arm
   a. These muscles connect the humerus to various regions of the pectoral girdle, ribs, and vertebral column.
   b. They include the coracobrachialis, pectoralis major, teres major, latissimus dorsi, supraspinatus, deltoid, subscapularis, infraspinatus, and teres minor.
6. Muscles that move the forearm
   a. These muscles connect the radius and ulna to the humerus or pectoral girdle.
   b. They include the biceps brachii, brachialis, brachioradialis, triceps brachii, supinator, pronator teres, and pronator quadratus.
7. Muscles that move the hand
   a. These muscles arise from the distal end of the humerus and from the radius and ulna.
   b. They include the flexor carpi radialis, flexor carpi ulnaris, palmaris longus, flexor digitorum profundus, extensor carpi radialis longus, extensor carpi radialis brevis, extensor carpi ulnaris, and extensor digitorum.
8. Muscles of the abdominal wall
   a. These muscles connect the rib cage and vertebral column to the pelvic girdle.
   b. They include the external oblique, internal oblique, transversus abdominis, and rectus abdominis.
9. Muscles of the pelvic outlet
   a. These muscles form the floor of the pelvic cavity and fill the space within the pubic arch.
   b. They include the levator ani, superficial transversus perinei, bulbospongiosus, and ischiocavernosus.
10. Muscles that move the thigh
    a. These muscles attach to the femur and to some part of the pelvic girdle.
    b. They include the psoas major, iliacus, gluteus maximus, gluteus medius, gluteus minimus, tensor fasciae latae, adductor longus, adductor magnus, and gracilis.
11. Muscles that move the leg
    a. These muscles connect the tibia or fibula to the femur or pelvic girdle.
    b. They include the biceps femoris, semitendinosus, semimembranosus, sartorius, and the quadriceps femoris group.
12. Muscles that move the foot
    a. These muscles attach the femur, tibia, and fibula to bones of the foot.
    b. They include the tibialis anterior, fibularis tertius, extensor digitorum longus, gastrocnemius, soleus, flexor digitorum longus, tibialis posterior, and fibularis longus.

Chapter Assessments

8.1 Introduction
1. The three types of muscle tissue are _____, _____, and _____. (p. 179)

8.2 Structure of a Skeletal Muscle
2. Describe the difference between a tendon and an aponeurosis. (p. 179)

3. Describe how connective tissue associates with skeletal muscle. (p. 179)
4. List the major parts of a skeletal muscle fiber, and describe the function of each part. (p. 179)
5. Describe a neuromuscular junction. (p. 182)

6. A neurotransmitter _____. (p. 182)

 a. binds actin filaments, causing them to slide

 b. diffuses across a synapse from a neuron to a muscle cell

 c. carries ATP across a synapse

 d. travels across a synapse from a muscle cell to a neuron

 e. is a contractile protein that is part of a skeletal muscle fiber

8.3 Skeletal Muscle Contraction

7. List the major events of muscle fiber contraction and relaxation. (p. 183)

8. Describe how ATP and creatine phosphate interact. (p. 185)

9. Describe how muscles obtain oxygen. (p. 186)

10. Describe how an oxygen debt may develop. (p. 186)

11. Explain how muscles may become fatigued. (p. 187)

12. Explain how skeletal muscle function affects the maintenance of body temperature. (p. 187)

8.4 Muscular Responses

13. Define *threshold stimulus*. (p. 187)

14. Sketch a myogram of a single muscular twitch, and identify the latent period, period of contraction, and period of relaxation. (p. 188)

15. Define *motor unit*. (p. 190)

16. Which of the following describes the addition of muscle fibers to take part in a contraction? (p. 190)

 a. summation

 b. recruitment

 c. tetany

 d. twitch

 e. relaxation

17. Explain how skeletal muscle stimulation produces a sustained contraction. (p. 190)

18. Distinguish between tetanic contraction and muscle tone. (p. 190)

8.5 Smooth Muscle

19. Distinguish between multiunit and visceral smooth muscle fibers. (p. 191)

20. Compare smooth and skeletal muscle contractions. (p. 191)

8.6 Cardiac Muscle

21. Make a table comparing contraction mechanisms of cardiac and skeletal muscle fibers. (p. 192)

8.7 Skeletal Muscle Actions

22. Distinguish between a muscle's origin and its insertion. (p. 192)

23. Define *prime mover, synergist,* and *antagonist.* (p. 194)

8.8 Major Skeletal Muscles

24. Match the muscles to their descriptions and functions. (pp. 194–207)

(1) Buccinator	A.	Inserted on coronoid process of mandible
(2) Epicranius	B.	Draws corner of mouth upward
(3) Orbicularis oris	C.	Can raise and adduct scapula
(4) Platysma	D.	Can pull head into an upright position
(5) Rhomboid major	E.	Raises eyebrow
(6) Splenius capitis	F.	Compresses cheeks
(7) Temporalis	G.	Extends over neck from chest to face
(8) Zygomaticus	H.	Closes lips
(9) Biceps brachii	I.	Extends forearm at elbow
(10) Brachialis	J.	Pulls shoulder back and downward
(11) Deltoid	K.	Abducts arm
(12) Latissimus dorsi	L.	Inserted on radial tuberosity
(13) Pectoralis major	M.	Pulls arm forward and across chest
(14) Pronator teres	N.	Rotates forearm medially
(15) Teres minor	O.	Inserted on coronoid process of ulna
(16) Triceps brachii	P.	Rotates arm laterally
(17) Biceps femoris	Q.	Inverts foot
(18) External oblique	R.	Member of quadriceps femoris group
(19) Gastrocnemius	S.	Plantar flexor of foot
(20) Gluteus maximus	T.	Compresses contents of abdominal cavity
(21) Gluteus medius	U.	Extends thigh
(22) Gracilis	V.	Hamstring muscle
(23) Rectus femoris	W.	Adducts thigh
(24) Tibialis anterior	X.	Abducts thigh

25. Which muscles can you identify in the bodies of these models? (pp. 194–207)

Integrative Assessments/Critical Thinking

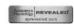

OUTCOMES 4.5, 8.3

1. As lactic acid and other substances accumulate in an active muscle, they stimulate pain receptors and the muscle may feel sore. How might the application of heat or substances that dilate blood vessels relieve such soreness?

OUTCOMES 5.3, 8.2

2. Discuss how connective tissue is part of the muscular system.

OUTCOMES 8.3, 8.4

3. A woman takes her daughter to a sports medicine specialist and requests that the specialist determine the percentage of fast- and slow-twitch fibers in the girl's leg muscles. The parent wants to know if the healthy girl should try out for soccer or cross-country running. Do you think this is a valid reason to test muscle tissue? Why or why not?

4. Following an injury to a nerve, the muscle it supplies with motor nerve fibers may become paralyzed. How would you explain to a patient the importance of moving the disabled muscles passively or contracting them using electrical stimulation?

OUTCOMES 8.5, 8.8

5. What steps might be taken to minimize atrophy of the skeletal muscles in patients confined to bed for prolonged times?

WEB CONNECTIONS

Visit the text website at **www.mhhe.com/shieress11** for additional quizzes, interactive learning exercises, and more.

APR

Anatomy & Physiology REVEALED includes cadaver photos that allow you to peel away layers of the human body to reveal structures beneath the surface. This program also includes animations, radiologic imaging, audio pronunciation, and practice quizzing. To learn more visit **www.aprevealed.com**.

9

Nervous System

Tennis Imagery   Spatial Navigation Imagery

Patient   SMA   PMC   PPC   PPA

Controls   SMA   PMC   PPC   PPA

X = -2   X = -24

"Islands of awareness" in the vegetative brain. The twenty-three-year-old had been in a persistent vegetative state for five months after sustaining traumatic brain injury in a car accident. She was awake, but apparently not aware, and unable to communicate in any way. To an observer, she had no sense of her own existence and did not react to sight or sound. But the young woman was aware—she just could not communicate. British researchers used functional MRI (fMRI), a form of neuroimaging that measures regional blood flow, to give the patient a way to communicate in response to a stimulus. In a preliminary experiment, fMRI tracked her response to speech. First she heard a sentence that made sense, and then a sentence that had the same cadence as the first but was all nonsense words. Her brain lit up in the speech-processing centers only when the sentence she heard had meaning. When she heard a sentence that included a homonym—a word that could have either of two meanings—an additional brain region lit up. Then the researchers asked her to imagine herself playing tennis and then walking through her house. Healthy individuals asked to do the same were controls. The young woman's brain and the control brains lit up in exactly the same areas. The researchers then devised a new set of experiments in which people in vegetative states who responded to "tennis" and "house" were told to use these imaginings as stand-ins for the words "yes" and "no." The test was to then ask the patients yes-no questions, and have them answer by imagining either being in a house or playing tennis, one meant "yes"

A woman who had suffered brain injury in a traffic accident and was in a persistent vegetative state was asked to imagine herself playing tennis and walking through the rooms of her home, while undergoing neuroimaging with functional MRI. Although she could not move or respond verbally, the patterns in which her brain lit up matched those of 12 healthy individuals as they completed the same tasks.

and one meant "no." The questions were highly specific about their pasts. When the patients imagined "tennis" or "house," the results were seen in the fMRI. They were correct every time, even after researchers switched the meanings from "tennis" signifying no and "house" yes to the opposite. This discovery of what the researchers call "islands of awareness" in the brain is changing the way that we view a persistent vegetative state. However, not all patients respond this way.

Learning Outcomes

After studying this chapter, you should be able to do the following:

9.1 Introduction

1. Distinguish between the two types of cells that compose nervous tissue. (p. 214)

2. Name the two major groups of nervous system organs. (p. 214)

9.2 General Functions of the Nervous System

3. Explain the general functions of the nervous system. (p. 215)

9.3 Neuroglia

4. State the functions of neuroglia in the central nervous system. (p. 216)

5. Distinguish among the types of neuroglia in the central nervous system. (p. 216)

6. Describe the Schwann cells of the peripheral nervous system. (p. 216)

9.4 Neurons

7. Describe the general structure of a neuron. (p. 216)

8. Explain how differences in structure and function are used to classify neurons. (p. 219)

9.5 The Synapse

9. Explain how information passes from one neuron to another. (p. 221)

9.6 Cell Membrane Potential

10. Explain how a membrane becomes polarized. (p. 222)

11. Describe the events that lead to the generation of an action potential. (p. 225)

9.7 Nerve Impulses

12. Compare nerve impulse conduction in myelinated and unmyelinated neurons. (p. 227)

9.8 Synaptic Transmission

13. Identify the changes in membrane potential associated with excitatory and inhibitory neurotransmitters. (p. 228)

 Learn   Practice   Assess

9.9 Impulse Processing

14. Describe the general ways in which the nervous system processes information. (p. 228)

9.10 Types of Nerves

15. Describe how peripheral nerves are classified. (p. 230)

9.11 Nerve Pathways

16. Describe the function of each part of a reflex arc, and name two reflex examples. (p. 231)

9.12 Meninges

17. Describe the coverings of the brain and spinal cord. (p. 232)

9.13 Spinal Cord

18. Describe the structure of the spinal cord and its major functions. (p. 234)

9.14 Brain

19. Name the major parts and functions of the brain. (pp. 236–246)

20. Distinguish among motor, sensory, and association areas of the cerebral cortex. (p. 238)

21. Describe the location, formation, and function of cerebrospinal fluid. (p. 240)

9.15 Peripheral Nervous System

22. List the major parts of the peripheral nervous system. (p. 246)

23. Name the cranial nerves, and list their major functions. (p. 246)

24. Describe the structure of a spinal nerve. (p. 249)

9.16 Autonomic Nervous System

25. Describe the functions of the autonomic nervous system. (p. 250)

26. Distinguish between the sympathetic and parasympathetic divisions of the autonomic nervous system. (p. 250)

27. Describe a sympathetic and a parasympathetic nerve pathway. (p. 251)

Anatomy & Physiology REVEALED®
aprevealed.com

Module 7: Nervous System

Aids to Understanding Words   *(Appendix A on page 564 has a complete list of Aids to Understanding Words.)*

ax- [axis] *ax*on: Cylindrical nerve fiber that carries impulses away from a neuron cell body.

dendr- [tree] *dendr*ite: Branched nerve cell process that serves as a receptor surface of a neuron.

funi- [small cord or fiber] *funi*culus: Major nerve tract or bundle of myelinated nerve cell axons in the spinal cord.

gangli- [a swelling] *gangli*on: Mass of neuron cell bodies.

-lemm [rind or peel] neuri*lemm*a: Sheath that surrounds the myelin of a nerve cell axon.

mening- [membrane] *mening*es: Membranous coverings of the brain and spinal cord.

moto- [moving] *moto*r neuron: Neuron that stimulates a muscle to contract or a gland to secrete.

peri- [around] *peri*pheral nervous system: Portion of the nervous system that consists of nerves branching from the brain and spinal cord.

plex- [interweaving] choroid *plex*us: Mass of specialized capillaries associated with spaces in the brain.

sens- [feeling] *sens*ory neuron: Neuron that conducts impulses into the brain or spinal cord.

syn- [together] *syn*apse: Junction between two neurons.

ventr- [belly or stomach] *ventr*icle: Fluid-filled space in the brain.

9.1 INTRODUCTION

Feeling, thinking, remembering, moving, and being aware of the world require activity from the nervous system. This vast collection of cells also helps coordinate all other body functions to maintain homeostasis and to enable the body to respond to changing conditions. Information from inside and outside the body is brought to the brain and spinal cord, which then stimulate responses from muscles and glands.

Recall from chapter 5 (p. 111) that nervous tissue consists of masses of nerve cells, or **neurons.** These cells are the main functional units of the nervous system and are specialized to react to physical and chemical changes in their surroundings (fig. 9.1). Neurons transmit information in the form of electrochemical changes, often called **nerve impulses,** which allow them to communicate with other neurons and with cells outside the nervous system.

Neurons typically have a rounded area called the **cell body,** and two types of extensions: dendrites and axons. **Dendrites,** which may be numerous, receive electrochemical messages. **Axons** are extensions that send information in the form of nerve impulses. Usually a neuron has only one axon. Figure 9.1 depicts these major parts of a neuron.

Nervous tissue also includes **neuroglia** that provide physical support, insulation, and nutrients for neurons. During development before birth, neuroglia release and relay signals that guide the differentiation of neurons from progenitor cells (see chapter 3, p. 71).

The organs of the nervous system can be divided into two groups. One group, consisting of the brain and spinal cord, forms the **central nervous system (CNS).** The other, composed of the nerves (bundles of axons) that connect the central nervous system to other body parts, is called the **peripheral nervous system (PNS)** (fig. 9.2). Together, these systems provide three general functions: sensory, integrative, and motor.

Practice

1. What are the two major types of cells that form nervous tissue?

2. What are the two major subdivisions of the nervous system?

9.2 GENERAL FUNCTIONS OF THE NERVOUS SYSTEM

The *sensory function* of the nervous system derives from **sensory receptors** (sen′so-re re-sep′torz) at the ends of peripheral neurons (see chapter 10, p. 263). These receptors gather information by detecting changes inside and outside the body. Sensory receptors monitor external environmental factors, such as light and sound intensities, and conditions of the body's internal environment, such as temperature and oxygen level.

Sensory receptors convert environmental information into nerve impulses, which are then transmitted over peripheral nerves to the central nervous system. There, the signals are integrated; that is, they are brought together, creating sensations, adding to memory, or helping produce thoughts that translate sensations into perceptions. As a result of this *integrative function,* we make conscious or subconscious decisions, and then we use *motor functions* to act on them.

Figure 9.1

Neurons are the structural and functional units of the nervous system (600×). The dark spots in the area surrounding the neuron are neuroglia. Note the dendrites and the single axon of the neuron.

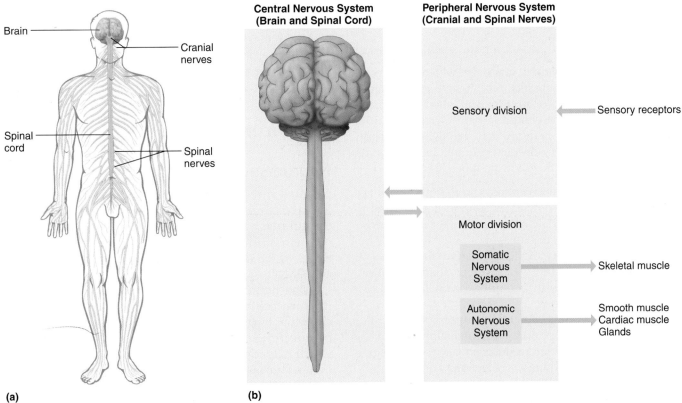

(a)

(b)

Figure 9.2

Nervous system. **(a)** The nervous system includes the central nervous system (brain and spinal cord) and the peripheral nervous system (cranial nerves and spinal nerves). **(b)** The nervous system receives information from sensory receptors and initiates responses through effector organs (muscles and glands).

The motor functions of the nervous system employ peripheral neurons, which carry impulses from the central nervous system to responsive structures called **effectors** (e-fek′torz). Effectors, which are outside the nervous system, include muscles and glands whose actions are either controlled or modified by nerve impulses.

The motor functions of the peripheral nervous system can be divided into two categories. Those that are under voluntary (conscious) control compose the **somatic nervous system,** which controls skeletal muscle. In contrast, the **autonomic nervous system** controls effectors that are involuntary, such as cardiac muscle, smooth muscle, and various glands.

The nervous system can detect changes outside and inside the body, make decisions based on the information received, and stimulate muscles or glands to respond. Typically these responses counteract the effects of the changes detected, and in this way the nervous system helps maintain homeostasis.

Practice

3. How do sensory receptors collect information?
4. How does the central nervous system integrate incoming information?
5. What are the two types of motor functions of the nervous system?

9.3 NEUROGLIA

Neurons cannot exist without neuroglia, which fill spaces, provide structural frameworks, produce the components of the electrical insulator **myelin** (mi′ĕ-lin), and carry on phagocytosis. In the central nervous system, neuroglia greatly outnumber neurons, and can divide, whereas neurons do not normally divide. Neuroglia in the central nervous system are of the following types (fig. 9.3):

1. **Microglial cells** are scattered throughout the central nervous system. They support neurons and phagocytize bacterial cells and cellular debris, and form scars in areas of damage.
2. **Oligodendrocytes** align along nerve fibers. They provide insulating layers of myelin, called a **myelin sheath** (mi′ĕ-lin shēth) around axons within the brain and spinal cord.
3. **Astrocytes,** commonly found between neurons and blood vessels, provide structural support, join parts by their abundant cellular processes, and help regulate the concentrations of nutrients and ions within the tissue. Astrocytes also form scar tissue that fills spaces following injury to the CNS.
4. **Ependymal cells** form an epithelial-like membrane that covers specialized brain parts (choroid plexuses) and form the inner linings that enclose spaces in the brain (ventricles) and spinal cord (central canal).

Neuroglia assemble in a special, protective way in the brain. Most capillaries (the smallest blood vessels) are "leaky," allowing small molecules to enter or leave the bloodstream. The cells that form capillaries in the brain, in contrast, are much more tightly connected, thanks partly to astrocytes. This specialized architecture creates a "blood–brain barrier" that shields delicate brain tissue from chemical fluctuations, blocking entry to many substances. Drug developers often have to invent ways to circumvent the barrier in delivering drugs to the brain. For example, experimental gene therapies for certain brain diseases deliver DNA through catheters placed in holes drilled into the skull.

The peripheral nervous system includes neuroglia called **Schwann cells.** They produce a myelin sheath around axons of myelinated neurons.

> Too few and too many neuroglia can harm health. Fast-growing gliomas are brain tumors consisting of rapidly dividing neuroglia (neurons do not divide). Immediately after a spinal cord injury, destruction of neuroglia strips axons of myelin. Subsequent overgrowth of neuroglia forms scars, which impede recovery of function.

Practice

6. List the functions of the cells that support neurons.
7. Distinguish among the types of neuroglia in the central nervous system.
8. What is the function of Schwann cells in the peripheral nervous system?

9.4 NEURONS

Neuron Structure

Neurons vary considerably in size and shape, but they all have common features. These include a cell body; the tubular, cytoplasm-filled dendrites, which conduct nerve impulses to the neuron cell body; and an axon, which conducts impulses away from the neuron cell body.

The neuron cell body consists of granular cytoplasm, a cell membrane, and organelles such as mitochondria, lysosomes, a Golgi apparatus, and a network of fine threads called **neurofibrils** (nu″ro-fi′brilz), which extends into the axon. Scattered throughout the cytoplasm are many membranous sacs called **chromatophilic substance** (Nissl bodies). These are similar to rough endoplasmic reticulum in other cells (fig. 9.4). Ribosomes attached to chromatophilic substance function in protein synthesis, as they do elsewhere. Near the center of the cell body is a large, spherical nucleus with a conspicuous nucleolus.

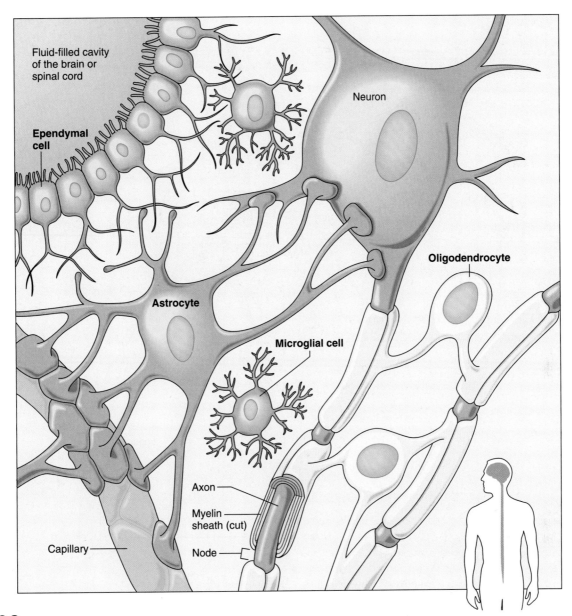

Figure 9.3

Types of neuroglia in the central nervous system include the oligodendrocyte, astrocyte, microglial cell, and ependymal cell. (Ependymal cells have cilia into early childhood. In adults, cilia remain only on ependymal cells in the ventricles of the brain.)

Dendrites are usually short and highly branched. These processes, together with the membrane of the cell body, are the neuron's main receptive surfaces with which axons from other neurons communicate.

In most neurons the axon arises from the cell body as a cone-shaped thickening called the *axon hillock.* Many mitochondria, microtubules, and neurofibrils are in the axon cytoplasm. An axon originates as a single structure but may give off side branches (collaterals). Its end may branch into many fine extensions that contact the receptive surfaces of other cells.

Larger axons of peripheral neurons are enclosed in *myelin sheaths* produced by Schwann cells (figs. 9.4 and 9.5). These cells wind tightly around axons, some-what like a bandage wrapped around a finger, coat-ing them with many layers of cell membrane that have little or no cytoplasm between them. The parts of the Schwann cells that contain most of the cytoplasm and the nuclei remain outside the myelin sheath and com-pose a **neurilemma** (nu″rĭ-lem′ah), or neurilemmal sheath, which surrounds the myelin sheath. Narrow gaps between Schwann cells are called **nodes of Ran-vier** (nō-dz uv ron′vee-ay) (fig. 9.5).

Axons with myelin sheaths are called *myelinated,* and those that lack sheaths are *unmyelinated.* Myelin is also in the CNS, where groups of myelinated axons appear white, and masses of such axons form the *white matter.* Unmyelinated axons and neuron cell bodies form *gray matter* in the CNS.

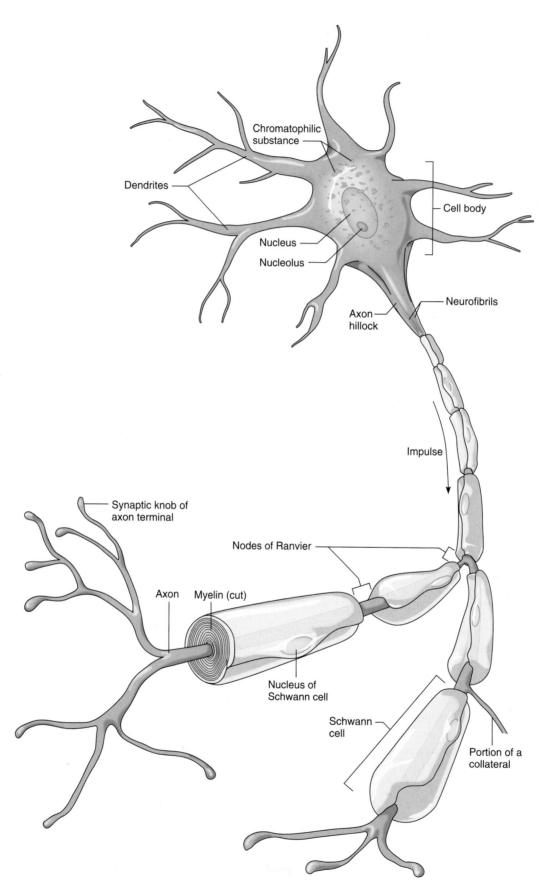

Figure 9.4

A common neuron.

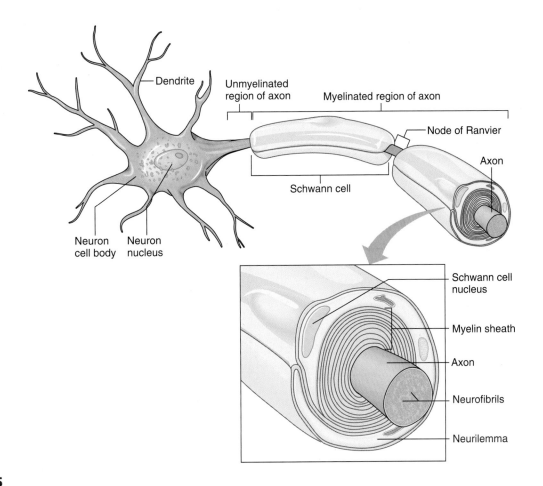

Figure 9.5

The portion of a Schwann cell that winds tightly around an axon forms a myelin sheath, and the cytoplasm and nucleus of the Schwann cell, remaining outside, form a neurilemma, or neurilemmal sheath.

Myelin begins to form on axons during the fourteenth week of prenatal development. Yet many of the axons in newborns are not completely myelinated. As a result, an infant's nervous system cannot function as effectively as that of an older child or adult. Infants' responses to stimuli are coarse and undifferentiated, and may involve the whole body. All myelinated axons begin to develop sheaths by the time a child starts to walk, and myelination continues into adolescence. Deficiencies of essential nutrients during the developmental years may limit myelin formation, which may impair nervous system function later in life.

When peripheral nerves are damaged, their axons can regenerate. The neurilemma plays an important role in this process. In contrast, CNS axons are myelinated by oligodendrocytes, which do not provide a neurilemma. Consequently, damaged CNS neurons usually do not regenerate.

The brain harbors small collections of neural stem cells that can divide to give rise to new neurons or neuroglia, depending upon their chemical surroundings. Neural stem cells are found in the hippocampus and near the brain's ventricles.

To picture the relative sizes of a typical neuron's parts, imagine that the cell body is the size of a tennis ball. The axon would then be a mile long and half an inch thick. The dendrites would fill a large bedroom.

Classification of Neurons

Neurons differ in the structure, size, and shape of their cell bodies. They also vary in the length and size of their axons and dendrites and in the number of connections they make with other neurons.

On the basis of structural differences, neurons are classified into three major groups (fig. 9.6). Each type of neuron is specialized to send a nerve impulse in one direction.

1. **Multipolar neurons** have many processes arising from their cell bodies. Only one process of each neuron is an axon; the rest are dendrites. Most neurons whose cell bodies lie within the brain or spinal cord are multipolar.

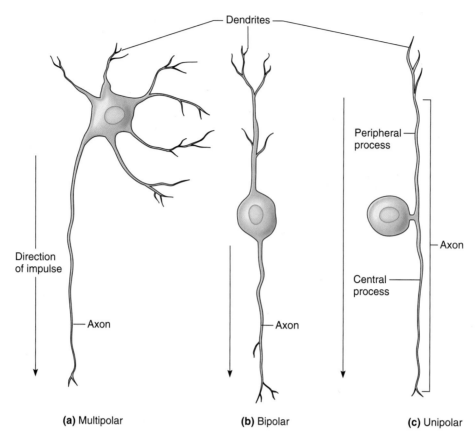

Figure 9.6

Structural types of neurons include **(a)** the multipolar neuron, **(b)** the bipolar neuron, and **(c)** the unipolar neuron.

2. **Bipolar neurons** have only two processes, one arising from each end of the cell body. These processes are structurally similar, but one is an axon and the other a dendrite. Neurons in specialized parts of the eyes, nose, and ears are bipolar.

3. **Unipolar neurons** have a single process extending from the cell body. A short distance from the cell body, this process divides into two branches, which really function as a single axon. One branch (the peripheral process) is associated with dendrites near a peripheral body part. The other branch (the central process) enters the brain or spinal cord. The cell bodies of some unipolar neurons aggregate in specialized masses of nervous tissue called **ganglia** (gang′gle-ah) (singular, *ganglion*), which are located outside the brain and spinal cord.

Neurons also vary in function. Different neurons may transmit impulses into the brain or spinal cord, transmit impulses from one area of the brain or spinal cord to another, or transmit impulses out of the brain or spinal cord. On the basis of functional differences, neurons are grouped as follows (fig. 9.7):

1. **Sensory neurons** (afferent neurons) transmit nerve impulses from peripheral body parts into the brain or spinal cord. Sensory neurons either have specialized

receptor ends at the tips of their dendrites, or they have dendrites that are closely associated with *receptor cells* in the skin or in sensory organs.

Changes that occur inside or outside the body stimulate receptor ends or receptor cells, triggering sensory nerve impulses. The impulses travel along the sensory neuron axons, which lead to the brain or spinal cord, where other neurons can process the impulses. Most sensory neurons are unipolar; some are bipolar.

2. **Interneurons** (also called *association* or *internuncial neurons*) lie entirely within the brain or spinal cord. They are multipolar and link other neurons. Interneurons transmit impulses from one part of the brain or spinal cord to another. That is, they may direct incoming sensory impulses to appropriate parts of the CNS for processing and interpreting. Other impulses are transferred to motor neurons. The cell bodies of some interneurons aggregate in specialized masses of nervous tissue called **nuclei** (singular, *nucleus*). Nuclei are similar to ganglia, but are within the central nervous system.

3. **Motor neurons** (efferent neurons) are multipolar and transmit impulses out of the brain or spinal cord to effectors. Motor impulses control muscle contraction and the secretions of glands.

Central Nervous System   **Peripheral Nervous System**

Cell body

Cell body

Axon (central process)

Interneurons

Axon

Axon

Dendrites

Sensory receptor

Axon (peripheral process)

Sensory (afferent) neuron

Motor (efferent) neuron

Effector (muscle or gland)

Axon terminal

Figure 9.7

Neurons are classified by function as well as structure. Sensory (afferent) neurons carry information into the central nervous system (CNS), interneurons are completely within the CNS, and motor (efferent) neurons carry instructions to the peripheral nervous system (PNS).

Neurons deprived of oxygen change shape as their nuclei shrink, and they eventually disintegrate. Oxygen deficiency can result from lack of blood flow (ischemia) through nervous tissue, an abnormally low blood oxygen concentration (hypoxemia), or toxins that prevent neurons from using oxygen by blocking aerobic respiration.

Practice

9. Distinguish between a dendrite and an axon.
10. Describe the components of a neuron.
11. Describe how a myelin sheath forms.
12. Explain why axons of peripheral nerves can regenerate, but axons of central nervous system nerves cannot.
13. Name three groups of neurons based on structure and three groups based on function.

9.5 THE SYNAPSE

The junction between any two communicating neurons is called a **synapse** (sin′aps). The neurons at a synapse are not in direct physical contact, but are separated by a gap called a *synaptic cleft*. Communication along a nerve pathway must cross these gaps (fig. 9.8).

When you receive a text message, the person writing the message is the sender and you are the receiver. Similarly, the neuron transmitting the impulse into the synapse is the sender, or *presynaptic neuron*. The neuron that receives input at the synapse is the receiver, or *postsynaptic neuron*. The process whereby this message crosses the synaptic cleft is called *synaptic transmission*. Clinical Application 9.1 discusses some factors that affect synaptic transmission.

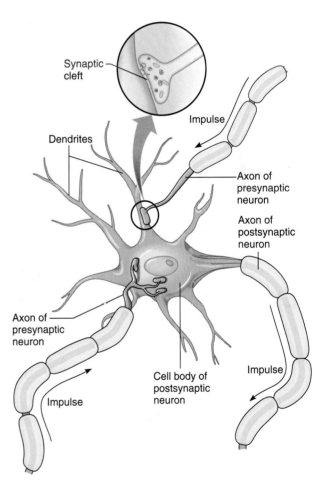

Figure 9.8

Synapses separate neurons. For an impulse to continue from one neuron to another, it must cross the synaptic cleft at a synapse. A synapse is usually between an axon and a dendrite or between an axon and a cell body.

Synaptic transmission is a one-way process carried out by biochemicals called **neurotransmitters.** The distal ends of axons have one or more extensions called *synaptic knobs,* absent in dendrites, which contain many membranous sacs, called *synaptic vesicles.* When a nerve impulse reaches a synaptic knob, some of the synaptic vesicles release neurotransmitter (figs. 9.9 and 9.10). The neurotransmitter diffuses across the synaptic cleft and reacts with specific receptors on the postsynaptic neuron membrane.

The action of neurotransmitter on a postsynaptic cell is either excitatory (turning a process on) or inhibitory (turning a process off). The net effect on the postsynaptic cell depends on the combined effect of the excitatory and inhibitory inputs from as few as one and as many as 10,000 presynaptic neurons.

9.6 CELL MEMBRANE POTENTIAL

The surface of a cell membrane (including a nonstimulated or *resting* neuron) is usually electrically charged, or *polarized,* with respect to the inside. This polarization arises from an unequal distribution of positive and negative ions across the membrane, and it is particularly important in the conduction of muscle and nerve impulses. A characteristic change in neuron membrane polarization and return to the resting state, called an *action potential,* forms a nerve impulse that is propagated along an axon.

Distribution of Ions

Because of the active transport of sodium and potassium ions, cells throughout the body have a greater concentration of sodium ions (Na^+) outside and a greater concentration of potassium ions (K^+) inside (see chapter 3, p. 65). The cytoplasm of these cells has many large, negatively charged particles, including phosphate ions (PO_4^{-3}), sulfate ions (SO_4^{-2}), and proteins, that cannot diffuse across the cell membranes.

Chapter 3 (p. 53) introduced cell membranes as selectively permeable phospholipid bilayers. Channels in the cell membranes partly determine the distribution of ions inside and outside of cells (see chapter 3, p. 54). Some channels are always open. Others, called "gated" channels, can be opened or closed. Furthermore, channels can be selective; that is, a channel may allow one kind of ion to pass through and exclude other kinds (fig. 9.11).

Potassium ions pass through cell membranes much more easily than sodium ions. This makes potassium ions a major contributor to membrane polarization. Calcium ions are less able to cross the resting cell membrane than either sodium ions or potassium ions, and have a special role in nerve function, described later on page 228.

Figure 9.9 AP|R

Action across a synapse. When a nerve impulse reaches the synaptic knob at the end of an axon, synaptic vesicles release a neurotransmitter that diffuses across the synaptic cleft and binds to specific receptors on the postsynaptic membrane.

Clinical Application 9.1

Factors Affecting Synaptic Transmission

Nerve impulses reaching synaptic knobs too rapidly can exhaust neurotransmitter supplies, and impulse conduction ceases until more neurotransmitters are synthesized. This happens during an epileptic seizure. Abnormal and too rapid impulses originate from certain brain cells and reach skeletal muscle fibers, stimulating violent contractions. In time, the synaptic knobs run out of neurotransmitters and the seizure subsides.

A drug called Dilantin (diphenylhydantoin) treats seizure disorders by blocking gated sodium channels, thereby limiting the frequency at which action potentials can occur. Many other drugs affect synaptic transmission. For example, caffeine in coffee, tea, and cola drinks stimulates nervous system activity by lowering the thresholds at synapses so that neurons are more easily excited. Antidepressants called "selective serotonin reuptake inhibitors" keep the neurotransmitter serotonin in synapses longer, compensating for a still little-understood deficit that presumably causes depression.

Resting Potential

Sodium and potassium ions follow the laws of diffusion discussed in chapter 3 (pp. 60–61) and show a net movement from high concentration to low concentration as permeabilities permit. Because a resting cell membrane is more permeable to potassium ions than to sodium ions, potassium ions diffuse out of the cell more rapidly than sodium ions can diffuse in (fig. 9.12*a*). Every millisecond, more positive charges leave the cell by diffusion than enter it. As a result, the outside of the cell membrane gains a slight surplus of positive charges, and the inside is left with a slight surplus of impermeant negative charges (fig. 9.12*b*).

The difference in electrical charge between two regions is called a *potential difference.* In a resting nerve cell, the potential difference between the region inside the membrane and the region outside the membrane is called a **resting potential.** As long as a nerve cell membrane is undisturbed, the membrane remains in this polarized state. At the same time, the cell continues to expend energy to drive the Na$^+$/K$^+$ "pumps" that actively transport

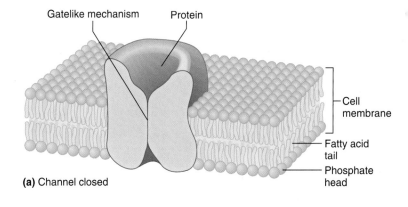

(a) Channel closed

(b) Channel open

Figure 9.11

Cell membrane polarization is necessary for nerve transmission, and depends upon the movements of ions through channels. A gatelike mechanism can **(a)** close or **(b)** open some of the channels in cell membranes through which ions pass.

Figure 9.10

This transmission electron micrograph of a synaptic knob shows abundant synaptic vesicles, which are filled with neurotransmitter molecules.

(a) If we imagine a cell before the membrane potential is established, concentration gradients are such that diffusion of potassium ions out of the cell exceeds diffusion of sodium ions into the cell, causing a net loss of positive charge from the cell.

(b) The net loss of positive charges from the inside of the cell has left the inside of the cell membrane slightly negative compared to the outside of the membrane, which is left slightly positive. This difference (an electrical "potential difference") is measured as −70 millivolts (mV) in a typical neuron, and is called the resting membrane potential.

(c) With the membrane potential established, sodium diffusion into the cell is now aided, and potassium diffusion opposed, by the negative charge on the inside of the membrane. As a result, slightly more sodium ions enter the cell than potassium ions leave, but the action of the sodium/potassium pump balances these movements, and as a result the concentrations of these ions, and the resting membrane potential, are maintained.

Figure 9.12

The resting potential. **(a)** Conditions that lead to the resting potential. **(b)** In the resting neuron, the inside of the membrane is negative relative to the outside. **(c)** The Na$^+$/K$^+$ pump maintains the concentration gradients for Na$^+$ and K$^+$ ions.

Q: *Constant activity of the Na$^+$/K$^+$ pump requires a constant supply of which substance?*

Answer can be found in Appendix E on page 568.

sodium and potassium ions in opposite directions. The pump maintains the concentration gradients responsible for diffusion of these ions in the first place (fig. 9.12c).

Potential Changes

Nerve cells are excitable; that is, they can respond to changes in their surroundings. Some nerve cells, for example, are specialized to detect changes in temperature, light, or pressure from outside the body. Many neurons respond to neurotransmitters from other neurons. Such changes (or stimuli) usually affect the resting potential in a particular region of a nerve cell membrane. If the membrane's resting potential decreases (as the inside of the membrane becomes less negative when compared to the outside), the membrane is said to be *depolarized* (fig. 9.13a).

Local potential changes are graded. This means that the magnitude of change in the resting potential is directly proportional to the intensity of the stimulus.

That is, if the membrane is being depolarized, then the greater the stimulus, the greater the depolarization. If neurons are depolarized sufficiently, the membrane potential reaches a level called the **threshold potential,** which is approximately –55 millivolts. If threshold is reached, an **action potential** results.

Action Potential

Recall that the axon arises from the cell body as a thickened region called the axon hillock. Functionally, this is referred to as the trigger zone. At the threshold potential, permeability changes at the trigger zone of the neuron being stimulated. Here, gated channels sensitive to changes in membrane potential, and highly selective for sodium ions, open and allow sodium to diffuse freely inward (figs. 9.13b and 9.14b). The negative electrical condition on the inside of the membrane aids this movement by attracting the positively-charged sodium ions.

As sodium ions diffuse inward, the membrane loses its negative electrical charge and becomes depolarized (more positive). At almost the same time, however, gated membrane channels open that allow potassium ions to pass through, and as these positive ions diffuse outward, the inside of the membrane becomes negatively charged

(a) If sodium or potassium channels open, more of that particular ion will cross the cell membrane, and the resting membrane potential will be altered. Here we see the effect of sodium channels opening in response to neurotransmitter. As sodium ions enter the cell, the membrane potential becomes more positive (or less negative), changing from –70 millivolts to –62 millivolts in this example. This change in a positive direction is called depolarization.

(b) If enough sodium ions enter the cell, the membrane potential depolarizes to a value called threshold, here shown to be –55 millivolts. A threshold depolarization causes another type of sodium channel to open. These channels are found along the axon of the neuron, especially near the beginning in an area known as the "trigger zone," and their opening is what triggers the action potential.

Figure 9.13

Action potentials. **(a)** A subthreshold depolarization will not result in an action potential. **(b)** Stimulation from multiple presynaptic neurons may cause the postsynaptic neuron to reach threshold, opening voltage-gated channels at the trigger zone.

once more (fig. 9.14*c*). The membrane potential may briefly become overly negative (*hyperpolarization*), but the membrane quickly returns to the resting potential (*repolarization*), and it remains in this state until stimulated again.

This rapid sequence of depolarization and repolarization, which takes about one-thousandth of a second (one millisecond), is the action potential (fig. 9.14*d*).

Because only a small fraction of the sodium and potassium ions move through the membrane during an action potential, action potentials can occur again and again, and resting potentials can be reestablished, before the original concentrations of these ions change significantly. Also, active transport across the membrane maintains the original concentrations of sodium and potassium ions on either side.

Figure 9.14 AP|R

Action potential. **(a)** At rest, the membrane potential is negative. **(b)** When the membrane reaches threshold, sodium channels open, some sodium (Na⁺) diffuses in, and the membrane is depolarized. **(c)** Soon afterward, potassium channels open, potassium (K⁺) diffuses out, and the membrane is repolarized. (For simplicity, negative ions are not shown.) **(d)** A recording of an action potential.

14. Describe the events that occur at a synapse.

15. Summarize how a nerve fiber becomes polarized.

16. List the major events of an action potential.

9.7 NERVE IMPULSES

An action potential in one region of a nerve cell membrane causes a bioelectric current to flow to adjacent regions of the membrane. This *local current* stimulates the adjacent membrane to its threshold level and triggers another action potential. This, in turn, stimulates the next adjacent region. In this way, a wave of action potentials moves down the axon to the end. This propagation of an action potential along a nerve axon is a nerve impulse (even though it is occurring in a single neuron) (fig. 9.15). Table 9.1 summarizes the events leading to a nerve impulse.

> The term "nerve impulse" is sometimes used to describe the wave of action potentials that propagates down the axon. Technically, this term is inaccurate, since the action potential occurs on a neuron, not on a nerve. However, we have retained some use of it here in the hope that familiarity will make the concept easier to grasp.

Impulse Conduction

An unmyelinated axon conducts an impulse along its entire length. A myelinated axon functions differently because myelin insulates and prevents almost all ion flow through the membrane it encloses. The myelin sheath would prevent a nerve impulse altogether if the sheath was continuous. However, nodes of Ranvier between Schwann cells interrupt the sheath (see fig. 9.4). Action potentials occur at these nodes, where the exposed axon membrane has sodium and potassium channels. In this case, the adjacent membrane that is brought to threshold is at the next node down the axon. A nerve impulse traveling along a myelinated axon thus appears to jump from node to node. This type of impulse conduction, termed saltatory, is many times faster than conduction on an unmyelinated axon.

The speed of a nerve impulse is proportional to the diameter of the axon—the greater the diameter, the faster the impulse. For example, an impulse on a relatively thick myelinated axon, such as that of a motor neuron associated with a skeletal muscle, might travel 120 meters per second. An impulse on a thin, unmyelinated axon, such as that of a sensory neuron associated with the skin, might move only 0.5 meter per second.

All-or-None Response

An action potential is not graded; rather it is an *all-or-none response*. That is, if a neuron responds at all, it responds completely. Thus, an action potential occurs

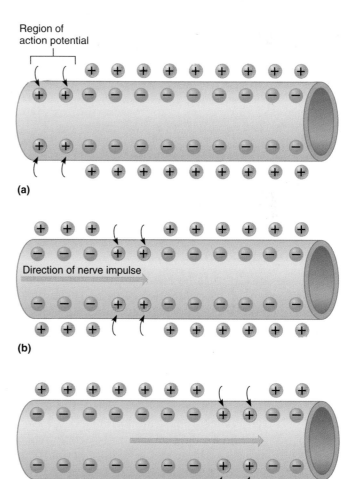

(a)

Region of action potential

(b)

Direction of nerve impulse

(c)

Figure 9.15 AP|R

A nerve impulse. **(a)** An action potential in one region stimulates the adjacent region, and **(b)** and **(c)** a wave of action potentials (a nerve impulse) moves along the axon.

Table 9.1	Events Leading to the Conduction of a Nerve Impulse

1. Neuron membrane maintains resting potential.

2. Threshold stimulus is received.

3. Sodium channels in the trigger zone of the neuron open.

4. Sodium ions diffuse inward, depolarizing the membrane.

5. Potassium channels in the membrane open.

6. Potassium ions diffuse outward, repolarizing the membrane.

7. The resulting action potential causes a local bioelectric current that stimulates adjacent portions of the membrane.

8. A wave of action potentials travels the length of the axon as a nerve impulse.

whenever a stimulus of threshold intensity or above is applied to an axon, and all action potentials propagated on that axon are of the same strength. A greater intensity of stimulation does not produce a stronger action potential; instead it produces more action potentials per second.

For a very short time following an action potential, a threshold stimulus will not trigger another action potential on that axon. This brief period, called the *refractory period,* limits the frequency of action potentials, and thus nerve impulses, along an axon. It also ensures that the nerve impulse proceeds in only one direction—down the axon, because the area upstream from where the action potential has just occurred is still in the refractory period from the previous action potential. Although a frequency of 700 impulses per second is possible, 100 impulses per second is more common.

Practice

17. What is the relationship between action potentials and nerve impulses?

18. Explain how impulse conduction differs in myelinated and unmyelinated nerve fibers.

19. Define *all-or-none response* as it relates to nerve impulse conduction.

9.8 SYNAPTIC TRANSMISSION

Neurotransmitters have various effects when they diffuse across the synaptic cleft and react with specific receptor molecules in the postsynaptic neuron membrane.

Excitatory and Inhibitory Actions

Neurotransmitters that increase postsynaptic membrane permeability to sodium ions will bring the postsynaptic membrane closer to threshold and may trigger nerve impulses. Such neurotransmitters are **excitatory.** Neurotransmitters that make it less likely that threshold will be reached are called **inhibitory,** because they decrease the chance that a nerve impulse will occur.

The synaptic knobs of a thousand or more neurons may communicate with the dendrites and cell body of a single postsynaptic neuron. Neurotransmitters released by some of these knobs have an excitatory action, while those from other knobs have an inhibitory action. The overall effect on the postsynaptic neuron depends on which presynaptic knobs are activated from moment to moment. If more excitatory than inhibitory neurotransmitters are released, the postsynaptic neuron's threshold may be reached, and an action potential triggered. Conversely, if most of the neurotransmitters released are inhibitory, threshold may not be reached.

Neurotransmitters

More than 100 neurotransmitters have been identified in the nervous system. Some neurons release only one, while others produce two or three. The neurotransmitters include *acetylcholine,* which stimulates skeletal muscle contractions (see chapter 8, p. 183); a group of compounds called *biogenic amines* (such as epinephrine, norepinephrine, dopamine, and serotonin), which form from modified amino acids; several *amino acids* (such as glycine, glutamic acid, aspartic acid, and gamma-aminobutyric acid—GABA); and more than 50 *neuropeptides,* which are short chains of amino acids. Acetylcholine and norepinephrine are excitatory. GABA and glycine are inhibitory. Neurotransmitters are usually synthesized in the cytoplasm of the synaptic knobs and stored in the synaptic vesicles. Table 9.2 lists some neurotransmitters and their actions.

When an action potential reaches the membrane of a synaptic knob, it increases the membrane's permeability to calcium ions by opening calcium ion channels in the membrane. Consequently, calcium ions diffuse inward, and in response some synaptic vesicles fuse with the membrane and release their contents into the synaptic cleft. After being released, some neurotransmitters are decomposed by enzymes. For example, the enzyme *acetylcholinesterase* decomposes acetylcholine and is present in the synapse and on the postsynaptic membrane of neuromuscular junctions, which control skeletal muscle contraction. Other neurotransmitters are transported back into the synaptic knob that released them (reuptake) or into nearby neurons or neuroglia. Decomposition or removal of neurotransmitters prevents continuous stimulation of postsynaptic neurons. Table 9.3 summarizes the events leading to the release of a neurotransmitter.

Practice

20. Distinguish between the actions of excitatory and inhibitory neurotransmitters.

21. What types of chemicals function as neurotransmitters?

22. What are possible fates of neurotransmitters?

9.9 IMPULSE PROCESSING

The way the nervous system processes and responds to nerve impulses reflects, in part, the organization of neurons and their axons in the brain and spinal cord.

Neuronal Pools

Neurons in the CNS are organized into **neuronal pools.** These are groups of neurons that make hundreds of synaptic connections with each other and work together to

Table 9.2	Some Neurotransmitters and Representative Actions		
Neurotransmitter	**Location**	**Major Actions**	
Acetylcholine	CNS	Controls skeletal muscle actions	
	PNS	Stimulates skeletal muscle contraction at neuromuscular junctions. May excite or inhibit at autonomic nervous system synapses	
Monoamines			
Norepinephrine	CNS	Creates a sense of feeling good; low levels may lead to depression	
	PNS	May excite or inhibit autonomic nervous system actions, depending on receptors	
Dopamine	CNS	Creates a sense of feeling good; deficiency in some brain areas is associated with Parkinson disease	
	PNS	Limited actions in autonomic nervous system; may excite or inhibit, depending on receptors	
Serotonin	CNS	Primarily inhibitory; leads to sleepiness; action is blocked by LSD, enhanced by selective serotonin reuptake inhibitor drugs	
Histamine	CNS	Release in hypothalamus promotes alertness	
Amino acids			
GABA	CNS	Generally inhibitory	
Glutamic acid	CNS	Generally excitatory	
Neuropeptides			
Substance P	PNS	Excitatory; pain perception	
Endorphins, enkephalins	CNS	Generally inhibitory; reduce pain by inhibiting substance P release	
Gases			
Nitric oxide	PNS	Vasodilation	
	CNS	May play a role in memory	

Table 9.3	Events Leading to the Release of a Neurotransmitter

1. Action potential passes along an axon and over the surface of its synaptic knob.
2. Synaptic knob membrane becomes more permeable to calcium ions, and they diffuse inward.
3. In the presence of calcium ions, synaptic vesicles fuse to synaptic knob membrane.
4. Synaptic vesicles release their neurotransmitter into synaptic cleft.

perform a common function. Each pool receives input from neurons (which may be part of other pools), and each pool generates output. Neuronal pools may have excitatory or inhibitory effects on other pools or on peripheral effectors.

Facilitation

As a result of incoming impulses and neurotransmitter release, a particular neuron of a neuronal pool may receive excitatory and inhibitory input. If the net effect of the input is excitatory, threshold may be reached, and an outgoing impulse triggered. If the net effect is excit-atory but subthreshold, an impulse is not triggered, but the neuron is more excitable to incoming stimulation than before, a state called **facilitation** (fah-sil″ĭ-ta′shun).

Convergence

Any single neuron in a neuronal pool may receive impulses from two or more incoming axons. Axons originating from different parts of the nervous system and leading to the same neuron exhibit **convergence** (kon-ver′jens) (fig. 9.16a).

Convergence makes it possible for impulses arriving from different sources to have an additive effect on a neuron. For example, if a neuron is facilitated by receiving subthreshold stimulation from one input neuron, it may reach threshold if it receives additional stimulation from a second input neuron. As a result, a nerve impulse may travel to a particular effector and evoke a response.

Incoming impulses often bring information from several sensory receptors that detect changes. Convergence allows the nervous system to collect a variety of kinds of information, process it, and respond to it in a special way.

Divergence

Impulses leaving a neuron of a neuronal pool often exhibit **divergence** (di-ver′jens) by passing into several

Figure 9.16

Impulse processing in neuronal pools. **(a)** Axons of neurons 1 and 2 converge to the cell body of neuron 3. **(b)** The axon of neuron 4 diverges to the cell bodies of neurons 5 and 6.

other output neurons (fig. 9.16*b*). For example, an impulse from one neuron may stimulate two others; each of these, in turn, may stimulate several others, and so forth. Divergence can amplify an impulse—that is, spread it to more neurons in the pool. As a result of divergence, an impulse originating from a single neuron in the CNS may be amplified so that impulses reach enough motor units within a skeletal muscle to cause forceful contraction (see chapter 8, p. 190). Similarly, an impulse originating from a sensory receptor may diverge and reach several different regions of the CNS, where the resulting impulses are processed and acted upon.

Practice

23. Define *neuronal pool*.
24. Distinguish between convergence and divergence.

9.10 TYPES OF NERVES

Recall from section 9.1 that nerves are bundles of axons. An axon is often referred to as a nerve fiber. Because of this, we will refer to the neuron processes that bring sensory information into the CNS as **sensory fibers,** or **afferent fibers.** In contrast, **motor fibers** or **efferent fibers** carry impulses from the CNS to effectors (muscles or glands). A nerve is a cordlike bundle (or group of bundles) of nerve fibers within layers of connective tissue (fig. 9.17).

The terminology used to describe muscle and nerve fibers is somewhat inconsistent. "Muscle fiber" refers to a muscle cell, whereas "nerve fiber" refers to an axon, which is part of a cell. However, names for the associated connective tissues are similar. Both muscle and nerve fibers are bundled into fascicles. Recall from figure 8.1 (p. 180) that epimysium, perimysium, and endomysium connective tissue separates muscle tissue into compartments. Similarly, a nerve is defined by an outer *epineurium,* with *perineurium* surrounding a nerve fascicle within the nerve, and *endoneurium* surrounding an individual nerve fiber.

Like neurons, nerves that conduct impulses to the brain or spinal cord are called **sensory nerves,** and those that carry impulses to muscles or glands are termed **motor nerves.** Most nerves include both sensory and motor fibers and are called **mixed nerves.**

Practice

25. What is a nerve?
26. How does a mixed nerve differ from a sensory nerve? From a motor nerve?

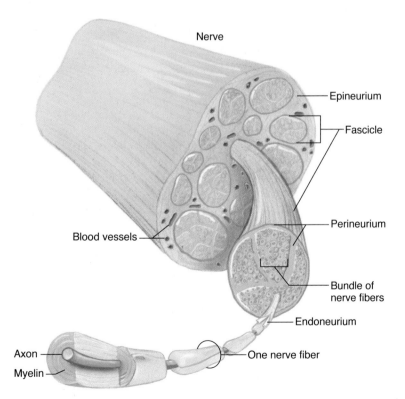

Figure 9.17

Connective tissue binds a bundle of nerve fibers, forming a fascicle. Many fascicles form a nerve.

9.11 NERVE PATHWAYS

Recall from section 9.5 that the routes nerve impulses follow as they travel through the nervous system are called *nerve pathways*. The simplest of these pathways includes only a few neurons and is called a **reflex** (re′fleks) **arc.** It constitutes the structural and functional basis for involuntary actions called **reflexes.**

Reflex Arcs

A reflex arc begins with a receptor at the end of a sensory (or afferent) neuron. This neuron usually leads to several interneurons in the CNS, which serve as a processing center, or *reflex center.* These interneurons can connect with interneurons in other parts of the nervous system. They also communicate with motor (or efferent) neurons, whose axons pass outward from the CNS to effectors, usually muscles or glands (fig. 9.18).

Reflex Behavior

Reflexes are automatic responses to changes (stimuli) within or outside the body. They help maintain homeostasis by controlling many involuntary processes, such as heart rate, breathing rate, blood pressure, and digestion. Reflexes also carry out the automatic actions of swallowing, sneezing, coughing, and vomiting.

The *patellar reflex* (knee-jerk reflex) is an example of a simple reflex involving a pathway of only two neurons—a sensory neuron communicating directly

Figure 9.18 AP|R

A reflex arc is the simplest nerve pathway. It involves a sensory neuron that sends a message to the CNS, and a motor neuron that sends the message from the CNS to a muscle or gland.

with a motor neuron. Striking the patellar ligament just below the patella initiates this reflex. The quadriceps femoris muscle group, which is attached to the patella by a tendon, is pulled slightly, stimulating stretch receptors in these muscles. These receptors, in turn, trigger impulses that pass along the axon of a sensory neuron into the spinal cord. Within the spinal cord, the sensory axon synapses with a motor neuron. An impulse is then triggered along the axon of the motor neuron and travels back to the quadriceps femoris group. The muscle group contracts in response, and the reflex is completed as the leg extends (fig. 9.19).

The patellar reflex helps maintain upright posture. If the knee begins to bend from the force of gravity when a person is standing still, the quadriceps femoris group is stretched, the reflex is triggered, and the leg straightens again.

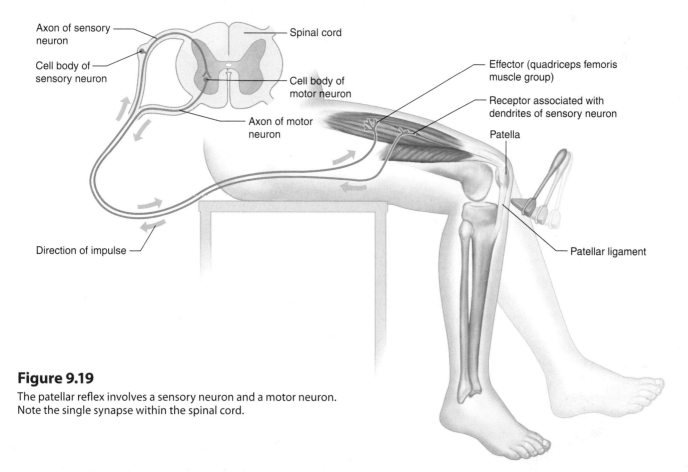

Figure 9.19

The patellar reflex involves a sensory neuron and a motor neuron. Note the single synapse within the spinal cord.

Another type of reflex, called a *withdrawal reflex,* occurs when a person unexpectedly touches a body part to something painful, such as stepping on a tack. This activates skin receptors and sends sensory impulses to the spinal cord. There, the impulses pass to the interneurons of a reflex center and are directed to motor neurons. The motor neurons transmit signals to flexor muscles in the injured part, and the muscles contract in response. At the same time, the antagonistic extensor muscles are inhibited, and the foot is rapidly and unconsciously withdrawn from the painful stimulus. Concurrent with the withdrawal reflex, other interneurons carry sensory impulses to the brain and the person becomes aware of the experience and may feel pain (fig. 9.20). A withdrawal reflex is protective because it may limit tissue damage caused by touching something harmful. Table 9.4 summarizes the parts of a reflex arc.

Reflexes provide information about the condition of the nervous system. An anesthesiologist may try to initiate a reflex in a patient being anesthetized to determine how well the anesthetic drug is affecting nerve functions. A neurologist may test reflexes to determine the location and extent of damage from a nervous system injury.

Practice

27. What is a nerve pathway?
28. List the parts of a reflex arc.
29. Define *reflex.*
30. List the actions that occur during a withdrawal reflex.

9.12 MENINGES

Bones, membranes, and fluid surround the organs of the CNS. The brain lies in the cranial cavity of the skull, and the spinal cord occupies the vertebral canal in the vertebral column. Layered membranes called **meninges** (mě-nin′jēz) (singular, *meninx*) lie between these bony coverings and the soft tissues of the CNS, protecting the brain and spinal cord (fig. 9.21*a*).

Figure 9.20
A withdrawal reflex involves a sensory neuron, an interneuron, and a motor neuron.

Table 9.4	Parts of a Reflex Arc	
Part	**Description**	**Function**
Receptor	Receptor end of a dendrite or a specialized receptor cell in a sensory organ	Senses specific type of internal or external change
Sensory neuron	Dendrite, cell body, and axon of a sensory neuron	Transmits nerve impulse from receptor into brain or spinal cord
Interneuron	Dendrite, cell body, and axon of a neuron within the brain or spinal cord	Conducts nerve impulse from sensory neuron to motor neuron
Motor neuron	Dendrite, cell body, and axon of a motor neuron	Transmits nerve impulse from brain or spinal cord out to effector
Effector	Muscle or gland	Responds to stimulation by motor neuron and produces reflex or behavioral action

(a)

(b)

Figure 9.21 AP|R

Meninges. **(a)** Membranes called meninges enclose the brain and spinal cord. **(b)** The meninges include three layers: dura mater, arachnoid mater, and pia mater.

The meninges have three layers—dura mater, arachnoid mater, and pia mater (fig. 9.21*b*). The **dura mater** (du′rah ma′ter) is the outermost layer. It is composed primarily of tough, white, fibrous connective tissue and contains many blood vessels and nerves. It attaches to the inside of the cranial cavity and forms the internal periosteum of the surrounding skull bones. In some regions, the dura mater extends inward between lobes of the brain and forms partitions that support and protect these parts.

The dura mater continues into the vertebral canal as a strong, tubular sheath that surrounds the spinal cord. It terminates as a blind sac below the end of the cord. The membrane around the spinal cord is not attached directly to the vertebrae but is separated by an *epidural space,* which lies between the dural sheath and the bony walls (fig. 9.22). This space contains loose connective and adipose tissues, which pad the spinal cord.

The **arachnoid mater** is a thin, weblike membrane without blood vessels that lies between the dura and pia

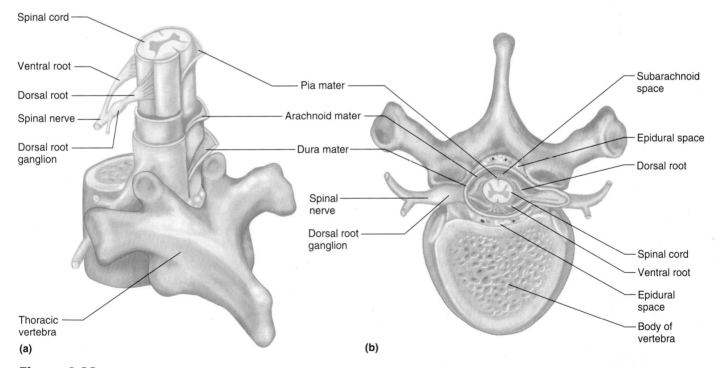

(a)

(b)

Figure 9.22

Meninges of the spinal cord. **(a)** The dura mater ensheaths the spinal cord. **(b)** Tissues forming a protective pad around the cord fill the epidural space between the dural sheath and the bone of the vertebra.

maters. Between the arachnoid and pia maters is a *sub-arachnoid space* that contains the clear, watery **cerebrospinal fluid (CSF).** The **pia mater** (pi′ah ma′ter) is very thin and contains many nerves and blood vessels that nourish underlying cells of the brain and spinal cord. This layer hugs the surfaces of these organs and follows their irregular contours, passing over high areas and dipping into depressions.

A blow to the head may break some blood vessels associated with the brain, and escaping blood may collect beneath the dura mater. Such a *subdural hematoma* increases pressure between the rigid bones of the skull and the soft tissues of the brain. Unless the accumulating blood is evacuated, compression of the brain may lead to functional losses or even death.

Practice

31. Describe the meninges.
32. State the location of cerebrospinal fluid.

9.13 SPINAL CORD

The **spinal cord** is a slender nerve column that passes downward from the brain into the vertebral canal. Although continuous with the brain, the spinal cord begins where nervous tissue leaves the cranial cavity at the level of the foramen magnum. The spinal cord tapers to a point and terminates near the intervertebral disc that separates the first and second lumbar vertebrae (fig. 9.23).

Structure of the Spinal Cord

The spinal cord consists of thirty-one segments, each of which gives rise to a pair of **spinal nerves.** These nerves (part of the peripheral nervous system) branch to various body parts and connect them with the CNS (see fig. 9.35).

In the neck region, a thickening in the spinal cord, called the *cervical enlargement,* supplies nerves to the upper limbs. A similar thickening in the lower back, the *lumbar enlargement,* gives off nerves to the lower limbs (fig. 9.23).

Two grooves, a deep *anterior median fissure* and a shallow *posterior median sulcus,* extend the length of the spinal cord, dividing it into right and left halves (fig. 9.24). A cross section of the cord reveals a core of gray matter within white matter. The pattern of gray matter roughly resembles a butterfly with its wings spread. The upper and lower wings of gray matter are called the *posterior horns* and *anterior horns,* respectively. Between them on either side in the thoracic and upper lumbar segments is a protrusion of gray matter called the *lateral horn.*

Neurons with large cell bodies located in the anterior horns give rise to motor fibers that pass out through spinal

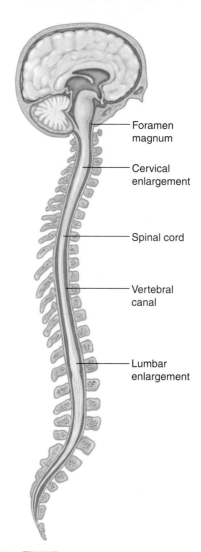

Figure 9.23 AP|R

The spinal cord begins at the level of the foramen magnum and ends near the intervertebral disc between the first and second lumbar vertebrae.

nerves to skeletal muscles. However, the majority of neurons in the gray matter of the spinal cord are interneurons.

Gray matter divides the white matter of the spinal cord into three regions on each side—the *anterior, lateral,* and *posterior funiculi* (fig. 9.24a). Each funiculus consists of longitudinal bundles of myelinated axons that comprise major neural pathways. In the central nervous system, such bundles of axons are called **tracts.**

A horizontal bar of gray matter in the middle of the spinal cord, the *gray commissure,* connects the wings of the gray matter on the right and left sides. This bar surrounds the **central canal,** which contains cerebrospinal fluid.

Functions of the Spinal Cord

The spinal cord has two major functions—conducting nerve impulses and serving as a center for spinal reflexes. The tracts of the spinal cord consist of axons that provide a two-way communication system between the brain and the body parts outside the nervous system. The tracts that carry sensory information to the brain are called **ascending tracts** (fig. 9.25); those that

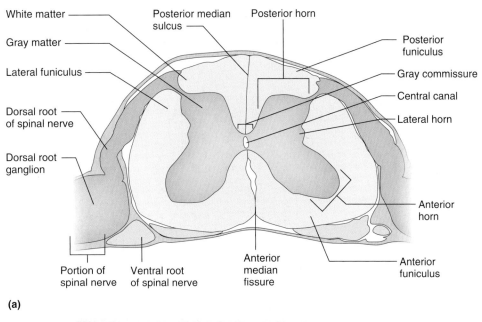

White matter —
Gray matter —
Posterior median sulcus —
Posterior horn
Lateral funiculus —
Dorsal root of spinal nerve —
Dorsal root ganglion —
Portion of spinal nerve
Ventral root of spinal nerve
Anterior median fissure
Posterior funiculus
Gray commissure
Central canal
Lateral horn
Anterior horn
Anterior funiculus

(a)

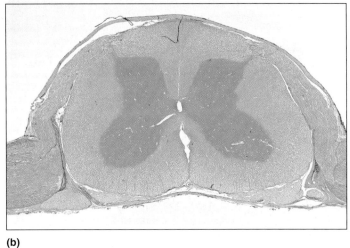

(b)

Figure 9.24 AP|R

The spinal cord. **(a)** A cross section of the spinal cord. **(b)** Identify the parts of the spinal cord in this micrograph (10×).

conduct motor impulses from the brain to muscles and glands are called **descending tracts** (fig. 9.26).

All the axons in a given tract typically originate from neuron cell bodies in the same part of the nervous system and terminate together in some other part. The names that identify tracts often reflect these common origins and terminations. For example, a *spinothalamic tract* begins in the spinal cord and carries sensory impulses associated with the sensations of pain, touch, and temperature to the thalamus of the brain. A *corticospinal tract* originates in the cortex of the brain and carries motor impulses downward through the spinal cord and spinal nerves. These impulses control skeletal muscle movements.

Corticospinal tracts are also called *pyramidal tracts* after the pyramid-shaped areas in the medulla oblongata of the brain through which they pass. Other descending tracts, called *extrapyramidal tracts,* control motor activities associated with maintaining balance and posture.

In addition to providing a pathway for tracts, the spinal cord functions in many reflexes, including the patellar and withdrawal reflexes described previously. These are called **spinal reflexes** because their reflex arcs pass through the spinal cord.

Some axons extend from the base of the spinal cord to the toes. If you stub your toe, a sensory message reaches the spinal cord in less than one-hundredth of a second.

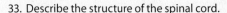

Practice

33. Describe the structure of the spinal cord.

34. Describe the general functions of the spinal cord.

35. Distinguish between an ascending and a descending tract.

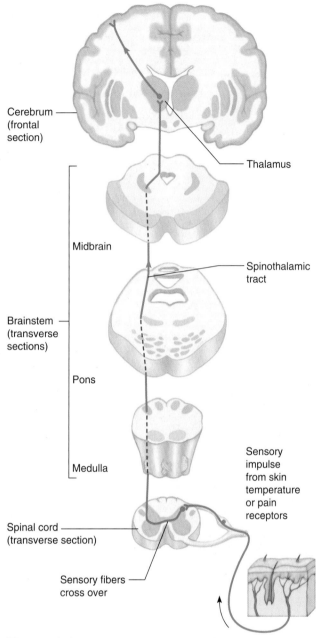

Figure 9.25

Ascending tracts. Sensory impulses originating in skin receptors cross over in the spinal cord and ascend to the brain. Other sensory tracts cross over in the medulla oblongata.

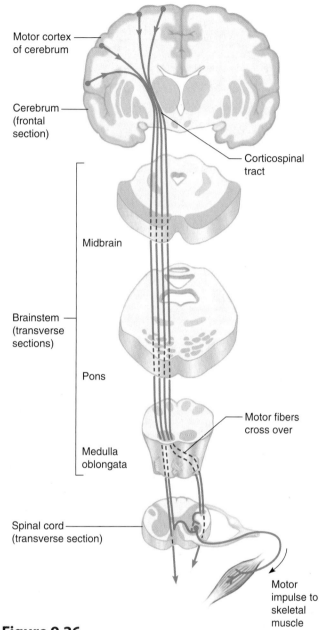

Figure 9.26

Descending tracts. Motor fibers of the corticospinal tract begin in the cerebral cortex, cross over in the medulla oblongata, and descend in the spinal cord. There, they synapse with neurons whose fibers lead to the spinal nerves that supply skeletal muscles.

9.14 BRAIN

The **brain** is composed of about 100 billion (10^{11}) multipolar neurons, which communicate with one another and with neurons in other parts of the nervous system. As figure 9.27 shows, the brain can be divided into four major portions—the cerebrum, the diencephalon, the brainstem, and the cerebellum. The *cerebrum,* the largest part, includes nerve centers associated with sensory and motor functions and provides higher mental functions, including memory and reasoning. The *diencephalon* also processes

sensory information. Nerve pathways in the *brainstem* connect various parts of the nervous system and regulate certain visceral activities. The *cerebellum* includes centers that coordinate voluntary muscular movements.

Structure of the Cerebrum

The **cerebrum** (ser′ĕ-brum) consists of two large masses called the left and right **cerebral hemispheres** (ser″ĕ-bral hem′ĭ-sfērz), which are essentially mirror images of each other. A broad, flat bundle of axons called the **corpus callosum** (kor′pus kah-lo′sum) connects the cere-

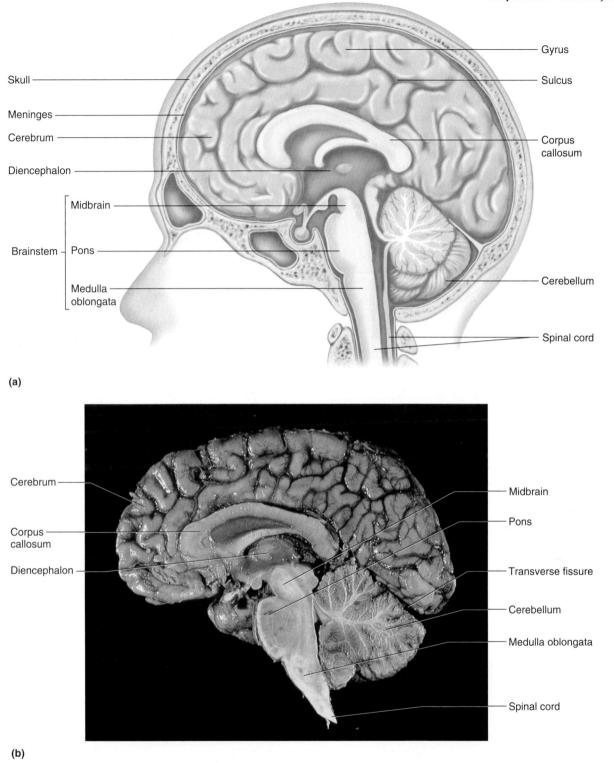

(a)

(b)

Figure 9.27 AP|R

The major portions of the brain are the cerebrum, the diencephalon, the brainstem, and the cerebellum.

bral hemispheres. A layer of dura mater (falx cerebri) separates them.

The surface of the cerebrum has many ridges (convolutions) or **gyri** (ji'ri) (singular, *gyrus*), separated by grooves. A shallow groove is called a **sulcus** (sul'kus), and a deep groove is called a **fissure.** The structural organization of these elevations and depressions is complex, but they form distinct patterns in all normal brains. For example, a *longitudinal fissure* separates the right and left cerebral hemispheres, a *transverse fissure* separates the cerebrum from the cerebellum, and several sulci divide each hemisphere into lobes.

The lobes of the cerebral hemispheres are named after the skull bones they underlie (fig. 9.28). They include:

1. **Frontal lobe** The frontal lobe forms the anterior part of each cerebral hemisphere. It is bordered posteriorly by a *central sulcus,* which extends from the longitudinal fissure at a right angle, and inferiorly by a *lateral sulcus,* which extends from the undersurface of the brain along its sides.
2. **Parietal lobe** The parietal lobe is posterior to the frontal lobe and separated from it by the central sulcus.
3. **Temporal lobe** The temporal lobe lies below the frontal and parietal lobes and is separated from them by the lateral sulcus.
4. **Occipital lobe** The occipital lobe forms the posterior part of each cerebral hemisphere and is separated from the cerebellum by a shelflike extension of dura mater (tentorium cerebelli). The boundary between the occipital lobe and the parietal and temporal lobes is not distinct.
5. **Insula** (in'su-lah) The insula is deep in the lateral sulcus and is covered by parts of the frontal, parietal, and temporal lobes. A *circular sulcus* separates the insula from the other lobes.

A thin layer of gray matter called the **cerebral cortex** (ser"ĕ-bral kor'teks) is the outermost part of the cerebrum. This layer covers the gyri and dips into the sulci and fissures. It contains nearly 75% of all the neuron cell bodies in the nervous system.

Just beneath the cerebral cortex is a mass of white matter that makes up the bulk of the cerebrum. This mass contains bundles of myelinated axons that connect neuron cell bodies of the cortex with other parts of the nervous system. Some of these fibers pass from one cerebral hemisphere to the other by way of the corpus callosum, and others carry sensory or motor impulses from parts of the cortex to nerve centers in the brain or spinal cord.

In a condition called lissencephaly, which means "smooth brain," sulci and gyri are absent. Lissencephaly is associated with mental retardation, developmental delay, and seizures.

Functions of the Cerebrum

The cerebrum provides higher brain functions. It has centers for interpreting sensory impulses arriving from sense organs and centers for initiating voluntary muscular movements. The cerebrum stores the information that constitutes memory and utilizes it to reason. Intelligence and personality also stem from cerebral activity.

Functional Regions of the Cerebral Cortex

Specific regions of the cerebral cortex perform specific functions. Although functions overlap among regions, the cortex can be divided into sensory, association, and motor areas.

Sensory areas located in several lobes of the cerebrum interpret impulses that arrive from sensory receptors, producing feelings or sensations. For example, sensations from all parts of the skin (cutaneous senses) arise in the anterior parts of the parietal lobes along

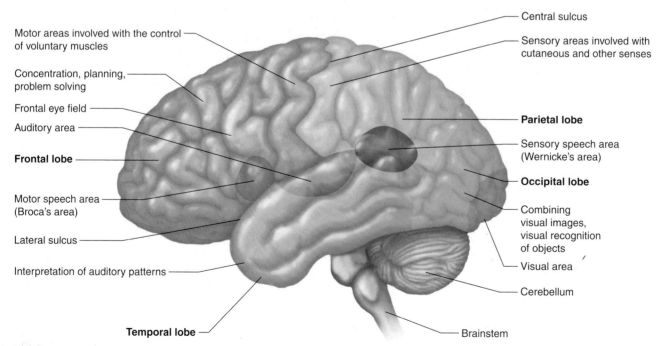

Motor areas involved with the control of voluntary muscles
Concentration, planning, problem solving
Frontal eye field
Auditory area
Frontal lobe
Motor speech area (Broca's area)
Lateral sulcus
Interpretation of auditory patterns
Temporal lobe

Central sulcus
Sensory areas involved with cutaneous and other senses
Parietal lobe
Sensory speech area (Wernicke's area)
Occipital lobe
Combining visual images, visual recognition of objects
Visual area
Cerebellum
Brainstem

Figure 9.28 AP|R

Some sensory, association, and motor areas of the left cerebral cortex.

the central sulcus (fig. 9.28). The posterior parts of the occipital lobes affect vision (visual area), and the temporal lobes contain the centers for hearing (auditory area). The sensory areas for taste are located near the bases of the central sulci along the lateral sulci, and the sense of smell arises from centers deep within the cerebrum.

Like motor fibers, sensory fibers cross over either in the spinal cord or in the brainstem (see fig. 9.25). Thus, the centers in the right cerebral hemisphere interpret impulses originating from the left side of the body, and vice versa.

Not all sensory areas are bilateral. The *sensory speech area* or *Wernicke's* (ver'nĭ-kēz) *area* includes parts of both the temporal and parietal lobes near the posterior end of the lateral sulcus, usually in the left hemisphere. This area receives and relays input from both the visual cortex and auditory cortex and is important for understanding written or spoken language.

Association areas are neither primarily sensory nor primarily motor. They connect with one another and with other brain structures. These areas analyze and interpret sensory experiences and oversee memory, reasoning, verbalizing, judgment, and emotion. Association areas occupy the anterior portions of the frontal lobes and are widespread in the lateral parts of the parietal, temporal, and occipital lobes (fig. 9.28).

The association areas of the frontal lobes control a number of higher intellectual processes. These include concentrating, planning, complex problem solving, and judging the possible consequences of behavior. Association areas of the parietal lobes help in understanding speech and choosing words to express thoughts and feelings.

Wernicke's area corresponds closely to a brain region that has been referred to as a "general interpretive area," near where the occipital, parietal, and temporal lobes meet. It plays a role in integrating visual, auditory, and other sensory information, and then interpretating a situation. For example, you hear a familiar voice, look up from your notes, see a friend from class, and realize that it is time for your study group.

The association areas of the temporal lobes and the regions of the posterior ends of the lateral sulcus store memory of visual scenes, music, and other complex sensory patterns. Association areas of the occipital lobes that are adjacent to the visual centers are important in analyzing visual patterns and combining visual images with other sensory experiences, as when one recognizes another person or an object.

The primary **motor areas** of the cerebral cortex lie in the frontal lobes, just in front of the central sulcus (fig. 9.28). The nervous tissue in these regions contains many large *pyramidal cells,* named for their pyramid-shaped cell bodies. These cells are also termed *upper motor neurons,* because of their location.

Impulses from the pyramidal cells travel downward through the brainstem and into the spinal cord on the corticospinal tracts (see fig. 9.26). Here they form synapses with *lower motor neurons* whose axons leave the spinal cord and reach skeletal muscle fibers. Most of the axons in these tracts cross over from one side of the brain to the other within the brainstem. As a result, the motor area of the right cerebral hemisphere generally controls skeletal muscles on the left side of the body, and vice versa.

In addition to the primary motor areas, certain other regions of the frontal lobe affect motor functions. For example, a region called the *motor speech area,* or *Broca's* (bro'kahz) *area,* is usually in the left hemisphere, just anterior to the primary motor cortex and superior to the lateral sulcus. This area generates the movements of muscles necessary for speech (fig. 9.28).

Above the motor speech area is a region called the *frontal eye field.* The motor cortex in this area controls voluntary movements of the eyes and eyelids. Another region just in front of the primary motor area controls the muscular movements of the hands and fingers that make skills such as writing possible.

Practice

36. List the major divisions of the brain.

37. Describe the cerebral cortex.

38. Describe the major functions of the cerebrum.

39. Locate the major functional regions of the cerebral cortex.

The effects of injuries to the cerebral cortex depend on the location and extent of the damage. For example, injury to the motor areas of one frontal lobe causes partial or complete paralysis on the opposite side of the body. Damage to the association areas of the frontal lobe may impair concentration on complex mental tasks, making a person appear disorganized and easily distracted. Damage to association areas of the temporal lobes may impair recognition of printed words or arranging words into meaningful thoughts.

Hemisphere Dominance

Both cerebral hemispheres participate in basic functions, such as receiving and analyzing sensory impulses, controlling skeletal muscles, and storing memory. However, in most persons, one side of the cerebrum is the **dominant hemisphere,** controlling the ability to use and understand language.

In most people the left hemisphere is dominant for the language-related activities of speech, writing, and reading, and for complex intellectual functions requiring verbal, analytical, and computational skills. In other persons, the right hemisphere is dominant for language-related abilities, or the hemispheres are equally dominant. Broca's area in the dominant hemisphere controls the muscles that function in speaking.

In addition to carrying on basic functions, the non-dominant hemisphere specializes in nonverbal functions, such as motor tasks that require orientation of the body in space, understanding and interpreting musical patterns, and nonverbal visual experiences. The nondominant hemisphere also controls emotional and intuitive thinking.

Nerve fibers of the corpus callosum, which connect the cerebral hemispheres, allow the dominant hemisphere to control the motor cortex of the nondominant hemisphere (see fig. 9.27). These fibers also transfer sensory information reaching the nondominant hemisphere to the dominant one, where the information can be used in decision making.

Deep within each cerebral hemisphere are several masses of gray matter called **basal nuclei** (often called basal ganglia) (fig. 9.29). They are the *caudate nucleus,* the *putamen,* and the *globus pallidus.* The basal nuclei produce the inhibitory neurotransmitter *dopamine.* The neurons of the basal nuclei interact with other brain areas, including the motor cortex, thalamus, and cerebellum. These interactions, through a combination of stimulation and inhibition, facilitate voluntary movement.

The signs of Parkinson disease and Huntington disease result from altered activity of basal nuclei neurons. In Parkinson disease, nearby neurons release less dopamine, and the basal nuclei become overactive, inhibiting movement. In Huntington disease, basal nuclei neurons gradually deteriorate, resulting in unrestrained movement.

Ventricles and Cerebrospinal Fluid

Interconnected cavities called **ventricles** lie within the cerebral hemispheres and brainstem (fig. 9.30). These spaces are continuous with the central canal of the spinal cord, and like it, they contain cerebrospinal fluid.

The largest ventricles are the *lateral ventricles* (first and second ventricles), which extend into the cerebral hemispheres and occupy parts of the frontal, temporal, and occipital lobes. A narrow space that constitutes the *third ventricle* is in the midline of the brain, beneath the corpus callosum. This ventricle communicates with the lateral ventricles through openings (interventricular foramina) in its anterior end. The *fourth ventricle* is in the brainstem just anterior to the cerebellum. A narrow canal, the *cerebral aqueduct,* connects it to the third ventricle and passes lengthwise through the brainstem. The fourth ventricle is continuous with the central canal of the spinal cord and has openings in its roof that lead into the subarachnoid space of the meninges.

Tiny, reddish, cauliflower-like masses of specialized capillaries from the pia mater, called **choroid plexuses** (plek′sus-ez), secrete cerebrospinal fluid (fig. 9.31). These structures project into the ventricles. Most of the cerebrospinal fluid is formed in the lateral ventricles. From there, it circulates slowly into the third and fourth ventricles and into the central canal of the spinal cord. Cerebrospinal fluid also enters the subarachnoid space of the meninges through the wall of the fourth ventricle near the cerebellum and completes its circuit by being reabsorbed into the blood.

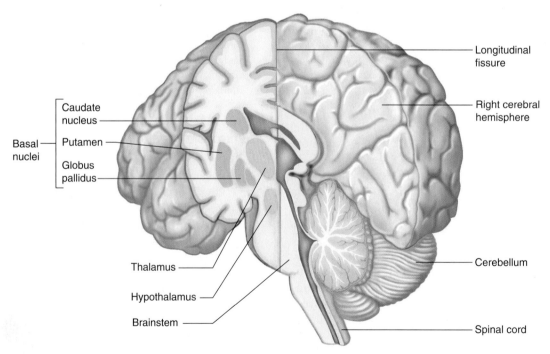

Figure 9.29 AP|R

A frontal (coronal) section of the left cerebral hemisphere (posterior view) reveals some of the basal nuclei.

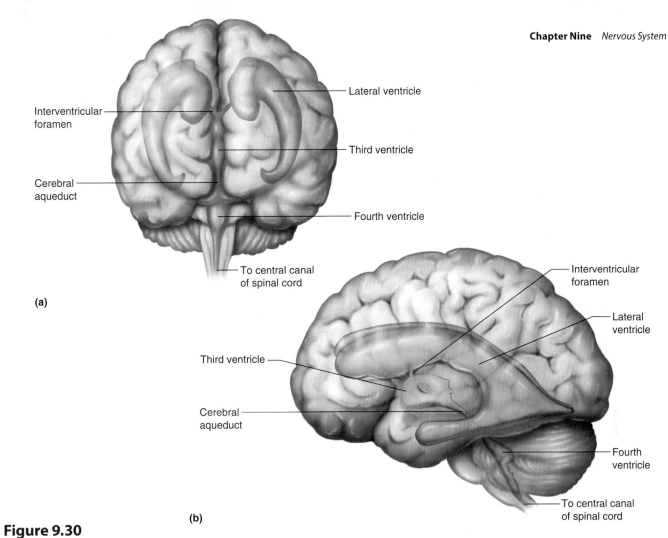

Interventricular foramen

Lateral ventricle

Third ventricle

Cerebral aqueduct

Fourth ventricle

To central canal of spinal cord

(a)

Interventricular foramen

Lateral ventricle

Third ventricle

Cerebral aqueduct

Fourth ventricle

To central canal of spinal cord

(b)

Figure 9.30

Ventricles in the brain. **(a)** Anterior view of the ventricles within the cerebral hemispheres and brainstem. **(b)** Lateral view.

Choroid plexuses of third ventricle

Third ventricle

Cerebral aqueduct

Fourth ventricle

Blood-filled dural space

Pia mater

Subarachnoid space

Arachnoid mater

Dura mater

Choroid plexus of fourth ventricle

Figure 9.31 AP|R

The choroid plexuses in the walls of the ventricles secrete cerebrospinal fluid. The fluid circulates through the ventricles and central canal, enters the subarachnoid space, and is reabsorbed into the blood.

Cerebrospinal fluid completely surrounds the brain and spinal cord because it occupies the subarachnoid space of the meninges. In effect, these organs float in the fluid, which supports and protects them by absorbing forces that might otherwise jar and damage them. Cerebrospinal fluid also maintains a stable ionic concentration in the CNS and provides a pathway to the blood for wastes.

> The fluid pressure in the ventricles normally remains relatively constant because cerebrospinal fluid is secreted and reabsorbed continuously and at equal rates. An infection, a tumor, or a blood clot can interfere with fluid circulation, increasing pressure in the ventricles and thus in the cranial cavity (intracranial pressure). This can injure the brain by forcing it against the rigid skull.
>
> A *lumbar puncture* (spinal tap) measures the pressure of cerebrospinal fluid. A very thin hollow needle is inserted into the subarachnoid space between the third and fourth or between the fourth and fifth lumbar vertebrae and an instrument called a *manometer* measures the pressure.

Practice

40. What is hemisphere dominance?
41. What are the major functions of the dominant hemisphere? The nondominant one?
42. Where are the ventricles of the brain?
43. Describe the circulation of cerebrospinal fluid.

Diencephalon

The **diencephalon** (di″en-sef′ah-lon) is located between the cerebral hemispheres and above the midbrain. It surrounds the third ventricle and is composed largely of gray matter. Within the diencephalon, a dense mass called the **thalamus** bulges into the third ventricle from each side (see figs. 9.29 and 9.32*b*). Another region of the diencephalon that includes many nuclei (masses of gray matter) is the **hypothalamus.** It lies below the thalamus and forms the lower walls and floor of the third ventricle.

Other parts of the diencephalon include: (1) the **optic tracts** and the **optic chiasma** that is formed by optic nerve fibers crossing over each other; (2) the **infundibulum,** a conical process behind the optic chiasma to which the pituitary gland attaches; (3) the **posterior pituitary gland,** which hangs from the floor of the hypothalamus; (4) the **mammillary bodies,** which appear as two rounded structures behind the infundibulum; and (5) the **pineal gland** (pin′e-al gland), a cone-shaped structure attached to the upper part of the diencephalon (see chapter 11, p. 309).

The thalamus is a central relay station for sensory impulses ascending from other parts of the nervous system to the cerebral cortex. It receives all sensory impulses (except those associated with the sense of smell) and channels them to the appropriate regions of the cortex for interpretation. In addition, all regions

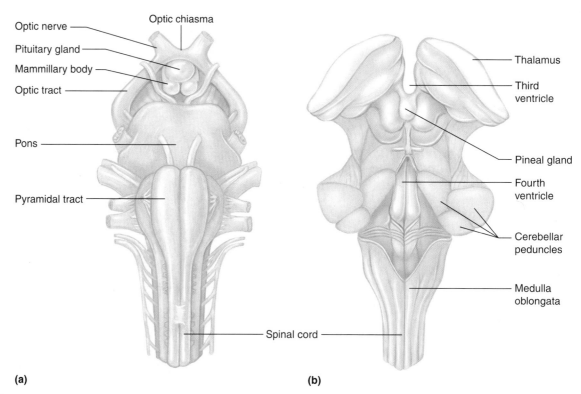

Optic nerve
Optic chiasma
Pituitary gland
Mammillary body
Optic tract
Pons
Pyramidal tract

Thalamus
Third ventricle
Pineal gland
Fourth ventricle
Cerebellar peduncles
Medulla oblongata

Spinal cord

(a)   (b)

Figure 9.32 AP|R

The brainstem. **(a)** Ventral view of the brainstem. **(b)** Dorsal view of the brainstem with the cerebellum removed, exposing the fourth ventricle.

Q: *What is the relative position of the fourth ventricle to the third ventricle?*

Answer can be found in Appendix E on page 568.

of the cerebral cortex can communicate with the thalamus by means of descending fibers. The cerebral cortex pinpoints the origin of sensory stimulation, and the thalamus produces a general awareness of certain sensations, such as pain, touch, and temperature.

Nerve fibers connect the hypothalamus to the cerebral cortex, thalamus, and other parts of the brainstem. The hypothalamus maintains homeostasis by regulating a variety of visceral activities and by linking the nervous and endocrine systems. The hypothalamus regulates:

1. Heart rate and arterial blood pressure
2. Body temperature
3. Water and electrolyte balance
4. Control of hunger and body weight
5. Control of movements and glandular secretions of the stomach and intestines
6. Production of neurosecretory substances that stimulate the pituitary gland to secrete hormones
7. Sleep and wakefulness

Structures in the general region of the diencephalon also control emotional responses. For example, regions of the cerebral cortex in the medial parts of the frontal and temporal lobes interconnect with a number of deep masses of gray matter, including the hypothalamus, thalamus, and basal nuclei. Together, these structures compose a complex called the **limbic system.**

The limbic system controls emotional experience and expression. It can modify the way a person acts by producing such feelings as fear, anger, pleasure, and sorrow. The limbic system recognizes upsets in a person's physical or psychological condition that might threaten life. By causing pleasant or unpleasant feelings about experiences, the limbic system guides a person into behavior that is likely to increase the chance of survival.

> A whiff of a certain scent may elicit vivid memories because sensory information from olfactory receptors (the sense of smell) also goes to the limbic system. Olfactory input to the limbic system is also why odors can alter mood. For example, the scent of just-mowed grass or an ocean breeze makes us feel good.

Brainstem

The **brainstem** is a bundle of nervous tissue that connects the cerebrum to the spinal cord. It consists of many tracts and several nuclei. The parts of the brainstem include the midbrain, pons, and medulla oblongata (figs. 9.27 and 9.32).

Midbrain

The **midbrain** is a short section of the brainstem between the diencephalon and the pons (see fig. 9.27). It contains bundles of myelinated axons that join lower parts of the brainstem and spinal cord with higher parts of the brain. Two prominent bundles of axons on the underside of the midbrain are the corticospinal tracts and are the main motor pathways between the cerebrum and lower parts of the nervous system.

The midbrain includes several masses of gray matter that serve as reflex centers. For example, the midbrain contains the centers for certain visual reflexes, such as those responsible for moving the eyes to view something as the head turns. It also contains the auditory reflex centers that enable a person to move the head to hear sounds more distinctly.

Pons

The **pons** (ponz) is a rounded bulge on the underside of the brainstem, where it separates the midbrain from the medulla oblongata (see fig. 9.27). The dorsal part of the pons consists largely of longitudinal nerve fibers, which relay impulses to and from the medulla oblongata and the cerebrum. The ventral part of the pons has large bundles of transverse nerve fibers, which transmit impulses from the cerebrum to centers in the cerebellum.

Several nuclei of the pons relay sensory impulses from peripheral nerves to higher brain centers. Other nuclei may contribute to the rhythm of breathing (see chapter 16, p. 456).

Medulla Oblongata

The **medulla oblongata** (mě-dul'ah ob"long-gah'tah) extends from the pons to the foramen magnum of the skull (see fig. 9.27). Its dorsal surface flattens to form the floor of the fourth ventricle, and its ventral surface is marked by the corticospinal tracts, most of whose fibers cross over at this level (see fig. 9.26).

All of the ascending and descending nerve fibers connecting the brain and spinal cord must pass through the medulla oblongata because of its location. As in the spinal cord, the white matter of the medulla oblongata surrounds a central mass of gray matter. Here, however, nerve fibers separate the gray matter into nuclei, some of which relay ascending impulses to the other side of the brainstem and then on to higher brain centers. Other nuclei in the medulla oblongata control vital visceral activities. These centers include:

1. **Cardiac center** Impulses originating in the cardiac center are transmitted to the heart on peripheral nerves, altering heart rate.
2. **Vasomotor center** Certain cells of the vasomotor center initiate impulses that travel to smooth muscles in the walls of certain blood vessels and stimulate them to contract. This constricts the blood vessels (vasoconstriction), maintaining blood pressure. Other cells of the vasomotor center produce the opposite effect—dilating blood vessels (vasodilation) and consequently dropping blood pressure.
3. **Respiratory center** The respiratory center acts to maintain the rhythm of breathing and adjusts the rate and depth of breathing.

Still other nuclei in the medulla oblongata are centers for the reflexes associated with coughing, sneezing, swallowing, and vomiting.

Reticular Formation

Scattered throughout the medulla oblongata, pons, and midbrain is a complex network of nerve fibers associated with tiny islands of gray matter. This network, the **reticular formation** (rĕ-tik′u-lar for-ma′shun) (reticular activating system), extends from the upper part of the spinal cord into the diencephalon. Its nerve fibers join centers of the hypothalamus, basal nuclei, cerebellum, and cerebrum with fibers in all the major ascending and descending tracts.

When sensory impulses reach the reticular formation, it responds by activating the cerebral cortex into a state of wakefulness. Without this arousal, the cortex remains unaware of stimulation and cannot interpret sensory information or carry on thought processes. Thus, decreased activity in the reticular formation results in sleep. If the reticular formation is injured so that it cannot function, the person remains unconscious and cannot be aroused, even with strong stimulation. This is called a comatose state. Barbiturate drugs, which dampen CNS activity, affect the reticular formation (see Clinical Application 9.2).

Practice

44. What are the major functions of the thalamus? The hypothalamus?
45. How may the limbic system influence behavior?
46. List the structures of the brainstem.
47. Which vital reflex centers are in the brainstem?
48. What is the function of the reticular formation?

Cerebellum

The **cerebellum** (ser″ĕ-bel′um) is a large mass of tissue located below the occipital lobes of the cerebrum and posterior to the pons and medulla oblongata (see fig. 9.27). It consists of two lateral hemispheres partially separated by a layer of dura mater (falx cerebelli) and connected in the midline by a structure called the *vermis*. Like the cerebrum, the cerebellum is composed primarily of white matter, with a thin layer of gray matter, the **cerebellar cortex,** on its surface.

The cerebellum communicates with other parts of the CNS by means of three pairs of nerve tracts called *cerebellar peduncles* (figs. 9.32 and 9.33). One pair (the inferior peduncles) brings sensory information concerning the position of the limbs, joints, and other body parts to the

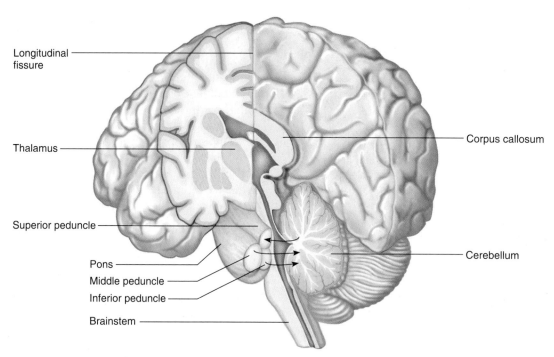

Longitudinal fissure

Thalamus

Superior peduncle

Pons

Middle peduncle

Inferior peduncle

Brainstem

Corpus callosum

Cerebellum

Figure 9.33

The cerebellum, which is located below the occipital lobes of the cerebrum, communicates with other parts of the nervous system by means of the cerebellar peduncles.

Clinical Application 9.2

Drug Abuse

Drug abuse is the chronic self-administration of a drug in doses high enough to cause *addiction*—a physical or psychological dependence in which the user is preoccupied with locating and taking the drug. Stopping drug use causes intense, unpleasant withdrawal symptoms. Prolonged and repeated abuse of a drug may also result in *drug tolerance,* in which the physiological response to a particular dose of the drug becomes less intense over time. Drug tolerance results as the drug increases synthesis of certain liver enzymes, which metabolize the drug more rapidly, so that the addict needs the next dose sooner. Drug tolerance also arises from physiological changes that lessen the drug's effect on its target cells. The most commonly abused drugs are CNS depressants ("downers"), CNS stimulants ("uppers"), hallucinogens, and anabolic steroids (see Clinical Application 8.1, p. 188).

CNS depressants include barbiturates, benzodiazepines, opiates, and cannabinoids. *Barbiturates* act uniformly throughout the brain, but the reticular formation is particularly sensitive to their effects. CNS depression occurs due to inhibited secretion of certain excitatory and inhibitory neurotransmitters. Effects range from mild calming of the nervous system (sedation) to sleep, loss of sensory sensations (anesthesia), respiratory distress, cardiovascular collapse, and death.

The *benzodiazepines,* such as diazepam, depress activity in the limbic system and the reticular formation. Low doses relieve anxiety, and higher doses cause sedation, sleep, or anesthesia. These drugs increase either the activity or release of the inhibitory neurotransmitter GABA. When benzodiazepines are metabolized, they may form other biochemicals that have depressing effects.

The *opiates* include heroin (which has no legal use in the United States), codeine, morphine, meperidine, and methadone. These drugs stimulate certain receptors (opioid receptors) in the CNS, and when taken in prescribed dosages, they sedate and relieve pain (analgesia). Opiates cause both physical and psychological dependence. Effects of overdose include a feeling of well-being (euphoria), respiratory distress, convulsions, coma, and possible death. On the other hand, these drugs are very important in treating chronic, severe pain. For example, cancer patients find pain relief with oxycodone, which is taken twice daily in a timed-release pill.

The *cannabinoids* include marijuana and hashish, both derived from the hemp plant. Hashish is several times more potent than marijuana. These drugs depress higher brain centers and release lower brain centers from the normal inhibitory influence of the higher centers. This induces an anxiety-free state, characterized by euphoria and a distorted perception of time and space. *Hallucinations* (sensory perceptions that have no external stimuli), respiratory distress, and vasomotor depression may occur with higher doses.

CNS stimulants include amphetamines and cocaine (including "crack"). These drugs have great abuse potential and may quickly produce psychological dependence. Cocaine, especially when smoked or inhaled, produces euphoria but may also change personality, cause seizures, and constrict certain blood vessels, leading to sudden death from stroke or cardiac arrhythmia. Cocaine's very rapid effect, and perhaps its addictiveness, reflect its rapid entry and metabolism in the brain. Cocaine arrives at the basal nuclei in four to six minutes and is mostly cleared within thirty minutes. The drug inhibits transporter molecules that remove dopamine from synapses after it is released. "Ecstasy" is a type of amphetamine.

Hallucinogens alter perceptions. They cause *illusions,* which are distortions of vision, hearing, taste, touch, and smell; *synesthesia,* such as "hearing" colors or "feeling" sounds; and hallucinations. The most commonly abused and most potent hallucinogen is lysergic acid diethylamide (LSD). LSD may act as an excitatory neurotransmitter. Persons under the influence of LSD may greatly overestimate their physical capabilities, such as believing they can fly off the top of a high building. Phencyclidine (PCP) is another commonly abused hallucinogen. Its use can lead to prolonged psychosis that may provoke assault, murder, and suicide.

---

cerebellum. Another pair (the middle peduncles) transmits signals from the cerebral cortex to the cerebellum concerning the desired positions of these parts. After integrating and analyzing this information, the cerebellum sends correcting impulses via a third pair (the superior peduncles) to the midbrain. These corrections are incorporated into motor impulses that travel downward through the pons,

medulla oblongata, and spinal cord in the appropriate patterns to move the body in the desired way.

The cerebellum is a reflex center for integrating sensory information concerning the position of body parts and for coordinating complex skeletal muscle movements. It also helps maintain posture. Damage to the cerebellum is likely to result in tremors, inaccurate movements of voluntary muscles, loss of muscle tone, a reeling walk, and loss of equilibrium.

A concussion is a temporary and reversible brain injury that results from the brain smashing against the cranium—as occurs commonly playing football. Helmets protect against skull fracture, but not concussions. Football players who repeatedly return to the game soon after suffering a concussion—as many do—risk neurological problems later in life, including dementia. Due to increasing recognition of this problem, among professional as well as high school and college football players, the National Football League has adopted new safety policies.

Practice

49. Where is the cerebellum located?
50. What are the major functions of the cerebellum?

9.15 PERIPHERAL NERVOUS SYSTEM

The peripheral nervous system (PNS) consists of nerves that branch from the CNS and connect it to other body parts. The PNS includes the cranial nerves, which arise from the brain, and the spinal nerves, which arise from the spinal cord.

The PNS can also be subdivided into the somatic and autonomic nervous systems. Generally, the **somatic** (so-mat'ik) **nervous system** consists of the cranial and spinal nerve fibers that connect the CNS to the skin and skeletal muscles; it oversees conscious activities. The **autonomic** (aw"to-nom'ik) **nervous system** includes fibers that connect the CNS to viscera, such as the heart, stomach, intestines, and glands; it controls unconscious activities. Table 9.5 outlines the subdivisions of the nervous system (see fig. 9.2).

Cranial Nerves

Twelve pairs of **cranial nerves** arise from the underside of the brain (fig. 9.34). The first pair begins in the cerebrum, the second in the diencephalon, and the rest in the brainstem. They pass from their sites of origin through foramina of the skull and lead to parts of the head, neck, and trunk.

Table 9.5	Subdivisions of the Nervous System
1. Central nervous system (CNS)	
a. Brain	
b. Spinal cord	
2. Peripheral nervous system (PNS)	
a. Cranial nerves arising from the brain and brainstem	
(1) Somatic fibers connecting to skin and skeletal muscles	
(2) Autonomic fibers connecting to viscera	
b. Spinal nerves arising from the spinal cord	
(1) Somatic fibers connecting to skin and skeletal muscles	
(2) Autonomic fibers connecting to viscera	

Most of the cranial nerves are mixed nerves containing both sensory and motor nerve fibers, but some of those associated with special senses, such as smell and vision, contain only sensory fibers. Other cranial nerves that affect muscles and glands are composed primarily of motor fibers.

Sensory fibers present in the cranial nerves have neuron cell bodies that are outside the brain, usually in groups called *ganglia*. On the other hand, motor neuron cell bodies are typically in the gray matter of the brain.

Numbers or names designate the cranial nerves. The numbers indicate the order in which the nerves arise from the front to the back of the brain, and the names describe their primary functions or the general distribution of their fibers (fig. 9.34).

The first pair of cranial nerves, the **olfactory nerves (I),** are associated with the sense of smell and contain axons only of sensory neurons. These bipolar neurons, located in the lining of the upper nasal cavity, serve as *olfactory receptor cells.* Axons from these receptors pass upward through the cribriform plates of the ethmoid bone, carrying impulses to the olfactory neurons in the *olfactory bulbs,* which are extensions of the cerebral cortex just beneath the frontal lobes (see fig. 10.4, p. 268). Sensory impulses are transmitted from the olfactory bulbs along *olfactory tracts* to cerebral centers, where they are interpreted. The result of this interpretation is the sensation of smell.

The second pair of cranial nerves, the **optic nerves (II),** lead from the eyes to the brain and are associated with vision. The sensory nerve cell bodies of these nerve fibers are in ganglion cell layers in the eyes, and their axons pass through the *optic foramina* of the orbits and continue into the visual nerve pathways of the brain (see chapter 10, pp. 286–287). Sensory impulses transmitted on the optic nerves are interpreted in the visual cortices of the occipital lobes.

The third pair of cranial nerves, the **oculomotor nerves (III),** arise from the midbrain and pass into the orbits of the eyes. One component of each nerve connects to the voluntary muscles that raise the eyelid and to four of the six muscles that move the eye. A second component of each oculomotor nerve is part of the autonomic nervous system and supplies involuntary muscles in the eyes. These muscles adjust the amount of light entering the eyes and focus the lenses.

The fourth pair of cranial nerves, the **trochlear nerves (IV),** arise from the midbrain and are the smallest cranial nerves. Each nerve carries motor impulses to a fifth voluntary muscle that moves the eye and is not innervated by the oculomotor nerve.

The fifth pair of cranial nerves, the **trigeminal nerves (V),** are the largest cranial nerves and arise from the pons. They are mixed nerves, with the sensory parts more extensive than the motor parts. Each sensory component includes three large branches, called the ophthalmic, maxillary, and mandibular divisions.

Olfactory bulb

Olfactory tract

Optic tract

Vestibulocochlear (VIII)

Hypoglossal (XII)

Vagus (X)

Olfactory (I)

Optic (II)

Oculomotor (III)

Trochlear (IV)

Trigeminal (V)

Abducens (VI)

Facial (VII)

Glossopharyngeal (IX)

Accessory (XI)

Figure 9.34 AP|R

The cranial nerves, except for the first pair, arise from the brainstem. They are identified by numbers indicating their order, by their function, or by the general distribution of their fibers.

The *ophthalmic division* of the trigeminal nerves consists of sensory fibers that carry impulses to the brain from the surface of the eyes, the tear glands, and the skin of the anterior scalp, forehead, and upper eyelids. The fibers of the *maxillary division* carry sensory impulses from the upper teeth, upper gum, and upper lip, as well as from the mucous lining of the palate and the skin of the face. The *mandibular division* includes both motor and sensory fibers. The sensory branches transmit impulses from the scalp behind the ears, the skin of the jaw, the lower teeth, the lower gum, and the lower lip. The motor branches supply the muscles of mastication and certain muscles in the floor of the mouth.

The sixth pair of cranial nerves, the **abducens nerves (VI),** are quite small and originate from the pons near the medulla oblongata. Each nerve enters the orbit of the eye and supplies motor impulses to the remaining muscle that moves the eye.

The seventh pair of cranial nerves, the **facial nerves (VII),** arise from the lower part of the pons and emerge on the sides of the face. Their sensory branches are associated with taste receptors on the anterior two-thirds of the tongue, and some of their motor fibers transmit impulses to the muscles of facial expression. Still other motor fibers of these nerves function in the autonomic nervous system and stimulate secretions from tear glands and salivary glands.

The eighth pair of cranial nerves, the **vestibulocochlear nerves (VIII),** are sensory nerves that arise from the medulla oblongata. Each of these nerves has two distinct parts—a vestibular branch and a cochlear branch.

The neuron cell bodies of the *vestibular branch* fibers are located in ganglia associated with parts of the inner ear. These parts contain the receptors involved with reflexes that help maintain equilibrium. The neu-

ron cell bodies of the *cochlear branch* fibers are located in the parts of the inner ear that house the hearing receptors. Impulses from these branches pass through the pons and medulla oblongata on their way to the temporal lobes, where they are interpreted.

The ninth pair of cranial nerves, the **glossopharyngeal nerves (IX),** are associated with the tongue and pharynx. These mixed nerves arise from the medulla oblongata, with predominantly sensory fibers. These sensory fibers carry impulses from the linings of the pharynx, tonsils, and posterior third of the tongue to the brain. Fibers in the motor component innervate muscles of the pharynx that function in swallowing.

The tenth pair of cranial nerves, the **vagus nerves (X),** originate in the medulla oblongata and extend downward through the neck into the chest and abdomen. These nerves are mixed, containing both somatic and autonomic branches, with autonomic fibers predominant. Certain somatic motor fibers carry impulses to muscles of the larynx that are associated with speech and swallowing. Autonomic motor fibers of the vagus nerves supply the heart and many smooth muscles and glands in the thorax and abdomen.

The eleventh pair of cranial nerves, the **accessory nerves (XI),** originate in the medulla oblongata and the spinal cord; thus, they have both cranial and spinal branches. Each *cranial branch* joins a vagus nerve and carries impulses to muscles of the soft palate, pharynx, and larynx. The *spinal branch* descends into the neck and supplies motor fibers to the trapezius and sterno-cleidomastoid muscles.

The twelfth pair of cranial nerves, the **hypoglossal nerves (XII),** arise from the medulla oblongata and pass into the tongue. They include motor fibers that carry impulses to muscles that move the tongue in speaking, chewing, and swallowing. Table 9.6 summarizes the functions of the cranial nerves.

Practice

51. Define *peripheral nervous system.*
52. Distinguish between somatic and autonomic nerve fibers.
53. Name the cranial nerves, and list the major functions of each.

The consequences of a cranial nerve injury depend on the injury's location and extent. Damage to one member of a nerve pair limits loss of function to the affected side, but injury to both nerves affects both sides. If a nerve is severed completely, functional loss is total; if the cut is incomplete, loss may be partial.

Spinal Nerves

Thirty-one pairs of **spinal nerves** originate from the spinal cord (fig. 9.35). They are mixed nerves that provide two-way communication between the spinal cord and parts of the upper and lower limbs, neck, and trunk.

Spinal nerves are not named individually, but are grouped according to the level from which they arise. Each nerve is numbered in sequence. On each vertebra the vertebral notches, the major parts of the intervertebral foramina, are associated with the inferior part of their respective vertebrae. For this reason, each spinal nerve, as it passes through the intervertebral foramen, is associated with the vertebra above it. The cervical spinal nerves are an exception, because spinal nerve C1 passes superior to the vertebra C1. Thus, although there are seven cervical vertebrae, there are eight pairs of *cervical nerves* (numbered C1 to C8). There are twelve pairs of *thoracic nerves* (numbered

Table 9.6	Functions of Cranial Nerves AP|R	
Nerve	**Type**	**Function**
I Olfactory	Sensory	Sensory fibers transmit impulses associated with the sense of smell.
II Optic	Sensory	Sensory fibers transmit impulses associated with the sense of vision.
III Oculomotor	Primarily motor	Motor fibers transmit impulses to muscles that raise eyelids, move eyes, adjust the amount of light entering the eyes, and focus lenses.
		Some sensory fibers transmit impulses associated with the condition of muscles.
IV Trochlear	Primarily motor	Motor fibers transmit impulses to muscles that move the eyes.
		Some sensory fibers transmit impulses associated with the condition of muscles.
V Trigeminal	Mixed	
Ophthalmic division		Sensory fibers transmit impulses from the surface of the eyes, tear glands, scalp, forehead, and upper eyelids.
Maxillary division		Sensory fibers transmit impulses from the upper teeth, upper gum, upper lip, lining of the palate, and skin of the face.
Mandibular division		Sensory fibers transmit impulses from the skin of the jaw, lower teeth, lower gum, and lower lip.
		Motor fibers transmit impulses to muscles of mastication and to muscles in the floor of the mouth.
VI Abducens	Primarily motor	Motor fibers transmit impulses to muscles that move the eyes.
		Some sensory fibers transmit impulses associated with the condition of muscles.
VII Facial	Mixed	Sensory fibers transmit impulses associated with taste receptors of the anterior tongue. Motor fibers transmit impulses to muscles of facial expression, tear glands, and salivary glands.
VIII Vestibulocochlear	Sensory	
Vestibular branch		Sensory fibers transmit impulses associated with the sense of equilibrium.
Cochlear branch		Sensory fibers transmit impulses associated with the sense of hearing.
IX Glossopharyngeal	Mixed	Sensory fibers transmit impulses from the pharynx, tonsils, posterior tongue, and carotid arteries.
		Motor fibers transmit impulses to muscles of the pharynx used in swallowing and to salivary glands.
X Vagus	Mixed	Somatic motor fibers transmit impulses to muscles associated with speech and swallowing; autonomic motor fibers transmit impulses to the heart, smooth muscles, and glands in the thorax and abdomen.
		Sensory fibers transmit impulses from the pharynx, larynx, esophagus, and viscera of the thorax and abdomen.
XI Accessory	Primarily motor	
Cranial branch		Motor fibers transmit impulses to muscles of the soft palate, pharynx, and larynx.
Spinal branch		Motor fibers transmit impulses to muscles of the neck and back.
XII Hypoglossal	Primarily motor	Motor fibers transmit impulses to muscles that move the tongue.

Posterior view

Cervical plexus
(C1–C4)

Brachial plexus
(C5–T1)

Musculocutaneous
nerve

Axillary nerve
Radial nerve
Median nerve
Ulnar nerve
Phrenic nerve

Intercostal
nerves

Cauda equina

Lumbosacral plexus
(T12–S5)

Femoral
nerve

Obturator nerve

Sciatic nerve

C1
C2
C3
C4
C5
C6
C7
C8
T1
T2
T3
T4
T5
T6
T7
T8
T9
T10
T11
T12
L1
L2
L3
L4
L5
S1
S2
S3
S4
S5
Co

Figure 9.35

The anterior branches of the spinal nerves in the thoracic region give rise to intercostal nerves. Those in other regions combine to form complex networks called plexuses. (Note that there are eight pairs of cervical nerves, one pair originating above the first cervical vertebra and the eighth pair originating below the seventh cervical vertebra.)

T1 to T12), five pairs of *lumbar nerves* (numbered L1 to L5), five pairs of *sacral nerves* (numbered S1 to S5), and one pair of *coccygeal nerves* (Co).

The adult spinal cord ends at the level between the first and second lumbar vertebrae. The lumbar, sacral, and coccygeal nerves descend beyond the end of the cord, forming a structure called the *cauda equina* (horse's tail).

Each spinal nerve emerges from the cord by two short branches, or *roots,* which lie within the vertebral column. The **dorsal root** (posterior or sensory root) can be identified by an enlargement called the *dorsal root ganglion* (see fig. 9.22a). This ganglion contains the cell bodies of the sensory neurons whose axons (peripheral processes) conduct impulses inward from the peripheral body parts.

The axons of these neurons (central processes) extend through the dorsal root and into the spinal cord, where they form synapses with dendrites of other neurons (see fig. 9.6). The **ventral root** (anterior or motor root) of each spinal nerve consists of axons from the motor neurons whose cell bodies are within the gray matter of the cord.

A ventral root and a dorsal root unite to form a spinal nerve, which extends outward from the vertebral canal through an *intervertebral foramen* (see fig. 7.17, p. 151). Just beyond its foramen, each spinal nerve divides into several parts.

Except in the thoracic region, the main parts of the spinal nerves combine to form complex networks called **plexuses** instead of continuing directly to peripheral body parts (fig. 9.35). In a plexus, spinal nerve axons are sorted and recombined so that axons that innervate a particular body part reach it in the same peripheral nerve, even though the axons originate from different spinal nerves.

Cervical Plexuses

The **cervical plexuses** lie deep in the neck on either side and form from the branches of the first four cervical nerves. Axons from these plexuses supply the muscles and skin of the neck. In addition, axons from the third, fourth, and fifth cervical nerves pass into the right and left **phrenic nerves,** which conduct motor impulses to the muscle fibers of the diaphragm.

Brachial Plexuses

Branches of the lower four cervical nerves and the first thoracic nerve give rise to the **brachial plexuses.** These networks of axons are deep within the shoulders between the neck and axillae (armpits). The major branches emerging from the brachial plexuses supply the muscles and skin of the arm, forearm, and hand, and include the **musculocutaneous, ulnar, median, radial,** and **axillary nerves.**

Lumbosacral Plexuses

The **lumbosacral plexuses** are formed on either side by the last thoracic nerve and the lumbar, sacral, and coccygeal nerves. These networks of axons extend from the lumbar region of the back into the pelvic cavity, giving rise to a number of motor and sensory axons associated with the muscles and skin of the lower abdominal wall, external genitalia, buttocks, thighs, legs, and feet. The major branches of these plexuses include the **obturator, femoral,** and **sciatic nerves.**

The anterior branches of the thoracic spinal nerves do not enter a plexus. Instead, they enter spaces between the ribs and become **intercostal nerves.** These nerves supply motor impulses to the intercostal muscles and the upper abdominal wall muscles. They also receive sensory impulses from the skin of the thorax and abdomen.

Practice

54. How are spinal nerves grouped?

55. Describe how a spinal nerve joins the spinal cord.

56. Name and locate the major nerve plexuses.

Spinal nerves may be injured in a variety of ways, including stabs, gunshot wounds, birth injuries, dislocations and fractures of the vertebrae, and pressure from tumors in surrounding tissues. For example, a sudden extension followed by flexion of the neck, called *whiplash,* can occur during rear-end automobile collisions and may stretch the superficial nerves of the cervical plexuses. Whiplash may cause continuing headaches and pain in the neck and skin, which the cervical nerves supply.

9.16 AUTONOMIC NERVOUS SYSTEM

The **autonomic nervous system** is the part of the PNS that functions independently (autonomously) and continuously without conscious effort. This system controls visceral functions by regulating the actions of smooth muscle, cardiac muscle, and glands. It regulates heart rate, blood pressure, breathing rate, body temperature, and other activities that maintain homeostasis. Parts of the autonomic nervous system respond to emotional stress and prepare the body to meet the demands of strenuous physical activity.

General Characteristics

Reflexes in which sensory signals originate from receptors in the viscera and the skin regulate autonomic activities. Axons carry these signals to centers in the brain or spinal cord. In response, motor impulses travel out from these centers on axons in cranial and spinal nerves. These axons typically lead to ganglia. The impulses they carry are integrated in these ganglia and relayed to effectors (muscles and glands) that respond by contracting, releasing secretions, or being inhibited. The integrative function of the ganglia provides the autonomic system with a degree of independence from the brain and spinal cord.

The autonomic nervous system includes two divisions—the **sympathetic** (sim″pah-thet′ik) and **parasympathetic** (par″ah-sim″pah-thet′ik) **divisions.** Some effectors are innervated by axons from each division. In such cases, impulses on one set of axons may activate an organ, while impulses on the other set inhibit it. Thus, the divisions may act antagonistically, alternately activating or inhibiting the actions of effectors.

The functions of the autonomic divisions are mixed; that is, each activates some organs and inhibits others. However, the divisions have important functional differences. The sympathetic division prepares the body for energy-expending, stressful, or emergency situations, as part of the *fight-or-flight* response. Conversely, the parasympathetic division is most active under ordinary, restful conditions. It also counterbalances the effects of the sympathetic division and restores the body to a resting state following a stressful experience. For example, during an emergency the sympathetic division increases heart rate; following the emergency, the parasympathetic division decreases heart rate.

Autonomic Neurons

The neurons of the autonomic nervous system are motor neurons. However, unlike the motor pathways of the somatic nervous system, which usually include a single neuron between the brain or spinal cord and a skeletal muscle, those of the autonomic system include two neurons (fig. 9.36). The cell body of the first, or preganglionic, neuron is located in the brain or spinal cord. Its axon, the

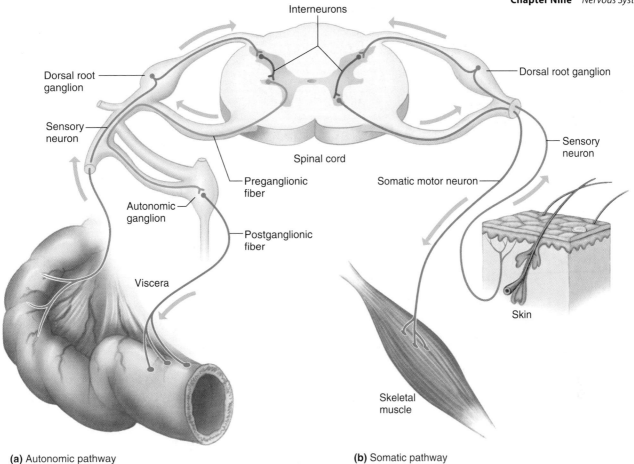

(a) Autonomic pathway   (b) Somatic pathway

Figure 9.36

Motor pathways. **(a)** Autonomic pathways include two neurons between the CNS and an effector. **(b)** Somatic pathways usually have a single neuron between the CNS and an effector. Note that in both cases the motor fibers pass through the ventral root of the spinal cord.

preganglionic fiber (pre″gang-gle-on′ik fi′ber), leaves the CNS and synapses with one or more neurons whose cell bodies are located in the PNS within an autonomic ganglion. The axon of such a second neuron, or postganglionic neuron, is called a **postganglionic fiber** (pō st″gang-gle-on′ik fi′ber), and it extends to a visceral effector.

Sympathetic Division

In the sympathetic division, the preganglionic fibers originate from neurons in the gray matter of the spinal cord (fig. 9.37). Their axons leave the cord through the ventral roots of spinal nerves in the first thoracic through the second lumbar segments. After traveling a short distance, these fibers leave the spinal nerves, and each enters a member of a chain of sympathetic ganglia (*paravertebral ganglia*). One of these sympathetic chains extends longitudinally along each side of the vertebral column.

In paravertebral ganglia, preganglionic fibers form synapses with second neurons. The axons of these neurons, the postganglionic fibers, typically return to spinal nerves and extend to visceral effectors.

Parasympathetic Division

The preganglionic fibers of the parasympathetic division arise from the brainstem and sacral region of the

spinal cord (fig. 9.38). From there, they lead outward in cranial or sacral nerves to ganglia located near or in various viscera. The relatively short postganglionic fibers continue from the ganglia to specific muscles or glands in these viscera.

Practice

57. Describe the parts of the autonomic nervous system.
58. Distinguish between the divisions of the autonomic nervous system.
59. Describe a sympathetic nerve pathway and a parasympathetic nerve pathway.

Autonomic Neurotransmitters

The preganglionic fibers of the sympathetic and parasympathetic divisions all secrete *acetylcholine* and are therefore called **cholinergic fibers** (ko″lin-er′jik fi′berz). The parasympathetic postganglionic fibers are also cholinergic. One exception, parasympathetic neurons that secrete nitric oxide, is described in chapter 19 (p. 512). However, most sympathetic postganglionic neurons

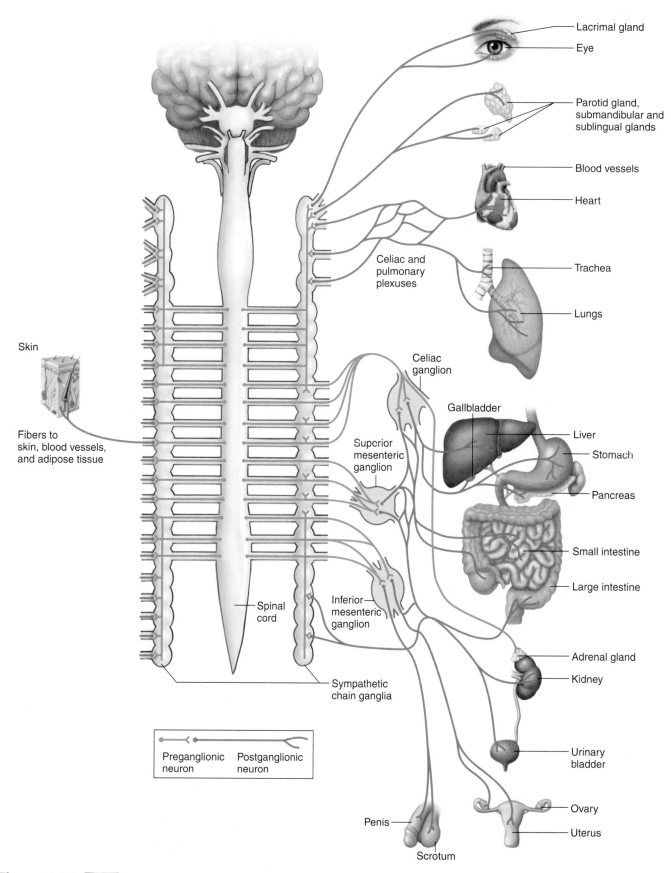

Lacrimal gland

Eye

Parotid gland, submandibular and sublingual glands

Blood vessels

Heart

Celiac and pulmonary plexuses

Trachea

Lungs

Skin

Celiac ganglion

Gallbladder

Liver

Fibers to skin, blood vessels, and adipose tissue

Superior mesenteric ganglion

Stomach

Pancreas

Small intestine

Large intestine

Spinal cord

Inferior mesenteric ganglion

Sympathetic chain ganglia

Adrenal gland

Kidney

Urinary bladder

Preganglionic neuron   Postganglionic neuron

Ovary

Penis

Uterus

Scrotum

Figure 9.37 AP|R

The preganglionic fibers of the sympathetic division of the autonomic nervous system arise from the thoracic and lumbar regions of the spinal cord (T1–L2). Note that the adrenal medulla is innervated directly by a preganglionic fiber.

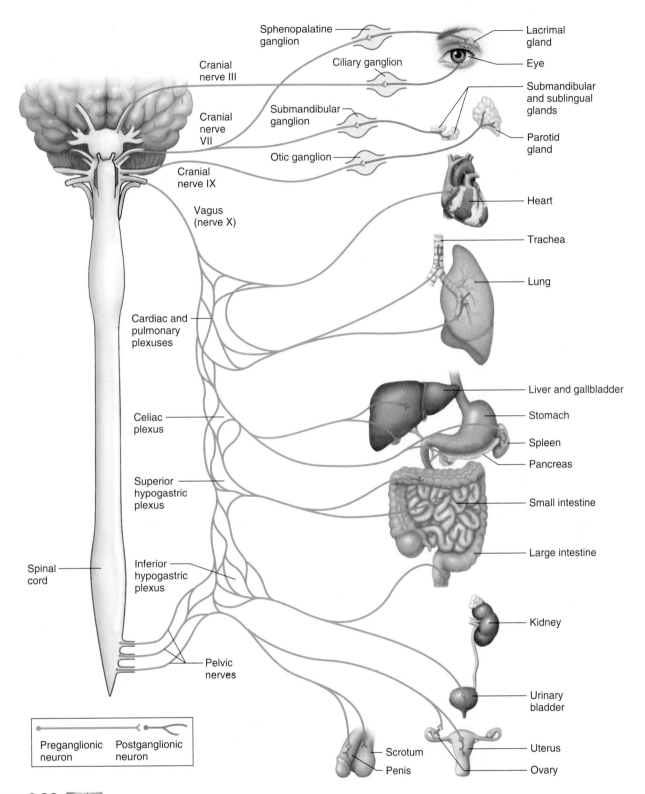

Figure 9.38 AP|R

The preganglionic fibers of the parasympathetic division of the autonomic nervous system arise from the brainstem and sacral region of the spinal cord.

secrete *norepinephrine* (noradrenalin) and are called **adrenergic fibers** (ad″ren-ur′jik fi′berz) (fig. 9.39). The different postganglionic neurotransmitters cause the different effects that the sympathetic and parasympathetic divisions have on their effector organs.

Most organs receive innervation from both sympathetic and parasympathetic divisions, usually with opposing actions. For example, parasympathetic activity increases activity of the digestive system, whereas sympathetic activity decreases it. Similarly, sympathetic stimulation increases heart rate, but parasympathetic action slows heart rate.

Some viscera are controlled primarily by one division or the other. That is, the divisions are not always actively antagonistic. For example, the sympathetic division regulates the diameter of most blood vessels, which lack parasympathetic innervation. Smooth muscles in the walls of these vessels are continuously stimulated and thus are in a state of partial contraction (sympathetic tone). Decreasing sympathetic stimulation increases (dilates) the diameter of the vessels, which relaxes their muscular walls. Conversely, increasing sympathetic stimulation constricts the vessels. Table 9.7 summarizes the effects of stimulation by adrenergic and cholinergic fibers on some visceral effectors.

Control of Autonomic Activity

The brain and spinal cord largely control the autonomic nervous system, despite the system's independence resulting from the integrative function of its ganglia. For example, control centers in the medulla oblongata for cardiac, vasomotor, and respiratory activities receive sensory impulses from viscera on vagus nerve fibers and use autonomic nerve pathways to stimulate motor responses in muscles and glands. Similarly, the hypothalamus helps regulate body temperature, hunger, thirst, and water and electrolyte balance by influencing autonomic pathways.

More complex centers in the brain, including the limbic system and the cerebral cortex, control the autonomic nervous system during emotional stress. These structures utilize autonomic pathways to regulate emotional expression and behavior.

Practice

60. Which neurotransmitters operate in the autonomic nervous system?

61. How do the divisions of the autonomic nervous system regulate visceral activities?

62. How are autonomic activities controlled?

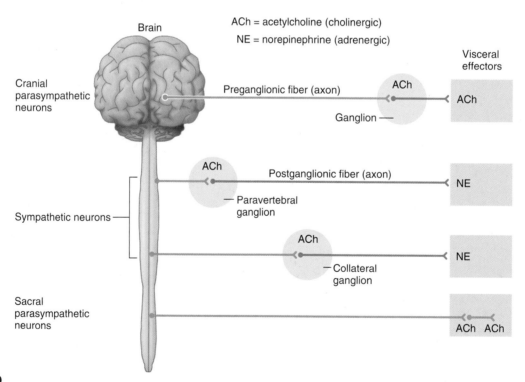

Figure 9.39

Most sympathetic fibers are adrenergic and secrete norepinephrine at the ends of the postganglionic fiber; parasympathetic fibers are cholinergic and secrete acetylcholine at the ends of the postganglionic fibers. Two arrangements of parasympathetic postganglionic fibers are seen in both the cranial and sacral portions. Similarly, sympathetic paravertebral and collateral ganglia are seen in both the thoracic and lumbar portions of the nervous system. (Note: This representation does not show dendrites.)

Nervous System

Neurons transmit impulses that allow body systems to communicate.

Integumentary System

Sensory receptors provide the nervous system with information about the outside world.

Skeletal System

Bones protect the brain and spinal cord and help maintain plasma calcium, which is important to neuron function.

Muscular System

Nerve impulses control movement and carry information about the position of body parts.

Endocrine System

The hypothalamus controls secretion of many hormones.

Cardiovascular System

Nerve impulses help control blood flow and blood pressure.

Lymphatic System

Stress may impair the immune response.

Digestive System

The nervous system can influence digestive function.

Respiratory System

The nervous system alters respiratory activity to control oxygen levels and blood pH.

Urinary System

Nerve impulses affect urine production and elimination.

Reproductive System

The nervous system plays a role in egg and sperm formation, sexual pleasure, childbirth, and nursing.

Table 9.7	Effects of Neurotransmitter Substances on Visceral Effectors or Actions	
Visceral Effector or Action	**Response to Adrenergic Stimulation (Sympathetic)**	**Response to Cholinergic Stimulation (Parasympathetic)**
Pupil of the eye	Dilation	Constriction
Heart rate	Increases	Decreases
Bronchioles of lungs	Dilation	Constriction
Muscles of intestinal wall	Slows peristaltic action	Speeds peristaltic action
Intestinal glands	Secretion decreases	Secretion increases
Blood distribution	More blood to skeletal muscles; less blood to digestive organs	More blood to digestive organs; less blood to skeletal muscles
Blood glucose concentration	Increases	Decreases
Salivary glands	Secretion decreases	Secretion increases
Tear glands	No action	Secretion
Muscles of gallbladder	Relaxation	Contraction
Muscles of urinary bladder	Relaxation	Contraction

Summary Outline

9.1 Introduction (p. 214)

1. Nervous tissue includes neurons, which are the structural and functional units of the nervous system, and neuroglia.
2. Organs of the nervous system are divided into the central and peripheral nervous systems.

9.2 General Functions of the Nervous System (p. 215)

1. Sensory functions involve receptors that detect internal and external changes.
2. Integrative functions collect sensory information and make decisions that motor functions carry out.
3. Motor functions stimulate effectors to respond.

9.3 Neuroglia (p. 216)

1. Neuroglia in the central nervous system include microglial cells, oligodendrocytes, astrocytes, and ependymal cells.
2. In the peripheral nervous system, Schwann cells form myelin sheaths.

9.4 Neurons (p. 216)

1. A neuron includes a cell body, dendrites, and an axon.
2. Dendrites and the cell body provide receptive surfaces.
3. A single axon arises from the cell body and may be enclosed in a myelin sheath and a neurilemma.
4. Classification of neurons
   a. Neurons are classified structurally as multipolar, bipolar, or unipolar.
   b. Neurons are classified functionally as sensory neurons, interneurons, or motor neurons.

9.5 The Synapse (p. 221)

A synapse is a junction between two neurons.
1. A presynaptic neuron carries an impulse into a synapse; a postsynaptic neuron responds.
2. Axons have synaptic knobs at their distal ends, which secrete neurotransmitters.
3. A neurotransmitter is released when a nerve impulse reaches the end of an axon.
4. A neurotransmitter reaching the postsynaptic neuron membrane is either excitatory or inhibitory.

9.6 Cell Membrane Potential (p. 222)

A cell membrane is usually polarized as a result of unequal ion distribution.
1. Distribution of ions
   a. Pores and channels in cell membranes that allow passage of some ions but not others set up differences in the concentrations of specific ions inside and outside a neuron.
   b. Potassium ions pass more easily through cell membranes than do sodium ions.
2. Resting potential
   a. A high concentration of sodium ions is outside a cell membrane, and a high concentration of potassium ions is inside.
   b. Many negatively charged ions are inside a cell.
   c. In a resting cell, more positive ions leave than enter, so the outside of the cell membrane develops a positive charge, while the inside develops a negative charge.
3. Potential changes
   a. Stimulation of a cell membrane affects the membrane's resting potential.
   b. When its resting potential becomes less negative, a membrane becomes depolarized.
   c. Potential changes are graded.
   d. Achieving threshold potential triggers an action potential.
4. Action potential
   a. At threshold, sodium channels open, and sodium ions diffuse inward, depolarizing the membrane.
   b. At almost the same time, potassium channels open, and potassium ions diffuse outward, repolarizing the membrane.
   c. This rapid sequence of depolarization and repolarization is an action potential.
   d. Many action potentials can occur in a neuron without disrupting the ion concentrations. Active transport contributes to maintaining these concentrations.

9.7 Nerve Impulses (p. 227)

A wave of action potentials is a nerve impulse.

1. Impulse conduction (action potential propagation)
   a. Unmyelinated axons conduct action potentials along their entire lengths.
   b. Myelinated axons conduct impulses more rapidly.
   c. Axons with larger diameters conduct impulses faster than those with smaller diameters.
2. All-or-none response
   a. An action potential occurs in an all-or-none manner whenever a stimulus of threshold intensity is applied to an axon.
   b. All of the action potentials triggered on an axon are of the same strength.

9.8 Synaptic Transmission (p. 228)

1. Excitatory and inhibitory actions
   a. Neurotransmitters that trigger nerve impulses are excitatory. Those that inhibit impulses are inhibitory.
   b. The net effect of synaptic knobs communicating with a neuron depends on which knobs are activated from moment to moment.
2. Neurotransmitters
   a. The nervous system produces many different neurotransmitters.
   b. Neurotransmitters include acetylcholine, biogenic amines, amino acids, and peptides.
   c. A synaptic knob releases neurotransmitters when an action potential increases membrane permeability to calcium ions.
   d. After being released, neurotransmitters are decomposed or removed from synaptic clefts.

9.9 Impulse Processing (p. 228)

How the nervous system processes and responds to nerve impulses reflects the organization of neurons in the brain and spinal cord.

1. Neuronal pools
   a. Neurons form pools in the central nervous system.
   b. Each pool receives impulses, processes them, and conducts impulses away.
2. Facilitation
   a. Each neuron in a pool may receive excitatory and inhibitory stimuli.
   b. A neuron is facilitated when it receives subthreshold stimuli and becomes more excitable.
3. Convergence
   a. Impulses from two or more incoming axons may converge on a single neuron.
   b. Convergence enables impulses from different sources to have an additive effect on a neuron.
4. Divergence
   a. Impulses leaving a pool may diverge by passing into several output neurons.
   b. Divergence amplifies impulses.

9.10 Types of Nerves (p. 230)

1. Nerves are cordlike bundles (fascicles) of nerve fibers (axons).
2. Nerves are sensory, motor, or mixed, depending on which type of axons they contain.

9.11 Nerve Pathways (p. 231)

A nerve pathway is the route an impulse follows through the nervous system.

1. A reflex arc usually includes a sensory neuron, a reflex center composed of interneurons, and a motor neuron.

2. Reflex behavior
   a. Reflexes are automatic, subconscious responses to changes.
   b. They help maintain homeostasis.
   c. Two neurons carry out the patellar reflex. It is therefore monosynaptic.
   d. Withdrawal reflexes are protective.

9.12 Meninges (p. 232)

1. Bone and meninges surround the brain and spinal cord.
2. The meninges are the dura mater, arachnoid mater, and pia mater.
3. Cerebrospinal fluid fills the space between the arachnoid and pia maters.

9.13 Spinal Cord (p. 234)

The spinal cord is a nerve column that extends from the brain into the vertebral canal.

1. Structure of the spinal cord
   a. Each of the spinal cord's thirty-one segments gives rise to a pair of spinal nerves (two pairs are associated with C1).
   b. The spinal cord has a cervical enlargement and a lumbar enlargement.
   c. A central core of gray matter lies within white matter.
   d. White matter consists of bundles of myelinated axons called tracts.
2. Functions of the spinal cord
   a. The spinal cord provides a two-way communication system between the brain and other body parts and serves as a center for spinal reflexes.
   b. Ascending tracts carry sensory impulses to the brain. Descending tracts carry motor impulses to muscles and glands.

9.14 Brain (p. 236)

The brain is subdivided into the cerebrum, diencephalon, brainstem, and cerebellum.

1. Structure of the cerebrum
   a. The cerebrum consists of two cerebral hemispheres connected by the corpus callosum.
   b. The cerebral cortex is a thin layer of gray matter near the surface.
   c. White matter consists of myelinated axons that connect neurons in the nervous system and communicate with other body parts.
2. Functions of the cerebrum
   a. The cerebrum provides higher brain functions.
   b. The cerebral cortex consists of sensory, association, and motor, areas.
   c. One cerebral hemisphere usually dominates for certain intellectual functions.
3. Ventricles and cerebrospinal fluid
   a. Ventricles are interconnected cavities within the cerebral hemispheres and brainstem.
   b. Cerebrospinal fluid fills the ventricles.
   c. The choroid plexuses in the walls of the ventricles secrete cerebrospinal fluid.
4. Diencephalon
   a. The diencephalon contains the thalamus, which is a central relay station for incoming sensory impulses, and the hypothalamus, which maintains homeostasis.
   b. The limbic system produces emotions and modifies behavior.
5. Brainstem
   a. The brainstem consists of the midbrain, pons, and medulla oblongata.
   b. The midbrain contains reflex centers associated with eye and head movements.

c. The pons transmits impulses between the cerebrum and other parts of the nervous system and contains centers that may help regulate breathing.

d. The medulla oblongata transmits all ascending and descending impulses and contains several vital and nonvital reflex centers.

e. The reticular formation filters incoming sensory impulses, arousing the cerebral cortex into wakefulness when significant impulses arrive.

6. Cerebellum

a. The cerebellum consists of two hemispheres.

b. It functions primarily as a reflex center for integrating sensory information required in the coordination of skeletal muscle movements and the maintenance of equilibrium.

9.15 Peripheral Nervous System (p. 246)

The peripheral nervous system consists of cranial and spinal nerves that branch from the brain and spinal cord to all body parts. It is subdivided into the somatic and autonomic systems.

1. Cranial nerves

a. Twelve pairs of cranial nerves connect the brain to parts in the head, neck, and trunk.

b. Most cranial nerves are mixed, but some are purely sensory, and others are primarily motor.

c. The names of the cranial nerves indicate their primary functions or the general distributions of their fibers.

d. Some cranial nerves are somatic, and others are autonomic.

2. Spinal nerves

a. Thirty-one pairs of spinal nerves originate from the spinal cord.

b. These mixed nerves provide a two-way communication system between the spinal cord and parts of the upper and lower limbs, neck, and trunk.

c. Spinal nerves are grouped according to the levels from which they arise, and they are numbered in sequence.

d. Each spinal nerve emerges by a dorsal and a ventral root.

e. Each spinal nerve divides into several branches just beyond its foramen.

f. Most spinal nerves combine to form plexuses in which nerve fibers are sorted and recombined so that those fibers associated with a particular part reach it together.

9.16 Autonomic Nervous System (p. 250)

The autonomic nervous system functions without conscious effort. It regulates the visceral activities that maintain homeostasis.

1. General characteristics

a. Autonomic functions are reflexes controlled from nerve centers in the brain and spinal cord.

b. The autonomic nervous system consists of two divisions—the sympathetic and the parasympathetic.

c. The sympathetic division responds to stressful and emergency conditions.

d. The parasympathetic division is most active under ordinary conditions.

2. Autonomic nerve fibers

a. Autonomic nerve fibers are motor fibers.

b. Sympathetic fibers leave the spinal cord and synapse in paravertebral ganglia.

c. Parasympathetic fibers begin in the brainstem and sacral region of the spinal cord and synapse in ganglia near viscera.

3. Autonomic neurotransmitters

a. Sympathetic and parasympathetic preganglionic fibers secrete acetylcholine.

b. Parasympathetic postganglionic fibers secrete acetylcholine. Sympathetic postganglionic fibers secrete norepinephrine.

c. The different effects of the autonomic divisions are due to the different neurotransmitters the postganglionic fibers release.

d. The two divisions usually have opposite actions.

4. Control of autonomic activity

a. The autonomic nervous system is somewhat independent.

b. Control centers in the medulla oblongata and hypothalamus utilize autonomic nerve pathways.

c. The limbic system and cerebral cortex control the autonomic system during emotional stress.

Chapter Assessments

9.1 Introduction

1. The general function of neurons is to _____ , whereas the general functions of neuroglia are to _____ . (p. 214)

2. Match the neuron part on the left to its description on the right. (p. 214)

(1) dendrite	A. A cell process that sends information
(2) axon	B. One of usually several cell processes that receive information
(3) cell body	C. The rounded part of a neuron

3. Explain the relationship between the CNS and the PNS. (p. 214)

9.2 General Functions of the Nervous System

4. List the general functions of the nervous system. (p. 215)

9.3 Neuroglia

5. Match the types of neuroglia to their functions. (p. 216)

(1) ependymal cells	A. Form a myelin sheath around peripheral nerves
(2) oligodendrocytes	B. Phagocytize cellular debris and bacteria
(3) astrocytes	C. Line inner parts of ventricles and spinal cord
(4) Schwann cells	D. Form scar tissue and regulate ion and nutrient concentrations in the CNS
(5) microglial cells	E. Form a myelin sheath around neurons in the CNS

9.4 Neurons

6. Describe three structures found in neurons that are also in other cell types, and describe two structures that are unique to neurons. (p. 216)

7. The part of a Schwann cell that contributes to the myelin sheath is the _____ , and the part that contributes to the neurilemma is the _____ . (p. 217)

8. Distinguish between myelinated and unmyelinated axons. (p. 217)

9. Distinguish among multipolar, bipolar, and unipolar neurons. (p. 219)

10. Distinguish among sensory neurons, interneurons, and motor neurons. (p. 220)

11. Distinguish between ganglia and nuclei. (p. 220)

9.5 The Synapse

12. Define *synapse*. (p. 221)

13. Explain how information passes from one neuron to another. (p. 222)

9.6 Cell Membrane Potential

14. Explain how a membrane becomes polarized. (p. 222)

15. Describe how ions associated with nerve cell membranes are distributed. (p. 222)

16. Define *resting potential*. (p. 223)

17. Explain the relationship between threshold potential and an action potential. (p. 225)

18. List the events that occur during an action potential. (p. 225)

9.7 Nerve Impulses

19. Choose the correct sequence of events along an axon: (p. 227)
   a. Resting potentials are propagated along a stimulated axon, causing a very small action potential.
   b. A threshold stimulus opens K^+ channels and the ions diffuse in, depolarizing the cell membrane. Then Na^+ channels open, Na^+ exits, and the cell membrane repolarizes, generating an action potential that stimulates adjacent cell membrane, forming the nerve impulse.
   c. A threshold stimulus opens Na^+ channels and the ions diffuse in, depolarizing the cell membrane. Then K^+ channels open, K^+ exits, and the cell membrane repolarizes, generating an action potential that stimulates adjacent cell membrane, forming the nerve impulse.
   d. A threshold stimulus opens Na^+ channels and the ions diffuse in, depolarizing the cell membrane. Then K^+ channels open, K^+ exits, and the cell membrane repolarizes, generating an action potential that inhibits adjacent cell membrane, forming the nerve impulse.
   e. Action potentials occur at different points along an axon, then join to form a very large action potential.

20. Explain why a myelin sheath covering an entire axon (with no nodes of Ranvier) would inhibit conduction of a nerve impulse. (p. 227)

21. "All-or-none" response in nerve impulse conduction means that _____ . (p. 227)

9.8 Synaptic Transmission

22. Distinguish between excitatory and inhibitory actions of neurotransmitters. (p. 228)

23. Neurotransmitters are synthesized in _____ and are stored in _____ . (p. 228)

24. Match the neurotransmitter to its description on the right. (p. 228)

(1) biogenic amine	A. Short chains of amino acids
(2) acetylcholine	B. A modified amino acid
(3) neuropeptide	C. An amino acid
(4) GABA	D. Stimulates skeletal muscle contraction

25. Explain what happens to neurotransmitters after they are released. (p. 228)

9.9 Impulse Processing

26. Describe the components of a neuronal pool. (p. 228)

27. "Facilitation in a neuronal pool" refers to _____ . (p. 228)

28. Distinguish between convergence and divergence in a neuronal pool. (p. 229)

9.10 Types of Nerves

29. Describe how sensory, motor, and mixed nerves differ. (p. 230)

9.11 Nerve Pathways

30. Distinguish between a reflex arc and a reflex. (p. 231)

31. Describe the components of a reflex arc and their functions. (p. 231)

32. List three body functions that reflexes control. (p. 231)

9.12 Meninges

33. Match each layer of the meninges to its description. (p. 233)

(1) dura mater	A. The thin, innermost layer, containing blood vessels and nerves
(2) arachnoid mater	B. The tough, outermost layer, consisting mostly of connective tissue
(3) pia mater	C. The lacy membrane, lacking blood vessels, sandwiched between the other two layers

9.13 Spinal Cord

34. Describe the structure of the spinal cord. (p. 234)

35. Distinguish between the ascending and descending tracts of the spinal cord. (p. 234)

9.14 Brain

36. Name the four major parts of the brain and describe their general functions. (p. 236)

37. The area of the brain that contains centers controlling visceral activities is the: (p. 236)
   a. cerebrum
   b. cerebellum
   c. brainstem
   d. diencephalon
   e. corpus callosum

38. The structure that connects the cerebral hemispheres is the _____ . (p. 236)

39. Distinguish between a sulcus and a fissure. (p. 237)

40. Relate the lobes of the cerebral hemispheres to the skull bones. (p. 238)

41. Locate the sensory, association, and motor areas of the cerebral cortex, and describe the general functions of each. (p. 238)

42. Define *hemisphere dominance*. (p. 239)

43. The function of the basal nuclei is to _____ . (p. 240)

44. Locate the ventricles in the brain. (p. 240)

45. Explain how cerebrospinal fluid is produced and how it functions. (p. 240)

46. The part of the diencephalon that regulates hunger, weight, water and electrolyte balance, sleep and wakefulness, temperature, arterial blood pressure, heart rate, production of substances that stimulate the pituitary gland, and movement and secretion in areas of the digestive tract is the: (p. 243)
 a. thalamus
 b. pineal gland
 c. infundibulum
 d. hypothalamus
 e. mammillary bodies

47. Define *limbic system*, and explain its functions. (p. 243)

48. The parts of the brainstem are the _____ , _____ , and _____ . (p. 243)

49. List the functions of the three parts of the brainstem. (p. 243)

50. Vomiting is controlled by: (p. 244)
 a. the reticular formation
 b. a nucleus in the medulla oblongata
 c. the midbrain
 d. the pons
 e. the thalamus

51. Describe what happens to the body when the reticular formation receives sensory impulses, and what happens when it does not receive stimulation. (p. 244)

52. Describe the functions of the cerebellum. (p. 245)

9.15 Peripheral Nervous System

53. Distinguish between the somatic nervous system and the autonomic nervous system. (p. 246)

54. Distinguish between cranial nerves and spinal nerves. (pp. 246, 248)

55. Match the cranial nerves to the body parts or functions that they affect. More than one nerve pair may correspond to the same structure or function. (pp. 246–248)

 (1) olfactory nerves (I)
 (2) optic nerves (II)
 (3) oculomotor nerves (III)
 (4) trochlear nerves (IV)
 (5) trigeminal nerves (V)
 (6) abducens nerves (VI)
 (7) facial nerves (VII)
 (8) vestibulocochlear nerves (VIII)
 (9) glossopharyngeal nerves (IX)
 (10) vagus nerves (X)
 (11) accessory nerves (XI)
 (12) hypoglossal nerves (XII)

 A. Vision
 B. Hearing and equilibrium
 C. Muscles of the larynx, pharynx, soft palate, sternocleidomastoid and trapezius muscles
 D. Heart, various smooth muscles and glands in the thorax and abdomen
 E. Taste, facial expressions, secretion of tears and saliva
 F. Sense of smell
 G. Tongue movements and swallowing
 H. Face and scalp
 I. Eye movements

56. Explain how the spinal nerves are classified and numbered. (p. 248)

57. Describe the structure of a spinal nerve. (p. 249)

58. Define *plexus*, and locate the major plexuses of the spinal nerves. (p. 249)

9.16 Autonomic Nervous System

59. Describe the general functions of the autonomic nervous system. (p. 250)

60. Distinguish between the sympathetic and parasympathetic divisions of the autonomic nervous system. (p. 250)

61. Distinguish between preganglionic and postganglionic neurons. (p. 251)

62. The effects of the sympathetic and parasympathetic autonomic divisions differ because _____ . (p. 251)

63. List two ways in which the CNS controls autonomic activities. (p. 254)

Integrative Assessments/Critical Thinking

OUTCOMES 3.4, 9.3, 9.4

1. State two reasons why rapidly growing brain cancers are composed of neuroglia rather than neurons.

OUTCOMES 9.3, 9.4, 9.7, 9.13, 9.14

2. In multiple sclerosis, nerve fibers in the CNS lose their myelin. Explain why this loss affects skeletal muscle function.

OUTCOMES 9.4, 9.5, 9.11, 9.13, 9.14

3. List four skills encountered in everyday life that depend on nervous system function, and list the part of the nervous system responsible for each.

OUTCOMES 9.11, 9.13

4. The biceps-jerk reflex is carried out by motor neurons that exit the spinal cord in the fifth spinal nerve (C5). The triceps-jerk reflex uses motor neurons in the seventh spinal nerve (C7). Describe how these reflexes might be tested to help pinpoint damage in a patient with a neck injury.

OUTCOMES 9.11, 9.14

5. Describe the roles of the cerebrum and cerebellum in athletics.

OUTCOMES 9.13, 9.14

6. Describe expected functional losses in a patient who has suffered injury to the right occipital lobe of the cerebral cortex compared to injury in the right temporal lobe of the cerebral cortex.

WEB CONNECTIONS

Visit the text website at **www.mhhe.com/shieress11** for additional quizzes, interactive learning exercises, and more.

APR

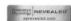

Anatomy & Physiology REVEALED includes cadaver photos that allow you to peel away layers of the human body to reveal structures beneath the surface. This program also includes animations, radiologic imaging, audio pronunciation, and practice quizzing. To learn more visit **www.aprevealed.com**.

10

The Senses

The sound of music. The band Nirvana and singer Tori Amos have each recorded the song "Smells Like Teen Spirit." In the original Nirvana version, Kurt Cobain's voice is loud and brash, as is the instrumentation; in contrast, Tori Amos's song is slow and subdued. Yet it is easy to tell that these are the same songs. What isn't easy is figuring out how the brain can tell this.

Some neurons in the auditory cortex sense a certain range of frequencies of incoming sound waves, but others are "pitch-sensitive," which means that they can recognize the same note, whether it comes from an oboe or an elephant. This property of sound, called pitch, is a vibration frequency from objects that vibrate periodically. The vibration is complex—plucking a string on an instrument vibrates the entire string, but also vibrates parts of it, creating a complex sound. Pitch-sensitive neurons recognize the "fundamental" vibration, which is the lowest one coming from the entire vibrating object, corresponding to plucking the entire string.

In experiments to identify and localize pitch-sensitive neurons, researchers placed electrodes over the auditory cortices of marmoset monkeys, who hear the same range of sounds as humans. When the monkeys listened to sounds that shared the fundamental vibration, even though different sources made the sounds, the same neurons fired action potentials. Moreover, the pitch-sensitive neurons in the monkey

Experiments in which monkeys listened to music suggest how the human brain processes pitch.

brains were in the same part of the auditory cortex that is damaged in humans who lose the ability to distinguish pitches after suffering a stroke. However, we don't yet know how the brain learns and matches the temporal combination of notes that make up a melody—which is how we perceive that Kurt Cobain and Tori Amos sang the same song. Presumably memory is part of the picture, which may explain why we can remember lyrics to a song many years after last hearing it but may not remember what we learned in a class just a day ago.

Learning Outcomes

After studying this chapter, you should be able to do the following:

10.1 Introduction
1. Distinguish between general senses and special senses. (p. 263)

10.2 Receptors, Sensations, and Perception
2. Name five kinds of receptors, and explain their functions. (p. 263)
3. Explain how a sensation arises. (p. 263)

10.3 General Senses
4. Describe the receptors associated with the senses of touch, pressure, temperature, and pain. (p. 264)
5. Describe how the sense of pain is produced. (p. 265)

10.4 Special Senses
6. Identify the locations of the receptors associated with the special senses. (p. 267)

10.5 Sense of Smell
7. Explain the relationship between the senses of smell and taste. (p. 267)
8. Explain the mechanism for smell. (p. 268)

10.6 Sense of Taste
9. Explain the mechanism for taste. (p. 270)

10.7 Sense of Hearing
10. Explain the function of each part of the ear. (p. 270)

10.8 Sense of Equilibrium
11. Distinguish between static and dynamic equilibrium. (p. 275)

10.9 Sense of Sight
12. Explain the function of each part of the eye. (p. 277)
13. Explain how the eye refracts light. (p. 284)
14. Describe the visual nerve pathway. (p. 286)

Module 7: Nervous System

 Learn   **Practice**   **Assess**

Aids to Understanding Words   *(Appendix A on page 564 has a complete list of Aids to Understanding Words.)*

choroid [skinlike] *choroid* coat: Middle, vascular layer of the eye.

cochlea [snail] *cochlea:* Coiled tube in the inner ear.

iris [rainbow] *iris:* Colored, muscular part of the eye.

labyrinth [maze] *labyrinth:* Complex system of connecting chambers and tubes of the inner ear.

lacri- [tears] *lacri*mal gland: Tear gland.

macula [spot] *macula* lutea: Yellowish spot on the retina.

olfact- [to smell] *olfact*ory: Pertaining to the sense of smell.

scler- [hard] *scler*a: Tough, outer protective layer of the eye.

tympan- [drum] *tympan*ic membrane: Eardrum.

vitre- [glass] *vitre*ous humor: Clear, jellylike substance within the eye.

10.1 INTRODUCTION

How dull life would be without sight and sound, smell and taste, touch and balance. Our senses are necessary not only for us to enjoy life, but to survive. They derive from structures called *sensory receptors* that detect environmental changes and trigger nerve impulses that travel on sensory pathways into the central nervous system for processing and interpretation. The body reacts with a particular feeling or sensation.

Sensory receptors vary greatly, but fall into two major categories. Receptors associated with the *general senses* are widely distributed throughout the skin and deeper tissues, and are structurally simple. Receptors of the second type are parts of complex, specialized sensory organs that provide the *special senses*.

10.2 RECEPTORS, SENSATIONS, AND PERCEPTION

Recall that all action potentials are the same (all-or-none). Our awareness of different sensory events therefore depends on receptors that respond to specific stimuli and the ability of different parts of the brain to interpret the resultant impulses.

Types of Receptors

Sensory receptors are diverse but share certain features. Each type of receptor is particularly sensitive to a distinct kind of environmental change and is much less sensitive to other forms of stimulation. Sensory receptors are categorized into five types according to their sensitivities: **Chemoreceptors** (ke″mo-re-sep′torz) are stimulated by changes in the concentration of certain chemicals; **pain receptors** (pān re-sep′torz) by tissue damage; **thermoreceptors** (ther′mo-re-sep′torz) by changes in temperature; **mechanoreceptors** (mek″ah-no-re-sep′torz) by changes in pressure or movement; and **photoreceptors** (fo″to-re-sep′torz) by light energy.

Sensations and Perception

A **sensation** occurs when sensory receptors reach threshold and the resulting action potentials cause the brain to become aware of that sensory event. A *perception* occurs when the brain interprets those sensory impulses. Thus, pain is a sensation, but realizing that you have just stepped on a tack is a perception. Because all the nerve impulses that travel away from sensory receptors into the central nervous system are alike, the resulting sensation depends on which region of the brain receives the impulse. For example, impulses reaching one region are always interpreted as sounds, and those reaching another are always sensed as touch. (Some receptors, such as those that measure oxygen levels in the blood, do not trigger sensations.)

At the same time that a sensation forms, the cerebral cortex causes the feeling to seem to come from the stimulated receptors. This process is called **projection** (pro-jek′shun) because the brain projects the sensation back to its apparent source. Projection allows a person to perceive the region of stimulation; this is how the eyes seem to see, and the ears seem to hear.

Sensory Adaptation

The brain must prioritize the sensory input it receives, or it would be overwhelmed by unimportant information. For example, until this sentence prompts you to think about it, you are probably unaware of the pressure of your clothing against your skin, or the background noise in the room. This ability of the nervous system to become less responsive to a maintained stimulus is called **adaptation** (ad″ap-ta′shun). It may result from receptors becoming unresponsive or inhibition along the central nervous system pathways leading to the respective sensory regions of the cerebral cortex.

Practice

1. List five general types of sensory receptors.

2. Explain how a perception is different from a sensation.

3. What is sensory adaptation?

10.3 GENERAL SENSES

General senses are widespread, and are associated with receptors in the skin, muscles, joints, and viscera. They include the senses of touch and pressure, temperature, and pain.

Touch and Pressure Senses

The senses of touch and pressure derive from three kinds of receptors (fig. 10.1). These receptors sense mechanical forces that deform or displace tissues. Touch and pressure receptors include:

1. **Free nerve endings** These receptors are common in epithelial tissues, where their free ends branch and extend between epithelial cells. They are responsible for the sensation of itching (discussed in the opening vignette to chapter 6, p. 116).
2. **Tactile (Meissner's) corpuscles** These are small, oval masses of flattened connective tissue cells in connective tissue sheaths. Two or more sensory fibers branch into each corpuscle and end in it as tiny knobs.

Tactile corpuscles are abundant in hairless areas of skin, such as the lips, fingertips, palms, soles, nipples, and external genital organs. They respond to the motion of objects that barely contact the skin, interpreting impulses from them as the sensation of light touch.

3. **Lamellated (Pacinian) corpuscles** These sensory bodies are relatively large structures composed of connective tissue fibers and cells, with a single fiber branch extending into each. They are common in the deeper dermal and subcutaneous tissues and in muscle tendons and joint ligaments. Lamellated corpuscles respond to heavy pressure and are associated with the sensation of deep pressure.

Temperature Senses

Temperature sensation depends on two types of free nerve endings in the skin. Those that respond to warmer temperatures are called *warm receptors,* and those that respond to colder temperatures are called *cold receptors.*

Warm receptors are most sensitive to temperatures above 25°C (77°F) and become unresponsive at

Figure 10.1

Touch and pressure receptors include **(a)** free ends of sensory nerve fibers, **(b)** tactile corpuscles (with 225× micrograph), and **(c)** lamellated corpuscles (with 50× micrograph).

temperatures above 45°C (113°F). Temperatures near and above 45°C stimulate pain receptors, producing a burning sensation.

Cold receptors are most sensitive to temperatures between 10°C (50°F) and 20°C (68°F). Temperatures below 10°C stimulate pain receptors, producing a freezing sensation.

Both warm and cold receptors adapt rapidly. Within about a minute of continuous stimulation, the sensation of warmth or cold begins to fade.

Sense of Pain

Certain receptors that consist of free nerve endings respond to painful stimuli. These receptors are widely distributed throughout the skin and internal tissues, except in the nervous tissue of the brain, which lacks pain receptors.

Pain receptors protect the body because tissue damage stimulates them. Pain sensation is usually perceived as unpleasant, and it signals a person to act to remove the stimulation. Pain receptors adapt poorly, if at all. Once a pain receptor is activated, even by a single stimulus, it may send impulses into the central nervous system for some time. Thus, pain may persist.

The way in which tissue damage stimulates pain receptors is poorly understood. Injuries likely promote the release of certain chemicals that build up and stimulate pain receptors. Deficiency of oxygen-rich blood (ischemia) in a tissue or stimulation of certain mechanoreceptors also triggers pain sensations. For example, the pain elicited during a muscle cramp stems from sustained contraction that squeezes capillaries and interrupts blood flow. Stimulation of mechanical-sensitive pain receptors also contributes to the sensation.

> Injuries to bones, tendons, or ligaments stimulate pain receptors that may also contract nearby skeletal muscles. The contracting muscles may become ischemic, which may trigger still other pain receptors.

Visceral Pain

As a rule, pain receptors are the only receptors in viscera whose stimulation produces sensations. Pain receptors in these organs respond differently to stimulation than those associated with surface tissues. For example, localized damage to intestinal tissue during surgical procedures may not elicit pain sensations, even in a conscious person. However, when visceral tissues are subjected to more widespread stimulation, such as when intestinal tissues are stretched or smooth muscles in intestinal walls undergo a spasm, a strong pain sensation may follow. Once again, the resulting pain seems to stem from stimulation of mechanoreceptors and from decreased blood flow accompanied by lower tissue oxygen concentration and accumulation of pain-stimulating chemicals via chemoreceptors.

Visceral pain may feel as if it is coming from a part of the body other than the part being stimulated, a phenomenon called **referred pain.** For example, pain originating in the heart may be referred to the left shoulder or left upper limb (fig. 10.2). Referred pain may arise from common nerve pathways that carry sensory impulses from skin areas as well as viscera. Pain impulses from the heart travel over the same nerve pathways as those from the skin of the left shoulder and left upper limb (fig. 10.3). Consequently, during a heart attack, the cerebral cortex may incorrectly interpret the source of the pain impulses as the left shoulder or upper limb, rather than the heart.

Practice

4. Describe the three types of touch and pressure receptors.
5. Describe the receptors that sense temperature.
6. What types of stimuli excite pain receptors?
7. What is referred pain?

Pain Nerve Fibers

The axons (fibers) that conduct impulses away from pain receptors are of two main types: acute pain fibers and chronic pain fibers. *Acute pain fibers* are myelinated. They conduct nerve impulses rapidly and are associated with the sensation of sharp pain, which typically originates from a restricted area of the skin and seldom continues after the pain-producing stimulus stops. *Chronic pain fibers* are unmyelinated. They conduct impulses more slowly and produce a dull, aching sensation that may be diffuse and difficult to pinpoint. Such pain may continue for some time after the original stimulus ceases. Acute pain is usually sensed as coming only from the skin; chronic pain is felt in deeper tissues as well.

An event that stimulates pain receptors usually triggers impulses on both acute and chronic pain fibers. This causes a dual sensation—a sharp, pricking pain, followed shortly by a dull, aching one. The aching pain is usually more intense and may worsen with time. Chronic pain can cause prolonged suffering.

Pain impulses that originate from the head reach the brain on sensory fibers of cranial nerves. All other pain impulses travel on the sensory fibers of spinal nerves, and they pass into the spinal cord by way of the dorsal roots of these spinal nerves. Within the spinal cord, neurons process pain impulses in the gray matter of the dorsal horn, and the impulses are transmitted to the brain. Here, most pain fibers terminate in the reticular formation (see chapter 9, p. 244). From there, other neurons conduct impulses to the thalamus, hypothalamus, and cerebral cortex.

Regulation of Pain Impulses

Awareness of pain arises when pain impulses reach the thalamus—that is, even before they reach the cerebral cortex. The cerebral cortex, however, determines pain

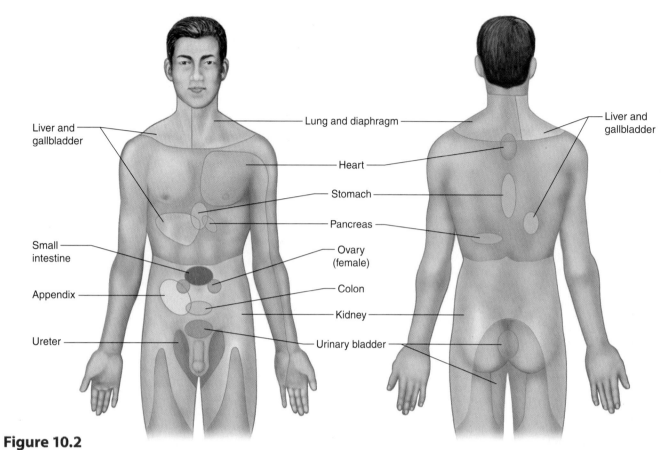

Figure 10.2

Referred pain feels as if it is coming from a different body part than the one being stimulated. Visceral pain may be felt at the surface regions indicated in the illustration.

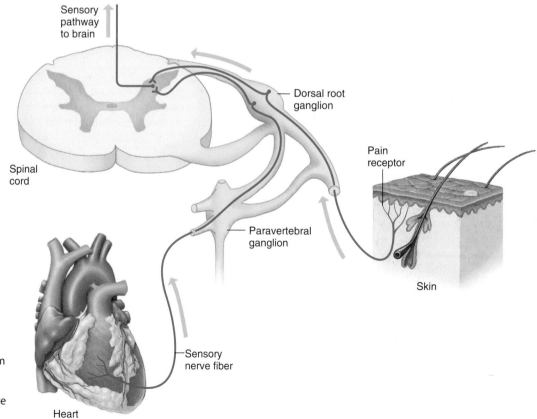

Figure 10.3

Pain originating in the heart may feel as if it is coming from the skin because sensory impulses from the heart and the skin follow common nerve pathways to the brain.

intensity, locates the pain source, and mediates emotional and motor responses to the pain.

Areas of gray matter in the midbrain, pons, and medulla oblongata regulate movement of pain impulses from the spinal cord. Impulses from special neurons in these brain areas descend in the lateral funiculus (see chapter 9, p. 234) to various levels of the spinal cord. These impulses stimulate the ends of certain nerve fibers to release biochemicals that can block pain signals by inhibiting presynaptic nerve fibers in the posterior horn of the spinal cord.

The inhibiting substances released in the posterior horn include neuropeptides called *enkephalins* and the biogenic amine *serotonin* (see chapter 9, p. 229). Enkephalins can suppress acute and chronic pain impulses and thus can relieve severe pain, much as morphine and other opiate drugs do. In fact, enkephalins bind to the same receptor sites on neuron membranes as does morphine. Serotonin stimulates other neurons to release enkephalins.

Endorphins are another group of neuropeptides with pain-suppressing, morphinelike actions. Endorphins are found in the pituitary gland and the hypothalamus. Enkephalins and endorphins are released in response to extreme pain and provide natural pain control.

Children who have *hereditary sensory and autonomic neuropathy* cannot feel pain, and as a result they inadvertently injure themselves. More common is *peripheral neuropathy,* in which the hands and/or feet become numb due to too few tactile corpuscles. In one study, people with normal pain sensation had an average of 12 corpuscles per square millimeter of skin, whereas people with peripheral neuropathy had fewer than 3. The most common causes of peripheral neuropathy are diabetes mellitus, cancer treatment, vitamin deficiency, and HIV infection.

Practice

8. Describe two types of pain fibers.
9. How do acute pain and chronic pain differ?
10. What parts of the brain interpret pain impulses?
11. How do neuropeptides help control pain?

10.4 SPECIAL SENSES

Special senses are those whose sensory receptors are in large, complex sensory organs in the head. These senses and their respective organs include the following:

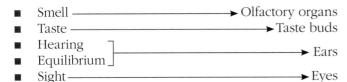

- Smell ──────────────▶ Olfactory organs
- Taste ──────────────▶ Taste buds
- Hearing ┐
- Equilibrium ┘──────────▶ Ears
- Sight ──────────────▶ Eyes

Clinical Application 10.1 describes a condition in which these assignments of senses to sense organs are enhanced.

10.5 SENSE OF SMELL

The sense of smell is associated with complex sensory structures in the upper region of the nasal cavity.

Olfactory Receptors

Smell (olfactory) receptors and taste receptors are chemoreceptors, which means that chemicals dissolved in liquids stimulate them. Smell and taste function closely together and aid in food selection because we usually smell food at the same time we taste it.

Olfactory Organs

The **olfactory organs,** which contain the olfactory receptors, are yellowish-brown masses of epithelium about the size of postage stamps that cover the upper parts of the nasal cavity, the superior nasal conchae, and a part of the nasal septum. **Olfactory receptor cells** are bipolar neurons surrounded by columnar epithelial cells (fig. 10.4). Hairlike cilia cover tiny knobs at the distal ends of these neurons' dendrites. In any particular such neuron, the cilia harbor many copies of one type of olfactory receptor protein. Chemicals called odorant molecules stimulate various sets of olfactory receptor proteins, and therefore various sets of olfactory receptor cells, to send a signal of a detected odor to the brain. Odorant molecules enter the nasal cavity as gases, but they must dissolve at least partially in the watery fluids that surround the cilia before receptors can detect them.

Olfactory Nerve Pathways

Stimulated olfactory receptor cells send impulses along their axons. These axons (which together form the first cranial nerves) synapse with neurons located in enlargements called **olfactory bulbs.** These structures lie on either side of the crista galli of the ethmoid bone (see fig. 7.14, p. 147). In the olfactory bulbs the impulses are analyzed, and as a result additional impulses travel along the **olfactory tracts** to the limbic system (see chapter 9, p. 243). The major interpreting areas for these impulses in the olfactory cortex lie within the temporal lobes and at the bases of the frontal lobes, anterior to the hypothalamus.

Humans smell the world using about 12 million olfactory receptor cells, but dogs have more than a billion such cells. Canines' excellent sense of smell is the basis of using service dogs to detect impending health problems in their owners. The dogs sense subtle odors that people emit when becoming ill in certain ways. Service dogs are used to sense imminent seizures, drops in blood glucose and heart rate, and lung, breast, and thyroid cancers. Experiments confirm that dogs are especially sensitive to odorant molecules on the skin or in the sweat of sick people.

Clinical Application 10.1

Synesthesia: Connected Senses

"The song was full of glittering orange diamonds."
"The paint smelled blue."
"The sunset was salty."
"The pickle tasted like a rectangle."

About 1 in 2,000 people have a condition called synesthesia ("joined sensation"), in which sensation and perception mix, so that the brain perceives a stimulus to one sense as coming from another. Most commonly, letters, numbers, or periods of time evoke specific colors. These associations are involuntary, are very specific, and persist over a lifetime. For example, a person might report that 3 is always mustard yellow, or Thursday a very dark, shiny brown.

Synesthesia runs in families, and geneticists have associated the condition with inheriting variants in any of four different genes. Female "synesthetes" outnumber males six to one. Creative individuals are overrepresented among those with the condition. They include musicians Syd Barrett, John Mayer, Tori Amos, and Franz Liszt, architect Frank Lloyd Wright,

and physicist Richard Feynman, who used to include the hues with which he visualized chemical equations on the chalkboard, to the amusement of his students. One of the co-authors of this book has it—to her, days are colors. The earliest recorded mention of synesthesia is an essay from John Locke in 1690. More and more people with synesthesia are recognizing that their peculiar talent has a name, thanks to Internet groups devoted to the condition.

Researchers hypothesize that mixed senses are present in all babies, but synesthesia develops in individuals who do not "prune" as many synapses as others as they age. (A loss of 20 billion synapses a day is normal for adults.) Imaging studies and animal experiments have localized the neurons that convey synesthetic connections to the general area where the temporal, parietal, and occipital lobes meet. Once sometimes referred to as a learning disability, synesthesia is instead now increasingly viewed as an enhancement to learning—and a fuller way of enjoying our sensual worlds.

Olfactory Stimulation

When odorant molecules bind to olfactory receptor proteins in olfactory receptor cell membranes, a biochemical pathway is activated that culminates in an influx of sodium ions, which may trigger an action potential if the depolarization reaches threshold. The action potentials from this and other olfactory receptor cells travel to the olfactory bulbs in the brain, where the sensation of smell arises.

The several hundred types of olfactory receptor cells can code for many thousands of odors when they signal the brain in groups. That is, an odorant molecule stimulates a distinct set of receptor types. An olfactory receptor cell has only one type of olfactory receptor, but that receptor can bind several types of odorant molecules. In addition, any one odorant molecule can bind several types of receptors. The brain interprets this binding information

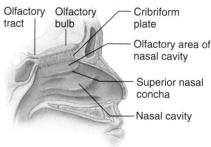

(a)   (b)

Figure 10.4 AP|R

Olfactory receptors convey the sense of smell. **(a)** Columnar epithelial cells support olfactory receptor cells, which have cilia at their distal ends. The actual olfactory receptors, which are proteins, are on the cilia. Binding of odorants to these receptors in distinctive patterns conveys the information that the brain interprets as an odor. **(b)** The olfactory area is associated with the superior nasal concha.

as a combinatorial olfactory code. In a simplified example, if there are ten odor receptors, parsley might stimulate receptors 3, 4, and 8, while chocolate might stimulate receptors 1, 5, and 10. Researchers identify olfactory and taste receptors by searching human genome sequence information for proteins that reside in cell membranes and are produced only in certain receptor cells. A person may have to sniff and force air up to the receptor areas to smell a faint odor because the olfactory organs are high in the nasal cavity above the usual pathway of inhaled air. Astronauts on the first space flights reportedly could not smell their food because they had to squeeze the food from tubes directly into their mouths—odorant molecules were not inhaled as they usually are.

The sense of smell adapts rapidly, but adaptation to one scent will not diminish sensitivity to new odors. For example, a person visiting a fish market might at first be acutely aware of the fishy smell, but then that odor fades. If a second person enters the fish market wearing a strong perfume, the person already there, who has become accustomed to the stinky fish, will nonetheless detect the flowery scent of the perfume.

> Partial or complete loss of smell is called *anosmia*. It may result from inflammation of the nasal cavity lining due to a respiratory infection, tobacco smoking, or using certain drugs, such as cocaine.

Practice

12. Where are olfactory receptors located?
13. Trace the pathway of an olfactory impulse from a receptor to the cerebrum.

10.6 SENSE OF TASTE

Taste buds are the special organs of taste (fig. 10.5). The 10,000 or so taste buds are located primarily on the surface of the tongue and are associated with tiny elevations called *papillae.* About 1,000 taste buds are scattered in the roof of the mouth and walls of the throat.

Taste Receptors

Each taste bud includes 50 to 150 modified epithelial cells which function as receptor cells, the **taste cells** (gustatory cells). Each taste cell is replaced on average every ten days. The taste bud also includes epithelial supporting cells. The entire structure is spherical, with an opening, the **taste pore,** on its free surface. Tiny projections called **taste hairs** protrude from the outer ends of the taste cells and extend from the taste pore. These taste hairs are the sensitive parts of the receptor cells.

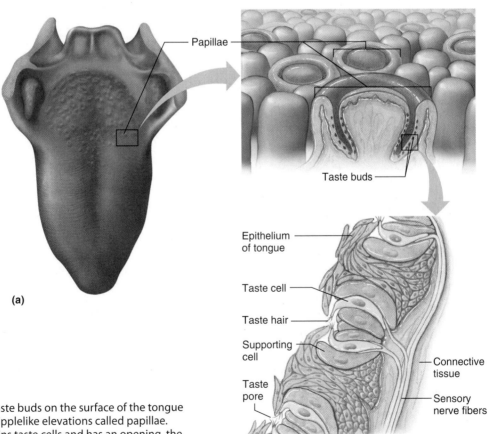

Figure 10.5

Taste receptors. **(a)** Taste buds on the surface of the tongue are associated with nipplelike elevations called papillae. **(b)** A taste bud contains taste cells and has an opening, the taste pore, at its free surface.

(a)

(b)

Interwoven among the taste cells and wrapped around them is a network of nerve fibers. Stimulation of a receptor cell triggers an impulse on a nearby nerve fiber, and the impulse then travels into the brain.

> Cats and dogs may be satisfied with less varied diets than humans because cats have only about 473 taste buds and dogs about 1,700.

A particular chemical must dissolve in the watery fluid surrounding the taste buds in order for it to be tasted. The salivary glands provide this fluid. Food molecules bind to specific receptor proteins embedded in taste hairs on the taste cells. The pattern of receptor types that bind food molecules and generate sensory impulses on nearby nerve fibers is interpreted as a particular taste sensation. Therefore, the chemical senses of smell and taste arise from molecules from the environment that bind receptors on neurons specialized as sensory receptors.

The taste cells in all taste buds appear alike microscopically, but are of at least five types. Each type is most sensitive to a particular kind of chemical stimulus, producing at least five primary taste (gustatory) sensations.

Taste Sensations

The five primary taste sensations are:

1. *Sweet,* such as table sugar
2. *Sour,* such as a lemon
3. *Salty,* such as table salt
4. *Bitter,* such as caffeine or quinine
5. *Umami* (a Japanese term meaning "delicious"), a response to certain amino acids and their chemical relatives, such as monosodium glutamate.

Some investigators recognize other taste sensations— *alkaline* and *metallic.*

A flavor results from one or a combination of the primary sensations. Experiencing flavors involves tasting, which reflects the concentrations of stimulating chemicals, as well as smelling and feeling the texture and temperature of foods. Furthermore, the chemicals in some foods—chili peppers and ginger, for instance—may stimulate pain receptors, which cause a burning sensation. In fact, the chemical in chili peppers that tastes "hot"—capsaicin—actually stimulates heat receptors.

Experiments indicate that each taste cell responds to one taste sensation only, with distinct receptors. Taste cells for each of the five taste sensations are in all areas of the tongue, but are distributed such that each sensation seems to arise most strongly from a particular region. The tip of the tongue is most sensitive to sweet stimuli, the margins of the tongue most sensitive to sourness, the back of the tongue more likely to detect bitter substances, and responsiveness to salt quite widely distributed.

Taste sensation, like the sense of smell, undergoes adaptation rapidly. Moving bits of food over the surface of the tongue to stimulate different receptors at different moments keeps us from losing taste due to sensory adaptation.

Taste Nerve Pathways

Sensory impulses from taste receptor cells in the tongue travel on fibers of the facial, glossopharyngeal, and vagus nerves into the medulla oblongata. From there, the impulses ascend to the thalamus and are directed to the gustatory cortex, in the parietal lobe of the cerebrum, along a deep part of the lateral sulcus (see fig. 9.28, p. 238).

> ### Practice
>
> 14. Why is saliva necessary for the sense of taste?
> 15. Name the five primary taste sensations.
> 16. Trace a sensory impulse from a taste receptor to the cerebral cortex.

10.7 SENSE OF HEARING

The organ of hearing, the ear, has outer, middle, and inner parts. The ear also functions in the sense of equilibrium.

Outer (External) Ear

The outer ear consists of three parts. The first is an outer, funnel-like structure called the **auricle** (aw'ri-kl) (pinna). The second is an S-shaped tube called the **external acoustic meatus** (me-a'tus), or external auditory canal, that leads inward through the temporal bone for about 2.5 centimeters (fig. 10.6).

The transmission of vibrations through matter produces sound. Vibrating strings or reeds produce the sounds of some musical instruments, and vibrating vocal folds in the larynx produce the voice. The auricle of the ear helps collect sound waves traveling through the air and directs them into the external acoustic meatus. The meatus terminates with the **eardrum** (tympanic membrane).

The eardrum is a semitransparent membrane covered by a thin layer of skin on its outer surface and by mucous membrane on the inside. It has an oval margin and is cone-shaped, with the apex of the cone directed inward. The attachment of one of the auditory ossicles (the malleus) maintains the eardrum's cone shape. Sound waves that enter the external acoustic meatus change the pressure on the eardrum, which moves back and forth in response and thus reproduces the vibrations of the sound-wave source.

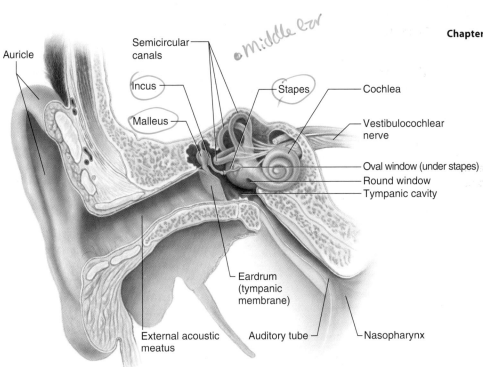

Semicircular
canals

•Middle ear

Auricle

Incus

Stapes

Cochlea

Malleus

Vestibulocochlear
nerve

Oval window (under stapes)
Round window
Tympanic cavity

Eardrum
(tympanic
membrane)

External acoustic
meatus

Auditory tube

Nasopharynx

Figure 10.6 AP|R

Major parts of the ear. The outer ear includes the auricle, external acoustic meatus, and eardrum. The middle ear includes the auditory ossicles (malleus, incus, and stapes) and the oval window. The inner ear includes the semicircular canals and the cochlea.

Q: *How do the action potentials generated along auditory pathways compare with those on taste and smell pathways?*

Answer can be found in Appendix E on page 568.

Middle Ear

The middle ear, or *tympanic cavity,* is an air-filled space in the temporal bone. It contains three small bones called **auditory ossicles** (aw′di-to″re os′i-klz): the *malleus,* the *incus,* and the *stapes* (fig. 10.7). Tiny ligaments attach them to the wall of the tympanic cavity, and they are covered by mucous membrane. These bones bridge the eardrum and the inner ear, transmitting vibrations between these parts. Specifically, the malleus attaches to the eardrum, and when the eardrum vibrates, the malleus vibrates in unison. The malleus causes the incus to vibrate, and the incus passes the movement on to the stapes. Ligaments hold the stapes to an opening in the wall of the tympanic cavity called the **oval window,** which leads into the inner ear. Vibration of the stapes at the oval window moves a fluid in the inner ear, which stimulates the hearing receptors.

The auditory ossicles help increase (amplify) the force of vibrations as they pass from the eardrum to the oval window, in addition to transmitting vibrations. The vibrational force concentrates as it moves from the outer to the inner ear because the ossicles transmit vibrations from the relatively large surface of the eardrum to a much smaller area at the oval window. As a result, the pressure (per square millimeter) that the stapes applies on the oval window is many times greater than the pressure that sound waves exert on the eardrum.

Auditory Tube

An **auditory tube** (aw′di-to″re tūb) (eustachian tube) connects each middle ear to the back of the nasal cavity (nasopharynx). This tube conducts air between the tympanic cavity and the outside of the body by way of

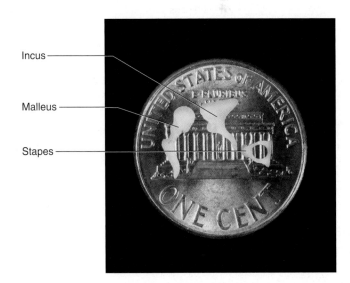

Incus

Malleus

Stapes

Figure 10.7

The auditory ossicles—the malleus, incus, and stapes—are bones that bridge the eardrum and the inner ear (2.5×) (see fig. 10.6). Comparison to a penny emphasizes their tiny size.

the nose and mouth. The auditory tube helps maintain equal air pressure on both sides of the eardrum, which is necessary for normal hearing.

The function of the auditory tube is noticeable during rapid changes in altitude. As a person moves from a high altitude to a lower one, air pressure on the outside of the eardrum increases. This may push the eardrum inward, impairing hearing. When the air pressure difference is great enough, air movement through the auditory tube equalizes the pressure on both sides of the eardrum, and the membrane moves back into its regular position. This produces a popping sound, which restores normal hearing.

> Mucous membrane infections of the throat may spread through the auditory tubes and cause middle ear infection because auditory tube mucous membranes connect directly with middle ear linings. Pinching a nostril when blowing the nose may force material from the throat up the auditory tube and into the middle ear.

Inner (Internal) Ear

The inner ear is a complex system of communicating chambers and tubes called a **labyrinth** (lab′ĭ-rinth). Each ear has two parts to the labyrinth—the *osseous labyrinth* and the *membranous labyrinth* (fig. 10.8). The osseous labyrinth is a bony canal in the temporal bone. The membranous labyrinth is a tube of similar shape that lies within the osseous labyrinth. Between the osseous and membranous labyrinths is a fluid called *perilymph,* which is secreted by cells in the wall of the bony canal. The membranous labyrinth contains another fluid, called *endolymph.*

The parts of the labyrinths include three **semicircular canals,** which provide a sense of equilibrium (discussed in section 10.8 on p. 275), and a **cochlea** (kok′le-ah), which functions in hearing. The cochlea has a bony core and a thin, bony shelf that winds around the core like the threads of a screw. The shelf divides the osseous labyrinth of the cochlea into upper and lower compartments. The upper compartment, called the

Osseous labyrinth

Perilymph

Membranous labyrinth

Endolymph

Osseous labyrinth (contains perilymph)

Membranous labyrinth (contains endolymph)

Semicircular canals

Utricle

Saccule

Vestibular nerve

Cochlear nerve

Scala chambers (cut) containing perilymph

Cochlear duct (cut) containing endolymph

Ampullae   Oval window   Round window   Maculae

Cochlea

Figure 10.8

A closer look at the inner ear. Perilymph separates the osseous labyrinth of the inner ear from the membranous labyrinth, which contains endolymph.

scala vestibuli, leads from the oval window to the apex of the spiral. The lower compartment, the *scala tympani,* extends from the apex of the cochlea to a membrane-covered opening in the wall of the inner ear called the **round window** (fig. 10.8).

The part of the membranous labyrinth within the cochlea is called the *cochlear duct.* It lies between the two bony compartments and ends as a closed sac at the apex of the cochlea. The cochlear duct is separated from the scala vestibuli by a *vestibular membrane* (Reissner's membrane) and from the scala tympani by a *basilar membrane* (fig. 10.9).

The basilar membrane has many thousands of stiff, elastic fibers, which lengthen from the base of the cochlea to its apex. Sound vibrations entering the perilymph at the oval window travel along the scala vestibuli

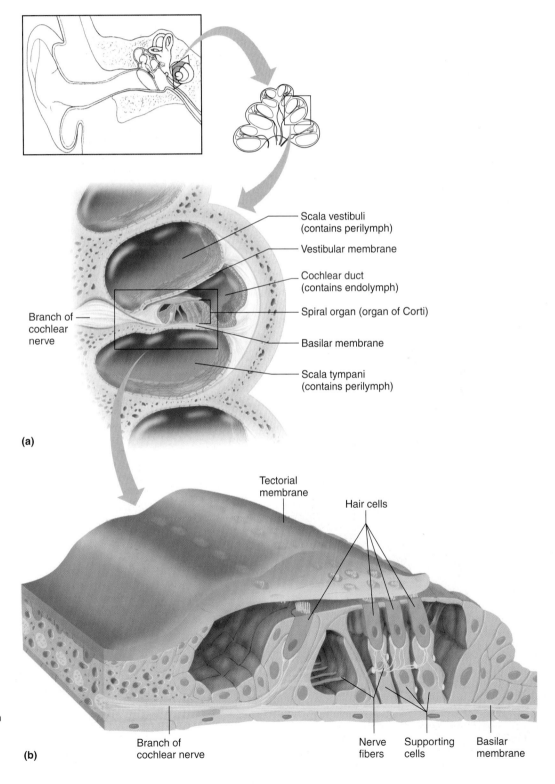

(a)

Scala vestibuli
(contains perilymph)

Vestibular membrane

Cochlear duct
(contains endolymph)

Spiral organ (organ of Corti)

Basilar membrane

Scala tympani
(contains perilymph)

Branch of
cochlear
nerve

Tectorial
membrane

Hair cells

Branch of
cochlear nerve

Nerve
fibers

Supporting
cells

Basilar
membrane

(b)

Figure 10.9 AP|R

The cochlea. **(a)** Cross section of the cochlea. **(b)** The spiral organ and the tectorial membrane.

and pass through the vestibular membrane and into the endolymph of the cochlear duct, where they move the basilar membrane.

After passing through the basilar membrane, the vibrations enter the perilymph of the scala tympani. Their forces are dissipated into the air in the tympanic cavity by movement of the membrane covering the round window.

The **spiral organ** (organ of Corti) contains the hearing receptors. It is located on the upper surface of the basilar membrane and stretches from the apex to the base of the cochlea (fig. 10.9). The receptor cells, called *hair cells,* are organized in rows and have many hairlike processes that project into the endolymph of the cochlear duct. Above these hair cells is a *tectorial membrane* attached to the bony shelf of the cochlea, passing over the receptor cells and contacting the tips of their hairs.

As sound vibrations pass through the inner ear, the hairs shear back and forth against the tectorial membrane, and the resulting mechanical deformation of the hairs stimulates the hair cells (figs. 10.9 and 10.10). Different hair cells have different sensitivities to the frequency (pitch) of sound waves. This enables us to hear sounds of different pitch simultaneously.

Hair cells are epithelial, but function somewhat like neurons. For example, when a hair cell is at rest, its membrane is polarized. When it is stimulated, selective ion channels open, depolarizing the membrane and making it more permeable to calcium ions. The hair cell has no axon or dendrites, but it has neurotransmitter-containing vesicles near its base. As calcium ions diffuse into the cell, some of these vesicles fuse with the cell membrane and release a neurotransmitter. The neurotransmitter stimulates the ends of nearby sensory neurons, and in response they transmit action potentials

along the cochlear branch of the vestibulocochlear nerve to the auditory cortex of the temporal lobe of the brain.

The ear of a young person with normal hearing can detect sound waves with frequencies ranging from 20 to 20,000 or more vibrations per second. The range of greatest sensitivity is 2,000–3,000 vibrations per second. Table 10.1 summarizes the steps of hearing.

Recall from chapter 9 (pp. 227–228) that action potentials are all-or-none. More-intense stimulation of the hair cells causes more action potentials per second to reach the auditory cortex, and we perceive a louder sound.

Units called *decibels* (dB) measure sound intensity on a logarithmic scale. The decibel scale begins at 0 dB, which is the intensity of the sound that is least perceptible by a normal human ear. A sound of 10 dB is 10 times as intense as the least perceptible sound; a sound of 20 dB is 100 times as intense; and a sound of 30 dB is 1,000 times as intense. A whisper has an intensity of about 40 dB, normal conversation measures 60–70 dB, and heavy traffic produces about 80 dB. A sound of 120 dB, common at a rock concert, produces discomfort, and a sound of 140 dB, such as that emitted by a jet plane at takeoff, causes pain. Frequent or prolonged exposure to sounds with intensities above 85 dB can damage hearing receptors and cause permanent hearing loss.

Auditory Nerve Pathways

The nerve fibers associated with hearing enter the auditory nerve pathways, which pass into the auditory cortices of the temporal lobes of the cerebrum, where

Figure 10.10

Scanning electron micrograph of hair cells in the spiral organ (3,800×).

Table 10.1	Steps in the Generation of Sensory Impulses from the Ear

1. Sound waves enter external acoustic meatus.

2. Waves of changing pressures cause eardrum to reproduce vibrations coming from sound wave source.

3. Auditory ossicles amplify and transmit vibrations to end of stapes.

4. Movement of stapes at oval window transmits vibrations to perilymph in scala vestibuli.

5. Vibrations pass through the vestibular membrane and enter endolymph of cochlear duct.

6. Different frequencies of vibration in endolymph stimulate different sets of receptor cells.

7. As a receptor cell depolarizes, its membrane becomes more permeable to calcium ions.

8. Inward diffusion of calcium ions causes vesicles at the base of the receptor cell to release neurotransmitter.

9. Neurotransmitter stimulates ends of nearby sensory neurons.

10. Sensory impulses are triggered on fibers of the cochlear branch of vestibulocochlear nerve.

11. Auditory cortex of temporal lobes interpret sensory impulses.

they are interpreted. On the way, some of these fibers cross over, so that impulses arising from each ear are interpreted on both sides of the brain. Consequently, damage to a temporal lobe on one side of the brain does not necessarily cause complete hearing loss in the ear on that side.

Several factors cause partial or complete hearing loss. Interference with the transmission of vibrations to the inner ear is called *conductive hearing loss.* Damage to the cochlea, auditory nerve, or auditory nerve pathways can cause *sensorineural hearing loss.* Conductive hearing loss may be due to plugging of the external acoustic meatus or to changes in the eardrum or auditory ossicles. For example, the eardrum may harden as a result of disease and become less responsive to sound waves, or disease or injury may tear or perforate the eardrum. Sensorineural hearing loss can be caused by loud sounds, tumors in the central nervous system, brain damage as a result of vascular accidents, or use of certain drugs.

Practice

17. How are sound waves transmitted through the outer, middle, and inner ears?
18. Distinguish between the osseous and membranous labyrinths.
19. Describe the spiral organ.

10.8 SENSE OF EQUILIBRIUM

The sense of equilibrium is really two senses—static equilibrium and dynamic equilibrium—that come from different sensory organs. The organs of **static equilibrium** (stat′ik e″kwĭ-lib′re-um) sense the position of the head, maintaining stability and posture when the head and body are still. When the head and body suddenly move or rotate, the organs of **dynamic equilibrium** (di-nam′ik e″kwĭ-lib′re-um) detect such motion and aid in maintaining balance.

Static Equilibrium

The organs of static equilibrium are in the **vestibule,** a bony chamber between the semicircular canals and the cochlea. The membranous labyrinth inside the vestibule consists of two expanded chambers—a **utricle** (u′trĭ-kl) and a **saccule** (sak′ūl) (see fig. 10.8).

Each of these chambers has a tiny structure called a **macula** (mak′u-lah). Maculae have many hair cells, which serve as sensory receptors. The hairs of the hair cells project into a mass of gelatinous material, which has grains of calcium carbonate (otoliths) embedded in it. These particles add weight to the gelatinous structure.

The head bending forward, backward, or to one side stimulates hair cells. Such movements tilt the gelatinous masses of the maculae, and as they sag in response to gravity, the hairs projecting into them bend. This action stimulates the hair cells, and they signal the neurons associated with them in a manner similar to that of hearing receptors. The resulting action potentials travel into the central nervous system on the vestibular branch of the vestibulocochlear nerve, informing the brain of the head's new position. The brain responds by sending motor impulses to skeletal muscles, which contract or relax to maintain balance (fig. 10.11).

Dynamic Equilibrium

The organs of dynamic equilibrium are the three semicircular canals in the labyrinth. They detect motion of the head and aid in balancing the head and body during sudden movement. These canals lie at right angles to each other, and each corresponds to a different anatomical plane (see fig. 10.8).

Suspended in the perilymph of the osseous portion of each semicircular canal is a membranous canal that ends in a swelling called an **ampulla** (am-pul′ah), which houses the sensory organs of the semicircular canals. Each of these organs, called a **crista ampullaris** (kris′tah am-pul′ar-is), contains a number of sensory hair cells and supporting cells (fig. 10.12). Like the hairs of the maculae, the hair cells extend upward into a dome-shaped, gelatinous mass called the *cupula.*

Rapid turns of the head or body stimulate the hair cells of the crista ampullaris (fig. 10.13). At such times, the semicircular canals move with the head or body, but the fluid inside the membranous canals remains stationary. This bends the cupula in one or more of the canals in a direction opposite that of the head or body movement, and the hairs embedded in it also bend. The stimulated hair cells signal their associated nerve fibers, sending impulses to the brain. The brain interprets these impulses as a movement in a particular direction.

Parts of the cerebellum are particularly important in interpreting impulses from the semicircular canals. Analysis of such information allows the brain to predict the consequences of rapid body movements, and by modifying signals to appropriate skeletal muscles, the cerebellum can maintain balance.

Other sensory structures aid in maintaining equilibrium. For example, certain mechanoreceptors (proprioceptors), particularly those associated with the joints of the neck, inform the brain about the position of body parts. In addition, the eyes detect changes in position that result from body movements. Such visual information is so important that even if the organs of equilibrium are damaged, a person may be able to maintain normal balance by keeping the eyes open and moving slowly.

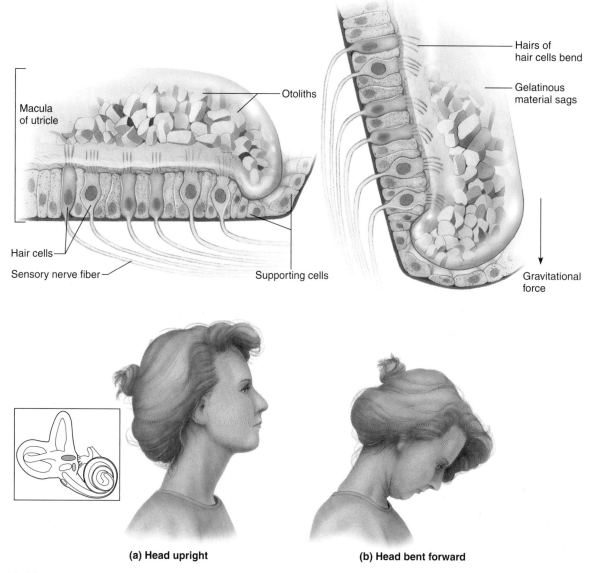

Figure 10.11

The maculae respond to changes in head position. **(a)** Macula of the utricle with the head in an upright position. **(b)** Macula of the utricle with the head bent forward.

The nausea, vomiting, dizziness, and headache of *motion sickness* arise from sensations that don't make sense. The eyes of a person reading in a moving car, for example, signal the brain that the person is stationary, because the print doesn't move. However, receptors in the skin detect bouncing, swaying, starting and stopping, as the inner ear detects movement. The contradiction triggers the symptoms. Similarly, in a passenger of an airplane flying through heavy turbulence, receptors in the skin and inner ear register the chaos outside, but the eyes focus on the immobile seats and surroundings.

To prevent or lessen the misery of motion sickness, focus on the horizon or an object in the distance ahead. Medications are available by pill (diphenhydramine and meclizine) and, for longer excursions, in a skin patch (scopalamine).

Practice

20. Distinguish between static and dynamic equilibrium.

21. Which structures provide the sense of static equilibrium? Of dynamic equilibrium?

22. How does sensory information from other receptors help maintain equilibrium?

10.9 SENSE OF SIGHT

The eye, the organ containing visual receptors, provides vision, with the assistance of *accessory organs*. These accessory organs include the eyelids and lacrimal apparatus, which protect the eye, and a set of extrinsic muscles, which move the eye.

(a) Head in still position

Figure 10.12

A crista ampullaris is located within the ampulla of each semicircular canal.

(b) Head rotating

Visual Accessory Organs

The eye, lacrimal gland, and associated extrinsic muscles are housed in the orbital cavity, or orbit, of the skull. Each orbit is lined with the periosteum of various bones, and also contains fat, blood vessels, nerves, and connective tissues.

Each **eyelid** has four layers—skin, muscle, connective tissue, and conjunctiva. The skin of the eyelid, which is the thinnest skin of the body, covers the lid's outer surface and fuses with its inner lining near the margin of the lid. The eyelids are moved by the *orbicularis oculi* muscle (see fig. 8.17*a*, p. 196), which acts as a sphincter and closes the lids when it contracts, and by the *levator palpebrae superioris* muscle, which raises the upper lid and thus helps open the eye (fig. 10.14). The **conjunctiva** is a mucous membrane that lines the inner surfaces of the eyelids and folds back to cover the anterior surface of the eyeball, except for its central portion (cornea).

The *lacrimal apparatus* consists of the **lacrimal gland,** which secretes tears, and a series of ducts that carry tears into the nasal cavity (fig. 10.15). The gland is located in the orbit and secretes tears continuously. The tears exit through tiny tubules and flow downward and medially across the eye.

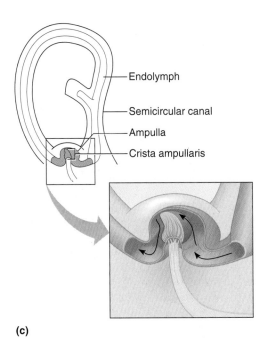

(c)

Figure 10.13

Equilibrium. **(a)** When the head is stationary, the cupula of the crista ampullaris remains upright. **(b)** and **(c)** When the head is moving rapidly, the cupula bends opposite the motion of the head, stimulating sensory receptors.

Tendon of levator palpebrae superioris

Superior rectus

Orbicularis oculi

Eyelid

Eyelash

Cornea

Conjunctiva

Inferior rectus

Figure 10.14

Sagittal section of the closed eyelids and anterior portion of the eye.

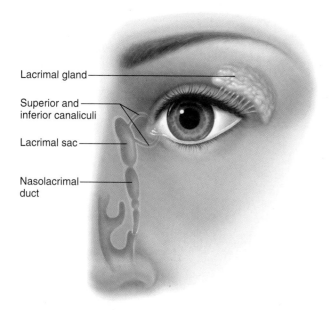

Lacrimal gland

Superior and inferior canaliculi

Lacrimal sac

Nasolacrimal duct

Figure 10.15

The lacrimal apparatus consists of a tear-secreting gland and a series of ducts.

Two small ducts (the superior and inferior canaliculi) collect tears, which flow into the *lacrimal sac,* located in a deep groove of the lacrimal bone, and then into the *nasolacrimal duct,* which empties into the nasal cavity. Secretion by the lacrimal gland moistens and lubricates the surface of the eye and the lining of the lids. Tears also have an enzyme (*lysozyme*) that kills bacteria, reducing the risk of eye infections.

The **extrinsic muscles** arise from the bones of the orbit and insert by broad tendons on the eye's tough outer surface. Six extrinsic muscles move the eye in various directions. Any given eye movement may utilize more than one extrinsic muscle, but each muscle is associated with one primary action. Figure 10.16 illustrates the locations of these extrinsic muscles, and table 10.2 lists their functions, as well as the functions of the eyelid muscles.

One eye deviating from the line of vision may result in double vision (diplopia). If this condition persists, the brain may suppress the image from the deviated eye. As a result, the turning eye may become blind (suppression amblyopia). Treating eye deviation early in life with exercises, eyeglasses, and surgery can prevent such monocular (one-eye) blindness.

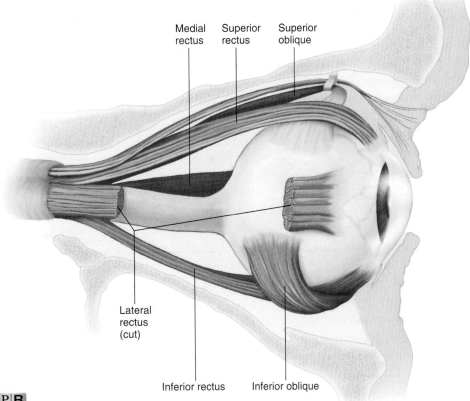

Medial rectus Superior rectus Superior oblique

Lateral rectus (cut)

Inferior rectus Inferior oblique

Figure 10.16 AP|R

Extrinsic muscles of the right eye (lateral view).

Practice

23. Explain how the eyelid moves.
24. Describe the conjunctiva.
25. What is the function of the lacrimal apparatus?

Structure of the Eye

The eye is a hollow, spherical structure about 2.5 centimeters in diameter. Its wall has three distinct layers—an outer (fibrous) layer, a middle (vascular) layer, and an inner (nervous) layer. The spaces within the eye are filled with fluids that support its wall and internal parts that help maintain its shape. Figure 10.17 shows the major parts of the eye.

Outer Layer

The anterior sixth of the outer layer bulges forward as the transparent **cornea** (kor′ne-ah), which is the window of the eye and helps focus entering light rays. The cornea is composed largely of connective tissue with a thin surface layer of epithelium. It is transparent because

Table 10.2	Muscles Associated with the Eyelids and Eyes	
Name	**Innervation**	**Function**
Muscles of the Eyelids		
Orbicularis oculi	Facial nerve (VII)	Closes eye
Levator palpebrae superioris	Oculomotor nerve (III)	Opens eye
Extrinsic Muscles of the Eyes		
Superior rectus	Oculomotor nerve (III)	Rotates eye upward and toward midline
Inferior rectus	Oculomotor nerve (III)	Rotates eye downward and toward midline
Medial rectus	Oculomotor nerve (III)	Rotates eye toward midline
Lateral rectus	Abducens nerve (VI)	Rotates eye away from midline
Superior oblique	Trochlear nerve (IV)	Rotates eye downward and away from midline
Inferior oblique	Oculomotor nerve (III)	Rotates eye upward and away from midline

Ciliary body
Suspensory ligaments
Iris
Lens
Pupil
Cornea
Aqueous humor
Anterior cavity — Anterior chamber / Posterior chamber

Lateral rectus
Retina
Choroid coat
Sclera
Vitreous humor
Fovea centralis
Optic nerve
Optic disc
Posterior cavity
Medial rectus

Figure 10.17 AP|R

Transverse section of the right eye (superior view).

Q: *What midline structure is the optic nerve in this figure angling towards?*

Answer can be found in Appendix E on page 568.

it contains few cells and no blood vessels, and the cells and collagenous fibers form unusually regular patterns.

Along its circumference, the cornea is continuous with the **sclera** (skle′rah), the white portion of the eye. The sclera makes up the posterior five-sixths of the outer layer and is opaque due to many large, disorganized, collagenous and elastic fibers. The sclera protects the eye and is an attachment for the extrinsic muscles. In the back of the eye, the **optic nerve** and certain blood vessels pierce the sclera. Clinical Application 10.2 describes headaches that include visual symptoms.

Worldwide, the most common cause of blindness is loss of transparency of the cornea. Each year, 40,000 corneal transplants are performed in the United States. Corneas are "immune privileged," not evoking an immune response, so anyone can donate to anyone. However, they do not work unless the transplanted tissue includes stem cells normally found in a layer of cells, called the limbus, which separates the cornea from the conjunctiva. (The cornea itself does not contain stem cells—their nuclei are so large that they would block light rays.) Researchers in England performed limbal cell transplants on patients with corneal damage in one eye. They took the stem cells from the patients' healthy eyes, culturing the limbal cells into a bluish, translucent gel, which they then applied to the affected eyes. Vision returned. Limbal stem cell transplants may prove to be more effective than cornea transplants, which have been done since 1905.

Middle Layer

The middle layer of the wall of the eye includes the choroid coat, ciliary body, and iris (fig. 10.17). The **choroid**

coat (ko′roid kōt), in the posterior five-sixths of the globe of the eye, is loosely joined to the sclera and is honeycombed with blood vessels, which nourish surrounding tissues. The choroid coat also has many pigment-producing melanocytes. The melanin that these cells produce absorbs excess light, which helps keep the inside of the eye dark.

The **ciliary body** (sil′e-er″e bod′e), which is the thickest part of the middle layer, extends forward from the choroid coat and forms an internal ring around the front of the eye. Within the ciliary body are many radiating folds called *ciliary processes* and groups of muscle fibers that constitute the *ciliary muscles*.

Many strong but delicate fibers, called *suspensory ligaments,* extend inward from the ciliary processes and hold the transparent **lens** in position (fig. 10.18). The distal ends of these fibers attach along the margin of a thin capsule that surrounds the lens. The body of the lens lies directly behind the iris and pupil and is composed of highly specialized epithelial cells called *lens fibers.* The cytoplasm of these cells is the transparent substance of the lens.

More than 90% of the proteins in a lens cell are lens crystallins, which aggregate into fibers. These proteins, along with the absence of organelles that scatter light (mitochondria, endoplasmic reticula, and nuclei), account for the transparency of the lens.

An eye disorder particularly common in older people is *cataract.* The lens or its capsule slowly becomes cloudy and opaque. Cataracts are treated on an outpatient basis with a laser. Without treatment, cataracts eventually cause blindness.

Clinical Application 10.2

Headache

Headaches are common. The cells of the nervous tissue in the brain lack pain receptors, but nearly all the other tissues of the head, including the meninges and blood vessels, are richly innervated, and can be the source of headache pain.

Many a *migraine* sufferer knows that an attack is imminent early in the morning, when an ominous dull throbbing begins, often on one side of the head. Migraine is more than head pain—the person feels unwell in a general sense, and an attack can be disabling, lasting from a few hours to several days. In a migraine, certain cranial blood vessels constrict, producing a localized cerebral blood deficiency. When vasodilation quickly follows, a severe headache results. Several types of drugs, such as the triptans, effectively relieve migraines, but they must be taken at the first sign of illness, and several drugs may need to be tried before an effective one is found. For some people, keeping a diary of conditions before an attack can reveal a trigger, such as eating chocolate or low-pressure weather systems.

In some individuals, migraine begins with an "aura" of shimmery bright lights in the peripheral vision. Accompanying the aura is often "photophobia," which is head pain when exposed to light. Photophobia arises from a group of brain neurons closely associated with the optic nerve (see figure 10.26). Completely blind migraine sufferers, whose optic nerves do not function, do not experience photophobia. But people who are "legally blind," with partially degenerated retinas but intact and functional optic nerves, do experience pain from light. Experiments on animals localized neurons near the optic nerve that trigger photophobia when stimulated.

Many headaches are associated with stressful life situations that cause fatigue, emotional tension, anxiety, or frustration. These conditions can trigger various physiological changes, such as prolonged contraction of the skeletal muscles in the forehead, sides of the head, or back of the neck, which stimulate pain receptors and produce a *tension headache*. More severe *vascular headaches* accompany constriction or dilation of the cranial blood vessels. For example, the throbbing headache of a "hangover" from drinking too much alcohol may be due to blood pulsating through dilated cranial vessels. Other causes of headaches include sensitivity to food additives, high blood pressure, increased intracranial pressure due to a tumor or to blood escaping from a ruptured vessel, decreased cerebrospinal fluid pressure following a lumbar puncture, and sensitivity to or withdrawal from certain drugs.

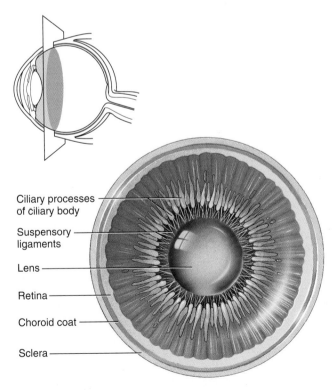

Ciliary processes of ciliary body

Suspensory ligaments

Lens

Retina

Choroid coat

Sclera

Figure 10.18

Lens and ciliary body viewed from behind.

The ciliary muscles and suspensory ligaments, along with the structure of the lens itself, enable the lens to adjust shape to facilitate focusing, a phenomenon called **accommodation** (ah-kom″o-da′shun). The lens is enclosed by a clear capsule composed largely of elastic fibers. This elastic nature keeps the lens under constant tension, and enables it to assume a globular shape. The suspensory ligaments attached to the margin of the capsule are also under tension. When they pull outward, flattening the capsule and the lens inside, the lens focuses on distant objects (fig. 10.19*a*). However, if the tension on the suspensory ligaments relaxes, the elastic lens capsule rebounds, and the lens surface becomes more convex—focused for viewing closer objects (fig. 10.19*b*).

The ciliary muscles control the actions of the suspensory ligaments in accommodation. For example, one set of these muscle fibers extends back from fixed points in the sclera to the choroid coat. When the fibers contract, the choroid coat is pulled forward, and the ciliary body shortens. This relaxes the suspensory ligaments, and the lens thickens in response (see fig. 10.19*b*). When the ciliary muscles relax, tension on the suspensory ligaments increases, and the lens becomes thinner and less convex again (see fig. 10.19*a*).

(a)

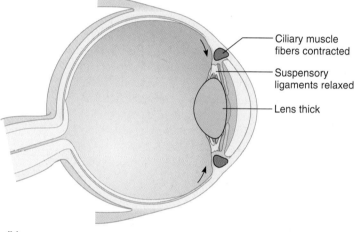

(b)

Figure 10.19

Accommodation. **(a)** The lens thins as the ciliary muscle fibers relax. **(b)** The lens thickens as the ciliary muscle fibers contract.

Practice

26. Describe the outer and middle layers of the eye.

27. What factors contribute to the transparency of the cornea?

28. How does the shape of the lens change during accommodation?

29. Why would reading for a long time cause eye fatigue, while looking at a distant scene is restful?

The **iris** (i′ris) is a thin diaphragm composed mostly of connective tissue and smooth muscle fibers. From the outside, the iris is the colored part of the eye. The iris extends forward from the periphery of the ciliary body and lies between the cornea and lens (see fig. 10.17). The iris divides the space (anterior cavity) separating these parts into an *anterior chamber* (between the cornea and the iris) and a *posterior chamber* containing the lens (between the iris and the vitreous body).

The epithelium on the inner surface of the ciliary body secretes a watery fluid called **aqueous humor** (a′kwe-us hu′mor) into the posterior chamber. The fluid circulates from this chamber through the **pupil** (pu′pil), a circular opening in the center of the iris, and into the anterior chamber. Aqueous humor fills the space between the cornea and lens, helps nourish these parts, and aids in maintaining the shape of the front of the eye. Aqueous humor leaves the anterior chamber through veins and a special drainage canal, the scleral venous sinus (canal of Schlemm), which is in the wall of the anterior chamber at the junction of the cornea and the sclera.

An eye disorder called *glaucoma* develops when aqueous humor forms faster than it is removed. As fluid accumulates in the anterior chamber of the eye, fluid pressure rises and is transmitted to all parts of the eye. In time, the building pressure squeezes shut blood vessels that supply the receptor cells of the retina. Cells that are robbed of nutrients and oxygen in this way may die, and permanent blindness can result.

When diagnosed early, glaucoma can usually be treated successfully with drugs, laser therapy, or surgery, all of which promote the outflow of aqueous humor. Since glaucoma in its early stages typically produces no symptoms, discovery of the condition usually depends on measuring intraocular pressure, using an instrument called a *tonometer*.

The smooth muscle fibers of the iris are organized into two groups, a *circular set* and a *radial set*. These muscles control the size of the pupil, through which light passes as it enters the eye. The circular set of muscle fibers acts as a sphincter. When the muscle fibers contract, the pupil gets smaller, and less light enters. Bright light stimulates the circular muscles to contract, which decreases the intensity of light entering the eye. Conversely, when the radial muscle fibers contract, the pupil's diameter increases, and more light enters (fig. 10.20). Dim light stimulates the radial muscles to contract, which dilates the pupil, allowing more light into the eye.

Inner Layer

The inner layer of the wall of the eye consists of the **retina** (ret′ĭ-nah), which contains the visual receptor cells (photoreceptors). The retina is a nearly transparent sheet of tissue continuous with the optic nerve in the back of the eye and extending forward as the inner lining of the eyeball. The retina ends just behind the margin of the ciliary body.

The retina is thin and delicate, but its structure is quite complex. It has a number of distinct layers, as figures 10.21 and 10.22 illustrate.

In the central region of the retina is a yellowish spot called the *macula lutea*. A depression in its center, called the **fovea centralis** (fo′ve-ah sen-tral′is), is in the region of the retina that produces the sharpest vision (see figs. 10.17 and 10.23).

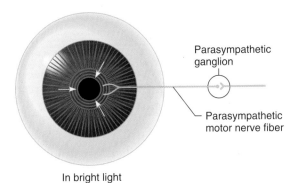

In dim light

Radially arranged
smooth muscle fibers
of the iris

Circularly arranged
smooth muscle fibers
of the iris

Pupil

In normal light

Parasympathetic
ganglion

Parasympathetic
motor nerve fiber

In bright light

Sympathetic
motor nerve
fiber

Figure 10.20

Dim light stimulates the radial muscles of the iris to contract, and the pupil dilates. Bright light stimulates the circular muscles of the iris to contract, and the pupil constricts.

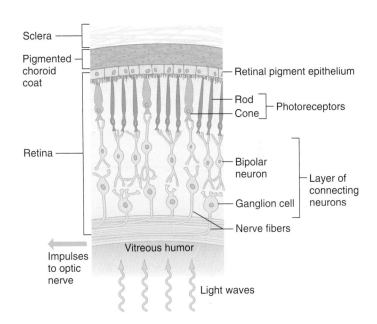

Figure 10.21

The retina consists of several cell layers. Light waves penetrate a layer of connecting neurons to impinge on the rods and cones, which are the photoreceptors. The pigmented epithelium absorbs stray light rays.

Figure 10.22

Retinal structure. Note the layers of cells and nerve fibers in this light micrograph of the retina (75×).

(a)

(b)

Figure 10.23

The retina. **(a)** Major features of the retina. **(b)** Nerve fibers leave the retina of the eye in the area of the optic disc (arrow) to form the optic nerve in this magnified view of the retina (53×).

The fovea centralis of the human eye has 150,000 cones per square millimeter. In contrast, a bird of prey's eye has about a million cones per square millimeter.

Just medial to the fovea centralis is an area called the **optic disc** (op'tik disk) (fig. 10.23). Here, nerve fibers from the retina leave the eye and join the optic nerve. A central artery and vein also pass through the optic disc. These vessels are continuous with the capillary networks of the retina, and along with vessels in the underlying choroid coat, they supply blood to the cells of the inner layer. Because the optic disc region lacks photoreceptors, it is commonly known as the *blind spot* of the eye.

The space bounded by the lens, ciliary body, and retina is the largest compartment of the eye and is called the *posterior cavity* (see fig. 10.17). It is filled with a transparent, jellylike fluid called **vitreous humor** (vit're-us hu'mor), which along with collagenous fibers forms the *vitreous body*. The vitreous body supports the internal parts of the eye and helps maintain its shape.

As a person ages, tiny, dense clumps of gel or deposits of crystal-like substances form in the vitreous humor. When these clumps cast shadows on the retina, the person sees small, moving specks in the field of vision, called *floaters*.

Practice

30. Explain the source of aqueous humor, and trace its path through the eye.
31. How does the pupil respond to changes in light intensity?
32. Describe the structure of the retina.

Light Refraction

When a person sees an object, either the object is giving off light, or it is reflecting light waves from another source. These light waves enter the eye, and an image of the object is focused on the retina. Focusing bends the light waves, a phenomenon called **refraction** (re-frak'shun).

Refraction occurs when light waves pass at an oblique angle from a medium of one optical density into a medium of a different optical density. This happens at the curved surface between the air and the cornea and at the curved surface of the lens itself. A lens with a *convex* surface (as in the eye) causes light waves to converge (fig. 10.24).

The convex surface of the cornea refracts light waves from outside objects. The convex surface of the lens and, to a lesser extent, the surfaces of the fluids in the chambers of the eye then refract the light again.

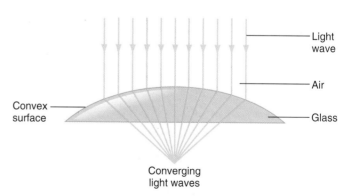

Figure 10.24

A lens with a convex surface causes light waves to converge. The lens of the eye functions the same way.

If eye shape is normal, light waves focus sharply on the retina, much as a motion picture image is focused on a screen for viewing. Unlike the motion picture image, however, the image that forms on the retina is upside down and reversed from left to right. The visual cortex interprets the image in its proper position.

Practice

33. What is refraction?
34. What parts of the eye provide refracting surfaces?

Photoreceptors

Photoreceptors are modified neurons of two distinct kinds, as figure 10.21 illustrates. One group, called *rods* (rodz), have long, thin projections at their ends, and provide black and white vision. The other group, *cones* (kōnz), have short, blunt projections, and provide color vision. There is only one type of rod, but three types of cones.

Rods and cones are in a deep part of the retina, where they project into an adjacent layer of retinal pigment epithelium (RPE) that absorbs light waves the photoreceptors do not absorb. With the pigment of the choroid coat, the RPE keeps light from reflecting off surfaces inside the eye (see fig. 10.22).

Photoreceptors are stimulated only when light reaches them. A light image focused on an area of the retina stimulates some photoreceptors, and impulses travel from them to the brain. However, the impulse leaving each activated photoreceptor provides only a fragment of the information required for the brain to interpret a complete scene.

Rods and cones provide different aspects of vision. Rods are hundreds of times more sensitive to light than cones and therefore can provide vision in dim light, without color. Cones detect color.

> A human eye has 125 million rods and 7 million cones. A cat has three types of cone cells, but sees mostly pastels. A dog has two types of cone cells, and its visual world is much like that of a person with colorblindness. Researchers corrected colorblindness in monkeys by introducing the genes for human cone pigments into their eyes.

Rods and cones also differ in the sharpness of the perceived images, or visual acuity. Cones provide sharp images, and rods provide more general outlines of objects. Rods give less precise images because nerve fibers from many rods converge, so that their impulses are transmitted to the brain on the same nerve fiber (fig. 10.25a). Thus, if a point of light stimulates a rod, the brain cannot tell which one of many receptors has been

Single sensory nerve fiber
(a)

Many sensory nerve fibers
(b)

(c)

Figure 10.25

Rods and cones are photoreceptors. **(a)** A single sensory nerve fiber transmits impulses from several rods to the brain. **(b)** Separate sensory nerve fibers transmit impulses from cones to the brain. **(c)** Scanning electron micrograph of rods and cones (1,350×).

stimulated. Convergence of impulses is less common among cones. When a cone is stimulated, the brain can pinpoint the stimulation more accurately (fig. 10.25*b*).

The fovea centralis, the area of sharpest vision, lacks rods but contains densely packed cones with few or no converging fibers (see fig. 10.17). Also in the fovea centralis, the overlying layers of the retina and the retinal blood vessels are displaced to the sides, more fully exposing photoreceptors to incoming light. Consequently, to view something in detail, a person moves the eyes so that the important part of an image falls on the fovea centralis.

Photopigments

Both rods and cones contain light-sensitive pigments that decompose when they absorb light energy. The light-sensitive biochemical in rods is called **rhodopsin** (ro-dop′sin), or *visual purple*. In the presence of light, rhodopsin molecules are broken down into a colorless protein called *opsin* and a yellowish substance called *retinal* (retinene) that is synthesized from vitamin A.

Decomposition of rhodopsin molecules activates an enzyme that initiates a series of reactions altering the permeability of the rod cell membrane. As a result, a complex pattern of nerve impulses originates in the retina. The impulses travel away from the retina along the optic nerve into the brain, where they are interpreted as vision.

In bright light, nearly all of the rhodopsin in the rods of the retina decomposes, greatly reducing rod sensitivity. In dim light, however, regeneration of rhodopsin from opsin and retinal is faster than rhodopsin breakdown. ATP provides the energy required for this regeneration (see chapter 4, p. 80).

Poor vision in dim light, called night blindness, results from vitamin A deficiency. Lack of the vitamin reduces the supply of retinal, rhodopsin production falls, and rod sensitivity is low. Supplementing the diet with vitamin A is used to treat night blindness.

In the inherited condition Leber congenital amaurosis, absence of an enzyme in the RPE prevents production of retinal from vitamin A, causing nightblindness that progresses to full blindness by early adulthood. Gene therapy on an eight-year-old boy and several others with this form of hereditary blindness restored their enzyme levels and their vision.

The light-sensitive pigments in cones, as in the rods, are composed of retinal and protein. In cones, however, three different opsin proteins, different from those found in rods, combine with retinal to form the three cone pigments. The three types of cones each contain one of these three photopigments.

The wavelength of light determines the color that the brain perceives from it. For example, the shortest wavelengths of visible light are perceived as violet, and the longest are perceived as red. One type of cone pigment (erythrolabe) is most sensitive to red light waves, another (chlorolabe) to green light waves, and a third (cyanolabe) to blue light waves. The color a person perceives depends on which set of cones or combination of sets the light in a given image stimulates. If all three sets of cones are stimulated, the person senses the light as white, and if none are stimulated, the person senses black. Different forms of colorblindness result from lack of different types of cone pigments.

Visual Nerve Pathways

Visual nerve pathways bring nerve impulses from the retina to the visual cortex, where they are perceived as vision. The pathways begin as the axons of the retinal neurons leave the eyes to form the *optic nerves* (fig. 10.26). Just anterior to the pituitary gland, these nerves give rise to the X-shaped *optic chiasma* (op′tik ki-az′mah), and within the chiasma, some of the fibers cross over. More specifically, the fibers from the nasal (medial) half of each retina cross over, but those from the temporal (lateral) sides do not. Thus, fibers from the

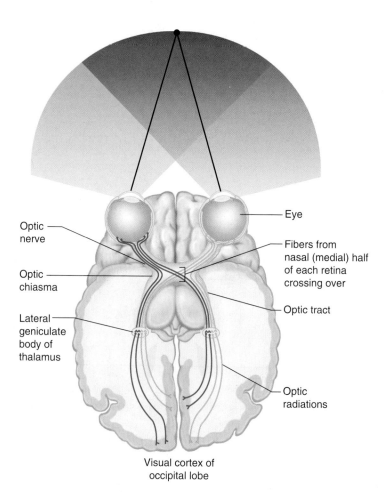

Optic nerve
Optic chiasma
Lateral geniculate body of thalamus
Eye
Fibers from nasal (medial) half of each retina crossing over
Optic tract
Optic radiations
Visual cortex of occipital lobe

Figure 10.26 AP|R

A visual pathway includes the optic nerve, optic chiasma, optic tract, and optic radiations.

nasal half of the left eye and the temporal half of the right eye form the *right optic tract,* and fibers from the nasal half of the right eye and the temporal half of the left eye form the *left optic tract.*

Just before the nerve fibers reach the thalamus, a few of them enter nuclei that function in various visual reflexes. Most of the fibers, however, enter the thalamus and synapse in its posterior portion (lateral geniculate body). From this region, the visual impulses enter nerve pathways called *optic radiations,* which lead to the visual cortex of the occipital lobes.

Practice

35. Distinguish between the rods and cones of the retina.
36. Explain the roles of visual pigments.
37. Trace a nerve impulse from the retina to the visual cortex.

Summary Outline

10.1 Introduction (p. 263)
Sensory receptors sense changes in their surroundings.

10.2 Receptors, Sensations, and Perception (p. 263)
1. Types of receptors
   a. Each type of receptor is most sensitive to a distinct type of stimulus.
   b. The major types of receptors are chemoreceptors, pain receptors, thermoreceptors, mechanoreceptors, and photoreceptors.
2. Sensations
   a. A sensation is the awareness that sensory stimulation has occurred.
   b. A particular part of the cerebral cortex interprets every impulse reaching it in a specific way.
   c. The cerebral cortex projects a sensation back to the region of stimulation.
3. Sensory adaptation may involve receptors becoming unresponsive or inhibition along the CNS pathways leading to the sensory regions of the cerebral cortex.

10.3 General Senses (p. 264)
General senses are associated with receptors in the skin, muscles, joints, and viscera.
1. Touch and pressure senses
   a. Free ends of sensory nerve fibers are receptors for the sensation of itching.
   b. Tactile corpuscles are receptors for the sensation of light touch.
   c. Lamellated corpuscles are receptors for the sensation of heavy pressure.
2. Temperature senses
   Temperature receptors include two sets of free nerve endings that are warm and cold receptors.
3. Sense of pain
   a. Pain receptors are free nerve endings that tissue damage stimulates.
   b. Visceral pain
      (1) Pain receptors are the only receptors in viscera that provide sensations.
      (2) Pain sensations produced from visceral receptors may feel as if they are coming from some other body part, called referred pain.
      (3) Visceral pain may be referred because sensory impulses from the skin and viscera travel on common nerve pathways.
   c. Pain nerve fibers
      (1) The two main types of pain fibers are acute pain fibers and chronic pain fibers.
      (2) Acute pain fibers conduct nerve impulses rapidly. Chronic pain fibers conduct impulses more slowly.
      (3) Pain impulses are processed in the gray matter of the spinal cord and ascend to the brain.
      (4) Within the brain, pain impulses pass through the reticular formation before being conducted to the cerebral cortex.
   d. Regulation of pain impulses
      (1) Awareness of pain occurs when pain impulses reach the thalamus.
      (2) The cerebral cortex determines pain intensity and locates its source.
      (3) Impulses descending from the brain stimulate neurons to release pain-relieving neuropeptides, such as enkephalins.

10.4 Special Senses (p. 267)
Special senses have receptors within large, complex sensory organs of the head.

10.5 Sense of Smell (p. 267)
1. Olfactory receptors
   a. Olfactory receptors are chemoreceptors that are stimulated by chemicals dissolved in liquid.
   b. Olfactory receptors function with taste receptors and aid in food selection.
2. Olfactory organs
   a. Olfactory organs consist of receptors and supporting cells in the nasal cavity.
   b. Olfactory receptor cells are bipolar neurons with cilia.
3. Olfactory nerve pathways
   Nerve impulses travel from the olfactory receptor cells through the olfactory nerves, olfactory bulbs, and olfactory tracts to interpreting centers in the temporal and frontal lobes of the cerebrum.
4. Olfactory stimulation
   a. Olfactory impulses may result when odorant molecules bind cell surface olfactory receptors on cilia of receptor cells. The binding pattern encodes a specific odor, which is interpreted in the brain.
   b. The sense of smell adapts rapidly.

10.6 Sense of Taste (p. 269)
1. Taste receptors
   a. Taste buds consist of taste (receptor) cells and supporting cells.
   b. Taste cells have taste hairs.
   c. Taste hair surfaces have receptors to which chemicals bind, stimulating nerve impulses.
2. Taste sensations
   a. The five primary taste sensations are sweet, sour, salty, bitter, and umami.
   b. Various taste sensations result from the stimulation of one or more types of taste receptors.
   c. A single taste receptor cell detects only one of the five tastes, but receptors corresponding to different tastes are scattered on the tongue.

3. Taste nerve pathways
   a. Sensory impulses from taste receptors travel on fibers of the facial, glossopharyngeal, and vagus nerves.
   b. These impulses are carried to the medulla oblongata and then ascend to the thalamus, from which they travel to the gustatory cortex in the parietal lobes.

10.7 Sense of Hearing (p. 270)

1. Outer ear
   The outer ear collects sound waves of vibrating objects.
2. Middle ear
   Auditory ossicles of the middle ear conduct sound waves from the eardrum to the oval window of the inner ear.
3. Auditory tube
   Auditory tubes connect the middle ears to the nasopharynx and help maintain equal air pressure on both sides of the eardrums.
4. Inner ear
   a. The inner ear is a complex system of connected tubes and chambers—the osseous and membranous labyrinths.
   b. The spiral organ contains hearing receptors that are stimulated by vibrations in the fluids of the inner ear.
   c. Different frequencies of vibrations stimulate different sets of receptor cells.
5. Auditory nerve pathways
   a. Auditory nerves carry impulses to the auditory cortices of the temporal lobes.
   b. Some auditory nerve fibers cross over, so that impulses arising from each ear are interpreted on both sides of the brain.

10.8 Sense of Equilibrium (p. 275)

1. Static equilibrium
   Static equilibrium maintains the stability of the head and body when they are motionless.
2. Dynamic equilibrium
   a. Dynamic equilibrium balances the head and body when they are moved or rotated suddenly.
   b. Other structures that help maintain equilibrium include the eyes and mechanoreceptors associated with certain joints.

10.9 Sense of Sight (p. 276)

1. Visual accessory organs
   Visual accessory organs include the eyelids, lacrimal apparatus, and extrinsic muscles of the eyes.
2. Structure of the eye
   a. The wall of the eye has an outer (fibrous), a middle (vascular), and an inner (nervous) layer.
      (1) The outer layer is protective, and its transparent anterior portion (cornea) refracts light entering the eye.
      (2) The middle layer is vascular and contains melanin that keeps the inside of the eye dark.
      (3) The inner layer contains the photoreceptors.
   b. The lens is a transparent, elastic structure. Ciliary muscles control its shape.
   c. The lens must thicken to focus on close objects.
   d. The iris is a muscular diaphragm that controls the amount of light entering the eye.
   e. Spaces within the eye are filled with fluids that help maintain its shape.
3. Light refraction
   The cornea and lens refract light waves to focus an image on the retina.
4. Photoreceptors
   a. Photoreceptors are rods and cones.
   b. Rods are responsible for colorless vision in dim light, and cones provide color vision.
5. Photopigments
   a. A light-sensitive pigment in rods decomposes in the presence of light and triggers a complex series of reactions that initiate nerve impulses.
   b. Color vision comes from three sets of cones containing different light-sensitive pigments.
6. Visual nerve pathways
   a. Nerve fibers from the retina form the optic nerves.
   b. Some fibers cross over in the optic chiasma.
   c. Most of the fibers enter the thalamus and synapse with others that continue to the visual cortex in the occipital lobes.

Chapter Assessments

10.1 Introduction

1. Distinguish between general senses and special senses. (p. 263)

10.2 Receptors, Sensations, and Perception

2. Match each sensory receptor to the type of stimulus to which it is likely to respond. (p. 263)

   (1) chemoreceptor   A. Approaching headlights
   (2) pain receptor   B. A change in blood pressure
   (3) thermoreceptor   C. The smell of roses
   (4) mechanoreceptor D. An infected tooth
   (5) photoreceptor   E. A cool breeze

3. Explain the difference between a sensation and a perception. (p. 263)

4. Explain the projection of a sensation. (p. 263)

5. You fill up the tub to take a hot bath, but the water is too hot to the touch. You try a second and third time, and within a few seconds it feels fine. Which of the following is the most likely explanation? (p. 263)
   a. The water has cooled down unusually quickly.
   b. Your ability to sense heat has adapted.
   c. Your nervous system is suddenly not functioning properly.
   d. Your ability to sense cold has adapted.
   e. All of the above.

10.3 General Senses

6. Describe the functions of free nerve endings, tactile corpuscles, and lamellated corpuscles. (p. 264)

7. Explain why pain may be referred, and provide an example. (p. 265)

10.4 Special Senses

8. Identify the location of the receptors for smell, taste, hearing, equilibrium, and sight. (p. 267)

10.5 Sense of Smell

9. Which two of the following are part of the olfactory organs? (p. 267)
   a. Olfactory receptors
   b. Columnar epithelial cells in the nasal mucosa
   c. The nose
   d. The brain
   e. The eyes

10. Trace a nerve impulse from an olfactory receptor to the interpreting center of the cerebrum. (p. 267)

10.6 Sense of Taste

11. Salivary glands are important in taste because (p. 270)
   a. they provide the fluid in which food molecules dissolve.
   b. the taste receptors are located in salivary glands.
   c. salivary glands are part of the brain.
   d. they lubricate the teeth.
   e. they produce enzymes to break down the food.

12. Name the five primary taste sensations. (p. 270)

13. Trace the pathway of a nerve impulse from a taste receptor to the interpreting center of the cerebrum. (p. 270)

10.7 Sense of Hearing

14. Match the ear area with the associated structure. (p. 270)
   (1) outer ear   A. Cochlea
   (2) middle ear   B. Eardrum
   (3) inner ear   C. Auditory ossicles

15. Trace the path of sound waves from the external acoustic meatus to the hearing receptors. (p. 270)

16. Describe the functions of the auditory ossicles. (p. 271)

17. The function of the auditory tube is to: (p. 272)
   a. equalize air pressure on both sides of the eardrum.
   b. transmit sound vibrations to the eardrum.
   c. contain the hearing receptors.
   d. connect the ears.
   e. provide for equilibrium.

18. Distinguish between the osseous and membranous labyrinths. (p. 272)

19. Describe the cochlea and its function. (p. 272)

20. Trace a nerve impulse from the spiral organ to the interpreting centers of the cerebrum. (pp. 274–275)

21. Which of the following best describes hearing receptor "hair cells"? (p. 274)
   a. They are neurons.
   b. They lack ion channels.
   c. They are epithelial, but function like neurons.
   d. They are built of the protein keratin.
   e. They are bipolar.

22. Explain how a hearing receptor stimulates a sensory neuron. (p. 274)

10.8 Sense of Equilibrium

23. Contrast static equilibrium and dynamic equilibrium. (p. 275)

24. Describe the organs of static and dynamic equilibrium and their functions. (p. 275)

10.9 Sense of Sight

25. Match the visual accessory organ with its function. (p. 277)
   (1) eyelid   A. Move the eye
   (2) conjunctiva   B. Covers the eye
   (3) lacrimal gland   C. Lines the eyelids
   (4) extrinsic muscles D. Produces tears

26. Name the three layers of the eye wall and describe the functions of each layer. (p. 279)

27. Explain why looking at a close object causes fatigue, in terms of how accommodation is accomplished. (p. 281)

28. Explain the mechanisms of pupil constriction and pupil dilation. (p. 282)

29. All of the following are compartments within the eye. In which one is vitreous humor found? (p. 282)
   a. Anterior chamber
   b. Posterior chamber
   c. Anterior cavity
   d. Posterior cavity
   e. All of the above

30. Distinguish between the fovea centralis and the optic disc. (p. 282)

31. Explain how light is focused on the retina. (p. 284)

32. Distinguish between rods and cones. (p. 285)

33. Explain why cone vision is generally more acute than rod vision. (p. 285)

34. Describe the function of rhodopsin (p. 286)

35. Explain why rod vision may be more important under dim light conditions. (p. 286)

36. Describe the relationship between light wavelength and color vision. (p. 286)

37. Trace a nerve impulse from the retina to the visual cortex. (p. 286)

Integrative Assessments/Critical Thinking

OUTCOMES 6.2, 9.14, 10.2, 10.9

1. PET (positron emission tomography) scans of the brains of people who have been blind since birth reveal high neural activity in the visual centers of the cerebral cortex when these people read Braille. However, when sighted individuals run their fingers over the raised letters of Braille, the visual centers do not show increased activity. Explain these experimental results.

OUTCOMES 6.2, 10.2, 10.3

2. Why are some serious injuries, like a bullet entering the abdomen, relatively painless, but others, such as a burn, considerably more painful?

OUTCOMES 10.2, 10.5

3. Loss of the sense of smell often precedes the major symptoms of Alzheimer disease and Parkinson disease. What additional information is needed to use this association to prevent or treat these diseases?

4. Describe how the taste of a medicine might be modified from sour to sweet, so that children would be more willing to take it.

OUTCOMES 10.2, 10.7, 10.8

5. People who are deaf due to cochlear damage do not suffer from motion sickness. Why not?

OUTCOMES 10.2, 10.8

6. Labyrinthitis is an inflammation of the inner ear. What symptoms would you expect in a patient with this disorder?

WEB CONNECTIONS

Visit the text website at **www.mhhe.com/shieress11** for additional quizzes, interactive learning exercises, and more.

APR

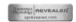

Anatomy & Physiology REVEALED includes cadaver photos that allow you to peel away layers of the human body to reveal structures beneath the surface. This program also includes animations, radiologic imaging, audio pronunciation, and practice quizzing. To learn more visit **www.aprevealed.com**.

11

Endocrine System

Smelly T-shirts. The endocrine system produces hormones, which are biochemicals that send messages in an individual. Less well understood are pheromones, which are chemical signals sent between individuals of a species. In insects and rodents, pheromones stimulate mating behavior. Experiments suggest that this may be the case with humans, too.

Mice and rats choose mates that are dissimilar to themselves with respect to a group of genes that provide immunity. Their sense of smell helps them discern appropriate mates. Biologists hypothesize that choosing mates based on scent may protect offspring in two ways—it prevents close relatives from mating, and it may team immune systems with different types of strengths.

Researchers have traced mouse social and mating behavior to receptors in the olfactory epithelium, in the nasal cavity. The receptors—called trace-amine-associated receptors—are attuned to molecules in mouse urine that direct social behavior. The genes that encode the receptors are also found in humans.

To test whether heterosexual humans use the sense of smell to respond to pheromones in mate selection as rodents do, researchers in Switzerland recruited forty-nine young women and forty-four young men. Each participant donated DNA, which was typed for genes that affect mating in rodents. The women used nasal spray for two weeks to clear their nasal passages. The men wore the same T-shirt on two consecutive days, using no deodorant or soap and avoiding contact with anything smelly that could linger. Each woman was then given three T-shirts from men genetically similar to her and three T-shirts from men genetically dissimilar to her, not knowing which shirts came from which men

The endocrine system produces hormones, which act within an individual. Humans may also produce pheromones, which affect other individuals and may play a role in mate selection, as they do in rodents and insects.

The women rated the shirts on intensity, pleasantness, and sexiness. Like the mice and rats, women preferred the sweaty T-shirts from the men least like them genetically.

Another, more specific experiment supported these findings. Women were given vials of fluid to sniff that either contained or did not contain a component of male sweat called androstadienone. Although they didn't know which samples they were sniffing, the women consistently reported mood elevation and sexual arousal when they smelled the sweat. In addition, their saliva had increased amounts of cortisol, a hormone that raises the blood sugar level, when they smelled androstadienone, suggesting that it might be a human pheromone. Despite the mounting scientific evidence for human pheromones, a definitive human pheromone has not yet been described.

Learning Outcomes

After studying this chapter, you should be able to do the following:

11.1 Introduction

1. Describe the secretions of the endocrine system. (p. 292)

2. Distinguish between paracrine and autocrine secretions. (p. 292)

3. Distinguish between endocrine and exocrine glands. (p. 292)

11.2 General Characteristics of the Endocrine System

4. Explain how the nervous and endocrine systems are alike and how they are different. (p. 292)

5. Describe the source of specificity of the endocrine system. (p. 293)

6. Name some functions of hormones. (p. 293)

11.3 Hormone Action

7. Explain how steroid and nonsteroid hormones affect target cells. (p. 294)

11.4 Control of Hormonal Secretions

8. Discuss how negative feedback mechanisms regulate hormonal secretions. (p. 296)

9. Explain how the nervous system controls secretion. (p. 297)

 Learn   **Practice**   **Assess**

**11.5–11.10 Pituitary Gland–Other
Endocrine Glands**

10. Name and describe the locations of the
major endocrine glands, and list the
hormones they secrete. (pp. 297–309)

11. Describe the functions of the hormones
that endocrine glands secrete.
(pp. 297–309)

12. Explain how the secretion of each hormone is regulated. (pp. 297–309)

11.11 Stress and Health

13. Describe how the body responds to
stress. (p. 311)

Module 8: Endocrine System

Aids to Understanding Words

(Appendix A on page 564 has a complete list of Aids to Understanding Words.)

-crin [to secrete] endo*crin*e: Pertaining to
internal secretions.

diuret- [to pass urine] *diuret*ic: Substance that
promotes urine production.

endo- [within] *endo*crine gland: Gland that
releases its secretion internally into a body
fluid.

exo- [outside] *exo*crine gland: Gland that releases
its secretion to the outside through a duct.

hyper- [above] *hyper*thyroidism: Condition
resulting from an above-normal secretion of
thyroid hormone.

hypo- [below] *hypo*thyroidism: Condition
resulting from a below-normal secretion of
thyroid hormone.

para- [beside] *para*thyroid glands: Set of glands
on the surface of the thyroid gland.

toc- [birth] oxy*toc*in: Hormone that stimulates
the uterine muscles to contract during
childbirth.

-tropic [influencing] adrenocortico*tropic*
hormone: Hormone that influences
secretions from the adrenal cortex.

11.1 INTRODUCTION

Regulating the functions of the human body to maintain
homeostasis is an enormous job. Two organ systems function in coordination to enable body parts to communicate with each other and to adjust constantly to changing
incoming signals. The nervous system is one biological
communication system; it utilizes nerve impulses carried
on nerve fibers. The other is the endocrine system.

The **endocrine system** includes cells, tissues, and
organs, collectively called endocrine glands, that secrete
substances called **hormones** (hor′mōnz) into the internal environment. The hormones diffuse from the interstitial fluid into the bloodstream, and eventually act on
cells called **target cells** some distance away.

Some glands secrete substances into the interstitial fluid, but because these secretions are rapidly broken down, they do not reach the bloodstream and are
not hormones by the traditional definition. However,
they do function similarly as messenger molecules and
are sometimes referred to as "local hormones." These
include **paracrine** secretions, which affect only neighboring cells, and **autocrine** secretions, which affect
only the secreting cell itself.

A different group of glands, called exocrine glands,
secrete outside the internal environment through tubes
or ducts that lead to the surface. Sweat, secreted by
sweat glands and reaching the surface of the skin, is one
example of an exocrine secretion (see chapter 5, p. 101).

Practice

1. What are the components of the endocrine system?

2. How do paracrine and autocrine secretions function
differently than traditionally defined hormones?

3. Distinguish between endocrine and exocrine glands.

11.2 GENERAL CHARACTERISTICS
OF THE ENDOCRINE SYSTEM

Both the endocrine system and the nervous system oversee cell-to-cell communication using chemical signals that
bind to receptor molecules. Table 11.1 summarizes some
similarities and differences between the nervous and
endocrine systems. In contrast to the nervous system, in
which neurons release neurotransmitter molecules into
synapses, the glandular cells of the endocrine system

Table 11.1	**A Comparison Between the Nervous System and the Endocrine System**	
	Nervous System	**Endocrine System**
Cells	Neurons	Glandular epithelium
Chemical signal	Neurotransmitter	Hormone
Specificity of response	Receptors on postsynaptic cell	Receptors on target cell
Speed of onset	Seconds	Seconds to hours
Duration of action	Very brief unless neuronal activity continues	May be brief or may last for days even if secretion ceases

release hormones into the bloodstream, which carries these messenger molecules everywhere (fig. 11.1). However, the endocrine system is no less precise, because only target cells can respond to a hormone. A hormone's target cells have specific receptors that other cells do not have. These receptors are proteins or glycoproteins with binding sites for a specific hormone.

Endocrine glands and their hormones help regulate metabolic processes. They control the rates of certain chemical reactions, aid in the transport of substances across cell membranes, and help regulate water and electrolyte balances. They also play vital roles in reproduction, development, and growth.

Specialized small groups of cells produce some hormones. However, the major endocrine glands are the pituitary gland, thyroid gland, parathyroid glands, adrenal glands, pancreas, pineal gland, reproductive glands (testes and ovaries), kidneys, and thymus (fig. 11.2).

Practice

4. Explain how the nervous and endocrine systems are alike and how they differ.

5. What determines whether a cell is a target cell for a particular hormone?

6. State some functions of hormones.

11.3 HORMONE ACTION

Most hormones are of two general types. They are either steroids (or steroidlike substances) synthesized from cholesterol, or they are amines, peptides, proteins, or glycoproteins synthesized from amino acids (table 11.2). Hormones can stimulate changes in target cells even in extremely low concentrations.

Table 11.2	Types of Hormones	
Type of Compound	**Formed From**	**Examples**
Steroids	Cholesterol	Estrogen, testosterone, aldosterone, cortisol
Amines	Amino acids	Norepinephrine, epinephrine
Peptides	Amino acids	Antidiuretic hormone, oxytocin, thyrotropin-releasing hormone
Proteins	Amino acids	Parathyroid hormone, growth hormone, prolactin
Glycoproteins	Protein and carbohydrate	Follicle-stimulating hormone, luteinizing hormone, thyroid-stimulating hormone

(a)

(b)

Figure 11.1

Chemical communication takes place in both the nervous system and the endocrine system. In both cases, cells respond to chemicals released from other cells. **(a)** Neurons release neurotransmitters into a synapse, affecting postsynaptic cells. **(b)** Glands release hormones into the bloodstream. Blood carries hormone molecules throughout the body, but only target cells respond.

Q: *What do postsynaptic cells and target cells have in common that allow them to respond to secreted chemicals?*

Answers can be found in Appendix E on page 568.

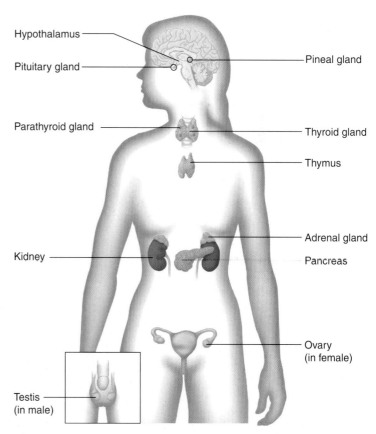

Figure 11.2

Locations of the major endocrine glands. The pituitary, thyroid, parathyroid, adrenal glands, and the pancreas are the main topics of this chapter. The functions of the other glands are described in more detail in subsequent chapters.

Steroid Hormones

Steroid molecules consist of complex rings of carbon and hydrogen atoms, and some oxygen atoms (see fig. 2.16, p. 43). Steroids differ according to the types and numbers of atoms attached to these rings and the ways they are joined.

Steroid hormones are insoluble in water. They are carried in the bloodstream weakly bound to plasma proteins in a way that allows them to be released in sufficient quantity in the vicinity of their target cells. However, unlike amine, peptide, and protein hormones, steroid hormones are soluble in lipids.

Steroid hormones can diffuse into cells relatively easily and may enter any cell in the body, because lipids make up the bulk of cell membranes and steroid molecules are lipid-soluble. When a steroid hormone molecule enters a target cell, the following events occur (fig. 11.3):

1. The lipid-soluble steroid hormone diffuses through the cell membrane.

2. The steroid hormone binds a specific protein molecule—the receptor for that hormone.
3. The resulting hormone-receptor complex binds in the nucleus to specific sequences of the target cell's DNA, activating transcription of specific genes into messenger RNA (mRNA) molecules.
4. The mRNA molecules leave the nucleus and enter the cytoplasm.
5. The mRNA molecules associate with ribosomes to direct the synthesis of specific proteins.

The newly synthesized proteins, which may be enzymes, transport proteins, or even hormone receptors, carry out the specific effects associated with the particular steroid hormone.

Nonsteroid Hormones

Nonsteroid hormones, such as amines, peptides, and proteins, usually bind receptors in target cell membranes. Each of these receptor molecules is a protein with a

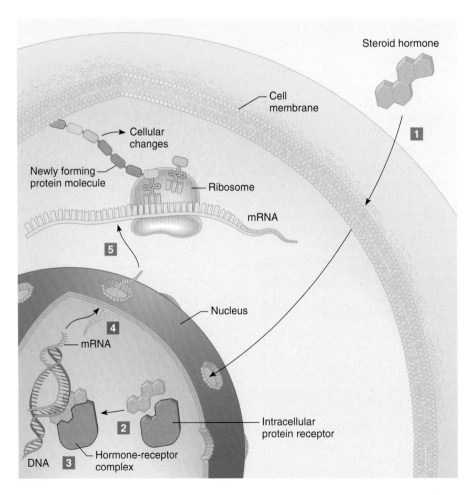

Figure 11.3 AP|R

Steroid hormones. **(1)** A steroid hormone crosses a cell membrane and **(2)** combines with a protein receptor, usually in the nucleus. **(3)** The hormone-receptor complex activates synthesis of specific messenger RNA (mRNA) molecules. **(4)** The mRNA molecules leave the nucleus and enter the cytoplasm **(5)**, where they guide synthesis of the encoded proteins.

binding site and an *activity site.* A hormone molecule delivers its message to its target cell by uniting with the binding site of its receptor. This combination stimulates the receptor's activity site to interact with other membrane proteins. The hormone that triggers this cascade of biochemical activity is called a *first messenger.* The biochemicals in the cell that induce changes in response to the hormone's binding are called *second messengers.* The entire process of chemical communication, from outside cells to inside, is called **signal transduction.**

The second messenger associated with one group of hormones is *cyclic adenosine monophosphate,* also called **cyclic AMP (cAMP)** (sī'klik ay em pee). This mechanism works as follows (fig. 11.4):

1. A hormone binds to its receptor.
2. The resulting hormone-receptor complex activates a membrane protein called a *G protein.*
3. The G protein activates an enzyme called *adenylate cyclase,* which is a membrane protein.
4. In the cytoplasm, activated adenylate cyclase catalyzes the formation of cAMP from ATP.

5. cAMP activates another set of enzymes, called protein kinases, which transfer phosphate groups from ATP to their substrate molecules, which are specific proteins in the cell. This action, called phosphorylation, alters the shapes of these substrate molecules, thereby activating them.

The activated proteins then alter various cellular processes, bringing about the characteristic effect of the hormone.

The type of membrane receptors present and the kinds of protein substrate molecules in a cell determine the cell's response to a hormone. Such responses to second messenger activation include altering membrane permeabilities, activating enzymes, promoting synthesis of certain proteins, stimulating or inhibiting specific metabolic pathways, moving the cell, and initiating secretion of hormones or other substances.

Another enzyme (phosphodiesterase) quickly inactivates cAMP, so that its action is short-lived. For this reason, a continuing response of a target cell requires a continuing signal from hormone molecules binding the target cell's membrane receptors.

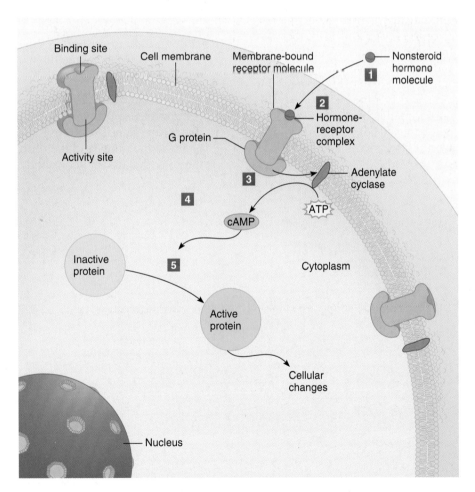

Figure 11.4 AP|R

Nonsteroid hormone action. **(1)** Body fluids carry nonsteroid hormone molecules to the target cell, where **(2)** they bind receptor molecules on the cell membrane. **(3)** This activates molecules of adenylate cyclase, which **(4)** catalyze conversion of ATP into cyclic adenosine monophosphate (cAMP). **(5)** The cAMP promotes a series of reactions leading to the cellular changes associated with the hormone's action.

A number of other second messengers work in much the same way. These include diacylglycerol (DAG) and inositol triphosphate (IP_3).

Abnormal or missing G proteins cause a variety of disorders, including colorblindness, precocious puberty, retinitis pigmentosa, and several thyroid problems.

Prostaglandins

A group of biochemicals called **prostaglandins** (pros″tah-glan′dinz) also regulates cells. Prostaglandins are lipids synthesized from a fatty acid (arachidonic acid) in cell membranes. A great variety of cells produce prostaglandins, including those of the liver, kidneys, heart, lungs, thymus, pancreas, brain, and reproductive organs. Prostaglandins usually act more locally than hormones, often affecting only the organ where they are produced.

Prostaglandins are potent and are present in very small amounts. They are not stored in cells, but rather synthesized just before release. They are rapidly inactivated.

Prostaglandins produce diverse and even opposite effects. Some prostaglandins, for example, relax smooth muscles in the airways of the lungs and in blood vessels, while others contract smooth muscles in the walls of the uterus and intestines. Prostaglandins stimulate hormone secretion from the adrenal cortex and inhibit secretion of hydrochloric acid from the stomach wall. They also influence the movements of sodium ions and water molecules in the kidneys, help regulate blood pressure, and have powerful effects on male and female reproductive physiology.

Practice

7. How does a steroid hormone promote cellular changes? How does a nonsteroid hormone do the same?
8. What is a second messenger?
9. What are prostaglandins?
10. What are the effects of prostaglandins?

11.4 CONTROL OF HORMONAL SECRETIONS

Hormones are continually excreted in the urine and broken down by various enzymes, primarily in the liver. Therefore, maintaining constant hormone levels requires ongoing hormone secretion. To increase or decrease the blood levels of a hormone, the body increases or decreases secretion. Not surprisingly, hormone secretion is precisely regulated.

Generally, hormone secretion is controlled in three ways, all of which use *negative feedback* (see chapter 1, p. 6). In each case, an endocrine gland or the system controlling it detects the concentration of the hormone the gland secretes, a process the hormone controls, or an action the hormone has on the internal environment (fig. 11.5). The three mechanisms of hormone control are described below.

1. The hypothalamus regulates the anterior pituitary gland's release of hormones that stimulate other endocrine glands to release hormones. Its location near the thalamus and the third ventricle allows the hypothalamus to constantly receive information

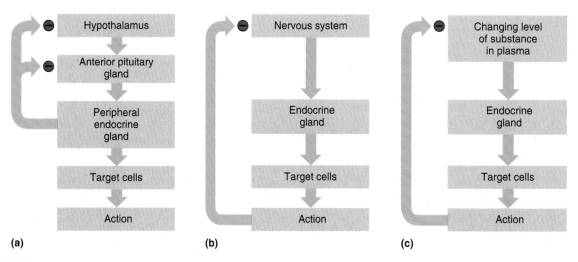

(a)   (b)   (c)

Figure 11.5

Control of the endocrine system occurs in three ways: **(a)** The hypothalamus and anterior pituitary stimulate other endocrine glands; **(b)** the nervous system stimulates a gland directly; or **(c)** changes in the internal environment stimulate glands directly.

(● indicates negative feedback inhibition.)

about the internal environment from neural connections and cerebrospinal fluid (fig. 11.5*a*).

2. The nervous system stimulates some glands directly. The adrenal medulla, for example, secretes its hormones in response to sympathetic nerve impulses (fig. 11.5*b*).

3. Another group of glands responds directly to changes in the composition of the internal environment. For example, when the blood glucose level rises, the pancreas secretes insulin, and when the blood glucose level falls, it secretes glucagon, as discussed later in the chapter (fig. 11.5*c*).

In each of these cases, as hormone levels rise in the blood and the hormone exerts its effects, negative feedback inhibits the system, and hormone secretion decreases. Then, as hormone levels in the blood decrease and the hormone's effects are no longer taking place, inhibition of the system is lifted, and secretion of that hormone increases again. As a result of negative feedback, hormone levels in the bloodstream remain relatively stable, tending to fluctuate slightly above and below an average value (fig. 11.6).

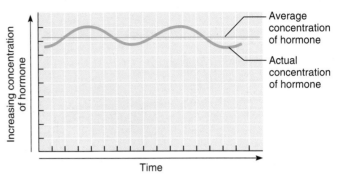

Figure 11.6
As a result of negative feedback, hormone concentrations remain relatively stable, although they may fluctuate slightly above and below average concentrations.

Practice

11. Explain three examples of control of hormonal secretion.

12. Describe a negative feedback system that controls hormone secretion.

11.5 PITUITARY GLAND

The **pituitary gland** (hypophysis) is located at the base of the brain, where a pituitary stalk (infundibulum) attaches it to the hypothalamus. The gland is about 1 centimeter in diameter and consists of an **anterior pituitary** (pĭ-tu′ĭ-tār″e), or anterior lobe, and a **posterior pituitary**, or posterior lobe (fig. 11.7).

> In the fetus, a narrow region develops between the anterior and posterior lobes of the pituitary gland. Called the *intermediate lobe* (pars intermedia), it produces melanocyte-stimulating hormone (MSH), which regulates the synthesis of melanin—the pigment in skin and in parts of the eyes and brain. This intermediate lobe is usually not present as a distinct structure in adults, .

The brain controls most of the pituitary gland's activities. For example, the posterior pituitary releases hormones when nerve impulses from the hypothalamus signal the axon terminals of neurosecretory cells in the posterior pituitary (fig. 11.8). On the other hand, **releasing hormones** (or release-inhibiting hormones) from the hypothalamus control secretion from the anterior pituitary (fig. 11.8). These hormones travel in a capillary network associated with the hypothalamus. The capillaries merge to form the **hypophyseal portal veins,** which pass downward along the pituitary stalk and give rise to a capillary network in the anterior pituitary. Thus, the hypothalamus releases substances that the blood carries directly to the anterior pituitary.

Upon reaching the anterior pituitary, each of the hypothalamic hormones acts on a specific population of cells. Some of the resulting actions are inhibitory, but most stimulate the anterior pituitary to release hormones that stimulate secretions from peripheral endocrine glands. In many of these cases, important negative feedback regulates hormone levels in the bloodstream.

Practice

13. Where is the pituitary gland located?

14. Explain how the hypothalamus controls the secretory activity of the posterior and anterior lobes of the pituitary gland.

Anterior Pituitary Hormones

The anterior pituitary is enclosed in a capsule of dense, collagenous connective tissue and consists largely of epithelial tissue organized in blocks around many thin-walled blood vessels. So far researchers have identified five types of secretory cells in this epithelium. Four of these cell types each secrete a different hormone—growth hormone (GH), prolactin (PRL), thyroid-stimulating hormone (TSH), and adrenocorticotropic hormone (ACTH). The fifth type of cell secretes both follicle-stimulating hormone (FSH) and luteinizing hormone (LH). (In males, luteinizing hormone had been known as interstitial cell stimulating hormone, or ICSH.)

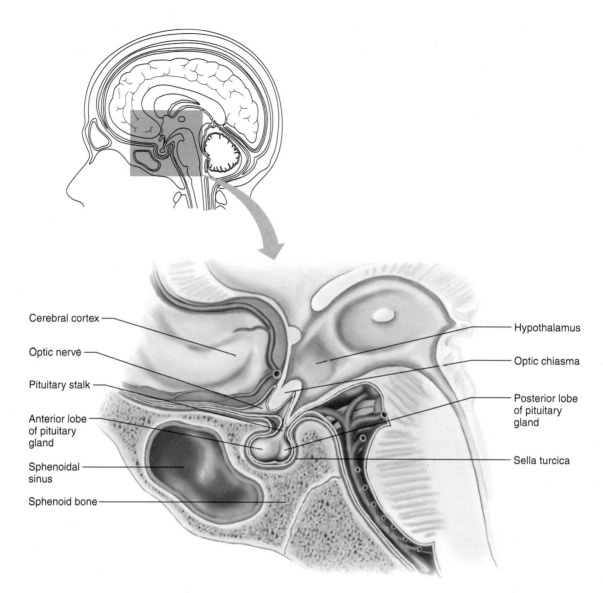

Figure 11.7   AP|R

The pituitary gland is attached to the hypothalamus and lies in the sella turcica of the sphenoid bone.

Growth hormone (GH) stimulates cells to increase in size and divide more frequently. It also enhances the movement of amino acids across cell membranes and speeds the rate at which cells utilize carbohydrates and fats. The hormone's effect on amino acids is important in stimulating growth.

Two hormones from the hypothalamus control GH secretion: *GH-releasing hormone (GHRH)* and *GH release-inhibiting hormone (GHIH)*. Nutritional state also influences control of GH. For example, more GH is released during periods of protein deficiency and abnormally low blood glucose concentration. Conversely, when blood protein and glucose concentrations increase, GH secretion decreases.

Insufficient secretion of growth hormone (GH) during childhood limits growth, causing pituitary dwarfism. Body parts are normally proportioned, and mental development is normal. Typically, hormone therapy can help.

Oversecretion of GH during childhood causes gigantism, in which height may exceed 8 feet. This rare condition is usually a result of a pituitary gland tumor, which may also cause oversecretion of other pituitary hormones. As a result, a person with gigantism often has several metabolic disturbances.

Acromegaly is the overproduction of growth hormone in adulthood. The many symptoms attesting to the wide effects of this hormone include enlarged heart, bones, thyroid gland, facial features, hands, feet, and head. Early symptoms include headache, joint pain, fatigue, and depression.

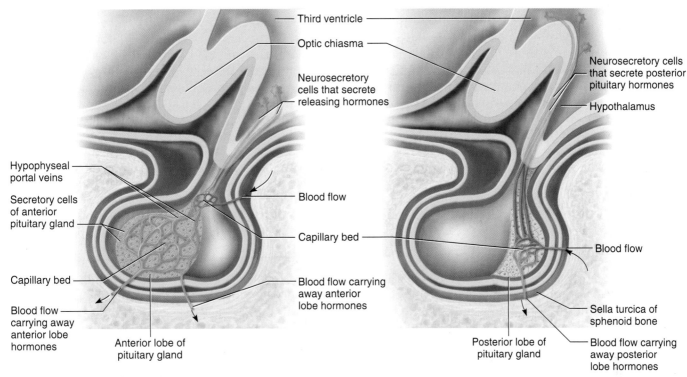

Figure 11.8 AP|R

Secretion of pituitary hormones. Releasing hormones from neurosecretory cells in the hypothalamus stimulate secretory cells of the anterior lobe of the pituitary gland to secrete hormones. Other neurosecretory cells in the hypothalamus release their hormones directly into capillaries of the posterior lobe of the pituitary gland.

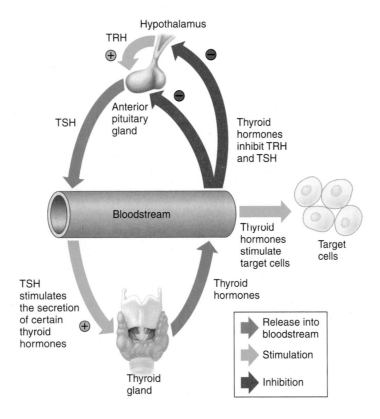

Figure 11.9

Thyrotropin-releasing hormone (TRH) from the hypothalamus stimulates the anterior pituitary gland to release thyroid-stimulating hormone (TSH), which stimulates the thyroid gland to release hormones. These thyroid hormones reduce the secretion of TSH and TRH by negative feedback. (⊕ = stimulation; ⊖ = inhibition)

Prolactin (pro-lak'tin) **(PRL)** stimulates and sustains a woman's milk production following the birth of an infant (see chapter 20, p. 552). No normal physiological role in human males has been firmly established, although PRL may help to maintain normal sperm production. In contrast, abnormally elevated levels of PRL can disrupt sexual function in both sexes.

Thyroid-stimulating hormone (TSH) controls thyroid gland secretions (described in section 11.6, on page 301). The hypothalamus partially regulates TSH secretion by producing *thyrotropin-releasing hormone (TRH)* (fig. 11.9). Circulating thyroid hormones inhibit release of TRH and TSH. As the blood concentration of thyroid hormones increases, secretion of TRH and TSH decreases.

Adrenocorticotropic hormone (ad-re″no-kor″te-ko-trō p'ik hor′mō n) **(ACTH)** controls the manufacture and secretion of certain hormones from the outer layer, or *cortex,* of the adrenal gland. (These hormones are discussed in section 11.8 on p. 304.) ACTH secretion is regulated in part by *corticotropin-releasing hormone (CRH),* which the hypothalamus releases in response to decreased concentrations of adrenal cortical hormones. Also, stress may increase ACTH secretion by stimulating the release of CRH.

Follicle-stimulating hormone (fol′ĭ-kl stim′u-la″ting hor′mōn) **(FSH)** and **luteinizing hormone** (lu′te-in-iz″ing hor′mōn) **(LH)** are *gonadotropins,* which means they exert their actions on the gonads, or reproductive organs. Gonads are the **testes** (tes′tēz) in the male and the **ovaries** (o′vah-rēz) in the female. (Chapter 19, pp. 513 and 515–522, discusses the functions of these gonadotropins and the ways they interact.)

Practice

15. How does growth hormone affect protein synthesis?

16. What is the function of prolactin?

17. How is secretion of thyroid-stimulating hormone regulated?

18. What is the function of adrenocorticotropic hormone?

19. What is a gonadotropin?

Posterior Pituitary Hormones

The posterior pituitary consists mostly of nerve fibers and neuroglia, unlike the anterior pituitary, which is composed primarily of glandular epithelial cells. The neuroglia support the nerve fibers, which originate in the hypothalamus.

Specialized neurons in the hypothalamus produce the two hormones associated with the posterior pituitary—**antidiuretic hormone** (an″tĭ-di″u-ret′ik hor′mōn) **(ADH)** and **oxytocin** (ok″sĭ-to′sin) **(OT)** (see fig. 11.8). These hormones travel down axons through the pituitary stalk to the posterior lobe, and are stored in vesicles (secretory granules) near the ends of the axons. Nerve impulses from the hypothalamus release the hormones into the blood. Thus, though synthesized in the hypothalamus, posterior pituitary hormones are named for where they enter the bloodstream.

A *diuretic* is a chemical that increases urine production, whereas an *antidiuretic* decreases urine formation. ADH produces an antidiuretic effect by reducing the volume of water the kidneys excrete. In this way, ADH regulates the water concentration of body fluids.

The hypothalamus regulates ADH secretion. Certain neurons in this part of the brain, called *osmoreceptors,* sense changes in the osmotic pressure of body fluids. Dehydration due to lack of water intake increasingly concentrates blood solutes. Osmoreceptors, sensing the resulting increase in osmotic pressure, signal the posterior pituitary to release ADH, which travels in the blood to the kidneys. As a result, the kidneys produce less urine, conserving water. On the other hand, drinking too much water dilutes body fluids, inhibiting ADH release. The kidneys excrete more dilute urine until the water concentration of body fluids returns to normal.

> If an injury or tumor damages any parts of the ADH-regulating mechanism, too little ADH may be synthesized or released, producing *diabetes insipidus*. An affected individual may produce as much as 25–30 liters of very dilute urine per day, and solute concentrations in body fluids rise.

OT contracts smooth muscles in the uterine wall and stimulates uterine contractions in the later stages of childbirth. Stretching of uterine and vaginal tissues late in pregnancy triggers OT release during childbirth. In the breast, OT contracts specialized cells (myoepithelial cells) associated with the milk-producing glands and their ducts. In lactating breasts, this action forces liquid from the milk glands into the milk ducts and ejects the milk from the breasts for breastfeeding. In addition, OT is an antidiuretic, but it is much weaker than ADH. Table 11.3 reviews the hormones of the pituitary gland.

> If a pregnant woman near or at her "due date" has certain signs of the approaching birth, but is not yet experiencing uterine contractions (labor pains), she may be given a form of oxytocin to stimulate contractions. Oxytocin may also be administered to the mother following childbirth to contract uterine muscles sufficiently to squeeze broken blood vessels closed, minimizing the risk of hemorrhage.

Practice

20. What is the function of antidiuretic hormone?

21. How is secretion of antidiuretic hormone controlled?

22. What effects does oxytocin produce in females?

Table 11.3	Hormones of the Pituitary Gland	
Hormone	**Action**	**Source of Control**
Anterior Lobe		
Growth hormone (GH)	Stimulates an increase in the size and division rate of body cells; enhances movement of amino acids across membranes	Growth hormone-releasing hormone and growth hormone release-inhibiting hormone from hypothalamus
Prolactin (PRL)	Sustains milk production after birth	Secretion restrained by prolactin release-inhibiting hormone and stimulated by prolactin-releasing factor from hypothalamus
Thyroid-stimulating hormone (TSH)	Controls secretion of hormones from thyroid gland	Thyrotropin-releasing hormone (TRH) from hypothalamus
Adrenocorticotropic hormone (ACTH)	Controls secretion of certain hormones from adrenal cortex	Corticotropin-releasing hormone (CRH) from hypothalamus
Follicle-stimulating hormone (FSH)	In females, responsible for the development of egg-containing follicles in ovaries and stimulates follicular cells to secrete estrogen; in males, stimulates production of sperm cells	Gonadotropin-releasing hormone from hypothalamus
Luteinizing hormone (LH)	Promotes secretion of sex hormones; plays a role in releasing an egg cell in females	Gonadotropin-releasing hormone from hypothalamus
*Posterior Lobe**		
Antidiuretic hormone (ADH)	Causes kidneys to conserve water; in high concentration, increases blood pressure	Hypothalamus in response to changes in water concentration in body fluids
Oxytocin (OT)	Contracts muscles in the uterine wall; contracts muscles associated with milk-secreting glands	Hypothalamus in response to stretching of uterine and vaginal walls and stimulation of breasts

*These hormones are synthesized in the hypothalamus, as explained in the text.

11.6 THYROID GLAND

The **thyroid gland** (thi′roid gland) is a very vascular structure that consists of two large lobes connected by a broad *isthmus* (is′mus). It is bilateral, just inferior to the larynx and anterior to the trachea (fig. 11.10 and reference plate 4, p. 26).

Structure of the Gland

A capsule of connective tissue covers the thyroid gland, which is made up of many secretory parts called *follicles*. The follicles have cavities that are lined with a single layer of cuboidal epithelial cells and filled with a clear, viscous substance called *colloid*. The follicular cells produce and secrete hormones that may be stored in the colloid or released into the blood in nearby capillaries.

Thyroid Hormones

The follicular cells of the thyroid gland synthesize two hormones—**thyroxine** (thi-rok′sin) (tetraiodothyronine), also known as T_4 because it contains four atoms of iodine, and **triiodothyronine** (tri″i-o″do-thi′ro-nēn), known as T_3 because it includes three atoms of iodine.

Thyroxine and triiodothyronine have similar actions, although triiodothyronine is five times more potent. These hormones help regulate the metabolism of carbohydrates, lipids, and proteins. They increase the rate at which cells release energy from carbohydrates, increase the rate of protein synthesis, and stimulate breakdown and mobilization of lipids. Thyroid hormones are the major factors determining how many calories the body must consume at rest in order to maintain life, which is known as the *basal metabolic rate (BMR)*. Thyroid hormones are required for normal growth and development, and are essential to nervous system maturation.

Up to 80% of the iodine in the body is in the thyroid gland.

Follicular cells require iodine salts (iodides) to produce thyroxine and triiodothyronine. Foods normally provide iodides, and after the iodides have been absorbed from the intestine, blood transports them to the thyroid gland. An efficient active transport mechanism moves the iodides into the follicular cells, where they are used to synthesize the hormones. The hypothalamus and anterior pituitary gland control the synthesis and release of thyroid hormones. Once in the blood, thyroxine and

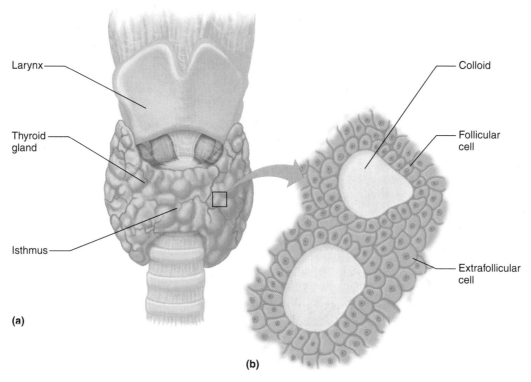

Larynx

Thyroid gland

Isthmus

(a)

Colloid

Follicular cell

Extrafollicular cell

(b)

Figure 11.10

Thyroid gland. **(a)** The thyroid gland consists of two lobes connected anteriorly by an isthmus. **(b)** Follicular cells secrete thyroid hormones. Extrafollicular cells secrete calcitonin.

triiodothyronine combine with proteins in the blood (plasma proteins) and are transported to body cells.

A third hormone of the thyroid gland, **calcitonin** (kal′sĭ-to′nin), is not referred to as "thyroid hormone" because it is produced by the thyroid's extrafollicular (other than follicle) cells. Along with parathyroid hormone (PTH) from the parathyroid glands, calcitonin regulates the concentrations of blood calcium and phosphate ions (see fig. 7.8, p. 141).

Blood concentration of calcium ions regulates calcitonin release. As this concentration increases, so does calcitonin secretion. Calcitonin inhibits the bone-resorbing activity of osteoclasts (see chapter 7, p. 136) and increases

the kidneys' excretion of calcium and phosphate ions—actions that lower the blood calcium and phosphate ion concentrations. Table 11.4 reviews the actions and controls of the thyroid hormones.

Practice

23. Where is the thyroid gland located?
24. Which hormones of the thyroid gland affect carbohydrate metabolism and protein synthesis?
25. How does the thyroid gland influence the concentrations of blood calcium and phosphate ions?

Table 11.4	Hormones of the Thyroid Gland AP R	
Hormone	**Action**	**Source of Control**
Thyroxine (T₄)	Increases rate of energy release from carbohydrates; increases rate of protein synthesis; accelerates growth; stimulates activity in nervous system	Thyroid-stimulating hormone from the anterior pituitary gland
Triiodothyronine (T₃)	Same as above, but five times more potent than thyroxine	Thyroid-stimulating hormone from the anterior pituitary gland
Calcitonin	Lowers blood calcium and phosphate ion concentrations by inhibiting release of calcium and phosphate ions from bones and by increasing excretion of these ions by kidneys	Blood calcium concentration

Thyroid disorders may produce underactivity (*hypothyroidism*) or overactivity (*hyperthyroidism*) of the glandular cells. A child with one form of hypothyroidism, called cretinism, may appear normal at birth because the mother provided enough thyroid hormones. When the infant's own thyroid gland does not produce enough of these hormones after birth, symptoms develop, including stunted growth, abnormal bone formation, retarded mental development, low body temperature, and sluggishness. Without receiving thyroid hormone within a month or so following birth, the child may suffer permanent mental retardation. Hypothyroidism is also common among older adults, producing fatigue and weight gain. Hyperthyroidism produces an elevated metabolic rate, restlessness, and overeating. The eyes protrude (exophthalmia) because of swelling in the tissues behind them, and the thyroid gland enlarges, producing a bulge in the neck called a *goiter*.

11.7 PARATHYROID GLANDS

The **parathyroid glands** (par″ah-thi′roid glandz) are on the posterior surface of the thyroid gland, as figure 11.11 shows. Usually, there are four parathyroid glands—a superior and an inferior gland associated with each of the thyroid's lateral lobes.

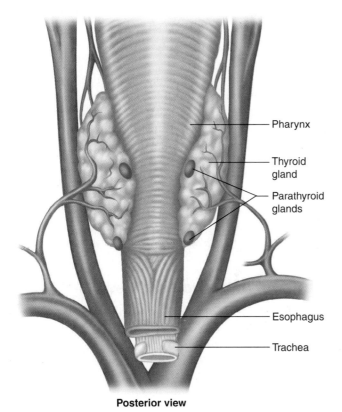

- Pharynx
- Thyroid gland
- Parathyroid glands
- Esophagus
- Trachea

Posterior view

Figure 11.11

The parathyroid glands are embedded in the posterior surface of the thyroid gland.

Structure of the Glands

A thin capsule of connective tissue covers each small, yellowish-brown parathyroid gland. The body of the gland consists of many tightly packed secretory cells closely associated with capillary networks.

Parathyroid Hormone

The parathyroid glands secrete **parathyroid hormone (PTH),** which increases blood calcium concentration and decreases blood phosphate ion concentration. PTH affects the bones, kidneys, and intestine.

The extracellular matrix of bone tissue is rich in mineral salts, including calcium phosphate (see chapter 7, p. 141). PTH inhibits the activity of osteoblasts and stimulates osteoclasts to resorb bone and release calcium and phosphate ions into the blood. At the same time, PTH causes the kidneys to conserve blood calcium and to excrete more phosphate ions in the urine. It also stimulates calcium absorption from food in the intestine, further increasing blood calcium concentration.

Negative feedback between the parathyroid glands and the blood calcium concentration regulates PTH secretion. As blood calcium concentration drops, more PTH is secreted; as blood calcium concentration rises, less PTH is released (fig. 11.12).

To summarize, calcitonin and PTH activities maintain stable blood calcium concentration. Calcitonin decreases an above-normal blood calcium concentration, while PTH increases a below-normal blood calcium concentration (see fig. 7.8, p. 141).

Practice

26. Where are the parathyroid glands?
27. How does parathyroid hormone help regulate concentrations of blood calcium and phosphate ions?

Injury to the parathyroids or their surgical removal can cause *hypoparathyroidism,* in which decreased PTH secretion reduces osteoclast activity. Although the bones remain strong, the blood calcium concentration decreases. The nervous system may become abnormally excitable, triggering spontaneous impulses. As a result, muscles may undergo tetanic contractions, possibly leading to respiratory failure and death.

A tumor in a parathyroid gland may cause *hyperparathyroidism,* which increases PTH secretion. This stimulates osteoclast activity, and as bone tissue is resorbed, the bones soften, deform, and more easily fracture spontaneously. In addition, excess calcium and phosphate released into body fluids may be deposited in abnormal places, causing new problems, such as kidney stones.

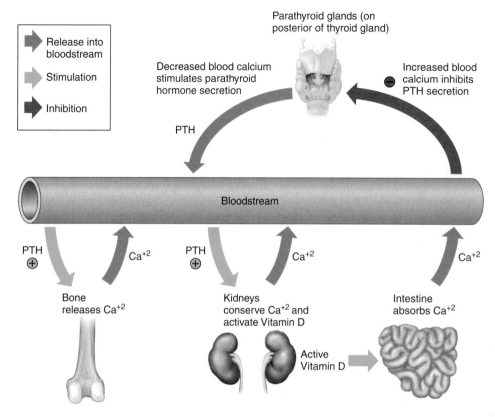

Release into
bloodstream

Stimulation

Inhibition

Parathyroid glands (on
posterior of thyroid gland)

Decreased blood calcium
stimulates parathyroid
hormone secretion

Increased blood
calcium inhibits
PTH secretion

PTH

Bloodstream

PTH
$\oplus$ Ca^{+2} PTH
$\oplus$ Ca^{+2} Ca^{+2}

Bone
releases Ca^{+2}

Kidneys
conserve Ca^{+2} and
activate Vitamin D

Intestine
absorbs Ca^{+2}

Active
Vitamin D

Figure 11.12 $\Lambda P|R$

Parathyroid hormone (PTH) stimulates bone to release calcium (Ca^{+2}) and the kidneys to conserve calcium. It indirectly stimulates the intestine to absorb calcium. The resulting increase in blood calcium concentration inhibits secretion of PTH by negative feedback. ($\oplus$ = stimulation; $\ominus$ = inhibition)

11.8 ADRENAL GLANDS

The **adrenal glands** (ah-dre′nal glandz) are closely associated with the kidneys (fig. 11.13 and reference plate 6, p. 28). A gland sits atop each kidney like a cap and is embedded in the mass of adipose tissue that encloses the kidney.

Structure of the Glands

Each adrenal gland is very vascular and consists of two parts: The central portion is the **adrenal medulla** (ah-dre′nal me-dul′ah), and the outer part is the **adrenal cortex** (ah-dre′nal kor′teks). These regions are not sharply divided, but they are functionally distinct glands that secrete different hormones.

The adrenal medulla consists of irregularly shaped cells organized in groups around blood vessels. These cells are intimately connected with the sympathetic division of the autonomic nervous system. Adrenal medullary cells are actually modified postganglionic neurons. Preganglionic autonomic nerve fibers lead to them from the central nervous system (see chapter 9, pp. 250–251).

The adrenal cortex, which makes up the bulk of the adrenal gland, is composed of closely packed masses of

epithelial cells, organized in layers. These layers form an outer (glomerulosa), middle (fasciculata), and inner (reticularis) zone of the cortex (fig. 11.13*b*). As in the adrenal medulla, the cells of the adrenal cortex are well supplied with blood vessels.

> **Practice**
>
> 28. Where are the adrenal glands?
> 29. Describe the two portions of an adrenal gland.

Hormones of the Adrenal Medulla

The cells of the adrenal medulla secrete two closely related hormones—**epinephrine** (ep″ĭ-nef′rin) (adrenalin) and **norepinephrine** (nor″ep-ĭ-nef′rin) (noradrenalin). These hormones have similar molecular structures and physiological functions. In fact, epinephrine, which makes up 80% of the adrenal medullary secretion, is synthesized from norepinephrine.

The effects of the adrenal medullary hormones resemble those of sympathetic neurons stimulating their effectors. Hormonal effects, however, last up to ten times longer because hormones are broken down more slowly

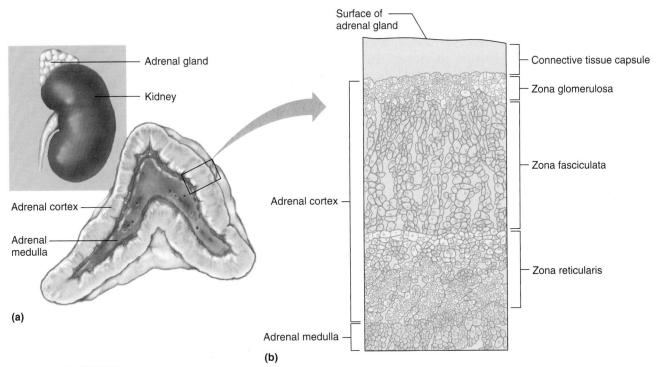

Figure 11.13 AP|R
Adrenal glands. **(a)** An adrenal gland consists of an outer cortex and an inner medulla. **(b)** The cortex consists of three layers, or zones, of cells.

than neurotransmitters. Epinephrine and norepinephrine increase heart rate, the force of cardiac muscle contraction, breathing rate, and blood glucose level. They also elevate blood pressure and decrease digestive activity.

Impulses arriving on sympathetic nerve fibers stimulate the adrenal medulla to release its hormones at the same time that sympathetic impulses are stimulating other effectors. These sympathetic impulses originate in the hypothalamus in response to stress. In this way, adrenal medullary secretions function with the sympathetic division of the autonomic nervous system in preparing the body for energy-expending action, sometimes called "fight-or-flight responses." Table 11.5 compares some of the effects of the adrenal medullary hormones.

Tumors in the adrenal medulla can increase hormonal secretion. Release of norepinephrine usually predominates, prolonging sympathetic responses—high blood pressure, increased heart rate, elevated blood sugar, and so forth. Surgical removal of the tumor corrects the condition.

Practice

30. Name the hormones the adrenal medulla secretes.

31. What effects do hormones from the adrenal medulla produce?

32. What stimulates release of hormones from the adrenal medulla?

Table 11.5	Comparative Effects of Epinephrine and Norepinephrine	
Part or Function Affected	**Epinephrine**	**Norepinephrine**
Heart	Increases rate and force of contraction	Increases rate and force of contraction
Blood vessels	Dilates vessels in skeletal muscle, decreasing resistance to blood flow	Increases blood flow to skeletal muscles, resulting from constriction of blood vessels in skin and viscera
Systemic blood pressure	Increases somewhat due to increased cardiac output	Increases greatly due to vasoconstriction
Airways	Dilates	Dilates slightly
Reticular formation of brain	Activates	Activates
Liver	Promotes breakdown of glycogen to glucose, increasing blood sugar concentration	Produces little effect on blood sugar concentration
Metabolic rate	Increases	Increases

Hormones of the Adrenal Cortex

The cells of the adrenal cortex produce more than thirty different steroids, including several hormones. Unlike the adrenal medullary hormones, without which a person can still survive, some adrenal cortical hormones are vital. Without them, a person usually dies within a week unless extensive electrolyte therapy is provided. The most important adrenal cortical hormones are aldosterone, cortisol, and certain sex hormones.

Aldosterone

Cells in the outer zone of the adrenal cortex synthesize **aldosterone** (al-dos′ter-ōn″). This hormone is a *mineralocorticoid* (min″er-al-o-kor′tĭ-koid) because it helps regulate the concentration of mineral electrolytes. More specifically, aldosterone causes the kidney to conserve sodium ions and excrete potassium ions. By conserving sodium ions, aldosterone stimulates water retention indirectly by osmosis, helping to maintain blood volume and blood pressure.

A decrease in the blood concentration of sodium ions or an increase in the blood concentration of potassium ions stimulates the cells that secrete aldosterone. The kidneys also indirectly stimulate aldosterone secretion if blood pressure falls (see chapter 17, p. 480).

Cortisol

Cortisol (kor′tĭ-sol) (hydrocortisone) is a *glucocorticoid* (gloo″ko-kor′tĭ-koid), which means it affects glucose metabolism. It is produced in the middle zone of the adrenal cortex and, like aldosterone, is a steroid. Cortisol also influences protein and fat metabolism.

The more important actions of cortisol include:

1. Inhibition of protein synthesis in tissues, increasing the blood concentration of amino acids.
2. Promotion of fatty acid release from adipose tissue, increasing the utilization of fatty acids as an energy source and decreasing the use of glucose.
3. Stimulation of liver cells to synthesize glucose from noncarbohydrates, such as circulating amino acids and glycerol, increasing the blood glucose concentration.

These actions of cortisol help keep blood glucose concentration within the normal range between meals, because a few hours without food can exhaust the supply of liver glycogen, a major source of glucose.

Negative feedback controls cortisol release. This is much like control of thyroid hormones, involving the hypothalamus, anterior pituitary gland, and adrenal cortex. The hypothalamus secretes corticotropin-releasing hormone (CRH) into the hypophyseal portal veins, which carry CRH to the anterior pituitary, stimulating it to secrete ACTH. In turn, ACTH stimulates the adrenal cortex to release cortisol. Cortisol inhibits the release of CRH and ACTH, and as concentrations of these fall, cortisol production drops (fig. 11.14).

The set point of the feedback mechanism controlling cortisol secretion changes from time to time, altering hormone output to meet the demands of changing conditions. For example, under stress—as from injury, disease, extreme temperature, or emotional upset—nerve impulses send the brain information concerning the stressful condition. In response, brain centers signal the hypothalamus to release more CRH, elevating cortisol concentration until the stress subsides (fig. 11.14).

Adrenal Sex Hormones

Cells in the inner zone of the adrenal cortex produce sex hormones. These hormones are male types (adrenal androgens), but some are converted to female hormones (estrogens) in the skin, liver, and adipose tissue. Adrenal sex hormones may supplement the supply of sex hormones from the gonads and stimulate early development of reproductive organs. Table 11.6 summarizes the characteristics of the adrenal cortical hormones.

> Hyposecretion of adrenal cortical hormones leads to *Addison disease*, a condition characterized by decreased blood sodium, increased blood potassium, low blood glucose concentration (hypoglycemia), dehydration, low blood pressure, and increased skin pigmentation. Without treatment with mineralocorticoids and glucocorticoids, Addison disease can be lethal in days because of severe disturbances in electrolyte balance.
>
> Hypersecretion of adrenal cortical hormones, which may be associated with an adrenal tumor or with the anterior pituitary oversecreting ACTH, causes *Cushing syndrome*. This condition alters carbohydrate and protein metabolism and electrolyte balance. For example, when mineralocorticoids and glucocorticoids are overproduced, blood glucose concentration remains high, depleting tissue protein. Also, too much sodium is retained, increasing tissue fluids, and the skin becomes puffy. At the same time, increase in adrenal sex hormone production may cause masculinizing effects in a female, such as beard growth and deepening of the voice.

Practice

33. Name the most important hormones of the adrenal cortex.
34. What is the function of aldosterone?
35. What actions does cortisol produce?
36. How are the blood concentrations of aldosterone and cortisol regulated?

11.9 PANCREAS

The **pancreas** (pan′kre-as) consists of two major types of secretory tissues. This organization reflects the pancreas's dual function as an exocrine gland that secretes

Figure 11.14

Negative feedback regulates cortisol secretion, similar to the regulation of thyroid hormone secretion (see fig. 11.9, p. 299). (⊕ = stimulation; ⊖ = inhibition)

Table 11.6	Hormones of the Adrenal Cortex AP|R	
Hormone	**Action**	**Factor Regulating Secretion**
Aldosterone	Helps regulate concentration of extracellular electrolytes by conserving sodium ions and excreting potassium ions	Electrolyte concentrations in body fluids
Cortisol	Decreases protein synthesis, increases fatty acid release, and stimulates glucose synthesis from noncarbohydrates	Corticotropin-releasing hormone from hypothalamus and adrenocorticotropic hormone from anterior pituitary
Adrenal androgens	Supplement sex hormones from the gonads; may be converted to estrogens in females	

digestive juice and an endocrine gland that releases hormones (fig. 11.15 and reference plate 6, p. 28).

Structure of the Gland

The pancreas is an elongated, somewhat flattened organ posterior to the stomach and behind the parietal peritoneum. A duct joins the pancreas to the duodenum (the first section of the small intestine) and transports pancreatic digestive juice to the intestine. The dual nature of the pancreas begins in the embryo. First, ducts form whose walls harbor progeni-

tor cells (see fig. 3.23, p. 72). Some of the progenitor cells divide to yield daughter cells that specialize as exocrine cells, and others divide to yield cells that differentiate into endocrine cells. The two functions are elaborated as the gland develops further.

The endocrine part of the pancreas consists of groups of cells that are closely associated with blood vessels. These groups form "islands" of cells called *pancreatic islets* (islets of Langerhans) (figs. 11.15 and 11.16). The pancreatic islets include two distinct types of cells—alpha cells, which secrete the hormone glucagon, and beta cells, which secrete the hormone insulin.

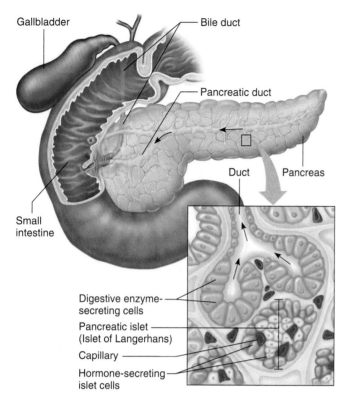

Figure 11.15

The hormone-secreting cells of the pancreas are grouped in clusters, or islets, that are in close proximity to blood vessels. Other pancreatic cells secrete digestive enzymes into ducts.

Q: *What term describes the secretions that enter the pancreatic duct?*

Answer can be found in Appendix E on page 568.

Figure 11.16

Light micrograph of a pancreatic islet (250×).

Clinical Application 11.1 discusses diabetes mellitus, a disorder that affects the beta cells. Chapter 15 (p. 413) discusses the digestive functions of the pancreas.

Hormones of the Pancreatic Islets

Glucagon (gloo'kah-gon) stimulates the liver to break down glycogen and convert certain noncarbohydrates, such as amino acids, into glucose, raising blood sugar concentration. Glucagon much more effectively elevates blood glucose than does epinephrine.

A negative feedback system regulates glucagon secretion. Low blood glucose concentration stimulates alpha cells to release glucagon. When blood glucose concentration rises, glucagon secretion falls. This control prevents hypoglycemia when glucose concentration is relatively low, such as between meals, or when glucose is used rapidly, such as during periods of exercise.

The main effect of **insulin** (in'su-lin) is exactly opposite that of glucagon. Insulin stimulates the liver to form glycogen from glucose and inhibits conversion of noncarbohydrates into glucose. Insulin also has the special effect of promoting facilitated diffusion (see chapter 3, p. 63) of glucose across cell membranes that

have insulin receptors, such as those of cardiac muscle, adipose tissue, and resting skeletal muscle. (Glucose uptake by exercising skeletal muscle does not require insulin.) These actions of insulin decrease blood glucose concentration. In addition, insulin secretion promotes transport of amino acids into cells, increases the rate of protein synthesis, and stimulates adipose cells to synthesize and store fat.

A negative feedback system sensitive to blood glucose concentration regulates insulin secretion. When blood glucose concentration is high, such as after a meal, beta cells release insulin. Insulin helps prevent too high a blood glucose concentration by promoting glycogen formation in the liver and entrance of glucose into adipose and muscle cells. When glucose concentration falls, such as between meals or during the night, insulin secretion decreases.

As insulin output decreases, less and less glucose enters adipose and muscle cells. Cells that lack insulin receptors, such as nerve cells, can then use the glucose that remains in the blood. At the same time that insulin is decreasing, glucagon secretion is increasing. Therefore, insulin and glucagon function in coordination to maintain a relatively stable blood glucose concentration, despite great variation in the amount of carbohydrates a person eats (fig. 11.17).

Nerve cells, including those of the brain, obtain glucose by a facilitated diffusion mechanism that does not require insulin, but rather depends only on the blood glucose concentration. For this reason, nerve cells are particularly sensitive to changes in blood glucose con-

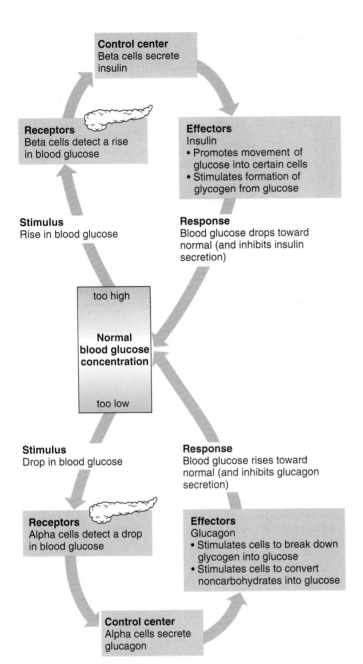

Figure 11.17 AP|R

Insulin and glucagon function together to help maintain a relatively stable blood glucose concentration. Negative feedback responding to blood glucose concentration controls the levels of both hormones.

centration, and conditions that cause such changes—oversecretion of insulin leading to decreased blood glucose, for example—are likely to alter brain functions.

Cancer cells that develop from nonendocrine tissues sometimes inappropriately synthesize and secrete great amounts of peptide hormones or peptide hormonelike chemicals. For example, in endocrine paraneoplastic syndrome, a person with a non-endocrine cancer overproduces ADH, ACTH, a PTH-like substance, or an insulin-like substance.

Practice

37. What is the endocrine portion of the pancreas called?

38. What is the function of glucagon?

39. What is the function of insulin?

40. How are glucagon and insulin secretion controlled?

41. Why are nerve cells particularly sensitive to changes in blood glucose concentration?

11.10 OTHER ENDOCRINE GLANDS

Other glands that produce hormones and thus are parts of the endocrine system include the pineal gland, the thymus, reproductive organs, and certain glands of the digestive tract, heart, and kidneys.

The **pineal gland** (pin′e-al gland) is a small structure located deep between the cerebral hemispheres, where it attaches to the upper part of the thalamus near the roof of the third ventricle (see fig. 11.2). The pineal gland secretes the hormone **melatonin** (mel″ah-to′nin) in response to light conditions outside the body. Nerve impulses originating in the retinas of the eyes send this information to the pineal gland. In the dark, nerve impulses from the eyes decrease, and melatonin secretion increases.

Melatonin acts on certain brain regions that function as a "biological clock," and may thereby help to regulate **circadian rhythms** (ser″kah-de′an rithmz), which are patterns of repeated activity associated with the environmental cycles of day and night. The changing levels of melatonin throughout the 24-hour day may enable the body to distinguish day from night. Circadian rhythms include the sleep–wake rhythm and seasonal cycles of fertility in many mammals. Clinical Application 11.2 discusses biological rhythms.

The fact that melatonin secretion responds to day length explains why traveling across several time zones produces the temporary insomnia of jet lag. Melatonin supplements are advertised as preventing jet lag, shift work disorder, and other sleeping disorders, but clinical trials for these applications are still in progress.

The mechanism of melatonin action is poorly understood, but the hormone inhibits the secretion of gonadotropins from the anterior pituitary and may help regulate the female reproductive cycle. It may also control the onset of puberty.

The **thymus** (thi′mus), which lies in the mediastinum posterior to the sternum and between the lungs, is relatively large in young children but shrinks with age (see

Clinical Application 11.1

Diabetes Mellitus

Diabetes mellitus is a metabolic disease that arises from lack of insulin or inability of cells to recognize it. Before insulin was isolated and its mechanism understood in 1921, a diagnosis of type 1 diabetes mellitus meant that a child had, at best, three years to live. By 1922, children began to be treated with insulin and they rapidly recovered (figure 11A). As people with diabetes began living longer by receiving insulin, the disease's long-term effects on other organs—such as the eyes, heart, kidneys, and peripheral nerves—became noticeable.

Insulin deficiency disturbs carbohydrate, protein, and fat metabolism. Because insulin helps glucose cross some cell membranes, movement of glucose into adipose and resting skeletal muscle cells becomes impaired in diabetes. At the same time, formation of glycogen, which is a long chain of glucose molecules, declines. As a result, blood sugar concentration rises (hyperglycemia). When it reaches a certain level, the kidneys begin to excrete the excess. Glucose in the urine (glycosuria) raises the urine's osmotic pressure, and too much water is excreted. Excess urine output causes dehydration and extreme thirst (polydipsia).

Diabetes mellitus also hampers protein and fat synthesis. Glucose-starved cells increasingly use proteins for energy, and as a result, tissues waste away as weight drops, hunger increases, exhaustion becomes overwhelming, children stop growing, and wounds do not heal. Changes in fat metabolism cause fatty acids and ketone bodies to accumulate in

Figure 11A

Before and after insulin treatment: The boy in his mother's arms is three years old but weighs only 15 pounds because of type 1 diabetes mellitus. The inset shows the same child after just two months of receiving insulin—his weight had doubled.

---

fig. 11.2). This gland secretes a group of hormones called **thymosins** (thi'mo-sinz) that affect the production and differentiation of certain white blood cells (lymphocytes). In this way, the thymus plays an important role in immunity, discussed in chapter 14 (p. 382).

The reproductive organs that secrete important hormones include the testes, which produce testosterone; the ovaries, which produce estrogens and progesterone; and the **placenta** (plah-sen'tah), which produces estrogens, progesterone, and gonadotropin. These glands and their secretions are discussed in chapter 19 (pp. 513 and 522) and chapter 20 (p. 542).

The digestive glands that secrete hormones are associated with the linings of the stomach and small intestine. Chapter 15 (pp. 412, 413 and 418) describes these structures and their secretions.

Other organs outside of the endocrine system produce hormones. The heart, for example, secretes *atrial natriuretic peptide,* a hormone that stimulates urinary sodium excretion (see chapter 17, p. 476). The kidneys secrete a red blood cell growth hormone called *erythropoietin* (see chapter 12, p. 321).

Practice

42. Where is the pineal gland located?
43. What is the function of the pineal gland?
44. Where is the thymus located?
45. Which reproductive organs secrete hormones?
46. Which other organs secrete hormones?

the blood, which lowers pH (acidosis). Dehydration and acidosis may harm brain cells, causing disorientation, coma, and, eventually, death.

There are two common forms of diabetes mellitus. These are type 1 (insulin-dependent or juvenile diabetes) and type 2 (non-insulin-dependent or maturity-onset diabetes).

Type 1 Diabetes Mellitus

Type 1 diabetes mellitus usually appears before age twenty. It is an autoimmune disease: the immune system destroys the beta cells of the pancreas (see chapter 14, pp. 394–395).

People with type 1 diabetes must carefully monitor their blood glucose levels. They do this in two ways. Every three or four months, a laboratory test checks the levels of hemoglobin molecules in the blood that bind glucose. This measurement is called "A1c," and should be between 6% and 7%. It provides a view of blood glucose level over the preceding months. The second type of test is called self-monitoring of blood glucose. A person uses a test kit to draw a drop of blood, applies it to a test strip, then uses a meter to read the concentration of glucose in the blood (in milligrams per deciliter). Normal plasma levels of glucose should range from 90 to 130 mg/dL before meals and less than 180 mg/dL one to two hours after meals. Most people with type 1 diabetes check their glucose this way two to four times a day.

Treatment for type 1 diabetes is still to give insulin, but delivery has improved so that treatment better mimics normal pancreatic function. Before 1978, for example, people with diabetes used insulin from pigs. Then genetically modified bacteria began to supply the human version of the hormone, to which allergy is far less likely. People with type 1 diabetes typically inject insulin several times a day, or receive the hormone from an implanted insulin pump. Delivery of insulin through nasal sprays or skin patches is still experimental. Replacing a diabetic pancreas with a healthy transplanted organ is too difficult to be practical—the surgery is complex, the supply of organs very limited, and immune rejection difficult to prevent or control. Instead, much research has focused on islet transplantation, which can decrease the need for insulin supplements for a few years.

Type 2 Diabetes Mellitus

About 85–90% of people with diabetes mellitus have type 2, in which the beta cells produce insulin but body cells lose the ability to recognize it. The condition usually develops gradually after age forty and has milder symptoms than type 1 diabetes. Most affected individuals are overweight when symptoms begin. Treatment includes controlling the diet, exercising, and maintaining a desirable body weight. Several oral drugs can help control glucose levels, which can delay the onset of diabetes-related complications.

People with either type of diabetes must monitor their blood glucose level at least daily, and do what they can to regulate it, to forestall complications, which include coronary artery disease, peripheral nerve damage, and retinal damage. Evidence suggests that these complications may begin even before blood glucose level indicates disease. The American Diabetes Association now recognizes "pre-diabetes" as blood glucose levels above the normal range but not yet indicative of type 2 diabetes. About 20 million people in the United States between the ages of forty and seventy-four fall into this category.

11.11 STRESS AND HEALTH

Survival depends on the maintenance of homeostasis. Therefore, factors that change the body's internal environment can threaten life. When the body senses danger, nerve impulses to the hypothalamus trigger physiological responses that preserve homeostasis. These responses include increased activity in the sympathetic division of the autonomic nervous system and increased secretion of adrenal and other hormones. A factor that can stimulate such a response is called a *stressor,* and the condition it produces in the body is called **stress.**

Types of Stress

Stressors include physical factors, such as exposure to extreme heat or cold, decreased oxygen concentration, infections, injuries, prolonged heavy exercise, and loud sounds. Stressors also include psychological factors, such as thoughts about real or imagined dangers, personal losses, and unpleasant social interactions. Feelings of anger, fear, grief, anxiety, depression, and guilt can also produce psychological stress. Sometimes, even pleasant stimuli, such as friendly social contact, feelings of joy and happiness, or sexual arousal, may be stressful.

Responses to Stress

Physiological responses to stress consist of reactions called the *stress response* or *general adaptation syndrome,* which is under hypothalamic control. These reactions proceed through two stages: the immediate "alarm" stage and the long-term "resistance" stage. Initially, the hypothalamus activates mechanisms that prepare the body for "fight or flight." These responses include raising

Clinical Application 11.2

Biological Rhythms

Biological rhythms are changes that systematically recur in organisms. In complex animals, they include the daily ebb and flow of biochemical levels in blood, reproductive cycles, and migration schedules. The period of any rhythm is the duration of one complete cycle. The frequency of a rhythm is the number of cycles per time unit. The study of biological rhythms is called *chronobiology.*

Three common types of rhythms in humans are ultradian, infradian, and circadian rhythms. *Ultradian rhythms* have periods shorter than 24 hours and include the cardiac cycle and the breathing cycle. Periods of *infradian rhythms,* such as the reproductive cycle, are longer than 24 hours. Periods of *circadian rhythms,* such as the sleep–wake cycle, are approximately 24 hours.

Both external (exogenous) and internal (endogenous) factors regulate human biological rhythms. Exogenous factors are environmental components, such as daily temperature changes and the light–dark cycle. Endogenous factors include "clock" genes. Many members of an extended family in Utah, for example, have "advanced sleep phase syndrome" due to a mutation in a gene called "period." The effect is striking—they promptly fall asleep at 7:30 each night and awaken suddenly at 4:30 A.M. The same gene alters sleep patterns in fruit flies and golden hamsters, used in chronobiology research. The mutation in all three species disrupts a signal that synchronizes the sleep–wake cycle to daily sunrise and sunset.

The cells that respond to environmental light and dark signals are located in a part of the brain called the suprachiasmatic

nuclei. Understanding how clock genes function may lead to new treatments for jet lag, insomnia, a form of advanced sleep phase syndrome that is common among older individuals, and people who work shifts longer than 12 hours, such as certain health-care professionals.

The sleep–wake cycle is the most obvious circadian rhythm in humans. It is largely controlled by the pattern of daylight and night, but under laboratory conditions of constant light or dark, the human body eventually follows an approximately 25-hour cycle. Other circadian rhythms in humans affect body temperature, cardiovascular functioning, and hormone secretion.

Body temperature is mostly endogenously regulated, but light exposure and physical activity help keep this rhythm on a 24- rather than 25-hour cycle. Body temperature is usually lowest between 4 and 6 A.M., and then increases and peaks between 5 and 11 P.M. It drops during the late evening hours and into the night.

Cardiovascular functioning is least efficient between 6 and 9 A.M.. Platelet cohesion, blood pressure, and pulse rate are typically highest 2 hours after awakening, which may explain why heart attacks and strokes are more likely to occur between 8 and 10 A.M. than at other times.

Plasma cortisol, for example, surges and peaks at about 6 A.M., and then gradually declines to its minimum level in late evening before increasing again in the early morning. Growth hormone secretion peaks during the night. Antidiuretic hormone secretion is greater at night, when it decreases urine formation.

---

blood concentrations of glucose, glycerol, and fatty acids; increasing heart rate, blood pressure, and breathing rate; dilating air passages; shunting blood from the skin and digestive organs to the skeletal muscles; and increasing epinephrine secretion from the adrenal medulla (fig. 11.18).

In the resistance response, the hypothalamus releases CRH, which stimulates the anterior pituitary to secrete ACTH, which increases cortisol secretion. Cortisol increases blood amino acid concentration, fatty acid release, and glucose formation from noncarbohydrates. Thus, while the alarm responses prepare the body for physical action to alleviate the stress, cortisol supplies cells with biochemicals required during stress (fig. 11.18).

Other hormones whose secretions increase with stress include glucagon, GH, and ADH. Glucagon and GH mobilize energy sources, such as glucose, glycerol, fatty acids, and amino acids. ADH stimulates the kid-

neys to retain water, which increases blood volume—particularly important if a person is bleeding or sweating heavily.

Increased cortisol secretion may be accompanied by a decrease in the number of certain white blood cells (lymphocytes), which lowers resistance to infectious diseases and some cancers. Also, excess cortisol production may raise the risk of developing high blood pressure, atherosclerosis, and gastrointestinal ulcers.

Practice

47. What is stress?
48. Distinguish between physical stress and psychological stress.
49. Describe the stress response.

Figure 11.18

Stress response. During stress, the hypothalamus helps prepare the body for "fight or flight" by triggering sympathetic impulses to various organs. It also stimulates epinephrine release, intensifying the sympathetic responses. The hypothalamus additionally secretes corticotropin-releasing hormone, which sets into motion more lasting responses to stress.

Summary Outline

11.1 Introduction (p. 292)

The endocrine and nervous systems maintain homeostasis.

1. The endocrine system is a network of glands that secrete hormones, which travel in the bloodstream and affect the functioning of target cells.
2. Paracrine secretions act locally, and autocrine secretions act on the cells that produce them.
3. Exocrine glands secrete through tubes or ducts.

11.2 General Characteristics of the Endocrine System (p. 292)

Like the nervous system, the endocrine system exerts precise effects in helping regulate metabolic processes.

11.3 Hormone Action (p. 293)

Endocrine glands secrete hormones that affect target cells with specific receptors. Hormones are very potent.

1. Chemically, hormones are steroids, amines, peptides, proteins, or glycoproteins.
2. Steroid hormones
   a. Steroid hormones enter a target cell and bind receptors, forming complexes in the nucleus.

b. These complexes activate specific genes, so that specific proteins are synthesized.
3. Nonsteroid hormones
   a. Nonsteroid hormones bind receptors in the target cell membrane.
   b. The hormone-receptor complex signals a G protein to stimulate a membrane protein, such as adenylate cyclase, to induce formation of second messenger molecules.
   c. A second messenger, such as cyclic adenosine monophosphate (cAMP), diacylglycerol (DAG), or inositol triphosphate (IP_3), activates protein kinases.
   d. Protein kinases activate protein substrate molecules, which in turn change a cellular process.
4. Prostaglandins
   a. Prostaglandins act on the cells of the organs that produce them.
   b. Prostaglandins are present in small amounts and have powerful hormonelike effects.

11.4 Control of Hormonal Secretions (p. 296)

The concentration of each hormone in body fluids is regulated.

1. Some endocrine glands secrete hormones in response to releasing hormones that the hypothalamus secretes.
2. Other glands secrete their hormones in response to nerve impulses.

Endocrine System

Integumentary System

Melanocytes produce skin pigment in response to hormonal stimulation.

Lymphatic System

Hormones stimulate lymphocyte production.

Skeletal System

Hormones act on bones to control calcium balance.

Digestive System

Hormones help control digestive system activity.

Muscular System

Hormones help increase blood flow to exercising muscles.

Respiratory System

Decreased oxygen causes hormonal stimulation of red blood cell production; red blood cells transport oxygen and carbon dioxide.

Nervous System

Neurons control the secretions of the anterior and posterior pituitary glands and the adrenal medulla.

Urinary System

Hormones act on the kidneys to help control water and electrolyte balance.

Cardiovascular System

Hormones are carried in the bloodstream; some have direct actions on the heart and blood vessels.

Reproductive System

Sex hormones play a major role in development of secondary sex characteristics, egg, and sperm.

Glands secrete hormones that have a variety of effects on cells, tissues, organs, and organ systems.

3. Some glands respond to levels of a substance in the bloodstream.
4. Negative feedback guides these control mechanisms.
    a. In a negative feedback system, a gland senses the concentration of a substance it regulates.
    b. When the concentration of the regulated substance reaches a certain point, it inhibits the gland.
    c. As the gland secretes less hormone, the amount of the controlled substance also decreases.
    d. Negative feedback systems maintain relatively stable hormone concentrations.

11.5 Pituitary Gland (p. 297)

The pituitary gland has an anterior lobe and a posterior lobe. The hypothalamus controls most pituitary secretions.

1. Anterior pituitary hormones
    a. The anterior pituitary secretes growth hormone (GH), prolactin (PRL), thyroid-stimulating hormone (TSH), adrenocorticotropic hormone (ACTH), follicle-stimulating hormone (FSH), and luteinizing hormone (LH).
    b. Growth hormone
        (1) GH stimulates cells to enlarge and divide more frequently.
        (2) GH-releasing hormone and GH release-inhibiting hormone from the hypothalamus control GH secretion.
    c. PRL stimulates and sustains milk production.
    d. Thyroid-stimulating hormone
        (1) TSH controls secretion of hormones from the thyroid gland.
        (2) The hypothalamus secretes thyrotropin-releasing hormone (TRH), which regulates TSH secretion.
    e. Adrenocorticotropic hormone
        (1) ACTH controls secretion of hormones from the adrenal cortex.
        (2) The hypothalamus secretes corticotropin-releasing hormone (CRH), which regulates ACTH secretion.
    f. FSH and LH are gonadotropins.
2. Posterior pituitary hormones
    a. The posterior lobe of the pituitary gland consists largely of neuroglia and nerve fibers.
    b. The hypothalamus produces the hormones of the posterior pituitary.
    c. Antidiuretic hormone (ADH)
        (1) ADH reduces the volume of water the kidneys excrete.
        (2) The hypothalamus regulates ADH secretion.
    d. Oxytocin (OT)
        (1) OT contracts muscles in the uterine wall.
        (2) OT also contracts cells that secrete and eject milk.

11.6 Thyroid Gland (p. 301)

The thyroid gland in the neck consists of two lobes.

1. Structure of the gland
    a. The thyroid gland consists of many follicles.
    b. The follicles are fluid-filled and store hormones.
2. Thyroid hormones
    a. Thyroxine and triiodothyronine increase the metabolic rate of cells, enhance protein synthesis, and stimulate lipid utilization.
    b. Calcitonin decreases blood calcium level and increases blood phosphate ion concentration.

11.7 Parathyroid Glands (p. 303)

The parathyroid glands are on the posterior surface of the thyroid gland.

1. Each parathyroid gland consists of secretory cells that are well supplied with capillaries.

2. Parathyroid hormone (PTH)
    a. PTH increases blood calcium level and decreases blood phosphate ion concentration.
    b. A negative feedback mechanism operates between the parathyroid glands and the blood.

11.8 Adrenal Glands (p. 304)

The adrenal glands are located atop the kidneys.

1. Structure of the glands
    a. Each gland consists of an adrenal medulla and an adrenal cortex.
    b. These parts are functionally distinct, and secrete different hormones.
2. Hormones of the adrenal medulla
    a. The adrenal medulla secretes epinephrine and norepinephrine, which have similar effects.
    b. Sympathetic impulses stimulate secretion of these hormones.
3. Hormones of the adrenal cortex
    a. The adrenal cortex produces several steroid hormones.
    b. Aldosterone is a mineralocorticoid that causes the kidneys to conserve sodium ions and water and to excrete potassium ions.
    c. Cortisol is a glucocorticoid that affects carbohydrate, protein, and fat metabolism.
    d. Adrenal sex hormones
        (1) These hormones are of the male type but may be converted to female hormones.
        (2) They may supplement the sex hormones the gonads produce.

11.9 Pancreas (p. 306)

The pancreas secretes digestive juices as well as hormones.

1. Structure of the gland
    a. The pancreas is attached to the small intestine.
    b. The pancreatic islets secrete glucagon and insulin.
2. Hormones of the pancreatic islets
    a. Glucagon stimulates the liver to produce glucose from glycogen and noncarbohydrates.
    b. Insulin moves glucose across some cell membranes, stimulates glucose and fat storage, and promotes protein synthesis.
    c. Nerve cells do not require insulin to obtain glucose.

11.10 Other Endocrine Glands (p. 309)

1. Pineal gland
    a. The pineal gland attaches to the thalamus.
    b. It secretes melatonin in response to varying light conditions.
    c. Melatonin may help regulate the female reproductive cycle by inhibiting gonadotropin secretion from the anterior pituitary.
2. Thymus
    a. The thymus lies behind the sternum and between the lungs.
    b. It secretes thymosins, which affect the production of certain lymphocytes that function in immunity.
3. Reproductive organs
    a. The testes secrete testosterone.
    b. The ovaries secrete estrogens and progesterone.
    c. The placenta secretes estrogens, progesterone, and gonadotropin.
4. Digestive glands
    Certain glands of the stomach and small intestine secrete hormones.
5. Other hormone-producing organs
    Other organs, such as the heart and the kidneys, also produce hormones.

11.11 Stress and Health (p. 311)

Stress occurs when the body responds to stressors that threaten the maintenance of homeostasis. Stress responses include increased activity of the sympathetic nervous system and increased secretion of adrenal hormones.

1. Types of stress
   a. Physical stress results from environmental factors that are harmful or potentially harmful to tissues.
   b. Psychological stress results from thoughts about real or imagined dangers.
2. Responses to stress
   a. Responses to stress maintain homeostasis.
   b. The hypothalamus controls the stress response.

Chapter Assessments

11.1 Introduction

1. Define *hormone* and *target cell*. (p. 292)

2. Contrast endocrine glands and exocrine glands. (p. 292)

11.2 General Characteristics of the Endocrine System

3. Compare and contrast the nervous and endocrine systems. (p. 292)

4. Explain the specificity of a hormone for its target cell. (p. 293)

5. Functions of hormones include which of the following? (p. 293)
   a. Control rates of certain chemical reactions
   b. Transport substances across cell membranes
   c. Help regulate water and electrolyte balances
   d. Play a role in reproduction
   e. All of the above

11.3 Hormone Action

6. List the steps of steroid hormone action. (p. 294)

7. List the steps in the action of most nonsteroid hormones. (p. 294)

8. Explain how prostaglandins are similar to hormones and how they are different. (p. 296)

11.4 Control of Hormonal Secretions

9. Draw diagrams of the three mechanisms by which hormone secretion is controlled, including negative feedback. (p. 296)

11.5 Pituitary Gland

10. Describe the location and structure of the pituitary gland. (p. 297)

11. Explain the two ways in which the brain controls pituitary gland activity. (p. 297)

12. Releasing hormones come from which one of the following? (p. 297)
   a. Thyroid gland
   b. Anterior pituitary gland
   c. Posterior pituitary gland
   d. Hypothalamus
   e. Parathyroid gland

13. List the hormones secreted by the anterior pituitary. (p. 297)

14. Match the following hormones with their actions. More than one hormone can correspond to the same function. (pp. 298–300)

   (1) growth hormone   A. Milk production
   (2) thyroid-stimulating hormone   B. Cell division
   (3) prolactin   C. Metabolic rate
   (4) adrenocorticotropic hormone   D. Exerts action on gonads
   (5) follicle-stimulating hormone   E. Controls secretion of adrenal cortex hormones
   (6) luteinizing hormone

15. Describe the control of growth hormone secretion. (p. 298)

16. Prolactin does which of the following? (p. 298)
   a. Stimulates breast milk secretion
   b. Stimulates breast milk production
   c. Inhibits breast milk secretion
   d. Inhibits breast milk production
   e. None of the above

17. Diagram the control of thyroid hormone secretion. (p. 298)

18. Describe the anatomical differences between the anterior and posterior lobes of the pituitary gland. (p. 300)

19. Describe the functions of the posterior pituitary hormones. (p. 300)

20. Under which of the following conditions would you expect an increase in antidiuretic hormone secretion? (p. 300)
   a. An individual ingests excess water.
   b. The posterior pituitary is removed from an individual because of a tumor.
   c. An individual is rescued after three days in the desert without food or water.
   d. An accident has destroyed part of an individual's hypothalamus in the area where the ADH-producing neurosecretory cells are located.
   e. An individual with normal kidneys is producing a large volume of urine.

11.6 Thyroid Gland

21. Describe the location and structure of the thyroid gland. (p. 301)

22. Match the hormones from the thyroid gland with their descriptions. (p. 301)

   (1) thyroxine   A. Most potent at controlling metabolism
   (2) triiodothyronine   B. Regulates blood calcium
   (3) calcitonin   C. Has four iodine atoms

23. List the source of control for each thyroid hormone. (p. 301)

11.7 Parathyroid Gland

24. Describe the location and structure of the parathyroid glands. (p. 303)

25. Explain the general function of parathyroid hormone. (p. 303)

26. Draw a diagram that shows how the secretion of parathyroid hormone is regulated. (p. 303)

11.8 Adrenal Glands

27. Distinguish between the adrenal medulla and the adrenal cortex. (p. 304)

28. Match the adrenal hormones with their source and actions. (pp. 304–306)

(1) cortisol   A. Cortex; sodium retention

(2) aldosterone   B. Cortex; fatty acid release

(3) epinephrine   C. Medulla; fight-or-flight response

29. Draw a diagram illustrating the regulation of cortisol secretion. (p. 306)

11.9 Pancreas

30. Describe the location and structure of the pancreas. (p. 307)

31. List the hormones secreted by the pancreatic islets, the type of cell that secretes each, and the actions of these hormones. (p. 307)

32. Draw a diagram that shows how the secretion of pancreatic hormones is regulated. (p. 308)

11.10 Other Endocrine Glands

33. Describe the location and general function of the pineal gland. (p. 309)

34. Describe the location and general function of the thymus. (p. 309)

35. Name five additional hormone-secreting organs. (p. 310)

11.11 Stress and Health

36. Define *stress*. (p. 311)

37. List the similarities and differences between the short-term alarm stage of stress and the long-term resistance stage. (p. 311)

Integrative Assessments/Critical Thinking

OUTCOMES 2.2, 11.3, 11.4, 11.5, 11.6

1. When reactor 4 at the Chernobyl Nuclear Power Station in Ukraine exploded at 1:23 P.M. on April 26, 1986, a great plume of radioactive isotopes erupted into the air and spread for thousands of miles. Most of the isotopes emitted immediately following the blast were of the element iodine. Which of the glands of the endocrine system would be most seriously—and immediately—affected by the blast, and how do you think this would become evident in the nearby population?

OUTCOMES 7.3, 11.4, 11.5

2. Growth hormone is administered to people who have pituitary dwarfism. Parents wanting their normal children to be taller have requested the treatment for them. Do you think this is a wise request? Why or why not?

OUTCOMES 11.4, 11.5

3. What hormone supplements would an adult whose anterior pituitary has been removed require?

OUTCOMES 11.4, 11.5, 11.6, 11.11

4. How might a patient with hyperthyroidism modify lifestyle to minimize the drain on body energy resources?

OUTCOMES 11.4, 11.8

5. The adrenal cortex of a patient who has lost a large volume of blood will increase secretion of aldosterone. What effect will this increased secretion have on the patient's blood concentrations of sodium and potassium ions?

OUTCOMES 11.4, 11.9

6. Why might oversecretion of insulin actually reduce glucose uptake by nerve cells?

WEB CONNECTIONS

Visit the text website at **www.mhhe.com/shieress11** for additional quizzes, interactive learning exercises, and more.

APR

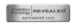

Anatomy & Physiology REVEALED includes cadaver photos that allow you to peel away layers of the human body to reveal structures beneath the surface. This program also includes animations, radiologic imaging, audio pronunciation, and practice quizzing. To learn more visit **www.aprevealed.com**.

12

Blood

Health-care workers wear personal protective equipment to shield themselves from body fluids containing disease-causing viruses.

Universal precautions. Blood can consist of more than cells, nutrients, proteins, and water—a single drop from an infected individual can harbor billions of viruses. In the wake of the AIDS epidemic, in 1988 the U.S. Centers for Disease Control and Prevention (CDC) devised "universal precautions," which are specific measures that health-care workers should take to prevent transmission of bloodborne infectious agents in the workplace. The CDC singled out HIV and the hepatitis B virus. The guidelines grew out of earlier suggestions for handling patients suspected to have been exposed to viruses. The term *universal* refers to the assumption that *any* patient may have been exposed to a pathogen that can be transmitted in a body fluid.

Attention to safety in the health-care setting can prevent transmission of infectious diseases. The World Health Organization estimates that 4–7% of new infections worldwide are transmitted via unsafe injections. Specific recommendations include:

- Use of personal protective equipment, such as gloves, goggles, and masks
- Engineering controls, such as fume hoods and sharps containers
- Work-practice controls, such as enforcing handwashing before and after performing procedures

Universal precautions were designed for, and work well in, preventing transmission of viral illnesses in settings that are already relatively safe, such as clinics. This isn't the case for outbreaks, natural disasters, and combat zones. For example, several pediatric nurses who aided neighbors infected with the Marburg virus in the isolated town of Uige in Angola, South Africa, in 2005 died from this hemorrhagic fever along with hundreds of others.

Headache, fever, vomiting, and diarrhea begin three to nine days after exposure to the virus. Then the person bleeds from all body openings, internally, and under the skin. Plummeting blood pressure kills most infected individuals within a week, and anyone contacting their blood is in danger of infection. Infected individuals must be isolated and not touched, but the scourge spreads because many family members become infected while tending their loved ones.

In the 2005 outbreak, contaminated medical equipment caused the rapid and deadly spread of the infection. Untrained clinic workers reused needles, and some people reused needles and intravenous equipment in their homes. However, even universal precautions might not have contained this outbreak because the infected body fluids were so copious. Universal precautions are critical for containing outbreaks under less dire circumstances.

Learning Outcomes

After studying this chapter, you should be able to do the following:

12.1 Introduction

1. Describe the general characteristics of blood, and discuss its major functions. (p. 319)
2. Distinguish among the formed elements and liquid portion of blood. (p. 319)

12.2 Blood Cells

3. Explain the significance of red blood cell counts. (p. 321)
4. Summarize the control of red blood cell production. (p. 321)
5. Distinguish among the five types of white blood cells, and give the function(s) of each type. (p. 325)

12.3 Blood Plasma

6. Describe the functions of each of the major components of plasma. (p. 328)

12.4 Hemostasis

7. Define *hemostasis,* and explain the mechanisms that help achieve it. (p. 330)
8. Review the major steps in blood coagulation. (p. 331)

 Learn **Practice** **Assess**

12.5 Blood Groups and Transfusions

9. Explain blood typing and how it is used to avoid adverse reactions following blood transfusions. (p. 334)

10. Describe how blood reactions may occur between fetal and maternal tissues. (p. 337)

Module 9: Cardiovascular System

Aids to Understanding Words

(Appendix A on page 564 has a complete list of Aids to Understanding Words.)

agglutin- [to glue together] *agglutin*ation: Clumping of red blood cells.

bil- [bile] *bil*irubin: Pigment excreted in the bile.

embol- [stopper] *embol*ism: Obstruction of a blood vessel.

erythr- [red] *erythr*ocyte: Red blood cell.

hema- [blood] *hema*tocrit: Percentage of red blood cells in a given volume of blood.

hemo- [blood] *hemo*globin: Red pigment responsible for the color of blood.

leuko- [white] *leuko*cyte: White blood cell.

-osis [abnormal condition] leuko*cytosis*: Condition in which white blood cells are overproduced.

-poie [make, produce] erythro*poie*tin: Hormone that stimulates the production of red blood cells.

-stasis [halt] hemo*stasis*: Arrest of bleeding from damaged blood vessels.

thromb- [clot] *thromb*ocyte: Blood platelet involved in the formation of a blood clot.

12.1 INTRODUCTION

Blood signifies life, and for good reason—it has many vital functions. This complex mixture of cells, cell fragments, and dissolved biochemicals transports nutrients, oxygen, wastes, and hormones; helps maintain the stability of the interstitial fluid; and distributes heat. The blood, heart, and blood vessels form the cardiovascular system and link the body's internal and external environments.

Blood is a type of connective tissue whose cells are suspended in a liquid extracellular matrix. Blood is vital in transporting substances between body cells and the external environment, thereby promoting homeostasis.

Whole blood is slightly heavier and three to four times more viscous than water. Its cells, which form mostly in red bone marrow, include red blood cells that transport gases and white blood cells that fight disease. Blood also contains cellular fragments called blood platelets that help control blood loss. Together, the cells and platelets are termed "formed elements" of the blood, in contrast to the liquid portion, which is called **plasma** (plaz'mah) (fig. 12.1).

A blood sample is usually about 45% red blood cells by volume. This percentage is called the **hematocrit (HCT).** The white blood cells and platelets account for less than 1% of blood volume. The remaining blood sample, about 55%, is the plasma, a clear, straw-colored liquid. Plasma is a complex mixture of water, amino acids, proteins, carbohydrates, lipids, vitamins, hormones, electrolytes, and cellular wastes (fig. 12.2).

Blood volume varies with body size, changes in fluid and electrolyte concentrations, and the amount of adipose tissue. An average-size adult has a blood volume of about 5 liters (5.3 quarts).

> Men have more blood than women. On average, men have 5–6 liters (1.500 gallons), compared to 4–5 liters (0.875 gallons) for women.

Practice

1. What are the major components of blood?
2. What factors affect blood volume?

12.2 BLOOD CELLS

Red Blood Cells

Red blood cells, or **erythrocytes** (ĕ-rith'ro-sītz), are biconcave discs. This shape is an adaptation for transporting gases; it increases the surface area through which gases can diffuse (fig. 12.3). The red blood cell's shape also places the cell membrane closer to oxygen-carrying **hemoglobin** in the cell.

Each red blood cell is about one-third hemoglobin by volume. This protein imparts the color of blood. When hemoglobin binds oxygen, the resulting *oxyhemoglobin* is bright red, and when oxygen is released, the resulting *deoxyhemoglobin* is darker. Blood rich in deoxyhemoglobin may appear bluish when it is viewed through blood vessel walls.

> Prolonged oxygen deficiency (hypoxia) causes *cyanosis*, in which the skin and mucous membranes appear bluish due to an abnormally high blood concentration of deoxyhemoglobin. Exposure to low temperature may also cause cyanosis, by constricting superficial blood vessels. This slows blood flow, allowing removal of more oxygen than usual from the blood flowing through the vessels.

Red blood cells have nuclei during their early stages of development, but the cells extrude the nuclei as they mature, providing more space for hemoglobin. Because they lack nuclei, mature red blood cells cannot synthesize proteins or divide. Because they also lack mitochondria, red blood cells produce ATP

White blood cells

Red blood cells

Platelets

Peripheral Blood Smear

Plasma

"Buffy coat" (white blood cells and platelets)

Red blood cells

Centrifuged Blood Sample

Figure 12.1 AP|R

Blood consists of a liquid portion called plasma and a solid portion (the formed elements) that includes red blood cells, white blood cells, and platelets. (Note: When blood components are separated by centrifugation, the white blood cells and platelets form a thin layer, called the "buffy coat," between the plasma and the red blood cells, which accounts for about 1% of the total blood volume.) Blood cells and platelets can be seen under a light microscope when a blood sample is smeared onto a glass slide.

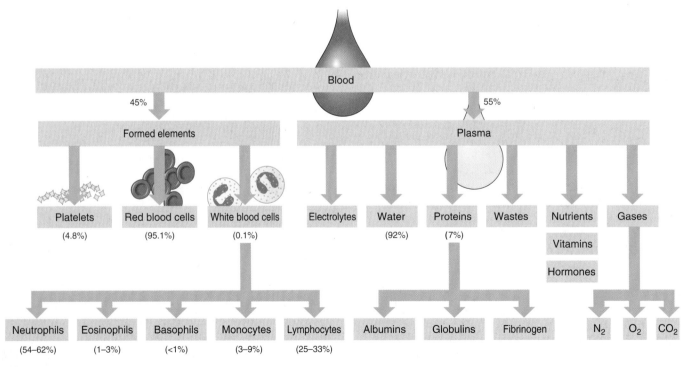

Blood

45% — Formed elements

55% — Plasma

Platelets (4.8%)   Red blood cells (95.1%)   White blood cells (0.1%)   Electrolytes   Water (92%)   Proteins (7%)   Wastes   Nutrients   Gases

Vitamins

Hormones

Neutrophils (54–62%)   Eosinophils (1–3%)   Basophils (<1%)   Monocytes (3–9%)   Lymphocytes (25–33%)   Albumins   Globulins   Fibrinogen N_2 O_2 CO_2

Figure 12.2

Blood composition. Blood is a complex mixture of formed elements in a liquid extracellular matrix, plasma.

Top view

7.5 micrometers

2.0 micrometers

Sectional view

(a)

(b)

Figure 12.3

Red blood cells. **(a)** The biconcave shape of a red blood cell makes it efficient at transporting gases. **(b)** Falsely colored scanning electron micrograph of human red blood cells (5,000×).

through glycolysis only and use none of the oxygen they carry.

Red Blood Cell Counts

The number of red blood cells in a microliter (μL or mcL or 1 mm^3) of blood is called the *red blood cell count* (*RBCC* or *RCC*). Although this number varies from time to time even in healthy individuals, the typical range for adult males is 4,600,000 to 6,200,000 cells per microliter, and that for adult females is 4,200,000 to 5,400,000 cells per microliter.

Because increasing the number of circulating red blood cells increases the blood's *oxygen-carrying*

capacity, changes in this number may affect health. For this reason, red blood cell counts are routinely consulted to help diagnose and evaluate the courses of various diseases.

Practice

3. Describe a red blood cell.

4. What is the function of hemoglobin?

5. How does a red blood cell change as it matures?

6. What is the typical red blood cell count for an adult male? For an adult female?

Red Blood Cell Production and Its Control

Recall from chapter 7 (p. 140) that red blood cell formation (erythropoiesis) initially occurs in the yolk sac, liver, and spleen. After birth, these cells are produced almost exclusively by tissue lining the spaces in bones, filled with red bone marrow. Figure 12.4 illustrates the stages in the formation of red blood cells from **hematopoietic** (he″mat-o-poi-et′ik) **stem cells** or *hemocytoblasts*.

The average life span of a red blood cell is 120 days. Many of these cells are removed from the circulation each day, and yet the number of cells in the circulating blood remains relatively stable. This observation suggests a homeostatic control of the rate of red blood cell production.

The combined surface area of all the red blood cells in the human body is roughly 2,000 times as great as the body's exterior surface.

The hormone **erythropoietin** (ĕ-rith″ro-poi′ĕ-tin) controls the rate of red blood cell formation through *negative feedback*. The kidneys, and to a lesser extent the liver, release erythropoietin in response to prolonged oxygen deficiency (fig. 12.5). At high altitudes, for example, where the amount of oxygen in the air is reduced, oxygen delivery to the tissues initially decreases. This drop in oxygen triggers the release of erythropoietin, which travels via the blood to the red bone marrow and stimulates red blood cell production.

After a few days, many newly formed red blood cells appear in the circulating blood. The increased rate of production continues until the number of erythrocytes in the circulation is sufficient to supply tissues with oxygen. When the availability of oxygen returns to normal, erythropoietin release decreases, and the rate of red blood cell production returns to normal as well. An excessive increase in red blood cells is *polycythemia*. This increases blood viscosity, slowing blood flow and impairing circulation.

Hematopoietic stem cell

Myeloid stem cell

Lymphoid stem cell

In red bone marrow

Proerythroblast

Megakaryoblast

Myeloblast

Monoblast

Lymphoblast
T cell
precursor

Lymphoblast
B cell
precursor

Erythroblast

Progranulocyte

Neutrophilic
myelocyte

Basophilic
myelocyte

Eosinophilic
myelocyte

Promonocyte

Prolymphocyte

Prolymphocyte

Normoblast

Megakaryocyte

Neutrophilic
band cell

Basophilic
band cell

Eosinophilic
band cell

In circulating blood

Reticulocyte

Thrombocytes
(platelets)

Neutrophil

Basophil

Eosinophil

Monocyte

T lymphocyte

B lymphocyte

Erythrocyte

Granulocytes

Agranulocytes

Activated in tissues
(some cells)

Macrophage

Plasma cell

(a)

(b)

Figure 12.4 AP|R

Blood cells. **(a)** Development of red blood cells, white blood cells,
and platelets from hematopoietic stem cells in bone marrow.
(b) Light micrograph of a hematopoietic stem cell (arrow) in red
bone marrow (500×).

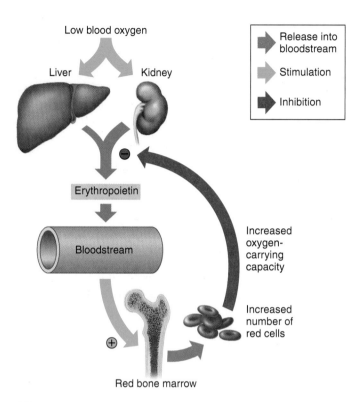

Figure 12.5

Low blood oxygen causes the kidneys and, to a lesser degree, the liver to release erythropoietin. Erythropoietin stimulates target cells in red bone marrow to increase the production of red blood cells that carry oxygen to tissues.

Dietary Factors Affecting Red Blood Cell Production

Availability of B-complex vitamins—*vitamin B₁₂* and *folic acid*—significantly influences red blood cell production. These vitamins are required for DNA synthesis, so they are necessary for the growth and division of cells. Cell division is rapid in blood-forming (hematopoietic) tissue, so this tissue is especially vulnerable to a deficiency of either of these vitamins. Hemoglobin synthesis and normal red blood cell production require iron. The small intestine absorbs iron slowly from food. The body reuses much of the iron released by the decomposition of hemoglobin from damaged red blood cells. Therefore, the diet need supply only small amounts of iron.

A deficiency of red blood cells, or a reduction in the amount of hemoglobin they contain, results in a condition called *anemia.* This reduces the oxygen-carrying capacity of the blood, and the affected person may appear pale and lack energy. A pregnant woman may become anemic if she doesn't eat iron-rich foods, because her blood volume increases due to fluid retention to accommodate the requirements of the fetus. This increased blood volume decreases the hematocrit.

In contrast to anemia, in an inherited disorder called hemochromatosis, the small intestine absorbs iron at ten times the normal rate. Iron builds up in organs, to toxic levels. Treatment is periodic blood removal, as often as every week following diagnosis. The blood is discarded.

In *sickle cell disease,* a single DNA base mutation changes one amino acid in the protein part of hemoglobin, causing hemoglobin to crystallize in a low-oxygen environment. This bends the red blood cells containing the abnormal hemoglobin into a sickle shape, which blocks circulation in small blood vessels, causing excruciating joint pain and damaging many organs.

Children with sickle cell disease are typically diagnosed at birth and receive antibiotics daily for years to prevent infection. Hospitalization for blood transfusions may be necessary for painful sickling "crises" of blocked circulation.

A drug, hydroxyurea, is used to activate production of a form of hemoglobin normally produced only in the fetus. The fetal hemoglobin slows sickling, which enables the red blood cells to reach the lungs—where fresh oxygen restores the cells' normal shapes. A bone marrow transplant or an umbilical cord stem cell transplant from a donor can completely cure sickle cell disease but have a 15% risk of fatality. The procedures are equally effective.

Practice

7. Where are red blood cells produced?

8. How is red blood cell production controlled?

9. Which vitamins are necessary for red blood cell production?

10. Why is iron required for the formation of red blood cells?

Destruction of Red Blood Cells

Red blood cells are elastic and flexible, and they readily bend as they pass through small blood vessels. With age, however, these cells become more fragile and may be damaged simply by passing through capillaries, particularly those in active muscles that must withstand contractile forces. **Macrophages** phagocytize and destroy damaged red blood cells, primarily in the liver and spleen. Recall from chapter 5, pages 102–103, that macrophages are large, phagocytic, wandering cells.

Hemoglobin molecules liberated from red blood cells break down into their four component polypeptide "globin" chains, each surrounding a *heme* group. The heme further decomposes into iron and a greenish pigment called **biliverdin.** The blood may transport the iron, combined with a protein, to the hematopoietic tissue in red bone marrow to be reused in synthesizing new hemoglobin. About 80% of the iron is stored in the liver in the form of an iron-protein complex. Biliverdin eventually is converted to an orange pigment called **bilirubin.** Biliverdin and bilirubin are excreted in the bile as bile pigments (see chapter 15, p. 418). Figure 12.6 summarizes the life cycle of a red blood cell.

In jaundice (icterus), accumulation of bilirubin turns the skin and eyes yellowish. Newborns can develop *physiologic jaundice* a few days after birth.

   Physiologic jaundice may be the result of immature liver cells that ineffectively excrete bilirubin into the bile. Treatment includes exposure to fluorescent light, which breaks down bilirubin in the tissues, and feedings that promote bowel movements. In hospital nurseries, babies being treated for physiologic jaundice lie under "bili lights," clad only in diapers and protective goggles. The healing effect of fluorescent light was discovered in the 1950s, when an astute nurse noted that jaundiced babies improved after sun exposure, except in the areas their diapers covered.

Practice

11. What happens to damaged red blood cells?
12. What are the products of hemoglobin breakdown?

White Blood Cells

White blood cells, or **leukocytes** (lu'ko-sītz), protect against disease. Leukocytes develop from hematopoietic stem cells in the red bone marrow (see fig. 12.4) in response to hormones, much as red blood cells form from precursors upon stimulation from erythropoietin. These hormones fall into two groups—**interleukins** and **colony-stimulating factors (CSFs).** Interleukins are numbered, while most colony-stimulating factors are named for the cell population they stimulate.

Blood transports white blood cells to sites of infection. White blood cells may then leave the bloodstream, as described on page 326.

Normally, five types of white blood cells are in circulating blood. They differ in size, the nature of their cytoplasm, the shape of the nucleus, and their staining characteristics, and are named for these distinctions. For example, leukocytes with granular cytoplasm are called **granulocytes,** whereas those without cytoplasmic granules are called **agranulocytes** (see fig. 12.4).

A typical granulocyte is about twice the size of a red blood cell. Members of this group include

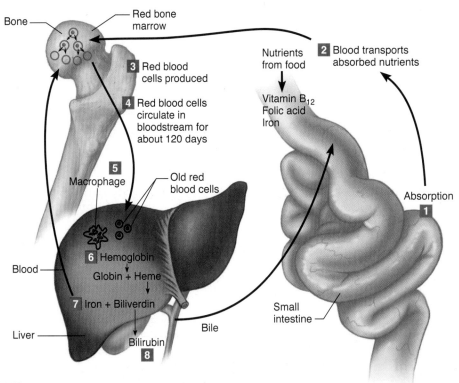

Figure 12.6 AP|R

Life cycle of a red blood cell. **(1)** The small intestine absorbs essential nutrients. **(2)** Blood transports nutrients to red bone marrow. **(3)** In the marrow, red blood cells arise from the division of less specialized progenitor cells. **(4)** Mature red blood cells are released into the bloodstream, where they circulate for about 120 days. **(5)** Macrophages destroy damaged red blood cells in the spleen and liver. **(6)** Hemoglobin liberated from red blood cells is broken down into heme and globin. **(7)** Iron from heme returns to red bone marrow and is reused. **(8)** Biliverdin and bilirubin are excreted in bile.

neutrophils, eosinophils, and basophils. Granulocytes develop in red bone marrow as do red blood cells, but have short life spans, averaging about 12 hours.

Neutrophils (nu′tro-filz) have fine cytoplasmic granules that appear light purple in neutral stain. The nucleus of an older neutrophil is lobed and consists of two to five sections (segments, so these cells are sometimes called *segs*) connected by thin strands of chromatin (fig. 12.7). Younger neutrophils are also called *bands* because their nuclei are C-shaped. Neutrophils account for 54–62% of the leukocytes in a typical blood sample from an adult.

Eosinophils (e″o-sin′o-filz) contain coarse, uniformly sized cytoplasmic granules that appear deep red in acid stain (fig. 12.8). The nucleus usually has only two lobes (termed bilobed). Eosinophils make up 1–3% of the total number of circulating leukocytes.

Basophils (ba′so-filz) are similar to eosinophils in size and in the shape of their nuclei, but they have fewer, more irregularly shaped cytoplasmic granules that become deep blue in basic stain (fig. 12.9). Basophils usually account for less than 1% of the circulating leukocytes.

The leukocytes of the agranulocyte group include monocytes and lymphocytes. Monocytes generally arise from red bone marrow. Lymphocytes are formed in the organs of the lymphatic system, as well as in the red bone marrow (see chapter 14, p. 382).

Monocytes (mon′o-sītz), the largest blood cells, are two to three times greater in diameter than red blood cells (fig. 12.10). Their nuclei vary in shape and are round, kidney-shaped, oval, or lobed. They usually make up 3–9% of the leukocytes in a blood sample and live for several weeks or even months.

Lymphocytes (lim′fo-sītz) are usually only slightly larger than red blood cells. A typical lymphocyte has a large, round nucleus surrounded by a thin rim of cytoplasm (fig. 12.11). These cells account for 25–33% of circulating leukocytes. Lymphocytes may live for years.

Practice

13. Which hormones are necessary for differentiation of white blood cells from hematopoietic stem cells in the red bone marrow?
14. Distinguish between granulocytes and agranulocytes.
15. List the five types of white blood cells, and explain how they differ from one another.

Functions of White Blood Cells

White blood cells protect against infection in various ways. Some leukocytes phagocytize bacterial cells in

the body, and others produce proteins (*antibodies*) that destroy or disable foreign particles.

Leukocytes can squeeze between the cells that form blood vessel walls. This movement, called *diapedesis,* allows the white blood cells to leave the circulation (fig. 12.12). Once outside the blood, they move through interstitial spaces using a form of self-propulsion called *amoeboid motion.*

The most mobile and active phagocytic leukocytes are neutrophils and monocytes. Monocytes leave the bloodstream and become *macrophages* that phagocytize bacteria, dead cells, and other debris in the tissues. Neutrophils cannot ingest particles much larger than bacterial cells, but monocytes can engulf large objects. Both of these phagocytes contain many *lysosomes,* which are organelles filled with digestive enzymes that break down organic molecules in captured bacteria, nutrients, and worn-out organelles. Neutrophils and monocytes may become so engorged with digestive products and bacterial toxins that they die.

Eosinophils are only weakly phagocytic, but they are attracted to and can kill certain parasites. Eosinophils also help control inflammation and allergic reactions by removing biochemicals associated with these reactions.

Basophils migrate to damaged tissues where they release *heparin,* which inhibits blood clotting, and *histamine,* which promotes inflammation, thus increasing blood flow to injured tissues. Basophils also play major roles in certain allergic reactions.

Lymphocytes are important in *immunity.* Some, for example, produce antibodies that attack specific foreign substances that enter the body. Chapter 14 (pp. 386–392) discusses immunity.

Practice

16. How do white blood cells fight infection?
17. How do white blood cells reach microorganisms that are outside blood vessels?
18. Which white blood cells are the most active phagocytes?

White Blood Cell Counts

The number of white blood cells in a microliter of human blood, called the *white blood cell count* (*WBCC* or *WCC*), normally is 4,000–11,000 cells. White blood cell counts are of clinical interest because their number may change in response to abnormal conditions. A total number of white blood cells exceeding 11,000 per microliter of blood constitutes **leukocytosis,** indicating acute infection, such as appendicitis. The white blood

Figure 12.7 AP|R
A neutrophil has a lobed nucleus with two to five segments (2,000×).

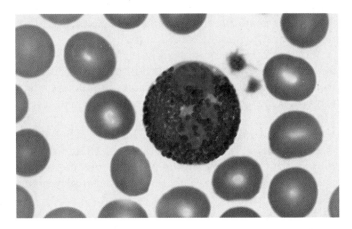

Figure 12.8 AP|R
An eosinophil has red-staining cytoplasmic granules (2,000×).

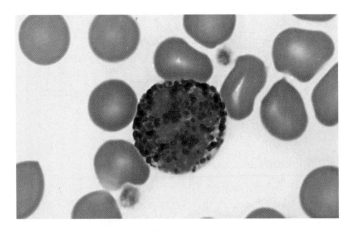

Figure 12.9 AP|R
A basophil has cytoplasmic granules that stain deep blue (2,000×).

Figure 12.10 AP|R
A monocyte is the largest of the blood cells (2,000×).

Figure 12.11 AP|R
A lymphocyte, the smallest of the white blood cells, has a large, round nucleus (2,000×).

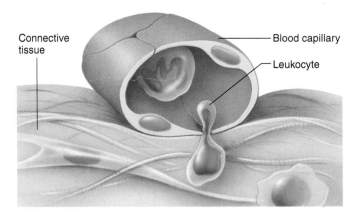

Figure 12.12
In a type of movement called diapedesis, leukocytes squeeze between the endothelial cells of a capillary wall and enter the tissue space outside the blood vessel.

Q: *What is a monocyte called once it has left the bloodstream and entered the tissues?*

Answer can be found in Appendix E on page 568.

cell count is greatly elevated in leukemia, as Clinical Application 12.1 describes.

A total white blood cell count below 4,000 per microliter of blood is called **leukopenia.** Such a deficiency may accompany typhoid fever, influenza, measles, mumps, chickenpox, AIDS, or poliomyelitis.

A *differential white blood cell count* (*DIFF*) lists percentages of the types of leukocytes in a blood sample. This test is useful because the relative proportions of white blood cells may change in particular diseases. For instance, the number of neutrophils usually increases during bacterial infections, and the number of eosinophils may increase during certain parasitic infections and allergic reactions. In AIDS, the numbers of a certain type of lymphocyte drop sharply.

Practice

19. What is the normal human white blood cell count?
20. Distinguish between leukocytosis and leukopenia.
21. What is a differential white blood cell count?

Blood Platelets AP|R

Platelets (plāt′letz), or **thrombocytes** (throm′bo-sītz), are not complete cells. They arise from very large cells in red bone marrow, called **megakaryocytes** (meg″ah-kar′e-o-sītz), that fragment like a shattered plate, releasing small sections of cytoplasm—platelets—into the circulation. The larger fragments of the megakaryocytes shrink and become platelets as they pass through blood vessels in the lungs. Megakaryocytes, and therefore platelets, develop from hematopoietic stem cells (see fig. 12.4) in response to the hormone **thrombopoietin** (throm″bo-poi′ĕ-tin).

Each platelet lacks a nucleus and is less than half the size of a red blood cell. It is capable of amoeboid movement and may live for about ten days. In normal blood, the *platelet count* varies from 130,000 to 360,000 per microliter. Platelets help close breaks in damaged blood vessels, as section 12.4 explains on page 331. Table 12.1 summarizes the characteristics of blood cells and platelets.

Practice

22. What is the normal blood platelet count?
23. What is the function of blood platelets?

12.3 BLOOD PLASMA

Plasma is the clear, straw-colored, liquid portion of the blood in which the cells and platelets are suspended. It is approximately 92% water and contains a complex

Table 12.1	Cellular Components of Blood		
Component	**Description**	**Number Present**	**Function**
Red blood cell (erythrocyte)	Biconcave disc without a nucleus; about one-third hemoglobin	4,200,000–6,200,000 per microliter	Transports oxygen and carbon dioxide
White blood cell (leukocyte)		4,000–11,000 per microliter	Destroys pathogenic microorganisms and parasites and removes worn cells
Granulocytes	About twice the size of red blood cells; cytoplasmic granules are present		
1. Neutrophil	Nucleus with two to five lobes; cytoplasmic granules stain light purple in neutral stain	54–62% of white blood cells present	Phagocytizes small particles
2. Eosinophil	Bilobed nucleus, cytoplasmic granules stain red in acid stain	1–3% of white blood cells present	Kills parasites and moderates allergic reactions
3. Basophil	Bilobed nucleus, cytoplasmic granules stain blue in basic stain	Less than 1% of white blood cells present	Releases heparin and histamine
Agranulocytes	Cytoplasmic granules are absent		
1. Monocyte	Two to three times larger than a red blood cell; nuclear shape varies from spherical to lobed	3–9% of white blood cells present	Phagocytizes large particles
2. Lymphocyte	Only slightly larger than a red blood cell; its nucleus nearly fills cell	25–33% of white blood cells present	Provides immunity
Platelet (thrombocyte)	Cytoplasmic fragment	130,000–360,000 per microliter	Helps control blood loss from broken vessels

Leukemia

When the twenty-three-year-old had a routine physical examination, she expected reassurance that her healthy lifestyle had indeed been keeping her healthy. After all, she felt great. What she got, a few days later, was a shock. Instead of having 4,000 to 11,000 white blood cells per microliter of blood, she had more than ten times that number—and many of the cells were cancerous. She had chronic myeloid leukemia (CML). Her red bone marrow was flooding her circulation with too many granulocytes, most of them poorly differentiated (figure 12A).

Another type of leukemia is lymphoid, in which the cancer cells are lymphocytes, produced in lymph nodes. Both myeloid and lymphoid leukemia can cause fatigue, headaches, nosebleeds and other bleeding, frequent respiratory infections, fever, bone pain, bruising, and other signs of slow blood clotting. The symptoms arise from the disrupted proportions of the blood's formed elements and their malfunction.

Immature white blood cells increase the risk of infection. Leukemic cells crowd out red blood cells and their precursors in the red marrow, causing anemia and resulting fatigue. Platelet deficiency (thrombocytopenia) slows clotting time, causing bruises and bleeding. Finally, spread of the cancer cells outside the marrow painfully weakens the surrounding bone. Eventually, without treatment, cancer cells spread outside the cardiovascular system, causing other tissues that would normally not produce white blood cells to do so.

Leukemia is also classified as acute or chronic. An acute condition appears suddenly, symptoms progress rapidly, and without treatment, death occurs in a few months. Chronic forms begin more slowly and may remain undetected for months or even years or, in rare cases, decades. Without treatment, life expectancy after symptoms develop is about three years.

Traditional cancer treatments destroy any cell that divides rapidly. A newer drug, Gleevec, targets only the cancer cells by nestling into ATP-binding sites on a type of enzyme called a tyrosine kinase, which blocks the message to divide. People with leukemia have other options. Bone marrow and stem cell transplants can cure the condition.

Another way that leukemia treatment is improving is refining diagnosis, based on identifying the proteins that leukemia cells produce. This information is used to predict which drugs are most likely to be effective, and which will cause intolerable side effects or not work in particular individuals. For example, some people with acute lymphoblastic leukemia (ALL), diagnosed on the basis of the appearance of the cancer cells in a blood smear, do not respond to standard chemotherapy. However, DNA microarray (also called DNA chip) technology revealed that the cells of patients who do not improve produce different proteins than the cancer cells of patients who do respond to the drugs used to treat ALL—the nonresponders have a different form of leukemia, called mixed-lineage leukemia. These patients respond to different drugs.

(a)

(b)

Figure 12A

Leukemia and blood cells. (**a**) Normal blood cells (700×). (**b**) Blood cells from a person with granulocytic leukemia, a type of myeloid leukemia (700×). Note the increased number of leukocytes.

mixture of organic and inorganic biochemicals. The functions of plasma constituents include transporting nutrients, gases, and vitamins; helping regulate fluid and electrolyte balance; and maintaining a favorable pH.

Plasma Proteins

Plasma proteins (plaz′mah pro′tēnz) are the most abundant of the dissolved substances (solutes) in plasma. These proteins remain in the blood and interstitial fluids, and ordinarily are not used as energy sources. The three main types of plasma proteins—albumins, globulins, and fibrinogen—differ in composition and function.

Albumins (al-bu′minz) are the smallest of the plasma proteins, yet account for about 60% of these proteins by weight. They are synthesized in the liver, and because they are so plentiful, albumins are an important determinant of the *osmotic pressure* of the plasma.

Recall from chapter 3 (p. 63) that the presence of an impermeant solute on one side of a selectively permeable membrane creates an osmotic pressure and that water always moves toward a greater osmotic pressure. Plasma proteins are too large to pass through the capillary walls, they are impermeant, and they create an osmotic pressure that holds water in the capillaries, despite blood pressure forcing water out of capillaries by filtration (see chapter 3, p. 64). The term *colloid osmotic pressure* is used to describe this osmotic effect due to the plasma proteins.

By maintaining the colloid osmotic pressure of plasma, albumins and other plasma proteins help regulate water movement between the blood and the tissues. In doing so, they help control blood volume, which, in turn, directly affects blood pressure (see chapter 13, p. 360).

If the concentration of plasma proteins falls, tissues swell, a condition called *edema*. This may result from starvation or a protein-deficient diet or from an impaired liver that cannot synthesize plasma proteins. As the concentration of plasma proteins drops, so does the colloid osmotic pressure, allowing fluid to accumulate in the interstitial spaces.

Globulins (glob′u-linz), which make up about 36% of the plasma proteins, can be further subdivided into *alpha, beta,* and *gamma globulins.* The liver synthesizes alpha and beta globulins, which have a variety of functions, including transport of lipids and fat-soluble vitamins. Lymphatic tissues produce the gamma globulins, which are a type of antibody (see chapter 14, p. 391).

Fibrinogen (fi-brin′o-jen), which constitutes about 4% of the plasma proteins, functions in blood coagulation, as discussed in section 12.4 on page 331. Synthesized in the liver, fibrinogen is the largest of the plasma

Table 12.2	Plasma Proteins		
Protein	Percentage of Total	Origin	Function
Albumin	60%	Liver	Helps maintain colloid osmotic pressure
Globulin	36%		
Alpha globulins		Liver	Transport lipids and fat-soluble vitamins
Beta globulins		Liver	Transport lipids and fat-soluble vitamins
Gamma globulins		Lymphatic tissues	Constitute a type of antibody
Fibrinogen	4%	Liver	Plays a key role in blood coagulation

proteins. Table 12.2 summarizes the characteristics of the plasma proteins.

Practice

24. List three types of plasma proteins.
25. How do albumins help maintain water balance between the blood and the tissues?
26. What are the functions of the globulins?
27. What is the role of fibrinogen?

Gases and Nutrients

The most important *blood gases* are oxygen and carbon dioxide. Plasma also contains a considerable amount of dissolved nitrogen, which ordinarily has no physiological function. Chapter 16 (pp. 459–461) discusses the blood gases and their transport.

The *plasma nutrients* include amino acids, simple sugars, nucleotides, and lipids absorbed from the digestive tract. For example, plasma transports glucose from the small intestine to the liver, where it may be stored as glycogen or converted to fat. If blood glucose concentration drops below the normal range, glycogen may be broken down into glucose, as described in chapter 11 (p. 308). Plasma also carries recently absorbed amino acids to the liver, where they can be used to manufacture proteins, or deaminated and used as an energy source (see chapter 15, p. 431).

Plasma lipids include fats (triglycerides), phospholipids, and cholesterol. Because lipids are not water-soluble and plasma is almost 92% water, these lipids are carried in the plasma by joining with proteins, forming lipoprotein complexes.

Nonprotein Nitrogenous Substances

Molecules that contain nitrogen atoms but are not proteins form a group called **nonprotein nitrogenous substances.** In plasma, this group includes amino acids, urea, uric acid, creatine (kre′ah-tin) and creatinine (kre-at′i-nin). Amino acids come from protein digestion and amino acid absorption. Urea and uric acid are products of protein and nucleic acid catabolism, respectively. Creatinine results from the metabolism of creatine. As discussed in chapter 8 (p. 185), creatine is present in *creatine phosphate* in muscle tissue as well as in the blood, where it stores energy in phosphate bonds.

Plasma Electrolytes

Blood plasma contains a variety of *electrolytes* that are absorbed from the intestine or released as by-products of cellular metabolism. They include sodium, potassium, calcium, magnesium, chloride, bicarbonate, phosphate, and sulfate ions. Sodium and chloride ions are the most abundant. Bicarbonate ions are important in maintaining the osmotic pressure and pH of plasma, and like other plasma constituents, they are regulated so that their blood concentrations remain relatively stable. Chapter 18 discusses these electrolytes in connection with water and electrolyte balance.

Practice

28. Which gases are in plasma?
29. Which nutrients are in plasma?
30. What is a nonprotein nitrogenous substance?
31. What are the sources of plasma electrolytes?

12.4 HEMOSTASIS

Hemostasis (he″mo-sta′sis) is the stoppage of bleeding, which is vitally important when blood vessels are damaged. Following an injury to the blood vessels, several actions may help to limit or prevent blood loss, including blood vessel spasm, platelet plug formation, and blood coagulation.

Blood Vessel Spasm

Cutting or breaking a small blood vessel stimulates the smooth muscles in its walls to contract, a phenomenon called **vasospasm,** and blood loss lessens almost immediately. Vasospasm may completely close the ends of a severed vessel.

Vasospasm may last only a few minutes, but the effect of the direct stimulation usually continues for about 30 minutes. By then, a *platelet plug* has formed, and blood is coagulating. Also, platelets release **serotonin,** which contracts smooth muscles in the blood vessel walls. This vasoconstriction further helps reduce blood loss.

Platelet Plug Formation

Platelets adhere to any rough surface, particularly to the collagen in connective tissue. When a blood vessel breaks, platelets adhere to the collagen underlying the endothelial lining of blood vessels. Platelets also adhere to each other, forming a platelet plug in the vascular break. A plug may control blood loss from a small break, but a larger break may require a blood clot to halt bleeding. Figure 12.13 shows the steps in platelet plug formation.

Practice

32. What is hemostasis?
33. How does a blood vessel spasm help control bleeding?
34. Describe the formation of a platelet plug.

Figure 12.13
Steps in platelet plug formation.

Blood Coagulation

Coagulation (ko-ag″u-la′shun), the most effective hemostatic mechanism, is the formation of a *blood clot.* Blood coagulation is complex and utilizes many biochemicals called *clotting factors.* Some of these factors promote coagulation, and others inhibit it. Whether or not blood coagulates depends on the balance between these two groups of factors. Normally, anticoagulants prevail, and the blood does not clot. However, as a result of injury (trauma), biochemicals that favor coagulation may increase in concentration, and the blood may coagulate.

The major event in blood clot formation is the conversion of the soluble plasma protein fibrinogen into insoluble threads of the protein **fibrin.** Formation of fibrin takes several steps. First, damaged tissues release *tissue thromboplastin,* initiating a series of reactions that results in the production of *prothrombin activator.* This series of changes requires calcium ions as well as certain proteins and phospholipids. As its name suggests, prothrombin activator acts on prothrombin (see figure 12.16).

Prothrombin is an alpha globulin that the liver continually produces and is thus a normal constituent of plasma. Prothrombin activator converts prothrombin into **thrombin,** which in turn catalyzes a reaction that joins fragments of fibrinogen into long threads of fibrin.

Once fibrin threads form, they stick to the exposed surfaces of damaged blood vessels, creating a meshwork that entraps blood cells and platelets (fig. 12.14). The resulting mass is a blood clot, which may block a vascular break and prevent further blood loss. The clear, yellow liquid that remains after the clot forms is called **serum.** Serum is plasma minus clotting factors.

The amount of prothrombin activator in the blood is directly proportional to the degree of tissue damage. Once a blood clot begins to form, it promotes additional clotting because thrombin also acts directly on blood clotting factors other than fibrinogen, causing prothrombin to form more thrombin. This is an example of a *positive feedback system,* in which the original action stimulates more of the action. Such a positive feedback mechanism produces unstable conditions and can operate for only a short time without disrupting the stable internal environment (see chapter 1, p. 7).

> Laboratory tests commonly used to evaluate the blood coagulation mechanisms include *prothrombin time (PT)* and *partial thromboplastin time (PTT).* These tests measure the time it takes for fibrin threads to form in a sample of plasma.

Normally, blood flow prevents formation of a massive clot by rapidly carrying excess thrombin away, keeping its concentration too low in any one place to promote further clotting. Consequently, blood usually coagulates in blood that is standing still (or moving slowly). Clotting ceases where a clot contacts circulating blood.

Fibroblasts (see chapter 5, p. 102) invade blood clots that form in ruptured vessels, producing fibrous connective tissue throughout, which helps strengthen and seal vascular breaks. Many clots, including those that form in tissues as a result of blood leakage (hematomas), disappear in time. This dissolution requires conversion of a plasma protein, *plasminogen,* to *plasmin,* a protein-splitting enzyme that can digest fibrin threads and other proteins associated with blood clots. Plasmin formation may dissolve a whole clot; however, clots that fill large blood vessels are seldom removed naturally.

A blood clot abnormally forming in a vessel is a **thrombus** (throm′bus). A clot that dislodges, or a fragment of a clot that breaks loose and is carried away by the blood flow, is called an **embolus** (em′bo-lus). Generally, emboli continue to move until they reach narrow places in vessels where they may lodge and block blood flow.

A blood clot forming in a vessel that supplies a vital organ, such as the heart (coronary thrombosis) or the brain (cerebral thrombosis), kills tissues the vessel serves (*infarction*) and may be fatal. A blood clot that travels and then blocks a vessel that supplies a vital organ, such as the lungs (pulmonary embolism), affects the portion of the organ the blocked blood vessel supplies. Genetics Connection 12.1 discusses several blood clotting disorders.

> Drugs based on "clot-busting" biochemicals can be lifesavers. *Tissue plasminogen activator* (tPA) may restore blocked coronary or cerebral circulation if given within three hours of a heart attack or stroke. A drug derived from bacteria called *streptokinase* may also be successful, for a fraction of the cost. Another plasminogen activator used as a drug is *urokinase,* an enzyme produced in certain kidney cells. Heparin and coumadin are drugs that interfere with clot formation, but do not dissolve clots.

Figure 12.14
Falsely-colored scanning electron micrograph of fibrin threads forming a blood clot (2,800×).

Genetics Connection 12.1

Coagulation Disorders

Genetic mutations cause several types of clotting disorders. Environmental factors can affect the severity of these conditions.

Hemophilia

Abnormalities of different clotting factors cause different forms of the bleeding disorder hemophilia. Symptoms include severe hemorrhage following minor injuries, frequent nosebleeds, large intramuscular hematomas, and blood in the urine. In the most common form, hemophilia A, factor VIII is deficient or absent. The gene for this clotting factor is on the X chromosome, so that primarily males are affected, because they do not have a second X chromosome to block the mutation's expression. One in 10,000 males has hemophilia A. Treatment is replacing factor VIII.

Hemophilia A has left its mark on history. A second-century B.C. Jewish document, the Talmud, states, "If she circumcised her first child and he died, and a second one also died, she must not circumcise her third child." Hemophilia A affected the royal families of England, Russia, Germany, and Spain. In 1985 hemophilia made history again when many patients who had received factor VIII pooled from donors contracted HIV infection. Today, the clotting factor is manufactured using recombinant DNA technology, so it is free of viruses.

von Willebrand Disease

The tendency to bleed and bruise easily may be a sign of von Willebrand disease, which is the most common hereditary coagulation disorder. One in 100 people inherits a mutation in any of four genes that encode the von Willebrand clotting factor, but only 1 in 10,000 individuals actually develops symptoms. It is equally likely to affect males as females.

Von Willebrand factor is a plasma protein secreted by the endothelial cells lining blood vessels. It enables platelets to adhere to damaged blood vessel walls, which is a key step preceding actual clotting. In von Willebrand disease, the mucous membranes of the gastrointestinal and urinary tracts can spontaneously bleed. Some people do not discover that they have von Willebrand disease until they bleed excessively following an injury. Usually no treatment is required. However, women who bleed very heavily during their menstrual periods may be advised to take oral contraceptives to diminish the flow, and a combination of factor VIII and von Willebrand factor can be taken before planned surgery to reduce the risk of uncontrolled bleeding.

Factor V Leiden

An inherited susceptibility plus other risk factors can cause dangerous clotting. This is the case for factor V Leiden, which is a mutation in the gene that encodes clotting factor V. This factor normally interacts with factor X to inactivate thrombin, which converts fibrinogen to fibrin to form a clot. A person with factor V Leiden has just one altered DNA base, which changes one amino acid in the clotting factor.

A forty-two-year-old woman who unknowingly had factor V Leiden went to an emergency department when she experienced shortness of breath, light-headedness, and difficulty walking. Her left leg was painful and had turned bluish-purple. An ultrasound scan revealed a clot in a leg vein extending from her hip to the calf. A careful clinical workup revealed several risk factors at play in the formation of the woman's dangerous blood clot:

Factor V Leiden (inherited tendency to form clots)
A very long car ride (stagnation of blood due to immobility)
Oral contraceptives (increased risk of clotting)
Following the grapefruit diet for the preceding three days (grapefruit inactivates the enzyme that breaks down the oral contraceptive)

Infusion of a "clot-buster" drug directly into the vein opened it up, and the woman recovered.

Abnormal clot formations are often associated with conditions that change the endothelial linings of vessels. For example, in *atherosclerosis* (ath″er-o″skle-ro′sis), accumulations of fatty deposits change arterial linings, sometimes initiating inappropriate clotting (fig. 12.15). Figure 12.16 summarizes the three primary hemostatic mechanisms: blood vessel spasm, platelet plug formation, and blood coagulation.

Practice

35. Review the major steps in blood clot formation.
36. What prevents the formation of massive clots throughout the cardiovascular system?
37. Distinguish between a thrombus and an embolus.

Lumen   Artery wall

(a)

Lumen   Plaque   Artery wall

(b)

Figure 12.15

Artery cross sections, falsely-colored light micrographs. **(a)** Normal artery (50×), and **(b)** the inner wall of an artery changed as a result of atherosclerosis (100×). Not only is blood flow impeded, but the uneven inner surface can snag platelets, triggering coagulation.

12.5 BLOOD GROUPS AND TRANSFUSIONS

Early attempts to transfer blood from one person to another produced varied results. Sometimes, the recipient improved. Other times, the recipient suffered a blood transfusion reaction in which the red blood cells clumped, obstructing vessels and producing great pain and organ damage.

Eventually, scientists determined that blood is of differing types and that only certain combinations of blood types are compatible. These discoveries led to the development of procedures for typing blood. Today, safe transfusions of whole blood depend on properly matching the blood types of donors and recipients.

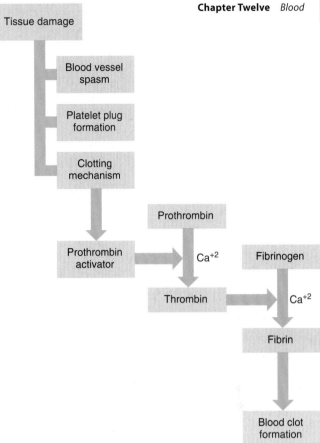

Figure 12.16

The three mechanisms of hemostasis: blood vessel spasm, platelet plug formation, and blood coagulation.

Q: *What is the major event in blood clot formation?*

Answer can be found in Appendix E on page 568.

Antigens and Antibodies

Agglutination is the clumping of red blood cells following a transfusion reaction. Red blood cell surface molecules called **antigens** (an′tĭ-jenz), also called *agglutinogens*, react with protein **antibodies** (an′tĭ-bod″ēz), also called *agglutinins,* carried in plasma.

Only a few of the many different antigens on red blood cell membranes can produce serious transfusion reactions. These include the antigens of the ABO group and those of the Rh group. Avoiding the mixture of certain kinds of antigens and antibodies prevents adverse transfusion reactions.

A mismatched blood transfusion quickly produces telltale signs of agglutination—anxiety, breathing difficulty, facial flushing, headache, and severe pain in the neck, chest, and lumbar area. Red blood cells burst, releasing free hemoglobin. Macrophages phagocytize the hemoglobin, converting it to bilirubin, which may sufficiently accumulate to cause the yellow skin of jaundice. Free hemoglobin in the kidneys may ultimately cause them to fail.

ABO Blood Group

The *ABO blood group* is based on the presence (or absence) of two major protein antigens on red blood cell membranes—antigen A and antigen B. A person's erythrocytes have on their surfaces one of four antigen combinations: only A, only B, both A and B, or neither A nor B. The resulting ABO blood type, because it reflects a protein combination, is inherited.

A person with only antigen A has *type A blood.* A person with only antigen B has *type B blood.* An individual with both antigen A and B has *type AB blood.* A person with neither antigen A nor B has *type O blood.* Thus, all people have one of four possible ABO blood types—A, B, AB, or O.

In the United States, the most common ABO blood types are O (47%) and A (41%). Rarer are type B (9%) and type AB (3%). These percentages vary in subpopulations and over time, reflect changes in the genetic structure of populations.

Certain antibodies that affect the ABO blood group are synthesized in the plasma about two to eight months following birth. Specifically, whenever antigen A is absent in red blood cells, an antibody called *anti-A* is produced, and whenever antigen B is absent, an antibody called *anti-B* is produced. Therefore, persons with type A blood have anti-B antibody in their plasma; those with type B blood have anti-A antibody;

Table 12.3	Antigens and Antibodies of the ABO Blood Group	
Blood Type	**Antigen**	**Antibody**
A	A	Anti-B
B	B	Anti-A
AB	A and B	Neither anti-A nor anti-B
O	Neither A nor B	Both anti-A and anti-B

those with type AB blood have neither antibody; and those with type O blood have both anti-A and anti-B antibodies (fig. 12.17 and table 12.3). The antibodies anti-A and anti-B are large and do not cross the placenta. Thus, a pregnant woman and her fetus may be of different ABO blood types, and agglutination in the fetus will not occur.

An antibody of one type will react with an antigen of the same type and clump red blood cells (fig. 12.18); therefore, such combinations must be avoided. The major concern in blood transfusion procedures is that the cells in the donated blood not clump due to antibodies in the recipient's plasma. For this reason, a person with type A (anti-B) blood must not receive blood of type B or AB, either of which would clump in the presence of anti-B in the recipient's type A blood. Likewise, a person with type B (anti-A) blood must not receive type A or AB blood, and a person with type O (anti-A and anti-B) blood must not receive type A, B, or AB blood.

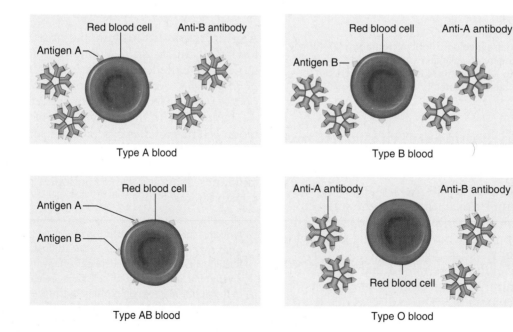

Figure 12.17

Different combinations of antigens and antibodies distinguish blood types. (Cells and antibodies not drawn to scale.)

Figure 12.18

Agglutination. **(a)** If red blood cells with antigen A are added to blood containing anti-A antibody, **(b)** the antibodies react with the antigens, causing clumping (agglutination). **(c)** Nonagglutinated blood (210×). **(d)** Agglutinated blood (220×). (Cells and antibodies in **(a)** and **(b)** not drawn to scale.)

Type AB blood lacks both anti-A and anti-B antibodies, so an AB person can receive a transfusion of blood of any other type. For this reason, type AB persons are sometimes called *universal recipients*. However, type A (anti-B) blood, type B (anti-A) blood, and type O (anti-A and anti-B) blood still contain antibodies (either anti-A and/or anti-B) that could agglutinate type AB cells if transfused rapidly. Consequently, even for

AB individuals, using donor blood of the same type as the recipient is best (table 12.4).

Type O blood lacks antigens A and B. Therefore, theoretically this type could be transfused into persons with blood of any other type. Individuals with type O blood are sometimes called *universal donors*. Type O blood, however, does contain both anti-A and anti-B antibodies. If type O blood is given to a person with blood type A, B, or AB, it should be transfused slowly so that the recipient's larger blood volume will dilute the donor blood, minimizing the chance of an adverse reaction.

Table 12.4	Preferred and Permissible Blood Types for Transfusions	
Blood Type of Recipient	Preferred Blood Type of Donor	If Preferred Blood Type Unavailable, Permissible Blood Type of Donor
A	A	O
B	B	O
AB	AB	A, B, O
O	O	No alternate types

Blood in the umbilical cord at birth is rich in stem cells that can be used to treat a variety of disorders, including leukemias, sickle cell disease and other hemoglobin abnormalities, and certain inborn errors of metabolism. The United States and the United Kingdom have public umbilical cord blood banks that provide stem cells for free. For many illnesses it is best to receive donor stem cells, because receiving one's own could reintroduce the disease.

Practice

38. Distinguish between antigens and antibodies.

39. What is the main concern when blood is transfused from one individual to another?

40. Why is a type AB person called a universal recipient?

41. Why is a type O person called a universal donor?

Rh Blood Group

The *Rh blood group* was named after the rhesus monkey in which it was first studied. In humans, this group includes several Rh antigens (factors). The most prevalent of these is *antigen D*. If the Rh antigen is present on the red blood cell membranes, the blood is said to be *Rh-positive*. Conversely, if the red blood cells lack Rh antigen, the blood is called *Rh-negative*.

Only 15% of the U.S. population is Rh-negative.

The presence (or absence) of Rh antigen is an inherited trait, as is ABO blood type. But unlike anti-A and anti-B, antibodies that react with Rh antigen (*anti-Rh antibodies*) do not appear spontaneously. Instead, they form only in Rh-negative persons in response to the presence of red blood cells with Rh antigens.

If an Rh-negative person receives a transfusion of Rh-positive blood, the Rh antigen stimulates the recipient to begin producing anti-Rh antibodies. Generally, this initial transfusion has no serious consequences, but if the Rh-negative person—who is now sensitized to Rh-positive blood—receives another transfusion of Rh-positive blood some months later, the donated red cells are likely to agglutinate.

A related condition may occur when an Rh-negative woman is pregnant with an Rh-positive fetus for the first time. Such a pregnancy may be uneventful; however, at birth (or if a miscarriage occurs), the placental membranes that separated the maternal blood from the fetal blood during the pregnancy tear, and some of the infant's Rh-positive blood cells may enter the maternal circulation. These Rh-positive cells may then stimulate the maternal tissues to begin producing anti-Rh antibodies.

If a woman who has already developed anti-Rh antibodies becomes pregnant with a second Rh-positive fetus, these antibodies, called hemolysins, cross the placental membrane and destroy the fetal red blood cells (fig. 12.19). The fetus then develops a condition called *erythroblastosis fetalis* or hemolytic disease of the fetus and newborn.

Erythroblastosis fetalis is extremely rare today because obstetricians carefully track Rh status. An Rh-negative woman who might carry an Rh-positive fetus is given an injection of a drug called RhoGAM. This injection is actually composed of anti-Rh antibodies, which bind to and shield any Rh-positive fetal cells that might contact the woman's cells and sensitize her immune system. RhoGAM must be given within 72 hours of possible contact with Rh-positive cells—including giving birth, terminating a pregnancy, miscarrying, or undergoing amniocentesis (a prenatal test in which a needle is inserted into the uterus).

Practice

42. What is the Rh blood group?

43. What are two ways that Rh incompatibility can arise?

Figure 12.19

Rh incompatibility. If a man who is Rh-positive and a woman who is Rh-negative conceive a child who is Rh-positive, the woman's body may manufacture antibodies that attack future Rh-positive offspring.

Rh-negative woman with Rh-positive fetus

Cells from Rh-positive fetus enter woman's bloodstream

Woman becomes sensitized— antibodies (◇) form to fight Rh-positive blood cells

In the next Rh-positive pregnancy, maternal antibodies attack fetal red blood cells

Summary Outline

12.1 Introduction (p. 319)

Blood is a type of connective tissue in which cells are suspended in a liquid extracellular matrix. It transports substances between body cells and the external environment, and helps maintain a stable internal environment.

1. Blood can be separated into formed elements and liquid portions.
   a. The formed elements portion is mostly red blood cells.
   b. The liquid plasma includes water, gases, nutrients, hormones, electrolytes, and cellular wastes.
2. Blood volume varies with body size, fluid and electrolyte balance, and adipose tissue content.

12.2 Blood Cells (p. 319)

1. Red blood cells
   a. Red blood cells are biconcave discs with shapes that increase surface area.
   b. Red blood cells contain hemoglobin, which combines loosely with oxygen.
2. Red blood cell counts
   a. The red blood cell count equals the number of cells per microliter of blood.
   b. The average count ranges from approximately 4 to 6 million cells per microliter of blood.
   c. Red blood cell count determines the oxygen-carrying capacity of the blood. It is used to diagnose and evaluate the courses of certain diseases.
3. Red blood cell production and its control
   a. Red bone marrow produces red blood cells.
   b. In health, the number of red blood cells remains relatively stable.
   c. Erythropoietin controls the rate of red blood cell formation by negative feedback.
4. Dietary factors affecting red blood cell production
   a. Availability of vitamin B_{12} and folic acid influences red blood cell production.
   b. Hemoglobin synthesis requires iron.
5. Destruction of red blood cells
   a. Macrophages in the liver and spleen phagocytize damaged red blood cells.
   b. Hemoglobin molecules decompose, and nearly all of the iron they contain is recycled.
   c. Biliverdin and bilirubin are pigments, released from the heme (iron) portion, excreted in bile.
6. White blood cells
   a. White blood cells develop from hematopoietic stem cells in red bone marrow, in response to interleukins and colony-stimulating factors.
   b. Granulocytes include neutrophils, eosinophils, and basophils.
   c. Agranulocytes include monocytes and lymphocytes.
7. Functions of white blood cells
   a. Neutrophils and monocytes phagocytize foreign particles.
   b. Eosinophils kill parasites and help control inflammation and allergic reactions.
   c. Basophils release heparin, which inhibits blood clotting, and histamine, which promotes inflammation, to increase blood flow to injured tissues.
   d. Lymphocytes are important in immunity.
8. White blood cell counts
   a. Normal total white blood cell counts vary from 4,000 to 11,000 cells per microliter of blood.
   b. The number of white blood cells may change in response to abnormal conditions, such as infections, emotional disturbances, or excessive loss of body fluids.
   c. A differential white blood cell count indicates the percentages of various types of leukocytes.
9. Blood platelets
   a. Blood platelets, which develop in the red bone marrow in response to thrombopoietin, are fragments of giant cells.
   b. The normal platelet count varies from 130,000 to 360,000 platelets per microliter of blood.
   c. Platelets help close breaks in blood vessels.

12.3 Blood Plasma (p. 327)

Plasma transports gases and nutrients, helps regulate fluid and electrolyte balance, and helps maintain stable pH.

1. Plasma proteins
   a. Plasma proteins remain in blood and interstitial fluids, and are not normally used as energy sources.
   b. Three major types exist.
      (1) Albumins help maintain the colloid osmotic pressure.
      (2) Globulins transport lipids and fat-soluble vitamins and include antibodies that provide immunity.
      (3) Fibrinogen functions in blood clotting.
2. Gases and nutrients
   a. Gases in plasma include oxygen, carbon dioxide, and nitrogen.
   b. Plasma nutrients include simple sugars, amino acids, and lipids.
      (1) The liver stores glucose as glycogen and releases glucose whenever blood glucose concentration falls.
      (2) Amino acids are used to synthesize proteins and are deaminated for use as energy sources.
      (3) Lipoproteins function in the transport of lipids.
3. Nonprotein nitrogenous substances
   a. Nonprotein nitrogenous substances are composed of molecules that contain nitrogen atoms but are not proteins.
   b. They include amino acids, urea, uric acid, creatine, and creatinine.
4. Plasma electrolytes
   a. Plasma electrolytes include ions of sodium, potassium, calcium, magnesium, chloride, bicarbonate, phosphate, and sulfate.
   b. Bicarbonate ions are important in maintaining the osmotic pressure and pH of plasma.

12.4 Hemostasis (p. 330)

Hemostasis is the stoppage of bleeding.

1. Blood vessel spasm
   a. Smooth muscles in blood vessel walls reflexly contract following injury.
   b. Platelets release serotonin, which stimulates vasoconstriction and helps maintain vessel spasm.
2. Platelet plug formation
   a. Platelets adhere to rough surfaces and exposed collagen.
   b. Platelets adhere to each other at injury sites and form platelet plugs in broken vessels.
3. Blood coagulation
   a. Blood clotting is the most effective means of hemostasis.
   b. Clot formation depends on the balance between factors that promote clotting and those that inhibit clotting.
   c. The basic event of coagulation is the conversion of soluble fibrinogen into insoluble fibrin.
   d. Biochemicals that promote clotting include prothrombin activator, prothrombin, and calcium ions.
   e. A thrombus is an abnormal blood clot in a vessel. An embolus is a clot or fragment of a clot that moves in a vessel.

12.5 Blood Groups and Transfusions (p. 333)

Blood can be typed on the basis of cell surface antigens.

1. Antigens and antibodies
   a. Agglutination is the clumping of red blood cells following a transfusion reaction.
   b. Red blood cell membranes may contain specific antigens, and blood plasma may contain antibodies against certain of these antigens.
2. ABO blood group
   a. Blood is grouped according to the presence or absence of antigens A and B.
   b. Mixing red blood cells that contain an antigen with plasma that contains the corresponding antibody results in an agglutination reaction or adverse transfusion reaction in a patient.
3. Rh blood group
   a. Rh antigen is present on the red blood cell membranes of Rh-positive blood. Rh antigen is absent in Rh-negative blood.
   b. An Rh-negative person exposed to Rh-positive blood produces anti-Rh antibodies in response to the presence of Rh antigens.
   c. Mixing Rh-positive red blood cells with plasma that contains anti-Rh antibodies agglutinates the positive cells.
   d. Anti-Rh antibodies in maternal blood may cross the placental tissues and react with the red blood cells of an Rh-positive fetus.

Chapter Assessments

12.1 Introduction

1. Major functions of blood include: (p. 319)
   a. nutrient, hormone, and oxygen transport.
   b. helping maintain the stability of interstitial fluid.
   c. heat distribution.
   d. waste transport
   e. all of the above.

2. Formed elements in blood are _____ , _____ , and _____ . (p. 319)

3. The liquid portion of blood is _____ . (p. 319)

12.2 Blood Cells

4. Describe a red blood cell. (p. 319)

5. Contrast oxyhemoglobin and deoxyhemoglobin. (p. 319)

6. Connect the significance of red blood cell counts with the function of red blood cells. (p. 321)

7. Describe the life cycle of a red blood cell, beginning with its production and ending with its destruction. (p. 321)

8. List dietary factors affecting red blood cell production. (p. 323)

9. Name five types of leukocytes, identifying which are granulocytes and which are agranulocytes, and list the major function(s) of each type. (p. 324)

10. _____ are fragments of megakaryocytes that function in _____ . (p. 327)

12.3 Blood Plasma

11. The most abundant component of plasma is: (p. 327)
    a. vitamins.   d. water.
    b. oxygen.   e. electrolytes.
    c. proteins.

12. Name three types of plasma proteins, and indicate the major function(s) of each type. (p. 328)

13. Name the gases and nutrients found in plasma. (p. 329)

14. Define *nonprotein nitrogenous substances,* and name those commonly present in plasma. (p. 329)

15. The most abundant plasma electrolytes are _____ and _____ . (p. 330)

12.4 Hemostasis

16. _____ is the stoppage of bleeding. (p. 330)

17. Explain how blood vessel spasm is stimulated following an injury. (p. 330)

18. Platelets adhering to form a plug may control blood loss from a _____ break, but a larger break may require a _____ to halt bleeding. (p. 330)

19. Describe the major steps leading to the formation of a blood clot. (p. 331)

20. Contrast thrombus and embolus. (p. 332)

12.5 Blood Groups and Transfusions

21. An individual with B antigens and anti-A antibodies is ABO blood type _____ . (p. 334)

22. Explain why the individual described in question 21 should not receive a transfusion with type AB blood. (p. 334)

23. Distinguish between Rh-positive and Rh-negative blood. (p. 335)

24. Describe *erythroblastosis fetalis,* and explain how this condition may develop. (p. 336)

Integrative Assessments/Critical Thinking

OUTCOMES 3.4, 12.2

1. If a patient with inoperable cancer is treated using a drug that reduces the rate of cell division, how might the patient's white blood cell count change? How might the patient's environment be modified to compensate for the effects of these changes?

OUTCOMES 8.3, 12.2

2. Erythropoietin is available as a drug. Why would athletes abuse it?

OUTCOME 12.2

3. How would you explain to a patient with leukemia, who has a greatly elevated white blood cell count, the importance of avoiding bacterial infections?

OUTCOMES 12.2, 12.5

4. Why can a person receive platelets donated by anyone, but must receive a particular type of whole blood?

OUTCOMES 12.3, 12.4

5. Why do patients with liver diseases commonly develop blood clotting disorders?

WEB CONNECTIONS

Visit the text website at **www.mhhe.com/shieress11** for additional quizzes, interactive learning exercises, and more.

APR

Anatomy & Physiology REVEALED includes cadaver photos that allow you to peel away layers of the human body to reveal structures beneath the surface. This program also includes animations, radiologic imaging, audio pronunciation, and practice quizzing. To learn more visit **www.aprevealed.com**.

13

Cardiovascular System

Cardiovascular defibrillators. A man rushing to catch a flight at a busy airport stops suddenly, looks about in confusion, and collapses. People gather around him, as a woman runs to a device mounted on a nearby wall. It is an automated external defibrillator (AED), and looks like a laptop computer. The woman learned how to use it in a cardiopulmonary resuscitation class. She brings it over to the man, opens it, and places electrode pads over the man's chest, as indicated in a drawing on the inner cover of the defibrillator. Then the device speaks: "Analyzing heart rhythm," it declares, as a computer assesses the heart rhythm. After a short pause, the device says, "Charging, stand clear," and then "Push button." The woman pushes the button, and the device delivers a shock to the man's chest. It assesses the heart rhythm again, and instructs the woman to deliver a second shock. Soon the man recovers, just as emergency technicians arrive.

The AEDs found in many public places can save the life of a person suffering sudden cardiac arrest. One study conducted at Chicago's O'Hare and Midway airports found that over a ten-month period, AEDs saved 64% of the people they were used on. Without defibrillation, only 5–7% of people survive sudden cardiac arrest. Each minute the odds of survival shrink by 10%, and after six minutes brain damage is irreversible.

Sudden cardiac arrest can result from an abnormally accelerated heartbeat or a chaotic and irregular contraction of the heart muscle (ventricular fibrillation). The bioelectrical malfunction that usually causes these conditions may result from an artery blocked with plaque or from buildup of scar tissue from a previous myocardial infarction

An implantable cardioverter defibrillator delivers a shock to a heart whose ventricles are contracting wildly, restoring a normal heartbeat.

(heart attack). Magnets in certain headphones can interfere with defibrillators and should be moved away from the person.

For people who know they have an inherited disorder that causes sudden cardiac arrest (by having suffered an event and having had genetic tests), a device called an implantable cardioverter defibrillator (ICD) can be placed under the skin of the chest in a one-hour procedure. Like the AED, the ICD monitors heart rhythm. When the telltale deviations of ventricular tachycardia or ventricular fibrillation begin, it delivers a shock, preventing cardiac arrest. Sometimes ICDs are offered to people who have had a previous myocardial infarction and have significant damage from it.

Learning Outcomes

After studying this chapter, you should be able to do the following:

13.1 Introduction

1. Name the structures composing the cardiovascular system. (p. 341)

13.2 Structure of the Heart

2. Distinguish between the coverings of the heart and the layers that compose the wall of the heart. (p. 342)

3. Identify and locate the major parts of the heart, and discuss the functions of each part. (p. 343)

4. Trace the pathway of blood through the heart and the vessels of coronary circulation. (p. 346)

13.3 Heart Actions

5. Describe the cardiac cycle and the cardiac conduction system. (p. 347)

6. Identify the parts of a normal ECG pattern, and discuss the significance of this pattern. (p. 351)

7. Explain control of the cardiac cycle. (p. 352)

13.4 Blood Vessels

8. Compare the structures and functions of the major types of blood vessels. (p. 354)

9. Describe how substances are exchanged between blood in capillaries and the tissue fluid surrounding body cells. (p. 357)

13.5 Blood Pressure

10. Explain how blood pressure is produced and controlled. (p. 359)

11. Describe the mechanisms that aid in returning venous blood to the heart. (p. 362)

 Learn Practice Assess

13.6 Paths of Circulation

12. Compare the pulmonary and systemic circuits of the cardiovascular system. (p. 363)

13.7–13.8 Arterial System–Venous System

13. Identify and locate the major arteries and veins. (pp. 363–371)

Module 9: Cardiovascular System

Aids to Understanding Words

(Appendix A on page 564 has a complete list of Aids to Understanding Words.)

brady- [slow] *brady*cardia: Abnormally slow heartbeat.

diastol- [dilation] *diastol*ic pressure: Blood pressure when the ventricle of the heart is relaxed.

-gram [something written] electrocardio*gram*: Recording of the electrical changes in the myocardium during a cardiac cycle.

papill- [nipple] *papill*ary muscle: Small mound of muscle projecting into a ventricle of the heart.

syn- [together] *syn*cytium: Mass of merging cells that act together.

systol- [contraction] *systol*ic pressure: Blood pressure resulting from a single ventricular contraction.

tachy- [rapid] *tachy*cardia: Abnormally fast heartbeat.

13.1 INTRODUCTION

The heart pumps 7,000 liters of blood through the body each day, contracting some 2.5 billion times in an average lifetime. This muscular pump forces blood through arteries, which connect to smaller-diameter vessels called arterioles. Arterioles branch into the tiniest tubes, the capillaries, which are sites of nutrient, electrolyte, gas, and waste exchange. Capillaries converge into venules, which in turn converge into veins that return blood to the heart, completing the closed system of blood circulation. These structures—the pump and its vessels—form the **cardiovascular system.**

The cardiovascular system has two closed pathways, or circuits, of blood flow. The **pulmonary** (pul'mo-ner″e) **circuit** sends oxygen-depleted (deoxygenated) blood to the lungs to pick up oxygen and unload carbon dioxide. The **systemic** (sis-tem′ik) **circuit** sends oxygen-rich (oxygenated) blood and nutrients to all body cells and removes wastes. Without circulation, tissues would lack a supply of oxygen and nutrients, and wastes would accumulate. Such deprived cells soon begin irreversible change, which quickly leads to their death. Figure 13.1 shows the general pattern of blood transport in the cardiovascular system.

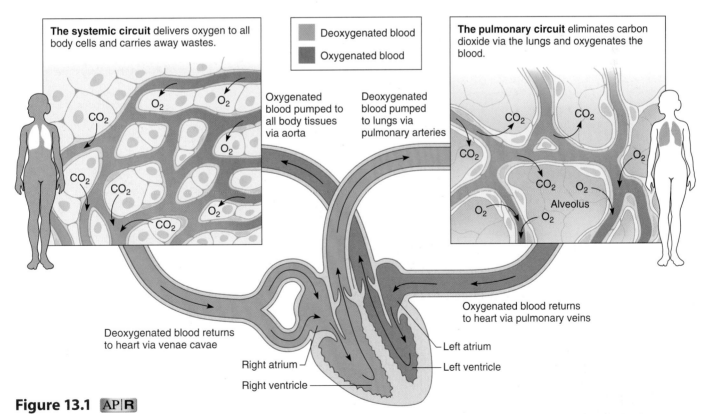

Figure 13.1 AP|R

The cardiovascular system transports blood between the body cells and organs such as the lungs, intestines, and kidneys that communicate with the external environment. Vessels in the pulmonary circuit carry blood from the heart to the lungs and back to the heart, replenishing oxygen and releasing the metabolic waste CO_2. Vessels of the systemic circuit supply all of the other cells.

13.2 STRUCTURE OF THE HEART

The heart is a hollow, cone-shaped, muscular pump. It lies within the thoracic cavity and rests on the diaphragm (fig. 13.2).

Size and Location of the Heart

Heart size varies with body size. An average adult's heart is about 14 centimeters long and 9 centimeters wide (fig. 13.3).

The heart is within the mediastinum, bordered laterally by the lungs, posteriorly by the vertebral column, and anteriorly by the sternum. The *base* of the heart, which attaches to several large blood vessels, lies beneath the second rib. The heart's distal end extends downward and to the left, terminating as a bluntly pointed *apex* at the level of the fifth intercostal space.

Coverings of the Heart

The **pericardium** (per″ĭ-kar′de-um) is a covering that encloses the heart and the proximal ends of the large blood vessels to which it attaches. The pericardium consists of an outer bag, the fibrous pericardium, composed of dense connective tissue. It is attached to the central part of the diaphragm, the posterior of the sternum, the vertebral column, and the large blood vessels emerging from the heart.

The fibrous pericardium surrounds a more delicate, double-layered sac. The innermost layer of this sac, the *visceral pericardium* (epicardium), covers the heart. At the base of the heart, the visceral pericardium turns back on itself to become the *parietal pericardium,* which forms the inner lining of the fibrous pericardium (figs. 13.2 and 13.4; see reference plate 3, p. 25). Between the parietal and visceral layers of the pericardium is a space, the *pericardial cavity,* that

Figure 13.2 AP|R

The heart is within the mediastinum and is enclosed by a layered pericardium.

Figure 13.3

Anterior view of a human heart. This photo is not life-size, so a proportionately reduced ruler has been included to help the student grasp the true size of the organ.

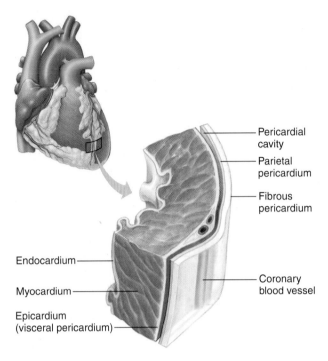

Figure 13.4

The heart wall has three layers: an endocardium, a myocardium, and an epicardium.

contains a small volume of serous fluid (fig. 13.4). This fluid reduces friction between the pericardial membranes as the heart moves within them.

> In *pericarditis,* inflammation of the pericardium due to viral or bacterial infection produces adhesions that attach the layers of the pericardium to each other. This condition is very painful and interferes with heart movements.

Practice

1. Where is the heart located?
2. Distinguish between the visceral pericardium and the parietal pericardium.

Wall of the Heart

The wall of the heart is composed of three distinct layers—an outer epicardium, a middle myocardium, and an inner endocardium (fig. 13.4). The **epicardium** (ep″ĭ-kar′de-um), which corresponds to the visceral pericardium, protects the heart by reducing friction. It is a serous membrane that consists of connective tissue beneath epithelium. Its deeper part typically contains adipose tissue, particularly along the paths of coronary

arteries and cardiac veins that carry blood through the myocardium.

The thick middle layer of the wall of the heart, or **myocardium** (mi″o-kar′de-um), consists mostly of cardiac muscle tissue that pumps blood out of the heart chambers. The muscle fibers are organized in planes, separated by connective tissue richly supplied with blood capillaries, lymph capillaries, and nerve fibers.

The inner layer of the wall of the heart, or **endocardium** (en″do-kar′de-um), consists of epithelium and connective tissue that contains many elastic and collagenous fibers. The endocardium also contains blood vessels and some specialized cardiac muscle fibers, called *Purkinje fibers,* described in section 13.3 on page 350. The endocardium is continuous with the inner linings of blood vessels attached to the heart.

Heart Chambers and Valves

Internally, the heart is divided into four hollow chambers—two on the left and two on the right (fig. 13.5). The upper chambers, called **atria** (a′tre-ah; singular, *atrium*), have thin walls and receive blood returning to the heart. Small, earlike projections called *auricles* extend anteriorly from the atria. The lower chambers, the **ventricles** (ven′trĭ-klz), receive blood from the atria and contract to force blood out of the heart into arteries.

A solid, wall-like **septum** separates the atrium and ventricle on the right side from their counterparts on the

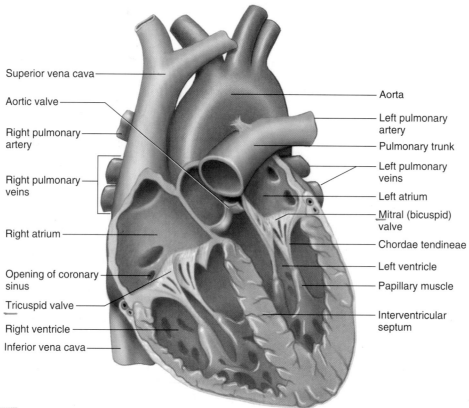

Superior vena cava

Aortic valve

Right pulmonary artery

Right pulmonary veins

Right atrium

Opening of coronary sinus

Tricuspid valve

Right ventricle

Inferior vena cava

Aorta

Left pulmonary artery

Pulmonary trunk

Left pulmonary veins

Left atrium

Mitral (bicuspid) valve

Chordae tendineae

Left ventricle

Papillary muscle

Interventricular septum

Figure 13.5 AP|R

Frontal section of the heart showing the connection between the left ventricle and the aorta, as well as the four hollow chambers. The plane of section does not show the pulmonary valve, and shows only parts of the tricuspid, mitral, and aortic valves.

left. As a result, blood from one side of the heart never mixes with blood from the other side (except in the fetus, see chapter 20, p. 550). An *atrioventricular valve* (AV valve), called the tricuspid on the right and the mitral on the left, ensures one-way blood flow between the atria and the ventricles.

The right atrium receives blood from two large veins—the *superior vena cava* and the *inferior vena cava*. A relatively smaller vein, the *coronary sinus,* also drains blood into the right atrium from the myocardium of the heart.

The large **tricuspid valve,** which has three tapered projections called *cusps* as its name implies, lies between the right atrium and the right ventricle (fig. 13.5). The valve permits blood to move from the right atrium into the right ventricle and prevents backflow.

Strong, fibrous strings called **chordae tendineae** (kor'de ten'di-ne) attach to the cusps of the tricuspid valve on the ventricular side. These strings originate from small mounds of cardiac muscle tissue, the **papillary muscles,** that project inward from the walls of the ventricle. The papillary muscles contract when the ventricle contracts. As the tricuspid valve closes, these muscles pull on the chordae tendineae and prevent the cusps from swinging back into the atrium.

The right ventricle has a thinner muscular wall than the left ventricle (fig. 13.5). This right chamber pumps blood a short distance to the lungs against a relatively

low resistance to blood flow. The left ventricle, on the other hand, must force blood to all the other parts of the body against a much greater resistance to flow.

When the muscular wall of the right ventricle contracts, the blood inside its chamber is put under increasing pressure, and the tricuspid valve closes passively. As a result, the only exit for the blood is through the *pulmonary trunk,* which divides to form the left and right *pulmonary arteries* that lead to the lungs. At the base of this trunk is a **pulmonary valve** with three cusps. This valve allows blood to leave the right ventricle and prevents backflow into the ventricular chamber (fig. 13.6).

The left atrium receives blood from the lungs through four *pulmonary veins*—two from the right lung and two from the left lung. Blood passes from the left atrium into the left ventricle through the **mitral valve** (shaped like a mitre, a type of headpiece), or bicuspid valve, which prevents blood from flowing back into the left atrium from the left ventricle (see fig. 13.5). As with the tricuspid valve, the papillary muscles and the chordae tendineae prevent the cusps of the mitral valve from swinging back into the left atrium during ventricular contraction.

When the left ventricle contracts, the mitral valve closes passively, and the only exit is through a large artery, the **aorta** (a-or'tah). At the base of the aorta is the **aortic valve** (a-or'tik valv), which has three cusps

(a)

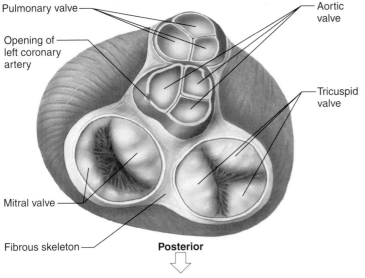

Pulmonary valve

Aortic valve

Opening of left coronary artery

Tricuspid valve

Mitral valve

Fibrous skeleton

Posterior

(b)

Figure 13.6

Heart valves. **(a)** Photograph of a transverse section through the heart, showing the four valves (superior view). **(b)** The skeleton of the heart consists of fibrous rings to which the heart valves are attached (superior view).

(fig. 13.6). The aortic valve opens and allows blood to leave the left ventricle as it contracts. When the ventricular muscles relax, this valve closes and prevents blood from backing up into the left ventricle (see fig. 13.5).

The mitral and tricuspid valves are called atrioventricular valves because they are between atria and ventricles. The pulmonary and aortic valves are called "semilunar" because of the half-moon shapes of their cusps. Table 13.1 summarizes the locations and functions of the heart valves.

Mitral valve prolapse (MVP) affects up to 6% of the U.S. population. In this condition, one (or both) of the cusps of the mitral valve stretches and bulges into the left atrium during ventricular contraction. The valve usually continues to function adequately, but sometimes blood regurgitates into the left atrium. Symptoms of MVP include chest pain, palpitations, fatigue, and anxiety.

Endocarditis, an inflammation of the endocardium due to an infection, appears as a plantlike growth on the mitral valve. Certain species of *Streptococcus* bacteria can damage the valve in this way. People with MVP are particularly susceptible to endocarditis. They take antibiotics before undergoing dental work to prevent *Streptococcus* infection in the mouth from migrating through the blood to the heart and causing infection.

Practice

3. Describe the layers of the heart wall.
4. Name and locate the four chambers of the heart.
5. Describe the function of each heart valve.

Skeleton of the Heart

Rings of dense connective tissue surround the pulmonary trunk and aorta at their proximal ends. These rings provide firm attachments for the heart valves and for muscle fibers; they also prevent the outlets of the atria and ventricles from dilating during contraction. The fibrous rings, together with other masses of dense

Table 13.1	Heart Valves	
Valve	**Location**	**Function**
Tricuspid valve	Opening between right atrium and right ventricle	Prevents blood from moving from right ventricle into right atrium during ventricular contraction
Pulmonary valve	Entrance to pulmonary trunk	Prevents blood from moving from pulmonary trunk into right ventricle during ventricular relaxation
Mitral (bicuspid) valve	Opening between left atrium and left ventricle	Prevents blood from moving from left ventricle into left atrium during ventricular contraction
Aortic valve	Entrance to aorta	Prevents blood from moving from aorta into left ventricle during ventricular relaxation

connective tissue in the part of the septum between the ventricles (interventricular septum), constitute the *skeleton of the heart* (fig. 13.6*b*).

Path of Blood Through the Heart

Blood that is low in oxygen and high in carbon dioxide enters the right atrium through the venae cavae and coronary sinus. As the right atrial wall contracts, the blood passes through the tricuspid valve and enters the chamber of the right ventricle (fig. 13.7). When the right ventricular wall contracts, the tricuspid valve closes, and blood moves through the pulmonary valve and into the pulmonary trunk and its branches (pulmonary arteries).

From the pulmonary arteries, blood enters the capillaries associated with the alveoli (microscopic air sacs) of the lungs. Gases are exchanged between blood in the capillaries (carbon dioxide or CO_2) and air in the alveoli (oxygen or O_2). The freshly oxygenated blood, low in carbon dioxide, returns to the heart through the pulmonary veins that lead to the left atrium.

The left atrial wall contracts, and blood moves through the mitral valve and into the chamber of the left ventricle. When the left ventricular wall contracts, the mitral valve closes, and blood moves through the aortic valve and into the aorta and its branches.

Blood Supply to the Heart

The first two branches of the aorta, called the right and left **coronary arteries,** supply blood to the tissues of the heart. Their openings lie just superior to the aortic valve (fig. 13.8). Blood flow through the coronary arteries increases during ventricular relaxation because the myocardial vessels are not compressed as happens in ventricular contraction, and the closed aortic valve does not block the openings to these vessels.

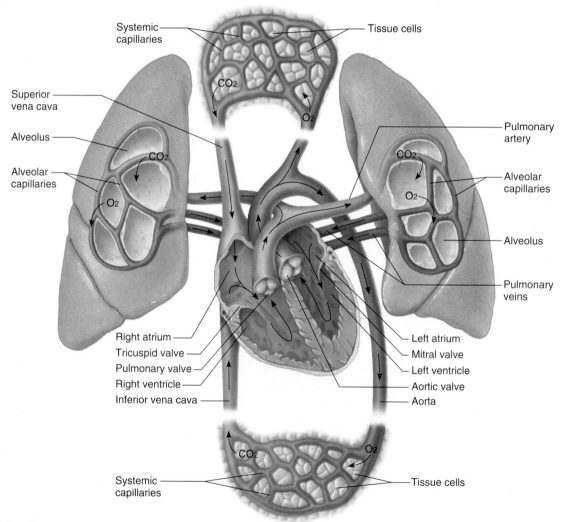

Figure 13.7   AP|R

The right ventricle forces blood to the lungs, whereas the left ventricle forces blood to all other body parts. (Structures are not drawn to scale.)

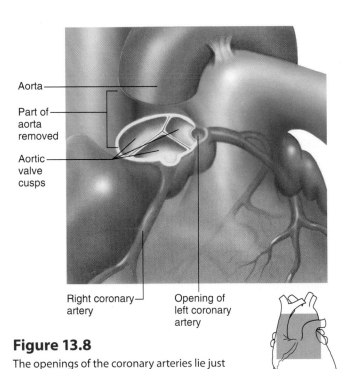

Aorta

Part of aorta removed

Aortic valve cusps

Right coronary artery

Opening of left coronary artery

Figure 13.8
The openings of the coronary arteries lie just superior to the aortic valve.

A thrombus or embolus that partially blocks or narrows a coronary artery branch causes a decrease in blood flow called *ischemia*. This deprives myocardial cells of oxygen, producing a painful condition called *angina pectoris*. The pain usually happens during physical activity, when oxygen demand exceeds oxygen supply. Pain lessens with rest. Emotional disturbance may also trigger angina pectoris.

Angina pectoris may cause a sensation of heavy pressure, tightening, or squeezing in the chest. The pain is usually felt behind the sternum or in the anterior part of the upper thoracic cavity, but may radiate to the neck, jaw, throat, left shoulder, left upper limb, back, or upper abdomen. Other symptoms include profuse perspiration (diaphoresis), shortness of breath (dyspnea), nausea, or vomiting.

A blood clot may completely obstruct a coronary artery (coronary thrombosis), killing part of the heart. This is a *myocardial infarction (MI)* (heart attack).

The heart must beat continually to supply blood to body tissues. To do this, myocardial cells require a constant supply of freshly oxygenated blood. Branches of the coronary arteries feed the many capillaries of the myocardium (fig. 13.9). The smaller branches of these arteries usually have connections (anastomoses) between vessels that provide alternate pathways for blood, called collateral circulation. These detours in circulation may supply oxygen and nutrients to the myocardium when a coronary artery is blocked.

Branches of the **cardiac veins,** whose paths roughly parallel those of the coronary arteries, drain blood that has passed through myocardial capillaries. As figure 13.9*b* shows, these veins join an enlarged vein

on the heart's posterior surface—the **coronary sinus**—which empties into the right atrium (see fig. 13.5).

In *heart transplantation,* the recipient's failing heart is removed, except for the posterior walls of the right and left atria and their connections to the venae cavae and pulmonary veins. The donor heart is prepared similarly and is attached to the atrial structures remaining in the recipient's thoracic cavity. Finally, the recipient's aorta and pulmonary arteries are connected to those of the donor heart.

Each year in the United States, more than 100,000 patients could benefit from a heart transplant, but only about 2,100 organs become available. A solution for some people in need of a heart is a left ventricular assist device, or LVAD. The patient wears a battery pack with controls that attach by a cable to the implanted device. LVADs, which began to be developed in the 1980s, are in their third generation. They are now smaller, lighter, with fewer touching parts. The patients with LVADs report improved quality of life—with more energy, better sleep, and improved mood—until they can receive their heart transplants.

Practice

6. Review the path of blood through the heart.
7. Which vessels supply blood to the myocardium?
8. How does blood return from the cardiac tissues to the right atrium?

13.3 HEART ACTIONS

The heart chambers function in a coordinated fashion. Their actions are regulated so that atria contract, called atrial **systole** (sis′to-le), while ventricles relax, called ventricular **diastole** (di-as′to-le); then ventricles contract (ventricular systole) while atria relax (atrial diastole). Then the atria and ventricles both relax for a brief interval. This series of events constitutes a complete heartbeat, or **cardiac cycle** (kar′de-ak si′kl).

Cardiac Cycle

During a cardiac cycle, pressure in the heart chambers rises and falls. These changes open and close the valves, like the wind blowing open and then closing a door. When pressure in the ventricles is low, early in diastole, the pressure difference between the atria and ventricles opens the AV valves. The ventricles fill. About 70% of the returning blood enters the ventricles prior to contraction, and ventricular pressure gradually increases. During atrial systole, the remaining 30% of returning blood is pushed into the ventricles, and ventricular pressure increases a bit more (fig. 13.10*a*). Then, as the

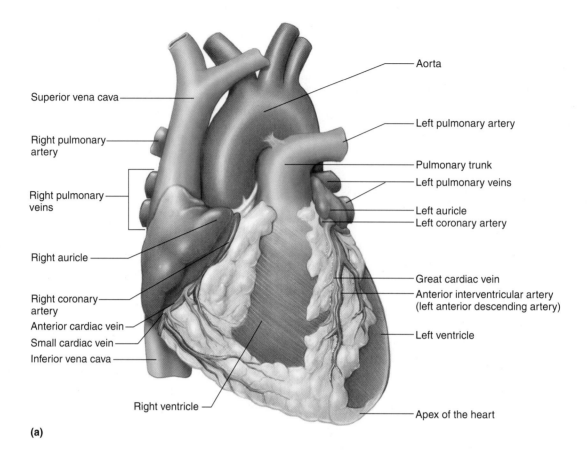

Superior vena cava

Right pulmonary artery

Right pulmonary veins

Right auricle

Right coronary artery

Anterior cardiac vein

Small cardiac vein

Inferior vena cava

Right ventricle

Aorta

Left pulmonary artery

Pulmonary trunk

Left pulmonary veins

Left auricle
Left coronary artery

Great cardiac vein

Anterior interventricular artery (left anterior descending artery)

Left ventricle

Apex of the heart

(a)

Aorta

Left pulmonary artery

Left pulmonary veins

Left auricle

Circumflex artery

Cardiac vein

Left ventricle

Superior vena cava

Right pulmonary artery

Right pulmonary veins

Left atrium

Right atrium

Inferior vena cava

Coronary sinus

Middle cardiac vein

Posterior interventricular artery

Right ventricle

Apex of the heart

(b)

Figure 13.9

Blood vessels associated with the surface of the heart. **(a)** Anterior view. **(b)** Posterior view.

ventricles contract, ventricular pressure rises sharply. As soon as the ventricular pressure exceeds the atrial pressure, the AV valves close. At the same time, the papillary muscles contract, and by pulling on the chordae tendineae, they prevent the cusps of the AV valves from bulging too far into the atria.

During ventricular systole, the AV valves remain closed. The atria are now relaxed, and pressure in the atria is quite low, even lower than venous pressure. As a result, blood flows into the atria from the large, attached veins. That is, as the ventricles are contracting, the atria are filling, already preparing for the next cardiac cycle (fig. 13.10*b*).

When ventricular pressure exceeds the pressure in the pulmonary trunk (right side) and aorta (left side), the pulmonary and aortic valves open; blood is ejected from each valve's respective ventricle into these arteries. As blood flows out of the ventricles, ventricular pressure begins to drop, dipping even further as the ventricles begin to relax. When ventricular pressure is lower than blood pressure in the aorta and pulmonary trunk, the pressure difference is reversed, and the semilunar valves close. The ventricles continue to relax, and as soon as ventricular pressure is less than atrial pressure, the AV valves open, and the ventricles begin to refill. Atria and ventricles are both relaxed for a brief interval. The graph in Appendix D (p. 567) summarizes some of the changes that occur during a cardiac cycle.

Heart Sounds

A heartbeat heard through a stethoscope sounds like *lubb-dupp*. These sounds are due to vibrations in the heart tissues associated with the valves closing.

The first part of a heart sound (*lubb*) originates during ventricular contraction, when the AV valves are closing. The second part (*dupp*) occurs during ventricular relaxation, when the pulmonary and aortic valves are closing.

Heart sounds provide information about the condition of the heart valves. For example, inflammation of the endocardium (endocarditis) may erode the edges of the valvular cusps, which then may not close completely. If blood leaks back through the valve, a sound called a *murmur* can be heard. The severity of a murmur depends on the degree of valvular damage. Many heart murmurs are harmless. Open heart surgery may repair or replace severely damaged valves.

Cardiac Muscle Fibers

Cardiac muscle fibers function much like those of skeletal muscles, but the fibers connect in branching networks. Stimulation to any part of the network sends impulses throughout the heart, which contracts as a unit.

A mass of merging cells that function as a unit is called a **functional syncytium** (funk'shun-al sin-sish'e-um). Two such structures are in the heart—in the atrial walls and in the ventricular walls. Parts of the heart's fibrous skeleton separate these masses of cardiac muscle fibers from each other, except for a small area in the right atrial floor. In this region, fibers of the cardiac conduction system connect the *atrial syncytium* and the *ventricular syncytium*.

Practice

9. Describe the pressure changes in the atria and ventricles during a cardiac cycle.
10. What causes heart sounds?
11. What is a functional syncytium?

Cardiac Conduction System

Throughout the heart are clumps and strands of specialized cardiac muscle tissue whose fibers contain only a few myofibrils. Instead of contracting, these areas

(a)

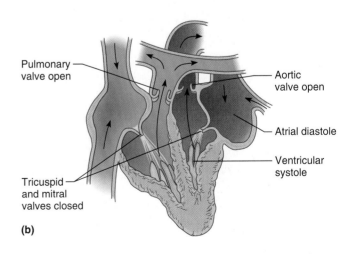

(b)

Figure 13.10 AP|R

A cardiac cycle. The atria **(a)** empty during atrial systole and **(b)** fill with blood during atrial diastole.

initiate and distribute impulses throughout the myocardium. They form the **cardiac conduction system** (kar'de-ak kon-duk'shun sis'tem), which coordinates the events of the cardiac cycle (fig. 13.11).

A key part of the cardiac conduction system is the **SA node,** or **sinoatrial node,** which is a small elongated mass of specialized cardiac muscle tissue located just beneath the epicardium in the right atrium near the opening of the superior vena cava. Its fibers are continuous with those of the atrial syncytium.

The cells of the SA node can reach threshold on their own, and their cell membranes contact one another. Without stimulation from nerve fibers or any other outside agents, the nodal cells initiate cardiac impulses that spread through preferential pathways into the surrounding myocardium and stimulate cardiac muscle fibers to contract.

SA node activity is rhythmic, initiating one cardiac impulse after another, seventy to eighty times a minute in an adult. Because it generates the heart's rhythmic contractions, the SA node is also called the **pacemaker.**

Figure 13.12 traces the path of a cardiac impulse. As a cardiac impulse travels from the SA node into the atrial syncytium, the right and left atria begin to contract almost simultaneously. The cardiac impulse does not pass directly into the ventricular syncytium, which is separated from the atrial syncytium by the fibrous skeleton of the heart. Instead, the cardiac impulse passes along fibers (junctional fibers) of the conduction system that lead to a mass of specialized cardiac muscle tissue called the **AV node,** or **atrioventricular node.** This

node is located in the inferior part of the septum that separates the atria (interatrial septum) and just beneath the endocardium. It provides the only normal conduction pathway between the atrial and ventricular syncytia.

The junctional fibers that conduct the cardiac impulse into the AV node have very small diameters, and because small fibers conduct impulses slowly, they delay transmission. The impulse is delayed further as it moves through the AV node. This allows more time for the atria to contract completely, enabling them to empty all their blood into the ventricles before they contract.

Once the cardiac impulse reaches the distal side of the AV node, it passes into a group of large fibers that make up the **AV bundle** (bundle of His). The AV bundle enters the upper part of the interventricular septum and divides into right and left bundle branches that lie just beneath the endocardium. About halfway down the septum, the branches give rise to enlarged **Purkinje fibers** (pur-kin'je fi'berz).

Purkinje fibers spread from the interventricular septum into the papillary muscles, which project inward from ventricular walls and then continue downward to the apex of the heart. There they curve around the tips of the ventricles and pass upward over the lateral walls of these chambers. Along the way, the Purkinje fibers give off many small branches, which become continuous with cardiac muscle fibers.

The muscle fibers in ventricular walls form irregular whorls. When impulses on the Purkinje fibers stimulate these muscle fibers, the ventricular walls contract with a twisting motion (fig. 13.13). This action squeezes blood out of the ventricular chambers and forces it into the aorta and pulmonary trunk.

Figure 13.11   AP|R

The cardiac conduction system coordinates the cardiac cycle.

Figure 13.12

Components of the cardiac conduction system.

(a)

(b)

Figure 13.13

The muscle fibers within the ventricular walls form patterns of whorls. The fibers of groups **(a)** and **(b)** surround both ventricles in these anterior views of the heart.

(a)

(b)

Figure 13.14

An electrocardiogram records electrical changes in the myocardium during a cardiac cycle. **(a)** A normal ECG. **(b)** In an ECG pattern, the P wave results from a depolarization of the atria, the QRS complex results from a depolarization of the ventricles, and the T wave results from a repolarization of the ventricles.

Q: *What two electrical events occur during the QRS complex?*

The answer can be found in Appendix E on page 568.

Practice

12. What types of tissues make up the cardiac conduction system?
13. How is a cardiac impulse initiated?
14. How is a cardiac impulse transmitted from the right atrium to the other heart chambers?

Electrocardiogram

An **electrocardiogram** (e-lek″tro-kar′de-o-gram″), or **ECG,** is a recording of the electrical changes in the myocardium during a cardiac cycle. (This pattern is generated as action potentials stimulate cardiac muscle fibers to contract, but it is not the same as individual action potentials.) These changes are detectable on the surface of the body because body fluids can conduct electrical currents.

To record an ECG, electrodes are placed on the skin and connected by wires to an instrument that responds to very weak electrical changes by moving a pen or stylus on a moving strip of paper. Up-and-down movements of the pen correspond to electrical changes in the myocardium. Because the paper moves past the pen at a known rate, the distance between pen deflections indicates the time between phases of the cardiac cycle.

As figure 13.14a illustrates, a normal ECG pattern includes several deflections, or *waves,* during each cardiac cycle. Between cycles, the muscle fibers remain polarized, with no detectable electrical changes, and the pen does not move but simply marks along the baseline. When the SA node triggers a cardiac impulse, atrial fibers depolarize, producing an electrical change. The pen moves, and at the end of the electrical change, returns to the base position. This first pen movement produces a *P wave,* corresponding to depolarization of the atrial fibers that will lead to contraction of the atria (fig. 13.14b).

When the cardiac impulse reaches the ventricular fibers, they rapidly depolarize. Because ventricular walls are thicker than those of the atria, the electrical change is greater, and the pen deflects more. When the electrical change ends, the pen returns to the baseline, leaving a mark called the *QRS complex.* This mark consists of a *Q wave,* an *R wave,* and an *S wave,* and corresponds to depolarization of ventricular fibers just prior to the contraction of the ventricular walls.

The electrical changes occurring as the ventricular muscle fibers repolarize slowly produce a *T wave* as the pen deflects again, ending the ECG pattern. The record of atrial repolarization seems to be missing from the pattern because atrial fibers repolarize at the same time that ventricular fibers depolarize. Thus, the QRS complex obscures the recording of atrial repolarization.

Physicians use ECG patterns to assess the heart's ability to conduct impulses. For example, the time period between the beginning of a P wave and the beginning of a QRS complex, called the *PQ interval* (or if the initial part of the QRS complex is upright, the PR interval), indicates the time for the cardiac impulse

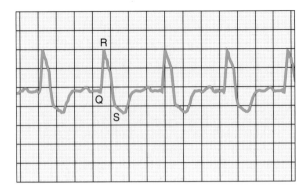

Figure 13.15
A prolonged QRS complex may result from damage to the AV bundle fibers.

to travel from the SA node through the AV node. Ischemia or other problems affecting the fibers of the AV conduction pathways can increase this PQ interval. Similarly, injury to the AV bundle can extend the QRS complex, because it may take longer for an impulse to spread throughout the ventricular walls (fig. 13.15).

Practice

15. What is an electrocardiogram?

16. Which cardiac events do the P wave, QRS complex, and T wave represent?

Regulation of the Cardiac Cycle

The volume of blood pumped changes to accommodate cellular requirements. For example, during strenuous exercise, skeletal muscles require more blood, and the heart rate increases in response. Changes in heart rate are often a response to factors that affect the SA node, such as the motor impulses carried on the parasympathetic and sympathetic nerve fibers (see chapter 9, p. 250).

The parasympathetic fibers that innervate the heart arise from neurons in the medulla oblongata (fig. 13.16). Most of these fibers branch to the SA and AV nodes. When the nerve impulses reach axon terminals, they secrete acetylcholine, which decreases SA and AV nodal activity. As a result, the heart rate decreases.

Parasympathetic fibers carry impulses continually to the SA and AV nodes, "braking" heart action. Consequently, parasympathetic activity can change heart rate in either direction. An increase in the impulses slows the heart rate, and a decrease in the impulses releases the parasympathetic "brake" and increases the heart rate.

Sympathetic fibers reach the heart and join the SA and AV nodes as well as other areas of the atrial and ventricular myocardium. Their axon terminals secrete norepinephrine in response to nerve impulses, which increases the rate and force of myocardial contractions.

Reflexes called *baroreceptor reflexes* involving the *cardiac control center* of the medulla oblongata maintain balance between the inhibitory effects of parasympathetic fibers and the excitatory effects of sympathetic fibers. This center receives sensory impulses from throughout the cardiovascular system and relays motor impulses to the heart in response. For example, receptors sensitive to stretch are located in certain regions of the aorta (aortic arch) and in the carotid arteries (carotid sinuses) (fig. 13.16). These receptors, called *baroreceptors* (pressoreceptors), can detect changes in blood pressure. Rising pressure stretches the receptors, and they signal the cardioinhibitor center in the medulla oblongata. In response, the medulla oblongata sends parasympathetic impulses to the heart, decreasing heart rate. This action helps lower blood pressure toward normal.

Impulses from the cerebrum or hypothalamus also influence the cardiac control center. Such impulses may decrease heart rate, as occurs when a person faints following an emotional upset, or they may increase heart rate during a period of anxiety.

Two other factors that influence heart rate are temperature change and certain ions. Rising body temperature increases heart action, which is why heart rate usually increases during fever. On the other hand, abnormally low body temperature decreases heart action.

The most important ions that influence heart action are potassium (K^+) and calcium (Ca^{+2}). In hyperkalemia, excess extracellular potassium ions decrease the rate and force of contractions. In hypokalemia, deficient extracellular potassium ions may cause a potentially life-threatening abnormal heart rhythm (arrhythmia).

In hypercalcemia, excess extracellular calcium ions increase heart actions, risking dangerously extended heart contraction. Conversely, in hypocalcemia, low extracellular calcium concentration depresses heart action.

An abnormally fast heartbeat, usually more than 100 beats per minute, is called *tachycardia*. Increase in body temperature, nodal stimulation by sympathetic fibers, certain drugs or hormones, heart disease, excitement, exercise, anemia, or shock can cause tachycardia.

Bradycardia means a slow heart rate, usually fewer than 60 beats per minute. Decreased body temperature, nodal stimulation by parasympathetic impulses, or certain drugs may cause bradycardia. It may also occur during sleep.

Practice

17. How do parasympathetic and sympathetic impulses help control heart rate?

18. How do changes in body temperature affect heart rate?

19. Describe the effects on the heart of abnormal concentrations of potassium and calcium ions.

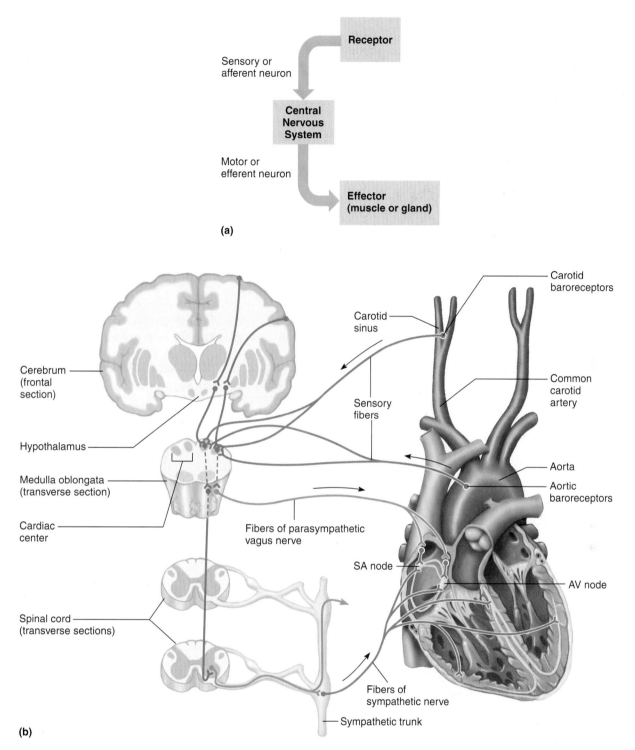

Figure 13.16 AP|R

Baroreceptor reflex. **(a)** Schematic of a general reflex arc. Note the similarity to figure 1.5 on page 6. **(b)** Autonomic nerve impulses alter the activities of the SA and AV nodes.

13.4 BLOOD VESSELS

The blood vessels form a closed circuit of tubes that carries blood from the heart to the body cells and back again. These vessels include arteries, arterioles, capillaries, venules, and veins.

The 62,000 or so miles of blood vessels in an average human body would stretch 2 1/2 times around the world if extended end-to-end.

Arteries and Arterioles

Arteries are strong, elastic vessels that are adapted for carrying blood away from the heart under relatively high pressure. These vessels subdivide into progressively thinner tubes and eventually give rise to finer, branched **arterioles** (ar-te′re-olz).

The wall of an artery consists of three distinct layers (fig. 13.17*a*). The innermost layer (*tunica interna*) is composed of a layer of simple squamous epithelium, called *endothelium,* that rests on a connective tissue membrane that is rich in elastic and collagenous fibers. Endothelium helps prevent blood clotting by providing a smooth surface that allows blood cells and platelets to flow through without being damaged and by secreting biochemicals that inhibit platelet aggregation. Endothelium also may help regulate local blood flow by secreting substances that dilate or constrict blood vessels. For example, endothelium releases the gas nitric oxide, which relaxes the smooth muscle of the vessel. Clinical Application 13.1 describes atherosclerosis, in which fatty deposits accumulate on the inner walls of arteries.

The middle layer (*tunica media*) makes up the bulk of the arterial wall. It includes smooth muscle fibers, which encircle the tube, and a thick layer of elastic connective tissue.

The outer layer (*tunica externa*) is relatively thin and chiefly consists of connective tissue with irregular elastic and collagenous fibers. This layer attaches the artery to the surrounding tissues.

If blood pressure dilates a weakened area of an artery wall, a pulsating sac called an *aneurysm* may form and enlarge. A sac that develops by a longitudinal splitting of the middle layer of the arterial wall is called a *dissecting aneurysm*. An aneurysm may cause symptoms by pressing on nearby organs, or it may rupture and cause great blood loss.

Aneurysms may also result from trauma, high blood pressure, infections, inherited disorders such as Marfan syndrome, or congenital defects in blood vessels. Common sites of aneurysms include the thoracic and abdominal aorta and an arterial circle at the base of the brain (circle of Willis). If areas of aortic weakening are detected before they balloon into aneurysms, the affected part of the vessel can be replaced with a synthetic or cadaver graft. A new technique grafts veins from the patient's own thighs, which are less likely to cause infection or be rejected by the immune system.

The sympathetic branches of the autonomic nervous system innervate smooth muscle in artery and arteriole walls. Impulses on these *vasomotor fibers* stimulate the smooth muscles to contract, reducing the diameter of the vessel. This is called **vasoconstriction** (vas″o-kon-strik′shun). If vasomotor impulses are inhibited, the muscle fibers relax, and the diameter of the vessel increases. This is called **vasodilation** (vas″o-di-la′shun). Changes in the diameters of arteries and arterioles greatly influence blood flow and blood pressure.

The walls of the larger arterioles have three layers, similar to those of arteries. These walls thin as arterioles

Figure 13.17 AP|R

Blood vessels. **(a)** The wall of an artery. **(b)** The wall of a vein.

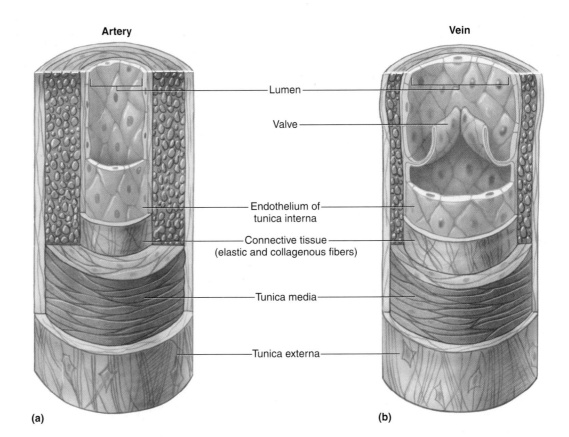

Artery   Vein

Lumen

Valve

Endothelium of tunica interna

Connective tissue (elastic and collagenous fibers)

Tunica media

Tunica externa

(a)   (b)

Clinical Application 13.1

Atherosclerosis

In the arterial disease *atherosclerosis* (ath"er-o-sklĕ-ro'sis), deposits of fatty materials, particularly cholesterol, form within and on the inner lining of the arterial walls. Such deposits, called *plaque,* protrude into the lumens of vessels and interfere with blood flow (fig. 13A). Furthermore, plaque often forms a surface texture that can initiate formation of a blood clot, increasing the risk of developing thrombi or emboli that cause decreased blood flow (ischemia) or tissue death (necrosis) downstream from the obstruction.

Plaque accumulation in coronary arteries may cause a *coronary thrombosis* or *coronary embolism.* Similar changes in brain arteries increase the risk of a *cerebral vascular accident (CVA),* or *stroke,* due to cerebral thrombosis, embolism, or hemorrhage. In addition, the walls of affected arteries may degenerate, losing their elasticity and becoming hardened, or *sclerotic.* In this stage of the disease a sclerotic vessel may rupture under the force of blood pressure.

Risk factors for developing atherosclerosis include a fatty diet, elevated blood pressure, tobacco smoking, obesity, and lack of physical exercise. Genetic factors may also increase the risk of developing atherosclerosis.

For millions of people who cannot control cholesterol with diet and exercise, drugs called statins can inhibit a liver enzyme that the body uses to synthesize cholesterol. Statins dramatically reduce LDL cholesterol, moderately increase HDL cholesterol, and regulate triglyceride levels, all of which lower the risk of atherosclerosis.

Several invasive treatments attempt to clear clogged arteries. In *percutaneous transluminal angioplasty,* a thin, plastic catheter with a tiny deflated balloon at its tip is passed through an incision in the skin into a large artery in the arm or thigh, guided through other arteries and into the lumen of the affected blood vessel. Once in position at the blockage, the balloon is inflated for several minutes, and presses the plaque against the arterial wall, widening the arterial lumen and restoring blood flow. Sometimes the catheter also introduces a stent, which is a coiled steel tube that opens as the balloon inflates, helping to keep the lumen clear. However, blockage can recur if the underlying cause of cholesterol buildup is not addressed.

Laser energy is also used to destroy atherosclerotic plaque and to channel through arterial obstructions to increase blood flow. In *laser angioplasty* the laser is passed through an incision in the skin and when it is in the lumen of an obstructed artery, light energy is transmitted through the optical fibers of the laser.

Another invasive procedure for treating arterial obstruction is *bypass graft surgery.* A surgeon uses part of a vein from the patient's lower limb or elsewhere to connect a healthy artery to the affected artery at a point beyond the obstruction. Blood from the healthy artery then bypasses the narrowed region of the affected artery, supplying the tissues

downstream. The vein is connected backward, so that its valves do not impede blood flow.

A new treatment for atherosclerosis is *fibroblast growth factor,* a body chemical given as a drug. It stimulates new blood vessels to grow in the heart, a process called angiogenesis.

(a)

(b)

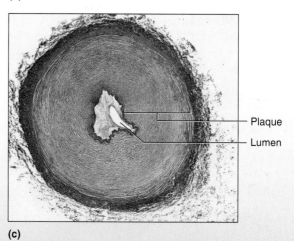

(c)

Figure 13A

Development of atherosclerosis. **(a)** Normal arteriole (100×). **(b)** and **(c)** Accumulation of plaque on the inner wall of the arteriole (*b* and *c* 100×).

approach capillaries. The wall of a very small arteriole consists only of an endothelial lining and some smooth muscle fibers, surrounded by a small amount of connective tissue (fig. 13.18).

Practice

20. Describe the wall of an artery.
21. What is the function of smooth muscle in the arterial wall?
22. How is the structure of an arteriole different from that of an artery?

Capillaries

Capillaries (kap′ĭ-lar″ēz), the smallest-diameter blood vessels, connect the smallest arterioles and the smallest venules. Capillaries are extensions of the inner linings of arterioles in that their walls are composed of endothelium (fig. 13.18). These thin walls form the semipermeable layer through which substances in the blood are exchanged for substances in the tissue fluid surrounding body cells.

The openings in capillary walls are thin slits where endothelial cells overlap (fig. 13.19). The sizes of these openings and, consequently, the permeability of the capillary wall vary from tissue to tissue. For example, the openings are smaller in capillaries of smooth, skeletal, and cardiac muscle than they are in capillaries associated with endocrine glands, the kidneys, and the lining of the small intestine.

Capillary density reflects tissues' rates of metabolism. Muscle and nerve tissues, which use abundant oxygen and nutrients, are richly supplied with capillaries. Tissues with slower metabolic rates have fewer capillaries, such as cartilage, or lack them entirely as in the cornea.

The spatial patterns of capillaries also differ in various body parts. For example, some capillaries pass directly from arterioles to venules, but others lead to highly branched networks (fig. 13.20).

Smooth muscles that encircle capillary entrances regulate blood distribution in capillary pathways. These muscles form *precapillary sphincters,* which may close a capillary by contracting or open it by relaxing (see fig. 13.18). A precapillary sphincter responds to the demands of the cells the capillary supplies. When these cells have low concentrations of oxygen and nutrients, the sphincter relaxes; when cellular requirements have been met, the sphincter may contract again. In this way, blood flow can follow different pathways through a tissue to meet the changing cellular requirements.

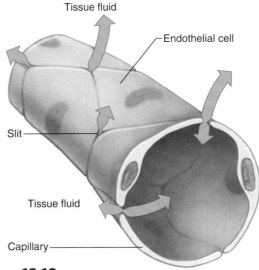

Figure 13.19
In capillaries, substances are exchanged between the blood and tissue fluid through openings (slits) separating endothelial cells.

Figure 13.20
Light micrograph of a capillary network (100×).

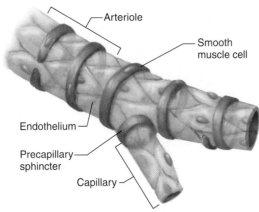

Figure 13.18
The smallest arterioles have only a few smooth muscle fibers in their walls. Capillaries lack these fibers.

Routing of blood flow to different parts of the body is due to vasoconstriction and vasodilation of arterioles and precapillary sphincters. During exercise, for example, blood enters the capillary networks of the skeletal muscles, where the cells have increased oxygen and nutrient requirements. At the same time, blood can bypass some of the capillary networks in the digestive tract tissues, where demand for blood is less immediate.

Practice

23. Describe a capillary wall.
24. What is the function of a capillary?
25. What controls blood flow into capillaries?

Exchanges in Capillaries

Gases, nutrients, and metabolic by-products are exchanged between the blood in capillaries and the tissue fluid surrounding body cells. The substances exchanged move through capillary walls by diffusion, filtration, and osmosis (see chapter 3, pp. 60–64).

Because blood entering systemic capillaries carries high concentrations of oxygen and nutrients, these substances diffuse through the capillary walls and enter the tissue fluid. Conversely, the concentrations of carbon dioxide and other wastes are generally greater in the tissues, and such wastes diffuse into the capillary blood.

Plasma proteins generally remain in the blood because they are too large to diffuse through the membrane pores or slitlike openings between the endothelial cells of most capillaries. Also, these bulky proteins are not soluble in the lipid parts of capillary cell membranes.

Whereas diffusion depends on concentration gradients, filtration with hydrostatic pressure forces molecules through a membrane. In capillaries, the blood pressure generated when ventricle walls contract provides the force for filtration.

Blood pressure also moves blood through the arteries and arterioles. This pressure decreases as the distance from the heart increases, because of friction (peripheral resistance) between the blood and the vessel walls. For this reason, blood pressure is greater in the arteries than in the arterioles, and greater in the arterioles than in the capillaries. Blood pressure is similarly greater at the arteriolar end of a capillary than at the venular end. Therefore, the filtration effect occurs primarily at the arteriolar ends of capillaries.

The presence of an impermeant solute on one side of a cell membrane creates an osmotic pressure. Plasma proteins trapped in the capillaries create an osmotic pressure that draws water into the capillaries. The term *colloid osmotic pressure* describes this osmotic effect due solely to the plasma proteins.

The effect of capillary blood pressure, which favors filtration, opposes the actions of the plasma colloid osmotic pressure, which favors reabsorption. At the arteriolar end of capillaries, the blood pressure is higher than the colloid osmotic pressure, so filtration predominates. At the venular end, the colloid osmotic pressure is essentially unchanged, but the blood pressure has decreased due to resistance through the capillary, so reabsorption predominates (fig. 13.21).

Net force at arteriolar end	
Outward force, including hydrostatic pressure	= 35 mm Hg
Inward force of osmotic pressure	= 24 mm Hg
Net outward pressure	= 11 mm Hg

Net force at venular end	
Outward force, including hydrostatic pressure	= 16 mm Hg
Inward force of osmotic pressure	= 24 mm Hg
Net inward pressure	= 8 mm Hg

Figure 13.21 AP|R

Water and other substances leave capillaries because of a net outward pressure at the capillaries' arteriolar ends. Water enters at the capillaries' venular ends because of a net inward pressure. Substances move in and out along the length of the capillaries according to their respective concentration gradients.

Q: *Which are the substances that do not leave the blood at the arteriolar end of the capillary and whose presence at the venular end of the capillary draws water by osmosis back into the capillary?*

The answer can be found in Appendix E on page 568.

Normally, more fluid leaves the capillaries than returns to them. Closed-ended vessels called lymphatic capillaries collect the excess fluid and return it through lymphatic vessels to the venous circulation. Chapter 14 (pp. 380–381) discusses this mechanism.

Unusual events may increase blood flow to capillaries, causing excess fluid to enter the spaces between tissue cells. This may occur in response to certain chemicals, such as *histamine,* that vasodilate the arterioles near capillaries and increase capillary permeability. Enough fluid may leak out of the capillaries to overwhelm lymphatic drainage. Affected tissues become swollen (edematous) and painful.

Practice

26. Which forces affect the exchange of substances between blood and tissue fluid?
27. Why is the fluid movement out of a capillary greater at its arteriolar end than at its venular end?

Venules and Veins

Venules (ven′ūlz) are the microscopic vessels that continue from the capillaries and merge to form **veins** (vānz). The veins, which carry blood back to the atria, follow pathways that roughly parallel those of the arteries.

The walls of veins are similar to those of arteries in that they are composed of three distinct layers (see fig. 13.17*b*). However, the middle layer of the venous wall is poorly developed compared to that of the arterial wall. Consequently, veins have thinner walls that have less smooth muscle and less elastic connective tissue than those of comparable arteries, but their lumens have a greater diameter (fig. 13.22).

Figure 13.22
Note the structural differences in these cross sections of a vein (left) and an artery (right) (60×).

Many veins, particularly those in the upper and lower limbs, have flaplike *valves,* which project inward from their linings. Valves are usually composed of two leaflets that close if blood begins to back up in a vein (fig. 13.23). These valves aid in returning blood to the heart because they open as long as the blood flow is toward the heart, but close if it is in the opposite direction.

Veins also function as blood reservoirs. For example, in hemorrhage accompanied by a drop in arterial blood pressure, sympathetic nerve impulses reflexly stimulate the muscular walls of the veins. The resulting venous constrictions help maintain blood pressure by returning more blood to the heart. This mechanism ensures a nearly normal blood flow even when as much as 25% of blood volume is lost. Table 13.2 summarizes the characteristics of blood vessels.

Varicose veins are abnormal and irregular dilations in superficial veins, particularly in the legs. This condition is usually associated with prolonged, increased back pressure in the affected vessels due to gravity, as when a person stands. Crossing the legs or sitting in a chair so that its edge presses against the area behind the knee can obstruct venous blood flow and aggravate varicose veins.

Increased venous back pressure stretches and widens the veins. The valves in these vessels do not change size, so they lose their abilities to block the backward flow of blood, and blood accumulates in the enlarged regions. Increased venous pressure is also accompanied by rising pressure in the venules and capillaries that supply the veins. Consequently, tissues in affected regions typically become edematous and painful.

Practice

28. How does the structure of a vein differ from that of an artery?
29. How does venous circulation help maintain blood pressure when hemorrhaging causes blood loss?

Figure 13.23
Venous valves **(a)** allow blood to move toward the heart, but **(b)** prevent blood from moving backward away from the heart.

Table 13.2	Characteristics of Blood Vessels	
Vessel	**Type of Wall**	**Function**
Artery	Thick, strong wall with three layers—an endothelial lining, a middle layer of smooth muscle and elastic connective tissue, and an outer layer of connective tissue	Carries blood under relatively high pressure from heart to arterioles
Arteriole	Thinner wall than an artery but with three layers; smaller arterioles have an endothelial lining, some smooth muscle tissue, and a small amount of connective tissue	Connects an artery to a capillary; helps control blood flow into a capillary by vasoconstricting or vasodilating
Capillary	Single layer of squamous epithelium	Allows nutrients, gases, and wastes to be exchanged between the blood and tissue fluid; connects an arteriole to a venule
Venule	Thinner wall than in an arteriole, less smooth muscle and elastic connective tissue	Connects a capillary to a vein
Vein	Thinner wall than an artery but with similar layers; the middle layer is more poorly developed; some have flaplike valves	Carries blood under relatively low pressure from a venule to the heart; valves prevent backflow of blood; serves as a blood reservoir

13.5 BLOOD PRESSURE

Blood pressure is the force blood exerts against the inner walls of blood vessels. Although this force occurs throughout the vascular system, the term *blood pressure* most commonly refers to pressure in arteries supplied by branches of the aorta (systemic arteries).

Contraction of the human heart can create enough pressure to squirt blood almost ten feet into the air.

Arterial Blood Pressure

Arterial blood pressure rises and falls in a pattern corresponding to the phases of the cardiac cycle. That is, contracting ventricles (ventricular systole) squeeze blood out and into the pulmonary trunk and aorta, which sharply increases the pressures in these arteries. The maximum pressure during ventricular contraction is called the **systolic pressure** (sis-tol'ik presh'ur). When the ventricles relax (ventricular diastole), the arterial pressure drops, and the lowest pressure that remains in the arteries before the next ventricular contraction is termed the **diastolic pressure** (di-a-stol'ik presh'ur). A sphygmomanometer is used to measure arterial blood pressure, reporting a fraction that is normally about 120/80. The upper number indicates the arterial systolic pressure in mm Hg (SP), and the lower number indicates the arterial diastolic pressure in mm Hg (DP). Figure 13.24 shows how these pressures decrease as distance from the left ventricle increases.

The surge of blood entering the arterial system during a ventricular contraction distends the elastic arterial walls, but the pressure begins to drop almost

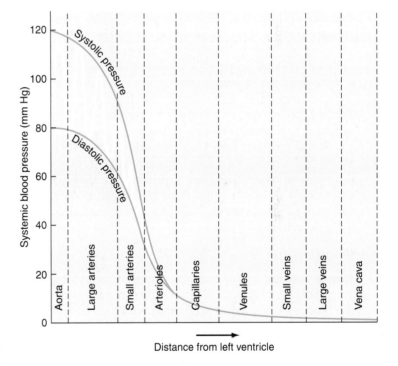

Figure 13.24

Blood pressure decreases as the distance from the left ventricle increases. Systolic pressure occurs during maximal ventricular contraction. Diastolic pressure occurs when the ventricles relax.

immediately as the contraction ends, and the arterial walls recoil. This alternate expanding and recoiling of the arterial wall can be felt as a *pulse* in an artery that runs close to the surface. The radial artery is commonly used to take a person's pulse. Other sites where an arterial pulse is easily detected include the carotid, brachial, and femoral arteries.

Factors That Influence Arterial Blood Pressure

Arterial blood pressure depends on a variety of factors. These include cardiac output, blood volume, peripheral resistance, and blood viscosity (fig. 13.25).

Cardiac Output

In addition to producing blood pressure by forcing blood into the arteries, heart action determines how much blood enters the arterial system with each ventricular contraction. The volume of blood discharged from the ventricle with each contraction is called the **stroke volume** and equals about 70 milliliters in an average-weight male at rest. The volume discharged from the ventricle per minute is called the **cardiac output.** It is calculated by multiplying the stroke volume by the heart rate in beats per minute (cardiac output = stroke volume × heart rate). Thus, if the stroke volume is 70 milliliters and the heart rate is 72 beats per minute, the cardiac output is 5,040 milliliters per minute.

Blood pressure varies with cardiac output. If either stroke volume or heart rate increases, so does cardiac output, and as a result, blood pressure initially rises. Conversely, if stroke volume or heart rate decreases, cardiac output decreases, and blood pressure also initially decreases.

Blood Volume

Blood volume equals the sum of the formed elements and plasma volumes in the vascular system. Although the blood volume varies somewhat with age, body size, and sex, it is usually about 5 liters for adults, or 8% of body weight in kilograms.

Blood pressure is normally directly proportional to blood volume in the cardiovascular system. Thus, any changes in blood volume can initially alter blood pressure. For example, if a hemorrhage reduces blood volume, blood pressure initially drops. If a transfusion restores normal blood volume, normal blood pressure may be reestablished. Blood volume can also fall if the fluid balance is upset, as happens in dehydration. Fluid replacement can reestablish normal blood volume and pressure.

> Black licorice contains a sweet-tasting compound called glycyrrhizic acid that can raise blood pressure if a person consumes it in excess.

Peripheral Resistance

Friction between the blood and the walls of blood vessels produces a force called **peripheral resistance** (pĕ-rif′er-al re-zis′tans), which hinders blood flow. Blood pressure must overcome this force if the blood is to continue flowing. Factors that alter the peripheral resistance change blood pressure. For example, contracting smooth muscles in arteriolar walls increase the peripheral resistance by constricting these vessels. Blood tends to back up into the arteries supplying the arterioles, and the arterial pressure rises. Dilation of arterioles has the opposite effect—peripheral resistance decreases, and arterial blood pressure drops in response.

Blood Viscosity

Viscosity (vis-kos′ĭ-te) is the ease with which the molecules of a fluid flow past one another. The greater the viscosity, the greater the resistance to flowing.

Blood cells and plasma proteins increase blood viscosity. The greater the blood's resistance to flowing, the greater is the force needed to move it through the vascular system. Thus, it is not surprising that blood pressure rises as blood viscosity increases and drops as viscosity decreases.

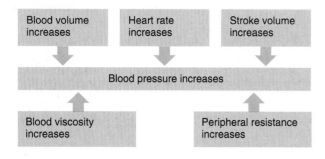

Figure 13.25
Some of the factors that influence arterial blood pressure.

Control of Blood Pressure

Blood pressure (BP) is determined by cardiac output (CO) and peripheral resistance (PR) according to this relationship:

$$BP = CO \times PR$$

Maintenance of normal arterial pressure therefore requires regulation of these two factors. For example, the volume of blood entering the ventricle affects cardiac output, depending on the volume of blood discharged from the ventricle (stroke volume). Entering blood mechanically stretches myocardial fibers in the ventricular wall. Within limits, the longer these fibers, the greater is the force with which they contract.

The relationship between fiber length (due to stretching of the cardiac muscle cell just before contraction) and force of contraction is called the *Frank-Starling law of the heart*. This relationship becomes important, for example, during exercise when much more blood returns to the heart from the veins. The more blood that enters the heart from the veins, the greater the ventricular distension, the stronger the contraction, the greater the stroke volume, and the greater the cardiac output. Conversely, the less blood that returns from the veins, the less the ventricle distends, the weaker the ventricular contraction, the lesser the stroke volume and cardiac output. This mechanism ensures that the volume of blood discharged from the heart is equal to the volume entering its chambers.

Baroreceptors are neurons in the walls of the aorta and carotid arteries that sense changes in blood pressure. If arterial pressure increases, nerve impulses travel from the baroreceptors to the cardiac center of the medulla oblongata. This center relays parasympathetic impulses to the SA node in the heart, and the heart rate decreases in response. As a result of this *cardioinhibitor reflex,* cardiac output falls, and blood pressure decreases toward the normal level (fig. 13.26). Conversely, decreasing arterial blood pressure initiates the *cardioaccelerator reflex,* which sends sympathetic impulses to the SA node. As a result, the heart beats faster, increasing cardiac output and arterial pressure. Other factors that increase heart rate and blood pressure include exercise, a rise in body temperature, and emotional responses, such as fear and anger. Clinical Application 13.2 discusses the effects of exercise on cardiovascular functioning.

Peripheral resistance also controls blood pressure. Changes in arteriole diameters regulate peripheral resistance. Because blood vessels with smaller diameters offer a greater resistance to blood flow, factors that cause arteriole vasoconstriction increase peripheral resistance, which raises blood pressure, and fac-

Figure 13.26

If blood pressure rises, baroreceptors initiate the cardioinhibitor reflex, which lowers the blood pressure.

tors causing vasodilation decrease peripheral resistance, lowering blood pressure.

The *vasomotor center* of the medulla oblongata continually sends sympathetic impulses to smooth muscles in the arteriole walls, keeping them in a state of tonic contraction. This action helps maintain the peripheral resistance associated with normal blood pressure. Because the vasomotor center responds to changes in blood pressure, it can increase peripheral resistance by increasing its outflow of sympathetic impulses, or it can decrease such resistance by decreasing its sympathetic outflow. In the latter case, the vessels vasodilate as sympathetic stimulation decreases.

Whenever arterial blood pressure suddenly increases, baroreceptors in the aorta and carotid arteries signal the vasomotor center, and the sympathetic outflow to the arterioles falls. The resulting vasodilation decreases peripheral resistance, and blood pressure lowers toward the normal level.

Certain chemicals, including carbon dioxide, oxygen, and hydrogen ions, also influence peripheral resistance by affecting precapillary sphincters and smooth muscle in arteriole walls. For example, increasing blood carbon dioxide, decreasing blood oxygen, and lowering blood pH relaxes smooth muscle in the systemic circulation. This increases local blood flow to tissues

Clinical Application 13.2

Exercise and the Cardiovascular System

The cardiovascular system adapts to exercise. The conditioned athlete experiences increases in heart pumping efficiency, blood volume, blood hemoglobin concentration, and the number of mitochondria in muscle fibers. All of these adaptations improve oxygen delivery to, and utilization by, muscle tissue.

An athlete's heart typically changes in response to these increased demands, and may enlarge 40% or more. Myocardial mass increases, the ventricular cavities expand, and the ventricle walls thicken. At rest stroke volume increases, and heart rate decreases, as does blood pressure. The lowest heart rate recorded in an athlete was 25 beats per minute! To a physician unfamiliar with a conditioned cardiovascular system, a trained athlete may appear abnormal. For example, a prolonged QT interval on an electrocardiogram for a long-time distance runner is not necessarily an indication of "long QT syndrome," as it is more likely to be in a less physically fit individual.

The cardiovascular system responds beautifully to a slow, steady buildup in exercise frequency and intensity. It does not react well to sudden demands, such as when a person who never exercises suddenly shovels snow. Sedentary people have a two- to sixfold increased risk of cardiac arrest while exercising than when not; people in shape have little or no excess risk while exercising.

For exercise to benefit the cardiovascular system, the heart rate must be elevated to 70–85% of its "theoretical maximum" for at least half an hour three times a week. You can calculate your theoretical maximum by subtracting your age from 220. If you are eighteen years old, your theoretical maximum is 202 beats per minute. Then, 70–85% of this value is 141–172 beats per minute. Some good activities for raising the heart rate are tennis, skating, skiing, handball, vigorous dancing, hockey, basketball, biking, and fast walking.

It is wise to consult a physician before starting an exercise program. People over age thirty are advised to have a stress test, which is an electrocardiogram taken while exercising. (The standard electrocardiogram is taken at rest.) An arrhythmia that appears only during exercise may indicate heart disease that has not yet produced symptoms.

The American Heart Association suggests that after a physical exam, a sedentary person wishing to start an exercise program begin with 30 minutes of activity (perhaps broken into two 15-minute sessions at first) at least five times per week.

---

with high metabolic rates, such as exercising skeletal muscles. In addition, epinephrine and norepinephrine vasoconstrict many systemic vessels, increasing peripheral resistance.

Hypertension, or high blood pressure, is persistently elevated systemic arterial pressure. It is one of the more common diseases of the cardiovascular system in industrialized nations.

High blood pressure with unknown cause is called *essential* (also primary or idiopathic) *hypertension.* Elevated pressure can be secondary, caused by another problem, such as kidney disease, high sodium intake, obesity, and psychological stress.

The consequences of prolonged, uncontrolled hypertension can be very serious. As the left ventricle works harder to pump sufficient blood, the myocardium thickens, enlarging the heart. If coronary blood vessels cannot support this overgrowth, parts of the heart muscle die and are replaced with fibrous tissue. Eventually, the enlarged and weakened heart is unable to function.

Exercising regularly, controlling body weight, reducing stress, and limiting sodium in the diet may control blood pressure. If not, drugs that are diuretics and/or inhibitors of sympathetic nerve activity may help.

Practice

36. What factors affect cardiac output?

37. What is the function of baroreceptors in the walls of the aorta and carotid arteries?

38. How does the vasomotor center control peripheral resistance?

Venous Blood Flow

Blood pressure decreases as blood moves through the arterial system and into the capillary networks, so that little pressure remains at the venular ends of capillaries (see fig. 13.24). Instead, blood flow through the venous system is only partly the direct result of heart action and depends on other factors, such as skeletal muscle contraction, breathing movements, and vasoconstriction of veins (*venoconstriction*).

Contracting skeletal muscles press on veins moving blood from one valve section to another (see fig. 13.23). This massaging action of contracting skeletal muscles helps push blood through the venous system toward the heart.

Respiratory movements also move venous blood. During inspiration, the pressure within the thoracic cavity falls as the diaphragm contracts and the rib cage moves upward and outward. At the same time, the pressure within the abdominal cavity increases as the diaphragm presses downward on the abdominal viscera. This action squeezes blood out of abdominal veins, forcing it into thoracic veins. During exercise, these respiratory movements act with skeletal muscle contractions to speed the return of venous blood to the heart.

Venoconstriction also returns venous blood to the heart. When venous pressure is low, sympathetic reflexes stimulate smooth muscles in the walls of veins to contract. The veins also provide a blood reservoir that can adapt its capacity to changes in blood volume. If some blood is lost and blood pressure falls, venoconstriction can force blood out of this reservoir. By maintaining venous return, venoconstriction helps to maintain blood pressure.

> ## Practice
> 39. What is the function of venous valves?
> 40. How do skeletal muscles and respiratory movements affect venous blood flow?
> 41. What factors stimulate venoconstriction?

13.6 PATHS OF CIRCULATION

Recall from figure 13.1 that the blood vessels can be divided into two major pathways. The *pulmonary circuit* or pulmonary circulation consists of vessels that carry blood from the heart to the lungs and back to the heart. The *systemic circuit* or systemic circulation carries blood from the heart to all other parts of the body and back again. The systemic circuit includes the coronary circulation.

The following sections describe the circulatory pathways of an adult. Chapter 20 (pp. 549–551) describes the somewhat different fetal pathways.

Pulmonary Circuit

Blood enters the pulmonary circuit as it leaves the right ventricle through the pulmonary trunk. The pulmonary trunk extends upward and posteriorly from the heart. About 5 centimeters above its origin, the pulmonary trunk divides into the right and left pulmonary arteries (see fig. 13.5), which penetrate the right and left lungs, respectively. After repeated divisions, the pulmonary arteries give rise to arterioles that continue into the capillary networks associated with the walls of the alveoli, where gas is exchanged between the blood and the air (see chapter 16, pp. 458–459).

From the pulmonary capillaries, blood enters the venules, which merge to form small veins, and these veins in turn converge to form still larger veins. Four pulmonary veins, two from each lung, return blood to the left atrium. This completes the vascular loop of the pulmonary circuit.

Systemic Circuit

Freshly oxygenated blood moves from the left atrium into the left ventricle. Contraction of the left ventricle forces this blood into the systemic circuit, which includes the aorta and its branches that lead to all the body tissues, as well as the companion system of veins that returns blood to the right atrium.

> ## Practice
> 42. Distinguish between the pulmonary and systemic circuits of the cardiovascular system.
> 43. Trace the path of blood through the pulmonary circuit from the right ventricle.

13.7 ARTERIAL SYSTEM

The **aorta** is the largest-diameter artery in the body. It extends upward from the left ventricle, arches over the heart to the left, and descends just anterior and to the left of the vertebral column. Figure 13.27 shows the aorta and its main branches.

Principal Branches of the Aorta

The first part of the aorta is called the *ascending aorta.* At its base are the three cusps of the aortic valve, and opposite each cusp is a swelling in the aortic wall called an **aortic sinus.** The right and left coronary arteries arise from two of these sinuses (see fig. 13.8).

Three major arteries originate from the *aortic arch* (arch of the aorta): the **brachiocephalic** (brāk"e-o-sĕ-fal'ik) **artery,** the left **common carotid** (kah-rot'id) **artery,** and the left **subclavian** (sub-kla've-an) **artery.**

The upper part of the *descending aorta* is left of the midline. It gradually extends medially and finally lies directly in front of the vertebral column at the level of the twelfth thoracic vertebra. The part of the descending aorta above the diaphragm is the **thoracic aorta.** It gives off many small branches to the thoracic wall and thoracic viscera.

Below the diaphragm, the descending aorta becomes the **abdominal aorta,** and it gives off branches to the abdominal wall and abdominal organs. Branches to abdominal organs include: the **celiac** (se'le-ak) **artery,** which gives rise to the *gastric, splenic,* and *hepatic arteries;* the **superior** (supplies small intestine and superior part of large intestine) and **inferior** (supplies inferior part of large intestine) **mesenteric** (mes"en-ter'ik) **arteries;** and the **suprarenal** (soo"prah-re'nal) **arteries, renal** (re'nal) **arteries,**

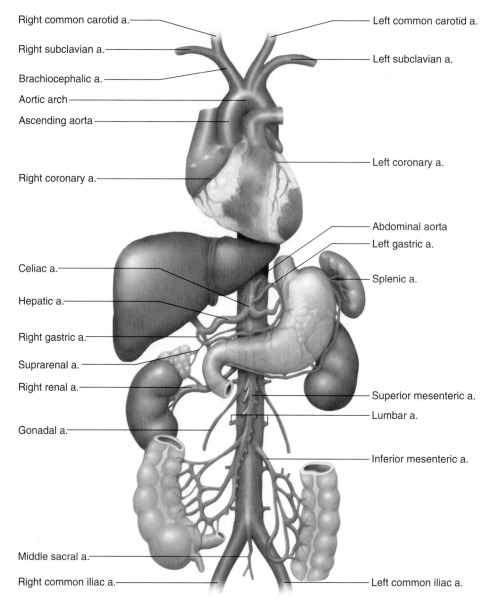

Right common carotid a.

Right subclavian a.

Brachiocephalic a.

Aortic arch

Ascending aorta

Right coronary a.

Left common carotid a.

Left subclavian a.

Left coronary a.

Abdominal aorta

Left gastric a.

Celiac a.

Splenic a.

Hepatic a.

Right gastric a.

Suprarenal a.

Right renal a.

Superior mesenteric a.

Lumbar a.

Gonadal a.

Inferior mesenteric a.

Middle sacral a.

Right common iliac a.

Left common iliac a.

Figure 13.27 AP|R

Major branches of the aorta. (*a.* stands for *artery.*)

and **gonadal** (go′nad-al) **arteries,** which supply blood to the adrenal glands, kidneys, and ovaries or testes, respectively. The abdominal aorta ends near the brim of the pelvis, where it divides into right and left **common iliac** (il′e-ak) **arteries.** These vessels supply blood to lower regions of the abdominal wall, the pelvic organs, and the lower extremities. Table 13.3 summarizes the major branches of the aorta.

Arteries to the Neck, Head, and Brain

Branches of the subclavian and common carotid arteries supply blood to structures in the neck, head, and brain (fig. 13.28). The main divisions of the subclavian artery to these regions include the vertebral and thyrocervical arteries. The common carotid artery communicates

with these regions by means of the internal and external carotid arteries.

The **vertebral arteries** pass upward through the foramina of the transverse processes of the cervical vertebrae and enter the skull through the foramen magnum. These vessels supply blood to the vertebrae and to their associated ligaments and muscles.

In the cranial cavity, the vertebral arteries unite to form a single *basilar artery.* This vessel passes along the ventral brainstem and gives rise to branches leading to the pons, midbrain, and cerebellum. The basilar artery ends by dividing into two *posterior cerebral arteries* that supply parts of the occipital and temporal lobes of the cerebrum. The posterior cerebral arteries also help form the **cerebral arterial circle** (circle of Willis) at the base of the brain, which connects the

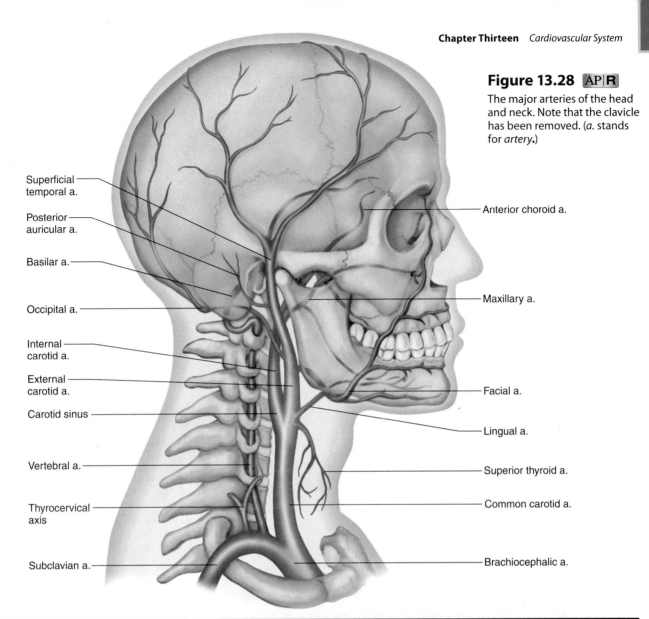

Superficial temporal a.

Posterior auricular a.

Basilar a.

Occipital a.

Internal carotid a.

External carotid a.

Carotid sinus

Vertebral a.

Thyrocervical axis

Subclavian a.

Anterior choroid a.

Maxillary a.

Facial a.

Lingual a.

Superior thyroid a.

Common carotid a.

Brachiocephalic a.

Figure 13.28 AP|R
The major arteries of the head and neck. Note that the clavicle has been removed. (*a.* stands for *artery.*)

Table 13.3	Major Branches of the Aorta	
Portion of Aorta	**Branch**	**General Regions or Organs Supplied**
Ascending aorta	Right and left coronary arteries	Heart
Arch of the aorta	Brachiocephalic artery	Right upper limb, right side of head
	Left common carotid artery	Left side of head
	Left subclavian artery	Left upper limb
Descending aorta		
Thoracic aorta	Bronchial artery	Bronchi
	Pericardial artery	Pericardium
	Esophageal artery	Esophagus
	Mediastinal artery	Mediastinum
	Posterior intercostal artery	Thoracic wall
Abdominal aorta	Celiac artery	Organs of upper digestive tract
	Phrenic artery	Diaphragm
	Superior mesenteric artery	Portions of small and large intestines
	Suprarenal artery	Adrenal gland
	Renal artery	Kidney
	Gonadal artery	Ovary or testis
	Inferior mesenteric artery	Lower portions of large intestine
	Lumbar artery	Posterior abdominal wall
	Middle sacral artery	Sacrum and coccyx
	Common iliac artery	Lower abdominal wall, pelvic organs, and lower limb

vertebral artery and internal carotid artery systems (fig. 13.29). The union of these systems provides alternate pathways for blood to circumvent blockages and reach brain tissues. It also equalizes blood pressure in the brain's blood supply.

The **thyrocervical** (thi″ro-ser′vĭ-kal) **arteries** are short vessels. At the thyrocervical axis, these vessels give off branches to the thyroid gland, parathyroid glands, larynx, trachea, esophagus, and pharynx, as well as to muscles in the neck, shoulder, and back.

The left and right *common carotid arteries* diverge into the internal and external carotid arteries. The **external carotid artery** courses upward on the side of the head, giving off branches to structures in the neck, face, jaw, scalp, and base of the skull. The **internal carotid artery** follows a deep course upward along the pharynx to the base of the skull. Entering the

cranial cavity, it provides the major blood supply to the brain. Near the base of the internal carotid arteries are enlargements called **carotid sinuses** that, like aortic sinuses, contain baroreceptors controlling blood pressure. Table 13.4 summarizes the major branches of the external and internal carotid arteries.

Arteries to the Shoulder and Upper Limb

The subclavian artery, after branching toward the neck, continues into the arm (fig. 13.30). It passes between the clavicle and the first rib, and becomes the axillary artery. The **axillary artery** supplies branches to structures in the axilla and chest wall and becomes the **brachial artery,** which courses along the humerus to the elbow. It gives rise to a *deep brachial artery* that curves posteriorly around the humerus and supplies the triceps

Figure 13.29 AP|R

The cerebral arterial circle (circle of Willis) is formed by the anterior and posterior cerebral arteries, which join the internal carotid arteries. (*a.* stands for *artery.*)

Table 13.4	Major Branches of the External and Internal Carotid Arteries	
Artery	**Branch**	**General Region or Organs Supplied**
External carotid artery	Superior thyroid artery	Larynx and thyroid gland
	Lingual artery	Tongue and salivary glands
	Facial artery	Pharynx, palate, chin, lips, and nose
	Occipital artery	Posterior scalp, meninges, and neck muscles
	Posterior auricular artery	Ear and lateral scalp
	Maxillary artery	Teeth, jaw, cheek, and eyelids
	Superficial temporal artery	Parotid salivary gland and surface of the face and scalp
Internal carotid artery	Ophthalmic artery	Eye and eye muscles
	Anterior choroid artery	Choroid plexus and brain
	Anterior cerebral artery	Frontal and parietal lobes of the brain

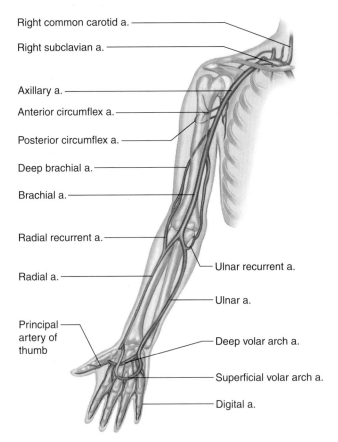

Right common carotid a.

Right subclavian a.

Axillary a.

Anterior circumflex a.

Posterior circumflex a.

Deep brachial a.

Brachial a.

Radial recurrent a.

Radial a.

Principal artery of thumb

Ulnar recurrent a.

Ulnar a.

Deep volar arch a.

Superficial volar arch a.

Digital a.

Figure 13.30

The major arteries to the shoulder and upper limb. (*a.* stands for *artery*.)

Q: *Blood from the brachial artery flows into which artery (arteries)?*

The answer can be found in Appendix E on page 568.

brachii. In the elbow, the brachial artery divides into an ulnar artery and a radial artery.

The **ulnar artery** leads downward on the ulnar side of the forearm to the wrist. Some of its branches supply the elbow joint, and some supply blood to muscles in the forearm.

The **radial artery** extends along the radial side of the forearm to the wrist, supplying the lateral muscles of the forearm. As the radial artery nears the wrist, it approaches the surface and provides a convenient vessel for taking the pulse (radial pulse).

At the wrist, the branches of the ulnar and radial arteries join to form a network of vessels. Arteries arising from this network supply blood to the hand.

Arteries to the Thoracic and Abdominal Walls

Blood reaches the thoracic wall through several vessels. The **internal thoracic artery,** a branch of the subclavian artery, gives off two *anterior intercostal*

(in″ter-kos′tal) *arteries* that supply the intercostal muscles and mammary glands. The *posterior intercostal arteries* arise from the thoracic aorta and enter the intercostal spaces. They supply the intercostal muscles, the vertebrae, the spinal cord, and the deep muscles of the back.

Branches of the *internal thoracic* and *external iliac arteries* provide blood to the anterior abdominal wall. Paired vessels originating from the abdominal aorta, including the *phrenic* and *lumbar arteries,* supply blood to structures in the posterior and lateral abdominal wall.

Arteries to the Pelvis and Lower Limb

The abdominal aorta divides to form the **common iliac** (il′e-ak) **arteries** at the level of the pelvic brim, and these vessels provide blood to the pelvic organs, gluteal region, and lower limbs (fig. 13.31). Each common iliac artery divides into an internal and an external branch. The **internal iliac artery** gives off many branches to pelvic muscles and visceral structures, as well as to the gluteal muscles and the external reproductive organs. The **external iliac artery** provides the main blood supply to the lower limbs. It passes downward along the brim of the pelvis and branches to supply the muscles and skin in the lower abdominal wall. Midway between the pubic symphysis and the anterior superior iliac spine of the ilium, the external iliac artery becomes the femoral artery.

The **femoral** (fem′or-al) **artery,** which approaches the anterior surface of the upper thigh, branches to muscles and superficial tissues of the thigh. These branches also supply the skin of the groin and the lower abdominal wall.

As the femoral artery reaches the proximal border of the space behind the knee, it becomes the **popliteal** (pop-lit′e-al) **artery.** Branches of this artery supply blood to the knee joint and to certain muscles in the thigh and calf. The popliteal artery diverges into the anterior and posterior tibial arteries.

The **anterior tibial artery** passes downward between the tibia and fibula, giving off branches to the skin and muscles in the anterior and lateral regions of the leg. This vessel continues into the foot as the *dorsalis pedis artery* (dorsal pedis artery), which supplies blood to the foot. The **posterior tibial artery,** the larger of the two popliteal branches, descends beneath the calf muscles, branching to the skin, muscles, and other tissues of the leg along the way.

Practice

44. Name the parts of the aorta.

45. Name the vessels that arise from the aortic arch.

46. Name the branches of the thoracic and abdominal aorta.

47. Which vessels supply blood to the head? To the upper limb? To the abdominal wall? To the lower limb?

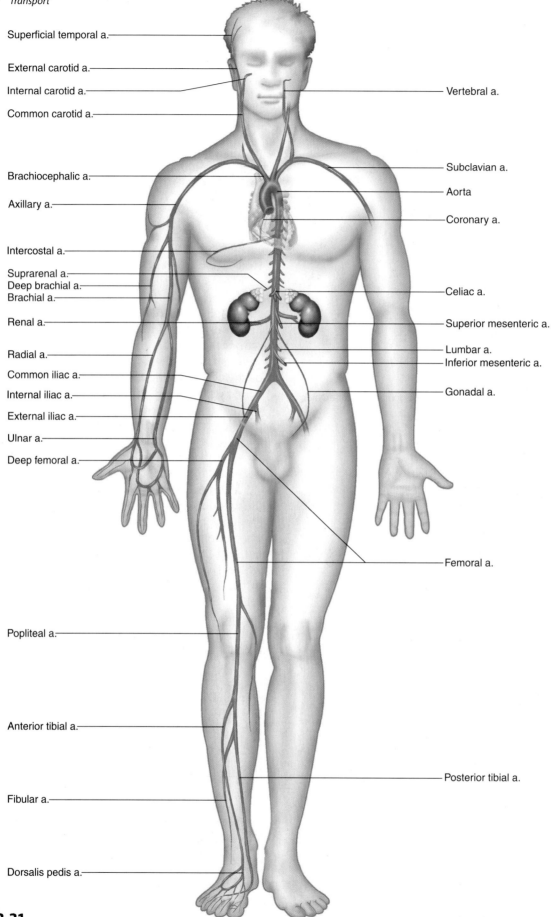

Superficial temporal a.

External carotid a.

Internal carotid a.

Common carotid a.

Vertebral a.

Brachiocephalic a.

Subclavian a.

Axillary a.

Aorta

Coronary a.

Intercostal a.

Suprarenal a.
Deep brachial a.
Brachial a.

Celiac a.

Renal a.

Superior mesenteric a.

Radial a.

Lumbar a.
Inferior mesenteric a.

Common iliac a.

Internal iliac a.

Gonadal a.

External iliac a.

Ulnar a.

Deep femoral a.

Femoral a.

Popliteal a.

Anterior tibial a.

Posterior tibial a.

Fibular a.

Dorsalis pedis a.

Figure 13.31

Major vessels of the arterial system. (*a.* stands for *artery.*)

13.8 VENOUS SYSTEM

Venous circulation returns blood to the heart after gases, nutrients, and wastes are exchanged between the blood and body cells.

Characteristics of Venous Pathways

The vessels of the venous system originate with the merging of the capillaries into venules, venules merge into small veins, and small veins meet to form larger ones. Unlike the arterial pathways, however, the vessels of the venous system are difficult to follow because they commonly connect in irregular networks. Many unnamed tributaries may join to form a large vein.

The pathways of larger veins are clearer. These veins typically parallel the courses of named arteries, and often bear the same names as their arterial counterparts. For example, the renal vein parallels the renal artery, and the common iliac vein accompanies the common iliac artery.

The veins that carry blood from the lungs and myocardium back to the heart have already been described. The veins from all the other parts of the body converge into two major pathways, the **superior** and **inferior venae cavae,** which lead to the right atrium.

Veins from the Brain, Head, and Neck

The **external jugular** (jug'u-lar) **veins** drain blood from the face, scalp, and superficial regions of the neck. These vessels descend on either side of the neck and empty into the *right* and *left subclavian veins* (fig. 13.32).

The **internal jugular veins,** which are somewhat larger than the external jugular veins, arise from numerous veins and venous sinuses of the brain and from deep veins in parts of the face and neck. They descend through the neck and join the subclavian veins. These unions of the internal jugular and subclavian veins form large **brachiocephalic veins** on each side. The vessels then merge and give rise to the superior vena cava, which enters the right atrium.

Veins from the Upper Limb and Shoulder

A set of deep veins and a set of superficial veins drain the upper limb. The deep veins generally parallel the arteries in each region and have similar names. Deep venous drainage of the upper limb begins in the digital veins that drain into pairs of radial veins and pairs of ulnar veins, which empty into a pair of brachial veins. The superficial veins connect in complex networks just beneath the skin. They also communicate with the deep vessels of the upper limb, providing many alternate

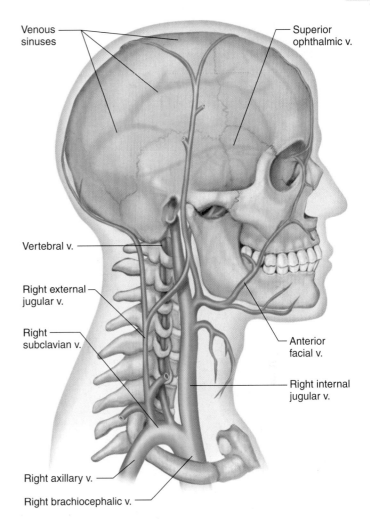

Figure 13.32
The major veins of the brain, head, and neck. Note that the clavicle has been removed. (*v.* stands for *vein.*)

pathways through which blood can leave the tissues (fig. 13.33). The main vessels of the superficial network are the basilic and cephalic veins.

The **basilic** (bah-sil'ik) **vein** ascends from the forearm to the middle of the arm, where it penetrates deeply and joins the *brachial vein.* The basilic and brachial veins merge, forming the *axillary vein.*

The **cephalic** (sĕ-fal'ik) **vein** courses upward from the hand to the shoulder. In the shoulder, it pierces the tissues and empties into the axillary vein. Beyond the axilla, the axillary vein becomes the subclavian vein.

> In the bend of the elbow, a *median cubital vein* ascends from the cephalic vein on the lateral side of the forearm to the basilic vein on the medial side. This large vein is usually visible. It is often used as a site for *venipuncture,* when it is necessary to remove a blood sample or to add fluids to blood.

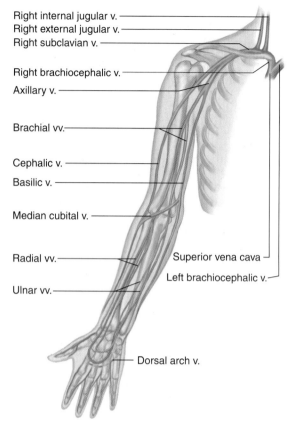

Right internal jugular v.
Right external jugular v.
Right subclavian v.
Right brachiocephalic v.
Axillary v.
Brachial vv.
Cephalic v.
Basilic v.
Median cubital v.
Radial vv.
Superior vena cava
Left brachiocephalic v.
Ulnar vv.
Dorsal arch v.

Figure 13.33

The major veins of the upper limb and shoulder. (*v.* stands for *vein, vv.* stands for *veins.*)

Q: *Blood from the brachial vein, basilic vein, and cephalic vein drains into which vein(s)?*

The answer can be found in Appendix E on page 568.

Veins from the Abdominal and Thoracic Walls

Tributaries of the brachiocephalic and azygos veins drain the abdominal and thoracic walls. For example, the *brachiocephalic vein* receives blood from the *internal thoracic vein,* which generally drains the tissues the internal thoracic artery supplies. Some *intercostal veins* also empty into the brachiocephalic vein.

The **azygos** (az′ĭ-gos) **vein** originates in the dorsal abdominal wall and ascends through the mediastinum on the right side of the vertebral column to join the superior vena cava. It drains most of the muscular tissue in the abdominal and thoracic walls.

Tributaries of the azygos vein include the *posterior intercostal veins* on the right side, which drain the intercostal spaces, and the *superior* and *inferior hemiazygos veins,* which receive blood from the posterior intercostal veins on the left. The right and left *ascending lumbar veins,* with tributaries that include vessels from the

lumbar and sacral regions, also connect to the azygos system.

Veins from the Abdominal Viscera

Most veins carry blood directly to the atria of the heart. Veins that drain the abdominal viscera are exceptions (fig. 13.34). They originate in the capillary networks of the stomach, intestines, pancreas, and spleen and carry blood from these organs through a **hepatic portal** (por′tal) **vein** to the liver. This unique venous pathway is called the **hepatic portal system.**

Tributaries of the hepatic portal vein include:

1. Right and left *gastric veins* from the stomach.
2. *Superior mesenteric vein* from the small intestine, ascending colon, and transverse colon.
3. *Splenic vein* from a convergence of several veins draining the spleen, the pancreas, and part of the stomach. Its largest tributary, the *inferior mesenteric vein,* brings blood upward from the descending colon, sigmoid colon, and rectum.

About 80% of the blood flowing to the liver in the hepatic portal system comes from capillaries in the stomach and intestines, and is oxygen-poor but nutrient-rich. As discussed in chapter 15 (pp. 415 and 417), the liver handles these nutrients in a variety of ways. It regulates blood glucose concentration by joining (polymerizing) excess glucose molecules into glycogen for storage, or by breaking down glycogen into glucose when blood glucose concentration drops below normal. The liver helps regulate blood concentrations of recently absorbed amino acids and lipids by modifying them into forms cells can use, by oxidizing them, or by changing them into storage forms. The liver also stores certain vitamins and detoxifies harmful substances. Blood in the hepatic portal vein nearly always contains bacteria that have entered through intestinal capillaries. Large *Kupffer cells* lining small vessels in the liver called hepatic sinusoids phagocytize these microorganisms, removing them from portal blood before it leaves the liver.

After passing through the hepatic sinusoids of the liver, blood in the hepatic portal system travels through a series of merging vessels into **hepatic veins.** These veins empty into the inferior vena cava, returning the blood to the general circulation.

Veins from the Lower Limb and Pelvis

Veins that drain blood from the lower limb are divided into deep and superficial groups, as is the case in the upper limb (fig. 13.35). The deep veins of the leg, such as the *anterior* and *posterior tibial veins,* are named for

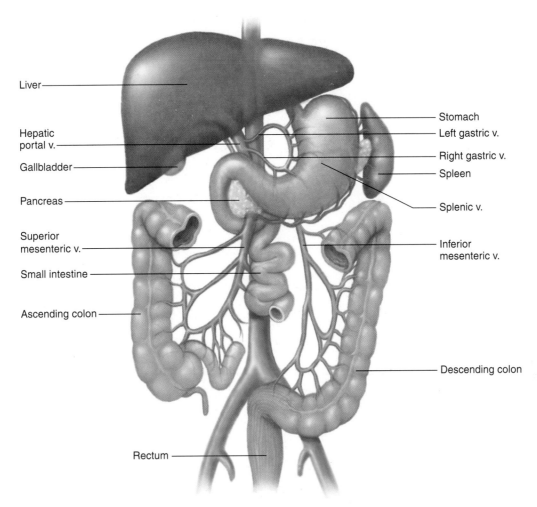

Figure 13.34 AP|R

Veins that drain the abdominal viscera. (*v.* stands for *vein.*)

the arteries they accompany. At the level of the knee, these vessels form a single trunk, the **popliteal vein.** This vein continues upward through the thigh as the **femoral vein,** which in turn becomes the **external iliac vein.**

The superficial veins of the foot, leg, and thigh connect to form a complex network beneath the skin. These vessels drain into two major trunks—the small and great saphenous veins. The **small saphenous (sah-fe′nus) vein** ascends along the back of the calf, enters the popliteal fossa, and joins the popliteal vein. The **great saphenous vein,** which is the longest vein in the body, ascends in front of the medial malleolus and extends upward along the medial side of the leg and thigh. In the thigh, it penetrates deeply and joins the femoral vein. Near its termination, the great saphenous vein receives tributaries from a number of vessels that drain the upper thigh, groin, and lower abdominal wall.

In addition to communicating freely with each other, the saphenous veins communicate extensively with the deep veins of the leg and thigh. Blood can thus return to the heart from the lower extremities by several routes.

In the pelvic region, vessels leading to the **internal iliac vein** carry blood away from the organs of the reproductive, urinary, and digestive systems. The internal iliac veins unite with the right and left external iliac veins to form the **common iliac veins.** These vessels, in turn, merge to produce the inferior vena cava.

Practice

48. Name the veins that return blood to the right atrium.

49. Which major veins drain blood from the head? From the upper limbs? From the abdominal viscera? From the lower limbs?

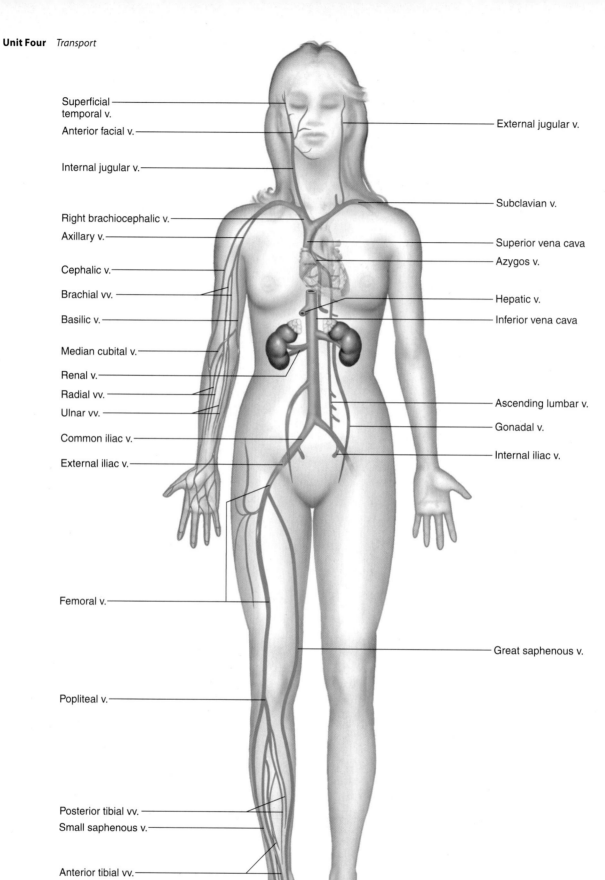

Superficial temporal v.

Anterior facial v.

Internal jugular v.

Right brachiocephalic v.

Axillary v.

Cephalic v.

Brachial vv.

Basilic v.

Median cubital v.

Renal v.

Radial vv.

Ulnar vv.

Common iliac v.

External iliac v.

Femoral v.

Popliteal v.

Posterior tibial vv.

Small saphenous v.

Anterior tibial vv.

External jugular v.

Subclavian v.

Superior vena cava

Azygos v.

Hepatic v.

Inferior vena cava

Ascending lumbar v.

Gonadal v.

Internal iliac v.

Great saphenous v.

Figure 13.35

Major vessels of the venous system. (*v.* stands for *vein, vv.* stands for *veins.*)

Cardiovascular System

Integumentary System

Changes in skin blood flow are important in temperature control.

Skeletal System

Bones help control plasma levels of calcium ions, which influence heart action.

Muscular System

Blood flow increases to exercising skeletal muscle, delivering oxygen and nutrients and removing wastes. Muscle actions help the blood circulate.

Nervous System

The brain depends on blood flow for survival. The nervous system helps control blood flow and blood pressure.

Endocrine System

Hormones are carried in the bloodstream. Some hormones directly affect the heart and blood vessels.

Lymphatic System

The lymphatic system returns tissue fluids to the bloodstream.

Digestive System

The digestive system breaks down nutrients into forms readily absorbed by the bloodstream.

Respiratory System

The respiratory system oxygenates the blood and removes carbon dioxide. Respiratory movements help the blood circulate.

Urinary System

The kidneys clear the blood of wastes and substances present in the body. The kidneys help control blood pressure and blood volume.

Reproductive System

Blood pressure is important in normal function of the sex organs.

The heart pumps blood through as many as 60,000 miles of blood vessels delivering nutrients to, and removing wastes from, all body cells.

Summary Outline

13.1 Introduction (p. 341)

The cardiovascular system, consisting of the heart and blood vessels, provides oxygen and nutrients to and removes wastes from body cells.

13.2 Structure of the Heart (p. 342)

1. Size and location of the heart
   a. The heart is about 14 centimeters long and 9 centimeters wide.
   b. It is located within the mediastinum and rests on the diaphragm.
2. Coverings of the heart
   a. A layered pericardium encloses the heart.
   b. The pericardial cavity is a space between the parietal and visceral layers of the pericardium.
3. Wall of the heart
   The wall of the heart has three layers—an epicardium, a myocardium, and an endocardium.
4. Heart chambers and valves
   a. The heart is divided into two atria and two ventricles.
   b. Right chambers and valves
      (1) The right atrium receives blood from the venae cavae and coronary sinus.
      (2) The tricuspid valve separates the right atrium from the right ventricle.
      (3) A pulmonary valve guards the base of the pulmonary trunk.
   c. Left chambers and valves
      (1) The left atrium receives blood from the pulmonary veins.
      (2) The mitral valve separates the left atrium from the left ventricle.
      (3) An aortic valve guards the base of the aorta.
5. Skeleton of the heart
   The skeleton of the heart consists of fibrous rings that enclose the bases of the pulmonary artery and aorta and masses of dense connective tissues in the septum between the ventricles.
6. Path of blood through the heart
   a. Blood low in oxygen and high in carbon dioxide enters the right side of the heart and is pumped into the pulmonary circulation.
   b. After blood is oxygenated in the lungs and some carbon dioxide is removed, it returns to the left side of the heart through the pulmonary veins.
7. Blood supply to the heart
   a. The coronary arteries supply blood to the myocardium.
   b. Blood returns to the right atrium through the cardiac veins and coronary sinus.

13.3 Heart Actions (p. 347)

1. Cardiac cycle
   a. The atria contract (atrial systole) while the ventricles relax (ventricular diastole). The ventricles contract (ventricular systole) while the atria relax (atrial diastole).
   b. Pressure within the chambers rises and falls in repeated cycles.
2. Heart sounds
   Heart sounds are due to the vibrations produced when the valves close.
3. Cardiac muscle fibers
   a. Cardiac muscle fibers connect to form a functional syncytium.
   b. If any part of the syncytium is stimulated, the whole structure contracts as a unit.
4. Cardiac conduction system
   a. This system initiates and conducts impulses throughout the myocardium.
   b. Impulses from the SA node pass slowly to the AV node. Impulses travel rapidly along the AV bundle and Purkinje fibers.
5. Electrocardiogram (ECG)
   a. An ECG records electrical changes in the myocardium during a cardiac cycle.
   b. The pattern contains several waves.
      (1) The P wave represents atrial depolarization.
      (2) The QRS complex represents ventricular depolarization.
      (3) The T wave represents ventricular repolarization.
6. Regulation of the cardiac cycle
   a. Physical exercise, body temperature, and the concentration of various ions affect heartbeat.
   b. Branches of sympathetic and parasympathetic nerve fibers innervate the SA and AV nodes.
   c. The cardiac control center in the medulla oblongata regulates autonomic impulses to the heart.

13.4 Blood Vessels (p. 353)

Blood vessels form a closed circuit of tubes that carry blood from the heart to body cells and back again.

1. Arteries and arterioles
   a. Arteries are adapted to carry blood under high pressure away from the heart.
   b. The walls of arteries and arterioles consist of layers of endothelium, smooth muscle, and connective tissue.
   c. Autonomic fibers innervate smooth muscle in vessel walls to produce vasoconstriction or vasodilation.
2. Capillaries
   a. Capillaries connect arterioles and venules.
   b. The capillary wall is a single layer of cells that forms a semipermeable membrane.
   c. Openings in capillary walls, where endothelial cells overlap, vary in size from tissue to tissue.
   d. Precapillary sphincters regulate capillary blood flow.
   e. Capillary blood and tissue fluid exchange gases, nutrients, and metabolic by-products.
      (1) Diffusion provides the most important means of transport.
      (2) Filtration, which is due to the hydrostatic pressure of blood, causes a net outward movement of fluid at the arteriolar end of a capillary.
      (3) Osmosis due to colloid osmotic pressure causes a net inward movement of fluid at the venular end of a capillary.
3. Venules and veins
   a. Venules continue from capillaries and merge to form veins.
   b. Veins carry blood to the heart.
   c. Venous walls are similar to arterial walls, but are thinner and contain less smooth muscle and elastic tissue.

13.5 Blood Pressure (p. 359)

Blood pressure is the force blood exerts against the inner walls of blood vessels.

1. Arterial blood pressure
   a. Arterial blood pressure rises and falls with the phases of the cardiac cycle.
   b. Systolic pressure is produced when the ventricle contracts. Diastolic pressure is the pressure in the arteries when the ventricle relaxes.
2. Factors that influence arterial blood pressure
   Arterial blood pressure increases as cardiac output, blood volume, peripheral resistance, or blood viscosity increases.

3. Control of blood pressure
   a. Blood pressure is controlled in part by the mechanisms that regulate cardiac output and peripheral resistance.
   b. The more blood that enters the heart, the stronger the ventricular contraction, the greater the stroke volume, and the greater the cardiac output.
   c. The baroreceptor reflexes involving the cardiac control center of the medulla oblongata regulate heart rate.
4. Venous blood flow
   a. Venous blood flow depends on skeletal muscle contraction, breathing movements, and venoconstriction.
   b. Many veins contain flaplike valves that prevent blood from backing up.

13.6 Paths of Circulation (p. 363)

1. Pulmonary circuit
   The pulmonary circuit consists of vessels that carry blood from the right ventricle to the lungs and back to the left atrium.
2. Systemic circuit
   a. The systemic circuit consists of vessels that lead from the left ventricle to the body cells (including those of the heart itself) and back to the heart.
   b. It includes the aorta and its branches.

13.7 Arterial System (p. 363)

1. Principal branches of the aorta
   a. The aorta is the largest artery with respect to diameter.
   b. Its major branches include the coronary, brachiocephalic, left common carotid, and left subclavian arteries.
   c. The branches of the descending aorta include the thoracic and abdominal groups.
   d. The abdominal aorta diverges into the right and left common iliac arteries.
2. Arteries to the neck, head, and brain
   These include branches of the subclavian and common carotid arteries.

3. Arteries to the shoulder and upper limb
   a. The subclavian artery passes into the upper limb, and in various regions is called the axillary and brachial artery.
   b. Branches of the brachial artery include the ulnar and radial arteries.
4. Arteries to the thoracic and abdominal walls
   a. Branches of the subclavian artery and thoracic aorta supply the thoracic wall.
   b. Branches of the abdominal aorta and other arteries supply the abdominal wall.
5. Arteries to the pelvis and lower limb
   The common iliac arteries supply the pelvic organs, gluteal region, and lower limbs.

13.8 Venous System (p. 369)

1. Characteristics of venous pathways
   a. Veins return blood to the heart.
   b. Larger veins usually parallel the paths of major arteries.
2. Veins from the brain, head, and neck
   a. Jugular veins drain these regions.
   b. Jugular veins unite with subclavian veins to form the brachiocephalic veins.
3. Veins from the upper limb and shoulder
   a. Sets of superficial and deep veins drain these regions.
   b. Deep veins parallel arteries with similar names.
4. Veins from the abdominal and thoracic walls
   Tributaries of the brachiocephalic and azygos veins drain these walls.
5. Veins from the abdominal viscera
   a. Blood from the abdominal viscera enters the hepatic portal system and is carried to the liver.
   b. From the liver, hepatic veins carry blood to the inferior vena cava.
6. Veins from the lower limb and pelvis
   a. Sets of deep and superficial veins drain these regions.
   b. The deep veins include the tibial veins, and the superficial veins include the saphenous veins.

Chapter Assessments

13.1 Introduction

1. The cardiovascular system includes the (p. 341)

   a. heart.   d. capillaries.

   b. arteries.   e. all of the above.

   c. veins.

13.2 Structure of the Heart

2. Describe the pericardium. (p. 342)

3. Compare the layers of the heart wall. (p. 343)

4. Draw a heart and label the chambers and valves. (p. 343)

5. Blood flows through the vena cavae and coronary sinus into the right atrium through the _____ , to the right ventricle through the pulmonary valve to the pulmonary trunk into the right and left _____ to the lungs, then through the pulmonary veins into the _____ through the mitral valve to the _____ and through the _____ to the aorta. (p. 344)

6. List the vessels through which blood flows from the aorta to the myocardium and back to the right atrium (p. 346)

13.3 Heart Actions

7. Describe a cardiac cycle, including the pressure changes in the atria and ventricles during the cardiac cycle. (p. 347)

8. Distinguish between the roles of the SA node and the AV node. (p. 350)

9. Explain how the cardiac conduction system controls the cardiac cycle. (p. 350)

10. Describe and explain the normal ECG pattern. (p. 351)

11. Discuss how the nervous system regulates the cardiac cycle. (p. 352)

13.4 Blood Vessels

12. Distinguish between an artery and an arteriole. (p. 354)

13. Explain control of vasodilation and vasoconstriction. (p. 354)

14. Describe the structure and function of a capillary. (p. 356)

15. Relate how diffusion functions in the exchange of substances between the blood and tissues. (p. 357)

16. Explain why water and dissolved substances leave the arteriolar end of a capillary and enter the venular end. (p. 357)

17. Distinguish between a vein and a venule. (p. 358)

13.5 Blood Pressure

18. Arterial blood pressure reaches its maximum when the ventricles contract. This point is called _____. (p. 359)

19. Name several factors that influence blood pressure, and explain how each produces its effect. (p. 360)

20. Describe the control of blood pressure. (p. 361)

21. Which of the following promote the flow of venous blood? (p. 362)

    a. skeletal muscle contraction   d. venoconstriction

    b. breathing   e. all of the above

    c. arterial blood pressure

13.6 Paths of Circulation

22. Distinguish between the pulmonary and systemic circuits of the cardiovascular system. (p. 363)

13.7–13.8 Arterial System–Venous System

23. Describe the aorta, and name its principal branches. (p. 363)

24. Discuss the relationship between the major venous pathways and the major arterial pathways to the head, upper limbs, abdominal viscera, and lower limbs. (p. 369)

Integrative Assessments/Critical Thinking

OUTCOMES 5.2, 5.3, 13.4

1. If you were asked to invent a blood vessel (artery, capillary, or vein) substitute, what materials might you use to build it? Include synthetic as well as natural materials.

OUTCOMES 5.5, 9.16, 13.2, 13.3

2. What structures and properties should an artificial heart have?

OUTCOMES 13.2, 13.4, 13.7

3. If a cardiologist inserts a catheter into a patient's right femoral artery, which arteries will the tube have to pass through in order to reach the entrance to the left coronary artery?

4. If a patient develops a blood clot in the femoral vein of the left lower limb and a portion of the clot breaks loose, where is the blood flow likely to carry the embolus? What symptoms are likely?

OUTCOMES 13.4, 13.5, 13.8

5. Cirrhosis of the liver, a disease commonly associated with alcoholism, obstructs blood flow through hepatic blood vessels. As a result, blood backs up and capillary pressure greatly increases in organs drained by the hepatic portal system. What effects might this increasing capillary pressure produce, and which organs would it affect?

WEB CONNECTIONS

Visit the text website at **www.mhhe.com/shieress11** for additional quizzes, interactive learning exercises, and more.

APR

Anatomy & Physiology REVEALED includes cadaver photos that allow you to peel away layers of the human body to reveal structures beneath the surface. This program also includes animations, radiologic imaging, audio pronunciations, and practice quizzing. To learn more visit **www.aprevealed.com.**

14

Lymphatic System and Immunity

Peanut allergy. The young woman went to the emergency department for sudden onset of difficulty breathing. She was also flushed and had vomited. An astute medical student taking a quick history from the woman's roommates discovered that she had just eaten cookies from a vending machine in their dorm. Suspecting that the cookies may have contained peanuts, the medical student alerted the attending physician, who treated the woman for suspected peanut allergy—giving oxygen, an antihistamine, a steroid drug, and epinephrine. She recovered.

Peanut allergy is common and on the rise, but only in certain countries. In the United States, 1% of children under the age of five and 2% of the population over ten years of age have had allergic reactions to peanuts. About 30,000 people react each year, and about 200 die. In Denmark and Norway, where peanuts are rarely eaten, peanut allergy is very rare.

Peculiarities of peanuts and our fondness for them may explain why peanut allergy prevalence is increasing. Three glycoproteins in peanuts are allergens, causing the misdirected immune response that is the allergy. These glycoproteins are highly concentrated in the peanut, and when ingested they disturb the intestinal lining in such a way that they enter the circulation rapidly, without being digested. Many allergens confront cells of the immune system beneath the intestinal lining.

Another factor in the increased incidence of peanut allergy is that people in the United States eat many peanuts. Virtually everyone has eaten a peanut by two years of age, usually in peanut butter. This is

Peculiarities of peanuts, combined with our fondness for them, sets the stage for allergy, a misdirected immune reaction.

sufficient exposure to set the stage for later allergy in genetically predisposed individuals. The young average age of first allergic reaction to peanuts—fourteen months—suggests that the initial exposure necessary to "prime" the immune system for future response happens through breast milk or in the uterus.

The dry roasting of peanuts in the United States may make the three glycoproteins that evoke the allergic response more active. In China, where peanuts are equally popular but are eaten boiled or fried, allergy is rare. However, children of Chinese immigrants in the United States have the same incidence of peanut allergy as other children in the United States, supporting the idea that method of preparation contributes to allergenicity.

Learning Outcomes

After studying this chapter, you should be able to do the following:

14.1 Introduction
1. Describe the general functions of the lymphatic system. (p. 378)

14.2 Lymphatic Pathways
2. Identify the locations of the major lymphatic pathways. (p. 379)

14.3 Tissue Fluid and Lymph
3. Describe how tissue fluid and lymph form, and explain the function of lymph. (p. 380)

14.4 Lymph Movement
4. Explain how lymphatic circulation is maintained. (p. 381)

14.5 Lymph Nodes
5. Describe a lymph node and its major functions. (p. 381)

14.6 Thymus and Spleen
6. Discuss the locations and functions of the thymus and spleen. (p. 382)

14.7 Body Defenses Against Infection
7. Distinguish between innate (nonspecific) and adaptive (specific) defenses. (p. 384)

14.8 Innate (Nonspecific) Defenses

8. List seven innate body defense mechanisms, and describe the action of each mechanism. (p. 384)

14.9 Adaptive (Specific) Defenses, or Immunity

9. Explain how two major types of lymphocytes are formed and activated, and how they function in immune mechanisms. (p. 386)

10. Discuss the actions of the five types of immunoglobulins. (p. 391)

11. Distinguish between primary and secondary immune responses. (p. 392)

12. Distinguish between active and passive immunity. (p. 393)

13. Explain how allergic reactions, tissue rejection reactions, and autoimmunity arise from immune mechanisms. (p. 393)

Module 10: Lymphatic System

Aids to Understanding Words   *(Appendix A on page 564 has a complete list of Aids to Understanding Words.)*

-gen [be produced] aller*gen*: Substance that evokes an allergic response

humor- [fluid] *humor*al immunity: Immunity resulting from antibodies in body fluids

immun- [free] *immun*ity: Resistance to (freedom from) a specific disease

inflamm- [set on fire] *inflamm*ation: Localized redness, heat, swelling, and pain in tissues

nod- [knot] *nod*ule: Small mass of lymphocytes surrounded by connective tissue

patho- [disease] *patho*gen: Disease-causing agent

14.1 INTRODUCTION

The **lymphatic** (lim-fat′ik) **system** is a vast collection of cells and biochemicals that travel in lymphatic vessels, and the organs and glands that produce them. The lymphatic system includes a network of vessels that assist in circulating body fluids, so it is closely associated with the cardiovascular system. Lymphatic vessels transport excess fluid away from interstitial spaces in most tissues and return it to the bloodstream (fig. 14.1). Without the lymphatic system, this fluid would accumulate in tissue spaces. Special lymphatic capillaries, called *lacteals* (lak′te-alz), are located in the lining of the small intestine, where they absorb digested fats and transport them to the venous circulation.

The lymphatic system has a second major function—it enables us to live in a world with different types of organisms, some of which take up residence in or on the human body and may cause infectious diseases. Cells and biochemicals of the lymphatic system launch both generalized and targeted attacks against "foreign" particles, enabling the body to destroy infectious microorganisms and viruses. This immunity against disease also protects against toxins and cancer cells. When the immune response is abnormal, persistent infection, cancer, autoimmune disorders, and allergies may result.

14.2 LYMPHATIC PATHWAYS

The **lymphatic pathways** begin as lymphatic capillaries. These tiny tubes merge to form larger lymphatic vessels, which in turn lead to larger vessels that unite with the veins in the thorax.

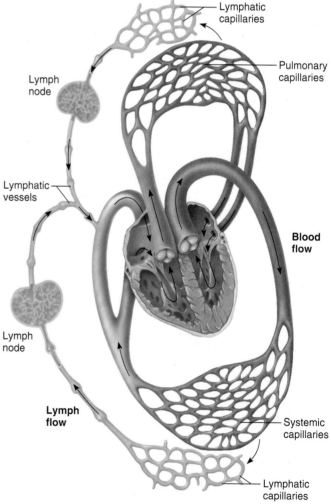

Figure 14.1 AP|R

Schematic representation of lymphatic vessels transporting fluid from interstitial spaces to the bloodstream.

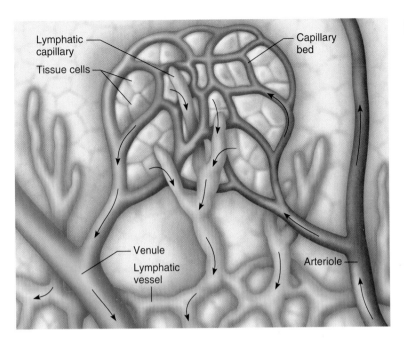

Figure 14.2
Lymphatic capillaries are microscopic, closed-ended tubes that originate in the interstitial spaces of most tissues.

Lymphatic Capillaries

Lymphatic capillaries are microscopic, closed-ended tubes (fig. 14.2). They extend into interstitial spaces, forming complex networks that parallel those of blood capillaries. Lymph capillaries are nearly everywhere there are blood capillaries, except in the central nervous system. The walls of lymphatic capillaries, like those of blood capillaries, are formed from a single layer of squamous epithelial cells. These thin walls allow tissue fluid (interstitial fluid) to enter lymphatic capillaries. Once inside lymphatic capillaries the fluid is called **lymph** (limf).

Lymphatic Vessels

The walls of **lymphatic vessels** are similar to those of veins, but thinner. Like veins, lymphatic vessels have flaplike valves that help prevent backflow of lymph (fig. 14.3).

The larger lymphatic vessels lead to specialized organs called **lymph nodes** (limf nōdz). After leaving the nodes, the vessels merge to form still larger lymphatic trunks.

Lymphatic Trunks and Collecting Ducts

Lymphatic trunks, which drain lymph from the lymphatic vessels, are named for the regions they serve. They join one of two **collecting ducts**—the thoracic duct or the right lymphatic duct (fig. 14.4a).

The **thoracic duct** is the larger and longer collecting duct. It receives lymph from the lower limbs and

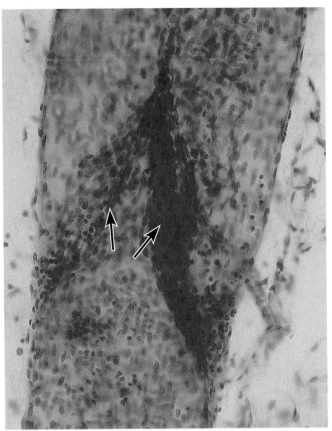

Figure 14.3
Light micrograph of the flaplike valve (arrows) within a lymphatic vessel (60×).

abdominal regions, left upper limb, and left side of the thorax, head, and neck, and empties into the left subclavian vein near the junction of the left jugular vein. The **right lymphatic duct** receives lymph from the right side of the head and neck, right upper limb, and right thorax, and empties into the right subclavian vein near the junction of the right jugular vein.

After leaving the two collecting ducts, lymph enters the venous system and becomes part of the plasma just before blood returns to the right atrium. Figure 14.5 summarizes the typical lymphatic pathway.

The skin has many lymphatic capillaries. Consequently, if the skin is broken or if something is injected into it (such as venom from a stinging insect), foreign substances can rapidly enter the lymphatic system.

Practice

1. What are the general functions of the lymphatic system?
2. Distinguish between the thoracic duct and the right lymphatic duct.

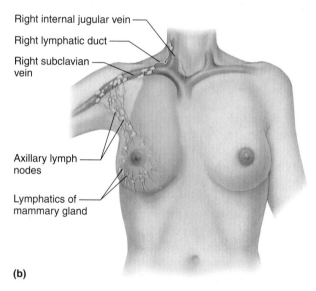

Figure 14.4

Lymphatic pathways. **(a)** The right lymphatic duct drains lymph from the upper right side of the body, whereas the thoracic duct drains lymph from the rest of the body. **(b)** Lymph drainage of the right breast illustrates a localized function of the lymphatic system. Surgery to treat breast cancer can disrupt this drainage, causing painful swelling (edema) in the arm.

Q: *Which lymphatic duct drains lymph from the right lower limb?*

Answer can be found in Appendix E on page 568.

Figure 14.5

The lymphatic pathway.

14.3 TISSUE FLUID AND LYMPH

Lymph is essentially tissue fluid (interstitial fluid) that has entered a lymphatic capillary. Thus, lymph formation depends upon tissue fluid formation.

Tissue Fluid Formation

Recall from chapter 13 (p. 357) that tissue fluid originates from blood plasma and is composed of water and dissolved substances that leave blood capillaries. Capillary blood pressure filters water and small molecules from the plasma. The resulting fluid is very similar in composition to the blood plasma (including nutrients, gases, and hormones), with the important exception of the plasma proteins, which are generally too large to pass through the capillary walls. The osmotic effect of these (called the *plasma colloid osmotic pressure*) helps draw fluid back into the capillaries by osmosis.

Lymph Formation and Function

Filtration from the plasma normally exceeds reabsorption, leading to the net formation of tissue fluid. This increases the tissue fluid hydrostatic pressure moving tissue fluid into lymphatic capillaries, forming lymph (see fig. 14.2). Lymph returns to the bloodstream most of the small proteins that the blood capillaries filtered. At the same time, lymph transports foreign particles, such as bacteria or viruses, to lymph nodes.

14.4 LYMPH MOVEMENT

The hydrostatic pressure of tissue fluid drives lymph into lymphatic capillaries. However, muscular activity largely influences the movement of lymph through the lymphatic vessels. Lymph, like venous blood, is under relatively low hydrostatic pressure and may not flow readily through the lymphatic vessels without help from contraction of skeletal muscles in the limbs, contraction of the smooth muscle in the walls of the larger lymphatic trunks, and pressure changes associated with breathing.

Contracting skeletal muscles compress lymphatic vessels. This squeezing action moves the lymph inside lymphatic vessels. Valves in these vessels prevent backflow, so lymph can only move toward a collecting duct. Additionally, the smooth muscle in the walls of larger lymphatic trunks can contract and compress the lymph inside, forcing the fluid onward.

Breathing aids lymph circulation by creating a relatively low pressure in the thoracic cavity during inhalation. At the same time, the contracting diaphragm increases the pressure in the abdominal cavity. Consequently, lymph is squeezed out of the abdominal vessels and forced into the thoracic vessels. Once again, valves in lymphatic vessels prevent lymph backflow.

The continuous movement of fluid from interstitial spaces into blood and lymphatic capillaries stabilizes the volume of fluid in these spaces. Conditions that interfere with lymph movement cause tissue fluid to accumulate within the interstitial spaces, producing *edema* (ĕ-de'mah), or swelling. This may happen when surgery removes lymphatic tissue, obstructing certain lymphatic vessels. For example, a surgeon removing a cancerous breast tumor may also remove nearby axillary lymph nodes to prevent associated lymphatic vessels from transporting cancer cells to other sites. Removing the lymphatic tissue can obstruct drainage from the upper limb, causing edema (see fig. 14.4*b*).

14.5 LYMPH NODES

Lymph nodes (lymph glands) are located along the lymphatic pathways. They contain large numbers of *lymphocytes* (B cells and T cells) and *macrophages* that fight invading microorganisms.

Structure of a Lymph Node

Lymph nodes vary in size and shape, but are usually less than 2.5 centimeters long and somewhat bean-shaped (figs. 14.6 and 14.7). Blood vessels and nerves join a lymph node through the indented region of the node,

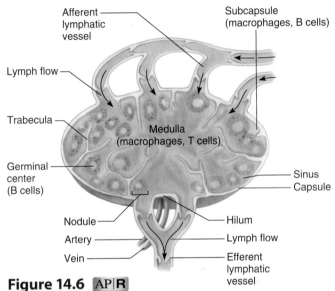

Figure 14.6 AP|R

A section of a lymph node.

Figure 14.7

Lymph enters and leaves a lymph node through lymphatic vessels.

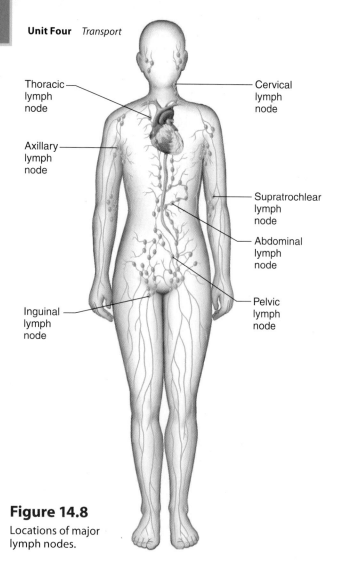

Figure 14.8

Locations of major lymph nodes.

Labels (clockwise from upper left):
Thoracic lymph node
Cervical lymph node
Axillary lymph node
Supratrochlear lymph node
Abdominal lymph node
Pelvic lymph node
Inguinal lymph node

called the **hilum.** The lymphatic vessels leading to a node (afferent vessels) enter separately at various points on its convex surface, but the lymphatic vessels leaving the node (efferent vessels) exit from the hilum.

A *capsule* of connective tissue encloses each lymph node and subdivides it into compartments. Masses of B cells and macrophages in the cortex, called **lymph nodules** (lymph follicles), are the functional units of the lymph node. The spaces within a node, called **lymph sinuses,** provide a complex network of chambers and channels through which lymph circulates.

Lymph nodules occur singly or in groups associated with the mucous membranes of the respiratory and digestive tracts. The *tonsils,* described in chapter 15 (pp. 404–405), are partially encapsulated lymph nodules. Aggregations of nodules called *Peyer's patches* are scattered throughout the mucosal lining of the distal portion of the small intestine.

Locations of Lymph Nodes

Lymph nodes are generally in groups or chains along the paths of the larger lymphatic vessels throughout the body, but are absent in the central nervous system. Figure 14.8 shows the locations of the major lymph nodes.

Functions of Lymph Nodes

Lymph nodes have two primary functions: (1) filtering potentially harmful particles from lymph before returning it to the bloodstream, and (2) monitoring body fluids (immune surveillance) provided by lymphocytes and macrophages. Along with red bone marrow, the lymph nodes are centers for lymphocyte production. Lymphocytes attack viruses, bacteria, and other parasitic cells that lymphatic vessels bring to the lymph nodes. Macrophages in the lymph nodes engulf and destroy foreign substances, damaged cells, and cellular debris.

> Superficial lymphatic vessels inflamed by bacterial infection appear as red streaks beneath the skin, a condition called *lymphangitis*. Inflammation of the lymph nodes, called *lymphadenitis*, often follows. In *lymphadenopathy*, affected lymph nodes enlarge and may be quite painful.

Practice

8. Distinguish between a lymph node and a lymph nodule.
9. What are the major functions of the lymph nodes?

14.6 THYMUS AND SPLEEN

Two other lymphatic organs whose functions are similar to those of the lymph nodes are the thymus and the spleen.

Thymus

The **thymus** (thī′mus) is a soft, bilobed gland enclosed in a connective tissue capsule and located anterior to the aorta and posterior to the upper part of the sternum (fig. 14.9*a*). The thymus is relatively large during infancy and early childhood, but shrinks after puberty and may be quite small in an adult. In elderly persons, adipose and connective tissues replace lymphatic tissue in the thymus.

Connective tissues extend inward from the surface of the thymus, subdividing it into *lobules* (fig. 14.9*b*). The lobules house many lymphocytes. Most of these cells (thymocytes) are inactive; however, some mature into *T lymphocytes,* which leave the thymus and provide immunity. Epithelial cells in the thymus secrete hormones called *thymosins,* which stimulate maturation of T lymphocytes.

> By age seventy years, the thymus is one-tenth the size it was at the age of ten, and the immune system is only 25% as powerful.

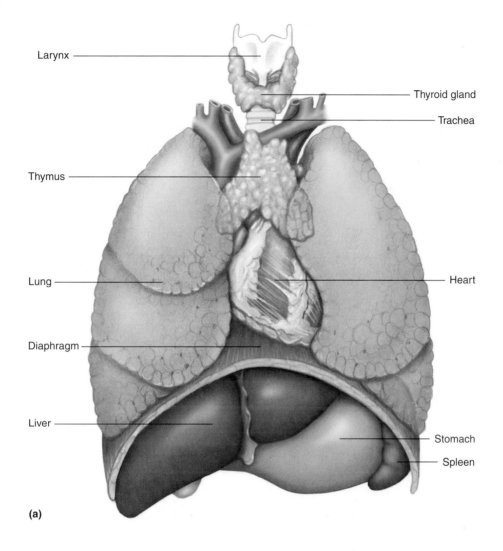

Larynx

Thyroid gland

Trachea

Thymus

Lung

Heart

Diaphragm

Liver

Stomach

Spleen

(a)

Connective tissue

Lobule

(b)

Figure 14.9

Thymus and spleen. **(a)** The thymus is bilobed and located between the lungs and superior to the heart. The spleen is located inferior to the diaphragm and posterior and lateral to the stomach. **(b)** A cross section of the thymus (15×). Note how the gland is subdivided into lobules.

Spleen

The **spleen** (splēn), the largest lymphatic organ, is in the upper left part of the abdominal cavity, just inferior to the diaphragm and posterior and lateral to the stomach (fig. 14.9*a*). The spleen resembles a large lymph node and is subdivided into lobules. However, unlike the sinuses of a lymph node, the spaces (venous sinuses) in the spleen are filled with blood instead of lymph.

The tissues within splenic lobules are of two types (fig. 14.10). The *white pulp* is distributed throughout the spleen in tiny islands. This tissue is composed of splenic nodules, which are similar to those in lymph nodes and are packed with lymphocytes. The *red pulp*, which fills the remaining spaces of the lobules, surrounds the venous sinuses. This pulp contains numerous red blood cells, which impart its color, plus many lymphocytes and macrophages.

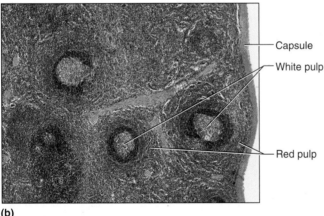

(b)

Figure 14.10 AP|R

Spleen. **(a)** The spleen resembles a large lymph node. **(b)** Light micrograph of the spleen (40×).

Blood capillaries in the red pulp are quite permeable. Red blood cells can squeeze through the pores in these capillary walls and enter the venous sinuses. The older, more fragile red blood cells may rupture during this passage, and the resulting cellular debris is removed by phagocytic macrophages in the splenic sinuses. These macrophages also engulf and destroy foreign particles, such as bacteria, that may be carried in the blood as it flows through the splenic sinuses. Thus, the spleen filters blood much as the lymph nodes filter lymph.

Practice

10. Why are the thymus and the spleen considered organs of the lymphatic system?

11. What are the major functions of the thymus and the spleen?

14.7 BODY DEFENSES AGAINST INFECTION

The presence and multiplication of a disease-causing agent, or **pathogen** (path′o-jen), may cause an **infection.** Pathogens include viruses, bacteria, fungi, and protozoans.

Viruses are not considered to be alive because they are not cells, and in order to reproduce they must use the host cell organelles to carry out protein synthesis. Viral structure is just a nucleic acid inside a coat of proteins, and perhaps glycoproteins.

Other pathogens are considered to be alive. A bacterium is a single, simple cell. A protozoan is a single, complex cell. A fungus may be single-celled, such as a yeast, or multicelled, such as a mold.

The human body can prevent the entry of pathogens or destroy them if they enter. Some mechanisms are quite general and protect against many types of pathogens, providing **innate (nonspecific) defense.** These mechanisms include species resistance, mechanical barriers, chemical barriers (enzyme action, interferon, and complement), natural killer cells, inflammation, phagocytosis, and fever. Other defense mechanisms are very precise, targeting specific pathogens and providing **adaptive (specific) defense,** or **immunity** (ĭ-mu′nĭ-te). Specialized lymphocytes that recognize foreign molecules (nonself antigens) in the body act against them. Innate and adaptive defense mechanisms work together to protect the body against infection. The innate defenses respond quite rapidly, while adaptive defenses develop more slowly.

14.8 INNATE (NONSPECIFIC) DEFENSES

Species Resistance

Species resistance refers to the fact that a given type of organism, or *species* (such as the human species, *Homo sapiens*), may be resistant to diseases that affect other species because its cells do not have receptors for the pathogen or its tissues do not provide the temperature or chemical environment that a particular pathogen requires. For example, other animal species are not infected by the infectious agents that cause measles, mumps, gonorrhea, and syphilis in humans.

A nonhuman species that can contract the same infectious diseases as humans can serve as a "model organism" to study the disease. This is the case for ferrets, which can be infected by the H1N1 influenza virus. The study of infected ferrets was instrumental in developing the H1N1 vaccine.

Mechanical Barriers

The skin and mucous membranes lining the passageways of the respiratory, digestive, urinary, and reproductive systems create **mechanical barriers** that prevent the entrance of infectious agents. As long as the skin and mucous membranes remain intact, many pathogens are unable to penetrate them. Hair traps infectious agents associated with the skin and mucous membranes, and sweat and mucus rinse away microorganisms. These barriers provide a *first line of defense.* The rest of the innate defenses discussed in this section are part of the *second line of defense.*

Chemical Barriers

Enzymes in body fluids provide a **chemical barrier** to pathogens. Gastric juice, for example, contains the protein-splitting enzyme pepsin and has a low pH due to the presence of hydrochloric acid (HCl) (see chapter 15, p. 411). The combined effect of pepsin and HCl is lethal to many pathogens that enter the stomach. Similarly, tears contain the enzyme lysozyme, which destroys certain bacteria on the eyes. The accumulation of salt from perspiration also kills certain bacteria on the skin.

Lymphocytes and fibroblasts produce hormonelike peptides called **interferons** in response to viruses or tumor cells. Once released from the virus-infected cell, interferon binds to receptors on uninfected cells, stimulating them to synthesize proteins that block replication of a variety of viruses. Thus, interferon's effect is nonspecific. Interferons also stimulate phagocytosis and enhance the activity of other cells that help resist infections and the growth of tumors.

Complement (kom'ple-ment) is a group of proteins, in plasma and other body fluids, that interact in an expanding series of reactions or cascade. Activation of complement stimulates inflammation, attracts phagocytes, and enhances phagocytosis.

Natural Killer (NK) Cells

Natural killer (NK) **cells** are a small population of lymphocytes different from the lymphocytes that provide adaptive (specific) defense mechanisms (discussed later in this chapter). NK cells defend the body against various viruses and cancer cells by secreting cytolytic ("cell-cutting") substances called **perforins** that lyse the cell membrane, destroying the infected cell. NK cells also secrete chemicals that enhance inflammation.

Inflammation

Inflammation (in"flah-ma'shun) is a tissue response to injury or infection, producing localized redness, swelling, heat, and pain. The redness is a result of blood vessel dilation that increases blood flow and volume in the affected tissues. This effect, coupled with an increase in the permeability of nearby capillaries, swells tissues (edema). The heat comes as blood enters from deeper body parts, which are generally warmer than the surface. Pain results from stimulation of nearby pain receptors.

Infected cells release chemicals that attract white blood cells to inflammation sites, where they phagocytize pathogens. Local heat speeds up phagocytic activity. In bacterial infections, the resulting mass of white blood cells, bacterial cells, and damaged tissue may form a thick fluid called **pus.**

Body fluids also collect in inflamed tissues. These fluids contain fibrinogen and other blood-clotting factors. Clotting forms a network of fibrin threads in the affected region. Later, fibroblasts may arrive and secrete matrix components until the area is enclosed in a connective tissue sac. This walling off of the infected area helps inhibit the spread of pathogens and toxins to adjacent tissues.

Phagocytosis

Phagocytosis removes foreign particles from the lymph as it moves from the interstitial spaces to the bloodstream. Phagocytes in the blood vessels and in the tissues of the spleen, liver, or bone marrow remove particles that reach the blood. Recall from chapter 12 (p. 326) that blood's most active phagocytic cells are *neutrophils* and *monocytes.* Chemicals released from injured tissues attract these cells (chemotaxis). Neutrophils engulf and digest smaller particles; monocytes phagocytize larger ones.

Monocytes that leave the bloodstream by diapedesis become *macrophages* (histiocytes), which may be *free or fixed* in various tissues. The fixed macrophages, which can divide and produce new macrophages, are found in the lymph nodes, spleen, liver, and lungs. Neutrophils, monocytes, and macrophages constitute the **mononuclear phagocytic system** (reticuloendothelial system).

Fever

Fever is body temperature elevated above an individual's normal temperature due to an elevated setpoint. It is part of the innate defense because it makes the body inhospitable to certain pathogens. Higher body temperature causes the liver and spleen to sequester iron, which reduces the level of iron in the blood. Bacteria and fungi require iron for normal metabolism, so their growth and reproduction in a fever-ridden body slows and may cease. Also, phagocytic cells attack more vigorously when the temperature rises. For these reasons, low-grade fever of short duration may be a desired natural response, not a symptom to be treated aggressively with medications.

Practice

12. What is an infection?
13. Explain seven innate (nonspecific) defense mechanisms.

14.9 ADAPTIVE (SPECIFIC) DEFENSES, OR IMMUNITY

The *third line of defense,* **immunity** (ĭ-mu'nĭ-te), is resistance to specific pathogens or to their toxins or metabolic by-products. Lymphocytes and macrophages that recognize and remember specific foreign molecules carry out adaptive immune responses.

Antigens

Antigens are proteins, polysaccharides, glycoproteins, or glycolipids, usually located on a cell's surface. Before birth, cells inventory the proteins and other large molecules in the body, learning to identify these as "self." The lymphatic system responds to "nonself," or foreign, antigens, but not normally to self antigens. Receptors on lymphocyte surfaces enable these cells to recognize foreign antigens.

The antigens that are most effective in eliciting an immune response are large and complex, with few repeating parts. Sometimes, a smaller molecule that cannot by itself stimulate an immune response combines with a larger one, which makes it able to do so. Such a small molecule is called a **hapten** (hap'ten). Stimulated lymphocytes react either to the hapten or to the larger molecule of the combination. Haptens are found in certain drugs such as penicillin, in household and industrial chemicals, in dust particles, and in animal dander.

Lymphocyte Origins

During fetal development (before birth), red bone marrow releases unspecialized precursors to lymphocytes into the circulation. About half of these cells reach the thymus, where they specialize into **T lymphocytes,** or **T cells.** Later, some of these T cells constitute 70–80% of the circulating lymphocytes in blood (fig. 14.11). Other T cells reside in lymphatic organs and are particularly abundant in the lymph nodes, thoracic duct, and white pulp of the spleen.

Other lymphocytes remain in the red bone marrow until they differentiate into **B lymphocytes,** or **B cells.** The blood distributes B cells, which constitute 20–30%

Figure 14.11
Scanning electron micrograph of a human circulating lymphocyte (7,000×).

of circulating lymphocytes. B cells settle in lymphatic organs along with T cells and are abundant in the lymph nodes, spleen, bone marrow, and intestinal lining (fig. 14.12). Table 14.1 compares the characteristics of T cells and B cells.

Practice

14. What is immunity?
15. What is the difference between an antigen and a hapten?
16. How do T cells and B cells originate?

T Cells and the Cellular Immune Response

A lymphocyte must be activated before it can respond to an antigen. T cell activation requires the presence of processed fragments of antigen attached to the surface of another type of cell, called an **antigen-presenting cell** (accessory cell). Macrophages, B cells, and several other cell types can be antigen-presenting cells.

T cell activation begins when a macrophage phagocytizes a bacterium, digesting it in its lysosomes. Some bacterial antigens exit the lysosomes and move to the macrophage's surface. Here, they are displayed on the cell membrane near certain protein molecules that are part of a group of proteins called the *major histocompatibility complex (MHC).* MHC antigens help T cells recognize that an antigen is foreign, not self.

Activated T cells interact directly with antigen-bearing cells. Such cell-to-cell contact is called the **cellular immune response,** or cell-mediated immunity. T cells

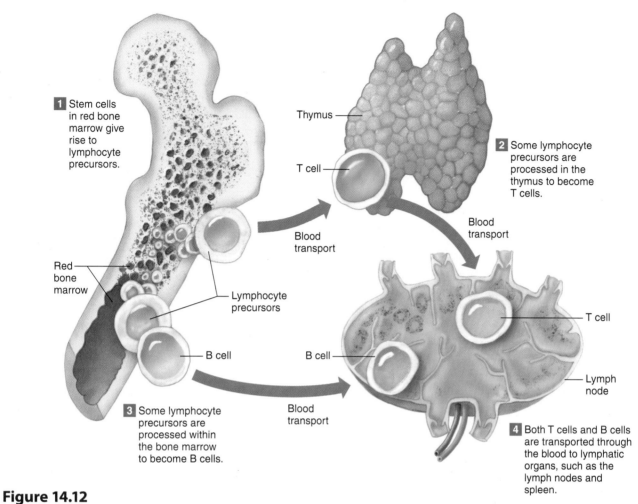

Figure 14.12

During fetal development, bone marrow releases unspecialized lymphocyte precursors, which after processing specialize as T cells (T lymphocytes) or B cells (B lymphocytes). Note that in the fetus, the medullary cavity contains red marrow.

Table 14.1	A Comparison of T Cells and B Cells	
Characteristic	**T Cells**	**B Cells**
Origin of Undifferentiated Cell	Red bone marrow	Red bone marrow
Site of Differentiation	Thymus	Red bone marrow
Primary Locations	Lymphatic tissues, 70–80% of the circulating lymphocytes	Lymphatic tissues, 20–30% of the circulating lymphocytes
Primary Functions	Provides cellular immune response in which T cells interact directly with the antigens or antigen-bearing agents to destroy them	Provides humoral immune response in which B cells interact indirectly, producing antibodies that destroy the antigens or antigen-bearing agents

(and some macrophages) also synthesize and secrete polypeptides called *cytokines* that enhance certain cellular responses to antigens. For example, *interleukin-1* and *interleukin-2* stimulate the synthesis of several cytokines by other T cells. In addition, interleukin-1 helps activate T cells, whereas interleukin-2 causes T cells to proliferate. Other cytokines, called *colony-stimulating factors (CSFs)*, stimulate leukocyte production in red bone marrow, and activate macrophages. T cells may also secrete toxins that kill their antigen-bearing target cells, growth-inhibiting factors that prevent target cell growth, or interferon that inhibits the proliferation of viruses and tumor cells. Several types of T cells have distinct functions.

A specialized type of T cell, called a *helper T cell,* becomes activated when its antigen receptor combines with displayed foreign antigen (fig. 14.13). Once activated, the helper T cell stimulates a B cell to produce antibodies that are specific for the displayed antigen.

Another type of T cell is a *cytotoxic T cell,* which recognizes and combines with nonself antigens that cancerous cells or virally infected cells display on their surfaces near certain MHC proteins. Cytokines from helper T cells activate the cytotoxic T cell. Next, the cytotoxic T cell proliferates, enlarging its **clone** (klōn) of cells, which is a group of genetically identical cells that descend from a single, original cell. Cytotoxic T cells then bind to the surfaces of antigen-bearing cells, where they release *perforin* protein that cuts porelike openings, destroying these cells. In this way, cytotoxic T cells continually monitor the body's cells, recognizing and eliminating tumor cells and cells infected with

viruses. Cytotoxic T cells provide much of the immune system's defense against HIV infection, discussed in Clinical Application 14.1. Unfortunately, the virus eventually kills these cells.

Some T cells do not respond to a foreign antigen on first exposure, but remain as *memory cells.* These memory cells immediately divide to yield more cytotoxic T cells and helper T cells upon subsequent exposure to the same antigen, often before symptoms arise.

Practice

17. How do T cells become activated?
18. What are some functions of cytokines?
19. Name three types of T cells.
20. How do cytotoxic T cells destroy cells bearing foreign antigens?

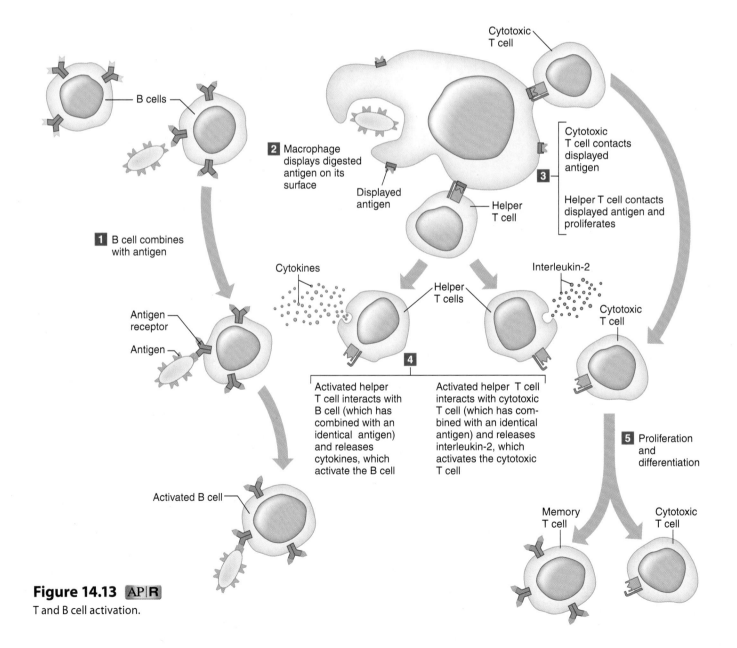

Figure 14.13 **AP|R**
T and B cell activation.

Immunity Breakdown: HIV/AIDS

In the early 1980s, health-care workers from large cities in the United States began seeing otherwise healthy young men who had rare infections and cancers. The conditions were opportunistic, taking advantage of a weakened immune system. An apparently new lethal disease emerged—acquired immune deficiency syndrome or AIDS. Today, 40 million people worldwide are infected with the virus that causes AIDS.

The human immunodeficiency virus (HIV), which causes AIDS, has RNA as its genetic material. The virus can be present for a decade or longer before symptoms begin, such as recurrent fever, weakness, and weight loss. However, a small percentage of people who are HIV positive have remained healthy, and people with rare mutations such that they lack the T cell receptors to which HIV binds cannot become infected at all.

Transmission of HIV requires contact with a body fluid containing abundant virus, such as blood or semen. Levels of the virus in sweat, tears, and saliva are so low that transmission is highly unlikely. Whether or not a person becomes infected depends on the amount of infected fluid contacted, the site of exposure in the body, and the individual's health and genetic background. Table 14A lists some of the ways that HIV infection can and cannot be spread.

HIV infection gradually shuts down the immune system. First, HIV enters macrophages, impairing this first line of defense. In these cells and later in helper T cells, the virus adheres with one of its surface proteins, called gp120, to two receptors on the host cell surface, called CD4 and CCR5. Once the virus enters the cell, a viral enzyme, reverse transcriptase, builds a DNA strand complementary to the viral RNA sequence. This newly formed viral DNA strand replicates to form a DNA double helix, which enters the cell's nucleus and inserts into a chromosome. The viral-produced DNA sequences are then transcribed and translated. The cell fills with viral pieces, which are assembled into complete new viral particles that burst from the cell.

Once the infected helper T cells are dying at a high rate, bacterial infections greatly increase because B cells aren't activated to produce antibodies. Much later in infection, HIV variants arise that bind receptors on cytotoxic T cells, killing them too. Loss of these cells renders the body very vulnerable to other infections and to cancers.

HIV replicates quickly, mutates often, and can hide, twisting and altering its surface features in ways that evade recognition and attack by antibodies or cytotoxic T cells. For several years, the bone marrow produces 2 billion new T and B cells a day, countering the million to billion new HIV particles that infected cells release daily. The population of HIV in a human host replicates and mutates so fast that within days of infection, viral variants can arise that resist the drugs used to treat HIV infection and AIDS.

Combining drugs that act in different ways minimizes the number of viruses (viral load) and delays symptom onset and progression, and also slows the spread of the virus in the population. The first drugs developed block viral replication. A second class of drugs, called protease inhibitors, prevent HIV from processing its proteins to a functional size. A third class of drugs, called entry inhibitors, either block the binding and/or fusing of HIV to T cell surfaces or keep the virus out even if it can bind to the cell. However, because viral RNA mutates at a rapid rate, viral variants usually emerge that resist these drugs. The drugs kill sensitive viruses, leaving only resistant ones, and the resistant ones take over. More than 200 drugs are also available to treat AIDS-associated opportunistic infections and cancers.

Developing a vaccine against HIV has been extremely difficult because components of the immune system recognize only small parts of the virus, and the virus often alters these very parts. Even without a vaccine, many people with HIV infection can keep viral levels low enough to stay healthy.

Table 14A	HIV Transmission
How HIV Is Transmitted	
Sexual contact, particularly anal intercourse, but also vaginal intercourse and oral sex	
Contaminated needles (intravenous drug use, injection of anabolic steroids, accidental needle stick in medical setting)	
During birth from infected mother	
Breast milk from infected mother	
Receiving infected blood or other tissue (precautions usually prevent this)	
How HIV Is Not Transmitted	
Casual contact (social kissing, hugging, handshakes)	
Objects (toilet seats, deodorant sticks, doorknobs)	
Mosquitoes	
Sneezing and coughing	
Sharing food	
Swimming in the same water	
Donating blood	

B Cells and the Humoral Immune Response

A B cell may become activated when it encounters an antigen whose molecular shape fits the shape of its antigen receptors. In response to the receptor-antigen combination, the B cell may divide repeatedly, expanding its clone. However, most of the time B cell activation requires T cell "help."

When an activated helper T cell encounters a B cell already combined with a foreign antigen identical to the one that activated the helper T cell, the helper cell releases certain cytokines that stimulate the B cell to proliferate, enlarging its clone of antibody-producing cells (fig. 14.14). The cytokines also attract

macrophages and leukocytes into inflamed tissues and help keep them there.

Some members of the activated B cell's clone differentiate further into memory cells. Like memory T cells, these memory B cells respond rapidly to subsequent exposure to a specific antigen.

Other members of the activated B cell's clone differentiate further into **plasma cells,** which produce and secrete large globular proteins called **antibodies** (an′tĭ-bod″ēz), or **immunoglobulins** (im″u-no-glob′u-linz), similar in structure to the antigen-receptor molecules on the original B cell's surface. Body fluids carry antibodies, which then react in various ways to destroy specific antigens or antigen-bearing particles.

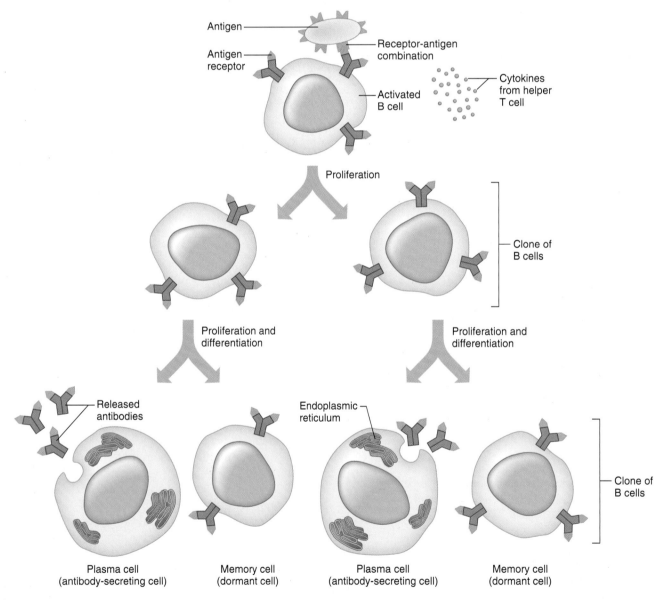

Figure 14.14

An activated B cell proliferates after stimulation by cytokines released from helper T cells. The B cell's clone enlarges. Some cells of the clone give rise to antibody-secreting plasma cells and others to dormant memory cells.

Table 14.2	Steps in Antibody Production

B Cell Activities

1. Antigen-bearing agents enter tissues.

2. B cell encounters an antigen that fits its antigen receptors.

3. Either alone or more often in conjunction with helper T cells, the B cell is activated. The B cell proliferates, enlarging its clone.

4. Some of the newly formed B cells differentiate further to become plasma cells.

5. Plasma cells synthesize and secrete antibodies whose molecular structure is similar to the activated B cell's antigen receptors.

T Cell Activities

1. Antigen-bearing agents enter tissues.

2. An accessory cell, such as a macrophage, phagocytizes the antigen-bearing agent, and the macrophage's lysosomes digest the agent.

3. Antigens from the digested antigen-bearing agents are displayed on the membrane of the accessory cell.

4. Helper T cell becomes activated when it encounters a displayed antigen that fits its antigen receptors.

5. Activated helper T cell releases cytokines when it encounters a B cell that has previously combined with an identical antigen-bearing agent.

6. Cytokines stimulate the B cell to proliferate.

7. Some of the newly formed B cells give rise to cells that differentiate into antibody-secreting plasma cells.

This antibody-mediated immune response is called the **humoral immune response** ("humoral" refers to fluid). Table 14.2 summarizes the steps leading to antibody production as a result of B cell and T cell activities.

A plasma cell secretes up to 2,000 identical antibodies per second.

An individual's B cells can produce an estimated 10 million to 1 billion different varieties of antibodies, each reacting against a specific antigen. The enormity and diversity of the antibody response defends against many pathogens.

Types of Antibodies

Antibodies (immunoglobulins) are soluble, globular proteins that constitute the *gamma globulin* fraction of plasma proteins (see chapter 12, p. 329). Of the five major types of immunoglobulins, the most abundant are immunoglobulin G, immunoglobulin A, and immunoglobulin M.

Immunoglobulin G (IgG) is in plasma and tissue fluids and is particularly effective against bacteria, viruses, and toxins. It also activates *complement.*

Immunoglobulin A (IgA) is commonly found in exocrine gland secretions. It is in breast milk, tears, nasal fluid, gastric juice, intestinal juice, bile, and urine.

Immunoglobulin M (IgM) is a type of antibody present in plasma in response to contact with certain antigens in foods or bacteria. The antibodies anti-A and anti-B, described in chapter 12 (pp. 334–335), are examples of IgM. IgM also activates complement.

Immunoglobulin D (IgD) is found on the surfaces of most B cells, especially those of infants. IgD is important in activating B cells.

Immunoglobulin E (IgE) is in exocrine secretions with IgA. It is associated with allergic reactions, which are described later in this section (see p. 394).

A newborn does not yet have its own antibodies, but does retain for a while IgG that passed through the placenta from the mother. These maternal antibodies protect the infant against some illnesses to which the mother is immune. At about the same time that the maternal antibody supply falls, the infant begins to manufacture its own antibodies. The newborn also receives IgA from colostrum, a substance secreted from the mother's breasts for the first few days after birth. Antibodies in colostrum protect against certain digestive and respiratory infections.

Practice

21. How are B cells activated?

22. How does the antibody response protect against diverse infections?

23. Which immunoglobulins are most abundant, and how do they differ from each other?

Antibody Actions

In general, antibodies react to antigens in three ways. Antibodies directly attack antigens, activate complement, or stimulate localized changes (inflammation) that help prevent the spread of pathogens or cells bearing foreign antigens.

In a direct attack, antibodies combine with antigens, causing them to clump (agglutinate) or to form insoluble substances (precipitate). Such actions make it easier for phagocytic cells to recognize and engulf the antigen-bearing agents and eliminate them. In other instances, antibodies cover the toxic portions of antigen molecules and neutralize their effects. However, under normal conditions, direct antibody attack is not as important as complement activation in protecting against infection.

When certain IgG or IgM antibodies combine with antigens, they expose reactive sites on antibody molecules. This triggers a series of reactions, leading to activation of the complement proteins, which in turn produce a variety of effects. These include: coating the antigen-antibody complexes (opsonization), making the complexes more susceptible to phagocytosis; attracting macrophages and neutrophils into the region (chemotaxis); rupturing membranes of foreign cells (lysis); clumping antigen-bearing cells; and altering the molecular structure of viruses, making them harmless (fig. 14.15).

Other proteins promote inflammation, which helps prevent the spread of infectious agents.

Practice

24. In what general ways do antibodies function?

25. How is complement activated?

26. What is the function of complement?

Immune Responses

Activation of B cells or T cells after first encountering the antigens for which they are specialized to react constitutes a **primary immune response.** During such a response, plasma cells release antibodies (IgM, followed by IgG) into the lymph. The antibodies are transported to the blood and then throughout the body, where they help destroy antigen-bearing agents. Production and release of antibodies continues for several weeks.

Following a primary immune response, some of the B cells produced during proliferation of the clone remain dormant as memory cells (see fig. 14.14). If the identical antigen is encountered again, clones of these memory cells enlarge, and can respond rapidly with IgG to the antigen to which they were previously sensitized. These memory B cells, along with the memory T cells, produce a **secondary immune response.**

As a result of a primary immune response, detectable concentrations of antibodies usually appear in the blood plasma five to ten days after exposure to antigens. If the identical antigen is encountered later in life, a secondary immune response may produce the same antibodies within a day or two (fig. 14.16). Although newly

Figure 14.15

Actions of the complement system.

- Opsonization— enhancing phagocytosis of antigens
- Chemotaxis— attracting macrophages and neutrophils
- Lysis— rupturing membranes of foreign cells
- Agglutination— clumping of antigen-bearing agents
- Neutralization— altering the molecular structure of viruses

(Complement system → Activation of complement proteins, triggered by Presence of antibodies combined with antigens)

Figure 14.16

A primary immune response occurs after the first exposure to an antigen. A secondary immune response occurs after subsequent exposure(s) to the same antigen. The break in the timeline represents a time lapse between exposure(s) to the antigen, when the time to develop antibodies begins once again.

Q: *Which immune response produces antibodies to a specific antigen more quickly, primary or secondary?*

Answer can be found in Appendix E on page 568.

formed antibodies may persist in the body for only a few months or years, memory cells live much longer.

Practical Classification of Immunity

Adaptive, also known as acquired, immunity can arise in response to natural events or be induced artificially by injection. Both naturally and artificially acquired immunities can be either active or passive. Active immunity results when the person produces an immune response (including memory cells) to the antigen; it is long-lasting. Passive immunity occurs when a person receives antibodies produced by another individual. Because the person does not produce an immune response, passive immunity is short-term, and the individual will be susceptible upon exposure to the antigen at some later date.

Naturally acquired active immunity occurs when a person exposed to a pathogen develops a disease. Resistance to that pathogen is the result of a primary immune response.

A **vaccine** (vak′sēn) is a preparation that produces another type of active immunity. A vaccine might consist of bacteria or viruses that have been killed or weakened so that they cannot cause a serious infection, or only molecules unique to the pathogens. A vaccine might also be a toxoid, which is a toxin from an infectious organism that has been chemically altered to destroy its dangerous effects. Whatever its composition, a vaccine includes the antigens that stimulate a primary immune response, but does not produce the severe symptoms of disease. A vaccine causes a person to develop *artificially acquired active immunity.*

> Vaccines stimulate active immunity against a variety of diseases, including typhoid fever, cholera, whooping cough, diphtheria, tetanus, polio, chickenpox, measles (rubeola), German measles (rubella), mumps, influenza, hepatitis A, hepatitis B, and bacterial pneumonia.

A person who has been exposed to infection may require protection against a disease-causing microorganism before active immunity has had the time to develop. An injection of antiserum (antibodies) may help. These antibodies may be obtained by separating gamma globulins from the plasma of persons who have already developed immunity against the particular disease. A gamma globulin injection provides *artificially acquired passive immunity.*

During pregnancy, certain antibodies (IgG) pass from the maternal blood into the fetal bloodstream. As a result, the fetus acquires limited immunity against pathogens that the pregnant woman has developed active immunities against. The fetus thus has *naturally acquired passive immunity,* which may last for six months to a year after birth. Table 14.3 summarizes the types of acquired immunity.

Practice

27. Distinguish between a primary and a secondary immune response.
28. Distinguish between active and passive immunity.

Allergic Reactions

An allergic reaction is an immune response to a nonharmful substance. Allergic reactions are similar to immune responses because they sensitize lymphocytes and provide antibodies that may combine with antigens. However, unlike normal immune responses, allergic reactions can damage tissues. Antigens that trigger allergic responses are called **allergens** (al′er-jenz).

Allergies can be classified by how quickly they follow exposure to the allergen. A *delayed-reaction allergy* results from repeated exposure of the skin to certain chemicals—commonly, household or industrial chemicals or ingredients of some cosmetics. After repeated contact, the presence of the foreign substance activates T cells, many of which collect in the skin. The T cells

Table 14.3	Practical Classification of Immunity	
Type	**Mechanism**	**Result**
Naturally acquired active immunity	Exposure to live pathogens	Stimulation of an immune response with symptoms of a disease
Artificially acquired active immunity	Exposure to a vaccine containing weakened or dead pathogens or their components	Stimulation of an immune response without the severe symptoms of a disease
Artificially acquired passive immunity	Injection of gamma globulin containing antibodies	Short-term immunity without stimulating an immune response
Naturally acquired passive immunity	Antibodies passed to fetus from pregnant woman with active immunity or to newborn through breast milk from a woman with active immunity	Short-term immunity for newborn without stimulating an immune response

and the macrophages they attract release chemical factors, which in turn cause eruptions and inflammation of the skin (dermatitis). This reaction is called *delayed* because it usually takes about 48 hours to occur. A delayed-reaction allergy may affect anyone. An example is seen in how some people react to poison ivy.

An *immediate-reaction allergy* occurs within minutes after contact with an allergen, and affects people who have an inherited tendency to overproduce IgE antibodies in response to certain antigens. IgE normally comprises a tiny fraction of plasma proteins. An immediate-reaction allergy activates B cells, which become sensitized when the allergen is first encountered. Subsequent exposures to the allergen trigger allergic reactions. In the initial exposure, IgE attaches to the cell membranes of widely distributed mast cells and basophils. Upon a subsequent allergen-antibody reaction, these cells release allergy mediators such as *histamine, prostaglandin D₂,* and *leukotrienes.* These substances cause a variety of physiological effects, including dilation of blood vessels, increased vascular permeability that swells tissues, contraction of bronchial and intestinal smooth muscles, and increased mucus production. The result is a severe inflammation reaction that is responsible for the symptoms of the allergy, such as hives, hay fever, asthma, eczema, or gastric disturbances.

> Anaphylactic shock is a severe form of immediate-reaction allergy in which mast cells release allergy mediators throughout the body. The entire body itches and breaks out in red hives. Vomiting and diarrhea may follow as the face, tongue, and larynx begin to swell, and breathing becomes difficult. An injection of epinephrine (adrenalin) and sometimes a tracheotomy (an incision into the windpipe to restore breathing) may save the person's life. Common triggers of anaphylactic shock are insect stings and allergy to penicillin. AP|R

Transplantation and Tissue Rejection

Transplantation of tissues or an organ, such as the skin, kidney, heart, or liver, from one person to another can replace a nonfunctional, damaged, or lost body part. However, the recipient's immune system may recognize the donor's cell surfaces as foreign and attempt to destroy the transplanted tissue, causing a **tissue rejection reaction.**

Tissue rejection resembles the cellular immune response against a foreign antigen. The greater the antigenic difference between the cell surface molecules (MHC antigens, discussed on page 386) of the recipient tissues and the donor tissues, the more rapid and severe the rejection reaction. Matching donor and recipient tissues can minimize the rejection reaction.

Immunosuppressive drugs are used to reduce rejection of transplanted tissues. These drugs interfere with the recipient's immune response by suppressing formation of antibodies or production of T cells, thereby dampening the humoral and cellular immune responses which may also cause the recipient to be more susceptible to infection. Immunosuppressive drugs were once given for the rest of the organ recipient's life. However, experience has shown that it is often better to begin immunosuppressive therapy before the transplant, and possibly end it after a few years, especially if the recipient has also received bone marrow stem cells from the donor. Lifelong immunosuppressants may disturb the process of the tissues from recipient and donor accepting each other.

> Donated organs must be transplanted quickly. How long can donated organs last outside the body?
>
> - A heart lasts three to five hours.
> - A liver lasts ten hours.
> - A kidney lasts twenty-four to forty-eight hours.

Autoimmunity

The immune system can turn against itself, and become unable to distinguish a particular self antigen from a nonself antigen, producing **autoantibodies** and cytotoxic T cells that attack and damage certain tissues and organs. This reaction against self is called **autoimmunity.** The specific nature of an autoimmune disorder reflects the cell types that are the target of the immune attack. Type 1 (insulin dependent) diabetes mellitus, rheumatoid arthritis, and systemic lupus erythematosus are autoimmune disorders. About 5% of the population have an autoimmune disorder.

Why might the immune system attack body tissues? Perhaps a virus, while replicating inside a human cell, takes proteins from the host cell's surface and incorporates them onto its own surface. When the immune system "learns" the surface of the virus in order to destroy it, it also learns to attack the human cells that normally bear the particular protein. Another explanation of autoimmunity is that somehow T cells never learn to distinguish self from nonself. A third possible route of autoimmunity is when a nonself antigen coincidentally resembles a self antigen. This is what happens when an infection by *Streptococcus* bacteria triggers inflammation of heart valves, as mentioned in chapter 13 (p. 345). Clinical Application 14.2 discusses how some disorders thought to be autoimmune may be the result of lingering fetal cells.

> ### Practice
>
> 29. How are allergic reactions and immune reactions similar yet different?
>
> 30. How does a tissue rejection reaction involve an immune response?
>
> 31. How is autoimmunity an abnormal functioning of the immune response?

Clinical Application 14.2

Persisting Fetal Cells and Autoimmunity

Some disorders thought to be autoimmune may have a stranger cause—fetal cells persisting in a woman's circulation, for decades. In response to an as yet unknown trigger, the fetal cells, perhaps "hiding" in a tissue such as skin, emerge, stimulating antibody production. The resulting antibodies and symptoms appear to be an autoimmune disorder. This situation of persisting fetal cells provoking an immune response is seen in a disorder called scleroderma, which means "hard skin."

Scleroderma, which typically begins between ages forty-five and fifty-five, is described as "the body turning to stone." Symptoms include fatigue, swollen joints, stiff fingers, and a masklike face. The hardening may affect blood vessels, the lungs, and the esophagus. Clues that scleroderma is a delayed response to persisting fetal cells include the following observations:

- It is much more common among women.
- Symptoms resemble those of graft-versus-host disease (GVHD), in which transplanted tissue produces chemicals that destroy the recipient's tissues. Antigens on cells in scleroderma lesions match those that cause GVHD.
- Mothers who have scleroderma and their sons have cell surfaces that are more similar than those of unaffected mothers and their sons. Perhaps the similarity of cell surfaces enabled the fetal cells to escape destruction by the woman's immune system. Female fetal cells probably have the same effect, but these cells cannot be distinguished from maternal cells by the presence of the Y chromosome. Perhaps other disorders considered autoimmune actually reflect an immune system response to lingering fetal cells.

Summary Outline

14.1 Introduction (p. 378)

The lymphatic system is closely associated with the cardiovascular system. It transports excess tissue fluid to the bloodstream, absorbs fats, and helps defend the body against disease-causing agents.

14.2 Lymphatic Pathways (p. 378)

1. Lymphatic capillaries
   a. Lymphatic capillaries are microscopic, closed-ended tubes that extend into interstitial spaces.
   b. They receive tissue fluid through their thin walls.
2. Lymphatic vessels
   a. Lymphatic vessels are formed by the merging of lymphatic capillaries.
   b. Lymphatic vessels have walls similar to those of veins, only thinner, and possess valves that prevent backflow of lymph.
   c. Larger lymphatic vessels lead to lymph nodes and then merge into lymphatic trunks.
3. Lymphatic trunks and collecting ducts
   a. Lymphatic trunks lead to two collecting ducts—the thoracic duct and the right lymphatic duct.
   b. Collecting ducts join the subclavian veins.

14.3 Tissue Fluid and Lymph (p. 380)

1. Tissue fluid formation
   a. Tissue fluid originates from plasma and includes water and dissolved substances that have passed through the blood capillary wall.
   b. It generally lacks large proteins, but some smaller proteins are filtered out of blood capillaries into interstitial spaces.
   c. As the protein concentration of tissue fluid increases, colloid osmotic pressure increases.

2. Lymph formation and function
   a. Increasing pressure within interstitial spaces forces some tissue fluid into lymphatic capillaries, and this fluid becomes lymph.
   b. Lymph returns protein molecules to the bloodstream and transports foreign particles to lymph nodes.

14.4 Lymph Movement (p. 381)

1. Lymph is under low pressure and may not flow readily without external aid.
2. Lymph is moved by the contraction of skeletal muscles, contraction of smooth muscle in the walls of large lymphatic trunks, and low pressure in the thorax created by breathing movements.

14.5 Lymph Nodes (p. 381)

1. Structure of a lymph node
   a. Lymph nodes are subdivided into nodules.
   b. Nodules contain masses of lymphocytes and macrophages.
2. Locations of lymph nodes
   Lymph nodes aggregate in groups or chains along the paths of larger lymphatic vessels.
3. Functions of lymph nodes
   a. Lymph nodes filter potentially harmful foreign particles from lymph.
   b. Lymph nodes are centers for the production of lymphocytes, and they also contain phagocytic cells.

14.6 Thymus and Spleen (p. 382)

1. Thymus
   a. The thymus, located anterior to the aorta and posterior to the upper part of the sternum, is composed of lymphatic tissue subdivided into lobules.
   b. It slowly shrinks after puberty.
   c. Some lymphocytes leave the thymus and provide immunity.

Lymphatic System

Integumentary System

The skin is a first line of defense against infection.

Cardiovascular System

The lymphatic system returns tissue fluid to the bloodstream. Lymph originates as tissue fluid, formed by the action of blood pressure.

Skeletal System

Cells of the immune system originate in the bone marrow.

Digestive System

Lymph plays a major role in the absorption of fats.

Muscular System

Muscle action helps pump lymph through the lymphatic vessels.

Respiratory System

Cells of the immune system patrol the respiratory system to defend against infection.

Nervous System

Stress may impair the immune response.

Urinary System

The kidneys control the volume of extracellular fluid, including lymph.

Endocrine System

Hormones stimulate lymphocyte production.

Reproductive System

Special mechanisms inhibit the female immune system in its attack of sperm as foreign invaders.

The lymphatic system is an important link between tissue fluid and the plasma; it also plays a major role in the response to infection.

2. Spleen
   a. The spleen, just inferior to the diaphragm and posterior and lateral to the stomach, resembles a large lymph node subdivided into lobules.
   b. Spaces within splenic lobules are filled with blood.
   c. The spleen contains many macrophages, which filter foreign particles and damaged red blood cells from blood.

14.7 Body Defenses Against Infection (p. 384)

The body has innate (nonspecific) and adaptive (specific) defenses against infection.

14.8 Innate (Nonspecific) Defenses (p. 384)

1. Species resistance
   Each species is resistant to certain diseases that may affect other species.
2. Mechanical barriers
   a. Mechanical barriers include the skin and mucous membranes, which block the entrance of some pathogens.
   b. Hair traps infectious agents; fluids such as tears, sweat, saliva, mucus, and urine wash away organisms before they can firmly attach.
3. Chemical barriers
   a. Enzymes in gastric juice and tears kill some pathogens.
   b. Interferons stimulate uninfected cells to synthesize antiviral proteins that stimulate phagocytosis, block proliferation of viruses, and enhance activity of cells that help resist infections and stifle tumor growth.
   c. Activation of complement proteins in plasma stimulates inflammation, attracts phagocytes, and enhances phagocytosis.
4. Natural killer cells
   Natural killer cells secrete perforins, which destroy cancer cells and cells infected with viruses.
5. Inflammation
   a. Inflammation is a tissue response to injury or infection, and includes localized redness, swelling, heat, and pain.
   b. Chemicals released by damaged tissues attract white blood cells to the site.
   c. Connective tissue may form a sac around injured tissue and thus block the spread of pathogens.
6. Phagocytosis
   a. The most active phagocytes in blood are neutrophils and monocytes. Monocytes give rise to macrophages, which can remain fixed in tissues.
   b. Phagocytic cells are associated with the linings of blood vessels in the red bone marrow, liver, spleen, lungs, and lymph nodes.
   c. Phagocytes remove foreign particles from tissues and body fluids.
7. Fever
   Higher body temperature and the resulting decrease in blood iron level and increase in phagocytic activity hamper infection.

14.9 Adaptive (Specific) Defenses, or Immunity (p. 386)

1. Antigens
   a. Before birth, body cells inventory "self" proteins and other large molecules.
   b. After inventory, lymphocytes develop receptors that allow them to differentiate between nonself (foreign) and self antigens.
   c. Haptens are small molecules that can combine with larger ones, becoming antigenic.

2. Lymphocyte origins
   a. Lymphocytes originate in red bone marrow and are released into the blood.
   b. Some reach the thymus, where they mature into T cells.
   c. Others, the B cells, mature in the red bone marrow.
   d. Both T cells and B cells reside in lymphatic tissues and organs.
3. T cells and the cellular immune response
   a. T cells are activated when an antigen-presenting cell displays a foreign antigen.
   b. When a macrophage acts as an accessory cell, it phagocytizes an antigen-bearing agent, digests the agent, and displays the antigens on its cell membrane in association with certain MHC proteins.
   c. T cells respond to antigens by cell-to-cell contact (cellular immune response).
   d. T cells secrete cytokines, such as interleukins, that enhance cellular responses to antigens.
   e. T cells may also secrete substances that are toxic to their target cells.
   f. A helper T cell becomes activated when it encounters displayed antigens for which it is specialized to react.
   g. Once activated, helper T cells stimulate B cells to produce antibodies.
   h. Cytotoxic T cells recognize foreign antigens on tumor cells and cells whose surfaces indicate that they are infected by viruses, then release perforin to destroy these cells.
   i. Memory T cells allow for immediate response to second and subsequent exposure to the same antigen.
4. B cells and the humoral immune response
   a. Sometimes a B cell is activated when it encounters an antigen that fits its antigen receptors or more often a B cell is activated when stimulated by a helper T cell.
   b. An activated B cell proliferates (especially when stimulated by a T cell), enlarging its clone.
   c. Some activated B cells differentiate further into memory B cells.
   d. Other activated B cells differentiate into antibody-producing plasma cells.
   e. Antibodies react against the antigen-bearing agent that stimulated their production (humoral immune response).
   f. An individual's diverse B cells defend against many pathogens.
5. Types of antibodies
   a. Antibodies are soluble proteins called immunoglobulins.
   b. The five major types of immunoglobulins are IgG, IgA, IgM, IgD, and IgE.
6. Antibody actions
   a. Antibodies directly attach to antigens, activate complement, or stimulate local tissue changes that are unfavorable to antigen-bearing agents.
   b. Direct attachment results in agglutination, precipitation, or neutralization.
   c. Activated complement proteins attract phagocytes, alter cells so that they become more susceptible to phagocytosis, and rupture foreign cell membranes (lysis).
7. Immune responses
   a. The first reaction to an antigen is called a primary immune response.
      (1) During this response, antibodies are produced for several weeks.
      (2) Some T cells and B cells remain dormant as memory cells.
   b. A secondary immune response occurs rapidly as a result of memory cells responding to subsequent exposure to an antigen.

8. Practical classification of immunity
   a. Naturally acquired immunity arises in the course of natural events, whereas artificially acquired immunity is the consequence of a medical procedure.
   b. Active immunity lasts much longer than passive immunity.
   c. A person who encounters a pathogen and has a primary immune response develops naturally acquired active immunity.
   d. A person who receives a vaccine containing a dead or weakened pathogen, or part of one, develops artificially acquired active immunity.
   e. A person who receives an injection of antibodies has artificially acquired passive immunity.
   f. When antibodies pass through a placental membrane from a pregnant woman to her fetus, the fetus develops naturally acquired passive immunity.
9. Allergic reactions
   a. Allergic reactions are excessive and misdirected immune responses that may damage tissue.
   b. Delayed-reaction allergy, which can occur in anyone and inflame the skin, results from repeated exposure to antigens.

c. Immediate-reaction allergy is an inborn ability to overproduce IgE.
   (1) Allergic reactions result from mast cells bursting and releasing allergy mediators such as histamine.
   (2) The released chemicals cause allergy symptoms such as hives, hay fever, asthma, eczema, or gastric disturbances.
10. Transplantation and tissue rejection
   a. A transplant recipient's immune system may react against the donated tissue in a tissue rejection reaction.
   b. Matching donor and recipient tissues and using immunosuppressive drugs can minimize tissue rejection.
   c. Immunosuppressive drugs may increase susceptibility to infection.
11. Autoimmunity
   a. In autoimmune disorders, the immune system manufactures autoantibodies that attack a person's own body tissues.
   b. Autoimmune disorders may result from a previous viral infection, faulty T cell development, or reaction to a nonself antigen that resembles a self antigen.

Chapter Assessments

14.1 Introduction

1. Explain the functions of the lymphatic system. (p. 378)

14.2 Lymphatic Pathways

2. Trace the general pathway of lymph from the interstitial spaces to the bloodstream. (p. 379)

14.3 Tissue Fluid and Lymph

3. Tissue fluid forms as a result of filtration from plasma exceeding _____ , whereas lymph forms due to increasing _____ in the tissue fluid. (p. 380)

4. Describe two functions of lymph. (p. 380)

14.4 Lymph Movement

5. Explain why physical exercise promotes lymphatic circulation. (p. 381)

14.5 Lymph Nodes

6. Draw a lymph node and label its parts. (p. 381)

7. Explain the functions of a lymph node. (p. 382)

14.6 Thymus and Spleen

8. Indicate the locations of the thymus and spleen. (p. 382)

9. Compare and contrast the functions of the thymus and spleen. (p. 382)

14.7 Body Defenses Against Infection

10. Defense mechanisms that prevent the entry of many types of pathogens and destroy them if they enter provide _____ (nonspecific) defense. Mechanisms that are very precise, targeting specific pathogens provide _____ (specific) defense. (p. 384)

14.8 Innate (Nonspecific) Defenses

11. Define *species resistance*. (p. 384)

12. Identify the barriers that provide the body's first line of defense against infectious agents. (p. 385)

13. Describe how enzymatic actions function as defense mechanisms. (p. 385)

14. Define *interferon,* and explain its action. (p. 385)

15. _____ is a group of plasma proteins that when activated stimulate inflammation, attract phagocytes, and enhance phagocytosis. (p. 385)

16. _____ _____ _____ are specialized lymphocytes that secrete perforins to lyse cell membranes of virus-infected cells. (p. 385)

17. List the major effects of inflammation, and explain why each occurs. (p. 385)

18. Identify the major phagocytic cells in blood and other tissues. (p. 385)

19. Discuss why low-grade fever of short duration may be a desired response to infection. (p. 385)

14.9 Adaptive (Specific) Defenses, or Immunity

20. Review the origin of T cells and B cells. (p. 386)

21. Explain the cellular immune response including the activation of T cells. (p. 386)

22. List three types of T cells, and describe the function of each in the immune response. (p. 387)

23. Explain the humoral immune response including the activation of B cells. (p. 390)

24. Explain the function of plasma cells. (p. 390)

25. Match the major types of immunoglobulins with their function and/or where each is found. (p. 391)

(1) associated with allergic reactions	A. IgA
(2) important in B cell activation, on surfaces of most B cells	B. IgM
(3) activates complement, anti-A and anti-B in blood	C. IgG
(4) effective against bacteria, viruses, toxins in plasma and tissue fluids	D. IgD
(5) found in exocrine secretions, including breast milk	E. IgE

26. Describe three ways in which an antibody's direct attack on an antigen helps remove that antigen. (p. 391)

27. List the various effects of complement activation. (p. 392)

28. Contrast a primary and a secondary immune response. (p. 392)

29. Match the practical classification of immunity with its example. (p. 393)

(1) naturally acquired active immunity	A. a breast-fed newborn
(2) artificially acquired active immunity	B. gamma globulin injection
(3) naturally acquired passive immunity	C. vaccination
(4) artificially acquired passive immunity	D. measles infection

30. List the major events leading to a delayed-reaction allergic response. (p. 393)

31. Describe how an immediate-reaction allergic response may occur. (p. 394)

32. Explain the relationship between tissue rejection and an immune response. (p. 394)

33. Explain the relationship between autoimmunity and an immune response. (p. 395)

Integrative Assessments/Critical Thinking

OUTCOMES 6.2, 14.2, 14.3, 14.4

1. Why is injecting a substance into the skin like injecting it into the lymphatic system?

OUTCOMES 14.2, 14.3, 14.4, 14.5

2. How can the removal of enlarged lymph nodes for microscopic examination aid in diagnosing certain diseases?

OUTCOME 14.9

3. The immune response is specific, diverse, and has memory. Give examples of each of these characteristics.

4. Some parents keep their preschoolers away from other children to prevent them from catching illnesses. How might these well-meaning parents actually be harming their children?

5. Why is a transplant consisting of fetal tissue less likely to provoke an immune rejection response than tissue from an adult?

6. Why does vaccination provide long-lasting protection against a disease, while gamma globulin (IgG) provides only short-term protection?

WEB CONNECTIONS

Visit the text website at **www.mhhe.com/shieress11** for additional quizzes, interactive learning exercises, and more.

APR

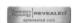

Anatomy & Physiology REVEALED includes cadaver photos that allow you to peel away layers of the human body to reveal structures beneath the surface. This program also includes animations, radiologic imaging, audio pronunciation, and practice quizzing. To learn more visit **www.aprevealed.com.**

15

Digestive System and Nutrition

The gut microbiome. Not all of the cells in an adult body are human—90% are microorganisms collectively called microflora. Terms from ecology are used to describe these microbes within us, including *community, ecosystem,* and *biome.* The Human Oral Microbiome Database, for example, lists more than 600 species that can live in the mouth. Each person has about 200 of these oral bacterial types.

To assess the microbial biome at the other end of the digestive tract, researchers analyzed DNA fragments in stool samples from a twenty-eight-year-old woman and a thirty-seven-year-old man, neither of whom had taken medications that could have affected the microflora. By comparing the DNA pieces to those of known microorganisms, the researchers discovered that the "distal gut microbial community" includes more than 6,800 species.

Researchers also tracked the formation and changing nature of the human gut microflora by classifying microbial DNA in a year's worth of stool collected daily from soiled diapers. Bacteria in the stool varied greatly from baby to baby at the onset, but by the babies' first birthdays, the gut communities were more alike and more closely resembled the microbial communities in adults.

The microorganisms that live in our large intestines are crucial to our health. They produce more than eighty types of enzymes that digest plant polysaccharides that our bodies cannot break down, and help process certain sugars. Our "gut" residents also synthesize essential vitamins and amino acids, and break down certain toxins and drugs.

We can use knowledge of our gut microbiome to improve health. For example, unusual bacterial communities can reflect disease. Specific microfloral profiles are associated with colorectal cancer, diar-

Several million microorganisms are normal residents of our digestive tracts. Escherichia coli, *pictured here (6,800×), produce vitamin K and if present in low numbers, will not cause diarrhea.*

rhea, inflammatory bowel disease, and peptic ulcers. A new focus of drug development is to target our microbial residents. Also, we can add bacteria to foods to prevent certain infections, an approach called probiotics. For example, certain *Lactobacillus* strains added to yogurt can help protect against *Salmonella* foodborne infection.

The numbers and types of microorganisms that live in our intestines vary somewhat from person to person, and these differences may be one reason some people can eat a great deal and not gain weight, yet others gain weight easily. Studies show that an item of food may yield different numbers of calories when eaten by different people. One investigation of the energy in one kind of cookie found that even though the package listed 110 calories, the cookie yielded anywhere from 90 to 110 calories, depending upon who ate it.

Learning Outcomes

After studying this chapter, you should be able to do the following:

15.1 Introduction

1. Describe the general functions of the digestive system. (p. 401)

2. Name the major organs of the digestive system. (p. 401)

15.2 General Characteristics of the Alimentary Canal

3. Describe the structure of the wall of the alimentary canal. (p. 401)

4. Explain how the contents of the alimentary canal are mixed and moved. (p. 403)

15.3 Mouth

5. Describe the functions of the structures associated with the mouth. (p. 403)

6. Describe how different types of teeth are adapted for different functions, and list the parts of a tooth. (p. 407)

15.4–15.10 Salivary Glands–Large Intestine

7. Identify the function of each enzyme secreted by the digestive organs. (pp. 408–422)

8. Describe how digestive secretions are regulated. (pp. 408–422)

9. Describe the mechanisms of swallowing and defecating. (pp. 409 and 427)

 Learn   **Practice**   **Assess**

10. Explain how the products of digestion are absorbed. (p. 422)

15.11 Nutrition and Nutrients

11. List the major sources of carbohydrates, lipids, and proteins. (p. 428)

12. Describe how cells utilize dietary carbohydrates, lipids, and proteins. (p. 429)

13. Identify the functions of each fat-soluble and water-soluble vitamin. (p. 432)

14. Identify the functions of each major mineral and trace element. (p. 433)

15. Describe an adequate diet. (p. 434)

Module 12: Digestive System

Aids to Understanding Words

(Appendix A on page 564 has a complete list of Aids to Understanding Words.)

aliment- [food] *aliment*ary canal: Tubelike part of the digestive system.

chym- [juice] *chym*e: Semifluid paste of food particles and gastric juice formed in the stomach.

decidu- [falling off] *decidu*ous teeth: Teeth shed during childhood.

gastr- [stomach] *gastr*ic gland: Part of the stomach that secretes gastric juice.

hepat- [liver] *hepat*ic duct: Duct that carries bile from the liver to the bile duct.

lingu- [tongue] *lingu*al tonsil: Mass of lymphatic tissue at the root of the tongue.

nutri- [nourish] *nutri*ent: Substance needed to nourish cells.

peri- [around] *peri*stalsis: Wavelike ring of contraction that moves material along the alimentary canal.

pyl- [gatekeeper] *pyl*oric sphincter: Muscle that serves as a valve between the stomach and small intestine.

vill- [hairy] *vill*i: Tiny projections of mucous membrane in the small intestine.

15.1 INTRODUCTION

Digestion (di-jest′yun) is the mechanical and chemical breakdown of foods and the absorption of the resulting nutrients by cells. *Mechanical digestion* breaks large pieces into smaller ones without altering their chemical composition. *Chemical digestion* breaks food into simpler chemicals. The organs of the **digestive system** carry out these processes.

The digestive system consists of the **alimentary canal** (al″i-men′tar-e kah-nal′), extending from the mouth to the anus, and several accessory organs, which secrete substances, into the canal, used in digestion. The alimentary canal includes the mouth, pharynx, esophagus, stomach, small intestine, large intestine, rectum, and anus; the accessory organs include the salivary glands, liver, gallbladder, and pancreas (fig. 15.1; see reference plates 4, 5, and 6, pp. 26–28). Overall, the digestive system is a tube, open at both ends, that has a surface area of 186 square meters. It supplies nutrients for body cells.

15.2 GENERAL CHARACTERISTICS OF THE ALIMENTARY CANAL

The alimentary canal is a muscular tube about 8 meters long that passes through the body's thoracic and abdominopelvic cavities (fig. 15.2). The structure of its wall, how it moves food, and its innervation are similar throughout its length.

Structure of the Wall

The wall of the alimentary canal consists of four distinct layers that are developed to different degrees from region to region. Although the four-layered structure persists throughout the alimentary canal, certain regions are specialized for particular functions. Beginning with the innermost tissues, these layers are (fig. 15.3):

1. **Mucosa** (mu-ko′sah), or **mucous membrane** (mu′kus mem′brān) Surface epithelium, underlying connective tissue, and a small amount of smooth muscle form this layer. In some regions, the mucosa is folded, with tiny projections that extend into the passageway, or **lumen** (lu′men), of the digestive tube. The folds increase the absorptive surface area. The mucosa also has glands that are tubular invaginations into which the lining cells secrete mucus and digestive enzymes. The mucosa protects the tissues beneath it and carries on secretion and absorption.

2. **Submucosa** (sub″mu-ko′sah) The submucosa consists of considerable loose connective tissue as well as glands, blood vessels, lymphatic vessels, and nerves. Its vessels nourish surrounding tissues and carry away absorbed materials.

3. **Muscular layer** This layer, which produces movements of the tube, consists of two coats of smooth muscle tissue. The fibers of the inner coat encircle the tube. When these *circular fibers* contract, the tube's diameter decreases. The fibers of the outer muscular coat run lengthwise. When these *longitudinal fibers* contract, the tube shortens.

4. **Serosa** (se′ro-sah), or **serous layer** (se′rus la′er) The layer of epithelium on the outside and the connective tissue beneath compose the serous layer, or outer covering, of the tube, also called the *visceral peritoneum*. The cells of the serosa protect underlying tissues and secrete serous fluid, which moistens and lubricates the tube's outer surface so that organs within the abdominal cavity slide freely against one another.

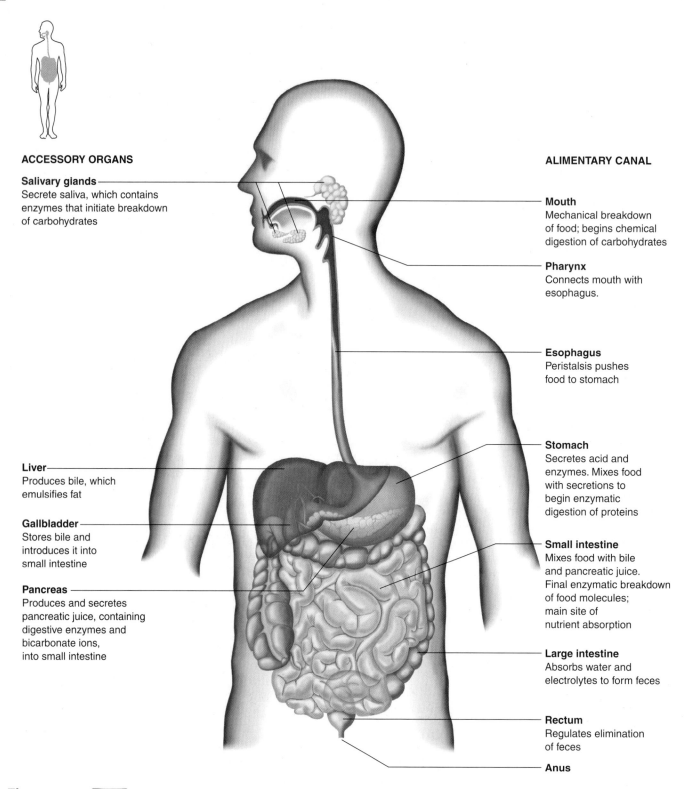

ACCESSORY ORGANS

Salivary glands
Secrete saliva, which contains
enzymes that initiate breakdown
of carbohydrates

Liver
Produces bile, which
emulsifies fat

Gallbladder
Stores bile and
introduces it into
small intestine

Pancreas
Produces and secretes
pancreatic juice, containing
digestive enzymes and
bicarbonate ions,
into small intestine

ALIMENTARY CANAL

Mouth
Mechanical breakdown
of food; begins chemical
digestion of carbohydrates

Pharynx
Connects mouth with
esophagus.

Esophagus
Peristalsis pushes
food to stomach

Stomach
Secretes acid and
enzymes. Mixes food
with secretions to
begin enzymatic
digestion of proteins

Small intestine
Mixes food with bile
and pancreatic juice.
Final enzymatic breakdown
of food molecules;
main site of
nutrient absorption

Large intestine
Absorbs water and
electrolytes to form feces

Rectum
Regulates elimination
of feces

Anus

Figure 15.1 AP|R
Organs of the digestive system.

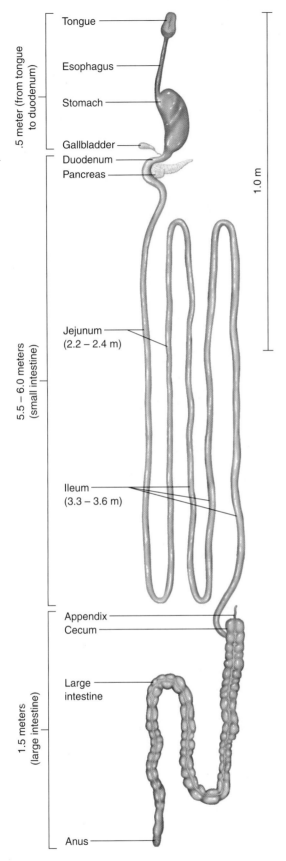

Figure 15.2

The alimentary canal is a muscular tube about 8 meters long.

Q: *What is the distance from the tongue to the duodenum in English units (inches)?*

Answer can be found in Appendix E on page 568.

Movements of the Tube

The motor functions of the alimentary canal are of two basic types—*mixing movements* and *propelling movements*. Mixing occurs when smooth muscles in small segments of the tube contract rhythmically (fig. 15.4*a*). For example, when the stomach is full, waves of muscular contractions move along its walls from one end to the other. These waves mix food with digestive juices that the mucosa secretes. In the small intestine, **segmentation** aids mixing movements by alternately contracting and relaxing the smooth muscle in nonadjacent segments of the organ. Because segmentation does not follow a set pattern, materials are not moved along the tract in one direction (fig. 15.4*b*).

Propelling movements include a wavelike motion called **peristalsis** (per″ĭ-stal′sis). During peristalsis, a ring of contraction occurs in the wall of the tube. At the same time, the muscular wall just ahead of the ring relaxes. As the peristaltic wave moves along, it pushes the tubular contents ahead of it (fig. 15.4*c*).

> **Practice**
> 1. Which organs constitute the digestive system?
> 2. Describe the wall of the alimentary canal.
> 3. Name the two basic types of movements in the alimentary canal.

15.3 MOUTH

The **mouth** receives food and begins digestion by mechanically breaking up solid particles into smaller pieces and mixing them with saliva. This action is called *mastication* (mas″tĭ-ka′shun). The lips, cheeks, tongue, and palate surround the mouth, which includes a chamber between the palate and tongue called the *oral cavity*, as well as a narrow space between the teeth, cheeks, and lips called the *vestibule* (fig. 15.5).

Cheeks and Lips

The **cheeks,** forming the lateral walls of the mouth, consist of outer layers of skin, pads of subcutaneous fat, muscles associated with expression and chewing, and inner linings of moist, stratified squamous epithelium. The **lips** are highly mobile structures that surround the mouth opening. They contain skeletal muscles and sensory receptors useful in judging the temperature and texture of foods. Blood vessels near lip surfaces impart a reddish hue.

Tongue

The **tongue** nearly fills the oral cavity when the mouth is closed. Mucous membrane covers the tongue, and a membranous fold called the **lingual frenulum**

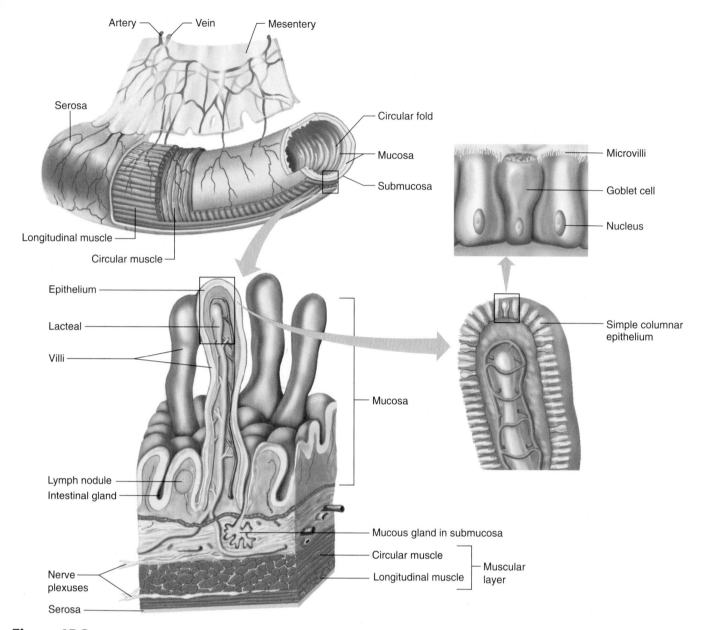

Figure 15.3

The wall of the small intestine, as in other portions of the alimentary canal, includes four layers: an inner mucosa, a submucosa, a muscular layer, and an outer serosa.

(ling'gwahl fren'u-lum) connects the midline of the tongue to the floor of the mouth.

The *body* of the tongue is mostly skeletal muscle. Muscular action mixes food particles with saliva during chewing and moves food toward the pharynx during swallowing. The tongue also helps move food underneath the teeth for chewing. Rough projections called **papillae** on the tongue surface provide friction, which helps handle food. These papillae also bear taste buds (see chapter 10, p. 269).

The posterior region, or *root,* of the tongue is anchored to the hyoid bone. It is covered with rounded masses of lymphatic tissue called **lingual tonsils** (ton'silz) (fig. 15.6).

Palate

The **palate** (pal'at) forms the roof of the oral cavity and consists of a bony anterior part (*hard palate*) and a muscular posterior part (*soft palate*). A muscular arch of the soft palate extends posteriorly and downward as a cone-shaped projection called the **uvula** (u'vu-lah).

In the back of the mouth, on either side of the tongue and closely associated with the palate, are masses of lymphatic tissue called **palatine** (pal'ah-tīn) **tonsils** (see figs. 15.5 and 15.6). These structures lie beneath the epithelial lining of the mouth and, like other lymphatic tissues, help protect the body against infection.

(a)

Digesting material

(b)

Movement of contents

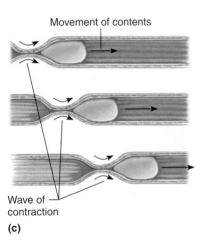

Wave of contraction

(c)

Figure 15.4

Movements through the alimentary canal. **(a)** Mixing movements occur when small segments of the muscular wall of the stomach rhythmically contract. **(b)** Segmentation mixes contents of the small intestine. **(c)** Peristaltic waves move the contents along the canal.

The palatine tonsils are common sites of infection, and become inflamed in *tonsillitis*. Infected tonsils may swell so greatly that they block the passageways of the pharynx and interfere with breathing and swallowing. When tonsillitis recurs and does not respond to antibiotic treatment, the tonsils may be surgically removed. Such tonsillectomies are done less often today than they were a generation ago because the tonsils' role in immunity is now recognized.

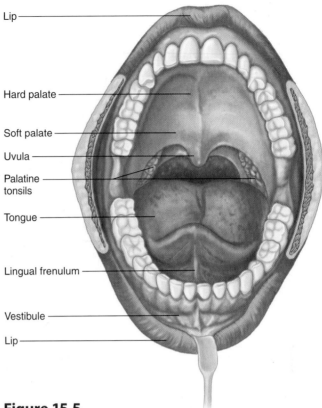

Lip
Hard palate
Soft palate
Uvula
Palatine tonsils
Tongue
Lingual frenulum
Vestibule
Lip

Figure 15.5

The mouth is adapted for ingesting food and beginning digestion, both mechanically and chemically.

Other masses of lymphatic tissue, called **pharyngeal** (fah-rin′je-al) **tonsils,** or *adenoids,* are on the posterior wall of the pharynx, above the border of the soft palate (fig. 15.6). Enlarged adenoids that block the passage between the nasal cavity and the pharynx may be surgically removed.

Practice

4. How does the tongue function as part of the digestive system?
5. Where are the tonsils located?

Teeth

Two different sets of **teeth** form during development. The first set, the *primary teeth* (deciduous teeth), usually erupt through the gums at regular intervals between the ages of six months and two to four years (fig. 15.7). There are twenty deciduous teeth—ten in each jaw.

The primary teeth are usually shed in the same order they erupted. After their roots are resorbed, then, the *secondary teeth* (permanent teeth) push the primary teeth out of their sockets. This secondary set consists of thirty-two teeth—sixteen in each jaw (fig. 15.8). The secondary teeth usually begin to appear at six years, but the set may not be complete until the third molars (wisdom teeth) emerge between seventeen and twenty-five years.

Frontal sinus

Nasal cavity

Hard palate

Vestibule

Tongue

Tooth

Lip

Hyoid bone

Larynx

Sphenoidal sinus

Pharyngeal tonsil

Opening of auditory tube

Soft palate

Nasopharynx

Oral cavity

Uvula

Palatine tonsil

Oropharynx

Lingual tonsil

Epiglottis

Laryngopharynx

Esophagus

Trachea

Figure 15.6 AP|R

Sagittal section of the mouth, nasal cavity, and pharynx.

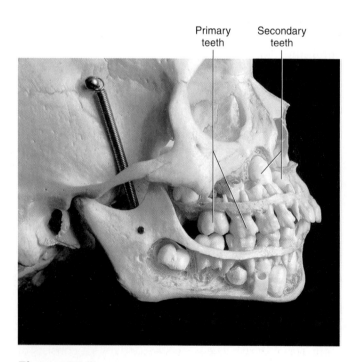

Primary teeth   Secondary teeth

Figure 15.7

This partially dissected child's skull reveals primary and developing secondary teeth in the maxilla and mandible.

Incisors

Canine (cuspid)

Premolars (bicuspids)

Molars

Molars

Premolars (bicuspids)

Canine (cuspid)

Incisors

Figure 15.8

The secondary teeth of the upper and lower jaws.

Clinical Application 15.1

Dental Caries

Sticky foods, such as caramel, lodge between the teeth and in the crevices of molars, feeding bacteria such as *Actinomyces, Streptococcus mutans*, and *Lactobacillus*. These microorganisms metabolize carbohydrates in the food, producing acid by-products that destroy tooth enamel and dentin. The bacteria also produce sticky substances that hold them in place.

If a person eats a candy bar but does not brush the teeth soon afterward, the acid-forming bacteria may decay tooth enamel, creating a condition called *dental caries*. Unless a dentist cleans and fills the resulting cavity that forms where enamel is destroyed, the damage will spread to the underlying dentin.

Dental caries can be prevented in several ways:

1. Brush and floss teeth regularly.
2. Have regular dental exams and cleanings.
3. Talk with your dentist about receiving a fluoride treatment. Fluoride is added to the water supply in many communities. Fluoride is incorporated into the enamel's chemical structure, strengthening it.
4. The dentist may apply a sealant to children's and adolescents' teeth where crevices might hold onto decay-causing bacteria. The sealant is a coating that keeps acids from eating away at tooth enamel.

Teeth begin mechanical digestion by breaking pieces of food into smaller pieces. This action increases the surface area of food particles, allowing digestive enzymes to react more effectively with the food molecules. Table 15.1 summarizes the number and types of teeth that appear during development and their functions.

Each tooth consists of two main parts—the *crown*, which projects beyond the gum (gingiva), and the *root*, which is anchored to the alveolar process of the jaw. Where these portions meet is called the *neck* of the tooth.

Glossy white *enamel* covers the crown. Enamel mainly consists of calcium salts and is the hardest substance in the body. Enamel damaged by abrasive action or injury is not replaced. Enamel also tends to wear away with age. Clinical Application 15.1 discusses tooth enamel destruction.

The bulk of a tooth beneath the enamel is *dentin*, a substance much like bone but somewhat harder. Dentin surrounds the tooth's central cavity (pulp cavity), which contains blood vessels, nerves, and connective tissue, collectively called *pulp*. Blood vessels and nerves reach this cavity through tubular *root canals* extending into the root.

A thin layer of bonelike material called *cementum*, surrounded by a *periodontal ligament*, encloses the root. This ligament contains blood vessels and nerves as well as bundles of thick collagenous fibers that pass between the cementum and the bone of the alveolar process, firmly attaching the tooth to the jaw (fig. 15.9).

Table 15.1	**Primary and Secondary Teeth**		
Type	**Number of Primary Teeth**	**Number of Secondary Teeth**	**Function**
Incisor			Bite off pieces of food
Central	4	4	
Lateral	4	4	
Canine (cuspid)	4	4	Grasp and tear food
Premolar (bicuspid)			Grind food particles
First	0	4	
Second	0	4	
Molar			Grind food particles
First	4	4	
Second	4	4	
Third	0	4	
Total	20	32	

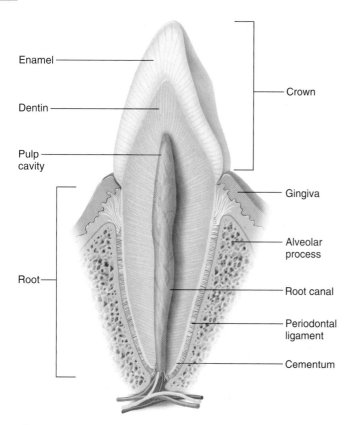

Enamel

Dentin

Pulp cavity

Root

Crown

Gingiva

Alveolar process

Root canal

Periodontal ligament

Cementum

Figure 15.9
A section of a tooth.

Extracted primary and wisdom teeth may one day provide stem cells that can be used to regenerate tooth roots and supporting periodontal ligaments. The stem cells are in the pulp and a region called the apical papilla. Dental researchers hope that these stem cells may be cultured to yield replacement teeth for people who do not have enough jawbone to support dental implants.

Practice

6. How do primary teeth differ from secondary teeth?
7. Describe the structure of a tooth.
8. Explain how a tooth is attached to the bone of the jaw.

15.4 SALIVARY GLANDS

The **salivary** (sal′ĭ-ver-e) **glands** secrete saliva. This fluid moistens food particles, helps bind them, and begins the chemical digestion of carbohydrates. Saliva is also a solvent, dissolving foods so that they can be tasted, and it helps cleanse the mouth and teeth.

Salivary Secretions

A salivary gland has two types of secretory cells—*serous cells* and *mucous cells*. Proportions of these cells vary in the different types of salivary glands. Serous cells produce a watery fluid that includes the digestive enzyme **salivary amylase** (am′ĭ-lās). This enzyme splits starch and glycogen molecules into disaccharides—the first step in the chemical digestion of carbohydrates. Mucous cells secrete a thick liquid called **mucus,** which binds food particles and lubricates the food during swallowing.

When a person sees, smells, tastes, or even thinks about appealing food, parasympathetic nerve impulses elicit the secretion of a large volume of watery saliva. Conversely, food that looks, smells, or tastes unpleasant inhibits parasympathetic activity and less saliva is produced. Swallowing may become difficult.

Major Salivary Glands

Three pairs of major salivary glands—the parotid, submandibular, and sublingual glands—and many minor ones are associated with the mucous membranes of the tongue, palate, and cheeks (fig. 15.10). The **parotid glands** (pah-rot′id glandz) are the largest of the major salivary glands. Each gland lies anterior and somewhat inferior to each ear, between the skin of the cheek and the masseter muscle. The parotid glands secrete a clear, watery fluid that is rich in salivary amylase.

The **submandibular** (sub″man-dib′u-lar) **glands** are located in the floor of the mouth on the inside surface of the lower jaw. The secretory cells of these glands are about equally serous and mucous. Consequently, the submandibular glands secrete a more viscous fluid than the parotid glands.

The **sublingual** (sub-ling′gwal) **glands,** the smallest of the major salivary glands, are on the floor of the mouth inferior to the tongue. Their secretory cells are primarily the mucous type, making their secretions thick and stringy.

Practice

9. What is the function of saliva?
10. What stimulates salivary glands to secrete saliva?
11. Where are the major salivary glands?

15.5 PHARYNX AND ESOPHAGUS

The pharynx is a cavity posterior to the mouth from which the tubular esophagus leads to the stomach (see fig. 15.1). The pharynx and the esophagus do not digest food, but both are important passageways whose muscular walls function in swallowing.

Figure 15.10 AP|R
Locations of the major salivary glands.

Structure of the Pharynx

The **pharynx** (far'inks) connects the nasal and oral cavities with the larynx and esophagus (see fig. 15.6). It has three parts:

1. The **nasopharynx** (na"zo-far'inks) communicates with the nasal cavity and provides a passageway for air during breathing. The auditory tubes, which connect the pharynx with the middle ears, open through the walls of the nasopharynx.
2. The **oropharynx** (o"ro-far'inks) is posterior to the soft palate and inferior to the nasopharynx. It is a passageway for food moving downward from the mouth and for air moving to and from the nasal cavity.
3. The **laryngopharynx** (lah-ring"go-far'inks), just inferior to the oropharynx, is a passageway to the esophagus.

Swallowing Mechanism

Swallowing has three stages. In the first stage, which is voluntary, food is chewed and mixed with saliva. Then the tongue rolls this mixture into a mass, or **bolus,** and forces it into the oropharynx.

The second stage of swallowing begins as food reaches the oropharynx and stimulates sensory receptors around the pharyngeal opening. This triggers the swallowing reflex, which includes the following actions:

1. The soft palate (including the uvula) raises, preventing food from entering the nasal cavity.
2. The hyoid bone and the larynx are elevated. A flaplike structure attached to the larynx, called the *epiglottis* (ep"i-glot'is), closes off the top of the larynx so that food is less likely to enter the trachea.
3. The tongue is pressed against the soft palate, sealing off the oral cavity from the nasopharynx.
4. The longitudinal muscles in the pharyngeal wall contract, pulling the pharynx upward toward the food.
5. Muscles in the laryngopharynx relax, opening the esophagus.
6. A peristaltic wave begins in the pharyngeal muscles and forces food into the esophagus.

The swallowing reflex momentarily inhibits breathing. Then, during the third stage of swallowing, peristalsis transports the food in the esophagus to the stomach.

Computer simulation studies show that each type of food requires an optimum range of number of chews to form a bolus. Eating raw carrots, for example, requires twenty to twenty-five chews.

Figure 15.11

Major regions of the stomach and its associated structures.

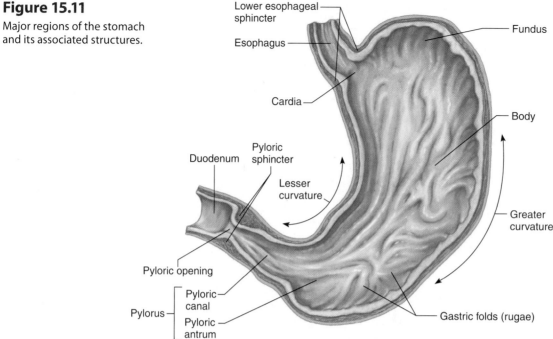

Esophagus

The **esophagus** (ĕ-sof'ah-gus), a straight, collapsible tube about 25 centimeters long, is a food passageway from the pharynx to the stomach (see figs. 15.1 and 15.6). The esophagus begins at the base of the laryngopharynx and descends posterior to the trachea, passing through the mediastinum. It penetrates the diaphragm through an opening, the *esophageal hiatus* (ĕ-sof"ah-je'al hi-a'tus), and is continuous with the stomach on the abdominal side of the diaphragm.

Mucous glands are scattered throughout the submucosa of the esophagus. Their secretions moisten and lubricate the tube's inner lining.

Just superior to the point where the esophagus joins the stomach, some of the circular smooth muscle fibers have increased muscle tone, forming the **lower esophageal sphincter,** or cardiac sphincter (fig. 15.11). These fibers usually remain contracted. They close the entrance to the stomach, preventing the stomach contents from regurgitating into the esophagus. When peristaltic waves reach the stomach, these muscle fibers temporarily relax and allow the swallowed food to enter.

In a *hiatal hernia,* part of the stomach protrudes through a weakened area of the diaphragm, through the esophageal hiatus, and into the thorax. Regurgitation (reflux) of gastric juice into the esophagus becomes more likely. This may inflame the esophageal mucosa, causing heartburn, difficulty in swallowing, or ulceration and blood loss. In response to the destructive action of gastric juice, columnar epithelium may replace the squamous epithelium that normally lines the esophagus (see chapter 5, page 98). This condition, called *Barrett's esophagus,* increases the risk of developing esophageal cancer.

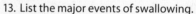

Practice

12. Describe the regions of the pharynx.

13. List the major events of swallowing.

14. What is the function of the esophagus?

15.6 STOMACH

The **stomach** is a J-shaped, pouchlike organ that hangs inferior to the diaphragm in the upper left portion of the abdominal cavity and has a capacity of about 1 liter or more (figs. 15.1 and 15.11; see reference plates 4 and 5, pp. 26–27). Thick folds (rugae) of mucosal and submucosal layers mark the stomach's inner lining and disappear when the stomach wall is distended. The stomach receives food from the esophagus, mixes the food with gastric juice, initiates protein digestion, carries on limited absorption, and moves food into the small intestine.

Parts of the Stomach

The stomach is divided into the cardia, fundus, body, and pylorus (fig. 15.11). The *cardia* is a small area near the esophageal opening. The *fundus,* which balloons superior to the cardia, is a temporary storage area. The dilated *body region,* which is the main part of the stomach, lies between the fundus and pylorus. The **pyloric canal** is a narrowing of the *pylorus* as it approaches the small intestine.

At the end of the pyloric canal the muscular wall thickens, forming a powerful circular muscle, the

pyloric sphincter. This muscle is a valve that controls gastric emptying.

Gastric Secretions

The mucous membrane that forms the inner lining of the stomach is thick. Its surface is studded with many small openings called *gastric pits* located at the ends of tubular **gastric glands** (gas′trik glandz) (fig. 15.12).

Gastric glands generally have three types of secretory cells. *Mucous cells* are in the necks of the glands, near the openings of the gastric pits. *Chief cells* and *parietal cells* are in the deeper parts of the glands. The chief cells secrete digestive enzymes, and the parietal cells release a solution containing hydrochloric acid. The products of the mucous cells, chief cells, and parietal cells together form **gastric juice** (gas′trik joos).

Pepsin (pep′sin) is by far the most important digestive enzyme in gastric juice. The chief cells secrete pepsin as the inactive enzyme precursor **pepsinogen** (pep-sin′o-jen). When pepsinogen contacts hydrochloric acid from the parietal cells, it breaks down rapidly, forming pepsin. Pepsin begins the digestion of nearly all types of dietary protein into polypeptides. This enzyme is most active in an acidic environment, which is provided by the hydrochloric acid in gastric juice.

The mucous cells of the gastric glands (*mucous neck cells*) and the mucous cells associated with the stomach's inner surface release a viscous, alkaline secretion that coats the inside of the stomach wall. This coating normally prevents the stomach from digesting itself.

Another component of gastric juice is **intrinsic factor** (in-trin′sik fak′tor), which the parietal cells secrete. Intrinsic factor helps the small intestine absorb vitamin B_{12}. Table 15.2 summarizes the components of gastric juice.

Practice

15. What are the secretions of the chief cells and parietal cells?

16. Which is the most important digestive enzyme in gastric juice?

17. Why doesn't the stomach digest itself?

Figure 15.12 AP|R

Lining of the stomach. **(a)** Gastric glands include mucous cells, parietal cells, and chief cells. **(b)** A light micrograph of cells associated with the gastric glands (60×).

Table 15.2	Major Components of Gastric Juice AP|R	
Component	**Source**	**Function**
Pepsinogen	Chief cells of the gastric glands	Inactive form of pepsin
Pepsin	Formed from pepsinogen in the presence of hydrochloric acid	A protein-splitting enzyme that digests nearly all types of dietary protein into polypeptides
Hydrochloric acid	Parietal cells of the gastric glands	Provides the acid environment needed for the production and action of pepsin
Mucus	Mucous cells	Provides a viscous, alkaline protective layer on the stomach's inner surface
Intrinsic factor	Parietal cells of the gastric glands	Aids in vitamin B_{12} absorption in the intestine

The 40 million cells that line the stomach's interior can secrete 2 to 3 quarts (about 2 to 3 liters) of gastric juice per day.

Regulation of Gastric Secretions

Gastric juice is produced continuously, but the rate varies considerably and is controlled both neurally and hormonally. When a person tastes, smells, or even sees appetizing food, or when food enters the stomach, parasympathetic impulses on the vagus nerves stimulate the release of acetylcholine (Ach) from nerve endings. This Ach stimulates gastric glands to secrete abundant gastric juice, which is rich in hydrochloric acid and pepsinogen. These parasympathetic impulses also stimulate certain stomach cells to release the peptide hormone **gastrin** (gas′trin), which increases the secretory activity of gastric glands (fig. 15.13).

Gastrin stimulates cell growth in the mucosa of the stomach and intestines, except where gastrin is produced. This cell growth helps replace mucosal cells damaged by normal stomach function, disease, or medical treatments.

As food moves into the small intestine, acid triggers sympathetic nerve impulses that inhibit gastric juice secretion. At the same time, proteins and fats in this region of the intestine cause the intestinal wall to release the peptide hormone **cholecystokinin** (ko″le-sis″to-ki′nin). This hormonal action decreases gastric motility as the small intestine fills with food.

An *ulcer* is an open sore in the skin or a mucous membrane resulting from localized tissue breakdown. Gastric ulcers form in the stomach, and duodenal ulcers form in the region of the small intestine nearest the stomach.

For many years, gastric and duodenal ulcers were attributed to stress and treated with medications to decrease stomach acid secretion. In 1982, two Australian researchers boldly suggested that stomach infection by the bacterium *Helicobacter pylori* causes gastric ulcers. When the medical community did not believe them, one of the researchers swallowed some bacteria, calling it "swamp water," to demonstrate their effect—and soon developed stomach pain (gastritis). Today, a short course of antibiotics, often combined with acid-lowering drugs, can be used to treat many gastric ulcers.

Gastric Absorption

Gastric enzymes begin breaking down proteins, but the stomach wall is not well adapted to absorbing digestive products. The stomach absorbs only small volumes of water and certain salts as well as certain lipid-soluble drugs. Alcohol, which is not a nutrient, is absorbed both in the small intestine and the stomach.

1 Parasympathetic preganglionic nerve fiber (in vagus nerve)

Release into bloodstream

Stimulation

4 Gastrin stimulates glands to release more gastric juice

2 Parasympathetic postganglionic impulses stimulate the release of gastric juice from gastric glands

3 Impulses stimulate the release of gastrin

Bloodstream

Figure 15.13 AP\|R

The secretion of gastric juice is regulated in part by parasympathetic nerve impulses that stimulate the release of gastric juice and gastrin.

Practice

18. What controls gastric juice secretion?

19. What is the function of cholecystokinin?

20. Which substances can the stomach absorb?

Mixing and Emptying Actions

Following a meal, the mixing movements of the stomach wall aid in producing a semifluid paste of food particles and gastric juice called **chyme** (kīm). Peristaltic waves push the chyme toward the pylorus of the stomach and, as chyme accumulates near the pyloric sphincter, this muscle begins to relax. Stomach contractions push chyme a little at a time into the small intestine.

The rate at which the stomach empties depends on the fluidity of the chyme and the type of food present. Liquids usually pass through the stomach rapidly, but solids remain until they are well mixed with gastric juice. Fatty foods may remain in the stomach from three to six hours; foods high in proteins move through more quickly; carbohydrates usually pass through faster than either fats or proteins.

As chyme enters the duodenum (the first part of the small intestine), accessory organs—the pancreas, liver, and gallbladder—add their secretions.

Vomiting results from a complex reflex that empties the stomach in the reverse of the normal direction. Irritation or distension in the stomach or intestines can trigger vomiting. Sensory impulses travel from the site of stimulation to the *vomiting center* in the medulla oblongata, and motor responses follow. These include taking a deep breath, raising the soft palate and thus closing the nasal cavity, closing the opening to the trachea (glottis), relaxing the circular muscle fibers at the base of the esophagus, contracting the diaphragm so it presses downward over the stomach, and contracting the abdominal wall muscles to increase pressure inside the abdominal cavity. As a result, the stomach is squeezed from all sides, forcing its contents upward and out through the esophagus, pharynx, and mouth.

Practice

21. How is chyme produced?
22. What factors influence how quickly chyme leaves the stomach?

15.7 PANCREAS

The **pancreas** was discussed as an endocrine gland in chapter 11 (pp. 306–309). It also has an exocrine function—secretion of a digestive fluid called **pancreatic juice** (pan″kre-at′ik joos).

Structure of the Pancreas

The pancreas is closely associated with the small intestine. It extends horizontally across the posterior abdominal wall, with its head in the C-shaped curve of the duodenum (part of the small intestine) and its tail against the spleen (figs. 15.1 and 15.14).

The cells that produce pancreatic juice, called *pancreatic acinar* (a′sĭ-nar) *cells,* make up the bulk of the pancreas. These cells cluster around tiny tubes into which they release their secretions. The smaller tubes unite to form larger ones, which in turn give rise to a *pancreatic duct* extending the length of the pancreas. The pancreatic duct usually connects with the duodenum at the same place where the bile duct from the liver and gallbladder joins the duodenum, although other connections may be present (fig. 15.14). A *hepatopancreatic sphincter* controls the movement of pancreatic juices into the duodenum.

Pancreatic Juice

Pancreatic juice contains enzymes that digest carbohydrates, fats, nucleic acids, and proteins. The carbohydrate-digesting enzyme **pancreatic amylase** splits molecules of starch or glycogen into disaccharides. The fat-digesting enzyme **pancreatic lipase** breaks triglyceride molecules into fatty acids and glycerol. Pancreatic juice also contains two **nucleases,** which are enzymes that break down nucleic acid molecules into nucleotides.

The protein-splitting (proteolytic) enzymes are **trypsin, chymotrypsin,** and **carboxypeptidase** (kar-bok″se-pep′tĭ-dās). These enzymes split the bonds between particular combinations of amino acids in proteins. No single enzyme can split all possible amino acid combinations, so several enzymes are necessary to completely digest protein molecules.

The proteolytic enzymes are stored in tiny cellular structures called *zymogen granules* (zi-mo′jen gran′ūlz). These enzymes, like gastric pepsin, are secreted in inactive forms. After the inactive forms of the proteolytic enzymes reach the small intestine, other enzymes activate them. For example, pancreatic cells release inactive *trypsinogen,* which is activated to trypsin when it contacts the enzyme **enterokinase** (en″ter-o-ki′nās) secreted by the mucosa of the small intestine.

Painful *acute pancreatitis* results from blockage of pancreatic juice release. Trypsin, activated as pancreatic juice accumulates, digests parts of the pancreas. Alcoholism, gallstones, certain infections, traumatic injuries, or the side effects of some drugs can cause pancreatitis.

Regulation of Pancreatic Secretion

The nervous and endocrine systems regulate release of pancreatic juice, as they do gastric and small intestinal secretions. For example, when parasympathetic impulses stimulate gastric juice secretion, other parasympathetic impulses stimulate the pancreas to release digestive enzymes. Also, as acidic chyme enters the duodenum, the duodenal mucous membrane releases the peptide hormone **secretin** (se-kre′tin) into the bloodstream (fig. 15.15). This hormone stimulates secretion of pancreatic juice that has a high concentration of bicarbonate ions.

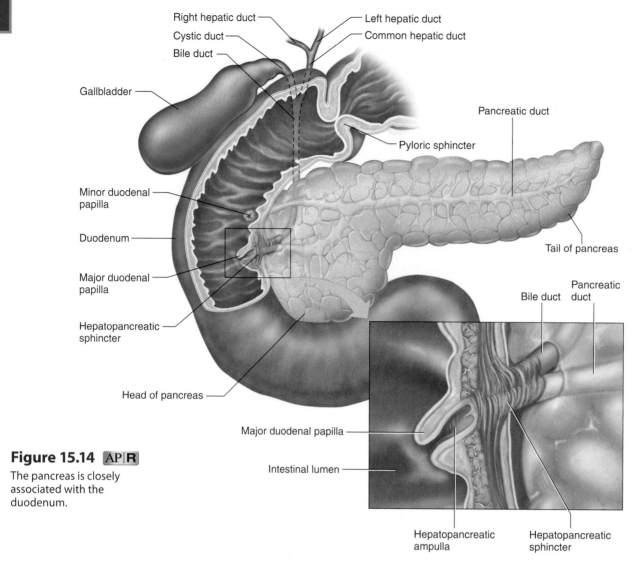

Right hepatic duct
Left hepatic duct
Cystic duct
Common hepatic duct
Bile duct
Gallbladder
Pancreatic duct
Pyloric sphincter
Minor duodenal papilla
Duodenum
Tail of pancreas
Major duodenal papilla
Pancreatic duct
Bile duct
Hepatopancreatic sphincter
Major duodenal papilla
Intestinal lumen
Head of pancreas
Hepatopancreatic ampulla
Hepatopancreatic sphincter

Figure 15.14 AP|R

The pancreas is closely associated with the duodenum.

Figure 15.15

Acidic chyme entering the duodenum from the stomach stimulates the release of secretin, which, in turn, stimulates the release of pancreatic juice.

1 Acidic chyme enters duodenum

4 Pancreatic juice rich in bicarbonate ions passes down pancreatic ducts to the duodenum

5 Bicarbonate ions neutralize acidic chyme

3 Secretin stimulates pancreas to secrete bicarbonate ions

2 Intestinal mucosa releases secretin into bloodstream

Bloodstream

Hormonal signals released into bloodstream

Stimulation of effector organ

These ions neutralize the acid in chyme and provide a favorable environment for digestive enzymes in the intestine.

Proteins and fats in chyme in the duodenum also stimulate the intestinal wall to release *cholecystokinin.* Like secretin, cholecystokinin travels via the bloodstream to the pancreas. Pancreatic juice secreted in response to cholecystokinin has a high concentration of digestive enzymes.

Practice

23. List the enzymes in pancreatic juice.
24. What are the functions of the enzymes in pancreatic juice?
25. What regulates secretion of pancreatic juice?

15.8 LIVER

The **liver** is in the upper right quadrant of the abdominal cavity, just inferior to the diaphragm. It is partially surrounded by the ribs, and extends from the level of the fifth intercostal space to the lower margin of the ribs. The reddish-brown liver is well supplied with blood vessels (see fig. 15.1 and reference plate 4, p. 26).

The average adult liver is the heaviest internal organ in the body. It weighs about 3 pounds.

Liver Structure

A fibrous capsule encloses the liver, and connective tissue divides the organ into a large *right lobe* and a smaller *left lobe* (fig. 15.16). Each lobe is separated into many tiny

hepatic lobules (hĕ-pat'ik lob'ulz), the liver's functional units (fig. 15.17). A lobule consists of many hepatic cells radiating outward from a *central vein.* Vascular channels called **hepatic sinusoids** separate platelike groups of these cells from each other. Blood from the digestive tract, which is carried in the *hepatic portal vein* (see chapter 13, p. 370), brings newly absorbed nutrients into the sinusoids and nourishes the hepatic cells (fig. 15.18).

Large phagocytic macrophages called *Kupffer cells* (koop'fer selz) are fixed to the inner linings of the hepatic sinusoids. They remove bacteria or other foreign particles that enter the blood in the portal vein through the intestinal wall. Blood passes from these sinusoids into the central veins of the hepatic lobules and exits the liver via the hepatic vein.

Within the hepatic lobules are many fine *bile canaliculi,* which carry secretions from hepatic cells to *bile ductules* (fig. 15.18). The ductules of neighboring lobules unite to form larger bile ducts which then converge to become the **hepatic ducts.** These ducts merge, in turn, to form the **common hepatic duct.**

Liver Functions

The liver carries on many important metabolic activities. Recall from chapter 11 (p. 308) that the liver plays a key role in carbohydrate metabolism by helping maintain concentration of blood glucose within the normal range. Liver cells responding to hormones such as insulin and glucagon lower the blood glucose level by polymerizing glucose to glycogen, and raise the blood glucose level by breaking down glycogen to glucose or by converting noncarbohydrates into glucose.

The liver's effects on lipid metabolism include oxidizing fatty acids at an especially high rate; synthesizing lipoproteins, phospholipids, and cholesterol; and

(a)

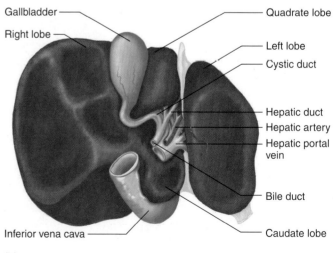
(b)

Figure 15.16
Lobes of the liver, viewed **(a)** anteriorly and **(b)** inferiorly.

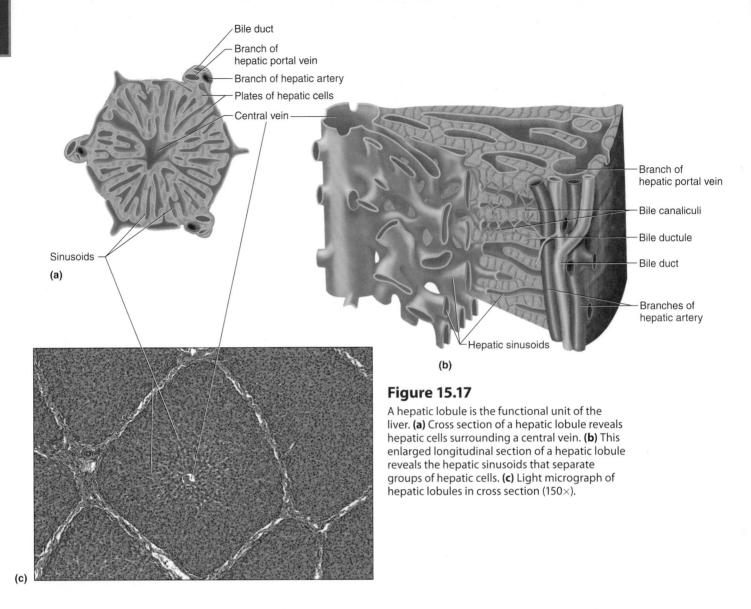

(a)

- Bile duct
- Branch of hepatic portal vein
- Branch of hepatic artery
- Plates of hepatic cells
- Central vein
- Sinusoids

- Branch of hepatic portal vein
- Bile canaliculi
- Bile ductule
- Bile duct
- Branches of hepatic artery
- Hepatic sinusoids

(b)

(c)

Figure 15.17

A hepatic lobule is the functional unit of the liver. **(a)** Cross section of a hepatic lobule reveals hepatic cells surrounding a central vein. **(b)** This enlarged longitudinal section of a hepatic lobule reveals the hepatic sinusoids that separate groups of hepatic cells. **(c)** Light micrograph of hepatic lobules in cross section (150×).

Figure 15.18 AP|R

The paths of blood and bile within a hepatic lobule.

- Bile duct
- Bile canaliculi
- Bile ductule
- Kupffer cell
- Hepatic cells
- Branch of hepatic portal vein
- Branch of hepatic artery
- Blood flow into liver
- Hepatic sinusoids
- Central vein (blood flow out of liver)

Hepatitis

Hepatitis is an inflammation of the liver. About half a million people develop hepatitis in the United States each year, and 6,000 die of the disease. Hepatitis has several causes, but the various types have similar symptoms.

Acute hepatitis may at first resemble the flu, producing mild headache, low fever, fatigue, lack of appetite, nausea and vomiting, and sometimes stiff joints. By the end of the first week, more distinctive symptoms arise: a rash, pain in the upper right quadrant of the abdomen, dark and foamy urine, and pale feces. The skin and sclera of the eyes turn yellow due to accumulating bile pigments (jaundice). Great fatigue may continue for two or three weeks, and then gradually the person begins to feel better. This is hepatitis in its most common, least dangerous acute guise. In a rare acute form called *fulminant hepatitis,* symptoms occur suddenly and severely, along with altered behavior and personality. Medical attention is necessary to prevent kidney or liver failure or coma.

Chronic hepatitis persists for more than six months. As many as 300 million people worldwide are hepatitis carriers. They do not have symptoms but can infect others. Five percent of carriers eventually develop liver cancer.

Only rarely does hepatitis result from alcoholism, autoimmunity, or the use of certain drugs. Usually, one of several types of viruses cause hepatitis. Viral types are distinguished by the route of infection, surface features, and whether the viral genetic material is DNA or RNA. Hepatitis B virus has DNA; the others have RNA. The viral types are classified as follows:

Hepatitis A spreads by contact with food or objects contaminated with virus-containing feces, including diapers. The course of hepatitis A is short and mild.

Hepatitis B spreads by contact with virus-containing body fluids, such as blood, saliva, or semen. It may be transmitted by blood transfusions, hypodermic needles, or sexual activity.

Hepatitis C accounts for about half of all known cases of hepatitis. This virus is primarily transmitted in blood—by sharing razors or needles, from pregnant woman to fetus, or through blood transfusions or use of blood products. As many as 60% of individuals infected with the hepatitis C virus suffer chronic symptoms.

Hepatitis D infection occurs only in the presence of the hepatitis B virus. It is blood-borne and associated with blood transfusions and intravenous drug use. About 20% of individuals infected with this virus die from the infection.

Hepatitis E virus is usually transmitted in water contaminated with feces. It most often affects visitors to developing nations.

Hepatitis G infection is rare but accounts for a significant percentage of cases of fulminant hepatitis. In people with healthy immune systems, the virus produces symptoms so mild that they may not be noticed.

Antibiotic drugs, which are effective against bacteria, are not helpful against viral hepatitis. Usually, the person must just wait out the symptoms. Hepatitis C, however, sometimes responds to a form of interferon, an immune system biochemical given as an injectable drug, and ribavirin, an antiviral drug. Several oral drugs to treat hepatitis C are in clinical trials.

---

converting parts of carbohydrate and protein molecules into fat molecules (see p. 423). The blood transports fats synthesized in the liver to adipose tissue for storage.

The most vital liver functions concern protein metabolism. They include deaminating amino acids; forming urea (see p. 431); synthesizing plasma proteins, such as clotting factors (see chapter 12, pp. 328–330); and converting certain amino acids into other amino acids.

The liver also stores many substances, including glycogen, iron, and vitamins A, D, and B_{12}. In addition, macrophages in the liver help destroy damaged red blood cells (see chapter 12, pp. 323–324) and phagocytize foreign antigens. The liver also removes toxic substances such as alcohol and certain other drugs from blood (detoxification) and secretes bile.

Many of these liver functions are not directly related to the digestive system and are, as indicated above, discussed in other chapters. Bile secretion, however, is important to digestion and is explained next in this chapter. Table 15.3 summarizes the major functions of the liver. Clinical Application 15.2 discusses viral infections of the liver.

> The liver is unlike most organs in that it can regenerate. Up to 75% of a liver can be destroyed and the organ regenerate and recover. For this reason, people can donate parts of their livers to people in liver failure, if the tissues of donor and recipient are compatible.

Practice

26. Locate the liver.
27. Describe a hepatic lobule.
28. Review liver functions.

Table 15.3	Major Functions of the Liver
General Function	**Specific Function**
Carbohydrate metabolism	Polymerizes glucose to glycogen; breaks down glycogen to glucose; converts noncarbohydrates to glucose
Lipid metabolism	Oxidizes fatty acids; synthesizes lipoproteins, phospholipids, and cholesterol; converts portions of carbohydrate and protein molecules into fats
Protein metabolism	Deaminates amino acids; forms urea; synthesizes plasma proteins; converts certain amino acids to other amino acids
Storage	Stores glycogen, iron, and vitamins A, D, and B_{12}
Blood filtering	Removes damaged red blood cells and foreign substances by phagocytosis
Detoxification	Removes toxins from blood
Secretion	Secretes bile

Composition of Bile

Bile (bīl) is a yellowish-green liquid continuously secreted from hepatic cells. In addition to water, bile contains *bile salts, bile pigments* (bilirubin and biliverdin), *cholesterol,* and *electrolytes.* Of these, bile salts are the most abundant and are the only bile substances that have a digestive function. Bile pigments are breakdown products of hemoglobin from red blood cells and are normally excreted in the bile (see chapter 12, pp. 323–324).

> Jaundice can have several causes. In *obstructive jaundice,* bile ducts are blocked (as with gallstones or tumors). In *hepatocellular jaundice,* the liver is diseased (as in cirrhosis or hepatitis). In *hemolytic jaundice,* red blood cells are destroyed too rapidly (as with a blood transfusion from an incompatible blood group or a blood infection such as malaria).

Gallbladder

The **gallbladder** (gawl'blad-er) is a pear-shaped sac in a depression on the liver's inferior surface. It connects to the **cystic duct** (sis'tik dukt), which in turn joins the common hepatic duct (figs. 15.1 and 15.19). The gallbladder is lined with epithelial cells and has a strong, muscular layer in its wall. The gallbladder stores bile between meals, reabsorbs water to concentrate bile, and contracts to release bile into the small intestine.

The common hepatic and cystic ducts join to form the *bile duct.* It leads to the duodenum (the proximal part of the small intestine) (figs. 15.14 and 15.19), where the *hepatopancreatic sphincter* guards its exit. This sphincter normally remains contracted, so as bile collects in the bile duct it backs up into the cystic duct. When this happens, bile flows into the gallbladder, where it is stored.

> Cholesterol in bile may precipitate and form crystals called *gallstones* under certain conditions (fig. 15.20). Gallstones in the bile duct may block bile flow into the small intestine and cause considerable pain. A surgical procedure called a *cholecystectomy* removes the gallbladder when gallstones are obstructive. The surgery can often be done with a laparoscope (small, lit probe) on an outpatient basis.

Regulation of Bile Release

Normally bile does not enter the duodenum until *cholecystokinin* stimulates the gallbladder to contract. The intestinal mucosa releases this hormone in response to proteins and fats in the small intestine. (Recall its action to stimulate pancreatic enzyme secretion, p. 415.) The hepatopancreatic sphincter usually remains contracted until a peristaltic wave in the duodenal wall approaches it. Then the sphincter relaxes, and bile is squirted into the duodenum (see fig. 15.19). Table 15.4 summarizes the hormones that help control digestion.

Functions of Bile Salts

Bile salts aid digestive enzymes. Bile salts affect *fat globules* (clumped molecules of fats) much like a soap or detergent would affect them. That is, bile salts break fat globules into smaller droplets, an action called

Table 15.4	Hormones of the Digestive Tract	
Hormone	**Source**	**Function**
Gastrin	Gastric cells, in response to food	Increases secretory activity of gastric glands
Cholecystokinin	Intestinal wall cells, in response to proteins and fats in the small intestine	Decreases secretory activity of gastric glands and inhibits gastric motility; stimulates pancreas to secrete fluid with a high digestive enzyme concentration; stimulates gallbladder to contract and release bile
Secretin	Cells in the duodenal wall, in response to acidic chyme entering the small intestine	Stimulates pancreas to secrete fluid with a high bicarbonate ion concentration

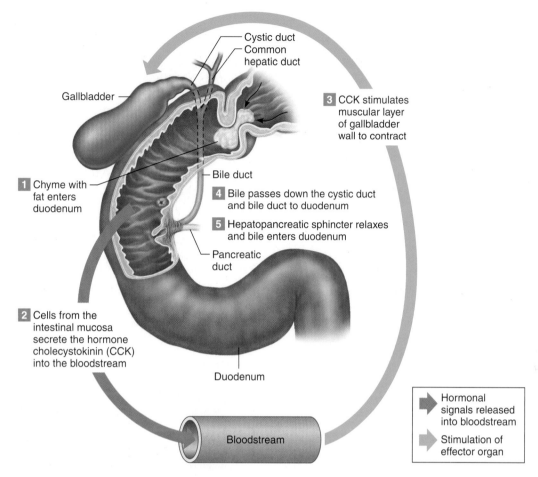

Figure 15.19

Fatty chyme entering the duodenum stimulates the gallbladder to release bile.

Q: *Which other organ, besides the gallbladder, responds to cholecystokinin stimulation, and what is the response of that organ to cholecystokinin stimulation?*

Answer can be found in Appendix E on page 568.

Figure 15.20

Falsely colored radiograph of a gallbladder that contains gallstones (arrow).

emulsification (e-mul″sĭ-fĭ-ka′shun), that greatly increases the total surface area of the fatty substance. The tiny fat droplets then mix with water. Fat-splitting enzymes (lipases) can then digest fat molecules more effectively.

Bile salts also enhance absorption of fatty acids, cholesterol, and the fat-soluble vitamins A, D, E, and K. Lack of bile salts results in poor lipid absorption and vitamin deficiencies.

Practice

29. Explain how bile forms.

30. Describe the function of the gallbladder.

31. How is secretion of bile regulated?

32. How do bile salts function in digestion?

15.9 SMALL INTESTINE

The **small intestine** is a tubular organ that extends from the pyloric sphincter to the beginning of the large intestine. With its many loops and coils, it fills much of the abdominal cavity (see fig. 15.1 and reference plates 4 and 5, pp. 26–27).

The small intestine receives secretions from the pancreas and liver. It also completes digestion of the nutrients in chyme, absorbs the products of digestion, and transports the residues to the large intestine.

Parts of the Small Intestine

The small intestine consists of three parts: the duodenum, the jejunum, and the ileum (figs. 15.21 and 15.22). The **duodenum** (du″o-de′num), about 25 centimeters long and 5 centimeters in diameter, lies posterior to the parietal peritoneum and is the most fixed portion of the small intestine. It follows a C-shaped path as it passes anterior to the right kidney and the upper three lumbar vertebrae.

The remainder of the small intestine is mobile and lies free in the peritoneal cavity. The proximal two-fifths of this portion of the small intestine is the **jejunum** (jě-joo′num), and the remainder is the **ileum** (il′e-um). A double-layered fold of peritoneal membrane called **mesentery** (mes′en-ter″e) suspends these parts from the posterior abdominal wall (figs. 15.21, 15.23, and

reference plate 5, p. 27). The mesentery supports the blood vessels, nerves, and lymphatic vessels that supply the intestinal wall. The jejunum and ileum are not distinctly separate parts; however, the diameter of the jejunum is usually greater than that of the ileum, and its wall is thicker, more vascular, and more active.

A filmy, double fold of peritoneal membrane called the *greater omentum* drapes like an apron from the stomach over the transverse colon and the folds of the small intestine (fig. 15.23 and reference plate 3, p. 25). If the wall of the alimentary canal becomes infected, cells from the omentum may adhere to the inflamed region, helping to wall off the area. This action prevents spread of the infection to the peritoneal cavity.

Structure of the Small Intestinal Wall

The inner wall of the small intestine throughout its length appears velvety due to many tiny projections of mucous membrane called **intestinal villi** (in-tes′tĭ-nal vil′i)

Figure 15.21 AP|R

The three parts of the small intestine are the duodenum, the jejunum, and the ileum.

Figure 15.22

Radiograph showing a normal small intestine containing a radiopaque substance that the patient ingested.

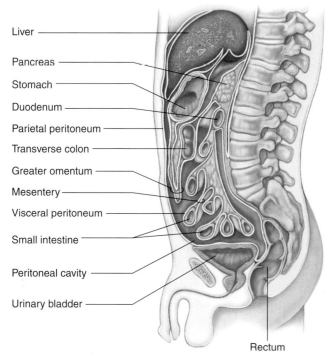

Figure 15.23 AP|R

Mesentery formed by folds of the peritoneal membrane suspends portions of the small intestine from the posterior abdominal wall.

(figs. 15.24 and 15.25; see fig. 15.3). These structures are most numerous in the duodenum and the proximal jejunum. They project into the lumen of the alimentary canal, contacting the intestinal contents. Villi greatly increase the surface area of the intestinal lining, aiding the absorption of digestive products.

Each villus consists of a layer of simple columnar epithelium and a core of connective tissue containing blood capillaries, a lymphatic capillary called a **lacteal,** and nerve fibers. Blood capillaries and lacteals carry away absorbed nutrients, and nerve fibers transmit impulses to stimulate or inhibit villus activities. Between the bases of adjacent villi are tubular **intestinal glands** that extend downward into the mucous membrane (figs. 15.24 and 15.25; see fig. 15.3).

The epithelial cells that form the lining of the small intestine are continually replaced. New cells form in the intestinal glands by mitosis and migrate outward onto the villus surface. When the migrating cells reach the tip of the villus, they are shed. This *cellular turnover* renews the small intestine's epithelial lining every three to six days. As a result, nearly one-quarter of the bulk of feces consists of dead epithelial cells from the small intestine.

Secretions of the Small Intestine

Mucus-secreting goblet cells are abundant throughout the mucosa of the small intestine. In addition, many specialized *mucus-secreting glands* in the submucosa

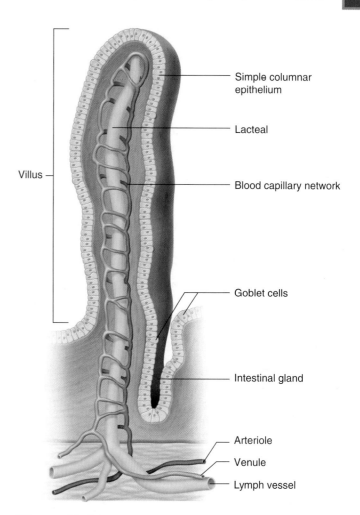

Figure 15.24

Structure of a single intestinal villus. The elongated shapes of intestinal villi dramatically increase the absorptive surface area of the small intestine.

Figure 15.25

Light micrograph of intestinal villi from the wall of the duodenum (50×).

in the proximal part of the duodenum secrete a thick, alkaline mucus in response to certain stimuli.

The intestinal glands at the bases of the villi secrete large volumes of a watery fluid, which brings digestive products to the villi. The fluid has a nearly neutral pH (6.5–7.5), and it lacks digestive enzymes. However, the epithelial cells of the intestinal mucosa have digestive enzymes embedded in the membranes of their microvilli on their luminal surfaces. These enzymes break down food molecules just before absorption takes place. The enzymes include **peptidases,** which split peptides into their constituent amino acids; **sucrase, maltase,** and **lactase,** which split the disaccharides sucrose, maltose, and lactose into the monosaccharides glucose, fructose, and galactose; and **intestinal lipase,** which splits fats into fatty acids and glycerol. Table 15.5 summarizes the sources and actions of the major digestive enzymes.

Regulation of Small Intestine Secretions

Goblet cells and intestinal glands secrete their products when chyme provides both mechanical and chemical stimulation. Distension of the intestinal wall activates the nerve plexuses within the wall and stimulates parasympathetic reflexes that also trigger release of small intestine secretions.

Practice

33. Describe the parts of the small intestine.
34. What is the function of an intestinal villus?
35. What is the function of the intestinal glands?
36. List the intestinal digestive enzymes.

Adults who have lactose intolerance do not produce sufficient lactase to adequately digest lactose, or milk sugar. Undigested lactose increases the osmotic pressure of the intestinal contents and draws water into the intestines. At the same time, intestinal bacteria metabolize undigested sugar, producing organic acids and gases. The overall result is bloating, intestinal cramps, and diarrhea. Taking lactase pills before eating dairy products may help avoid these symptoms. Infants with lactose intolerance may drink formula based on soybeans.

Genetic evidence suggests that lactose intolerance may be the "normal" condition, with ability to digest lactose the result of a mutation that occurred recently in our evolutionary past and became advantageous when the advent of agriculture brought dairy foods to human populations. The trait of ability to digest lactose has increased in parallel to increased use of dairy foods at least three times in history, in different populations.

Table 15.5	Summary of the Major Digestive Enzymes	
Enzyme	**Source**	**Digestive Action**
Salivary amylase	Salivary glands	Begins carbohydrate digestion by breaking down starch and glycogen to disaccharides
Pepsin	Gastric chief cells	Begins protein digestion
Pancreatic amylase	Pancreas	Breaks down starch and glycogen into disaccharides
Pancreatic lipase	Pancreas	Breaks down fats into fatty acids and glycerol
Proteolytic enzymes (a) Trypsin (b) Chymotrypsin (c) Carboxypeptidase	Pancreas	Breaks down proteins or partially digested proteins into peptides
Nucleases	Pancreas	Breaks down nucleic acids into nucleotides
Peptidase	Intestinal mucosal cells	Breaks down peptides into amino acids
Sucrase, maltase, lactase	Intestinal mucosal cells	Breaks down disaccharides into monosaccharides
Intestinal lipase	Intestinal mucosal cells	Breaks down fats into fatty acids and glycerol
Enterokinase	Intestinal mucosal cells	Converts trypsinogen into trypsin

Absorption in the Small Intestine

Villi greatly increase the surface area of the intestinal mucosa, making the small intestine the most important absorbing organ of the alimentary canal. So effective is the small intestine in absorbing digestive products, water, and electrolytes that very little absorbable material reaches its distal end.

Carbohydrate digestion begins in the mouth with the activity of salivary amylase, and is completed in the small intestine by enzymes from the intestinal mucosa and pancreas. Villi absorb the resulting monosaccharides, which enter blood capillaries. Simple sugars are absorbed by facilitated diffusion or active transport (see chapter 3, pp. 63 and 65)

Protein digestion begins in the stomach as a result of pepsin activity, and is completed in the small intestine by enzymes from the intestinal mucosa and the pancreas. During this process, large protein molecules are broken down into amino acids, which are then actively transported into the villi and carried away by the blood.

Fat molecules are digested almost entirely by enzymes from the pancreas and intestinal mucosa. The resulting fatty acids and glycerol molecules diffuse into villi epithelial cells (fig. 15.26 *(1)*). The endoplasmic reticula of the cells use the fatty acids to resynthesize fat molecules similar to those previously digested (fig. 15.26 *(2)*). These fats are encased in protein to form *chylomicrons* (fig. 15.26 *(3)*), which make their way to the lacteals of the villi (fig. 15.26 *(4)*). Lymph in the lacteals and other lymphatic vessels carries chylomicrons to the bloodstream, as discussed in chapter 14 on page 379 (fig. 15.26 *(5)*). Some fatty acids with very short carbon chains may be absorbed directly into the blood capillary of a villus without being changed back into fat.

Chylomicrons transport dietary fats to muscle and adipose cells. Similarly, VLDL (very low-density lipoprotein with a high concentration of triglycerides) molecules, produced in the liver, transport triglycerides synthesized from excess dietary carbohydrates. As VLDL molecules reach adipose cells, an enzyme, *lipoprotein lipase,* catalyzes reactions that unload their triglycerides,

converting the VLDL to LDL (low-density lipoproteins). Because most of the triglycerides have been removed, LDL molecules have a higher cholesterol content than do the original VLDL molecules. Cells in the peripheral tissues remove LDL from plasma by receptor-mediated endocytosis, thus obtaining a supply of cholesterol (see chapter 3, p. 66).

While LDL delivers cholesterol to tissues, HDL (high-density lipoprotein with a high concentration of protein and a low concentration of lipids) removes cholesterol from tissues. The liver produces the basic HDL framework and secretes the HDL molecules into the bloodstream. As they circulate, the HDL molecules pick up cholesterol from peripheral tissues and return to the liver, where they enter liver cells by receptor-mediated endocytosis. The liver disposes of the cholesterol it obtains in this manner by secreting it into bile or by using it to synthesize bile salts.

The intestine reabsorbs much of the cholesterol and bile salts in bile, which are then transported back to the liver, and the secretion-reabsorption cycle repeats. During each cycle, some of the cholesterol and bile salts escape reabsorption, reach the large intestine, and are eliminated with the feces.

In addition to the products of carbohydrate, protein, and fat digestion, the intestinal villi absorb electrolytes by diffusion and active transport and water by osmosis. Table 15.6 summarizes the absorption process.

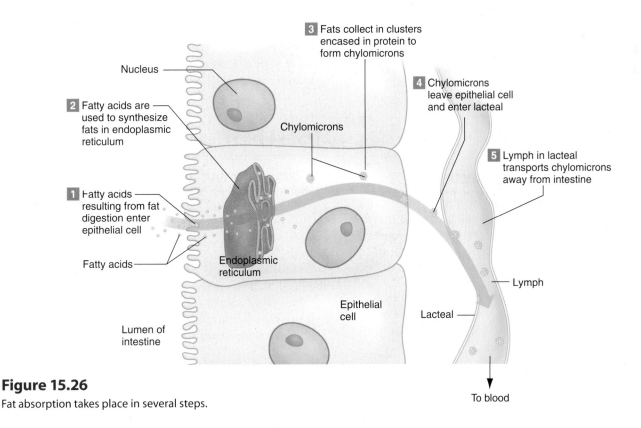

Figure 15.26

Fat absorption takes place in several steps.

Table 15.6	Intestinal Absorption of Nutrients	
Nutrient	**Absorption Mechanism**	**Means of Transport**
Monosaccharides	Facilitated diffusion and active transport	Blood in capillaries
Amino acids	Active transport	Blood in capillaries
Fatty acids and glycerol	Facilitated diffusion of glycerol; diffusion of fatty acids into cells	
	(a) Most fatty acids are resynthesized into fats and incorporated in chylomicrons for transport	Lymph in lacteals
	(b) Some fatty acids with relatively short carbon chains are transported without being changed back into fats	Blood in capillaries
Electrolytes	Diffusion and active transport	Blood in capillaries
Water	Osmosis	Blood in capillaries

Practice

37. Which substances resulting from digestion of carbohydrate, protein, and fat molecules does the small intestine absorb?

38. Describe how fatty acids are absorbed and transported.

In *malabsorption,* the small intestine digests, but does not absorb, some nutrients. Symptoms of malabsorption include diarrhea, weight loss, weakness, vitamin deficiencies, anemia, and bone demineralization. Causes of malabsorption include surgical removal of a portion of the small intestine, obstruction of lymphatic vessels due to a tumor, or interference with the production and release of bile as a result of liver disease.

Another cause of malabsorption is a reaction to *gluten,* called *celiac disease.* Gluten is a composite of two types of proteins that are found in certain grains, such as wheat, barley, and rye. In celiac disease, microvilli are damaged, and in severe cases, villi may be destroyed. These effects reduce the surface area of the small intestine, preventing adequate absorption of some nutrients. Grocery stores sell many gluten-free products.

Movements of the Small Intestine

The small intestine carries on mixing movements and peristalsis, like the stomach. The major mixing movement is segmentation, in which periodic small, ringlike contractions cut chyme into segments and move it back and forth. Segmentation also slows the movement of chyme through the small intestine.

Weak peristaltic waves propel chyme short distances through the small intestine. Consequently, chyme moves slowly through the small intestine, taking from three to ten hours to travel its length.

If the small intestine wall becomes overdistended or irritated, a strong *peristaltic rush* may pass along the organ's entire length. This movement sweeps the contents of the small intestine into the large intestine so quickly that water, nutrients, and electrolytes that would normally be absorbed are not. The result is *diarrhea,* characterized by more frequent defecation and watery stools. Prolonged diarrhea causes imbalances in water and electrolyte concentrations.

At the distal end of the small intestine, the **ileocecal** (il″e-o-se′kal) **sphincter** joins the small intestine's ileum to the large intestine's cecum (fig. 15.27). Normally, this sphincter remains constricted, preventing the contents of the small intestine from entering the large intestine, and the contents of the large intestine from backing up into the ileum. However, after a meal, a gastroileal reflex increases peristalsis in the ileum and relaxes the sphincter, forcing some of the contents of the small intestine into the cecum.

Practice

39. Describe the movements of the small intestine.

40. What stimulus relaxes the ileocecal sphincter?

15.10 LARGE INTESTINE

The **large intestine** is so named because its diameter is greater than that of the small intestine. This part of the alimentary canal is about 1.5 meters long. It begins in the lower right side of the abdominal cavity, where the ileum joins the cecum. From there, the large intestine ascends on the right side, crosses obliquely to the left, and descends into the pelvis. At its distal end, it opens to the outside of the body as the anus (see fig. 15.1).

The large intestine absorbs water and electrolytes from chyme remaining in the alimentary canal. It also forms and stores feces.

Parts of the Large Intestine

The large intestine consists of the cecum, colon, rectum, and anal canal (figs. 15.27 and 15.28; see reference plates 4 and 5, pp. 26–27). The **cecum,** at the beginning of the

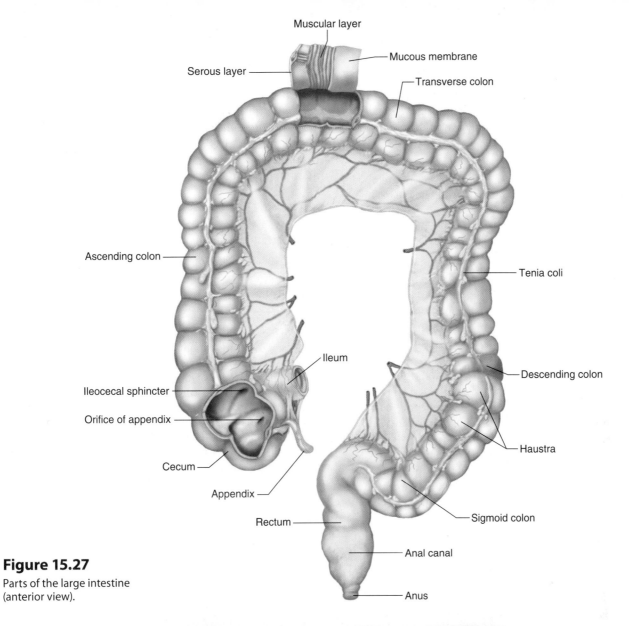

Figure 15.27

Parts of the large intestine (anterior view).

Figure 15.28

Radiograph of the large intestine containing a radiopaque substance that the patient ingested.

large intestine, is a dilated, pouchlike structure that hangs slightly below the ileocecal opening. Projecting downward from it is a closed end, narrow tube containing lymphatic tissue called the **appendix.** The human appendix has no known digestive function.

> In *appendicitis,* the appendix becomes inflamed and infected. Surgery is often required to remove the appendix before it ruptures. If it does rupture, this may allow contents of the large intestine to enter the abdominal cavity and cause a serious infection of the peritoneum called *peritonitis.*

The **colon** is divided into four parts—the ascending, transverse, descending, and sigmoid colons. The **ascending colon** begins at the cecum and continues upward against the posterior abdominal wall to a point just inferior to the liver. There, it turns sharply to the left and becomes the **transverse colon.** The transverse colon is the longest and most movable part of the large intestine. It is suspended by a fold of peritoneum and sags in the middle below the stomach. As the transverse colon approaches the spleen, it turns abruptly downward and becomes the **descending colon.** At the brim of the pelvis, the descending colon makes an S-shaped curve called the **sigmoid colon** and then becomes the rectum.

The **rectum** lies next to the sacrum and generally follows its curvature. The peritoneum firmly attaches the rectum to the sacrum, and the rectum ends about 5 centimeters below the tip of the coccyx, where it becomes the anal canal (see fig. 15.27).

The last 2.5–4.0 centimeters of the large intestine form the **anal canal** (fig. 15.29). The mucous membrane in the canal is folded into six to eight longitudinal *anal columns.* At its distal end, the canal opens to the outside as the **anus.** Two sphincter muscles guard the anus—an *internal anal sphincter muscle,* composed of smooth muscle under involuntary control, and an *external anal sphincter muscle,* composed of skeletal muscle under voluntary control.

Practice

41. What is the general function of the large intestine?
42. Describe the parts of the large intestine.

> *Hemorrhoids* are enlarged and inflamed branches of the rectal vein in the anal columns that cause intense itching, sharp pain, and sometimes bright red bleeding. The hemorrhoids may be internal or bulge out of the anus. Causes of hemorrhoids include anything that puts prolonged pressure on the delicate rectal tissue, including obesity, pregnancy, constipation, diarrhea, and liver disease.
>
> Eating more fiber-rich foods and drinking lots of water can usually prevent or cure hemorrhoids. Warm soaks in the tub, cold packs, and careful wiping of painful areas also help, as do external creams and ointments. Surgery—with a scalpel or a laser—can remove severe hemorrhoids.

Structure of the Large Intestinal Wall

The wall of the large intestine is composed of the same types of tissues as other parts of the alimentary canal but also has some unique features. The large intestinal wall lacks the villi characteristic of the small intestine, and the layer of longitudinal muscle fibers is not uniformly distributed throughout the large intestinal wall. Instead, the fibers form three distinct bands (teniae coli) that extend the entire length of the colon (see fig. 15.27). These bands exert tension lengthwise on the wall, creating a series of pouches (haustra).

> People over fifty years of age or with a family history of colorectal cancer should have a screening of the large intestine, performed with a fiberoptic colonoscope. Under sedation, this flexible lit tube is inserted into the rectum, and polyps and tumors identified and removed. Computed tomographic colonography (popularly called a virtual colonoscopy) requires the same preparatory bowel cleansing, but does not require sedation and is not invasive, and is faster and less costly. However, if a lesion is detected, fiberoptic colonoscopy must be used to remove the suspicious tissue.

Functions of the Large Intestine

The large intestine has little or no digestive function, in contrast to the small intestine, which secretes digestive

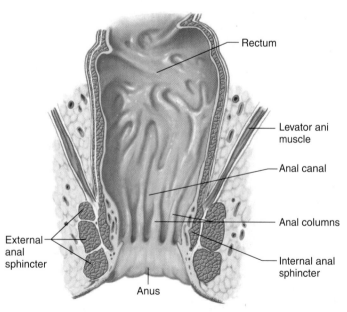

Figure 15.29

The rectum and the anal canal are at the distal end of the alimentary canal.

Lumen of large intestine

Goblet cells

Figure 15.30

Light micrograph of the large intestinal mucosa (560×).

Q: *Note the many goblet cells in the mucosa of the large intestine. Why are there so many of these cells in the large intestinal wall as opposed to in the small intestinal wall?*

Answer can be found in Appendix E on page 568.

enzymes and absorbs the products of digestion. However, the mucous membrane that forms the large intestine's inner lining contains many tubular glands. Structurally, these glands are similar to those of the small intestine, but they are composed almost entirely of goblet cells (fig. 15.30). Consequently, mucus is the large intestine's only significant secretion.

Mucus secreted into the large intestine protects the intestinal wall against the abrasive action of the materials passing through it. Mucus also binds particles of fecal matter, and its alkalinity helps control the pH of the large intestinal contents.

Chyme entering the large intestine contains materials that the small intestine did not digest or absorb. It also contains water, electrolytes, mucus, and bacteria. The large intestine normally absorbs water (although most water is absorbed in the small intestine) and electrolytes in the proximal half of the tube. Substances that remain in the tube become feces and are stored in the distal part of the large intestine.

The many bacteria that normally inhabit the large intestine, called *intestinal flora,* break down some of the molecules that escape the actions of human digestive enzymes. For instance, cellulose, a complex carbohydrate in food of plant origin, passes through the alimentary canal almost unchanged, but colon bacteria can break down cellulose and use it as an energy source. These bacteria, in turn, synthesize certain vitamins, such as K, B_{12}, thiamine, and riboflavin, which the intestinal mucosa absorbs. Bacterial actions in the large intestine may produce intestinal gas (flatus).

Practice

43. How does the structure of the large intestine differ from that of the small intestine?

44. Which substances does the large intestine absorb?

Movements of the Large Intestine

The movements of the large intestine—mixing and peristalsis—are similar to those of the small intestine, although usually slower. Also, peristaltic waves of the large intestine happen only two or three times each day. These waves produce *mass movements* in which a large section of the intestinal wall constricts vigorously, forcing the intestinal contents toward the rectum. Typically, mass movements follow a meal as a result of the gastrocolic reflex initiated in the small intestine. Irritations of the intestinal mucosa also can trigger such movements. For instance, a person with an inflamed colon (colitis) may experience frequent mass movements. Clinical Application 15.3 discusses inflammatory bowel disease.

A person can usually initiate a *defecation reflex* by holding a deep breath and contracting the abdominal wall muscles. This action increases internal abdominal pressure and forces feces into the rectum. As the rectum fills, its wall distends, triggering the defecation reflex that stimulates peristaltic waves in the descending colon. The internal anal sphincter relaxes. At the same time, other reflexes involving the sacral region of the spinal cord strengthen the peristaltic waves, lower the diaphragm, close the glottis, and contract the abdominal wall muscles. These actions further increase internal abdominal pressure and squeeze the rectum. The external anal sphincter is signaled to relax, and the feces are forced to the outside. Contracting the external anal sphincter allows voluntary inhibition of defecation.

Feces

Feces (fe′sēz) include materials not digested or absorbed, plus water, electrolytes, mucus, shed intestinal cells, and bacteria. Usually, feces are about 75% water, and their color derives from bile pigments altered by bacterial action. Feces' pungent odor results from a variety of compounds that bacteria produce.

Practice

45. How does peristalsis in the large intestine differ from peristalsis in the small intestine?

46. List the major events that occur during defecation.

47. Describe the composition of feces.

Clinical Application 15.3

Inflammatory Bowel Disease

Inflammatory bowel disease (IBD) is a group of disorders that affect about a million people in the United States. The most common ages of onset are between ten and thirty years, and between fifty and sixty years. IBD includes ulcerative colitis and Crohn disease. These disorders differ by the site and extent of inflammation and ulceration of the intestines. Both produce abdominal cramps and diarrhea.

IBD is usually the result of an inflammatory response, such as to infection, that does not stop. About 10–20% of people with IBD have relatives with the condition. Mutations in two specific genes cause severe IBD in young children. The genes normally encode receptors for interleukin 10, a cytokine that halts the inflammatory response. The mutation impairs the receptors so that inflammation is not shut off. Painfully swollen intestines result.

Ulcerative colitis affects the mucosa and submucosa of the distal large intestine and the rectum. In about 25% of cases, the disease extends no farther than the rectum. Bouts of bloody diarrhea and cramps may last for days or weeks, and vary in frequency. The severe diarrhea leads to weight loss and electrolyte imbalances and may develop into colon cancer or affect other organs, including the skin, eyes, or liver. The inflamed and ulcerous tissue is continuous.

Crohn disease is more extensive than ulcerative colitis, extending into the small and large intestines and penetrating all tissue layers. In contrast to the uniformity of ulcerative colitis, affected parts of intestine in Crohn disease are interspersed with unaffected areas, producing a "cobblestone" effect after many years. Rarely, the disease affects more proximal structures of the gastrointestinal tract. The diarrhea is often not bloody, and complications such as cancer are rare.

Several classes of drugs are used to control IBD. They affect different parts of the gastrointestinal tract and/or different parts of the immune response that contribute to the excess inflammation. Surgery may be necessary if drug therapy is ineffective or if cancer develops.

15.11 NUTRITION AND NUTRIENTS

Nutrition is the study of nutrients and how the body utilizes them. **Nutrients** (nu'tre-ents) include carbohydrates, lipids, proteins, vitamins, minerals, and water (see chapter 2, pp. 41–44, and chapter 4, p. 80). Carbohydrates, lipids, and proteins are called **macronutrients** because they are required in large amounts. They provide energy as well as other specific functions. Vitamins and minerals are required in much smaller amounts and are therefore called **micronutrients.** They do not directly provide energy, but make possible the biochemical reactions that extract energy from macronutrient molecules.

Macronutrients provide potential energy that can be expressed in calories, which are units of heat. A **calorie** (kal'o-re) is the amount of heat required to raise the temperature of a gram of water by 1° Celsius. The calorie used to measure food energy is 1,000 times greater. This larger calorie (Cal) is technically a kilocalorie, but nutritional studies commonly refer to it simply as a calorie. As a result of cellular oxidation, 1 gram of carbohydrate or 1 gram of protein yields on average about 4.1 calories, and 1 gram of fat about 9.5 calories (more than twice as much chemical energy as carbohydrates or proteins).

Foods provide nutrients, and digestion breaks nutrients down to sizes that can be absorbed and transported in the bloodstream. Nutrients that human cells cannot synthesize, such as certain amino acids, are called **essential nutrients.**

Practice

48. Identify and distinguish among macronutrients and micronutrients.

49. How is food energy measured?

Carbohydrates

Carbohydrates are organic compounds and include the sugars and starches. The energy held in their chemical bonds is used primarily to power cellular processes.

Carbohydrate Sources

Carbohydrates are ingested in a variety of forms, including starch from grains and vegetables; glycogen from meats; disaccharides from milk sugar, cane sugar, beet sugar, and molasses; and monosaccharides from honey and fruits (see chapter 2, pp. 41–42). Digestion breaks down complex carbohydrates into monosaccharides, which are small enough to be absorbed into the bloodstream.

Cellulose is a complex carbohydrate that is abundant in food—it gives celery its crunch and lettuce its crispness. Humans cannot digest cellulose, so the portion of it that is not broken down by intestinal flora passes through the alimentary canal largely unchanged. In this way, cellulose provides bulk (also called fiber or roughage) against which the muscular wall of the digestive system can push, easing the movement of intestinal contents.

Carbohydrate Use

The monosaccharides absorbed from the digestive tract include *fructose, galactose,* and *glucose.* Liver enzymes catalyze reactions that convert fructose and galactose into glucose, which is the carbohydrate form most commonly oxidized for cellular fuel (see chapter 4, pp. 80–82).

Many cells obtain energy by oxidizing fatty acids. Some cells, however, such as neurons, require a continuous supply of glucose for survival. Even a temporary decrease in the glucose supply may seriously impair nervous system function. Consequently, the body requires a minimum of carbohydrates. If foods do not provide an adequate carbohydrate supply, the liver may convert some noncarbohydrates, such as amino acids from proteins, into glucose. The requirement for glucose has physiological priority over the requirement to synthesize proteins from available amino acids.

Carbohydrates have functions other than providing energy. Some excess glucose is polymerized to form *glycogen,* which is stored in the liver and muscles. When glucose is required to supply energy, it can be mobilized rapidly by breaking down glycogen. However, the body can store only a certain amount of glycogen, so excess glucose is usually converted into fat and stored in adipose tissue. To obtain energy, the body first metabolizes glucose, then glycogen stores, and finally fats and proteins.

Cells use carbohydrates as starting materials for synthesizing such vital biochemicals as the five-carbon sugars *ribose* and *deoxyribose,* required for production of the nucleic acids RNA and DNA. Carbohydrates are also required to synthesize the disaccharide *lactose* (milk sugar) when the breasts are actively secreting milk.

Carbohydrate Requirements

Carbohydrates provide the primary fuel source for cellular processes, so the need for carbohydrates varies with individual energy expenditure. Physically active individuals require more fuel than those who are sedentary. The minimal requirement for carbohydrates in the human diet is unknown. It is estimated, however, that

an intake of at least 125 to 175 grams daily is necessary to spare protein (that is, to avoid protein breakdown) and to avoid metabolic disorders resulting from excess fat use.

Practice

50. List several common sources of carbohydrates.
51. Explain the importance of cellulose in the diet.
52. Explain why the requirement for glucose has priority over protein synthesis.
53. Why do daily requirements for carbohydrates vary from person to person?

Lipids

Lipids are organic compounds that include fats, oils, and fatlike substances such as phospholipids and cholesterol. They supply energy for cellular processes and help build structures, such as cell membranes. The most common dietary lipids are the fats called *triglycerides.* Recall from chapter 2 (p. 42) that a triglyceride molecule consists of a glycerol and three fatty acids.

Lipid Sources

Triglycerides are found in plant- and animal-based foods. Saturated fats are mainly found in foods of animal origin, such as meats, eggs, milk, and lard, as well as in palm and coconut oils. Unsaturated fats are in seeds, nuts, and plant oils. Monounsaturated fats, such as those in olive, peanut, and canola oils, are the healthiest. Saturated fats in excess are a risk factor for cardiovascular disease.

Cholesterol is abundant in liver and egg yolk and, to a lesser extent, in whole milk, butter, cheese, and meats. It is not present in foods of plant origin.

Lipid Use

Many foods contain phospholipids, cholesterol, or triglycerides. Lipids provide a variety of physiological functions; however, fats mainly supply energy. Before a triglyceride molecule can release energy, it must undergo hydrolysis (breakdown in the presence of water) as part of digestion, releasing the constituent fatty acids and glycerol. After being absorbed, these products are transported in lymph to the blood, then on to tissues. Figure 15.31 shows that some of the fatty acid portions can react to form molecules of acetyl coenzyme A by a series of reactions called **beta oxidation** (ba′tah ok″sĭ-da′shun). Excess acetyl coenzyme A can be converted into compounds called *ketone bodies,* such as acetone, which later may be changed back to acetyl coenzyme A. In either case, the resulting acetyl coenzyme A can be oxidized in the citric acid cycle. The glycerol parts of the triglyceride

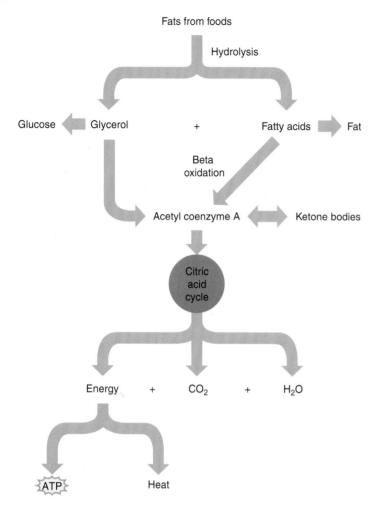

Figure 15.31

The body digests fats from foods into glycerol and fatty acids, which may enter catabolic pathways and provide energy.

molecules can also enter metabolic pathways leading to the citric acid cycle, or they can be used to synthesize glucose. Fatty acid molecules released from fat hydrolysis can combine to form fat molecules and be stored in adipose tissue.

The liver can convert fatty acids from one form to another, but it cannot synthesize certain fatty acids, called **essential fatty acids.** *Linoleic acid,* for example, is required for phospholipid synthesis, which in turn is necessary for constructing cell membranes and transporting circulating lipids. Good sources of linoleic acid include corn, cottonseed, and soy oils. Another essential fatty acid is *linolenic acid.*

The liver regulates circulating lipids by using free fatty acids to synthesize triglycerides, phospholipids, and lipoproteins that may then be released into the bloodstream. Because lipids are less dense than proteins, as the proportion of lipids in a lipoprotein increases, the density of the particle decreases. Conversely, as the pro-

portion of lipids decreases, the density increases. Lipoproteins are classified on the basis of their densities, which reflect their composition. *Very-low-density lipoproteins* (VLDL) have a relatively high concentration of triglycerides. *Low-density lipoproteins* (LDL) have a relatively high concentration of cholesterol and are the major cholesterol-carrying lipoproteins. *High-density lipoproteins* (HDL) have a relatively high concentration of protein and a lower concentration of lipids.

In addition to regulating circulating lipids, the liver controls the total amount of cholesterol in the body by synthesizing cholesterol and releasing it into the blood, or by removing cholesterol from the blood and excreting it into bile. The liver also uses cholesterol to produce bile salts. Cholesterol is not an energy source. It provides structural material for cell and organelle membranes and it furnishes starting material for the synthesis of certain sex hormones and adrenal cortex hormones.

Adipose tissue stores excess triglycerides. If the blood lipid concentration drops (in response to fasting, for example), some of these triglycerides are hydrolyzed into free fatty acids and glycerol, and then released into the bloodstream.

Lipid Requirements

The amounts and types of fats required for health vary with individuals' habits and goals. Because linoleic acid is an essential fatty acid, nutritionists recommend that formula-fed infants receive 3% of their energy intake in the form of linoleic acid to prevent deficiency conditions. Fat intake must be sufficient to carry fat-soluble vitamins. Lipids provide flavor to food, which is one reason why adhering to a very low-fat diet is difficult. The USDA and American Heart Association recommend that lipid intake not exceed 30% of calories.

Practice

54. Which fatty acids are essential nutrients?

55. What is the liver's role in the use of lipids?

56. What are the functions of cholesterol?

Proteins

Proteins are polymers of amino acids. They are involved in a wide variety of functions.

Protein Sources

Foods rich in proteins include meats, fish, poultry, cheese, nuts, milk, eggs, and cereals. Legumes, including beans and peas, contain less protein. The cells of an adult can synthesize all but eight required amino acids, and the cells of a child can produce all but ten. Amino acids that the body can synthesize are termed nonessential; those that it cannot synthesize and must be obtained in the diet are **essential amino acids.** Table 15.7 lists the amino acids in foods and indicates those that are essential.

All twenty types of amino acids must be present in the body at the same time for growth and tissue repair to occur. If just one type of essential amino acid is missing from the diet, normal protein synthesis cannot take place, because many proteins include all twenty types of amino acids.

Proteins are classified as complete or incomplete on the basis of the amino acid types they provide. **Complete proteins,** such as those in milk, meats, and eggs, have adequate amounts of the essential amino acids. **Incomplete proteins,** such as *zein* in corn, which has too little of the essential amino acids tryptophan and lysine, are unable by themselves to maintain human tissues or to support normal growth and development. A protein in wheat called *gliadin* is a **partially complete protein** because it has very little of the amino acid lysine.

> Many plant proteins have too little of one or more essential amino acids to provide adequate nutrition for a person. However, combining appropriate plant foods can provide an adequate diversity of dietary amino acids. For example, beans are low in methionine but have enough lysine. Rice lacks lysine but has enough methionine. A meal of beans and rice provides enough of both types of amino acids.

Protein Use

Proteins include enzymes that control metabolic rates, clotting factors, the keratin of skin and hair, elastin and collagen of connective tissue, plasma proteins that regulate water balance, the muscle components actin and myosin, certain hormones, and the antibodies that protect against infection.

Proteins may also supply energy after digestion breaks them down into amino acids. The liberated amino acids are transported to the liver, where they undergo *deamination,* losing their nitrogen-containing ($-NH_2$) groups (see fig. 2.17, p. 44). These $-NH_2$ groups subsequently react to form the waste *urea* (u-re′ah), which is excreted in urine.

Depending upon the particular amino acids involved, the remaining deaminated parts are decomposed in one of several pathways (fig. 15.32). Some of these pathways lead to formation of acetyl coenzyme A, and others lead more directly to the steps of the citric acid cycle. As energy is released from the cycle, some of it is captured in ATP molecules. If energy is not required immediately, the deaminated parts of the amino acids may react to form glucose or fat molecules in other metabolic pathways.

Figure 15.32

The body digests proteins from foods into amino acids, but must deaminate these smaller molecules before they can be used as energy sources.

Table 15.7	Amino Acids in Foods
Alanine	Leucine (e)
Arginine (ch)	Lysine (e)
Asparagine	Methionine (e)
Aspartic acid	Phenylalanine (e)
Cysteine	Proline
Glutamic acid	Serine
Glutamine	Threonine (e)
Glycine	Tryptophan (e)
Histidine (ch)	Tyrosine
Isoleucine (e)	Valine (e)

Eight essential amino acids (e) cannot be synthesized by human cells and must be provided in the diet. Two additional amino acids (ch) are essential in growing children.

Protein Requirements

Proteins supply essential amino acids. They also provide nitrogen and other elements for the synthesis of nonessential amino acids and certain nonprotein nitrogenous substances. The amount of dietary protein individuals require varies according to body size, metabolic rate, and other factors, such as activity level. Bodybuilders, for example, require more protein to help heal small muscle tears that result from weight lifting.

For an average adult, nutritionists recommend a daily protein intake of about 0.8 grams per kilogram of body weight. Another way to estimate desirable protein intake is to divide weight in pounds by two. Most people should consume 60 to 150 grams of protein a day. For a pregnant woman, the recommendation adds 30 grams of protein per day. Similarly, a nursing mother requires an additional 20 grams of protein per day to maintain a high level of milk production.

Practice

57. Which foods are rich sources of proteins?

58. Why are some amino acids called essential?

59. Distinguish between complete and incomplete proteins.

60. List some examples of proteins associated with the body.

61. How does dietary protein provide energy?

Vitamins

Vitamins (vi′tah-minz) are organic compounds (other than carbohydrates, lipids, and proteins) required in small amounts for normal metabolism, that cells cannot synthesize in adequate amounts. Therefore, they are essential nutrients that must come from foods.

Vitamins are classified on the basis of solubility. Some are soluble in fats (or fat solvents) and others are soluble in water. *Fat-soluble vitamins* are vitamins A, D, E, and K; *water-soluble vitamins* are the B vitamins and vitamin C.

Fat-Soluble Vitamins

Fat-soluble vitamins dissolve in fats, and therefore they associate with lipids and respond to the same factors that affect lipid absorption. For example, bile salts in the intestine promote absorption of these vitamins. Fat-soluble vitamins accumulate in various tissues, which is why excess intake can lead to overdose conditions. For example, too much beta carotene, a vitamin A precursor, can tinge the skin orange. Fat-soluble vitamins resist the effects of heat; therefore, cooking and food processing usually do not destroy them. Table 15.8 lists the fat-soluble vitamins and their characteristics, functions, sources, and recommended daily allowances (RDAs) for adults.

Water-Soluble Vitamins

The water-soluble vitamins include the B vitamins and vitamin C. The **B vitamins** are several compounds that are essential for normal cellular metabolism. They help oxidize carbohydrates, lipids, and proteins. The B vitamins are in many of the same foods, so they are referred to as the *vitamin B complex*. Members of this group differ chemically and functionally. Cooking and food processing destroy some of them.

Vitamin C (ascorbic acid) is one of the least stable vitamins and is fairly widespread in plant foods. It is necessary for collagen production, the conversion of folacin to folinic acid, and the metabolism of certain

Table 15.8	Fat-Soluble Vitamins		
Vitamin	**Characteristics**	**Functions**	**Sources and RDA* for Adults**
Vitamin A	Exists in several forms; synthesized from carotenes; stored in liver; stable in heat, acids, and bases; unstable in light	An antioxidant necessary for synthesis of visual pigments, mucoproteins, and mucopolysaccharides; for normal development of bones and teeth; and for maintenance of epithelial cells	Liver, fish, whole milk, butter, eggs, leafy green vegetables, yellow and orange vegetables and fruits; RDA = 4,000–5,000 IU[†]
Vitamin D	A group of steroids; resistant to heat, oxidation, acids, and bases; stored in liver, skin, brain, spleen, and bones	Promotes absorption of calcium and phosphorus; promotes development of teeth and bones	Produced in skin exposed to ultraviolet light; in milk, egg yolk, fish liver oils, fortified foods; RDA = 400 IU
Vitamin E	A group of compounds; resistant to heat and visible light; unstable in presence of oxygen and ultraviolet light; stored in muscles and adipose tissue	An antioxidant; prevents oxidation of vitamin A and polyunsaturated fatty acids; may help maintain stability of cell membranes	Oils from cereal seeds, salad oils, margarine, shortenings, fruits, nuts, and vegetables; RDA = 30 IU
Vitamin K	Exists in several forms; resistant to heat, but destroyed by acids, bases, and light; stored in liver	Required for synthesis of prothrombin, which functions in blood clotting	Leafy green vegetables, egg yolk, pork liver, soy oil, tomatoes, cauliflower; RDA = 55–70 μg

*RDA = recommended daily allowance.

[†]IU = international unit.

amino acids. Vitamin C also promotes iron absorption and synthesis of certain hormones from cholesterol. Table 15.9 lists the water-soluble vitamins and their characteristics, functions, sources, and RDAs for adults.

> Sailors on English ships ate limes to protect them from scurvy (vitamin C deficiency). American ships carried cranberries for the same purpose.

Practice

62. What are vitamins?

63. How are vitamins classified?

64. How do bile salts affect the absorption of fat-soluble vitamins?

65. List the fat-soluble and water-soluble vitamins.

Minerals

Dietary **minerals** (min′er-alz) are inorganic elements essential in human metabolism. Plants usually extract these elements from soil, and humans obtain them from plant foods or from animals that have eaten plants.

Characteristics of Minerals

Minerals contribute about 4% of body weight and are most concentrated in the bones and teeth. Minerals are usually incorporated into organic molecules. For example, phosphorus is found in phospholipids, iron in hemoglobin, and iodine in thyroxine. However, some minerals are part of inorganic compounds, such as the calcium phosphate of bone. Other minerals are free ions, such as sodium, chloride, and calcium ions in blood.

Minerals are parts of the structural materials of all body cells. They also constitute parts of enzyme molecules, contribute to the osmotic pressure of body fluids,

Table 15.9	Water-Soluble Vitamins		
Vitamin	**Characteristics**	**Functions**	**Sources and RDA* for Adults**
Thiamine (vitamin B$_1$)	Destroyed by heat and oxygen, especially in alkaline environment	Part of coenzyme required to oxidize carbohydrates; coenzyme required for ribose synthesis	Lean meats, liver, eggs, whole-grain cereals, leafy green vegetables, legumes; RDA =1.5 mg
Riboflavin (vitamin B$_2$)	Stable to heat, acids, and oxidation; destroyed by bases and ultraviolet light	Part of enzymes and coenzymes required for oxidation of glucose and fatty acids and for cellular growth	Meats, dairy products, leafy green vegetables, whole-grain cereals; RDA = 1.7 mg
Niacin (nicotinic acid) (vitamin B$_3$)	Stable to heat, acids, and bases; converted to niacinamide by cells; synthesized from tryptophan	Part of coenzymes required for oxidation of glucose and synthesis of proteins, fats, and nucleic acids	Liver, lean meats, peanuts, legumes; RDA = 20 mg
Pantothenic acid (vitamin B$_5$)	Destroyed by heat, acids, and bases	Part of coenzyme A required for oxidation of carbohydrates and fats	Meats, whole-grain cereals, legumes, milk, fruits, vegetables; RDA = 10 mg
Vitamin B$_6$	Group of three compounds; stable to heat and acids; destroyed by oxidation, bases, and ultraviolet light	Coenzyme required for synthesis of proteins and certain amino acids, for conversion of tryptophan to niacin, for production of antibodies, and for nucleic acid synthesis	Liver, meats, bananas, avocados, beans, peanuts, whole-grain cereals, egg yolk; RDA = 2 mg
Cyanocobalamin (vitamin B$_{12}$)	Complex, cobalt-containing compound; stable to heat; inactivated by light, strong acids, and strong bases; absorption regulated by intrinsic factor from gastric glands; stored in liver	Part of coenzyme required for synthesis of nucleic acids and for metabolism of carbohydrates; plays role in myelin synthesis; needed for normal red blood cell production	Liver, meats, milk, cheese, eggs; RDA = 3–6 μg
Folacin (folic acid)	Occurs in several forms; destroyed by oxidation in acid environment or by heat in alkaline environment; stored in liver, where it is converted into folinic acid	Coenzyme required for metabolism of certain amino acids and for DNA synthesis; promotes production of normal red blood cells	Liver, leafy green vegetables, whole-grain cereals, legumes; RDA = 0.4 mg
Biotin	Stable to heat, acids, and light; destroyed by oxidation and bases	Coenzyme required for metabolism of amino acids and fatty acids, and for nucleic acid synthesis	Liver, egg yolk, nuts, legumes, mushrooms; RDA = 0.3 mg
Ascorbic acid (vitamin C)	Chemically similar to monosaccharides; stable in acids but destroyed by oxidation, heat, light, and bases	Required for collagen production, conversion of folacin to folinic acid, and metabolism of certain amino acids; promotes absorption of iron and synthesis of hormones from cholesterol	Citrus fruits, tomatoes, potatoes, leafy green vegetables; RDA = 60 mg

*RDA = recommended daily allowance.

and play vital roles in nerve impulse conduction, muscle fiber contraction, blood coagulation, and maintenance of the pH of body fluids.

Major Minerals

Calcium and *phosphorus* account for nearly 75% by weight of the mineral elements in the body. Therefore, they are termed **major minerals.** Other major minerals, each of which accounts for 0.05% or more of the body weight, include potassium, sulfur, sodium, chlorine, and magnesium. Table 15.10 lists the distribution, functions, sources, and adult RDAs of major minerals.

> The human body has enough phosphorus to make 2,000 match tips.

Trace Elements

Trace elements are essential minerals found in minute amounts, each making up less than 0.005% of adult body weight. They include iron, manganese, copper, iodine, cobalt, zinc, fluorine, selenium, and chromium. Table 15.11 lists the distribution, functions, sources, and adult RDAs of the trace elements.

> A human body has enough iron to make a small nail.

Practice

66. What are minerals?

67. What are the major functions of minerals?

68. Distinguish between a major mineral and a trace element.

69. Name the major minerals and trace elements.

Adequate Diets

An adequate diet provides sufficient energy (calories), essential fatty acids, essential amino acids, vitamins, and minerals to support optimal growth and to maintain and repair body tissues. Individual requirements for nutrients vary greatly with age, sex, growth rate, physical activity, and level of stress, as well as with genetic and other environmental factors. Therefore, designing a diet that is adequate for everyone is impossible. Diagrams called food guide pyramids are used to organize foods according to suggested relative amounts, often in serving sizes. Figure 15.33 depicts one such food pyramid, developed by the U.S. Department of Agriculture.

Table 15.10	Major Minerals		
Mineral	**Distribution**	**Functions**	**Sources and RDA* for Adults**
Calcium (Ca)	Mostly in the inorganic salts of bones and teeth	Structure of bones and teeth; essential for neurotransmitter release, muscle fiber contraction, and blood coagulation; increases permeability of cell membranes; activates certain enzymes	Milk, milk products, leafy green vegetables; RDA = 800 mg
Phosphorus (P)	Mostly in the inorganic salts of bones and teeth	Structure of bones and teeth; component in nearly all metabolic reactions; in nucleic acids, many proteins, some enzymes, and some vitamins; in cell membrane, ATP, and phosphates of body fluids	Meats, cheese, nuts, whole-grain cereals, milk, legumes; RDA = 800 mg
Potassium (K)	Widely distributed; tends to be concentrated inside cells	Helps maintain intracellular osmotic pressure and regulate pH; required for nerve impulse conduction	Avocados, dried apricots, meats, nuts, potatoes, bananas; RDA = 2,500 mg
Sulfur (S)	Widely distributed; abundant in skin, hair, and nails	Essential part of various amino acids, thiamine, insulin, biotin, and mucopolysaccharides	Meats, milk, eggs, legumes; No RDA established
Sodium (Na)	Widely distributed; mostly in extracellular fluids and bound to inorganic salts of bone	Helps maintain osmotic pressure of extracellular fluids; regulates water movement; plays a role in nerve impulse conduction; regulates pH and transport of substances across cell membranes	Table salt, cured ham, sauerkraut, cheese, graham crackers; RDA = 2,500 mg
Chlorine (Cl)	Closely associated with sodium (as chloride); most highly concentrated in cerebrospinal fluid and gastric juice	Helps maintain osmotic pressure of extracellular fluids; regulates pH; maintains electrolyte balance; forms hydrochloric acid; aids transport of carbon dioxide by red blood cells	Same as for sodium; No RDA established
Magnesium (Mg)	Abundant in bones	Required in metabolic reactions in mitochondria that produce ATP; plays a role in the breakdown of ATP to ADP	Milk, dairy products, legumes, nuts, leafy green vegetables; RDA = 300–350 mg

*RDA = recommended daily allowance.

Table 15.11	Trace Elements		
Trace Element	**Distribution**	**Functions**	**Sources and RDA* for Adults**
Iron (Fe)	Primarily in blood; stored in liver, spleen, and bone marrow	Part of hemoglobin molecule; catalyzes vitamin A formation; incorporated into a number of enzymes	Liver, lean meats, dried apricots, raisins, enriched whole-grain cereals, legumes, molasses; RDA = 10–18 mg
Manganese (Mn)	Most concentrated in liver, kidneys, and pancreas	Activates enzymes required for fatty acid and cholesterol synthesis, urea formation, and normal functioning of the nervous system	Nuts, legumes, whole-grain cereals, leafy green vegetables, fruits; RDA = 2.5–5 mg
Copper (Cu)	Most highly concentrated in liver, heart, and brain	Essential for hemoglobin synthesis, bone development, melanin production, and myelin formation	Liver, oysters, crabmeat, nuts, whole-grain cereals, legumes; RDA = 2–3 mg
Iodine (I)	Concentrated in thyroid gland	Essential component for synthesis of thyroid hormones	Food content varies with soil content in different geographic regions; iodized table salt; RDA = 0.15 mg
Cobalt (Co)	Widely distributed	Component of cyanocobalamin; needed for synthesis of several enzymes	Liver, lean meats, milk; No RDA established
Zinc (Zn)	Most concentrated in liver, kidneys, and brain	Component of enzymes involved in digestion, respiration, bone metabolism, liver metabolism; necessary for normal wound healing and maintaining skin integrity	Meats, cereals, legumes, nuts, vegetables; RDA = 15 mg
Fluorine (F)	Primarily in bones and teeth	Component of tooth structure (enamel)	Fluoridated water; RDA = 1.5–4.0 mg
Selenium (Se)	Concentrated in liver and kidneys	Component of certain enzymes	Lean meats, fish, cereals; RDA = 0.05–2.00 mg
Chromium (Cr)	Widely distributed	Essential for use of carbohydrates	Liver, lean meats, wine; RDA = 0.05–2.00 mg

*RDA = recommended daily allowance.

Activity
Activity is represented by the steps and the person climbing them, as a reminder of the importance of daily physical activity.

Moderation
Moderation is represented by the narrowing of each food group from bottom to top. The wider base stands for foods with little or no solid fats or added sugars. These should be selected more often. The narrower top area stands for foods containing more added sugars and solid fats. The more active you are, the more of these foods can fit into your diet.

Personalization
Personalization is shown by the person on the steps and the slogan.

Proportionality
Proportionality is shown by the different widths of the food group bands. The widths suggest how much food a person should choose from each group. The widths are just a general guide, not exact proportions.

Variety
Variety is symbolized by the 6 color bands representing the 5 food groups of the Pyramid and oils. This illustrates that foods from all groups are needed each day for good health.

Gradual Improvement
Gradual improvement is encouraged by the slogan. It suggests that individuals can benefit from taking small steps to improve their diet and lifestyle each day.

MyPyramid.gov
STEPS TO A HEALTHIER YOU

USDA U.S. Department of Agriculture
Center for Nutrition Policy and Promotion
April 2005 CNPP-16
USDA is an equal opportunity provider and employer.

GRAINS | VEGETABLES | FRUITS | OILS | MILK | MEAT & BEANS

Figure 15.33
The USDA food pyramid symbolizes an individual approach to healthy eating and physical exercise.

Clinical Application 15.4

Eating Extremes: Undereating and Overeating

In the United States, many people are either underweight or overweight. Two-thirds of people over the age of twenty are overweight or obese, yet fashion models and celebrities grow ever more skeletal.

Being thin is not a disease, but abnormal eating behavior may be. Anorexia nervosa is self-imposed starvation. The sufferer perceives herself or himself as overweight and eats barely enough to survive, losing as much as 25% of her or his body weight. Anorexia leads to low blood pressure, slowed or irregular heartbeat, constipation, and constant chills. In the female, menstruation may stop as body fat level plunges. Hair becomes brittle, the skin dries out, and soft, pale, fine body hair called lanugo, normally seen only on a fetus, grows to preserve body heat. A person with anorexia may be hospitalized so that intravenous feeding can prevent sudden death from heart failure due to an electrolyte imbalance. Psychotherapy and nutritional counseling may help identify and remedy the underlying cause of the abnormal eating behavior. Despite these interventions, 15–21% of people with anorexia die from the disease.

In bulimia, a person eats large amounts of food, and then gets rid of the thousands of extra calories by vomiting, taking laxatives, or exercising frantically. The binge-and-purge cycle is very hard to break, even with psychotherapy and nutritional counseling.

Because body weight reflects energy balance, excess food means, ultimately, excess weight. Being overweight or obese raises the risk of developing hypertension, diabetes, stroke, gallstones, sleep apnea, and certain cancers. The body strains to support the extra weight—miles of blood vessels are required to nourish additional body mass. Usually, being overweight stems from overeating and inactivity.

Weight loss requires eating less and exercising more. A safe goal is to lose one pound of fat per week. A pound of fat contains 3,500 calories of energy. Shedding that pound therefore requires eating 500 fewer calories per day or exercising off 500 calories each day, or some other combination. Certain weight-loss drugs inhibit the function of pancreatic lipase, preventing the digestion and absorption of about a third of dietary fat.

For people with BMIs above 40, or above 35 in addition to an obesity-related disorder, bariatric surgery can lead to great weight loss. Two types of procedures are done. In laparoscopic adjustable gastric banding, a silicone band ties off part of the stomach, limiting its capacity to hold food. The band can be inflated or deflated in a doctor's office by adding or removing saline. The second type of bariatric surgery is gastric bypass, in which part of the stomach is stapled shut, forming a pouch that is surgically connected to the jejunum, bypassing the duodenum. Both procedures lead to decreased hunger, very reduced food intake, and some decrease in the absorption of nutrients. A special diet, liquid at first, must be followed. Many patients who have had bariatric surgery report improvement in or disappearance of type 2 diabetes, back pain, arthritis, varicose veins, sleep apnea, and hypertension. However, now that some people have lived many years following bariatric surgery, some of them have regained the weight by gradually increasing their food intake yet again.

Malnutrition (mal″nu-trish′un) is poor nutrition that results from a lack of nutrients or a failure to use them. This condition may be due to either *undernutrition,* producing the symptoms of deficiency diseases, or to *overnutrition,* arising from excess nutrient intake.

A variety of factors can lead to malnutrition. For example, a deficiency condition may stem from lack of food or from poor-quality food. On the other hand, malnutrition may result from overeating or from taking too many vitamin supplements.

A measurement called the body mass index (BMI) is used to determine whether a person is of adequate weight, overweight, or obese. To calculate your BMI, divide your weight in kilograms (a kilogram equals 2.2 pounds) by your height in meters squared (one foot equals about .3 meters). Figure 15.34 interprets the BMI. Clinical Application 15.4 discusses the effects of undereating and overeating.

Practice

70. What is an adequate diet?

71. Which factors influence individual needs for nutrients?

72. What causes malnutrition?

Weight in pounds

Developed by the National Center for Health Statistics in collaboration with the
National Center for Chronic Disease Prevention and Health Promotion

Figure 15.34

Body mass index (BMI). BMI equals weight/height^2, with weight measured in kilograms and height measured in meters. This chart provides a shortcut—the calculations have been done and converted to the English system of measurement. The uncolored squares indicate lower than healthy weight according to this index.

Q: *What is your body mass index? Based on if you are lower than healthy weight, healthy weight, overweight, or obese, what might you do with regards to your diet and activity level to achieve/maintain healthy weight?*

Answer can be found in Appendix E on page 568.

Summary Outline

15.1 Introduction (p. 401)

Digestion mechanically and chemically breaks down foods and absorbs the products. The digestive system consists of an alimentary canal and several accessory organs.

15.2 General Characteristics of the Alimentary Canal (p. 401)

Regions of the alimentary canal perform specific functions.

1. Structure of the wall
   The wall consists of four layers—the mucosa, submucosa, muscular layer, and serosa.
2. Movements of the tube
   Motor functions include mixing and propelling movements.

15.3 Mouth (p. 403)

The mouth receives food and begins digestion.

1. Cheeks and lips
   a. Cheeks consist of outer layers of skin, pads of fat, muscles associated with expression and chewing, and inner linings of epithelium.
   b. Lips are highly mobile and have sensory receptors.

2. Tongue
   a. The tongue's rough surface handles food and has taste buds.
   b. Lingual tonsils are on the root of the tongue.
3. Palate
   a. The palate includes hard and soft portions.
   b. Palatine tonsils are located on either side of the tongue in the back of the mouth.
4. Teeth
   a. There are twenty primary and thirty-two secondary teeth.
   b. Teeth begin mechanical digestion by breaking food into smaller pieces, increasing the surface area exposed to digestive actions.
   c. Each tooth consists of a crown and root, and is composed of enamel, dentin, pulp, nerves, and blood vessels.
   d. A periodontal ligament attaches a tooth to the alveolar process.

15.4 Salivary Glands (p. 408)

Salivary glands secrete saliva, which moistens food, helps bind food particles, begins chemical digestion of carbohydrates, makes taste possible, and helps cleanse the mouth.

1. Salivary secretions
   Salivary glands include serous cells that secrete digestive enzymes and mucous cells that secrete mucus.
2. Major salivary glands
   a. The parotid glands secrete saliva rich in amylase.
   b. The submandibular glands produce viscous saliva.
   c. The sublingual glands primarily secrete mucus.

15.5 Pharynx and Esophagus (p. 408)

The pharynx and esophagus are important passageways.

1. Structure of the pharynx
   The pharynx is divided into a nasopharynx, oropharynx, and laryngopharynx.
2. Swallowing mechanism
   Swallowing occurs in three stages:
   a. Food is mixed with saliva and forced into the oropharynx.
   b. Involuntary reflex actions move the food into the esophagus.
   c. Peristalsis transports food to the stomach.
3. Esophagus
   a. The esophagus passes through the diaphragm and joins the stomach.
   b. Circular muscle fibers at the distal end of the esophagus help prevent regurgitation of food from the stomach.

15.6 Stomach (p. 410)

The stomach receives food, mixes it with gastric juice, carries on a limited amount of absorption, and moves food into the small intestine.

1. Parts of the stomach
   a. The stomach is divided into cardia, fundus, body, and pylorus.
   b. The pyloric sphincter is a valve between the stomach and small intestine.
2. Gastric secretions
   a. Gastric glands secrete gastric juice.
   b. Gastric juice contains pepsin (begins chemical digestion of proteins), hydrochloric acid, and intrinsic factor.
3. Regulation of gastric secretions
   a. Parasympathetic impulses and the hormone gastrin enhance gastric secretion.
   b. Food in the small intestine reflexly inhibits gastric secretions.
4. Gastric absorption
   The stomach wall may absorb a few substances, such as water and other small molecules.

Digestive System

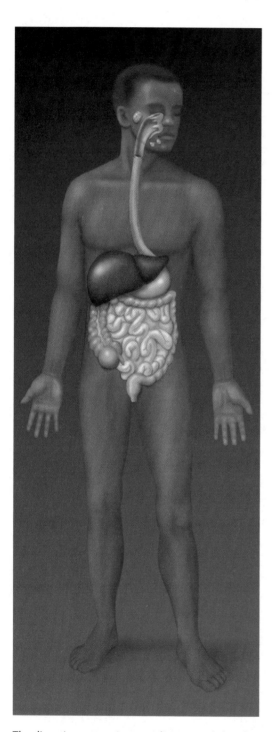

The digestive system ingests, digests, and absorbs nutrients for use by all body cells.

Integumentary System

Vitamin D activated in the skin plays a role in absorption of calcium from the digestive tract.

Cardiovascular System

The bloodstream carries absorbed nutrients to all body cells.

Skeletal System

Bones are important in mastication. Calcium absorption is necessary to maintain bone matrix.

Lymphatic System

The lymphatic system plays a major role in the absorption of fats.

Muscular System

Muscles are important in mastication, swallowing, and the mixing and moving of digestion products through the gastrointestinal tract.

Respiratory System

The digestive system and the respiratory system share common anatomical structures.

Nervous System

The nervous system can influence digestive system activity.

Urinary System

The kidneys and liver work together to activate vitamin D.

Endocrine System

Hormones can influence digestive system activity.

Reproductive System

In a woman, nutrition is essential for conception and normal development of an embryo and fetus.

5. Mixing and emptying actions
   a. Mixing movements help produce chyme. Peristaltic waves move chyme into the pylorus.
   b. The muscular wall of the pylorus regulates chyme movement into the small intestine.
   c. The rate of emptying depends on the fluidity of chyme and the type of food present.

15.7 Pancreas (p. 413)

1. Structure of the pancreas
   a. The pancreas produces pancreatic juice that is secreted into a pancreatic duct.
   b. The pancreatic duct leads to the duodenum.
2. Pancreatic juice
   a. Pancreatic juice contains enzymes that can split carbohydrates, fats, nucleic acids, and proteins.
   b. Pancreatic juice has a high bicarbonate ion concentration that helps neutralize chyme and causes intestinal contents to be alkaline.
3. Hormones regulate pancreatic secretion
   a. Secretin stimulates the release of pancreatic juice with a high bicarbonate ion concentration.
   b. Cholecystokinin stimulates the release of pancreatic juice with a high concentration of digestive enzymes.

15.8 Liver (p. 415)

1. Liver structure
   a. The lobes of the liver consist of hepatic lobules, the functional units of the gland.
   b. Bile canals carry bile from hepatic lobules to hepatic ducts.
2. Liver functions
   a. The liver metabolizes carbohydrates, lipids, and proteins; stores some substances; filters blood; destroys toxins; and secretes bile.
   b. Bile is the only liver secretion that directly affects digestion.
3. Composition of bile
   a. Bile contains bile salts, bile pigments, cholesterol, and electrolytes.
   b. Only the bile salts have digestive functions.
4. Gallbladder
   a. The gallbladder stores bile between meals.
   b. A sphincter muscle controls release of bile from the bile duct.
5. Regulation of bile release
   a. Cholecystokinin from the small intestine stimulates bile release.
   b. The sphincter muscle at the base of the bile duct relaxes as a peristaltic wave in the duodenal wall approaches.
6. Functions of bile salts
   Bile salts emulsify fats and aid in the absorption of fatty acids, cholesterol, and certain vitamins.

15.9 Small Intestine (p. 420)

The small intestine receives secretions from the pancreas and liver, completes nutrient digestion, absorbs the products of digestion, and transports the residues to the large intestine.

1. Parts of the small intestine
   The small intestine consists of the duodenum, jejunum, and ileum.
2. Structure of the small intestinal wall
   a. The wall is lined with villi that greatly increase the surface area and aid in mixing and absorption.
   b. Intestinal glands are located between the villi.
3. Secretions of the small intestine
   a. Secretions include mucus and digestive enzymes.
   b. Digestive enzymes embedded in the surfaces of microvilli split molecules of sugars, proteins, and fats into simpler forms.

4. Regulation of small intestine secretions
   Secretion is stimulated by gastric juice, chyme, and reflexes stimulated by distension of the small intestinal wall.
5. Absorption in the small intestine
   a. Blood capillaries in the villi absorb monosaccharides and amino acids.
   b. Blood capillaries in the villi also absorb water and electrolytes.
   c. Fat molecules with longer chains of carbon atoms enter the lacteals of the villi.
   d. Fatty acids with relatively short carbon chains enter blood capillaries of the villi.
6. Movements of the small intestine
   a. Movements include mixing by segmentation and peristalsis.
   b. The ileocecal sphincter controls movement of the intestinal contents from the small intestine into the large intestine.

15.10 Large Intestine (p. 424)

The large intestine reabsorbs water and electrolytes, and forms and stores feces.

1. Parts of the large intestine
   a. The large intestine consists of the cecum, colon, rectum, and anal canal.
   b. The colon is divided into ascending, transverse, descending, and sigmoid portions.
2. Structure of the large intestinal wall
   a. The large intestinal wall resembles the wall in other parts of the alimentary canal.
   b. The large intestinal wall has a unique layer of longitudinal muscle fibers arranged in distinct bands.
3. Functions of the large intestine
   a. The large intestine has little or no digestive function.
   b. It secretes mucus.
   c. The large intestine absorbs water and electrolytes.
   d. The large intestine forms and stores feces.
4. Movements of the large intestine
   a. Movements are similar to those in the small intestine.
   b. Mass movements occur two to three times each day.
   c. A reflex stimulates defecation.
5. Feces
   a. Feces consist of water, undigested material, electrolytes, mucus, and bacteria.
   b. The color of feces is due to bile pigments that have been altered by bacterial actions.

15.11 Nutrition and Nutrients (p. 428)

Nutrition is the study of nutrients and how the body utilizes them. The macronutrients (carbohydrates, lipids, and proteins) are required in large amounts. The micronutrients (vitamins and minerals) are required in smaller amounts. Calories measure potential energy in foods.

1. Carbohydrates
   a. Carbohydrate sources
      (1) Starch, glycogen, disaccharides, and monosaccharides are carbohydrates.
      (2) Cellulose is a polysaccharide that human enzymes cannot digest.
   b. Carbohydrate use
      (1) Oxidation releases energy from glucose.
      (2) Excess glucose is stored as glycogen or combined to produce fat.
      (3) Carbohydrates supply energy and are also part of nucleic acids and milk.

c. Carbohydrate requirements
  (1) Humans survive with a wide range of carbohydrate intakes.
  (2) Excess carbohydrates may lead to weight gain.
2. Lipids
  a. Lipid sources
    (1) Foods of plant and animal origin provide triglycerides.
    (2) Foods of animal origin provide dietary cholesterol.
  b. Lipid use
    (1) The liver and adipose tissue control triglyceride metabolism.
    (2) Linoleic acid and linolenic acid are essential fatty acids.
    (3) Lipids supply energy and are used to build membranes and steroid hormones.
  c. Lipid requirements
    (1) The amounts and types of lipids needed for health are unknown.
    (2) Fat intake must be sufficient to carry fat-soluble vitamins.
3. Proteins
  a. Protein sources
    (1) Proteins are mainly obtained from meats, dairy products, cereals, and legumes.
    (2) Complete proteins contain adequate amounts of all the essential amino acids.
    (3) Incomplete proteins lack adequate amounts of one or more essential amino acids.
  b. Protein use
    Proteins serve as structural materials, function as enzymes, and provide energy.
  c. Protein requirements
    Proteins and amino acids must supply essential amino acids and nitrogen for the synthesis of nitrogen-containing molecules.

4. Vitamins
  a. Fat-soluble vitamins
    (1) These include vitamins A, D, E, and K.
    (2) They are carried in lipids and are influenced by the same factors that affect lipid absorption.
    (3) They resist the effects of heat; thus, they are not destroyed by cooking or food processing.
  b. Water-soluble vitamins
    (1) This group includes the B vitamins and vitamin C.
    (2) B vitamins make up a group (the vitamin B complex) and oxidize carbohydrates, lipids, and proteins.
    (3) Cooking or processing food destroys some water-soluble vitamins.
5. Minerals
  a. Characteristics of minerals
    (1) Most minerals are in the bones and teeth.
    (2) Minerals are usually incorporated into organic molecules; some occur in inorganic compounds or as free ions.
    (3) They serve as structural materials, function in enzymes, and play vital roles in metabolism.
  b. Major minerals include calcium, phosphorus, potassium, sulfur, sodium, chlorine, and magnesium.
  c. Trace elements include iron, manganese, copper, iodine, cobalt, zinc, fluorine, selenium, and chromium.
6. Adequate diets
  a. An adequate diet provides sufficient energy and essential nutrients to support optimal growth, maintenance, and repair of tissues.
  b. Individual requirements vary so greatly that designing a diet that is adequate for everyone is not possible. Food guide pyramids can help to personalize diets.
  c. Malnutrition is poor nutrition due to lack of food or failure to use available food.

Chapter Assessments

15.1 Introduction

1. Functions of the digestive system include (p. 401)
   a. mechanical breakdown of foods.
   b. chemical breakdown of foods.
   c. breaking large pieces into smaller ones without altering their chemical composition.
   d. breaking food into simpler chemicals.
   e. all of the above

2. List the major parts of the alimentary canal, then separately list the accessory organs of the digestive system. (p. 401)

15.2 General Characteristics of the Alimentary Canal

3. Contrast the composition of the layers of the wall of the alimentary canal. (p. 401)

4. Distinguish between mixing and propelling movements. (p. 403)

15.3 Mouth

5. Discuss the functions of the mouth and its parts. (p. 403)

6. Distinguish between primary and secondary teeth. (p. 405)

7. The teeth that are best adapted for grasping and tearing food are the (p. 407)
   a. incisors.
   b. canines.
   c. premolars.
   d. molars.
   e. bicuspids.

8. Describe the structure of a tooth. (p. 407)

15.4–15.10 Salivary Glands–Large Intestine

9. Match the organ or gland with the enzyme(s) it secretes. Enzymes may be used more than once. An organ or gland may secrete more than one enzyme. (pp. 408–422)

(1) salivary glands (serous cells)
(2) stomach (chief cells)
(3) pancreas (acinar cells)
(4) small intestine (mucosal cells)

A. peptidase
B. amylase
C. nuclease
D. lipase
E. pepsin
F. trypsin, chymotrypsin, carboxypeptidase
G. sucrase, maltase, lactase

10. Match the enzyme(s) with its (their) function(s). (pp. 408–422)

(1) peptidase
(2) amylase
(3) nuclease
(4) lipase
(5) pepsin
(6) trypsin, chymotrypsin, carboxypeptidase
(7) sucrase, maltase, lactase

A. Begins protein digestion
B. Breaks fats into fatty acids and glycerol
C. Breaks down proteins into peptides
D. Breaks down starch into disaccharides
E. Breaks down peptides into amino acids
F. Breaks down nucleic acids into nucleotides
G. Breaks down disaccharides into monosaccharides

11. List the steps in swallowing. (p. 409)

12. Explain the stimulus for and response of the parasympathetic nervous system in digestion. (p. 412)

13. Explain how hormones control the secretions and/or release of secretions from the stomach, pancreas, and gallbladder. (pp. 412, 413, 415, 418)

14. Discuss absorption of amino acids, monosaccharides, glycerol, fatty acids, electrolytes, and water from substances in the small and large intestines. (pp. 422–424 and 427)

15. List the steps in defecating. (p. 427)

15.11 Nutrition and Nutrients

16. Identify dietary sources of carbohydrates, lipids, and proteins. (pp. 428–431)

17. Explain how cells use carbohydrates, lipids, and proteins for the normal functioning of the body. (pp. 429–431)

18. Match the vitamins with their general functions, and indicate if the vitamin is fat-soluble or water-soluble. Functions may be used more than once, and more than one function may be applied to a vitamin. (pp. 432–433)

(1) vitamin A
(2) vitamin B$_1$ (thiamine)
(3) vitamin B$_2$ (riboflavin)
(4) vitamin B$_3$ (niacin)
(5) vitamin B$_5$ (pantothenic acid)
(6) vitamin B$_6$
(7) vitamin B$_{12}$ (cyanocobalamin)
(8) folacin
(9) biotin
(10) vitamin C (ascorbic acid)
(11) vitamin D
(12) vitamin E
(13) vitamin K

A. Part of coenzyme A in oxidation of carbohydrates
B. Required for ribose synthesis
C. Necessary for synthesis of visual pigments
D. Required for synthesis of prothrombin
E. Required to produce collagen
F. Required to synthesize nucleic acids
G. Promotes red blood cell production
H. Plays a role in myelin synthesis
I. Antioxidant, helps stabilize cell membranes
J. Promotes development of teeth and bones
K. Required to produce antibodies
L. Required for cellular growth
M. Part of coenzymes to synthesize proteins, fats, and nucleic acids

19. Match the minerals/elements with their functions, and indicate whether each is a major mineral or a trace element required for nutrition. Functions may be used more than once, and more than one function may be applied to a mineral or trace element. (pp. 434–435)

(1) calcium
(2) chlorine
(3) chromium
(4) cobalt
(5) copper
(6) fluorine
(7) iodine
(8) iron
(9) magnesium
(10) manganese
(11) phosphorus
(12) potassium
(13) selenium
(14) sodium
(15) sulfur
(16) zinc

A. Essential for the use of carbohydrates
B. Component of certain enzymes
C. Component of tooth enamel
D. Component of teeth and bones
E. Helps maintain intracellular osmotic pressure
F. Essential part of certain amino acids
G. Helps maintain extracellular fluid osmotic pressure
H. Necessary for normal wound healing
I. Component of cyanocobalamin
J. Essential for synthesis of thyroid hormones
K. Required in metabolic reactions associated with ATP production
L. Component of hemoglobin molecules
M. Essential for hemoglobin synthesis and melanin production
N. Required for cholesterol synthesis and urea formation

20. Define *adequate diet*. (p. 434)

Integrative Assessments/ Critical Thinking

OUTCOMES 11.9, 15.8, 15.11

1. Why does blood sugar concentration stay relatively stable in a person whose diet is low in carbohydrates?

OUTCOMES 15.6, 15.11

2. How would removal of 95% of the stomach (subtotal gastrectomy) to treat severe ulcers or cancer affect digestion and absorption? How would the patient's eating habits have to be altered? Why? Do you think that people should have this type of surgery to treat life-threatening obesity?

OUTCOMES 15.7, 15.8

3. Why might a person with inflammation of the gallbladder (cholecystitis) also develop inflammation of the pancreas (pancreatitis)?

OUTCOME 15.11

4. Examine the label information on the packages of a variety of dry breakfast cereals. Which types of cereals provide adequate sources of vitamins and minerals? Which major nutrients are lacking in these cereals?

WEB CONNECTIONS

Visit the text website at **www.mhhe.com/shieress11** for additional quizzes, interactive learning exercises, and more.

APR

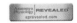

Anatomy & Physiology REVEALED includes cadaver photos that allow you to peel away layers of the human body to reveal structures beneath the surface. This program also includes animations, radiologic imaging, audio pronunciation, and practice quizzing. To learn more visit **www.aprevealed.com.**

16

Respiratory System

The dangers of secondhand smoke. Evidence is mounting that exposure to environmental tobacco smoke (ETS)—also called secondhand smoke—is as dangerous as actually smoking. ETS has two sources: *sidestream smoke* from lit cigarettes, cigars, or pipes, and *mainstream smoke* that smokers exhale. Smoke contains more than 4,000 chemicals, many of which are irritants, carcinogens, mutagens, or systemic or developmental toxins. More than sixty carcinogens are in tobacco smoke. Toxins include benzene, formaldehyde, vinyl chloride, ammonia, arsenic, and cyanide.

Exposure to ETS likely kills an estimated 53,000 nonsmokers in the United States each year. That includes 46,000 heart disease deaths and 3,400 lung cancer deaths.

ETS is also responsible for 150,000 to 300,000 cases of lower respiratory infection (bronchitis or pneumonia) in children under eighteen months of age; worsening of asthma in children; ear infections in children; and low birth weight. Even short exposures are dangerous to anyone. Just a half hour of breathing someone else's smoke can activate platelets, damage endothelium, decrease coronary artery blood flow, and decrease heart rate variability—all changes that set the stage for cardiovascular disease. Smokers are further harmed by secondhand smoke. A study of newsagents in Italy who smoke, and who sell news-

The danger of secondhand smoke, long debated, is now widely accepted, and new rules and regulations attempt to limit exposure.

papers from inside booths, showed that exposure to smoke from their customers adds the equivalent of smoking three extra cigarettes a day.

The only way to decrease exposure to ETS is to eliminate it. Dividing a space into smoking and nonsmoking areas does not work, because the smoke lingers for hours—this is why airplanes, many restaurants, and many workplaces are smoke-free. The *U.S. Surgeon General's Report on the Health Consequences of Involuntary Exposure to Tobacco Smoke* concluded that there is "no risk-free level of exposure to secondhand smoke."

Learning Outcomes

After studying this chapter, you should be able to do the following:

16.1 Introduction
1. Identify the general functions of the respiratory system. (p. 443)

16.2 Organs of the Respiratory System
2. Locate the organs of the respiratory system. (p. 443)
3. Describe the functions of each organ of the respiratory system. (p. 443)

16.3 Breathing Mechanism
4. Explain the mechanisms of inspiration and expiration. (p. 450)
5. Define each of the respiratory volumes and capacities. (p. 455)

16.4 Control of Breathing
6. Locate the respiratory areas in the brainstem and explain how they control breathing. (p. 456)
7. Discuss how various factors affect the respiratory areas. (p. 456)

16.5 Alveolar Gas Exchanges
8. Describe the structure and function of the respiratory membrane. (p. 459)
9. Explain how air and blood exchange gases. (p. 459)

16.6 Gas Transport
10. List the ways blood transports oxygen and carbon dioxide. (p. 460)

aprevealed.com

Module 11: Respiratory System

 Learn   **Practice**   **Assess**

Aids to Understanding Words   *(Appendix A on page 564 has a complete list of Aids to Understanding Words.)*

alveol- [small cavity] *alveol*us: Microscopic air sac within a lung.

bronch- [windpipe] *bronch*us: Primary branch of the trachea.

cric- [ring] *cric*oid cartilage: Ring-shaped mass of cartilage at the base of the larynx.

epi- [upon] *epi*glottis: Flaplike structure that partially covers the opening into the larynx during swallowing.

hemo- [blood] *hemo*globin: Pigment in red blood cells that transports oxygen and carbon dioxide.

16.1 INTRODUCTION

Cells require oxygen to break down nutrients to release energy and produce ATP, and must excrete the carbon dioxide that results. Obtaining oxygen and removing carbon dioxide are the primary functions of the **respiratory system.** It includes tubes that remove (filter) particles from incoming air and transport air into and out of the lungs, as well as microscopic air sacs where gases are exchanged. The respiratory organs also entrap particles from incoming air, help control the temperature and water content of the air, produce vocal sounds, and participate in the sense of smell and the regulation of blood pH.

The entire process of gas exchange between the atmosphere and cells is called **respiration** (res″pĭ-ra′shun). The events of respiration include: (1) movement of air into and out of the lungs—commonly called breathing or *ventilation;* (2) gas exchange between the blood and the air in the lungs (external respiration); (3) gas transport in blood between the lungs and body cells; and (4) gas exchange between the blood and the cells (internal respiration). The process of oxygen utilization and carbon dioxide production at the cellular level is called *cellular respiration.*

16.2 ORGANS OF THE RESPIRATORY SYSTEM

The organs of the respiratory system can be divided into two groups, or tracts. Those in the *upper respiratory tract* include the nose, nasal cavity, paranasal sinuses, and pharynx. Those in the *lower respiratory tract* include the larynx, trachea, bronchial tree, and lungs (fig. 16.1; see reference plates 3, 4, 5, and 6, pp. 25–28).

Nose

Bone and cartilage support the **nose** internally. Its two *nostrils* are openings through which air can enter and leave the nasal cavity. Many internal hairs guard the nostrils, preventing entry of large particles carried in the air.

Nasal Cavity

The **nasal cavity** is a hollow space behind the nose (fig. 16.1). The **nasal septum,** composed of bone and cartilage, divides the nasal cavity into right and left parts. **Nasal conchae** are bones and bone processes that curl out from the lateral walls of the nasal cavity on each side, dividing the cavity into passageways (fig. 16.2). Nasal conchae also support the mucous membrane that lines the nasal cavity and help increase its surface area.

The mucous membrane has pseudostratified ciliated epithelium that is rich in mucus-secreting goblet

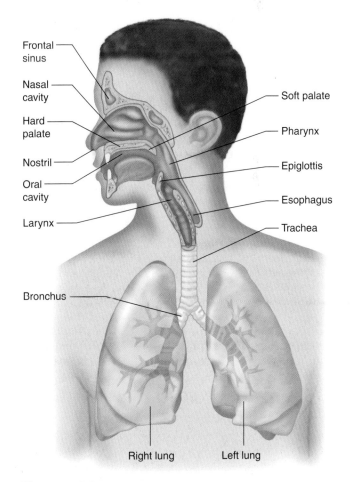

Figure 16.1

Organs and associated structures of the respiratory system.

Figure 16.2 AP|R

Major structures associated with the respiratory tract in the head and neck.

cells (see chapter 5, p. 98). It also includes an extensive network of blood vessels. As air passes over the mucous membrane, heat leaves the blood and warms the air, adjusting the air's temperature to that of the body. In addition, incoming air is moistened as water evaporates from the mucous lining. The sticky mucus that the mucous membrane secretes entraps dust and other small particles entering with the air.

> The nasal septum is usually straight at birth, but it can bend as the result of a birth injury. With age, the septum bends toward one side or the other. Such a *deviated septum* may obstruct the nasal cavity, making breathing difficult.

As the cilia of the epithelial lining move, they push a thin layer of mucus and entrapped particles toward the pharynx, where the mucus is swallowed (fig. 16.3). In the stomach, gastric juice destroys microorganisms in the mucus.

> A spore of the bacterium that causes anthrax is only half a micrometer wide. When spores are coated with powder to create a "bioweapon," they are still small enough to bypass the hairs and mucus in the nose, reaching the lungs, where they can cause inhalation anthrax. The bacteria release a toxin that causes death.

Paranasal Sinuses

Recall from chapter 7 (pp. 144–148) that the **paranasal sinuses** are air-filled spaces within the *frontal, ethmoid, sphenoid,* and *maxillary bones* of the skull and opening into the nasal cavity. Mucous membranes line the sinuses and are continuous with the lining of the nasal cavity. The paranasal sinuses reduce the weight of the skull and are resonant chambers that affect the quality of the voice.

> A painful sinus headache can result from blocked drainage caused by an infection or allergic reaction.

Practice

1. What is respiration?
2. Which organs constitute the respiratory system?
3. What are the functions of the mucous membrane that lines the nasal cavity?
4. Where are the paranasal sinuses?
5. What are the functions of the paranasal sinuses?

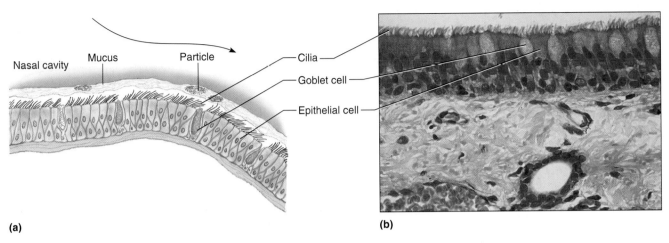

Figure 16.3

Mucus movement in the respiratory tract. **(a)** Cilia move mucus and trapped particles from the nasal cavity to the pharynx. **(b)** Micrograph of ciliated epithelium in the respiratory tract (275×).

Pharynx

The **pharynx,** or throat, is behind the oral cavity, the nasal cavity, and the larynx (see fig. 16.1). It is a passageway for food moving from the oral cavity to the esophagus and for air passing between the nasal cavity and the larynx. The pharynx also helps produce the sounds of speech. Chapter 15 (p. 409) describes the subdivisions of the pharynx—the nasopharynx, oropharynx, and laryngopharynx, which are shown in figure 16.2.

Larynx

The **larynx** (lar′inks) is an enlargement in the airway at the top of the trachea and below the pharynx. It conducts air in and out of the trachea and prevents foreign objects from entering the trachea. It also houses the *vocal cords.*

The larynx is composed of a framework of muscles and cartilages bound by elastic tissue. The largest of the cartilages are the *thyroid* ("Adam's apple"), *cricoid,* and *epiglottic cartilages* (fig. 16.4).

Inside the larynx, two pairs of horizontal *vocal folds* composed of muscle tissue and connective tissue with a covering of mucous membrane extend inward from the lateral walls. The upper folds are called *false vocal cords* because they do not produce sounds (fig. 16.5a). Muscle fibers within these folds help close the airway during swallowing.

The lower folds of muscle tissue and elastic fibers are the *true vocal cords.* Air forced between the vocal cords causes them to vibrate from side to side, which generates sound waves. Changing the shapes of the pharynx and oral cavity and using the tongue and lips transform these sound waves into words.

Contracting or relaxing muscles that alter the tension on the vocal cords controls the pitch (musical tone) of a sound. Increasing tension raises pitch, and decreasing tension lowers pitch. The intensity (loudness) of a sound reflects the force of air passing through the vocal folds. Stronger blasts of air produce louder sound; weaker blasts produce softer sound.

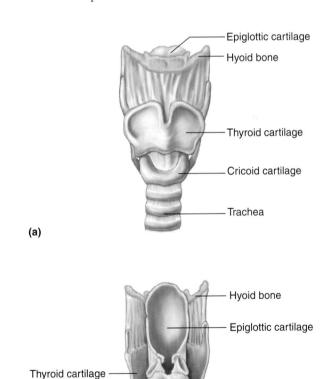

Figure 16.4 APR

Larynx. **(a)** Anterior and **(b)** posterior views of the larynx.

Damage to the nerves (recurrent laryngeal nerves) that supply the laryngeal muscles can alter the quality of a person's voice. These nerves pass through the neck as parts of the vagus nerves, and they can be injured by trauma or surgery to the neck or thorax. Nodules or other growths on the margins of the vocal folds that interfere with the free flow of air can also cause vocal problems. Surgery can remove such lesions.

During normal breathing, the vocal cords are relaxed and the opening between them, called the **glottis** (glot'is), is a triangular slit. However, when food or liquid is swallowed, muscles in the false vocal cords close the glottis, which prevents food or liquid from entering the trachea (fig. 16.5).

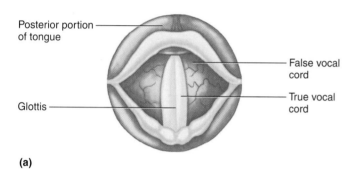

(a)

Posterior portion of tongue

False vocal cord

True vocal cord

Glottis

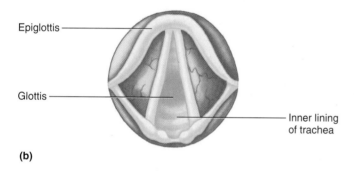

(b)

Epiglottis

Glottis

Inner lining of trachea

(c)

Figure 16.5

The vocal cords as viewed from above with the glottis **(a)** closed and **(b)** open. **(c)** Photograph of the glottis and vocal folds.

The epiglottic cartilage is the central part of a flap-like structure called the **epiglottis.** This structure usually stands upright and allows air to enter the larynx. During swallowing, however, the larynx rises, and the epiglottis presses downward to partially cover the opening into the larynx. This helps prevent foods and liquids from entering the air passages (see chapter 15, p. 409).

Laryngitis—hoarseness or lack of voice—occurs when the mucous membrane of the larynx becomes inflamed and swollen, due to an infection or an irritation from inhaled vapors, and prevents the vocal cords from vibrating as freely as before. Laryngitis is usually mild, but may be dangerous if swollen tissues obstruct the airway and interfere with breathing. Inserting a tube (endotracheal tube) into the trachea through the nose or mouth can restore the passageway until the inflammation subsides.

Practice

6. Describe the structure of the larynx.

7. How do the vocal cords produce sounds?

8. What is the function of the glottis? The epiglottis?

Trachea

The **trachea** (tra'ke-ah), or windpipe, is a flexible cylindrical tube about 2.5 centimeters in diameter and 12.5 centimeters in length (fig. 16.6). It extends downward anterior to the esophagus and into the thoracic cavity, where it splits into right and left bronchi.

A ciliated mucous membrane with many goblet cells lines the trachea's inner wall. This membrane filters incoming air and moves entrapped particles upward into the pharynx, where the mucus can be swallowed.

Within the tracheal wall are about twenty C-shaped pieces of hyaline cartilage, one above the other. The open ends of these incomplete rings are directed posteriorly, and smooth muscle and connective tissues fill the gaps between the ends. These cartilaginous rings prevent the trachea from collapsing and blocking the airway. The soft tissues that complete the rings in the back allow the nearby esophagus to expand as food moves through it to the stomach.

Bronchial Tree

The **bronchial tree** (brong'ke-al tre) consists of branched airways leading from the trachea to the microscopic air sacs in the lungs (fig. 16.7). Its branches begin with the right and left **main (primary) bronchi,** which arise from the trachea at the level of the fifth thoracic vertebra.

Each primary bronchus divides into lobar (secondary) bronchi a short distance from its origin. The lobar

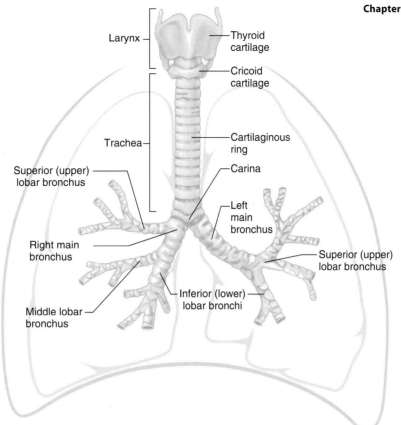

Figure 16.6

The trachea conducts air between the larynx and the bronchi.

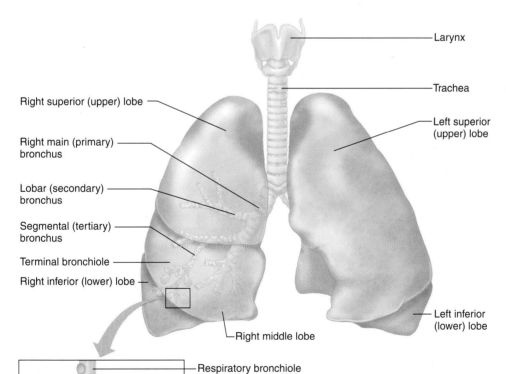

Figure 16.7

The bronchial tree consists of the passageways that connect the trachea and the alveoli. The alveolar ducts and alveoli are enlarged to show their locations.

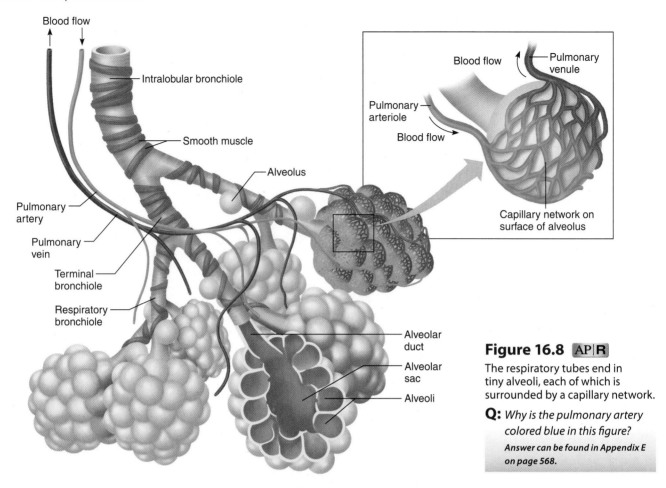

Figure 16.8 AP|R

The respiratory tubes end in tiny alveoli, each of which is surrounded by a capillary network.

Q: *Why is the pulmonary artery colored blue in this figure?*

Answer can be found in Appendix E on page 568.

Figure 16.9 AP|R

Light micrograph of alveoli (250×).

bronchi branch into segmental (tertiary) bronchi, and then into increasingly finer tubes. Among these smaller tubes are **bronchioles** that continue to branch, giving rise to terminal bronchioles, respiratory bronchioles, and finally to very thin tubes called **alveolar ducts.** These ducts lead to thin-walled outpouchings called **alveolar sacs.** Alveolar sacs lead to smaller micro-

scopic air sacs called **alveoli** (al-ve′o-li; singular, alveolus), which lie within capillary networks (figs. 16.8 and 16.9).

The structure of a bronchus is similar to that of the trachea, but the tubes that branch from it have less cartilage in their walls, and the bronchioles lack cartilage. As the cartilage diminishes, a layer of smooth muscle surrounding the tube becomes more prominent. This muscular layer persists even in the smallest bronchioles, but only a few muscle fibers are in the alveolar ducts.

The branches of the bronchial tree are air passages whose mucous membranes filter incoming air and distribute the air to alveoli throughout the lungs. The Genetics Connection entitled "Cystic Fibrosis" discusses what happens when the mucus formed is extremely thick.

The alveoli provide a large surface area of thin simple squamous epithelial cells through which gases are easily exchanged. Oxygen diffuses from the alveoli into the blood in nearby capillaries, and carbon dioxide diffuses from the blood into the alveoli (fig. 16.10).

Combined, two adult lungs have about 300 million alveoli, providing a total surface area nearly half the size of a tennis court.

Cystic Fibrosis

"Woe to that child which when kissed on the forehead tastes salty. He is bewitched and soon must die." So went a seventeenth-century British saying about a child with cystic fibrosis (CF). Until recently, salty skin, foul stools, and poor weight gain ("failure to thrive") were typically the first symptoms of CF. Today most new cases are detected before birth, using genetic tests. The disease, inherited from two carrier parents, affects about 30,000 people in the United States and 70,000 worldwide. It isn't known how many people have mild forms of the disease, merely with symptoms of frequent respiratory infection. More than 1,000 mutations can cause CF, so severity varies widely.

In 1938, physicians first described CF as a defect in channels leading from certain glands. This causes formation of extremely thick, sticky mucus, which encourages infections by microorganisms not otherwise common in the lungs. A clogged pancreas prevents digestive juices from reaching the intestines and thus impairs absorption of nutrients.

In the 1930s, life expectancy for a child with CF was five years, but by 1960 it became possible to treat the symptoms. Antibiotics control the respiratory infections, and daily "bronchial drainage" exercises shake the stifling mucus free from the lungs of infants. Older children and adults wear a vibrating vest for half-hour stretches two to four times a day to shake the mucus free. Some people multitask, taking daily antibiotics in a nebulizer as they wear the vest. Digestive enzymes mixed into soft foods enhance nutrient absorption.

The gene that is mutant in CF normally encodes a protein called the "cystic fibrosis transmembrane regulator," or CFTR for short. It is an ion channel that controls chloride transport out of cells. In severe CF, the chloride channel is missing one crucial amino acid, and is so deformed that it fails to function. The abnormal handling of chloride ions thickens the mucus. Organs become clogged.

Discovery of the most common CFTR mutation in 1989 enabled development of more targeted treatments. Some drugs allow more chloride to leave the cells lining the lungs. Two new drugs, still experimental, are small molecules that escort abnormal CFTR protein to the cell surface, where it apparently functions. The drugs act as "correctors," saving the errant CFTR proteins from being dismantled before they can reach the cell surface.

Life with severe CF is difficult. One little girl did not mind the twice-daily vibrating vest, or even the feeding tube she needed at night to pack in nutrients. But she hated the measures to avoid respiratory infections, especially in summertime. She had to stay away from hoses, which harbor lung-loving *Pseudomonas* bacteria. Bonfires or cookouts could expose her to lung-clogging particulates in the air. She couldn't even go into a pool—too little chlorine would invite bacterial infections, and too much would irritate her lungs. But unlike children of a generation ago, her disease is controlled enough that she will likely live well into adulthood.

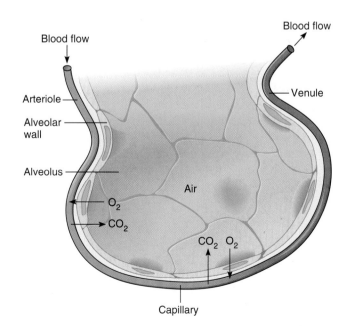

Figure 16.10 AP|R

Oxygen (O_2) diffuses from air within the alveolus into the capillary, while carbon dioxide (CO_2) diffuses from blood within the capillary into the alveolus.

Practice

9. What is the function of the cartilaginous rings in the tracheal wall?

10. Describe the bronchial tree.

11. Predict the direction of diffusion of gases between alveoli and alveolar capillaries.

Lungs

The **lungs** are soft, spongy, cone-shaped organs in the thoracic cavity (see fig. 16.1 and reference plates 4 and 5, pp. 26–27). The mediastinum separates the right and left lungs medially, and the diaphragm and thoracic cage enclose them.

Each lung occupies most of the thoracic space on its side. A bronchus and some large blood vessels suspend each lung in the cavity. These tubular structures enter the lung on its medial surface. A layer of serous membrane, the **visceral pleura** (vis′er-al ploo′rah), firmly attaches to each lung surface and folds back to become the **parietal pleura** (pah-ri′ĕ-tal ploo′rah). The parietal pleura, in turn, forms part of the mediastinum and lines the inner wall of the thoracic cavity (fig. 16.11).

No significant space exists between the visceral and parietal pleurae, but the potential space between them is called the **pleural cavity** (ploo′ral kav′ĭ-te). It has a thin film of serous fluid that lubricates adjacent pleural surfaces, reducing friction as they move against one another during breathing. This fluid also helps hold the pleural membranes together, as explained in section 16.3.

The right lung is larger than the left one and is divided into three lobes. The left lung has two lobes (see figs. 16.1 and 16.7).

A major branch of the bronchial tree supplies each lobe. A lobe also has connections to blood and lymphatic vessels and lies within connective tissues. Thus, a lung includes air passages, alveoli, blood vessels, connective tissues, lymphatic vessels, and nerves. Table 16.1 summarizes the characteristics of the major parts of the respiratory system.

Practice

12. Where are the lungs located?

13. What is the function of serous fluid in the pleural cavity?

14. What types of structures make up a lung?

16.3 BREATHING MECHANISM

Breathing, or ventilation, is the movement of air from outside the body into and out of the bronchial tree and alveoli. The actions providing these air movements are termed **inspiration** (in″spĭ-ra′shun), or inhalation, and **expiration** (ek″spĭ-ra′shun), or exhalation.

Inspiration

Atmospheric pressure, the pressure of the air around us, provides the force that moves air into the lungs. At sea level, this pressure is sufficient to support a column of mercury about 760 millimeters (mm) high in a tube. Thus, normal air pressure is equal to 760 mm of mercury (Hg).

Air pressure is exerted on all surfaces in contact with the air, and because people breathe air, the inside surfaces of their lungs also are subjected to pressure. The pressures on the inside of the lungs and alveoli and on the outside of the thoracic wall are about the same.

Figure 16.11

The potential spaces between the pleural membranes, called the left and right pleural cavities, are shown here as actual spaces.

Table 16.1	Parts of the Respiratory System	
Part	**Description**	**Function**
Nose	Part of face centered above mouth, in and below space between eyes	Nostrils provide entrance to nasal cavity; internal hairs begin to filter incoming air
Nasal cavity	Hollow space behind nose	Conducts air to pharynx; mucous lining filters, warms, and moistens incoming air
Paranasal sinuses	Hollow spaces in certain skull bones	Reduce weight of skull; serve as resonant chambers
Pharynx	Chamber behind nasal cavity, oral cavity, and larynx	Passageway for air moving from nasal cavity to larynx and for food moving from oral cavity to esophagus
Larynx	Enlargement at top of trachea	Passageway for air; prevents foreign objects from entering trachea; houses vocal cords
Trachea	Flexible tube that connects larynx with bronchial tree	Passageway for air; mucous lining continues to filter particles from incoming air
Bronchial tree	Branched tubes that lead from trachea to alveoli	Conducts air from trachea to alveoli; mucous lining continues to filter incoming air
Lungs	Soft, cone-shaped organs that occupy a large portion of the thoracic cavity	Contain air passages, alveoli, blood vessels, connective tissues, lymphatic vessels, and nerves of the lower respiratory tract

Pressure and volume are related in an opposite (inverse) way. For example, if we pull back on the plunger of a syringe, the volume inside the barrel increases, decreasing the air pressure inside. Atmospheric pressure then pushes outside air into the syringe. In contrast, if we push on the plunger of a syringe, the volume inside the syringe is reduced, but the pressure inside increases, forcing air out into the atmosphere. The movement of air into and out of the lungs occurs in much the same way.

If the pressure inside the lungs and alveoli decreases, atmospheric pressure will push outside air into the airways. That is what happens during normal inspiration. Impulses carried on the phrenic nerves, which are associated with the cervical plexuses (see chapter 9, p. 250), stimulate muscle fibers in the dome-shaped *diaphragm* below the lungs to contract. The diaphragm moves downward, the thoracic cavity enlarges, and the pressure in the alveoli falls to about 2 mm Hg below that of atmospheric pressure. In response, atmospheric pressure forces air into the airways (fig. 16.12).

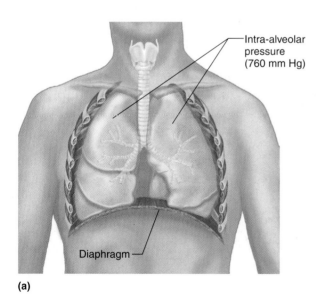

Intra-alveolar pressure (760 mm Hg)

Diaphragm

(a)

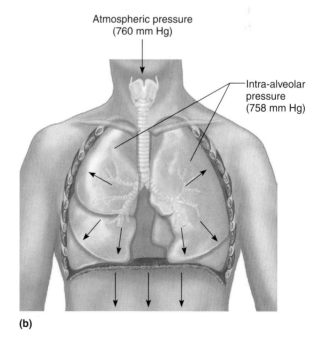

Atmospheric pressure (760 mm Hg)

Intra-alveolar pressure (758 mm Hg)

(b)

Figure 16.12 AP|R

Normal inspiration. **(a)** Prior to inspiration, the intra-alveolar pressure is 760 mm Hg. **(b)** The intra-alveolar pressure decreases to about 758 mm Hg as the thoracic cavity enlarges, and atmospheric pressure forces air into the airways.

While the diaphragm is contracting and moving downward, the *external (inspiratory) intercostal muscles* between the ribs may be stimulated to contract. This moves the ribs and the sternum upward and outward, enlarging the thoracic cavity even more. As a result, the pressure inside is reduced further, and the greater atmospheric pressure forces even more air into the airways.

Lung expansion in response to movements of the diaphragm and chest wall depends on movements of the pleural membranes. Any separation of the pleural membranes decreases pressure in the intrapleural space, holding these membranes together. In addition, only a thin film of serous fluid separates the parietal pleura on the inner wall of the thoracic cavity from the visceral pleura attached to the surface of the lungs. The water molecules in this fluid greatly attract the pleural membranes and each other, helping to adhere the moist surfaces of the pleural membranes, much as a wet coverslip sticks to a microscope slide. As a result of these factors, when the external intercostal muscles move the thoracic wall upward and outward, the parietal pleura moves too, and the visceral pleura follows it. This helps expand the lung in all directions.

The moist pleural membranes play a role in expanding the lungs, but the moist inner surfaces of the alveoli have the opposite effect. Here, the attraction of water molecules creates a force called **surface tension** that makes it difficult to inflate the alveoli and may actually collapse them. Certain alveolar cells, however, synthesize a mixture of lipids and proteins called **surfactant** (ser-fak'tant). It is secreted continuously into alveolar air spaces and reduces the alveoli's tendency to collapse, especially when lung volumes are low. Surfactant makes it easier for inspiratory efforts to inflate the alveoli.

> Surfactant is particularly important in the minutes after birth, when the newborn's lungs fully inflate for the first time. Premature infants may suffer respiratory distress syndrome if they do not produce sufficient surfactant. To help many of these newborns survive, physicians inject synthetic surfactant into the tiny lungs through an endotracheal tube. A ventilator machine especially geared to an infant's size assists breathing.

If a person needs to take a deeper than normal breath, the diaphragm and external intercostal muscles contract more forcefully. Additional muscles, such as the pectoralis minor and the sternocleidomastoid, can also pull the thoracic cage farther upward and outward, enlarging the thoracic cavity and decreasing internal pressure (fig. 16.13).

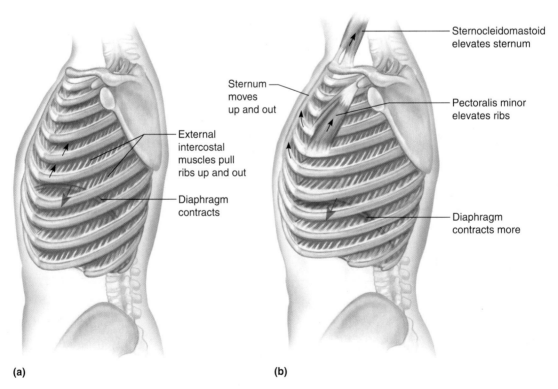

(a)   (b)

Figure 16.13

Maximal inspiration. **(a)** Shape of the thorax at the end of normal inspiration. **(b)** Shape of the thorax at the end of maximal inspiration, aided by contraction of the sternocleidomastoid and pectoralis minor muscles.

The first breath is the toughest. A newborn must use twenty times the energy to take the first breath as for subsequent breaths.

Expiration

The forces for expiration come from the *elastic recoil* of tissues and from surface tension. The lungs and thoracic wall contain considerable elastic tissue, which stretches with lung expansion during inspiration. Also, the diaphragm lowering compresses the abdominal organs beneath it. As the diaphragm and external intercostal muscles relax following inspiration, the elastic tissues cause the lungs and thoracic cage to recoil and return to their original shapes. Similarly, the abdominal organs spring back into their previous shapes, pushing the diaphragm upward (fig. 16.14*a*). At the same time, the surface tension that develops between the moist surfaces of the alveolar linings decreases the diameters of the alveoli. Each of these factors increases alveolar pressure about 1 mm Hg above atmospheric pressure, so that the air inside the lungs is forced out through respiratory passages. Thus, normal resting expiration is a passive process.

Low pressure and wet surfaces hold the visceral and parietal pleural membranes together, so no actual space normally exists in the pleural cavity between them. However, a puncture in the thoracic wall admits atmospheric air into the pleural cavity and creates a real space between the membranes. This condition, called *pneumothorax,* may collapse the lung on the affected side because of the lung's elasticity. A collapsed lung is called *atelectasis.*

If a person needs to exhale more air than normal, the posterior *internal (expiratory) intercostal muscles* can be contracted (fig. 16.14*b*). These muscles pull the ribs and sternum downward and inward, increasing the pressure in the lungs. Also, the *abdominal wall muscles,* including the external and internal obliques, transversus abdominis, and rectus abdominis, can squeeze the abdominal organs inward (see fig. 8.19, p. 199). Thus, the abdominal wall muscles can increase pressure in the abdominal cavity and force the diaphragm still higher against the lungs. These actions squeeze additional air out of the lungs. Clinical Application 16.1 discusses problems in breathing associated with two common diseases, emphysema and lung cancer.

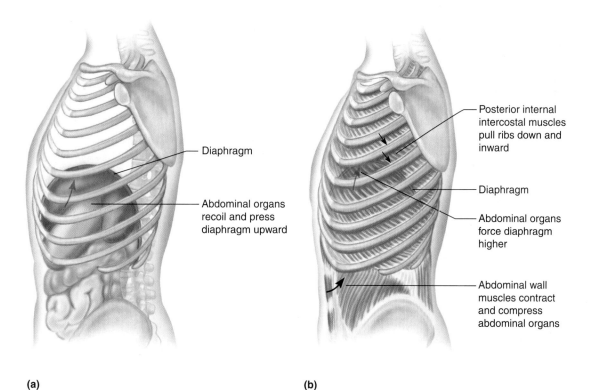

(a)   (b)

Figure 16.14 AP|R

Expiration. **(a)** Normal resting expiration is due to elastic recoil of the lung tissues and the abdominal organs. **(b)** Contraction of the abdominal wall muscles and the posterior internal intercostal muscles aids maximal expiration.

Clinical Application 16.1

Emphysema and Lung Cancer

Emphysema is a progressive, degenerative disease that destroys alveolar walls. As a result, clusters of small air sacs merge into larger chambers, which greatly decreases the surface area of the respiratory membrane, thereby reducing the volume of gases that can be exchanged through the membrane. Alveolar walls lose some of their elasticity, and capillary networks associated with the alveoli diminish (fig. 16A).

Loss of tissue elasticity in the lungs makes it increasingly difficult for a person with emphysema to exhale, because normal expiration involves the passive elastic recoil of inflated lungs.

Emphysema may develop in response to prolonged exposure to respiratory irritants, such as those in tobacco smoke and polluted air. The disease may also result from an inherited enzyme deficiency. Emphysema is a type of chronic obstructive pulmonary disease (COPD).

Lung cancer, like other cancers, is the uncontrolled division of abnormal cells that rob normal cells of nutrients and oxygen, eventually crowding them out. Some cancerous growths in the lungs result secondarily from cancer cells that have spread (metastasized) from other parts of the body, such as the breasts, intestines, liver, or kidneys. Cancers that begin in the lungs are called *primary pulmonary cancers*. These may arise from epithelia, connective tissue, or blood cells. The most common form originates from epithelium in a bronchiole (fig. 16B) and is called *bronchogenic carcinoma*. This type of cancer is a response to irritation, such as prolonged exposure to tobacco smoke. Susceptibility to primary pulmonary cancers may be inherited.

Cancer cells divide to form tumor masses that obstruct air passages and reduce gas exchange. Bronchogenic carcinoma can spread quickly, establishing secondary cancers in the lymph nodes, liver, bones, brain, or kidneys. Lung cancer is treated with surgery, ionizing radiation, and drugs, but the survival rate is low.

(a)

(b)

Figure 16A

Comparison of lung tissues. **(a)** Normal lung tissue. **(b)** As emphysema develops, alveoli merge, forming larger chambers (100×).

(a)

(b)

(c)

Figure 16B

About 95% of lung cancers start in the lining (epithelium) of a bronchiole. **(a)** The normal lining shows **(4)** columnar cells with **(2)** hairlike cilia, **(3)** goblet cells that secrete **(1)** mucus, and **(5)** basal cells from which new columnar cells arise. **(6)** A basement membrane separates the epithelial cells from **(7)** the underlying connective tissue. **(b)** In the first stage of lung cancer, the basal cells divide repeatedly. The goblet cells secrete excess mucus, and the cilia are less efficient in moving the heavy mucus secretion. **(c)** Continued division of basal cells displaces the columnar and goblet cells. The basal cells penetrate the basement membrane and invade the deeper connective tissue.

Air movements other than breathing are called *nonrespiratory movements.* They are used to clear air passages, as in coughing and sneezing, or to express emotion, as in laughing and crying.

Nonrespiratory movements usually result from *reflexes*, although sometimes they are initiated voluntarily. A *cough*, for example, can be produced through conscious effort or may be triggered by a foreign object in an air passage.

Coughing involves taking a deep breath, closing the glottis, and forcing air upward from the lungs against the closure. Then the glottis is suddenly opened, and a blast of air is forced upward from the lower respiratory tract. Usually, this rapid rush of air removes the substance that triggered the reflex.

A *sneeze* is much like a cough, but it clears the upper respiratory passages rather than the lower ones. This reflex is usually initiated by a mild irritation in the lining of the nasal cavity, and in response, a blast of air is forced up through the glottis. This time, the air is directed into the nasal passages by depressing the uvula, thus closing the opening between the pharynx and the oral cavity. A sneeze can propel a particle out of the nose at 200 miles per hour.

Laughing involves taking a breath and releasing it in a series of short expirations. *Crying* consists of very similar movements. It may be necessary to note a person's facial expression to distinguish laughing from crying.

A *hiccup* is caused by sudden inspiration due to a spasmodic contraction of the diaphragm while the glottis is closed. Air striking the vocal folds generates the sound of the hiccup. We do not know the function, if any, of hiccups.

Yawning is familiar to everyone, yet its significance and how it is contagious remain poorly understood. Evidence points away from a role in increasing oxygen intake, as had long been thought.

Practice

15. Describe the events in inspiration.

16. How does expansion of the chest wall expand the lungs during inspiration?

17. Which forces cause normal expiration?

Respiratory Air Volumes and Capacities

Different intensities in breathing move different volumes of air in or out of the lungs. *Spirometry* is a test that measures such air volumes. Three distinct **respiratory volumes** (re-spi′rah-to″re vol′ūmz) can be measured using spirometry and a fourth (residual volume) cannot.

One inspiration plus the following expiration is called a **respiratory cycle.** The volume of air that enters (or leaves) during a single respiratory cycle is termed the **tidal volume.** About 500 milliliters (mL) of air enter during a normal, resting inspiration. Approximately the same volume leaves during a normal, resting expiration. Thus, the **resting tidal volume** is about 500 mL (fig. 16.15).

During forced inspiration, air in addition to the resting tidal volume enters the lungs. This extra volume is the **inspiratory reserve volume** (complemental air), and at maximum, it equals about 3,000 mL.

During forced expiration, the lungs can expel up to about 1,100 mL of air beyond the resting tidal volume. This volume is called the **expiratory reserve volume** (supplemental air). However, even after the most forceful expiration, about 1,200 mL of air remains in the lungs. This is called the **residual volume** and can only be measured using special gas dilution techniques.

Because of the residual volume and the expiratory reserve volumes, newly inhaled air always mixes with air already in the lungs. This prevents the oxygen and carbon dioxide concentrations in the lungs from fluctuating greatly with each breath.

Combining two or more of the respiratory volumes yields four **respiratory capacities** (re-spi′rah-to″re kah-pas′ĭ-tēz). Combining the inspiratory reserve volume (3,000 mL) with the tidal volume (500 mL) and the expiratory reserve volume (1,100 mL) gives the **vital capacity** (4,600 mL). This is the maximum volume of air a person can exhale after taking the deepest breath possible.

The tidal volume (500 mL) plus the inspiratory reserve volume (3,000 mL) gives the **inspiratory capacity** (3,500 mL). This is the maximum volume of air a person can inhale following a resting expiration. Similarly, the expiratory reserve volume (1,100 mL) plus

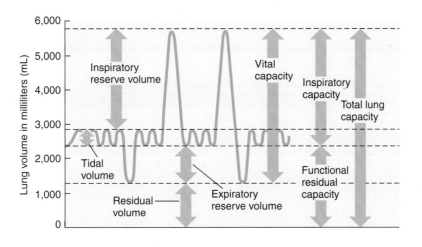

Figure 16.15

Respiratory volumes and capacities.

Q: *During inspiration, which way would the pen move in this figure?*

Answer can be found in Appendix E on page 568.

the residual volume (1,200 mL) equals the **functional residual capacity** (2,300 mL). This is the volume of air that remains in the lungs following a resting expiration.

The vital capacity plus the residual volume equals the **total lung capacity** (about 5,800 mL). This total varies with age, sex, and body size.

Some of the air that enters the respiratory tract during breathing does not reach the alveoli. This volume (about 150 mL) remains in the passageways of the trachea, bronchi, and bronchioles. Because gas is not exchanged through the walls of these passages, this air is said to occupy *anatomic dead space.* Table 16.2 summarizes the respiratory air volumes and capacities.

> An instrument called a *spirometer* measures respiratory air volumes, except residual volume, which requires a special technique. Such measurements are used to evaluate the courses of emphysema, pneumonia, and lung cancer, conditions in which functional lung tissue is lost. Spirometry may also be used to track the progress of diseases such as bronchial asthma that obstruct air passages.

Practice

18. What is tidal volume?

19. Distinguish between inspiratory and expiratory reserve volumes.

20. How is vital capacity determined?

21. How is total lung capacity calculated?

16.4 CONTROL OF BREATHING

Normal breathing is a rhythmic, involuntary act that continues even when a person is unconscious. The respiratory muscles, however, are also under voluntary control. (Take a deep breath and consider this!)

Respiratory Areas

Groups of neurons in the brainstem form the **respiratory areas,** which control both inspiration and expiration. The components of the respiratory areas are widely scattered throughout the pons and medulla oblongata. Two parts of the respiratory areas are of special interest: the respiratory center of the medulla and the respiratory group of the pons (fig. 16.16).

The **medullary respiratory center** includes two bilateral groups of neurons that extend throughout the length of the medulla oblongata. They are called the ventral respiratory group and the dorsal respiratory group.

Current evidence suggests that the basic rhythm of breathing arises from the *ventral respiratory group.* The *dorsal respiratory group* stimulates the inspiratory muscles, primarily the diaphragm. The dorsal respiratory group also helps process sensory information regarding the respiratory system and may play a role in certain cardiopulmonary reflexes that affect respiratory rhythm (fig. 16.17).

Neurons in another part of the brainstem, the pons, form the *pontine respiratory group* (formerly called the *pneumotaxic center*). They may contribute to the rhythm of breathing by limiting inspiration.

Practice

22. Where are the respiratory areas?

23. Describe how the respiratory areas maintain a normal breathing pattern.

24. Explain how the breathing pattern may change.

Factors Affecting Breathing

The respiratory areas affect breathing rate and depth, and so do certain chemicals in body fluids, the degree to which lung tissues stretch, a person's emotional state,

Table 16.2	Respiratory Air Volumes and Capacities	
Name	**Volume***	**Description**
Tidal volume (TV)	500 mL	Volume moved in or out of lungs during respiratory cycle
Inspiratory reserve volume (IRV)	3,000 mL	Volume that can be inhaled during forced breathing in addition to tidal volume
Expiratory reserve volume (ERV)	1,100 mL	Volume that can be exhaled during forced breathing in addition to tidal volume
Residual volume (RV)	1,200 mL	Volume that remains in lungs even after maximal expiration
Inspiratory capacity (IC)	3,500 mL	Maximum volume of air that can be inhaled following exhalation of tidal volume: IC = TV + IRV
Functional residual capacity (FRC)	2,300 mL	Volume of air that remains in the lungs following exhalation of tidal volume: FRC = ERV + RV
Vital capacity (VC)	4,600 mL	Maximum volume of air that can be exhaled after taking the deepest breath possible: VC = TV + IRV + ERV
Total lung capacity (TLC)	5,800 mL	Total volume of air that the lungs can hold: TLC = VC + RV

*Values are typical for a tall, young adult.

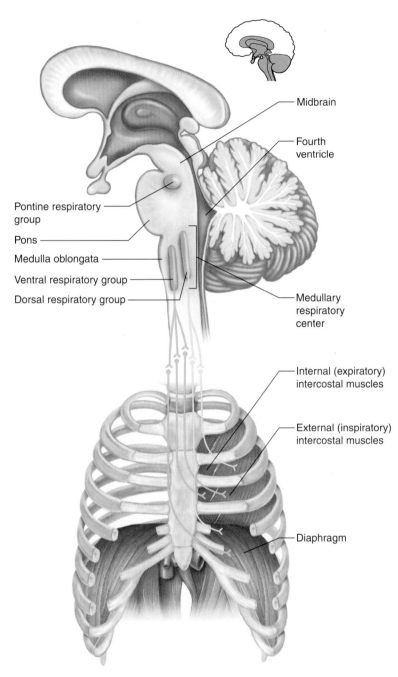

Figure 16.16

The respiratory areas are located in the pons and the medulla oblongata.

Figure 16.17

The medullary respiratory center and the pontine respiratory group control breathing.

Adding carbon dioxide to air can stimulate the rate and depth of breathing. Ordinary air is about 0.04% carbon dioxide. Inhaling air containing 4% carbon dioxide usually doubles the breathing rate.

and level of physical activity (Clinical Application 16.2). For example, *chemosensitive areas* (central chemoreceptors) in the ventral part of the medulla oblongata near the origins of the vagus nerves sense changes in the cerebrospinal fluid (CSF) levels of carbon dioxide and hydrogen ions. If either level rises, the central chemoreceptors signal the respiratory areas, and respiratory rate and tidal volume increase. As a result of the increased ventilation, more carbon dioxide is exhaled, the blood and CSF levels of these chemicals fall, and breathing rate decreases.

Low blood oxygen has little direct effect on the central chemoreceptors associated with the respiratory areas. Instead, *peripheral chemoreceptors* in specialized structures called the *carotid bodies* and the *aortic bodies* sense changes in blood oxygen levels. Peripheral chemoreceptors are in the walls of certain large arteries (the carotid arteries and the aorta) in the neck and thorax (fig. 16.18). Stimulated peripheral chemoreceptors transmit impulses to the respiratory areas, increasing the breathing rate. However, blood oxygen levels must be very low to trigger this mechanism. Thus, oxygen plays only a minor role in the control of normal respiration.

An *inflation reflex* helps regulate the depth of breathing. This reflex occurs when stretched lung tissues stimulate stretch receptors in the visceral pleura, bronchioles, and alveoli. The sensory impulses of this reflex travel via the vagus nerves to the pontine respiratory group and shorten the duration of inspiratory movements. This action prevents overinflation of the lungs during forceful breathing.

Emotional upset can alter the normal breathing pattern. Fear and pain typically increase the breathing rate. Conscious control of breathing is also possible because the respiratory muscles are voluntary.

A person can voluntarily stop breathing for a very short time. If breathing stops, blood levels of carbon dioxide and hydrogen ions rise, and oxygen levels fall.

Clinical Application 16.2

Exercise and Breathing

Moderate to heavy physical exercise greatly increases the volume of oxygen the skeletal muscles use. For example, a young man at rest utilizes about 250 mL of oxygen per minute, but maximal exercise may require 3,600 mL of oxygen per minute.

As oxygen utilization increases, the volume of carbon dioxide produced also increases. Because decreased blood oxygen and increased blood carbon dioxide concentrations stimulate the respiratory areas, exercise would be expected to increase breathing rate. Studies reveal, however, that blood oxygen and carbon dioxide concentrations usually do not change during exercise.

The cerebral cortex and sensory structures called *proprioceptors* that are associated with muscles and joints cause much of the increased breathing rate during vigorous exercise. Specifically, whenever the cerebral cortex signals skeletal muscles to contract, it also transmits stimulating impulses to the respiratory areas. Muscular movements stimulate proprioceptors, triggering a *joint reflex* that sends impulses to the respiratory center, increasing breathing rate.

When breathing rate increases during exercise, increased blood flow is also required to power skeletal muscles. Thus, physical exercise taxes both the cardiovascular and respiratory systems. If either of these systems fails to keep pace with cellular demands, the person feels shortness of breath. This feeling usually reflects an inability of the heart and blood vessels to move enough blood between the lungs and cells, rather than the respiratory system's inability to provide enough air.

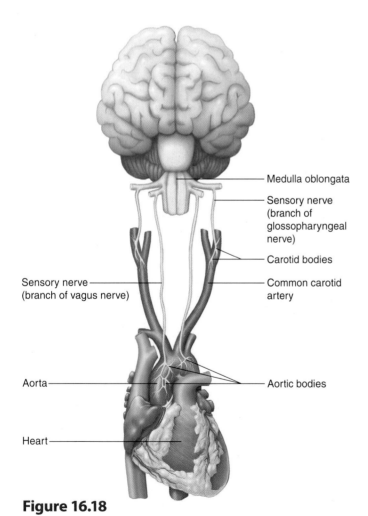

Medulla oblongata

Sensory nerve (branch of glossopharyngeal nerve)

Carotid bodies

Common carotid artery

Sensory nerve (branch of vagus nerve)

Aorta

Aortic bodies

Heart

Figure 16.18

Decreased blood oxygen concentration stimulates peripheral chemoreceptors in the carotid and aortic bodies.

These changes (primarily the increased carbon dioxide) stimulate the chemoreceptors, and soon the urge to inhale overpowers the desire to hold the breath.

A person can increase breath-holding time by breathing rapidly and deeply enough in advance to exhale excessive carbon dioxide. This action, called **hyperventilation** (hi″per-ven″tĭ-la′shun), lowers the blood carbon dioxide level. Following hyperventilation, it takes longer for carbon dioxide to rise to the level that stimulates the respiratory areas. (*Note:* Prolonging breath-holding in this way can cause abnormally low blood oxygen levels. Hyperventilation should never be used to help hold the breath while swimming because the person may lose consciousness underwater and drown.)

Interference with the oxygen supply to the brain causes fainting. A person who is emotionally upset may hyperventilate, become dizzy, and lose consciousness. This condition is due to a lowered carbon dioxide concentration followed by a rise in pH (alkalosis), a localized vasoconstriction of cerebral arterioles, and resulting decreased blood flow to nearby brain cells.

Practice

25. Which chemical factors affect breathing?

26. Describe the inflation reflex.

27. How does hyperventilation decrease the respiratory rate?

16.5 ALVEOLAR GAS EXCHANGES

The parts of the respiratory system discussed so far conduct air in and out of air passages. The alveoli carry on the vital process of exchanging gases between the air and the blood.

Alveoli

Alveoli are microscopic air sacs clustered at the distal ends of the narrowest respiratory tubes, the alveolar ducts (see fig. 16.8). Each alveolus consists of a tiny space within a thin wall that separates it from adjacent alveoli.

Respiratory Membrane

The wall of an alveolus is simple squamous epithelium. In close association with an alveolus is a dense network of capillaries, which also have walls of simple squamous epithelium. Thin, fused basement membranes separate the layers of these flattened cells, and in the spaces between the cells are elastic and collagenous fibers that support the alveolar wall. At least two thicknesses of epithelial cells and a layer of fused basement membranes separate the air in an alveolus from the blood in a capillary. These layers constitute the **respiratory membrane** (re-spi′rah-to″re mem′brān) across which blood and alveolar air exchange gases (fig. 16.19).

If all of the capillaries that surround the alveoli were unwound and laid end to end, they would extend for about 620 miles.

Diffusion Across the Respiratory Membrane

Recall from chapter 3 (pp. 60–62) that molecules diffuse from regions where they are in higher concentration toward regions where they are in lower concentration. For gases, it is more useful to think of diffusion from regions of higher pressure toward regions of lower pressure. The pressure of a gas determines the rate at which it diffuses from one region to another.

Measured by volume, ordinary air is about 78% nitrogen, 21% oxygen, and 0.04% carbon dioxide. Air also has traces of other gases that have little or no physiological importance.

In a mixture of gases such as air, each gas accounts for a portion of the total pressure the mixture produces. The amount of pressure each gas contributes is called the **partial pressure** (par′shal presh′ur) of that gas and is proportional to its concentration. For example, because air is 21% oxygen, oxygen accounts for 21% of the atmospheric pressure (21% of 760 mm Hg), or 160 mm Hg. Thus, the partial pressure of oxygen, symbolized P_{O_2}, in atmospheric air is 160 mm Hg. Similarly, the partial pressure of carbon dioxide (P_{CO_2}) in air is 0.3 mm Hg.

Gas molecules from the air may enter, or dissolve in, a liquid. This is what happens when carbon dioxide is added to a carbonated beverage, or when inspired gases dissolve in the blood in the alveolar capillaries.

When a mixture of gases dissolves in blood, the resulting concentration of each gas is proportional to its partial pressure. Each gas diffuses between blood and its surroundings from areas of higher partial pressure to areas of lower partial pressure until the partial pressures in the two regions reach equilibrium. For example, the P_{CO_2} in capillary blood is 45 mm Hg, but the P_{CO_2} in alveolar air is 40 mm Hg. Because of the difference in these partial pressures, carbon dioxide diffuses from blood, where its partial pressure is higher, across the respiratory membrane and into alveolar air (fig. 16.20). When blood leaves the lungs, its P_{CO_2} is 40 mm Hg, which is the same as the P_{CO_2} of alveolar air. Similarly, the P_{O_2} of capillary blood is 40 mm Hg, but that of alveolar air is 104 mm Hg. Thus, oxygen diffuses from alveolar air into blood, and blood leaves the lungs with a P_{O_2} of 104 mm Hg. (Because of the large volume of air always in the lungs, as long as breathing continues, alveolar P_{O_2} stays relatively constant at 104 mm Hg.)

Cell of alveolar wall   Surfactant-secreting cell   Fluid with surfactant   Macrophage

Alveolus

Respiratory membrane

Capillary lumen

Cell of capillary wall

Figure 16.19
The respiratory membrane consists of the wall of the alveolus and the wall of the capillary.

Figure 16.20

Gases are exchanged between alveolar air and capillary blood because of differences in partial pressures.

A number of factors affect diffusion across the respiratory membrane. More surface area, shorter distance, greater solubility of gases, and a steeper partial pressure gradient all favor increased diffusion. Thus, diseases that harm the respiratory membrane, such as pneumonia, or diseases that reduce the surface area for diffusion, such as emphysema, may require increased P_{O_2} for treatment.

The respiratory membrane is normally so thin that certain soluble chemicals other than carbon dioxide may diffuse into alveolar air and be exhaled. This is why breath analysis can reveal alcohol in the blood or acetone on the breath of a person who has untreated diabetes mellitus. Breath analysis may also detect substances associated with kidney failure, certain digestive disturbances, and liver disease.

Practice

28. Describe the structure of the respiratory membrane.
29. What is the partial pressure of a gas?
30. Which force moves oxygen and carbon dioxide across the respiratory membrane?

16.6 GAS TRANSPORT

Blood transports oxygen and carbon dioxide between the lungs and the cells. As these gases enter blood, they dissolve in the liquid portion (plasma) or combine chemically with blood components.

Oxygen Transport

Almost all the oxygen (over 98%) that blood transports binds the iron-containing protein **hemoglobin** (he″mo-glo′bin) in red blood cells. The remainder of the oxygen dissolves in plasma.

In the lungs, where the P_{O_2} is relatively high, oxygen dissolves in blood and combines rapidly with the iron atoms of hemoglobin, forming **oxyhemoglobin** (ok″si-he″mo-glo′bin) (fig. 16.21*a*). The chemical bonds between oxygen and hemoglobin molecules are unstable. In regions of body cells where the P_{O_2} decreases, oxyhemoglobin molecules release oxygen, which diffuses into nearby cells that have depleted their oxygen supplies in cellular respiration (fig. 16.21*b*).

Several other factors affect how much oxygen oxyhemoglobin releases. More oxygen is released as the blood concentration of carbon dioxide increases, as blood becomes more acidic, or as blood temperature increases. This explains why more oxygen is released to skeletal muscles during physical exercise. The increased muscular activity and oxygen utilization increase carbon dioxide concentration, decrease pH, and raise temperature. Less active cells receive proportionately less oxygen.

A deficiency of O_2 reaching the tissues is called **hypoxia.** It may be a response to decreased arterial P_{O_2} (*hypoxemia*), diminished ability of the blood to transport O_2 (anemic hypoxia), inadequate blood flow (ischemic hypoxia), or a defect at the cellular level (histotoxic hypoxia), such as in cyanide poisoning.

Practice

31. How is oxygen transported from the lungs to cells?
32. What stimulates blood to release oxygen to tissues?

Carbon Dioxide Transport

Blood flowing through capillaries gains carbon dioxide because tissues have a relatively high P_{CO_2}. Blood transports carbon dioxide to the lungs in one of three forms: as carbon dioxide dissolved in plasma, as part of a com-

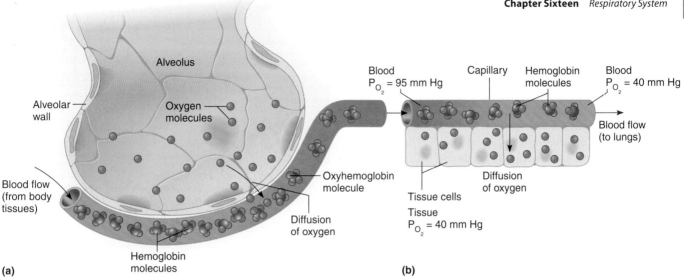

(a)

(b)

Figure 16.21 AP|R

Blood transports oxygen. **(a)** Oxygen molecules, entering the blood from the alveolus, bond to hemoglobin, forming oxyhemoglobin. **(b)** In the regions of the body cells, oxyhemoglobin releases oxygen. Much oxygen is still bound to hemoglobin at the P_{O_2} of systemic venous blood.

pound formed by bonding to hemoglobin, or as a bicarbonate ion (fig. 16.22).

The amount of carbon dioxide that dissolves in plasma is determined by its partial pressure. The higher the P_{CO_2} of the tissues, the more carbon dioxide will go into solution. However, only about 7% of the carbon dioxide that blood transports is in this form.

Unlike oxygen, which binds the iron atoms (part of the "heme" part) of hemoglobin molecules, carbon dioxide bonds with the amino groups ($-NH_2$) of the "globin" or protein portion of these molecules. Conse-

quently, oxygen and carbon dioxide do not compete for binding sites, and a hemoglobin molecule can transport both gases at the same time.

Carbon dioxide loosely bonds with hemoglobin, forming **carbaminohemoglobin** (kar-bam″ĭ-no-he″mo-glo′bin). This molecule decomposes readily in regions of low P_{CO_2}, releasing its carbon dioxide. Transporting carbon dioxide this way is theoretically quite effective, but carbaminohemoglobin forms slowly. Only about 23% of the carbon dioxide that blood transports is in this form.

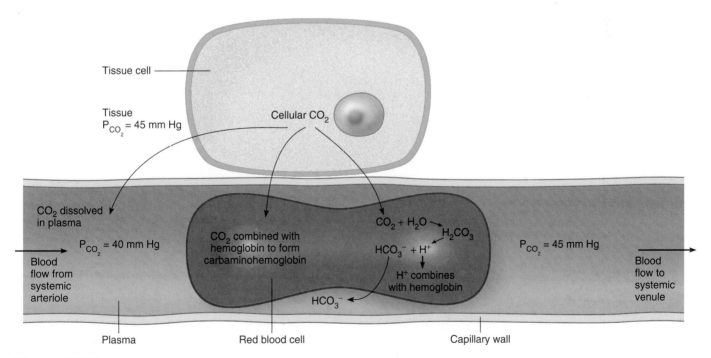

Figure 16.22

Carbon dioxide produced by cells is transported in the blood plasma in a dissolved state, bound to hemoglobin, or in the form of bicarbonate ions (HCO_3^-).

The most important carbon dioxide transport mechanism forms **bicarbonate ions** (HCO_3^-). Carbon dioxide reacts with water to form carbonic acid (H_2CO_3):

$$CO_2 + H_2O \longrightarrow H_2CO_3$$

This reaction occurs slowly in plasma, but much of the carbon dioxide diffuses into red blood cells. These cells have the enzyme **carbonic anhydrase** (kar-bon′ik an-hi′drās), which speeds the reaction between carbon dioxide and water. The resulting carbonic acid then dissociates, releasing hydrogen ions (H^+) and bicarbonate ions (HCO_3^-):

$$H_2CO_3 \longrightarrow H^+ + HCO_3^-$$

Most of the hydrogen ions bind hemoglobin molecules quickly, and thus do not accumulate and greatly change blood pH. The bicarbonate ions diffuse out of red blood cells and enter the plasma. Nearly 70% of the carbon dioxide that blood transports is in this form.

When blood passes through the capillaries of the lungs, its dissolved carbon dioxide diffuses into alveoli in response to the relatively low P_{CO_2} of alveolar air (fig. 16.23). At the same time, hydrogen ions and bicarbonate ions in red blood cells recombine to form carbonic acid, and under the influence of carbonic anhydrase, the carbonic acid quickly breaks down to yield carbon dioxide and water:

$$H^+ + HCO_3^- \longrightarrow H_2CO_3 \longrightarrow CO_2 + H_2O$$

Carbaminohemoglobin also releases its carbon dioxide, and carbon dioxide continues to diffuse out of the blood until the P_{CO_2} of the blood and that of alveolar air are in equilibrium. Table 16.3 summarizes transport of blood gases.

Table 16.3	Gases Transported in Blood	
Gas	**Reaction Involved**	**Substance Transported**
Oxygen	1–2% dissolves in plasma; 98–99% combines with iron atoms of hemoglobin molecules	Oxyhemoglobin
Carbon dioxide	About 7% dissolves in plasma	Carbon dioxide
	About 23% combines with amino groups of hemoglobin molecules	Carbamino-hemoglobin
	About 70% reacts with water to form carbonic acid; the carbonic acid then dissociates to release hydrogen ions and bicarbonate ions	Bicarbonate ions

The normal percentage of hemoglobin molecules that bind carbon monoxide (CO) in people who do not smoke is 2%. In people with CO poisoning, levels may exceed 20%. Since CO binds with the same heme part of hemoglobin as oxygen, the increased CO binding prevents oxygen from binding. This starves tissues of oxygen and causes chest pain, shortness of breath, fatigue, confusion, an irregular pulse, and abnormal heart rhythm.

Practice

33. Describe three forms in which blood can transport carbon dioxide from cells to the lungs.
34. How can hemoglobin carry oxygen and carbon dioxide at the same time?
35. How is carbon dioxide released from blood into the lungs?

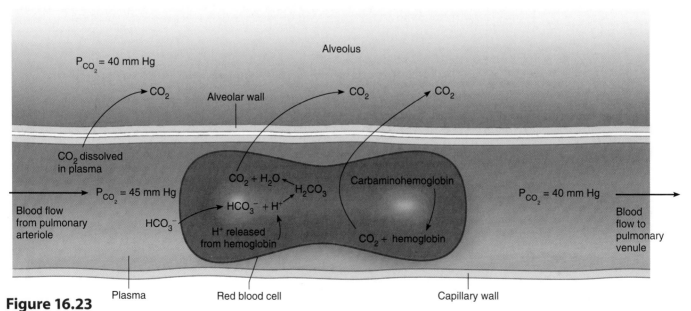

Figure 16.23
In the lungs, carbon dioxide diffuses from the blood into the alveoli.

Respiratory System

Integumentary System

Stimulation of skin receptors may alter respiratory rate.

Cardiovascular System

As the heart pumps blood through the lungs, the lungs oxygenate the blood and excrete carbon dioxide.

Skeletal System

Bones provide attachments for muscles involved in breathing.

Lymphatic System

Cells of the immune system patrol the lungs and defend against infection.

Muscular System

The respiratory system eliminates carbon dioxide produced by exercising muscles.

Digestive System

The digestive system and respiratory system share openings to the outside.

Nervous System

The brain controls the respiratory system.

Urinary System

The kidneys and the respiratory system work together to maintain blood pH. The kidneys compensate for water lost through breathing.

Endocrine System

Hormonelike substances control the production of red blood cells that transport oxygen and carbon dioxide.

Reproductive System

Respiration increases during sexual activity. Fetal gas exchange begins before birth.

The respiratory system provides oxygen for the internal environment and excretes carbon dioxide.

Summary Outline

16.1 Introduction (p. 443)

The respiratory system includes tubes that remove particles from incoming air and transport air to and from the lungs and the air sacs where gases are exchanged. Respiration is the entire process of gas exchange between the atmosphere and body cells.

16.2 Organs of the Respiratory System (p. 443)

The organs of the respiratory system can be divided into two groups. The upper respiratory tract includes the nose, nasal cavity, paranasal sinuses, and pharynx; the lower respiratory tract includes the larynx, trachea, bronchial tree, and lungs.

1. Nose
   a. Bone and cartilage support the nose.
   b. The nostrils are openings for air.
2. Nasal cavity
   a. Nasal conchae divide the nasal cavity into passageways and help increase the surface area of the mucous membrane.
   b. The mucous membrane filters, warms, and moistens incoming air.
   c. Ciliary action carries particles trapped in mucus to the pharynx, where they are swallowed.
3. Paranasal sinuses
   a. The paranasal sinuses are spaces in the bones of the skull that open into the nasal cavity.
   b. Mucous membrane lines the sinuses.
4. Pharynx
   a. The pharynx is behind the nasal cavity, oral cavity, and larynx.
   b. It is a passageway for air and food.
5. Larynx
   a. The larynx conducts air and helps prevent foreign objects from entering the trachea.
   b. It is composed of muscles and cartilages and is lined with mucous membrane.
   c. The larynx contains the vocal cords, which vibrate from side to side and produce sounds when air passes between them.
   d. The glottis and epiglottis help prevent foods and liquids from entering the trachea.
6. Trachea
   a. The trachea extends into the thoracic cavity anterior to the esophagus.
   b. It divides into right and left main bronchi.
7. Bronchial tree
   a. The bronchial tree consists of branched air passages that lead from the trachea to the air sacs.
   b. Alveoli are at the distal ends of the narrowest tubes, the alveolar ducts.
8. Lungs
   a. The mediastinum separates the left and right lungs, and the diaphragm and thoracic cage enclose them.
   b. The visceral pleura attaches to the surface of the lungs. The parietal pleura lines the thoracic cavity.
   c. Each lobe of the lungs is composed of alveoli, blood vessels, and supporting tissues.

16.3 Breathing Mechanism (p. 450)

Changes in the size of the thoracic cavity accompany inspiration and expiration.

1. Inspiration
   a. Atmospheric pressure provides the force that moves air into the lungs.
   b. Inspiration occurs when the pressure inside alveoli decreases.
   c. Pressure within alveoli decreases when the diaphragm moves downward and the thoracic cage moves upward and outward.
   d. Surface tension aids lung expansion.
2. Expiration
   a. Elastic recoil of tissues and surface tension within alveoli provide the forces of expiration.
   b. Thoracic and abdominal wall muscles aid expiration.
3. Respiratory air volumes and capacities
   a. One inspiration followed by one expiration is a respiratory cycle.
   b. The amount of air that moves in (or out) during a single respiratory cycle is the tidal volume.
   c. Additional air that can be inhaled is the inspiratory reserve volume. Additional air that can be exhaled is the expiratory reserve volume.
   d. Residual volume remains in the lungs after a maximal expiration.
   e. The vital capacity is the maximum amount of air a person can exhale after taking the deepest breath possible.
   f. The inspiratory capacity is the maximum volume of air a person can inhale following exhalation of the tidal volume.
   g. The functional residual capacity is the volume of air that remains in the lungs after a person exhales the tidal volume.
   h. The total lung capacity equals the vital capacity plus the residual volume.

16.4 Control of Breathing (p. 456)

Normal breathing is rhythmic and involuntary.

1. Respiratory areas
   a. The respiratory areas are in the brainstem and include parts of the medulla oblongata and pons.
   b. The medullary respiratory center includes two groups of neurons.
      (1) The ventral respiratory group gives rise to the basic rhythm of breathing.
      (2) The dorsal respiratory group stimulates inspiratory muscles.
   c. The pontine respiratory group may contribute to the rhythm of breathing by limiting inspiration.
2. Factors affecting breathing
   a. Chemicals, stretching of lung tissues, emotional state, and exercise affect breathing.
   b. Chemosensitive areas (central chemoreceptors) are associated with the respiratory center.
      (1) Blood levels of carbon dioxide and hydrogen ions influence the central chemoreceptors.
      (2) Stimulation of these receptors increases breathing rate.
   c. Peripheral chemoreceptors are in the walls of certain large arteries.
      (1) These chemoreceptors sense low oxygen levels.
      (2) When oxygen levels are low, breathing rate increases.
   d. Overstretching lung tissues triggers an inflation reflex.
      (1) This reflex shortens the duration of inspiratory movements.
      (2) The inflation reflex prevents overinflation of the lungs during forceful breathing.
   e. Hyperventilation decreases blood carbon dioxide levels, but *this is very dangerous when done before swimming underwater.*

16.5 Alveolar Gas Exchanges (p. 459)

Gas exchange between air and blood occurs in alveoli.

1. Alveoli
   Alveoli are tiny air sacs clustered at the distal ends of alveolar ducts.

2. Respiratory membrane
   a. This membrane consists of alveolar and capillary walls.
   b. Blood and alveolar air exchange gases across this membrane.
3. Diffusion across the respiratory membrane
   a. The partial pressure of a gas is proportional to the concentration of that gas in a mixture or the concentration dissolved in a liquid.
   b. Gases diffuse from regions of higher partial pressure toward regions of lower partial pressure.
   c. Carbon dioxide diffuses from blood into alveolar air. Oxygen diffuses from alveolar air into blood.

16.6 Gas Transport (p. 460)

Blood transports gases between the lungs and cells.

1. Oxygen transport
   a. Blood mainly transports oxygen in combination with hemoglobin molecules.

   b. The resulting oxyhemoglobin is unstable and releases its oxygen in regions where the P_{O_2} is low.
   c. More oxygen is released as the plasma P_{CO_2} increases, as blood becomes more acidic, and as blood temperature increases.
2. Carbon dioxide transport
   a. Carbon dioxide may be carried dissolved in plasma, bound to hemoglobin, or as a bicarbonate ion.
   b. Most carbon dioxide is transported in the form of bicarbonate ions.
   c. The enzyme carbonic anhydrase speeds the reaction between carbon dioxide and water to form carbonic acid.
   d. Carbonic acid dissociates to release hydrogen ions and bicarbonate ions.

Chapter Assessments

16.1 Introduction

1. List the general functions of the respiratory system. (p. 443)

16.2 Organs of the Respiratory System

2. Which one of the following is the beginning of the lower respiratory tract? (p. 443)

   a. nostril   d. larynx
   b. nasal cavity   e. oral cavity
   c. pharynx

3. Explain how the nose and nasal cavity filter the incoming air. (p. 443)

4. Identify the locations of the major paranasal sinuses. (p. 444)

5. Match the following structures with their descriptions: (p. 445–449)

   (1) true vocal cords   A. Serous membrane on lungs
   (2) false vocal cords   B. Contains the vocal cords
   (3) larynx   C. Vibrate to make sound
   (4) visceral pleura   D. Air sacs
   (5) alveoli   E. Muscular folds

6. Name and describe the locations of the larger cartilages of the larynx. (p. 445)

16.3 Breathing Mechanism

7. Explain how inspiration and expiration depend on pressure changes. (p. 450)

8. Compare the muscles used in a resting inspiration with those in a forced inspiration. (p. 450)

9. Define *surface tension* and explain how it aids breathing. (p. 452)

10. Define *surfactant* and explain its function. (p. 452)

11. Compare the muscles used (if any) in a resting expiration with those in a forced expiration. (p. 453)

12. Distinguish between the vital capacity and the total lung capacity. (p. 455)

16.4 Control of Breathing

13. Describe the location of the respiratory areas and name the major components. (p. 456)

14. Chemosensitive areas in the medulla oblongata are most sensitive to levels of _____. (p. 457)

    a. nitrogen
    b. oxygen
    c. carbon dioxide
    d. sodium

15. Describe the function of the chemoreceptors in the carotid and aortic bodies. (p. 457)

16. Describe the inflation reflex. (p. 457)

17. Hyperventilation is which one of the following? (p. 458)

    a. any decrease in breathing
    b. a decrease in breathing that brings in oxygen too slowly
    c. an increase in breathing that eliminates carbon dioxide too quickly
    d. an increase in breathing that has no effect on blood gases
    e. any increase in breathing

16.5 Alveolar Gas Exchanges

18. Define *respiratory membrane* and indicate its function. (p. 459)

19. Explain the relationship between the partial pressure of a gas and diffusion of that gas. (p. 459)

20. Summarize the exchange of oxygen and carbon dioxide across the respiratory membrane. (p. 459)

16.6 Gas Transport

21. Identify how blood transports oxygen. (p. 460)

22. List three factors that increase the release of oxygen from hemoglobin. (p. 460)

23. Identify the three ways blood transports carbon dioxide. (p. 460)

Integrative Assessments/Critical Thinking

OUTCOME 16.2

1. Why does breathing through the mouth dry out the throat?

OUTCOMES 16.2, 16.3

2. It is below 0°F outside, but the dedicated runner bundles up and hits the road anyway. "You're crazy," shouts a neighbor. "Your lungs will freeze." Why is the well-meaning neighbor wrong?

OUTCOMES 16.2, 16.3, 16.4

3. Emphysema reduces the lungs' capacity to recoil elastically. Which respiratory air volumes does emphysema affect?

OUTCOMES 16.3, 16.4, 16.5

4. When a woman is very close to delivering a baby, she may hyperventilate. Breathing into a paper bag regulates her breathing. How does this action return her breathing to normal?

OUTCOMES 16.3, 16.5, 16.6

5. Why were the finishing times of endurance events rather slow at the 1968 Olympics, held in 2,200-meter-high Mexico City?

OUTCOMES 16.4, 16.5

6. If a person has stopped breathing and is receiving pulmonary resuscitation, would it be better to administer pure oxygen or a mixture of oxygen and carbon dioxide? Why?

WEB CONNECTIONS

Visit the text website at **www.mhhe.com/shieress11** for additional quizzes, interactive learning exercises, and more.

APR

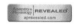

Anatomy & Physiology REVEALED includes cadaver photos that allow you to peel away layers of the human body to reveal structures beneath the surface. This program also includes animations, radiologic imaging, audio pronunciation, and practice quizzing. To learn more, visit **www.aprevealed.com**.

17

Urinary System

Spinach tainted with Escherichia coli strain 0157:H7 caused food poisoning in at least 199 people in late summer, 2006. Several died of hemolytic uremic syndrome.

Hemolytic uremic syndrome. In late summer of 2006, 199 people became infected with *Escherichia coli* strain 0157:H7 from eating raw spinach, mostly in sandwiches or salads. Two-year-old Kyle had drunk spinach that his mother had mixed into a fruit drink. When the story was reported in the news, many people commented on the irony of becoming ill from eating or drinking such a healthy food, but three years later another outbreak of *E. coli*-related food poisoning swept the country, traced this time to tainted, raw, chocolate chip cookie dough.

E. coli is a common bacterium that can cause severe illness when it produces a poison called shigatoxin. The illness begins with sharp abdominal pain and bloody diarrhea. Intensifying pain sends many victims to hospitals. For about 10% of them, the condition worsens to hemolytic uremic syndrome (HUS), which develops as the bloodstream transports the toxin to the kidneys, where the toxin destroys the microscopic capillaries that normally prevent proteins and blood cells from being excreted. With the capillaries compromised, proteins and blood cells, as well as damaged kidney cells, appear in the urine. HUS causes acute kidney failure. Often blood clots around the sites of the damaged capillaries, and new cells can form—after weeks of hospitalization, the person recovers. In some cases, the kidney damage may be permanent. For some not so lucky, such as two-year-old Kyle, HUS is deadly.

Past *E. coli* outbreaks were associated with eating undercooked hamburger, drinking unpasteurized apple cider, or exposure to the bacteria at petting zoos. People who became ill after visiting petting zoos had ingested the bacteria from not washing their hands after close contact with excrement. All of these infection routes came from a single type of source—manure. Epidemiologists used DNA tests to trace the spinach that had sickened people in twenty-two states to a single facility in California that had processed spinach contaminated with runoff from a nearby cattle ranch. Most cases of *E. coli* poisoning, however, are caused by toxin that gets into hamburger meat from cows' intestines. But the tainted raw cookie dough remains a mystery. Epidemiologists traced the telltale bacteria and their toxin to a vat of the dough in Virginia. The illness affected seventy-two people in thirty states.

Learning Outcomes

After studying this chapter, you should be able to do the following:

17.1 Introduction

1. List the general functions of the organs of the urinary system. (p. 468)

17.2 Kidneys

2. Describe the locations and structure of the kidneys. (p. 468)

3. List the functions of the kidneys. (p. 469)

4. Trace the pathway of blood through the major vessels in a kidney. (p. 469)

5. Describe a nephron, and explain the functions of its major parts. (p. 470)

17.3 Urine Formation

6. Explain how glomerular filtrate is produced, and describe its composition. (p. 473)

7. Explain the factors that affect the rate of glomerular filtration and how this rate is regulated. (p. 475)

8. Discuss the role of tubular reabsorption in urine formation. (p. 477)

9. Define *tubular secretion,* and explain its role in urine formation. (p. 479)

17.4 Urine Elimination

10. Describe the structure of the ureters, urinary bladder, and urethra. (p. 481)

11. Explain the process and control of micturition. (p. 483)

Module 13: Urinary System

Aids to Understanding Words   *(Appendix A on page 564 has a complete list of Aids to Understanding Words.)*

calyc- [small cup] major *calyc*es: Cuplike divisions of the renal pelvis.

cort- [covering] renal *cort*ex: Shell of tissues surrounding the inner kidney.

detrus- [to force away] *detrus*or muscle: Muscle within the bladder wall that expels urine.

glom- [little ball] *glom*erulus: Cluster of capillaries within a renal corpuscle.

mict- [to pass urine] *mict*urition: Process of expelling urine from the urinary bladder.

nephr- [pertaining to the kidney] *nephr*on: Functional unit of a kidney.

papill- [nipple] renal *papill*ae: Small elevations that project into a renal calyx.

trigon- [triangle] *trigon*e: Triangular area on the internal floor of the urinary bladder.

17.1 INTRODUCTION

Cells produce a variety of wastes that are toxic if they accumulate. Body fluids, such as blood and lymph, carry wastes from the tissues that produce them, while other structures remove wastes from the blood and transport them to the outside. The respiratory system removes carbon dioxide from the blood, and the *urinary system* removes certain salts and nitrogenous wastes. The urinary system also helps maintain the normal concentrations of water and electrolytes in body fluids, regulates the pH and volume of body fluids, and helps control red blood cell production and blood pressure.

The urinary system consists of a pair of kidneys, which remove substances from blood, form urine, and help regulate certain metabolic processes; a pair of tubular ureters, which transport urine from the kidneys; a saclike urinary bladder, which stores urine; and a tubular urethra, which conveys urine to the outside of the body. Figure 17.1 and reference plate 6 (p. 28) show these organs.

17.2 KIDNEYS

A **kidney** is a reddish-brown, bean-shaped organ with a smooth surface. An adult kidney is about 12 centimeters long, 6 centimeters wide, and 3 centimeters thick, and is enclosed in a tough, fibrous capsule (fig. 17.2).

Location of the Kidneys

The kidneys lie on either side of the vertebral column in a depression high on the posterior wall of the abdominal cavity. The upper and lower borders of the kidneys are generally at the levels of the twelfth thoracic and third lumbar vertebrae, respectively. The left kidney is usually 1.5–2.0 centimeters higher than the right one.

The kidneys are positioned **retroperitoneally** (ret″ro-per″ĭ-to-ne′alē), which means they are behind the parietal peritoneum and against the deep muscles of the back. Connective tissue and masses of adipose

tissue surround the kidneys and hold them in position (see fig. 1.11, p. 11).

Kidney Structure

The lateral surface of each kidney is convex, but its medial side is deeply concave. The resulting medial depression leads into a hollow chamber called the **renal sinus.** The entrance to this sinus is the *hilum,* and through it pass blood vessels, nerves, lymphatic vessels, and the ureter (see fig. 17.1).

The superior end of the ureter expands to form a funnel-shaped sac called the **renal pelvis** (re′nal pel′vis) inside the renal sinus. The pelvis is divided into two or three tubes, called *major calyces* (singular, *calyx*), and these in turn are divided into several *minor calyces* (fig. 17.2*a*).

Figure 17.1 AP|R

The urinary system includes the kidneys, ureters, urinary bladder, and urethra. Note the relationship of these structures to the major blood vessels.

Figure 17.2 AP|R

The kidney. **(a)** Longitudinal section of a kidney. **(b)** A renal pyramid containing nephrons. **(c)** A single nephron.

A series of small elevations called *renal papillae* project into the renal sinus from its wall. Tiny openings that lead into a minor calyx pierce each projection.

Each kidney has two distinct regions—an inner medulla and an outer cortex. The **renal medulla** (re′nal mĕ-dul′ah) is composed of conical masses of tissue called *renal pyramids* and appears striated. The **renal cortex** (re′nal kor′teks) forms a shell around the medulla and dips into the medulla between renal pyramids, forming *renal columns.* The granular appearance of the cortex is due to the random organization of tiny tubules associated with the **nephrons** (nef′ronz), which are the kidney's functional units (fig. 17.2*b, c*).

Practice

1. Where are the kidneys located?
2. Describe kidney structure.
3. Name the kidney's functional unit.

Kidney Functions

The primary function of the kidneys is to help maintain homeostasis by regulating the composition (including pH) and the volume of the extracellular fluid. They accomplish this by removing metabolic wastes from the blood and diluting them with water and electrolytes to form urine, which they then excrete.

The kidneys have several other important functions:

- Secreting the hormone erythropoietin (see chapter 12, p. 321) to help control the rate of red blood cell production.
- Playing a role in the activation of vitamin D.
- Helping to maintain blood volume and blood pressure by secreting the enzyme renin.

Renal Blood Vessels

The **renal arteries,** which arise from the abdominal aorta, supply blood to the kidneys. These arteries transport a large volume of blood. When a person is at rest, the renal arteries usually carry 15–30% of the total cardiac output into the kidneys.

A renal artery enters a kidney through the hilum and gives off several branches, called *interlobar arteries,* which pass between the renal pyramids. At the junction between the medulla and the cortex, the interlobar arteries branch, forming a series of incomplete arches, the *arcuate arteries,* which in turn give rise to *cortical radiate arteries (interlobular arteries).* The final branches of the cortical radiate arteries, called **afferent arterioles** (af′er-ent ar-te′re-ōlz), lead to the nephrons (figs. 17.3 and 17.4).

Venous blood returns through a series of vessels that correspond generally to arterial pathways. The **renal vein** then joins the inferior vena cava as it courses through the abdominal cavity (see fig. 17.1).

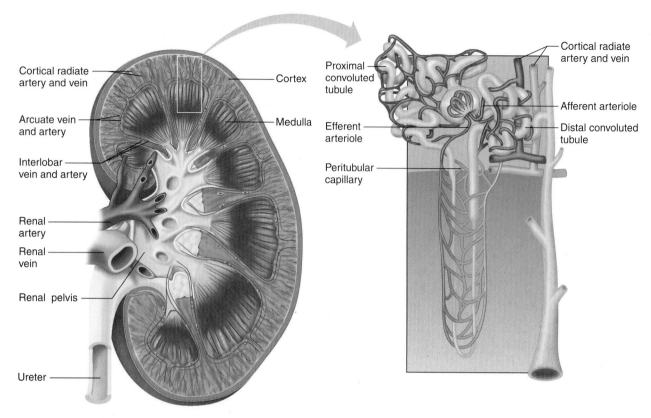

Figure 17.3 AP|R

Main branches of the renal artery and renal vein.

Figure 17.4

Scanning electron micrograph of a cast of the renal blood vessels associated with glomeruli (200×). From *Tissues and Organs: A Text-Atlas of Scanning Electron Microscopy,* by R. G. Kessel and R. H. Kardon, © 1979 W. H. Freeman and Company, all rights reserved.

A kidney transplant can help patients with end-stage renal disease. This procedure requires a kidney from a living or recently deceased donor whose tissues are antigenically similar (histocompatible) to those of the recipient. A surgeon places the kidney in the depression on the medial surface of the right or left ilium (iliac fossa). The surgeon then connects the renal artery and vein of the donor kidney to the recipient's iliac artery and vein, respectively, and the ureter of the donor kidney to the dome (apex) of the recipient's urinary bladder.

Nephrons

Nephron Structure

A kidney contains about one million nephrons. Each nephron consists of a **renal corpuscle** (re'nal kor'pusl) and a **renal tubule** (re'nal tu'būl) (see fig. 17.2c). Fluid flows through renal tubules on its way out of the body.

A renal corpuscle is composed of a tangled cluster of blood capillaries called a **glomerulus** (glo-mer'u-lus). Glomerular capillaries filter fluid, which is the first step in urine formation. A thin-walled, saclike structure called a **glomerular capsule** (glo-mer'u-lar kap'sūl) surrounds the glomerulus (figs. 17.5, 17.6, 17.7). The glomerular capsule, which is an expansion at the proximal

(a) Renal tubules Glomerular capsule Glomerulus Renal corpuscle

(b) Collecting ducts Epithelial cell Blood vessel

Figure 17.5

Microscopic view of the kidney. **(a)** Light micrograph of a section of the human renal cortex (220×). **(b)** Light micrograph of the renal medulla (80×).

end of a renal tubule, receives the fluid filtered at the glomerulus. The renal tubule leads away from the glomerular capsule and coils into a part of the nephron called the *proximal convoluted tubule.*

The proximal convoluted tubule dips toward the renal pelvis, where it becomes the *descending limb of the nephron loop* (loop of Henle). The tubule then curves back toward its renal corpuscle and forms the *ascending limb of the nephron loop.* The ascending limb returns to the region of the renal corpuscle, where it coils tightly again and is called the *distal convoluted tubule.*

Distal convoluted tubules from several nephrons merge in the renal cortex to form a *collecting duct* (technically not part of the nephron), which in turn passes into the renal medulla and enlarges as other distal convoluted tubules join it. The resulting tube empties into a minor calyx through an opening in a renal papilla. Figure 17.6 summarizes the structure of a nephron and its associated blood vessels.

Practice

4. List the general functions of the kidneys.

5. Trace the blood supply to the nephron.

6. Name the parts of a nephron.

Blood Supply of a Nephron

The cluster of capillaries that forms a glomerulus arises from an afferent arteriole. After passing through the glomerular capillaries, blood (minus any filtered fluid) enters an **efferent arteriole** (ef′er-ent ar-te′re-ōl) (see fig. 17.4). This is different from entering a venule, the usual circulatory route.

The efferent arteriole branches into a complex, freely interconnecting network of capillaries, called the **peritubular capillary** (per″ĭ-tu′bu-lar kap′ĭ-ler″e) **system,** that surrounds the renal tubule (see figs. 17.4 and 17.6). Blood in the peritubular capillary system is under low pressure. After flowing through the capillary network, the blood rejoins blood from other branches of the peritubular capillary system and enters the venous system of the kidney.

Juxtaglomerular Apparatus

Near its end, the ascending limb of the nephron loop passes between and contacts afferent and efferent arterioles. At the point of contact, the epithelial cells of the distal tubule are quite narrow and densely packed. These cells form a structure called the *macula densa.*

In the walls of the arterioles near their attachments to the glomerulus are large smooth muscle cells called *juxtaglomerular cells.* With cells of the macula densa, they constitute the **juxtaglomerular apparatus**

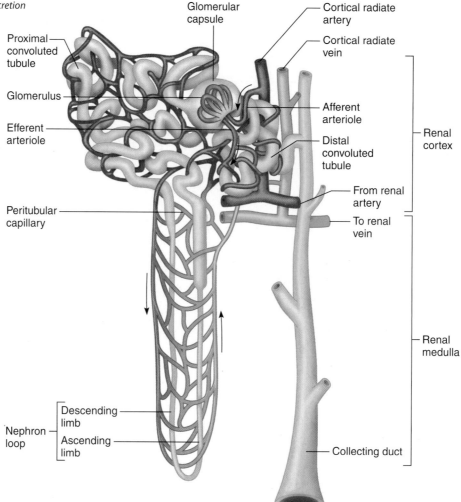

Figure 17.6 AP|R

Structure of a nephron and associated blood vessels. Arrows indicate the direction of blood flow.

(juks″tah-glo-mer′u-lar ap″ah-ra′tus), or juxtaglomerular complex (fig. 17.7). Its role in the control of renin secretion is described on page 476.

Practice

7. Describe the system of blood vessels associated with a nephron.
8. Which structures form the juxtaglomerular apparatus?

17.3 URINE FORMATION

The process of urine formation begins when the glomerular capillaries filter plasma, a process called **glomerular filtration.** Recall from chapter 13 (p. 357) that the force of blood pressure promotes filtration at capillaries throughout the body, but most of this fluid is reabsorbed into the bloodstream by the colloid osmotic pressure of the plasma (fig. 17.8a). Nephrons take filtration and reabsorption to another level, using two capillary beds working in series, the glomerulus and the peritubular capillaries. The first capillary bed, the glomerulus, is specialized only to filter. Instead of forming interstitial fluid, the filtered fluid (filtrate) moves into the renal tubule, where much of it is destined to become urine (fig. 17.8b).

Glomerular filtration produces about 180 liters of fluid, more than four times the total body water, every 24 hours. This is many times more than the amount filtered at capillaries elsewhere in the body. However, this is not your daily urine output! Glomerular filtration could not continue for very long unless most of this filtered fluid were returned to the bloodstream. This is accomplished by **tubular reabsorption** (tu′bu-lar re-absorp′shun), which moves substances from the tubular fluid back into the blood within the peritubular capillaries. In contrast, **tubular secretion** (tu′bu-lar se-kre′shun), the reverse process, moves substances from the blood within the peritubular capillary system into the renal tubule (fig. 17.8b).

Therefore, in addition to glomerular filtration, two other processes contribute to urine formation. In tubular reabsorption, the kidney selectively reclaims just the right amounts of substances, such as water, electrolytes, and glucose, that the body requires. Waste and substances that are in excess exit the body. In tubular secretion, some substances that the body must excrete, such as hydrogen ions and certain toxins, are removed even faster than through filtration alone.

The final product of these three processes— glomerular filtration, tubular reabsorption, and tubular secretion—is **urine.** The following relationship

Figure 17.7
Juxtaglomerular apparatus.
(a) Location of the juxtaglomerular apparatus. **(b)** Enlargement of a section of the juxtaglomerular apparatus, which consists of the macula densa and the juxtaglomerular cells.

(b)

determines the amount of any given substance excreted in the urine:

Amount filtered at the glomerulus
− Amount reabsorbed by the tubule
+ Amount secreted by the tubule
——————————————————
= Amount excreted in the urine

As the kidneys selectively excrete waste products and excess materials in the urine, they contribute to homeostasis by maintaining the composition of the internal environment.

Glomerular Filtration

Urine formation begins when water and certain dissolved substances are filtered out of glomerular capillaries and into glomerular capsules (fig. 17.9*a*). This filtration is similar to filtration at the arteriolar ends of other capillaries. However, many tiny openings (fenes-

trae) in glomerular capillary walls make glomerular capillaries much more permeable than capillaries in other tissues, even though cells called *podocytes* cover these capillaries and help make them impermeable to plasma proteins (fig. 17.9*b*, see fig. 17.7*b*).

The glomerular capsule receives the resulting **glomerular filtrate,** which is similar in composition to the filtrate that becomes tissue fluid elsewhere in the body. That is, glomerular filtrate is mostly water and the same components as blood plasma, except for the large protein molecules. Table 17.1 shows the relative concentrations of some substances in plasma, glomerular filtrate, and urine.

Filtration Pressure

The afferent arterioles have diameters larger than arterioles elsewhere in the body, allowing blood to enter the glomerular capillaries more easily. The efferent arterioles

(a) In most systemic capillaries, filtration predominates at the arteriolar end and osmotic reabsorption predominates at the venular end.

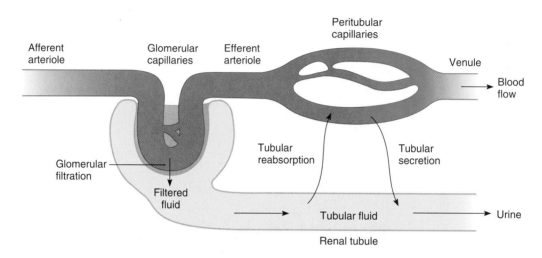

(b) In the kidneys, the glomerular capillaries are specialized for filtration. The renal tubule is specialized to control movements of substances back into the blood of the peritubular capillaries (tubular reabsorption) or from the blood into the renal tubule (tubular secretion).

Figure 17.8

Compared to most capillary beds in the (a) systemic circulation, those in the (b) kidneys are highly specialized for filtration.

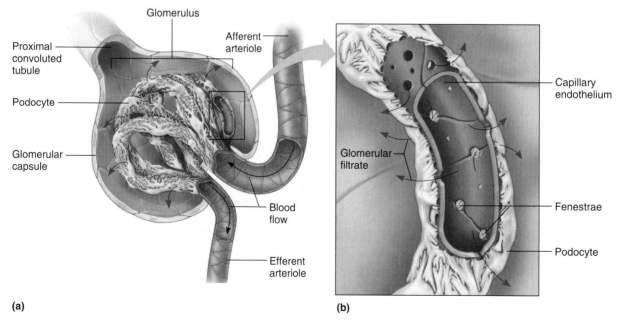

(a)

(b)

Figure 17.9

Glomerular filtration. (a) The first step in urine formation is filtration of substances out of glomerular capillaries and into the glomerular capsule. (b) Glomerular filtrate passes through the fenestrae of the capillary endothelium.

Q: *Vasoconstriction of which blood vessel in this figure would decrease glomerular filtration?*

Answer can be found in Appendix E on page 568.

Table 17.1	Relative Concentrations of Substances in the Plasma, Glomerular Filtrate, and Urine		
CONCENTRATIONS (mEq/L)			
Substance	Plasma	Glomerular Filtrate	Urine
Sodium (Na$^+$)	142	142	128
Potassium (K$^+$)	5	5	60
Calcium (Ca^{+2})	4	4	5
Magnesium (Mg^{+2})	3	3	15
Chloride (Cl$^-$)	103	103	134
Bicarbonate (HCO$_3^-$)	27	27	14
Sulfate (SO$_4^{-2}$)	1	1	33
Phosphate (PO$_4^{-3}$)	2	2	40
CONCENTRATIONS (mg/100 mL)			
Substance	Plasma	Glomerular Filtrate	Urine
Glucose	100	100	0
Urea	26	26	1,820
Uric acid	4	4	53

Note: mEq/L = milliequivalents per liter.

Net Outward Pressure

Outward force, glomerular hydrostatic pressure	= +60 mm
Inward force of plasma colloid osmotic pressure	= −32 mm
Inward force of capsular hydrostatic pressure	= −18 mm
Net filtration pressure	= +10 mm

Figure 17.10

Normally the glomerular net filtration pressure is positive, causing filtration. The forces involved include the hydrostatic and osmotic pressure of the plasma and the hydrostatic pressure of the fluid in the glomerular capsule.

have relatively narrow diameters, which tend to back up blood into the glomerular capillaries. The differences in the diameters raise the blood pressure in the glomerular capillaries. As in other capillaries, the hydrostatic pressure of blood forces substances through the glomerular capillary wall. The colloid osmotic pressure of plasma in the glomerulus and the hydrostatic pressure inside the glomerular capsule oppose this movement. An increase in either of these pressures reduces filtration. The net pressure forcing substances out of the glomerulus is the **net filtration pressure,** and it is normally always positive, favoring filtration at the glomerulus (fig. 17.10).

> If arterial blood pressure plummets, as can occur during *shock,* glomerular hydrostatic pressure may fall below the level required for filtration. At the same time, epithelial cells of the renal tubules may not receive sufficient nutrients to maintain their high metabolic rates. As a result, cells die (tubular necrosis), impairing renal functions. Such changes can cause renal failure.

Filtration Rate

The glomerular filtration rate (GFR), the most commonly measured index of kidney function, is directly proportional to net filtration pressure. Consequently, factors that affect glomerular hydrostatic pressure, glomerular plasma osmotic pressure, or hydrostatic pressure in the glomerular capsule also affect filtration rate. For example, any change in the diameters of the afferent and efferent arterioles changes glomerular hydrostatic pressure, also altering the glomerular filtration rate.

The afferent arteriole, which delivers blood to the glomerulus, may constrict in response to sympathetic nerve impulses. Blood flow diminishes, filtration pressure decreases, and filtration rate drops. On the other hand, if the efferent arteriole (which transports blood from the glomerulus) constricts, blood backs up into the glomerulus, net filtration pressure increases, and filtration rate rises. Vasodilation of these vessels causes opposite effects.

In capillaries, the plasma colloid osmotic pressure that attracts water inward (see chapter 12, p. 329) opposes the blood pressure that forces water and dissolved substances outward. During filtration through the capillary wall, proteins remaining in the plasma raise colloid osmotic pressure within the glomerular capillary. As this pressure rises, filtration decreases. Conversely, conditions that decrease plasma colloid osmotic pressure, such as a decrease in plasma protein concentration, increase the filtration rate.

> In *glomerulonephritis,* the glomerular capillaries are inflamed and become more permeable to proteins, which appear in the glomerular filtrate and in urine (proteinuria). At the same time, the protein concentration in blood plasma decreases (hypoproteinemia), and this decreases plasma colloid osmotic pressure. As a result, less tissue fluid moves into the capillaries, and edema develops.

The hydrostatic pressure in the glomerular capsule sometimes changes because of an obstruction, such as a stone in a ureter or an enlarged prostate gland pressing on the urethra. If this occurs, fluids back up into renal tubules and raise the hydrostatic pressure in the glomerular capsule. Because any increase in capsular pressure opposes glomerular filtration, the filtration rate may decrease significantly.

At rest, the kidneys receive about 25% of the cardiac output, and about 20% of the blood plasma is filtered as it flows through the glomerular capillary. This means that in an average adult, the GFR for the nephrons of both kidneys is about 125 milliliters per minute, or 180,000 milliliters (180 liters, or nearly 45 gallons) in 24 hours. Only a small fraction is excreted as urine. Instead, most of the fluid that passes through the renal tubules is reabsorbed and reenters the plasma.

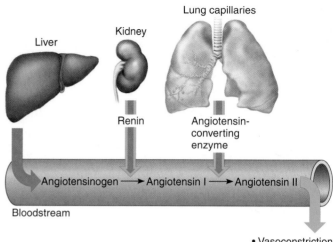

Practice

9. Which processes form urine?

10. Which forces affect net filtration pressure?

11. Which factors influence the rate of glomerular filtration?

Figure 17.11

The formation of angiotensin II in the bloodstream involves several organs and results in multiple actions that conserve sodium and water.

Regulation of Filtration Rate

The glomerular filtration rate is usually relatively constant. To help maintain homeostasis, however, the glomerular filtration rate may increase when body fluids are in excess and decrease when the body must conserve fluid.

Sympathetic nervous system reflexes that respond to changes in blood pressure and blood volume can alter the glomerular filtration rate. If blood pressure or volume drops sufficiently, afferent arterioles vasoconstrict, decreasing the glomerular filtration rate. This helps ensure that less urine forms when the body must conserve water. Conversely, vasodilation of afferent arterioles increases the glomerular filtration rate to counter increased blood volume or blood pressure.

Another mechanism to control filtration rate involves the enzyme *renin*. Juxtaglomerular cells secrete renin in response to three types of stimuli: (1) when special cells in the afferent arteriole sense a drop in blood pressure; (2) in response to sympathetic stimulation; and (3) when the macula densa (see fig. 17.7) senses decreased numbers of chloride, potassium, and sodium ions reaching the end of the ascending limb of the nephron loop. Once in the bloodstream, renin reacts with the plasma protein *angiotensinogen* to form *angiotensin I.* A second enzyme (*angiotensin-converting enzyme,* or ACE) in the lungs and in plasma quickly converts angiotensin I to *angiotensin II.*

Angiotensin II carries out a number of actions that help maintain sodium balance, water balance, and blood pressure (fig. 17.11). Angiotensin II vasoconstricts the efferent arteriole, which causes blood to back up into the glomerulus, raising glomerular capillary hydrostatic pressure. This important action helps minimize the decrease in glomerular filtration rate when systemic blood pressure is low. Angiotensin II has a major effect on the kidneys by stimulating secretion of the adrenal hormone aldosterone, which stimulates tubular reabsorption of sodium.

The heart secretes another hormone, atrial natriuretic peptide (ANP), when blood volume increases. ANP increases sodium excretion by a number of mechanisms, including increasing the glomerular filtration rate.

Elevated blood pressure (hypertension) is sometimes associated with excessive release of renin, followed by increased formation of the vasoconstrictor angiotensin II. Patients with this form of high blood pressure often take a drug called an *angiotensin-converting enzyme inhibitor.* These "ACE inhibitors" prevent the formation of angiotensin II by inhibiting the action of the enzyme that converts angiotensin I into angiotensin II.

Practice

12. What is the function of the macula densa?

13. How does renin help regulate filtration rate?

Tubular Reabsorption

Comparing the composition of glomerular filtrate entering the renal tubule with that of urine leaving the tubule reveals that the fluid changes as it passes through the tubule (see table 17.1). For example, glucose is present in glomerular filtrate but absent in urine. In contrast, urea and uric acid are much more concentrated in urine than in glomerular filtrate. Such changes in fluid composition are largely the result of tubular reabsorption, the process by which filtered substances are returned to the bloodstream. In this process, substances are transported out of the tubular fluid, through the epithelium of the renal tubule, and into the interstitial fluid. These substances then diffuse into the peritubular capillaries (fig. 17.12a).

Tubular reabsorption returns substances to the internal environment. The term *tubular* is used because the epithelial cells that make up the renal tubules control this process. In tubular reabsorption, substances must first cross the cell membrane facing the inside of the tubule and then cross the cell membrane facing the interstitial fluid.

The basic rules for movements across cell membranes apply to tubular reabsorption. Substances moving down a concentration gradient must be lipid-soluble, or there must be a carrier molecule or channel for that substance in the renal tubular cells. Active transport, requiring ATP, may move substances uphill against a concentration gradient.

Peritubular capillary blood is under relatively low pressure because it has already passed through two arterioles. Also, the walls of the peritubular capillaries are more permeable than other capillaries. Finally, because fluid is lost through glomerular filtration, the plasma protein concentration in the peritubular capillaries is relatively high. All of these factors enhance the rate of fluid reabsorption from the renal tubule.

Tubular reabsorption occurs throughout the renal tubule, but most of it takes place in the proximal convoluted part. The epithelial cells here have many microscopic projections called *microvilli* that form a "brush border" on their free surfaces. These tiny extensions greatly increase the surface area exposed to glomerular filtrate and enhance reabsorption.

Segments of the renal tubule are adapted to reabsorb specific substances, using particular modes of transport. Active transport, for example, reabsorbs glucose through the walls of the proximal convoluted tubule. Water is then reabsorbed by osmosis through the epithelium of the proximal convoluted tubule. However, parts of the distal convoluted tubule and collecting duct may be almost impermeable to water. This is a

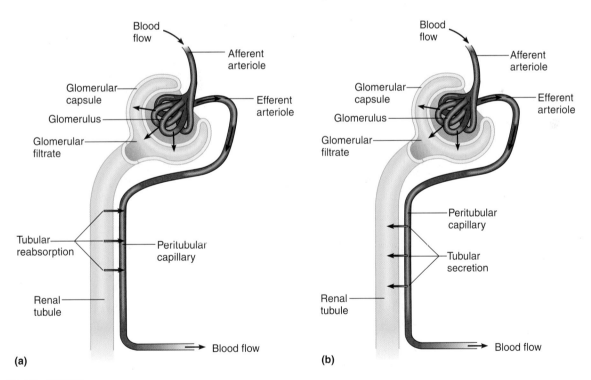

(a) **(b)**

Figure 17.12 AP|R

Two processes in addition to glomerular filtration contribute to form urine. **(a)** Tubular reabsorption transports substances from the glomerular filtrate into the blood within the peritubular capillary. **(b)** Tubular secretion transports substances from the blood within the peritubular capillary into the renal tubule.

Q: *Which of the three processes (glomerular filtration, tubular reabsorption, tubular secretion), if increased for a substance, would reduce urinary excretion of that substance?*

Answer can be found in Appendix E on page 568.

characteristic important in regulating urine concentration and volume, described on page 480.

Active transport utilizes carrier molecules in cell membranes (see chapter 3, p. 65). These carriers transport certain molecules across the membrane, release them, and then repeat the process. However, such a mechanism has a *limited transport capacity;* that is, it can transport only a certain number of molecules in a given time because the number of carriers is limited.

Usually, carrier molecules are able to transport all of the glucose in glomerular filtrate. But when the plasma glucose concentration increases to a critical level, called the *renal plasma threshold,* more glucose molecules are in the filtrate than can be actively transported. As a result, some glucose remains in the tubular fluid and is excreted in urine.

Amino acids enter the glomerular filtrate and are reabsorbed in the proximal convoluted tubule. Three active transport mechanisms reabsorb different groups of amino acids whose members have similar structures. Normally, only a trace of amino acids remains in urine.

Glomerular filtrate is nearly free of protein except for traces of albumin, a small protein that is taken up by endocytosis through the brush border of epithelial cells lining the proximal convoluted tubule. Once these proteins are inside an epithelial cell, they are broken down to amino acids, which then move into the blood of the peritubular capillary.

The epithelium of the proximal convoluted tubule reabsorbs other substances, including creatine; lactic, citric, uric, and ascorbic (vitamin C) acids; and phosphate, sulfate, calcium, potassium, and sodium ions. Active transport mechanisms with limited transport capacities reabsorb these chemicals. However, these substances usually do not appear in urine until glomerular filtrate concentration exceeds a particular substance's threshold.

Sodium and Water Reabsorption

Substances that remain in the renal tubule become more concentrated as water is reabsorbed from the filtrate. Water reabsorption occurs passively by osmosis, primarily in the proximal convoluted tubule, and is closely associated with the active reabsorption of sodium ions. It increases if sodium reabsorption increases, and decreases if sodium reabsorption decreases.

Active transport (the sodium pump) reabsorbs about 70% of sodium ions in the proximal segment of the renal tubule. As these positively charged ions (Na^+) move through the tubular wall, negatively charged ions, including chloride ions (Cl^-), phosphate ions (PO_4^{-3}), and bicarbonate ions (HCO_3^-), accompany them. These negatively charged ions move because of the elec-

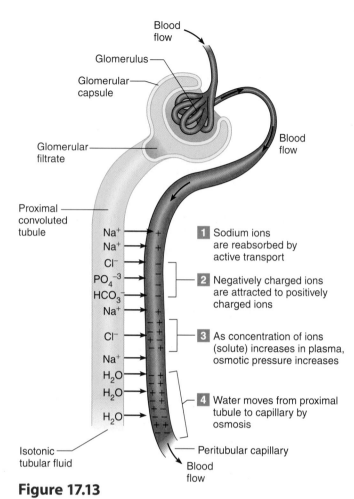

Figure 17.13

In the proximal portion of the renal tubule, osmosis reabsorbs water in response to active transport reabsorbing sodium and other solutes.

trochemical attraction between particles of opposite charge. This is a form of *passive transport* because it does not require direct expenditure of cellular energy (fig. 17.13).

As active transport moves more sodium ions out of the proximal tubule, along with (passively) various negatively charged ions, the concentration of solutes within the peritubular blood increases. Since water moves across cell membranes from regions of lesser solute concentration toward regions of greater solute concentration, water moves by osmosis from the renal tubule into the peritubular capillary. Movement of solutes and water into the peritubular capillary greatly reduces the fluid volume within the renal tubule. The end of the proximal convoluted tubule is in osmotic equilibrium, and the remaining tubular fluid is isotonic (fig. 17.13).

Active transport continues to reabsorb sodium ions as the tubular fluid moves through the nephron loop, the distal convoluted tubule, and the collecting duct. Water is absorbed passively by osmosis in various segments of the renal tubule. As a result, almost all the sodium ions and water that enter the renal tubule as part of the glomerular filtrate are reabsorbed before urine is excreted.

Tubular Secretion

About 20% of the plasma flowing through the kidneys is filtered in the glomeruli, and approximately 80% escapes filtration and continues on through the peritubular capillaries. In tubular secretion, certain substances move from the plasma of blood in the peritubular capillary into the fluid of the renal tubule. As a result, the amount of a particular chemical excreted in the urine may exceed the amount filtered from the plasma in the glomerulus (fig. 17.12*b*). As in the case of tubular reabsorption, the term *tubular* refers to control by the epithelial cells that make up the renal tubules.

Active transport mechanisms similar to those that function in reabsorption secrete some substances. Secretory mechanisms, however, transport substances in the opposite direction. For example, the epithelium of the proximal convoluted segment actively secretes certain organic compounds, including penicillin, creatinine, and histamine, into the tubular fluid.

Hydrogen ions are also actively secreted throughout the entire renal tubule. Secretion of hydrogen ions is important in regulating the pH of body fluids, as chapter 18 (p. 499) explains.

Most potassium ions in the glomerular filtrate are actively reabsorbed in the proximal convoluted tubule, but some may be secreted in the distal segment and collecting duct. During this process, active reabsorption of sodium ions from the tubular fluid results in a negative electrical charge within the tubule. Because positively charged potassium ions (K^+) and hydrogen ions (H^+) are attracted to negatively charged regions, these ions move passively through the tubular epithelium and enter the tubular fluid (fig. 17.14). Potassium ions are also secreted by active processes.

To summarize, urine forms as a result of the following:

- Glomerular filtration of materials from blood plasma.
- Reabsorption of substances, including glucose; water; creatine; amino acids; lactic, citric, and uric acids; and phosphate, sulfate, calcium, potassium, and sodium ions.
- Secretion of substances, including penicillin, histamine, phenobarbital, hydrogen ions, ammonia, and potassium ions.

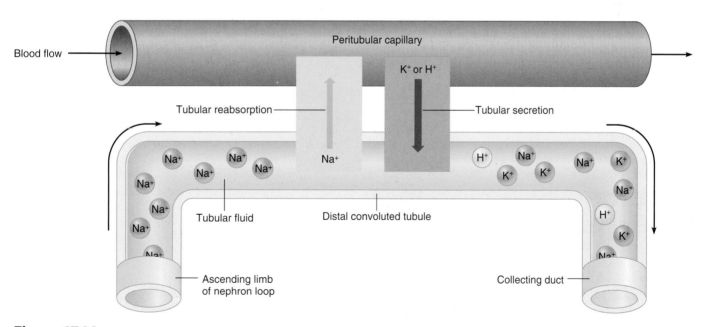

Figure 17.14

In the distal convoluted tubule, potassium ions (and hydrogen ions) may be passively secreted in response to the active reabsorption of sodium ions.

Regulation of Urine Concentration and Volume

The hormones aldosterone and ADH (antidiuretic hormone) may stimulate additional reabsorption of sodium and water, respectively. The changes in sodium and water excretion in response to these hormones are the final adjustments the kidney makes to maintain a constant internal environment.

The adrenal glands secrete aldosterone in response to changes in the blood concentrations of sodium and potassium ions (see chapter 11, p. 306). Aldosterone stimulates the distal convoluted tubule to reabsorb sodium and secrete potassium. Angiotensin II is another important stimulator of aldosterone secretion.

Neurons in the hypothalamus produce ADH, which the posterior pituitary releases in response to a decreasing water concentration in blood or a decrease in blood volume. When ADH reaches the kidney, it increases the water permeability of the epithelial linings of the distal convoluted tubule and collecting duct, and water moves rapidly out of these segments by osmosis—that is, water is reabsorbed. Urine volume falls, and soluble wastes and other substances become more concentrated, which minimizes loss of body fluids when dehydration is likely.

If body fluids have excess water, ADH secretion decreases. As blood levels of ADH drop, the epithelial linings of the distal segment and collecting duct become less permeable to water, less water is reabsorbed, and urine is more dilute, excreting the excess water. Table 17.2 summarizes the role of ADH in urine production.

Urea and Uric Acid Excretion

Urea (u-re′ah) is a by-product of amino acid catabolism. Consequently, its plasma concentration reflects the amount of protein in the diet. Urea enters the renal tubule by filtration. About 80% of it is reabsorbed, and the remainder is excreted in urine.

Uric acid is a product of the metabolism of certain organic bases in nucleic acids. Active transport reabsorbs all the uric acid normally present in glomerular filtrate, but a small amount is secreted into the renal tubule and is excreted in urine. Table 17.3 summarizes

Table 17.2	Role of ADH in Regulating Urine Concentration and Volume
1. Concentration of water in blood decreases.	
2. Increase in osmotic pressure of body fluids stimulates osmoreceptors in hypothalamus of brain.	
3. Hypothalamus signals posterior pituitary to release ADH.	
4. Blood carries ADH to kidneys.	
5. ADH causes distal convoluted tubules and collecting ducts to increase water reabsorption by osmosis.	
6. Urine becomes concentrated, and urine volume decreases.	

some specific functions of the nephron segments and the collecting duct.

Excess uric acid may precipitate in the plasma and be deposited as crystals in joints, causing the inflammation and extreme pain of gout, particularly in the digits and especially in the great toe. Gout has had an interesting history. Hippocrates mentioned it. King Charles I of Spain gave up his vast empire in 1556 due to the painful condition. In 2006, Spanish researchers confirmed the diagnosis by detecting uric acid deposits in the terminal joint of a finger that, for reasons unknown, had been preserved in a small box apart from the rest of the king. Today, gout is treated with drugs that inhibit uric acid reabsorption or block an enzyme in the biosynthetic pathway for uric acid. Limiting foods rich in uric acid, such as organ meats and seafood, and drinking more to dilute urine can help. Gout is inherited, but an attack may not occur until the person eats the offending foods.

Practice

20. How does the hypothalamus regulate urine concentration and volume?

21. Explain how urea and uric acid are excreted.

Urine Composition

Urine composition reflects the volumes of water and amounts of solutes that the kidneys must eliminate from the body or retain in the internal environment to maintain homeostasis. The kidney is able to accomplish this because it handles each of the substances discussed above independently. Urinary excretion of some solutes may increase while that of other solutes may decrease. Urine composition differs considerably from time to time because of variations in dietary intake and physical activity. About 95% water, urine usually contains urea and uric acid. It may also have a trace of amino acids and a variety of electrolytes, whose concentrations vary directly with amounts in the diet (see table 17.1).

The volume of urine produced is usually between 0.6 and 2.5 liters per day, depending on fluid intake, environmental temperature and relative humidity of the surrounding air, as well as the person's emotional condition, respiratory rate, and body temperature. Urine output of 50–60 milliliters per hour is normal; output of less than 30 milliliters per hour may indicate kidney failure.

Glucose, proteins, ketones, and blood cells are not usually found in urine. These unusual urinary components may reflect certain normal circumstances, or disease. For example, glucose in urine may follow a large intake of carbohydrates, precede giving birth, or may be due to diabetes mellitus (see Clinical Application 11.1, p. 310). Proteins may appear following vigorous physical exercise, and ketones may appear after a prolonged fast.

Table 17.3	Functions of Nephron Components
Part	**Function**
Renal corpuscle	
Glomerulus	Filtration of water and dissolved substances from plasma
Glomerular capsule	Receives glomerular filtrate
Renal tubule	
Proximal convoluted tubule	Reabsorption of glucose; amino acids; creatine; lactic, uric, citric, and ascorbic acids; phosphate, sulfate, calcium, potassium, and sodium ions by active transport
	Reabsorption of water by osmosis
	Reabsorption of chloride ions and other negatively charged ions by electrochemical attraction
	Active secretion of substances such as penicillin, histamine, creatinine, and hydrogen ions
Descending limb of nephron loop	Reabsorption of water by osmosis
Ascending limb of nephron loop	Reabsorption of sodium, potassium, and chloride ions by active transport
Distal convoluted tubule	Reabsorption of sodium ions by active transport
	Reabsorption of water by osmosis
	Secretion of hydrogen and potassium ions both actively and passively by electrochemical attraction
Collecting duct	Reabsorption of water by osmosis

Note: Although the collecting duct is not anatomically part of the nephron, it is included here because of its functional importance.

Practice

22. List the normal constituents of urine.
23. Which factors affect urine volume?

17.4 URINE ELIMINATION

After urine forms in the nephrons, it passes from the collecting ducts through openings in the renal papillae and enters the calyces of the kidney (see fig. 17.2). From there, it passes through the renal pelvis, and a ureter conveys it to the urinary bladder (see fig. 17.1 and reference plate 6, p. 28). The urethra passes urine to the outside.

Ureters

Each **ureter** (u-re′ter) is a tube about 25 centimeters long that begins as the funnel-shaped renal pelvis. It descends behind the parietal peritoneum and runs parallel to the vertebral column. In the pelvic cavity, each ureter courses forward and medially, joining the urinary bladder from underneath.

The ureter wall has three layers. The inner layer, or *mucous coat,* is continuous with the linings of the renal tubules and the urinary bladder. The middle layer, or *muscular coat,* consists largely of smooth muscle fibers. The outer layer, or *fibrous coat,* is connective tissue (fig. 17.15).

The muscular walls of the ureters propel the urine. Muscular peristaltic waves, originating in the renal pelvis, force urine along the length of the ureter. When a peristaltic wave reaches the urinary bladder, a jet of urine spurts into the urinary bladder. A flaplike fold of mucous membrane covers the opening through which urine enters the bladder. This fold acts as a valve, allowing urine to enter the bladder from the ureter but preventing it from backing up.

Practice

24. Describe the structure of a ureter.
25. How is urine moved from the renal pelvis to the urinary bladder?
26. What prevents urine from backing up from the urinary bladder into the ureters?

Urinary Bladder

The **urinary bladder** is a hollow, distensible, muscular organ that stores urine and forces it into the urethra (see fig. 17.1 and reference plate 6, p. 28). The bladder is in the pelvic cavity, behind the symphysis pubis and beneath the parietal peritoneum.

The pressure of surrounding organs alters the bladder's somewhat spherical shape. When empty, the inner

Kidney Stones

Kidney stones, which are usually composed of uric acid, calcium oxalate, calcium phosphate, or magnesium phosphate, can form in the collecting ducts and renal pelvis (fig. 17A). Such a stone passing into a ureter causes sudden, severe pain that begins in the region of the kidney and radiates into the abdomen, pelvis, and lower limbs. It may also cause nausea and vomiting, and blood in the urine.

About 60% of kidney stones pass from the body on their own. Other stones were once removed surgically but are now shattered with intense sound waves. In this procedure, called *extracorporeal shock-wave lithotripsy (ESWL),* the patient is placed in a stainless steel tub filled with water. A spark-gap electrode produces underwater shock waves, and a reflector concentrates and focuses the shock-wave energy on the stones. The resulting sandlike fragments then leave in urine.

The tendency to form kidney stones is inherited, particularly the stones that contain calcium, which account for more than half of all cases. Eating calcium-rich foods does not increase the risk, but taking calcium supplements can. People who have calcium oxalate stones can reduce the risk of recurrence by avoiding specific foods: chocolate, coffee, wheat bran, cola, strawberries, spinach, nuts, and tea. Other causes of kidney stones include excess vitamin D, blockage of the urinary tract, or a complication of a urinary tract infection.

Figure 17A
This kidney stone is small, held against this fingertip, but it is large enough to cause severe pain.

It is very helpful for a physician to analyze the composition of the stones, because certain drugs can prevent recurrence. Stones can be obtained during surgery or by the person, using a special collection device.

Figure 17.15
Cross section of a ureter (75×).

— Mucous coat
— Lumen
— Muscular coat
— Fibrous coat
— Adipose tissue

wall of the bladder forms many folds, but as the bladder fills with urine, the wall becomes smoother. At the same time, the superior surface of the bladder expands upward into a dome.

Tissue engineers create replacement urinary bladders for children and teens who have certain birth defects. A postage-stamp-size sample of bladder tissue from a patient is expanded in a dish to 1.5 billion cells, which are seeded onto synthetic, three-dimensional domes. After the cells form confluent layers, the domes are surgically attached to the lower parts of the patient's bladder, after removing the upper parts. The synthetic parts degenerate, leaving a new bladder built from the patient's own cells.

The internal floor of the bladder includes a triangular area called the *trigone,* which has an opening at each of its three angles (fig. 17.16*a*). Posteriorly, at the base of the trigone, the openings are those of the ureters. Anteriorly, at the apex of the trigone, a short, funnel-shaped extension called the *neck* of the bladder contains the opening into the urethra.

The wall of the urinary bladder has four layers. The inner layer, or *mucous coat,* includes several layers of transitional epithelial cells. The thickness of this tissue

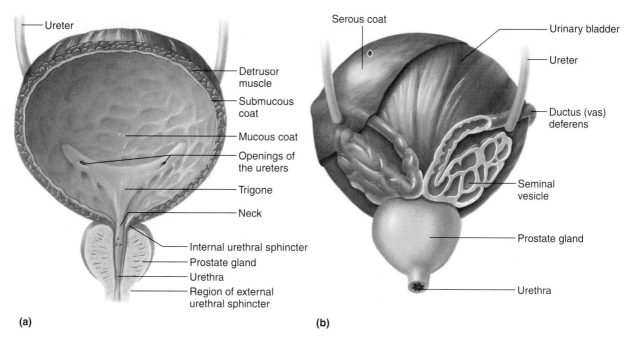

Ureter

Serous coat

Urinary bladder

Ureter

Detrusor muscle

Submucous coat

Mucous coat

Openings of the ureters

Trigone

Neck

Internal urethral sphincter

Prostate gland

Urethra

Region of external urethral sphincter

Ductus (vas) deferens

Seminal vesicle

Prostate gland

Urethra

(a)

(b)

Figure 17.16

A urinary bladder in a male. **(a)** Longitudinal section. **(b)** Posterior view.

changes as the bladder expands and contracts. During distension, the tissue may be only two or three cells thick; during contraction, it may be five or six cells thick (see fig. 5.8, p. 100).

The second layer of the bladder wall is the *submucous coat*. It consists of connective tissue and has many elastic fibers. The third layer of the bladder wall, or *muscular coat*, is composed primarily of coarse bundles of smooth muscle fibers. These bundles are interlaced in all directions and at all depths, and together they comprise the **detrusor muscle** (de-truz'or mus'l). The part of the detrusor muscle that surrounds the neck of the bladder forms an *internal urethral sphincter*. Sustained contraction of this muscle prevents the bladder from emptying until pressure in the bladder increases to a certain level. The detrusor muscle is innervated with parasympathetic nerve fibers that function in the micturition reflex, discussed in the next section.

The outer layer of the bladder wall, or *serous coat*, consists of the parietal peritoneum. This layer is only on the bladder's upper surface. Elsewhere, the outer coat is connective tissue.

Inflammation of the urinary bladder, called *cystitis,* is more common in women than in men because the female urethral pathway is shorter. Infectious agents, such as bacteria, may ascend from the urinary bladder into the ureters because the linings of these structures are continuous. Inflammation of the ureter is called *ureteritis.*

Practice

27. Describe the trigone of the urinary bladder.
28. Describe the structure of the bladder wall.
29. What kind of nerve fibers supply the detrusor muscle?

Micturition

Micturition (mik″tu-rish′un), or urination, is the process that expels urine from the urinary bladder. In micturition, the detrusor muscle contracts, as do muscles in the abdominal wall and pelvic floor. At the same time, muscles in the thoracic wall and diaphragm do not contract. Micturition also requires relaxation of the *external urethral sphincter*. This muscle, which is part of the urogenital diaphragm described in chapter 8 (pp. 201–203), surrounds the urethra about 3 centimeters from the urinary bladder and is composed of voluntary skeletal muscle tissue.

Distension of the bladder wall as it fills with urine stimulates stretch receptors, triggering the micturition reflex. The *micturition reflex center* is in the spinal cord. When sensory impulses from the stretch receptors signal the reflex center, parasympathetic motor impulses travel to the detrusor muscle, which contracts rhythmically in response. A sensation of urgency accompanies this action.

The urinary bladder may hold as much as 600 milliliters of urine before stimulating pain receptors, but the urge to urinate usually begins when it contains about

150 milliliters. As urine volume increases to 300 milliliters or more, the sensation of fullness intensifies, and contractions of the bladder wall become more powerful. When these contractions are strong enough to force the internal urethral sphincter open, another reflex signals the external urethral sphincter to relax, and the bladder can empty.

The external urethral sphincter is composed of skeletal muscle. Therefore it is under conscious control, and typically it is contracted until a person decides to urinate. Nerve centers in the brainstem and cerebral cortex that can partially inhibit the micturition reflex aid this control. When a person decides to urinate, the external urethral sphincter relaxes and the micturition reflex is no longer inhibited. Nerve centers in the pons and the hypothalamus of the brain facilitate the micturition reflex. Then the detrusor muscle contracts, and urine is excreted through the urethra. Within a few moments, the neurons of the micturition reflex undergo adaptation, the detrusor muscle relaxes, and the urinary bladder begins to fill with urine again.

Damage to the spinal cord above the sacral region destroys the sensation of fullness and the voluntary control of urination. However, if the micturition reflex center and its sensory and motor fibers are uninjured, micturition may continue to occur reflexly. In this case, the urinary bladder collects urine until its walls stretch enough to trigger a micturition reflex, and the detrusor muscle contracts in response. This condition is called an *automatic bladder.*

Urethra

The **urethra** (u-re′thrah) is a tube that conveys urine from the urinary bladder to the outside (see fig. 17.1 and reference plate 7, p. 29). Its wall is lined with mucous membrane and has a thick layer of smooth muscle tissue, whose fibers are generally directed longitudinally. The urethral wall also has abundant mucous glands, called *urethral glands,* which secrete mucus into the urethral canal (fig. 17.17).

In a female, the urethra is about 4 centimeters long. Its opening, the *external urethral orifice* (urinary

Figure 17.17
Cross section through the urethra (10×).

meatus) is anterior to the vaginal opening and posterior to the clitoris. In a male, the urethra functions as part of both the urinary system and the reproductive system and extends from the urinary bladder to the tip of the penis.

In a urinary tract infection (UTI), urination is frequent, painful, scant, and may be bloody, with accompanying pelvic pain. Usually pathogenic bacteria in the urinary tract remain outside the cells, and they are easily killed with antibiotic drugs or prevented from attaching to ureter lining cells by exposure to compounds in cranberry and blueberry juices that the person drinks. However, certain bacteria, such as *Escherichia coli,* enter the lining cells, persisting inside the cells and causing recurrent UTIs.

Practice

30. Describe micturition.
31. How is it possible to consciously inhibit the micturition reflex?
32. Describe the structure of the urethra.

Urinary System

Integumentary System

The urinary system compensates for water loss due to sweating. The kidneys and skin both play a role in vitamin D production.

Cardiovascular System

The urinary system controls blood volume. Blood volume and blood pressure play a role in determining water and solute excretion.

Skeletal System

The kidneys and bone tissue work together to control plasma calcium levels.

Lymphatic System

The kidneys control extracellular fluid volume and composition (including lymph).

Muscular System

Muscle tissue controls urine elimination from the bladder.

Digestive System

The kidneys compensate for fluids lost by the digestive system.

Nervous System

The nervous system influences urine production and elimination.

Respiratory System

The kidneys and the lungs work together to control the pH of the internal environment.

Endocrine System

The endocrine system influences urine production.

Reproductive System

The urinary system in males shares organs with the reproductive system. The kidneys compensate for fluids lost from the male and female reproductive systems.

The urinary system controls the composition of the internal environment.

Summary Outline

17.1 Introduction (p. 468)

The urinary system consists of the kidneys, ureters, urinary bladder, and urethra.

17.2 Kidneys (p. 468)

1. Location of the kidneys
   a. The kidneys are high on the posterior wall of the abdominal cavity.
   b. They are behind the parietal peritoneum.
2. Kidney structure
   a. A kidney has a hollow renal sinus.
   b. The ureter expands into the renal pelvis.
   c. Renal papillae project into the renal sinus.
   d. Each kidney is divided into a medulla and a cortex.
3. Kidney functions
   a. The kidneys maintain homeostasis by removing metabolic wastes from blood and excreting them.
   b. They also help regulate red blood cell production; blood volume and blood pressure; and the volume, composition, and pH of body fluids.
4. Renal blood vessels
   a. Arterial blood flows through the renal artery, interlobar arteries, arcuate arteries, cortical radiate arteries, and afferent arterioles to the nephrons.
   b. Venous blood returns through a series of vessels that correspond to the arterial pathways.
5. Nephrons
   a. Nephron structure
      (1) A nephron is the functional unit of the kidney.
      (2) It consists of a renal corpuscle and a renal tubule.
         (a) The corpuscle consists of a glomerulus and a glomerular capsule.
         (b) Segments of the renal tubule include the proximal convoluted tubule, nephron loop (descending and ascending limbs), and distal convoluted tubule, which empties into a collecting duct.
         (3) The collecting duct (technically not part of a nephron) empties into the minor calyx of the renal pelvis.
   b. Blood supply of a nephron
      (1) The glomerular capillary receives blood from the afferent arteriole and passes it to the efferent arteriole.
      (2) The efferent arteriole gives rise to the peritubular capillary system, which surrounds the renal tubule.
   c. Juxtaglomerular apparatus
      (1) The juxtaglomerular apparatus is at the point of contact between the last portion of the ascending limb of the nephron loop and the afferent and efferent arterioles.
      (2) It consists of the macula densa and juxtaglomerular cells.

17.3 Urine Formation (p. 472)

Nephrons remove wastes from blood and regulate water and electrolyte concentrations. Urine is the end product.

1. Glomerular filtration
   a. Urine formation begins when water and dissolved materials filter out of glomerular capillaries.
   b. Glomerular capillaries are much more permeable than the capillaries in other tissues.
   c. The composition of the filtrate is similar to that of tissue fluid.
2. Filtration pressure
   a. Filtration is due mainly to hydrostatic pressure inside glomerular capillaries.
   b. The osmotic pressure of plasma and the hydrostatic pressure in the glomerular capsule also affect filtration.
   c. Net filtration pressure is the net force moving material out of the glomerulus and into the glomerular capsule.
3. Filtration rate
   a. Rate of filtration varies with filtration pressure.
   b. Filtration pressure changes with the diameters of the afferent and efferent arterioles.
   c. As colloid osmotic pressure in the glomerulus increases, filtration rate decreases.
   d. As hydrostatic pressure in a glomerular capsule increases, filtration rate decreases.
   e. The kidneys produce about 125 milliliters of glomerular fluid per minute, most of which is reabsorbed.
4. Regulation of filtration rate
   a. Glomerular filtration rate (GFR) remains relatively constant, but may increase or decrease as required.
   b. Increased sympathetic nerve activity can decrease glomerular filtration rate.
   c. When the macula densa senses decreased amounts of chloride, potassium, and sodium ions in the last part of the ascending limb of the nephron loop, it causes juxtaglomerular cells to release renin.
   d. This triggers a series of changes leading to vasoconstriction of afferent and efferent arterioles, which may affect glomerular filtration rate, and aldosterone secretion, which stimulates tubular sodium reabsorption.
5. Tubular reabsorption
   a. Substances are selectively reabsorbed from glomerular filtrate.
   b. The peritubular capillary's permeability adapts it for reabsorption.
   c. Most reabsorption occurs in the proximal tubule, where epithelial cells have microvilli.
   d. Different modes of transport reabsorb various substances in particular segments of the renal tubule.
      (1) Active transport reabsorbs glucose and amino acids.
      (2) Osmosis reabsorbs water.
   e. Active transport mechanisms have limited transport capacities.
6. Sodium and water reabsorption
   a. Substances that remain in the filtrate are concentrated as water is reabsorbed.
   b. Active transport reabsorbs sodium ions.
   c. As positively charged sodium ions move out of the filtrate, negatively charged ions follow them.
   d. Water is passively reabsorbed by osmosis.
7. Tubular secretion
   a. Secretion transports substances from plasma in the peritubular capillaries to the renal tubular fluid.
   b. Various organic compounds are secreted actively.
   c. Potassium and hydrogen ions are secreted both actively and passively.
8. Regulation of urine concentration and volume
   a. Most sodium is reabsorbed before urine is excreted.
   b. Antidiuretic hormone increases the permeability of the distal convoluted tubule and collecting duct, promoting water reabsorption.

9. Urea and uric acid excretion
   a. Diffusion passively reabsorbs 80% of the urea.
   b. Active transport reabsorbs uric acid. Some uric acid is secreted into the renal tubule.
10. Urine composition
    a. Urine is about 95% water, and it also usually contains urea and uric acid.
    b. Urine contains varying amounts of electrolytes and may contain a trace of amino acids.
    c. Urine volume varies with fluid intake, certain environmental factors, and a person's emotional condition, respiratory rate, and body temperature.

17.4 Urine Elimination (p. 481)

1. Ureters
   a. The ureter extends from the kidney to the urinary bladder.
   b. Peristaltic waves in the ureter force urine to the urinary bladder.
2. Urinary bladder
   a. The urinary bladder stores urine and forces it through the urethra during micturition.
   b. The openings for the ureters and urethra are at the three angles of the trigone.
   c. A portion of the detrusor muscle forms an internal urethral sphincter.
3. Micturition
   a. Micturition expels urine.
   b. Micturition contracts the detrusor muscle and relaxes the external urethral sphincter.
   c. Micturition reflex
      (1) Distension stimulates stretch receptors in the bladder wall.
      (2) The micturition reflex center in the spinal cord sends parasympathetic motor impulses to the detrusor muscle.
      (3) As the bladder fills, its internal pressure increases, forcing the internal urethral sphincter open.
      (4) A second reflex relaxes the external urethral sphincter unless voluntary control maintains its contraction.
      (5) Nerve centers in the cerebral cortex and brainstem aid control of urination.
4. Urethra
   The urethra conveys urine from the urinary bladder to the outside.

Chapter Assessments

17.1 Introduction

1. Name and identify the general functions of the organs of the urinary system. (p. 468)

17.2 Kidneys

2. Explain why the kidneys are said to be retroperitoneal. (p. 468)
3. Describe the external and internal structure of a kidney. (p. 468)
4. Identify the functions of the kidneys. (p. 469)
5. List in correct order the vessels through which blood passes as it travels from the renal artery to the renal vein. (pp. 469 and 471)
6. Distinguish between a renal corpuscle and a renal tubule. (p. 470)
7. Name in correct order the parts of the nephron through which fluid passes from the glomerulus to the collecting duct. (p. 470)
8. Describe the location and structure of the juxtaglomerular apparatus. (p. 471)

17.3 Urine Formation

9. Which one of the following is abundant in blood plasma, but only in small amounts in glomerular filtrate? (p. 475)
   a. sodium ions
   b. water
   c. glucose
   d. protein
   e. potassium ions
10. Define *net filtration pressure*. (p. 475)
11. Explain how the diameters of the afferent and efferent arterioles affect the rate of glomerular filtration. (p. 475)
12. Explain how changes in the osmotic pressure of blood plasma affect the glomerular filtration rate. (p. 475)
13. Explain how the hydrostatic pressure of a glomerular capsule affects the rate of glomerular filtration. (p. 475)
14. Describe two mechanisms by which the body regulates filtration rate. (p. 476)
15. Discuss how tubular reabsorption is selective. (p. 477)
16. Explain how the peritubular capillary is adapted for reabsorption. (p. 477)
17. Explain how epithelial cells of the proximal convoluted tubule are adapted for reabsorption. (p. 477)
18. Explain why active transport mechanisms have limited transport capacities. (p. 478)
19. Define *renal plasma threshold*. (p. 478)
20. Explain how amino acids and proteins are reabsorbed. (p. 478)
21. Describe the effect of sodium reabsorption on the reabsorption of negatively charged ions. (p. 478)
22. Explain how sodium reabsorption affects water reabsorption. (p. 478)
23. Explain how potassium ions may be secreted passively. (p. 479)
24. The major action of ADH in the kidneys is to: (p. 480)
    a. increase water absorption by the proximal convoluted tubule.
    b. increase glomerular filtration rate.
    c. increase water reabsorption by the collecting duct.
    d. increase potassium's excretion.
    e. increase sodium's excretion.
25. Compare the processes that reabsorb urea and uric acid. (p. 480)
26. List the common constituents of urine and their sources. (p. 480)
27. Identify some of the factors that affect the volume of urine produced daily. (p. 480)

17.4 Urine Elimination

28. Describe the structure and function of a ureter. (p. 481)

29. Explain how the muscular wall of the ureter helps move urine. (p. 481)

30. Describe the structure and location of the urinary bladder. (p. 481)

31. Define *detrusor muscle.* (p. 483)

32. Distinguish between the internal and external urethral sphincters. (p. 483)

33. Describe the micturition reflex. (p. 483)

34. Which of the following involves skeletal muscle? (p. 484).

   a. contraction of the internal urethral sphincter

   b. contraction of the external urethral sphincter

   c. ureteral peristalsis

   d. detrusor muscle contraction

   e. all of the above

Integrative Assessments/Critical Thinking

OUTCOMES 3.3, 17.1, 17.2, 17.3

1. Imagine you are adrift at sea. Why will you dehydrate more quickly if you drink seawater instead of fresh water to quench your thirst?

2. Why are people following high-protein diets advised to drink large volumes of water?

OUTCOMES 13.5, 17.2, 17.3

3. If blood pressure plummets in a patient in shock as a result of a severe injury, how would you expect urine production to change? Why?

OUTCOMES 17.2, 17.3

4. Why may protein in the urine be a sign of kidney damage? What structures in the kidney are probably affected?

5. An infant is born with narrowed renal arteries. What effect will this condition have on urine volume?

OUTCOMES 17.2, 17.4

6. Why do urinary tract infections frequently accompany sexually transmitted diseases?

WEB CONNECTIONS

Visit the text website at **www.mhhe.com/shieress11** for additional quizzes, interactive learning exercises, and more

APR

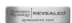

Anatomy & Physiology REVEALED includes cadaver photos that allow you to peel away layers of the human body to reveal structures beneath the surface. This program also includes animations, radiologic imaging, audio pronunciations, and practice quizzing. To learn more, visit **www.aprevealed.com**.

18

Water, Electrolyte, and Acid-Base Balance

An extremely high environmental temperature can challenge the body's ability to maintain a normal body temperature.

Heatstroke can kill. Heatstroke, a form of hyperthermia, is the result of extreme environmental heat that can be quickly fatal. It occurs when the body is exposed to a heat index (heat considering humidity) of more than 105°F. Under these conditions, evaporation of sweat becomes less efficient at cooling the body, and the body temperature may exceed 106°F, causing organ failure.

The symptoms of heatstroke happen in a sequence. First come headache, dizziness, and exhaustion. Sweating is profuse, then stops, as the skin becomes dry, hot, and red. Respiratory rate rises, and the pulse may race up to 180 beats per minute. If there is no cooling with fluids, water applied to the skin, fanning, and removal of clothing, neurological symptoms may begin. These include disorientation, hallucinations, and odd behavior. Kidney failure and/or heart arrhythmia may prove fatal.

During heat waves, the very young and the very old are more susceptible to heatstroke, because their temperature control mechanisms may be compromised. However, heatstroke also affects two groups of young, otherwise healthy individuals—athletes who work out in extreme heat and soldiers deployed to hot climates. In the Persian Gulf in July, the temperatures may soar to 122°F at midday, dipping down only to about 100°F at night. Some cases of unexpected sudden death in soldiers in Iraq and Afghanistan were due to heatstroke. For this reason, military officials insist that soldiers carry drinking water with them at all times and drink throughout the day, whether they feel thirsty or not.

Clues to the cause of heatstroke have come from mice bred to have a version of a disease that affects humans, called malignant hyperthermia.

In this condition, a patient receiving general anesthesia experiences increased heart rate, whole-body muscle rigidity, and a spike in body temperature up to 112°F. Affected mice die upon exposure to anesthesia—and also when subjected to 105°F heat. The gene encodes a receptor that activates calcium entry into skeletal muscles. In the mouse mutant, under intense heat, too many calcium ions enter cells, causing uncontrolled contraction. In addition, calcium ions move from the receptor area and free radicals are released, which causes oxidative damage. Adding an antioxidant drug to the animals' water dampened their reaction to heat.

The experiments done with mice that were genetically susceptible to heatstroke suggest that people, particularly those who have had malignant hyperthermia, might be predisposed to react dangerously to heat, too. Researchers are using the mice as models to test treatments for heatstroke.

Learning Outcomes

After studying this chapter, you should be able to do the following:

18.1 Introduction
1. Explain water and electrolyte balance. (p. 490)

18.2 Distribution of Body Fluids
2. Explain body fluid distribution in compartments. (p. 490)

18.3 Water Balance
3. List the routes by which water enters and leaves the body, and explain how water intake and output are regulated. (p. 492)

18.4 Electrolyte Balance
4. Explain how electrolytes enter and leave the body, and describe how electrolyte intake and output are regulated. (p. 493)

18.5 Acid-Base Balance
5. List the major sources of hydrogen ions in the body. (p. 497)
6. Distinguish between strong and weak acids and bases. (p. 497)

7. Explain how chemical buffer systems, the respiratory center, and the kidneys keep the pH of body fluids constant. (p. 498)

18.6 Acid-Base Imbalances
8. Describe the causes and consequences of an increase or decrease in body fluid pH. (p. 500)

 Learn   Practice Assess

Aids to Understanding Words   *(Appendix A on page 564 has a complete list of Aids to Understanding Words.)*

de- [separation from] *de*hydration: Removal of water from the cells or body fluids.

extra- [outside] *extra*cellular fluid: Fluid outside of the body cells.

im- [not] *im*balance: Condition in which factors are not in equilibrium.

intra- [within] *intra*cellular fluid: Fluid in body cells.

neutr- [neither one nor the other] *neutr*al: Solution that is neither acidic nor basic.

18.1 INTRODUCTION

Two types of substances that are important in maintaining homeostasis in the body are water and **electrolytes,** which are molecules that release ions in water. To maintain homeostasis, the quantities of water and electrolytes must be in balance; that is, the amounts entering the body must equal the amounts leaving it. Therefore, the body requires mechanisms to (1) replace lost water and electrolytes, and (2) excrete any excess.

Water balance and electrolyte balance are interdependent because electrolytes are dissolved in the water of body fluids. Consequently, anything that alters the concentrations of the electrolytes will alter the concentration of the water either by adding solutes to it or by removing solutes from it. Likewise, anything that changes the concentration of water will change concentrations of the electrolytes by concentrating or diluting them.

> A human being is 60% water by weight. Losing one-fifth of that water volume can be fatal.

18.2 DISTRIBUTION OF BODY FLUIDS

Body fluids are not uniformly distributed. Instead, they occupy regions, or *compartments,* of different volumes that contain fluids of varying compositions. The movement of water and electrolytes between these compart-

ments is regulated to stabilize both the distribution and the composition of body fluids.

Fluid Compartments

The body of an average adult female is about 52% water by weight, and that of an average male is about 63% water. The reason for this difference is that females generally have more adipose tissue, which contains little water. Males generally have more muscle tissue, which contains a great deal of water. Water in the adult human body (about 40 liters), with its dissolved electrolytes, is distributed into two major compartments: an intracellular fluid compartment and an extracellular fluid compartment.

The **intracellular** (in″trah-sel′u-lar) **fluid compartment** includes all the water and electrolytes that cell membranes enclose. In other words, intracellular fluid is the fluid inside cells, and in an adult it accounts for about 63% by volume of total body water.

The **extracellular** (ek″strah-sel′u-lar) **fluid compartment** includes all the fluid outside of cells—in tissue spaces (interstitial fluid), blood vessels (plasma), and lymphatic vessels (lymph). Epithelial layers separate a specialized fraction of extracellular fluid from other extracellular fluids. This **transcellular** (trans-sel′u-lar) **fluid** includes *cerebrospinal fluid* of the central nervous system, *aqueous* and *vitreous humors* of the eyes, *synovial fluid* of the joints, *serous fluid* in the body cavities, and fluid *secretions* of the exocrine glands. The fluids of the extracellular compartment constitute about 37% by volume of total body water (fig. 18.1).

Figure 18.1

Cell membranes separate fluid in the intracellular compartment from fluid in the extracellular compartment. Approximately two-thirds of the water in the body is inside cells.

Body Fluid Composition

Extracellular fluids generally are similar in composition, including high concentrations of sodium, chloride, calcium, and bicarbonate ions and lesser concentrations of potassium, magnesium, phosphate, and sulfate ions. The blood plasma fraction of extracellular fluid has considerably more protein than does either interstitial fluid or lymph.

Intracellular fluid has high concentrations of potassium, phosphate, and magnesium ions. It includes a greater concentration of sulfate ions and lesser concentrations of sodium, chloride, calcium, and bicarbonate ions than does extracellular fluid. Intracellular fluid also has a greater concentration of protein than plasma. Figure 18.2 shows these relative concentrations.

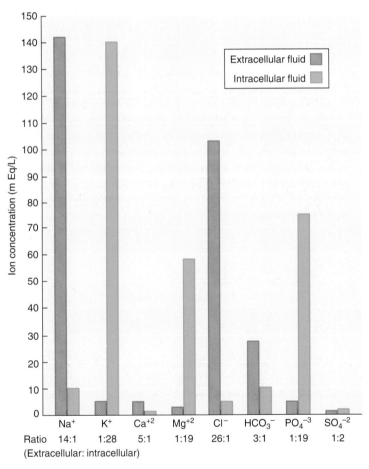

Figure 18.2

Extracellular fluids have relatively high concentrations of sodium (Na^+), calcium (Ca^{+2}), chloride (Cl^-), and bicarbonate (HCO_3^-) ions. Intracellular fluid has relatively high concentrations of potassium (K^+), magnesium (Mg^{+2}), phosphate (PO_4^{-3}), and sulfate (SO_4^{-2}) ions.

Q: *According to the graph, which positively charged intracellular fluid ion is in the highest concentration?*

Answer can be found in Appendix E on page 568.

Movement of Fluid Between Compartments

Two major factors regulate the movement of water and electrolytes from one fluid compartment to another: *hydrostatic pressure* and *osmotic pressure* (fig. 18.3). For example, as chapter 13 explained (pp. 357–358), fluid leaves the plasma at the arteriolar ends of capillaries and enters the interstitial spaces because of the net outward force of hydrostatic pressure (blood pressure). Fluid returns to the plasma from the interstitial spaces at the venular ends of capillaries because of the net inward force of *colloid osmotic pressure* due to the plasma proteins. Likewise, as chapter 14 mentioned (p. 380), fluid leaves the interstitial spaces and enters the lymph capillaries due to the hydrostatic pressure of the interstitial fluid. The circulation of lymph returns interstitial fluid to the plasma.

Pressures similarly control fluid movement between the intracellular and extracellular compartments. Because hydrostatic pressure within the cells and surrounding interstitial fluid is ordinarily equal and stable, a

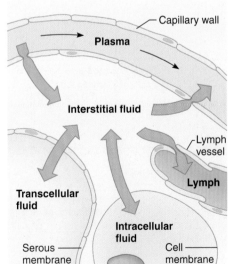

Figure 18.3

Net movements of fluids between compartments result from differences in hydrostatic and osmotic pressures.

change in intracellular or extracellular osmotic pressure is the likely cause of any net fluid movement.

The sodium ion concentration in extracellular fluids is especially high. A decrease in extracellular sodium ion concentration causes a net movement of water from the extracellular compartment into the intracellular compartment by osmosis. The cell swells. Conversely, if the extracellular sodium ion concentration increases, cells shrink as they lose water by osmosis.

Practice

5. Which factors control the movement of water and electrolytes from one fluid compartment to another?

6. How does the sodium ion concentration of body fluids affect the net movement of water between the compartments?

18.3 WATER BALANCE

Water balance exists when water intake equals water output. Homeostasis requires control of both water intake and water output. Clinical Application 18.1 discusses water balance disorders, including dehydration, edema, and water intoxication.

Water Intake

The volume of water gained each day varies among individuals. An average adult living in a moderate envi-

ronment takes in about 2,500 milliliters. Probably 60% is obtained from drinking water or beverages, and another 30% comes from moist foods. The remaining 10% is a by-product of the oxidative metabolism of nutrients (chapter 4, p. 80) and is called **water of metabolism** (fig. 18.4*a*).

> The kangaroo rat is a desert rodent that does not have to drink water. It can survive on the water of metabolism alone.

Regulation of Water Intake

The primary regulator of water intake is thirst. The intense feeling of thirst derives from the osmotic pressure of extracellular fluids and a *thirst center* in the hypothalamus of the brain. As the body loses water, the osmotic pressure of extracellular fluids increases. Such a change stimulates *osmoreceptors* in the thirst center, which cause the person to feel thirsty and to seek water.

Thirst is a homeostatic mechanism, normally triggered when total body water decreases by as little as 1%. Drinking distends the stomach wall, triggering nerve impulses that inhibit the thirst mechanism. In this way, drinking stops even before the swallowed water is absorbed, preventing the person from drinking more than is required to replace the volume lost, avoiding development of an imbalance.

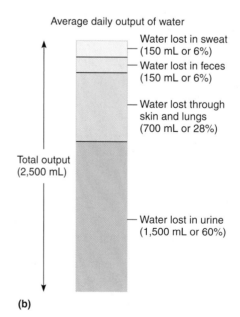

Figure 18.4

Water balance. **(a)** Major sources of body water. **(b)** Routes by which the body loses water.

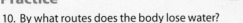
Water Output

Water normally enters the body only through the mouth, but it can be lost by a variety of routes. These include obvious losses in urine, feces, and sweat (sensible perspiration), as well as evaporation of water from the skin (insensible perspiration) and from the lungs during breathing.

If an average adult takes in 2,500 milliliters of water each day, then 2,500 milliliters must be eliminated to maintain water balance. Of this volume, perhaps 60% is lost in urine, 6% in feces, and 6% in sweat. About 28% is lost by evaporation from the skin and lungs (fig. 18.4*b*). These percentages vary with such environmental factors as temperature and relative humidity and with physical exercise.

Regulation of Water Output

The primary means of regulating water output is urine production. The distal convoluted tubules of the nephrons and collecting ducts are the effectors of the mechanism that regulates urine volume. The epithelial linings in these structures remain relatively impermeable to water unless antidiuretic hormone (ADH) is present. ADH increases the permeability of the distal convoluted tubule and collecting duct, thereby increasing water reabsorption and reducing urine production. In the absence of ADH, less water is reabsorbed and more urine is produced (see chapter 17, p. 480).

Diuretics are chemicals that promote urine production. They act in different ways. Alcohol and certain narcotic drugs promote urine formation by inhibiting ADH release. Caffeine inhibits the reabsorption of sodium ions or other solutes in parts of the renal tubules. As a consequence, the osmotic pressure of the tubular fluid increases, reducing osmotic reabsorption of water and increasing urine volume.

18.4 ELECTROLYTE BALANCE

Electrolyte balance exists when the quantities of electrolytes the body gains equal those lost. Homeostatic mechanisms maintain electrolyte balance.

Electrolyte Intake

The electrolytes of greatest importance to cellular functions dissociate to release sodium, potassium, calcium, magnesium, chloride, sulfate, phosphate, bicarbonate, and hydrogen ions. These electrolytes are obtained primarily from foods, but they may also be found in drinking water and other beverages. In addition, some electrolytes are by-products of metabolic reactions.

Regulation of Electrolyte Intake

Ordinarily, responding to hunger and thirst provides sufficient electrolytes. A severe electrolyte deficiency may produce a *salt craving,* which is a strong desire to eat salty foods.

Electrolyte Output

The body loses some electrolytes by perspiring, with more lost in sweat on warmer days and during strenuous exercise. Varying amounts of electrolytes are lost in the feces. The greatest electrolyte output is a result of kidney function and urine production. The kidneys alter electrolyte output to maintain balance.

Regulation of Electrolyte Output

Precise concentrations of positively charged ions, such as sodium (Na^+), potassium (K^+), and calcium (Ca^{+2}), are required for nerve impulse conduction, muscle fiber contraction, and maintenance of cell membrane potential. *Sodium ions* account for nearly 90% of positively charged ions in extracellular fluids. The kidneys and the hormone aldosterone regulate these ions. Aldosterone, which the adrenal cortex secretes, increases sodium ion reabsorption in the distal convoluted tubules of the kidneys' nephrons and collecting ducts.

Aldosterone also regulates potassium ions. A rising potassium ion concentration directly stimulates the adrenal cortex to secrete aldosterone. This hormone enhances tubular reabsorption of sodium ions and at

Clinical Application 18.1

Water Balance Disorders

Dehydration, water intoxication, and edema are among the more common disorders that involve a water imbalance in body fluids.

Dehydration

In *dehydration,* water output exceeds water intake. Dehydration may develop following excessive sweating or as a result of prolonged water deprivation accompanied by continued water output. The extracellular fluid becomes more concentrated, and water leaves cells by osmosis (fig. 18A). Dehydration may also accompany prolonged vomiting or diarrhea that depletes body fluids.

During dehydration, the skin and mucous membranes of the mouth feel dry, and body weight drops. Severe hyper-

thermia may develop as the body's temperature-regulating mechanism falters due to lack of water for sweat.

Infants are more likely to become dehydrated because their kidneys are less efficient at conserving water than those of adults. Elderly people are also especially susceptible to developing water imbalances because the sensitivity of their thirst mechanism decreases with age, and physical disabilities may make it difficult for them to obtain adequate fluids.

The treatment for dehydration is to replace the lost water and electrolytes. If only water is replaced, the extracellular fluid will become more dilute than normal. This may produce a condition called water intoxication.

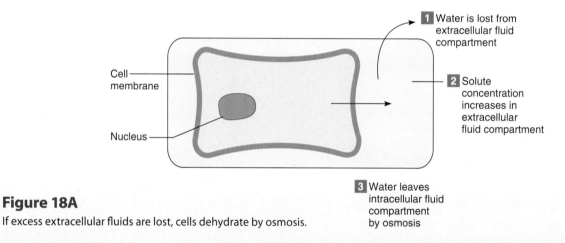

Figure 18A

If excess extracellular fluids are lost, cells dehydrate by osmosis.

Figure 18B

If excess water is added to the extracellular fluid compartment, cells gain water by osmosis.

Water Intoxication

Until recently, runners were advised to drink as much fluid as they could, particularly in long events. But the death of a young woman in the 2002 Boston Marathon from low blood sodium (*hyponatremia*) due to excessive water intake leading to *water intoxication* inspired further study and a reevaluation of this advice. Researchers from Harvard Medical School studied 488 runners from the race and found that 13% of them had developed hyponatremia. Tendency to develop the condition was associated with longer race time, high or low body mass index, and significant weight gain during the race. Drinking sports drinks instead of water does not make a difference because these beverages are mostly water.

In recognition of the possibility of hyponatremia, USA Track and Field, the national governing body for the sport, advises runners how to determine exactly how much water to drink during a one-hour training run. The goal is to replace exactly what is lost.

Edema

Edema is an abnormal accumulation of extracellular fluid in the interstitial spaces (fig. 18B). Causes include a decrease in the plasma protein concentration (*hypoproteinemia*), obstructions in lymphatic vessels, increased venous pressure, and increased capillary permeability.

Hypoproteinemia may result from failure of the liver to synthesize plasma proteins; kidney disease (glomerulonephritis) that damages glomerular capillaries, allowing proteins to enter the urine; or starvation, in which amino acid intake is insufficient to support synthesis of plasma proteins.

In each of these instances, the plasma protein concentration is decreased, which decreases plasma colloid osmotic pressure, reducing the normal return of tissue fluid to the venular ends of capillaries. Consequently, tissue fluid accumulates in the interstitial spaces.

Edema may result from *lymphatic obstructions* due to surgery or parasitic infections of lymphatic vessels, as discussed in chapter 14 (p. 381). Back pressure develops in the lymphatic vessels, which interferes with the normal movement of tissue fluid into them. At the same time, proteins that the lymphatic circulation ordinarily removes accumulate in interstitial spaces, raising the osmotic pressure of interstitial fluid. This effect draws still more fluid into the interstitial spaces.

If blood outflow from the liver into the inferior vena cava is blocked, the venous pressure in the liver and portal blood vessels greatly increases. As a result, fluid with a high protein concentration is exuded from the surfaces of the liver and intestine into the peritoneal cavity. This elevates the osmotic pressure of the abdominal fluid, which in turn draws more water into the peritoneal cavity by osmosis. This condition, called *ascites,* is painful.

Edema may also result from increased capillary permeability accompanying *inflammation*. Recall that inflammation is a response to tissue damage and usually releases chemicals such as histamine from damaged cells. Histamine causes vasodilation and increased capillary permeability. As a result, excess fluid leaks out of the capillary and enters the interstitial spaces. Table 18A summarizes the factors that result in edema.

Table 18A	Factors Associated with Edema	
Factor	**Cause**	**Effect**
Low plasma protein concentration	Liver disease and failure to synthesize proteins; kidney disease and loss of proteins in urine; lack of proteins in diet due to starvation	Plasma colloid osmotic pressure decreases; less fluid enters venular ends of capillaries by osmosis
Obstruction of lymphatic vessels	Surgical removal of portions of lymphatic pathways; certain parasitic infections	Back pressure in lymph vessels interferes with movement of fluid from interstitial spaces into lymph capillaries
Increased venous pressure	Venous obstructions or faulty venous valves	Back pressure in veins increases capillary filtration and interferes with return of fluid from interstitial spaces into venular ends of capillaries
Inflammation	Tissue damage	Capillaries become abnormally permeable; fluid leaks from plasma into interstitial spaces

Clinical Application 18.2

Sodium and Potassium Imbalances

Extracellular fluids usually have high sodium ion concentrations, and intracellular fluid usually has a high potassium ion concentration. Renal regulation of sodium is closely related to that of potassium, because active reabsorption of sodium (under the influence of aldosterone) is accompanied by tubular secretion (and excretion) of potassium. Therefore, conditions resulting from sodium ion imbalance often also involve potassium ion imbalance.

Such disorders include:

1. *Low sodium concentration (hyponatremia)* Possible causes of sodium deficiencies include prolonged sweating, vomiting, or diarrhea; renal disease in which sodium is inadequately reabsorbed; adrenal cortex disorders in which aldosterone secretion is insufficient to promote sodium reabsorption (Addison disease); and drinking too much water. One possible effect of hyponatremia is the development of hypotonic extracellular fluid that promotes water movement into cells by osmosis, producing symptoms of water intoxication.

2. *High sodium concentration (hypernatremia)* Possible causes of elevated sodium concentration include excess water loss by evaporation (despite decreased sweating, as may occur during high fever), or increased water loss accompanying diabetes insipidus. In one form of dia-

betes insipidus, the secretion of antidiuretic hormone (ADH) is insufficient for renal tubules and collecting ducts to maintain water balance. Hypernatremia may disturb the central nervous system, causing confusion, stupor, and coma.

3. *Low potassium concentration (hypokalemia)* Possible causes of potassium deficiency include the release of excess aldosterone by the adrenal cortex (Cushing syndrome), which increases renal excretion of potassium; use of diuretic drugs that promote potassium excretion; kidney disease; and prolonged vomiting or diarrhea. Possible effects of hypokalemia include muscular weakness or paralysis, respiratory difficulty, and severe cardiac disturbances, such as atrial or ventricular arrhythmias.

4. *High potassium concentration (hyperkalemia)* Possible causes of elevated potassium concentration include renal disease, which decreases potassium excretion; use of drugs that promote renal conservation of potassium; the release of insufficient aldosterone by the adrenal cortex (Addison disease); or a shift of potassium from intracellular to extracellular fluid, a change that accompanies an increase in plasma hydrogen ion concentration (acidosis). Possible effects of hyperkalemia include paralysis of the skeletal muscles and severe cardiac disturbances, such as cardiac arrest.

the same time causes tubular secretion of potassium ions (fig. 18.5). Conditions caused by an imbalance of these ions are discussed in Clinical Application 18.2.

Recall from chapter 11 (p. 303) that the calcium ion concentration dropping below normal directly stimulates the parathyroid glands to secrete parathyroid hormone. This hormone returns the concentration of calcium in extracellular fluids toward normal.

Generally the regulatory mechanisms that control positively charged ions secondarily control the concentrations of negatively charged ions. For example, renal tubules passively reabsorb chloride ions (Cl^-), the most abundant negatively charged ions in extracellular fluids, in response to active tubular reabsorption of sodium ions. That is, the negatively charged chloride ions are electrically attracted to positively charged sodium ions and accompany them as they are reabsorbed (see chapter 17, p. 475).

Active transport mechanisms with limited transport capacities partially regulate some negatively charged ions, such as phosphate ions (PO_4^{-3}) and sulfate ions (SO_4^{-2}). Therefore, if extracellular phosphate ion

Figure 18.5

If potassium ion concentration increases, the kidneys conserve sodium ions and excrete potassium ions.

concentration is low, renal tubules reabsorb phosphate ions. On the other hand, if the renal plasma threshold is exceeded, excess phosphate is excreted in urine.

Practice

15. How does aldosterone regulate sodium and potassium ion concentration?

16. How is calcium regulated?

17. What mechanism regulates the concentrations of most negatively charged ions?

18.5 ACID-BASE BALANCE

Electrolytes that dissociate in water and release hydrogen ions are called **acids,** and electrolytes that release ions that combine with hydrogen ions are called **bases,** as chapter 2 discussed (p. 39). Maintenance of homeostasis depends on controlling the concentrations of acids and bases in body fluids.

Sources of Hydrogen Ions

Most of the hydrogen ions in body fluids originate as by-products of metabolic processes, although the digestive tract may directly absorb some hydrogen ions. The major metabolic sources of hydrogen ions include the following (fig. 18.6):

1. **Aerobic respiration of glucose**   This process produces carbon dioxide and water. Carbon dioxide diffuses out of the cells and reacts with the water in the extracellular fluids to form *carbonic acid,* which then ionizes to release hydrogen ions and bicarbonate ions:

$$H_2CO_3 \longrightarrow H^+ + HCO_3^-$$

2. **Anaerobic respiration of glucose**   Anaerobically metabolized glucose produces *lactic acid,* which adds hydrogen ions to body fluids.

3. **Incomplete oxidation of fatty acids**   This process produces *acidic ketone bodies,* which increase hydrogen ion concentration.

4. **Oxidation of amino acids containing sulfur**   This process yields *sulfuric acid* (H_2SO_4), which ionizes to release hydrogen ions.

5. **Breakdown (hydrolysis) of phosphoproteins and nucleic acids**   Phosphoproteins and nucleic acids contain phosphorus. Their oxidation produces *phosphoric acid* (H_3PO_4), which ionizes to release hydrogen ions.

The acids resulting from metabolism vary in strength. Therefore, their effects on the hydrogen ion concentration of body fluids vary.

Practice

18. Distinguish between an acid and a base.

19. What are the major sources of hydrogen ions in the body?

Strengths of Acids and Bases

Acids that ionize to release hydrogen ions more completely are *strong acids,* and those that ionize to release hydrogen ions less completely are *weak acids.* For example, the hydrochloric acid (HCl) of gastric juice is a strong acid, but the carbonic acid (H_2CO_3) produced when carbon dioxide reacts with water is weak.

Bases release ions, such as hydroxide ions (OH^-), which can combine with hydrogen ions and thereby lower their own concentration. Thus, sodium hydroxide (NaOH), which releases hydroxide ions, and sodium bicarbonate (NaHCO$_3$), which releases bicarbonate ions (HCO_3^-), are bases. Strong bases dissociate to release more OH^- or its equivalent than do weak bases. Often the negative ions themselves are called bases. For example, HCO_3^- acting as a base combines with H^+ from the strong acid HCl to form the weak acid carbonic acid (H_2CO_3).

Figure 18.6

Some of the metabolic processes that provide hydrogen ions.

Regulation of Hydrogen Ion Concentration

Chemical buffer systems, the respiratory center in the brainstem, and the nephrons in the kidneys regulate hydrogen ion concentration in body fluids. The pH scale is used to measure hydrogen ion concentration.

Chemical Buffer Systems

Chemical buffer systems, in all body fluids, consist of chemicals that combine with excess acids or bases. More specifically, the chemical components of a buffer system can combine with strong acids to convert them into weak acids. Likewise, these buffers can combine with strong bases to convert them into weak bases. Such actions help minimize pH changes in body fluids. The three most important chemical buffer systems in body fluids are:

1. **Bicarbonate buffer system** The bicarbonate buffer system, which is present in both intracellular and extracellular fluids, uses the bicarbonate ion (HCO_3^-), acting as a weak base, and carbonic acid (H_2CO_3), acting as a weak acid. In the presence of excess hydrogen ions, bicarbonate ions combine with hydrogen ions to form carbonic acid, thereby minimizing any increase in the hydrogen ion concentration of the body fluids:

$$H^+ + HCO_3^- \longrightarrow H_2CO_3$$

On the other hand, if conditions are basic or alkaline, carbonic acid dissociates to release bicarbonate ions and hydrogen ions:

$$H_2CO_3 \longrightarrow H^+ + HCO_3^-$$

It is important to remember that even though this reaction releases bicarbonate ions, the increase of free hydrogen ions at equilibrium is what minimizes the shift toward a more alkaline pH.

2. **Phosphate buffer system** The phosphate buffer system also operates in both intracellular and extracellular body fluids. However, it is particularly important in the control of hydrogen ion concentrations in the tubular fluid of the nephrons and in urine. This buffer system consists of two phosphate ions, monohydrogen phosphate (HPO_4^{-2}) and dihydrogen phosphate ($H_2PO_4^-$). Under acidic conditions, monohydrogen phosphate ions react with hydrogen ions to produce dihydrogen phosphate:

$$H^+ + HPO_4^{-2} \longrightarrow H_2PO_4^-$$

On the other hand, if conditions are basic or alkaline, dihydrogen phosphate ions release hydrogen ions:

$$H_2PO_4^- \longrightarrow H^+ + HPO_4^{-2}$$

3. **Protein buffer system** The protein buffer system consists of the plasma proteins, such as albumins, and certain proteins in cells, including the hemoglobin of red blood cells. As described in chapter 2 (pp. 43–44), proteins are chains of amino acids. Some of these amino acids have freely exposed amino groups ($-NH_2$). When the solution pH falls, these amino groups can accept hydrogen ions:

$$-NH_2 + H^+ \longrightarrow -NH_3^+$$

Some amino acids of a protein also have freely exposed *carboxyl groups* ($-COOH$). When the solution pH rises, these carboxyl groups can ionize, releasing hydrogen ions:

$$-COOH \longrightarrow -COO^- + H^+$$

Therefore, protein molecules can function as bases by accepting hydrogen ions into their amino groups or as acids by releasing hydrogen ions from their carboxyl groups. This special property allows protein molecules to operate as an acid-base buffer system, minimizing changes in pH.

Table 18.1 summarizes the actions of the three major chemical buffer systems.

Neurons are particularly sensitive to changes in the pH of body fluids. If the interstitial fluid becomes more alkaline than normal (alkalosis), neurons become more excitable and seizures may result. Conversely, acidic conditions (acidosis) depress neuron activity, reducing the level of consciousness.

Table 18.1	Chemical Buffer System	
Buffer System	**Constituents**	**Actions**
Bicarbonate system	Bicarbonate ion (HCO_3^-)	Combines with a hydrogen ion in the presence of excess acid
	Carbonic acid (H_2CO_3)	Releases a hydrogen ion in the presence of excess base
Phosphate system	Monohydrogen phosphate ion (HPO_4^{-2})	Combines with a hydrogen ion in the presence of excess acid
	Dihydrogen phosphate ion ($H_2PO_4^-$)	Releases a hydrogen ion in the presence of excess base
Protein system (and amino acids)	$-NH_2$ group of an amino acid or protein	Combines with a hydrogen ion in the presence of excess acid
	$-COOH$ group of an amino acid or protein	Releases a hydrogen ion in the presence of excess base

Practice

20. What is the difference between a strong acid or base and a weak acid or base?
21. How does a chemical buffer system help regulate the pH of body fluids?
22. List the major chemical buffer systems of the body.

Respiratory Excretion of Carbon Dioxide

The respiratory center in the brainstem helps regulate the hydrogen ion concentrations in the body fluids by controlling the rate and depth of breathing (see chapter 16, pp. 456–457). Figure 18.7 traces this process. Specifically, if body cells increase their production of carbon dioxide, as occurs during periods of physical exertion, carbonic acid production increases. As the carbonic acid dissociates, the concentration of hydrogen ions increases, and the pH of the internal environment drops. Such an increasing concentration of carbon dioxide in the central nervous system and the subsequent increase in hydrogen ion concentration in the cerebrospinal fluid stimulate chemosensitive areas in the respiratory center.

In response, the respiratory center increases the depth and rate of breathing, so that the lungs excrete more carbon dioxide. Hydrogen ion concentration in body fluids returns toward normal because the released carbon dioxide comes from carbonic acid:

$$H_2CO_3 \longrightarrow CO_2 + H_2O$$

Conversely, if body cells are less active, concentrations of carbon dioxide and hydrogen ions in body fluids

remain low. As a result, breathing rate and depth stay closer to resting levels.

Renal Excretion of Hydrogen Ions

Nephrons help regulate the hydrogen ion concentration of body fluids by excreting hydrogen ions in urine. Recall from chapter 17 (p. 479) that epithelial cells lining certain segments of the renal tubules secrete hydrogen ions into the tubular fluid.

Time Course of Hydrogen Ion Regulation

The various regulators of hydrogen ion concentration operate at different rates. Chemical buffers can convert strong acids or bases into weak acids or bases almost immediately. For this reason, these chemical buffer systems are called the body's *first line of defense* against shifts in pH.

Physiological buffer systems, such as the respiratory and renal mechanisms, function more slowly and constitute the *second line of defense* against shifts in pH. The respiratory mechanism may require several minutes to begin resisting a change in pH, and the renal mechanism may require one to three days to regulate a changing hydrogen ion concentration. Figure 18.8 compares the actions of chemical buffers and physiological buffers.

Figure 18.7

An increase in carbon dioxide production increases carbon dioxide elimination.

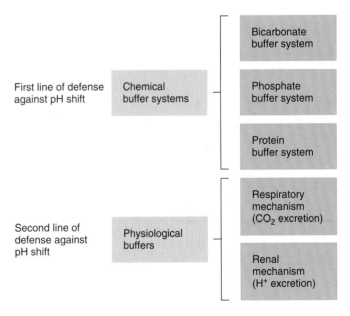

Figure 18.8

Chemical buffers act rapidly, whereas physiological buffers may require several minutes to several days to begin resisting a change in pH.

Q: *How does respiratory excretion of CO_2 buffer the pH of body fluids?*

Answer can be found in Appendix E on page 568.

Practice

23. How does the respiratory system help regulate acid-base balance?

24. How do the kidneys respond to excess hydrogen ions?

25. How do the rates at which chemical and physiological buffer systems act differ?

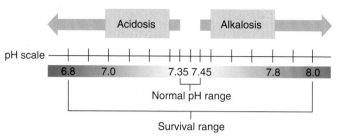

Figure 18.9

If the pH of arterial blood drops to 6.8 or rises to 8.0 for more than a few hours, the person usually cannot survive.

18.6 ACID-BASE IMBALANCES

Chemical and physiological buffer systems ordinarily maintain the hydrogen ion concentration of body fluids within very narrow pH ranges. Abnormal conditions may disturb the acid-base balance. For example, the pH of arterial blood is normally 7.35–7.45. A pH value below 7.35 produces *acidosis*. A pH above 7.45 produces *alkalosis*. Such shifts in the pH of body fluids can be life-threatening. A person usually cannot survive if the pH drops to 6.8 or rises to 8.0 for more than a few hours (fig. 18.9).

Acidosis results from an accumulation of acids or loss of bases, both of which cause abnormal increases in the hydrogen ion concentrations of body fluids. Conversely, alkalosis results from a loss of acids or an accumulation of bases accompanied by a decrease in hydrogen ion concentrations (fig. 18.10).

Acidosis

The two major types of acidosis are *respiratory acidosis* and *metabolic acidosis*. Factors that increase carbon dioxide levels, also increasing the concentration of carbonic acid (the respiratory acid), cause respiratory acidosis. Metabolic acidosis is due to an abnormal accumulation of any other acids in the body fluids or to loss of bases, including bicarbonate ions.

Respiratory acidosis may be due to hindered pulmonary ventilation, which increases carbon dioxide concentration. This may result from the following conditions:

1. Injury to the respiratory center of the brainstem, decreasing rate and depth of breathing.
2. Obstruction in air passages that interferes with air movement into alveoli.
3. Diseases that decrease gas exchange, such as pneumonia, or those that reduce the surface area of the respiratory membrane, such as emphysema.

Figure 18.11 summarizes the factors that can lead to respiratory acidosis. Any of these conditions can increase the level of carbonic acid and hydrogen ions in body fluids, lowering pH. Chemical buffers, such as hemoglobin, may resist this shift in pH. At the same time, rising concentrations of carbon dioxide and hydrogen ions stimulate the respiratory center, increasing the breathing rate and depth and thereby lowering the

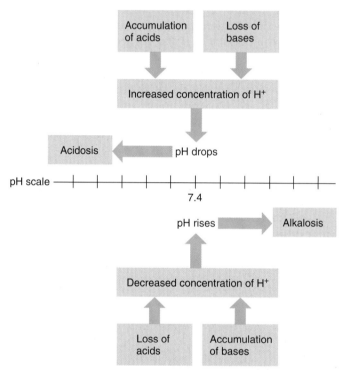

Figure 18.10

Acidosis results from accumulation of acids or loss of bases. Alkalosis results from loss of acids or accumulation of bases.

Figure 18.11

Some of the factors that lead to respiratory acidosis.

carbon dioxide concentration. Also, the kidneys may begin to excrete more hydrogen ions. Eventually, these chemical and physiological buffers return the pH of the body fluids to normal. The acidosis is thus *compensated*.

The symptoms of respiratory acidosis result from depression of central nervous system function. They include drowsiness, disorientation, stupor, labored breathing, and cyanosis. In *uncompensated acidosis,* the person may become comatose and die.

Metabolic acidosis is due to either accumulation of nonrespiratory acids or loss of bases. Factors that may lead to this condition include the following:

1. Kidney disease that reduces glomerular filtration so that the kidneys fail to excrete acids produced in metabolism (uremic acidosis).
2. Prolonged vomiting that loses the alkaline contents of the upper intestine and the stomach contents. (Losing only the stomach contents produces metabolic alkalosis.) Vomiting can empty not only the stomach, but also the first foot or so of the intestine.
3. Prolonged diarrhea in which excess alkaline intestinal secretions are lost (especially in infants).
4. Diabetes mellitus, in which some fatty acids react to produce ketone bodies, such as *acetoacetic acid, beta-hydroxybutyric acid,* and *acetone.* Normally these molecules are scarce and cells oxidize them as energy sources. However, if fats are being utilized at an abnormally high rate, as may occur in diabetes mellitus, ketone bodies may accumulate faster than they can be oxidized and as a result spill over into the urine (ketonuria); in addition, the lungs may release acetone, which is volatile and imparts a fruity odor to the breath. More seriously, the accumulation of acetoacetic acid and beta-hydroxybutyric acid may lower pH (ketoacidosis). These acids may also combine with bicarbonate ions in the urine. Excess bicarbonate ions are excreted, interfering with the function of the bicarbonate acid-base buffer system.

Figure 18.12 summarizes the factors leading to metabolic acidosis. In each case, pH is lowered. Countering this lower pH are chemical buffer systems, which accept excess hydrogen ions; the respiratory center, which increases breathing rate and depth; and the kidneys, which excrete more hydrogen ions.

Alkalosis

The two major types of alkalosis are *respiratory alkalosis* and *metabolic alkalosis.* Respiratory alkalosis results from excessive loss of carbon dioxide and consequent loss of carbonic acid. Metabolic alkalosis is due to excessive loss of hydrogen ions or gain of bases.

Respiratory alkalosis develops as a result of *hyperventilation* (described in chapter 16, pp. 457–458), in which too much carbon dioxide is lost, decreasing car-

Figure 18.12

Some of the factors that lead to metabolic acidosis.

bonic acid and hydrogen ion concentrations. Hyperventilation may be a response to anxiety or may accompany fever or poisoning from salicylates, such as aspirin. At high altitudes, hyperventilation may be a response to low oxygen pressure. Musicians can hyperventilate when providing the large volume of air needed to play sustained passages on wind instruments. In each case, rapid, deep breathing depletes carbon dioxide, and the pH of body fluids increases. Figure 18.13 illustrates the factors leading to respiratory alkalosis.

Chemical buffers, such as hemoglobin, that release hydrogen ions resist the increase in pH. The lower

Figure 18.13

Some of the factors that lead to respiratory alkalosis.

levels of carbon dioxide and hydrogen ions stimulate the respiratory center to a lesser degree. This inhibits hyperventilation, thereby reducing further carbon dioxide loss. At the same time, the kidneys decrease their secretion of hydrogen ions, and the urine becomes alkaline as bases are excreted.

The symptoms of respiratory alkalosis include light-headedness, agitation, dizziness, and tingling sensations. In severe cases, impulses may be triggered spontaneously on peripheral nerves, and muscles may respond with tetanic contractions (see chapter 8, p. 190).

Metabolic alkalosis results from a great loss of hydrogen ions or from a gain in bases, both accompanied by a rise in the pH of the blood (alkalemia). This condition may occur following gastric drainage (lavage), prolonged vomiting in which the stomach contents are lost, or the use of certain diuretic drugs. Gastric juice is acidic, so its loss leaves body fluids more basic. Metabolic alkalosis may also develop as a result of ingesting too much antacid, such as sodium bicarbonate to relieve the symptoms of indigestion. Symptoms of metabolic alkalosis include a decrease in the breathing rate and depth, which in turn results in an increased concentration of carbon dioxide in the blood. Figure 18.14 illustrates the factors leading to metabolic alkalosis.

Practice

26. What is the difference between a respiratory acid-base disturbance and a metabolic disturbance?
27. How do the symptoms of acidosis compare with those of alkalosis?

Figure 18.14
Some of the factors that lead to metabolic alkalosis.

Summary Outline

18.1 Introduction (p. 490)
Maintenance of water and electrolyte balance requires that the quantities of these substances entering the body equal the quantities leaving it. Altering the water balance affects the electrolyte balance.

18.2 Distribution of Body Fluids (p. 490)
1. Fluid compartments
   a. The intracellular fluid compartment includes the fluids and electrolytes that enclose cell membranes.
   b. The extracellular fluid compartment includes all the fluids and electrolytes outside cell membranes.
2. Body fluid composition
   a. Extracellular fluids have high concentrations of sodium, chloride, calcium, and bicarbonate ions, with less potassium, magnesium, phosphate, and sulfate ions. Plasma contains more protein than does either interstitial fluid or lymph.
   b. Intracellular fluid contains high concentrations of potassium, magnesium, and phosphate ions. It also has a greater concentration of sulfate ions and lesser concentrations of sodium, chloride, calcium, and bicarbonate ions than does extracellular fluid.
3. Movement of fluid between compartments
   a. Hydrostatic and osmotic pressure regulate fluid movements.
      (1) Hydrostatic pressure forces fluid out of plasma, and colloid osmotic pressure returns fluid to plasma.
      (2) Hydrostatic pressure drives fluid into lymph vessels.
      (3) Osmotic pressure regulates fluid movement in and out of cells.
   b. Sodium ion concentrations are especially important in regulating fluid movement.

18.3 Water Balance (p. 492)
1. Water intake
   a. Most water comes from consuming liquids or moist foods.
   b. Oxidative metabolism produces some water.
2. Regulation of water intake
   a. Thirst is the primary regulator of water intake.
   b. Drinking and the resulting stomach distension inhibit thirst.
3. Water output
   Water is lost in urine, feces, and sweat, and by evaporation from the skin and lungs.
4. Regulation of water output
   The distal convoluted tubules of the nephrons and collecting ducts are the effectors of the control system that regulate water output.

18.4 Electrolyte Balance (p. 493)
1. Electrolyte intake
   a. The electrolytes of greatest importance to cellular functions in body fluids dissociate to release ions of sodium, potassium, calcium, magnesium, chloride, sulfate, phosphate, bicarbonate, and hydrogen.
   b. These ions are obtained in foods and beverages or as by-products of metabolic processes.
2. Regulation of electrolyte intake
   a. Food and drink usually provide sufficient electrolytes.
   b. A severe electrolyte deficiency may produce a salt craving.
3. Electrolyte output
   a. Electrolytes are lost through perspiration, feces, and urine.
   b. Quantities lost vary with temperature and physical exercise.
   c. Most electrolytes are lost as a result of kidney function.

4. Regulation of electrolyte output
   a. Concentrations of sodium, potassium, and calcium ions in body fluids are particularly important.
   b. The adrenal cortex secretes aldosterone to regulate sodium and potassium ions.
   c. Parathyroid hormone regulates calcium ions.
   d. The mechanisms that control positively charged ions secondarily regulate negatively charged ions.

18.5 Acid-Base Balance (p. 497)

Acids are electrolytes that ionize to release hydrogen ions. Bases release ions that combine with hydrogen ions. Body fluid pH must remain within a certain range.

1. Sources of hydrogen ions
   a. Aerobic respiration of glucose produces carbonic acid.
   b. Anaerobic respiration of glucose produces lactic acid.
   c. Incomplete oxidation of fatty acids releases acidic ketone bodies.
   d. Oxidation of sulfur-containing amino acids produces sulfuric acid.
   e. Hydrolysis of phosphoproteins and nucleic acids produces phosphoric acid.
2. Strengths of acids and bases
   a. Acids vary in the extent to which they ionize to release ions.
      (1) Strong acids, such as hydrochloric acid, ionize more completely.
      (2) Weak acids, such as carbonic acid, ionize less completely.
   b. Bases also vary in strength.

3. Regulation of hydrogen ion concentration
   a. Chemical buffer systems
      (1) Buffer systems convert strong acids into weaker acids or strong bases into weaker bases.
      (2) They include the bicarbonate buffer system, phosphate buffer system, and protein buffer system.
      (3) Buffer systems minimize pH changes.
   b. The respiratory center controls the rate and depth of breathing to regulate pH.
   c. The kidneys excrete hydrogen ions to regulate pH.
   d. Chemical buffers act more rapidly. Physiological buffers act more slowly.

18.6 Acid-Base Imbalances (p. 500)

1. Acidosis
   a. Respiratory acidosis results from increased levels of carbon dioxide and carbonic acid.
   b. Metabolic acidosis results from accumulation of other acids or loss of bases.
2. Alkalosis
   a. Respiratory alkalosis results from loss of carbon dioxide and carbonic acid.
   b. Metabolic alkalosis results from loss of hydrogen ions or gain of bases.

Chapter Assessments

18.1 Introduction

1. Explain how water balance and electrolyte balance are interdependent. (p. 490)

18.2 Distribution of Body Fluids

2. Water and electrolytes enclosed by cell membranes constitute the _____ . (p. 490)
   a. transcellular fluid
   b. intracellular fluid
   c. extracellular fluid
   d. lymph
   e. plasma
3. Explain how the fluids in the compartments differ in composition. (p. 491)
4. Describe how fluid movements between the compartments are controlled. (p. 491)

18.3 Water Balance

5. Prepare a list of sources of normal water gain and loss to illustrate how the input of water equals the output of water. (p. 492)
6. Define *water of metabolism*. (p. 492)
7. Explain how water intake is regulated. (p. 492)
8. Explain how the kidneys regulate water output. (p. 492)

18.4 Electrolyte Balance

9. Electrolytes in body fluids of importance to cellular functions include _____ . (p. 493)
   a. sodium
   b. potassium
   c. calcium
   d. chloride
   e. all of the above
10. Explain how electrolyte intake is regulated. (p. 493)
11. List the routes by which electrolytes leave the body. (p. 493)
12. Explain how the adrenal cortex functions to regulate electrolyte balance. (p. 493)
13. Describe the role of the parathyroid glands in regulating electrolyte balance. (p. 496)

18.5 Acid-Base Balance

14. List five sources of hydrogen ions in body fluids, and name an acid that originates from each source. (p. 497)
15. _____ ionize to release hydrogen ions more completely. An example is hydrochloric acid. (p. 497)
16. _____ ionize to release fewer hydroxide ions. (p. 497)
17. Explain how the bicarbonate and phosphate buffer systems resist pH changes. (p. 498)

18. Explain why a protein has both acidic and basic properties. (p. 498)

19. Explain how the respiratory system and the kidneys function in the regulation of acid-base balance. (p. 499)

18.6 Acid-Base Imbalances

20. Distinguish between respiratory and metabolic acid-base imbalances. (p. 500)

21. Explain how the body compensates for acid-base imbalances. (p. 500)

Integrated Assessments/Critical Thinking

OUTCOMES 13.4, 13.5, 14.3, 18.2

1. If the right ventricle of a patient's heart is failing, increasing the systemic venous pressure, what changes might occur in the patient's extracellular fluid compartments?

OUTCOMES 15.2, 15.6, 15.9, 18.4, 18.6

2. Radiation therapy may damage the mucosa of the stomach and intestines. What effect might this have on the patient's electrolyte balance?

OUTCOMES 15.9, 15.10, 18.5, 18.6

3. After eating an undercooked hamburger, a twenty-five-year-old male developed diarrhea due to infection with a strain of *Escherichia coli* that produces a shigatoxin. How would this affect his blood pH, urine pH, and respiratory rate?

OUTCOMES 16.4, 18.5, 18.6

4. A student hyperventilates and is disoriented just before an exam. Is this student likely to be experiencing acidosis or alkalosis? How will the body compensate in an effort to maintain homeostasis?

WEB CONNECTIONS

Visit the text website at **www.mhhe.com/shieress11** for additional quizzes, interactive learning exercises, and more.

APR

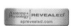

Anatomy & Physiology REVEALED includes cadaver photos that allow you to peel away layers of the human body to reveal structures beneath the surface. This program also includes animations, radiologic imaging, audio pronunciations, and practice quizzing. To learn more visit **www.aprevealed.com**.

19

Reproductive Systems

Selling eggs. The ad in the student newspaper seemed too good to be true—the fee for donating a few eggs would pay nearly a semester's tuition. Intrigued, the young woman submitted a health history, had a checkup, and a month later received a call. A young couple struggling with infertility sought an egg donor. They'd chosen Sherrie because, with her strawberry-blond hair, she looked like Linda, the woman whose cancer had left her unable to conceive. The donor eggs would be fertilized in a laboratory dish (*in vitro*) with sperm from Linda's partner, Ted, and then implanted in Linda's uterus.

For two weeks Sherrie injected herself in the thigh with a drug that acts like gonadotropin-releasing hormone, suppressing release of an egg from an ovary (ovulation). When daily hormone checks indicated that her endocrine system was in sync with Linda's, Sherrie began giving herself shots twice a day at the back of the hip. This second drug mimicked follicle-stimulating hormone, and it caused several ovarian follicles to mature. Finally, injections of luteinizing hormone brought the eggs close to being released. Then, at a health-care facility, Sherrie received pain medication and light sedation. A needle inserted through her vaginal wall retrieved a dozen of the mature eggs as they became accessible on the surface of her ovary.

Four *in vitro* fertilized ova formed. Two of them divided a few times and were implanted into Linda's uterus. The other fertilized ova were frozen for possible later use. The preparation and procedure weren't too painful. Sherrie had felt a dull aching the last day, and felt bloated for a few days after the egg retrieval, but she did not

Falsely colored, scanning electron micrograph of an egg being released from the surface of an ovary.

experience bleeding, infection, cramping, or mood swings. Nor did she develop a complication in which too many eggs mature, leaking fluid from blood vessels into the abdomen. About 6% of egg donors develop the syndrome, which can cause infertility, kidney failure, and even death. Future risks, however, are uncertain, because eggs haven't been collected long enough to know the long-term consequences. Case reports point to ovary scarring and possibly cancer. Another side effect that Sherrie had not fully considered was how she would feel afterward. Although she was happy to have helped the couple and to have paid her tuition, she would always wonder about the twins her eggs had become.

Learning Outcomes

After studying this chapter, you should be able to do the following:

19.1 Introduction

1. State the general functions of the male and female reproductive systems. (p. 506)

19.2 Organs of the Male Reproductive System

2. Describe the general functions of each part of the male reproductive system. (p. 506)

3. Outline the process of spermatogenesis. (p. 507)

4. Describe semen production and exit from the body. (p. 511)

19.3 Hormonal Control of Male Reproductive Functions

5. Explain how hormones control the activities of the male reproductive organs and the development of male secondary sex characteristics. (p. 513)

19.4 Organs of the Female Reproductive System

6. Describe the general functions of each part of the female reproductive system. (p. 516)

7. Outline the process of oogenesis. (p. 517)

19.5 Hormonal Control of Female Reproductive Functions

8. Explain how hormones control the activities of the female reproductive organs

 Learn   **Practice**   **Assess**

and the development of female secondary sex characteristics. (p. 522)

9. Describe the major events of a reproductive cycle. (p. 522)

19.6 Mammary Glands

10. Review the structure of the mammary glands. (p. 525)

19.7 Birth Control

11. Describe several methods of birth control, including the relative effectiveness of each method. (p. 526)

19.8 Sexually Transmitted Infections

12. List the general symptoms of sexually transmitted infections. (p. 530)

Module 14: Reproductive System

Aids to Understanding Words

(Appendix A on page 564 has a complete list of Aids to Understanding Words.)

andr- [man] *andr*ogens: Male sex hormones.

ejacul- [to shoot forth] *ejacul*ation: Expulsion of semen from the male reproductive tract.

fimb- [fringe] *fimb*riae: Irregular extensions on the margin of the infundibulum of the uterine tube.

follic- [small bag] *follic*le: Ovarian structure that contains an egg.

genesis- [origin] spermato*genesis*: Formation of sperm cells.

labi- [lip] *labi*a minora: Flattened, longitudinal folds that extend along the margins of the female vestibule.

mens- [month] *mens*es: Monthly flow of blood from the female reproductive tract.

mons- [an eminence] *mons* pubis: Rounded elevation overlying the pubic symphysis in a female.

puber- [adult] *puber*ty: The time in life when a person becomes able to reproduce.

19.1 INTRODUCTION

The male and female reproductive systems are connected series of organs and glands that produce and nurture sex cells (gametes) and transport them to sites of fertilization. Male sex cells are **sperm.** Female sex cells are eggs, or **oocytes** (o-o-sitz), which in Latin means "egg cells."

Sex cells have one set of genetic instructions, carried on 23 chromosomes, compared to two sets on 46 chromosomes in other cells. When sex cells join at fertilization, the amount of genetic information held in 46 chromosomes is restored. Some of the reproductive organs secrete hormones vital to the development and maintenance of secondary sex characteristics and the regulation of reproductive physiology.

19.2 ORGANS OF THE MALE REPRODUCTIVE SYSTEM

The *primary sex organs* (gonads) of the male reproductive system are the two testes, in which sperm cells and the male sex hormones are formed. The *accessory sex organs* of the male reproductive system are the internal and external reproductive organs (fig. 19.1; reference plates 3 and 4, pp. 25–26).

Testes

The **testes** (tes'tēz; sing., *testis*) are ovoid structures about 5 centimeters in length and 3 centimeters in diameter. Both testes are within the cavity of the saclike *scrotum.*

Structure of the Testes

A tough, white, fibrous capsule encloses each testis. Along the capsule's posterior border, the connective tissue thickens and extends into the testis, forming thin septa that divide the testis into about 250 *lobules.*

Each lobule contains one to four highly coiled, convoluted **seminiferous tubules** (se"mĭ-nif'er-us too'būlz), each approximately 70 centimeters long uncoiled. These tubules course posteriorly and unite to form a complex network of channels that give rise to several ducts that join a tube called the *epididymis.* The epididymis is coiled on the outer surface of the testis and continues to become the *ductus deferens* (fig. 19.2*a*).

A specialized stratified epithelium with **spermatogenic** (sper"mah-to-jen'ik) **cells** (germ cells), which give rise to sperm cells, lines the seminiferous tubules. Other specialized cells, called **interstitial cells** (cells of Leydig), lie in the spaces between the seminiferous tubules (fig. 19.2*b, c*). Interstitial cells produce and secrete male sex hormones.

The epithelial cells of the seminiferous tubules can give rise to *testicular cancer,* a common cancer in young men. In most cases the first sign is a painless testis enlargement or a scrotal mass attached to a testis. If a biopsy (tissue sample) reveals cancer cells, surgery is performed to remove the affected testis (orchiectomy). Radiation and/or chemotherapy often prevents the cancer from recurring.

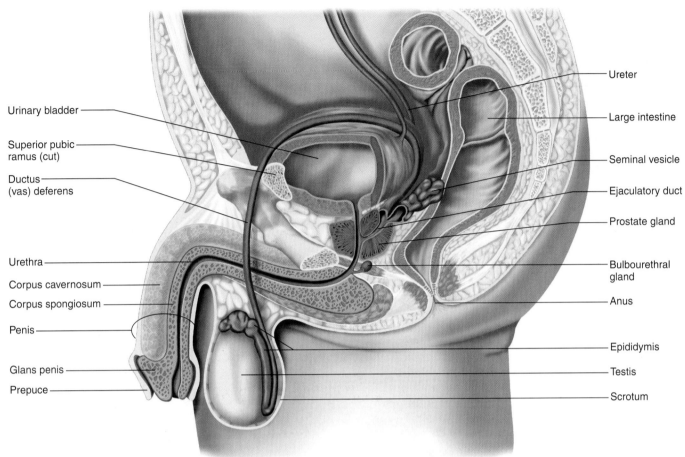

Figure 19.1 AP|R

Male reproductive organs (sagittal view). The paired testes are the primary sex organs, and the other reproductive structures, both internal and external, are accessory sex organs.

Practice

1. Describe the structure of a testis.
2. Where in the testes are the sperm cells produced?
3. Which cells produce male sex hormones?

Formation of Sperm Cells

The epithelium of the seminiferous tubules consists of sustentacular cells (Sertoli cells) and spermatogenic cells. Sustentacular cells provide support for the spermatogenic cells, and also nourish and regulate them.

Spermatogenesis occurs continually in a male, starting at puberty. The resulting sperm cells collect in the lumen of each seminiferous tubule, and then pass to the epididymis, where they accumulate and mature.

In the male embryo, the undifferentiated spermatogenic cells are called **spermatogonia** (sper″mah-to-go′ne-ah). Each spermatogonium contains 46 chromosomes (23 pairs) in its nucleus, the usual number for human body cells. Beginning during embryonic development, hormones stimulate spermatogonia to undergo mitosis (see chapter 3, p. 69). Each cell division gives rise to two new cells, one of which (type A) maintains the supply of undifferentiated cells, the other of which (type B) differentiates, becoming a *primary spermatocyte*. Sperm cell production or **spermatogenesis** (sper″mah-to-jen′ĕ-sis) is arrested at this stage.

At puberty the primary spermatocytes then reproduce by a special type of cell division called **meiosis** (mi-o′sis) (fig. 19.3). Meiosis includes two successive divisions, called the *first* and *second meiotic divisions*. The first meiotic division (meiosis I) separates homologous chromosome pairs. Homologous pairs are the same, gene for gene. They may not be identical, however, because a gene may have variants, and the chromosome that comes from the person's mother may carry a different variant for the corresponding gene from the father's homologous chromosome. Before meiosis I, each homologous chromosome is replicated, so it consists of two complete DNA strands called *chromatids*. The chromatids of a replicated chromosome attach at regions called *centromeres*. Each chromatid has the complete genetic information associated with that chromosome.

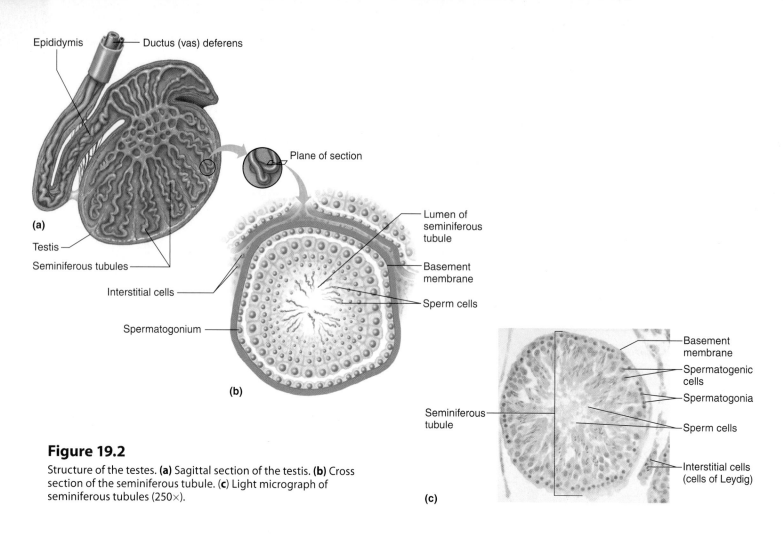

Figure 19.2

Structure of the testes. **(a)** Sagittal section of the testis. **(b)** Cross section of the seminiferous tubule. **(c)** Light micrograph of seminiferous tubules (250×).

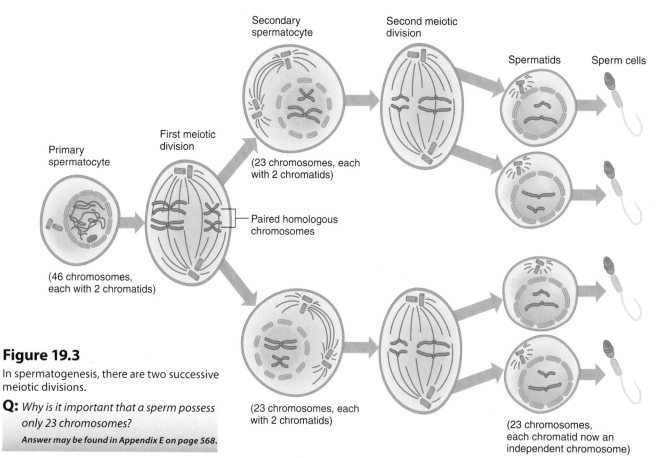

Figure 19.3

In spermatogenesis, there are two successive meiotic divisions.

Q: *Why is it important that a sperm possess only 23 chromosomes?*

Answer may be found in Appendix E on page 568.

Each of the cells that undergoes the second meiotic division (meiosis II) begins with one member of each homologous pair, a condition termed **haploid** (hap′loyd). That is, a haploid cell has one set of chromosomes. This second division separates the chromatids, producing cells that are still haploid, but whose chromosomes are no longer in the replicated form. After meiosis II, each of the chromatids is an independent chromosome (fig. 19.3). Consequently, for each primary spermatocyte that undergoes meiosis, four sperm cells, with 23 chromosomes in each of their nuclei, form.

During spermatogenesis, each primary spermatocyte divides to form two *secondary spermatocytes.* Each of these cells, in turn, divides to form two *spermatids,* which mature into sperm cells (fig. 19.4).

Figure 19.4 AP|R

Spermatogonium mitosis results in spermatogonia (type A) that continue the germ cell line and spermatogonia (type B) that give rise to primary spermatocytes. The spermatocytes, in turn, give rise to sperm cells by meiosis. Changes in chromosome number and structure are represented by a single pair of chromosomes. Note that as the cells approach the lumen, they mature.

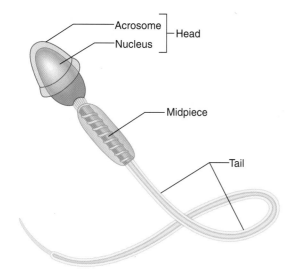

Figure 19.5

Parts of a mature sperm cell.

Structure of a Sperm Cell

A mature sperm cell is a tiny, tadpole-shaped structure about 0.06 millimeters long. It consists of a flattened head, a cylindrical midpiece (body), and an elongated tail (fig. 19.5; see fig. 3.9*b*, p. 59).

The oval *head* of a sperm cell is primarily composed of a nucleus and contains highly compacted chromatin consisting of 23 chromosomes. A small protrusion at its anterior end, called the *acrosome,* contains enzymes that help the sperm cell penetrate an egg cell during fertilization. (Chapter 20, pp. 538–539, describes this process.)

The *midpiece* of a sperm cell has a central, filamentous core and many mitochondria organized in a spiral. The *tail* (flagellum) consists of several microtubules enclosed in an extension of the cell membrane. The mitochondria provide ATP for the tail's lashing movement that propels the sperm cell through fluid.

> **Practice**
>
> 4. Explain the function of supporting cells in the seminiferous tubules.
> 5. Review the events of spermatogenesis.
> 6. Describe the structure of a sperm cell.

Male Internal Reproductive Organs

The internal accessory organs of the male reproductive system are specialized to nurture and transport sperm cells. These structures include the two epididymides, two ductus deferentia, two ejaculatory ducts, and the urethra, as well as the two seminal vesicles, prostate gland, and two bulbourethral glands.

Epididymides

The **epididymides** (ep″ĭ-di-dy′mides; sing., *epididymis*) are tightly coiled, threadlike tubes about 6 meters long (see figs. 19.1 and 19.2). Each epididymis is connected to ducts within a testis. It emerges from the top of the testis, descends along the posterior surface of the testis, and then courses upward to become the ductus deferens.

Immature sperm cells reaching the epididymis are nonmotile. As rhythmic peristaltic contractions help move these cells through the epididymis, the cells mature. Following this aging process, the sperm cells can move independently and fertilize egg cells. However, they usually do not "swim" until after ejaculation.

Ductus Deferentia

The **ductus deferentia** (duk′tus def″er-en′sha; sing., *ductus deferens*), also called vasa deferentia, are muscular tubes about 45 centimeters long (see fig. 19.1). Each passes upward along the medial side of a testis and through a passage in the lower abdominal wall (inguinal canal), enters the pelvic cavity, and ends behind the urinary bladder. Just outside the prostate gland, the ductus deferens unites with the duct of a seminal vesicle to form an **ejaculatory duct,** which passes through the prostate gland and empties into the urethra.

Seminal Vesicles

The **seminal vesicles** are convoluted, saclike structures about 5 centimeters long. Each attaches to the ductus deferens near the base of the urinary bladder (see fig. 19.1). The glandular tissue lining the inner wall of a seminal vesicle secretes a slightly alkaline fluid. This fluid helps regulate the pH of the tubular contents as sperm cells travel to the outside. Seminal vesicle secretions also contain *fructose,* a monosaccharide that provides energy to sperm cells, and *prostaglandins* (see chapter 11, p. 296), which stimulate muscular contractions within the female reproductive organs, aiding the movement of sperm cells toward the egg cell.

> **Practice**
>
> 7. Describe the structure of the epididymis.
> 8. Trace the path of the ductus deferens.
> 9. What is the function of a seminal vesicle?

Prostate Gland

The **prostate** (pros′tāt) **gland** is a chestnut-shaped structure about 4 centimeters across and 3 centimeters thick that surrounds the proximal part of the urethra, just inferior to the urinary bladder (see fig. 19.1). It is enclosed in connective tissue and composed of many branched tubular glands whose ducts open into the urethra.

Prostate Cancer

Each year in the United States, nearly 200,000 men receive a diagnosis of prostate cancer, and more than 25,000 die of the disease. The diagnostic process typically begins with a rectal exam in which a health care practitioner feels an enlargement of the prostate gland, and a blood test to detect elevated levels of a biomarker called prostate-specific antigen (PSA) (see Clinical Application 2.2, page 47). Normally, secretory epithelium in the prostate gland releases PSA, which liquefies the ejaculate. When cancer cells accumulate, more PSA is produced, and it enters capillaries in the prostate. If the elevated PSA is still present on a test a few months later, the next step is a biopsy that samples cells from several parts of the gland. If cancer is detected, it is assigned a two-digit number, called a Gleason score, which indicates how differentiated (specialized) the cancer cells are. The less specialized the cancer cells, the more advanced the disease. Imaging technologies can be used to assess whether the cancer has spread beyond the prostate capsule.

In the past, most men with a diagnosis of prostate cancer were treated quickly, based on the premise that early treatment lowers the fatality rate. The two most common treatments are surgery to remove the prostate gland, and radiation. Both can cause urinary incontinence and/or erectile dysfunction, which may improve with time.

Another option for men diagnosed with prostate cancer is "active surveillance," which means to regularly monitor the disease, with PSA tests twice a year and a biopsy every one to two years, and only treat if the condition worsens. Active surveillance grew out of several very large studies reported in 2009 that showed that for men diagnosed with early stage prostate cancer, it is safe to delay treatment until and if the disease progresses. For many men, treatment never becomes necessary. Considering the possible serious adverse effects of surgery or radiation, active surveillance may be the safer option for some men.

The challenge in deciding when to treat a cancer is to identify those patients in whom the cancer is likely to progress. This is becoming possible using gene expression profiling, introduced in the Genetics Connection 4.1 on pages 90–91. For prostate cancer, the activities of six specific genes change dramatically when the disease spreads. By combining PSA screening with the still experimental gene expression approach, detection rises from about 65% with PSA alone to more than 90% with both. In response to the findings for prostate cancer, researchers are now looking at whether active surveillance might help people with other types of cancer decide when it is best to undergo treatment.

The prostate gland secretes a thin, milky fluid with an alkaline pH. This secretion neutralizes the fluid containing sperm cells, which is acidic from accumulation of metabolic wastes that stored sperm cells produce. Prostatic fluid also enhances the motility of sperm cells and helps neutralize the acidic secretions of the vagina. Clinical Application 19.1 discusses prostate cancer.

> The prostate gland is small in boys, begins to grow in early adolescence, and reaches adult size several years later. Usually the gland does not grow again until age fifty, when in about half of all men it enlarges enough to press on the urethra. This produces a feeling of pressure because the bladder cannot empty completely and the man feels the need to urinate frequently.

Bulbourethral Glands

The two **bulbourethral** (bul″bo-u-re′thral) **glands** (Cowper's glands) are each about a centimeter in diameter and are inferior to the prostate gland (see fig. 19.1). Bulbourethral glands are composed of many tubes whose epithelial linings secrete a mucuslike fluid in response to sexual stimulation. This fluid lubricates the end of the penis in preparation for sexual intercourse. However, females secrete most of the lubricating fluid for sexual intercourse.

Semen

Semen (se′men) is the fluid the male urethra conveys to the outside during ejaculation. It consists of sperm cells from the testes and secretions of the seminal vesicles, prostate gland, and bulbourethral glands. Semen is slightly alkaline (pH about 7.5), and it includes prostaglandins and nutrients.

The volume of semen released at one time varies from 2 to 5 milliliters. The average number of sperm cells in the fluid is about 120 million per milliliter. If a man can ejaculate at least 60 million sperm per milliliter, he is likely to eventually be able to father a child.

Sperm cells are nonmotile while in the ducts of the testis and epididymis, but begin to swim as they mix with accessory gland secretions. Sperm cells cannot fertilize an egg cell until they enter the female reproductive tract. Here, they undergo *capacitation*, which weakens the sperm cells' acrosomal membranes.

Practice

10. Where is the prostate gland located?
11. What are the functions of the prostate gland's secretion?
12. What are the components of semen?

Male External Reproductive Organs AP|R

The male external reproductive organs are the scrotum, which encloses two testes, and the penis. The urethra passes through the penis.

Scrotum

The **scrotum** is a pouch of skin and subcutaneous tissue that hangs from the lower abdominal region posterior to the penis (see fig. 19.1). A medial septum divides the scrotum into two chambers, each of which encloses a testis. Each chamber also contains a serous membrane, which covers the testis and helps it move smoothly within the scrotum. The scrotum protects and helps regulate the temperature of the testes. These factors are important to sex cell production.

Exposure to cold stimulates the smooth muscle fibers in the wall of the scrotum to contract, the scrotal skin to wrinkle, and the testes to move closer to the pelvic cavity, where they can absorb heat. Exposure to warmth stimulates the smooth muscle to relax and the scrotum to hang loosely, providing an environment 3°C (about 5°F) below body temperature, which is more conducive to sperm production and survival.

Penis

The **penis** is a cylindrical organ that conveys urine and semen through the urethra to the outside (see fig. 19.1). During erection, it enlarges and stiffens, enabling it to be inserted into the vagina during sexual intercourse.

The *body,* or shaft, of the penis has three columns of erectile tissue—a pair of dorsally located *corpora cavernosa* and a single, ventral *corpus spongiosum.* A tough capsule of dense connective tissue surrounds each column. Skin, a thin layer of subcutaneous tissue, and a layer of connective tissue enclose the penis.

The corpus spongiosum, through which the urethra extends, enlarges at its distal end to form a sensitive, cone-shaped **glans penis.** The glans covers the ends of the corpora cavernosa and bears the urethral opening (external urethral orifice). The skin of the glans is very thin and hairless, and contains sensory receptors for sexual stimulation. A loose fold of skin called the *prepuce* (foreskin) originates just posterior to the glans and extends anteriorly to cover it as a sheath. A surgical procedure called *circumcision* removes the prepuce.

Practice

13. Describe the structure of the penis.
14. What is circumcision?

Erection, Orgasm, and Ejaculation

During sexual stimulation, parasympathetic nerve impulses from the sacral part of the spinal cord release nitric oxide (NO), which dilates the arteries leading into the penis, increasing blood flow into erectile tissues. At the same time, the increasing pressure of arterial blood entering the vascular spaces of erectile tissue compresses the veins of the penis, reducing the flow of venous blood away from the organ. Consequently, blood accumulates in erectile tissues, and the penis swells and elongates, producing an **erection.**

The culmination of sexual stimulation is **orgasm** (or'gazm), a pleasurable feeling of physiological and psychological release. Emission and ejaculation are events that accompany male orgasm.

Emission (e-mish'un) is the movement of sperm cells from the testes and secretions from the prostate gland and seminal vesicles into the urethra, where they mix to form semen. Emission is a response to sympathetic nerve impulses from the spinal cord, which stimulate peristaltic contractions in smooth muscles in the walls of the testicular ducts, epididymides, ductus deferentia, and ejaculatory ducts. At the same time, other sympathetic impulses stimulate rhythmic contractions of the seminal vesicles and prostate gland.

As the urethra fills with semen, sensory impulses pass into the sacral part of the spinal cord. In response, motor impulses are transmitted from the spinal cord to certain skeletal muscles at the base of the penile erectile columns, rhythmically contracting them. This increases the pressure in the erectile tissues and aids in forcing the semen through the urethra to the outside, a process called **ejaculation** (e-jak"u-la'shun). At the time of ejaculation, the posterior pituitary gland releases a burst of oxytocin, which stimulates contractions of the epididymides, seminiferous tubules, and prostate gland, aiding the release of sperm.

The sequence of events during emission and ejaculation is coordinated so that the fluid from the bulbourethral glands is expelled first. This is followed by the release of fluid from the prostate gland, the passage of sperm cells, and finally the ejection of fluid from the seminal vesicles.

Immediately after ejaculation, sympathetic impulses constrict the arteries that supply the erectile tissue, reducing inflow of blood. Smooth muscles in the walls of the vascular spaces partially contract again, and the veins of the penis carry excess blood out of these

Table 19.1	Functions of the Male Reproductive Organs
Organ	**Function**
Testis	
Seminiferous tubules	Produce sperm cells
Interstitial cells	Produce and secrete male sex hormones
Epididymis	Promotes sperm cell maturation; stores sperm cells; conveys sperm cells to ductus deferens
Ductus deferens	Conveys sperm cells to ejaculatory duct
Seminal vesicle	Secretes an alkaline fluid containing nutrients and prostaglandins that helps neutralize the acidic components of semen
Prostate gland	Secretes an alkaline fluid that helps neutralize semen's acidity and enhances sperm cell motility
Bulbourethral gland	Secretes fluid that lubricates end of penis
Scrotum	Encloses, protects, and regulates temperature of testes
Penis	Conveys urine and semen to outside of body; inserted into vagina during sexual intercourse; the glans penis is richly supplied with sensory nerve endings associated with feelings of pleasure during sexual stimulation

spaces. The penis gradually returns to its flaccid state. Table 19.1 summarizes the functions of the male reproductive organs. Clinical Application 19.2 discusses male infertility.

> Spontaneous emission and ejaculation are common in sleeping adolescent males. They are caused by changes in hormonal concentrations that accompany adolescent development and sexual maturation.

Practice

15. What controls blood flow into penile erectile tissues?

16. Distinguish among orgasm, emission, and ejaculation.

17. Review the events associated with emission and ejaculation.

19.3 HORMONAL CONTROL OF MALE REPRODUCTIVE FUNCTIONS

The hypothalamus, anterior pituitary gland, and testes secrete hormones that control male reproductive functions. These hormones initiate and maintain sperm cell production and oversee the development and maintenance of male secondary sex characteristics.

Hypothalamic and Pituitary Hormones

The male body is reproductively immature before ten years of age. It is childlike, with spermatogenic cells undifferentiated. Then, a series of changes leads to development of a reproductively functional adult. The hypothalamus controls many of these changes.

Recall from chapter 11 (p. 301) that the hypothalamus secretes gonadotropin-releasing hormone (GnRH), which enters the blood vessels leading to the anterior pituitary gland. In response, the anterior pituitary secretes the **gonadotropins** (go-nad″o-trōp′inz) *luteinizing hormone (LH)* and *follicle-stimulating hormone (FSH)*. LH, which in males has also been called interstitial cell stimulating hormone (ICSH), promotes development of interstitial cells of the testes, and they in turn secrete male sex hormones. FSH stimulates the sustentacular cells of the seminiferous tubules to respond to the effects of the male sex hormone *testosterone*. Then, in the presence of FSH and testosterone, these cells stimulate spermatogenic cells to undergo spermatogenesis, giving rise to sperm cells (fig. 19.6). The sustentacular cells also secrete a hormone called *inhibin,* which inhibits the anterior pituitary gland by negative feedback. This action prevents oversecretion of FSH.

Male Sex Hormones

Male sex hormones are termed **androgens** (an′dro-jenz). Interstitial cells of the testes produce most of them, but small amounts are synthesized in the adrenal cortex (see chapter 11, p. 306). **Testosterone** (tes-tos′tĕ-rōn) is the most important androgen. It loosely attaches to plasma proteins for secretion and transport in the blood.

Testosterone secretion begins during fetal development and continues for several weeks following birth; then it nearly ceases during childhood. Between the ages of thirteen and fifteen, a young man's androgen production usually increases rapidly. This phase in development, when an individual becomes reproductively functional, is **puberty** (pu′ber-te). After puberty, testosterone secretion continues throughout the life of a male.

Clinical Application 19.2

Male Infertility

Male infertility—the inability of sperm cells to fertilize an egg cell—has several causes. If, during fetal development, the testes do not descend into the scrotum, the higher temperature of the abdominal cavity or inguinal canal causes the developing sperm cells in the seminiferous tubules to degenerate. Certain diseases, such as mumps, may inflame the testes (orchitis), impairing fertility by destroying cells in the seminiferous tubules.

The quality and the quantity of sperm cells are essential factors in the ability of a man to father a child. If a sperm head is misshapen, if a sperm cell cannot swim, or if there are too few sperm cells, completing the journey to the well-protected egg cell may be impossible. However, all men make some abnormal sperm cells.

Computer-aided sperm (or semen) analysis (CASA) is standardizing and expanding criteria for normalcy in human semen and the sperm it contains. For this analysis, a man abstains from intercourse for two to three days and then provides a sperm sample. The man also provides information about his reproductive history and possible exposure to toxins. The CASA system captures images with a digital camera and analyzes and integrates information on sperm cell density, motility, and morphology. The result is a "spermiogram." Figure 19A shows a CASA of normal sperm cells, depicting different swimming patterns as they travel. Table 19A lists the components of a semen analysis.

Table 19A	Sperm Cell Analysis		
Characteristic	Density (million cells/ milliliter)	Motile	Normal morphology
Fertile	>48	>12%	>63%
Borderline fertile	13.5–48	>12%	>63%
Infertile	<13.5	<9%	<32%

(a) (b) (c) (d)

Figure 19A

A computer tracks sperm cell movements. In semen, sperm cells swim in a straight line **(a)**, but as they are activated by biochemicals in the woman's body, their trajectories widen **(b)**. The sperm cells in **(c)** are in the mucus of a woman's cervix, and the sperm cells in **(d)** are attempting to digest through the structures surrounding an egg cell.

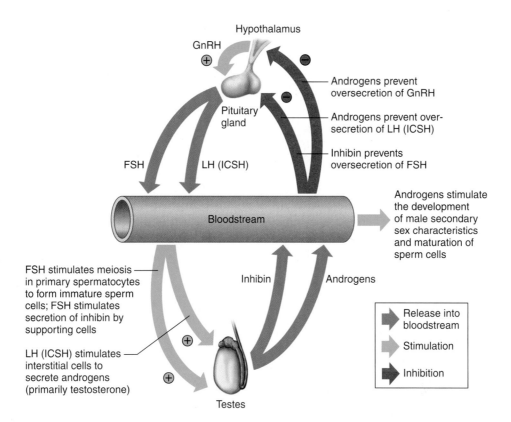

Figure 19.6

The hypothalamus controls maturation of sperm cells and development of male secondary sex characteristics. Negative feedback among the hypothalamus, the anterior lobe of the pituitary gland, and the testes controls the concentration of testosterone in the male body.

Actions of Testosterone

During puberty, testosterone stimulates enlargement of the testes and accessory organs of the reproductive system, as well as development of *male secondary sex characteristics,* which are special features associated with the adult male body. Secondary sex characteristics in the male include:

1. Increased growth of body hair, particularly on the face, chest, axillary region, and pubic region. Sometimes, hair growth on the scalp slows.
2. Enlargement of the larynx and thickening of the vocal folds, with lowering of the pitch of the voice.
3. Thickening of the skin.
4. Increased muscular growth, broadening of the shoulders, and narrowing of the waist.
5. Thickening and strengthening of the bones.

Testosterone also increases the rate of cellular metabolism and red blood cell production. For this reason, the average number of red blood cells in a microliter of blood is usually greater in males than in females. Testosterone stimulates sexual activity by affecting certain parts of the brain.

Regulation of Male Sex Hormones

The extent to which male secondary sex characteristics develop is directly related to the amount of testosterone

that interstitial cells secrete. The hypothalamus regulates testosterone output through negative feedback (fig. 19.6).

An increasing blood testosterone concentration inhibits the hypothalamus, and hypothalamic stimulation of the anterior pituitary gland by GnRH decreases. As the pituitary gland's secretion of LH falls in response, testosterone release from the interstitial cells decreases.

As the blood testosterone concentration drops, the hypothalamus becomes less inhibited, and it once again stimulates the anterior pituitary gland to release LH. Increasing LH secretion then causes interstitial cells to release more testosterone, and the blood testosterone concentration increases.

Testosterone level decreases somewhat during and after the *male climacteric,* which is a decline in sexual function associated with aging. At any given age, the testosterone concentration in the male body is regulated to remain relatively constant.

Practice

18. Which hormone initiates the changes associated with male sexual maturity?
19. Describe several male secondary sex characteristics.
20. List the functions of testosterone.
21. Explain how the secretion of male sex hormones is regulated.

19.4 ORGANS OF THE FEMALE REPRODUCTIVE SYSTEM

The organs of the female reproductive system produce and maintain the female sex cells, the egg cells (or oocytes); transport these cells to the site of fertilization; provide a favorable environment for a developing offspring; move the offspring to the outside; and produce female sex hormones. The *primary sex organs* (gonads) of this system are the two ovaries, which produce the female sex cells and sex hormones. The *accessory sex organs* of the female reproductive system are the internal and external reproductive organs (fig. 19.7; reference plates 5 and 6, pp. 27–28).

Ovaries

The two **ovaries** are solid, ovoid structures, each about 3.5 centimeters long, 2 centimeters wide, and 1 centimeter thick. The ovaries lie in shallow depressions in the lateral wall of the pelvic cavity (fig. 19.7).

Ovary Structure

Ovarian tissues are divided into two indistinct regions—an inner *medulla* and an outer *cortex*. The ovarian medulla is composed of loose connective tissue and has many blood vessels, lymphatic vessels, and nerve fibers. The ovarian cortex consists of more compact tissue and has a granular appearance due to tiny masses of cells called *ovarian follicles.*

A layer of cuboidal epithelium covers the ovary's free surface. Just beneath this epithelium is a layer of dense connective tissue.

Practice

22. What are the primary sex organs of the female?

23. Describe the structure of an ovary.

Primordial Follicles

During prenatal (before birth) development of a female, small groups of cells in the outer region of the ovarian cortex form several million **primordial follicles.** Each follicle is a single, large cell, called a *primary oocyte,* surrounded by epithelial cells called *follicular cells.*

Early in development, the primary oocytes begin to undergo meiosis, but the process soon halts and does not continue until the individual reaches puberty. Once the primordial follicles appear, no new ones form. Instead, the number of oocytes in the ovary steadily

Figure 19.7 AP|R

Female reproductive organs (sagittal view). The paired ovaries are the primary sex organs, and the other reproductive structures, both internal and external, are accessory sex organs.

declines as many of the oocytes degenerate. Of the several million oocytes that formed in the embryo, only a million or so remain at birth, and perhaps 400,000 are present at puberty. Of these, probably fewer than 400 or 500 oocytes will be released from the ovary during a female's reproductive life.

Oogenesis

Oogenesis (o″o-jen′ĕ-sis) is the process of egg cell formation. Beginning at puberty, some primary oocytes are stimulated to continue meiosis. Like sperm cells, the resulting cells have one-half as many chromosomes (23) in their nuclei as their parent cells.

Unlike a primary spermatocyte, when a primary oocyte divides, the cytoplasm is distributed unequally. One of the resulting cells, called a *secondary oocyte* (egg cell), is large, and the other, called the *first polar body,* is small (fig. 19.8). The large secondary oocyte represents a future *egg cell* (ovum) that can be fertilized by uniting

with a sperm cell. If this happens, the secondary oocyte divides unequally to produce a tiny *second polar body* and a large fertilized egg cell, or a **zygote** (zi′gōt).

The polar bodies have no further function and soon degenerate. Their role in reproduction allows for production of an egg cell that has the massive amounts of cytoplasm and the abundant organelles required to carry the zygote through the first few cell divisions, yet the right number of chromosomes.

The largest cell in the human body is the egg cell. The smallest cell is the sperm.

Practice

24. Describe the major events of oogenesis.
25. What is the function of polar body formation?

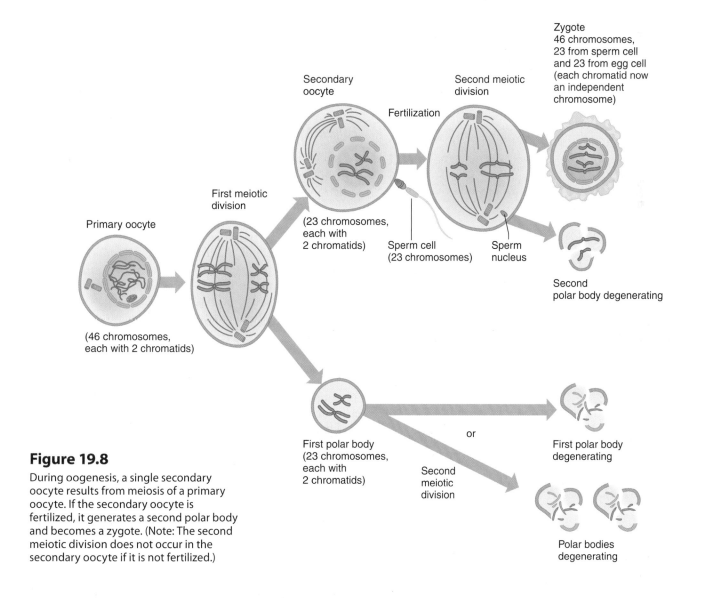

Figure 19.8

During oogenesis, a single secondary oocyte results from meiosis of a primary oocyte. If the secondary oocyte is fertilized, it generates a second polar body and becomes a zygote. (Note: The second meiotic division does not occur in the secondary oocyte if it is not fertilized.)

Primary oocyte
(46 chromosomes, each with 2 chromatids)

First meiotic division

Secondary oocyte
(23 chromosomes, each with 2 chromatids)

Fertilization

Sperm cell (23 chromosomes)

Sperm nucleus

Second meiotic division

Zygote
46 chromosomes, 23 from sperm cell and 23 from egg cell (each chromatid now an independent chromosome)

Second polar body degenerating

First polar body (23 chromosomes, each with 2 chromatids)

Second meiotic division

or

First polar body degenerating

Polar bodies degenerating

Corpus albicans

Corpus luteum

Ovulation

Time

Time

Uterine tube

Secondary oocyte

Zona pellucida

Corona radiata

Primordial follicle

Primary follicle

Ovary

Time

Follicular cells

Time

Time

First polar body

Primary oocyte

Follicular fluid

Figure 19.9 AP|R

In an ovary, as a follicle matures, a developing primary oocyte enlarges and becomes surrounded by follicular cells and fluid. Eventually, the mature follicle ruptures, releasing the secondary oocyte and layers of surrounding follicular cells.

Fluid-filled antrum

Corona radiata

Zona pellucida

Secondary oocyte

Figure 19.10

Light micrograph of a maturing follicle (250×).

Q: *Name the structure formed by the follicular cells attached to and surrounding the secondary oocyte.*

Answer may be found in Appendix E on page 568.

Follicle Maturation

At puberty, the anterior pituitary gland secretes increased amounts of FSH, and the ovaries enlarge in response. With each reproductive cycle, some of the primordial follicles mature into **primary follicles** (pri′ma-re fol′ĭ-klz). Figure 19.9 traces the maturation of a follicle in an ovary.

During maturation, a primary oocyte enlarges, and surrounding follicular cells proliferate by mitosis. These follicular cells organize into layers, and soon a cavity (*antrum*) appears in the cellular mass. A clear *follicular fluid* fills the cavity and bathes the primary oocyte. The enlarging fluid-filled cavity presses the primary oocyte to one side.

In time, the mature follicle reaches a diameter of 10 millimeters or more and bulges outward on the ovary surface, like a blister. The secondary oocyte within the mature follicle is a large, spherical cell, surrounded by a layer of glycoprotein called the zona pellucida and attached to a mantle of follicular cells (corona radiata) (fig. 19.10). Processes from these follicular cells extend through the zona pellucida and supply nutrients to the secondary oocyte. Although as many as twenty primary follicles may begin maturing at any one time, one follicle usually outgrows the others. Typically, only the dominant follicle fully develops, and the other follicles degenerate.

Ovulation

As a follicle matures, its primary oocyte undergoes oogenesis, giving rise to a secondary oocyte and a first polar body. The process called **ovulation** (o''vu-la′shun) releases the secondary oocyte and first polar body with one or two surrounding layers of follicular cells from the mature follicle.

Release of LH from the anterior pituitary gland triggers ovulation, which rapidly swells the mature follicle and weakens its wall. Eventually, the wall ruptures, and follicular fluid and the secondary oocyte ooze from the ovary's surface (see fig. 19.9).

After ovulation, the secondary oocyte and surrounding follicular cells are usually propelled to the opening of a nearby uterine tube (fig. 19.11). If the oocyte is not fertilized within hours, it degenerates.

Practice

26. What changes occur in a follicle and its oocyte during maturation?

27. What causes ovulation?

28. What happens to an oocyte following ovulation?

Female Internal Reproductive Organs

The internal accessory organs of the female reproductive system include a pair of uterine tubes, a uterus, and a vagina.

Uterine Tubes

The **uterine tubes** (fallopian tubes or oviducts) open near the ovaries (fig. 19.11). Each tube is about 10 centimeters long and passes medially to the uterus, penetrates its wall, and opens into the uterine cavity.

Near each ovary, a uterine tube expands, forming a funnel-shaped *infundibulum* (in''fun-dib′u-lum), which partially encircles the ovary. Fingerlike extensions called *fimbriae* (fim′bre) fringe the infundibulum margin. Although the infundibulum generally does not touch the ovary, one of the larger extensions connects directly to the ovary.

Simple columnar epithelial cells, some *ciliated,* line the uterine tube. The epithelium secretes mucus, and the cilia beat toward the uterus. These actions help draw the secondary oocyte and expelled follicular fluid into the infundibulum following ovulation. Ciliary action and peristaltic contractions of the uterine tube's muscular layer help transport the secondary oocyte

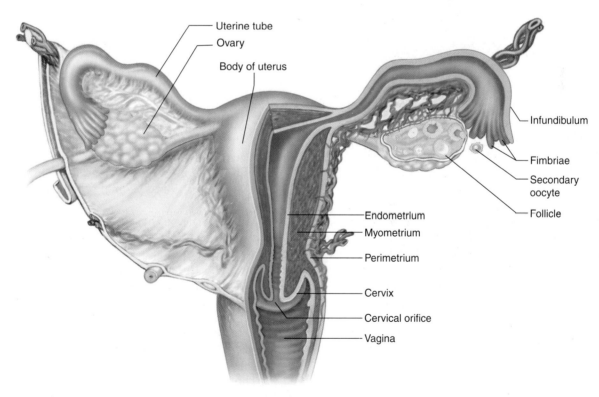

Figure 19.11

The funnel-shaped infundibulum of the uterine tube partially encircles the ovary (posterior view).

down the uterine tube. Fertilization occurs in the uterine tube.

Uterus

The **uterus** receives the embryo that develops from an egg cell fertilized in the uterine tube, and sustains its development. It is a hollow, muscular organ shaped somewhat like an inverted pear.

The size of the uterus changes greatly during pregnancy. In its nonpregnant, adult state, the uterus is about 7 centimeters long, 5 centimeters wide (at its broadest point), and 2.5 centimeters in diameter. The uterus is located medially in the anterior part of the pelvic cavity, superior to the vagina, and usually bends forward over the urinary bladder.

The upper two-thirds, or *body*, of the uterus has a dome-shaped top called the *fundus* (fig. 19.11). The uterine tubes enter the top of the uterus at its broadest part. The lower third of the uterus is called the **cervix.** This tubular part extends downward into the upper part of the vagina. The cervix surrounds the opening called the *cervical orifice*, through which the uterus opens to the vagina

The uterine wall is thick and has three layers (figs. 19.11 and 19.12). The **endometrium** (en″do-me′tre-um),

the inner mucosal layer, is covered with columnar epithelium and contains abundant tubular glands. The **myometrium** (mi″o-me′tre-um), a thick, middle, muscular layer, consists largely of bundles of smooth muscle fibers. During the monthly female reproductive cycles and during pregnancy, the endometrium and myometrium change extensively. The **perimetrium** (per-ĭ-me′tre-um) is an outer serosal layer that covers the body of the uterus and part of the cervix.

> During pregnancy, the uterus expands to 500 times its normal volume.

> A precancerous condition, called carcinoma *in situ* (CIS), affects the cervix. It is detected using a *Pap (Papanicolaou) smear test* that identifies cellular changes. Government public health guidelines advise testing annually starting three years after becoming sexually active or by age twenty-one. At age thirty, testing can switch to every three years if no abnormalities have been found. Most cases of CIS do not progress to cancer, and those that do can usually be removed.

Vagina

The **vagina** is a fibromuscular tube, about 9 centimeters long, extending from the uterus to the outside of the body (see fig. 19.7). It conveys uterine secretions, receives the erect penis during sexual intercourse, and provides an open channel for the offspring to exit the body during birth.

The vagina extends upward and back into the pelvic cavity. It is posterior to the urinary bladder and urethra, anterior to the rectum, and attached to these structures by connective tissues.

A thin membrane of connective tissue and stratified squamous epithelium called the **hymen** partially covers the *vaginal orifice*. A central opening of varying size allows uterine and vaginal secretions to pass to the outside.

The vaginal wall has three layers. The inner *mucosal layer* is stratified squamous epithelium. This layer lacks mucous glands; the mucus in the lumen of the vagina comes from uterine glands and from vestibular glands at the mouth of the vagina.

The middle *muscular layer* mainly consists of smooth muscle fibers. A thin band of striated muscle at the lower end of the vagina helps close the vaginal opening. Another voluntary muscle (bulbospongiosus) is primarily responsible for closing this orifice.

The outer *fibrous layer* consists of dense connective tissue interlaced with elastic fibers. It attaches the vagina to surrounding organs.

Lumen

Endometrium

Myometrium

Perimetrium

Figure 19.12 AP|R

Light micrograph of the uterine wall (10×).

Practice

29. How is a secondary oocyte moved along a uterine tube?
30. Describe the structure of the uterus.
31. Describe the structure of the vagina.

Female External Reproductive Organs

The external accessory organs of the female reproductive system include the labia majora, labia minora, clitoris, and the vestibular glands. These structures that surround the openings of the urethra and vagina compose the **vulva** (see fig. 19.7).

Labia Majora

The **labia majora** (singular, *labium majus*) enclose and protect the other external reproductive organs. The labia majora correspond to the scrotum in males and are composed of rounded folds of adipose tissue and a thin layer of smooth muscle, covered by skin.

The labia majora lie close together. A cleft that includes the urethral and vaginal openings separates the labia longitudinally. At their anterior ends, the labia merge to form a medial, rounded elevation of adipose tissue called the *mons pubis,* which overlies the pubic symphysis.

Labia Minora

The **labia minora** (singular, *labium minus*) are flattened, longitudinal folds between the labia majora (see fig. 19.7). They are composed of connective tissue richly supplied with blood vessels, giving a pinkish appearance. Posteriorly, the labia minora merge with the labia majora, while anteriorly, they converge to form a hoodlike covering around the clitoris.

Clitoris

The **clitoris** (klit′o-ris) is a small projection at the anterior end of the vulva between the labia minora (see fig. 19.7). It is usually about 2 centimeters long and 0.5 centimeters in diameter, including a portion embedded in surrounding tissues. The clitoris corresponds to the penis in males and has a similar structure. It is composed of two columns of erectile tissue called *corpora cavernosa.* At its anterior end, a small mass of erectile tissue forms a glans, which is richly supplied with sensory nerve fibers.

Vestibule

The labia minora enclose a space called the **vestibule.** The vagina opens into the posterior portion of the vesti-

bule, and the urethra opens in the midline, just anterior to the vagina and about 2.5 centimeters posterior to the glans of the clitoris.

A pair of **vestibular glands,** corresponding to the bulbourethral glands in males, lie one on either side of the vaginal opening. Beneath the mucosa of the vestibule on either side is a mass of vascular erectile tissue called the *vestibular bulb.*

Practice

32. What is the male counterpart of the labia majora? Of the clitoris?
33. Which structures are within the vestibule?

Erection, Lubrication, and Orgasm

Erectile tissues in the clitoris and around the vaginal entrance respond to sexual stimulation. Following such stimulation, parasympathetic nerve impulses from the sacral part of the spinal cord release the vasodilator nitric oxide (NO), which dilates the arteries associated with the erectile tissues. As a result, blood inflow increases, erectile tissues swell, and the vagina expands and elongates.

If sexual stimulation is sufficiently intense, parasympathetic impulses stimulate the vestibular glands to secrete mucus into the vestibule. This moistens and lubricates the tissues surrounding the vestibule and the lower end of the vagina, facilitating insertion of the penis into the vagina.

The clitoris is abundantly supplied with sensory nerve fibers, which are especially sensitive to local stimulation. Such stimulation culminates in orgasm.

Just prior to orgasm, the tissues of the outer third of the vagina engorge with blood and swell. This increases the friction on the penis during intercourse. Orgasm initiates a series of reflexes involving the sacral and lumbar parts of the spinal cord, and in response, the muscles of the perineum and the walls of the uterus and uterine tubes contract rhythmically. These contractions help transport sperm cells through the female reproductive tract toward the upper ends of the uterine tubes. Table 19.2 summarizes the functions of the female reproductive organs.

Practice

34. What events result from parasympathetic stimulation of the female reproductive organs?
35. What changes occur in the vagina just prior to and during orgasm?

Table 19.2	Functions of the Female Reproductive Organs
Organ	**Function**
Ovary	Produces oocytes and female sex hormones
Uterine tube	Conveys secondary oocyte toward uterus; site of fertilization; conveys developing embryo to uterus
Uterus	Protects and sustains embryo during pregnancy
Vagina	Conveys uterine secretions to outside of body; receives erect penis during sexual intercourse; provides open channel for offspring during birth process
Labia majora	Enclose and protect other external reproductive organs
Labia minora	Form margins of vestibule; protect openings of vagina and urethra
Clitoris	Produces feelings of pleasure during sexual stimulation due to abundant sensory nerve endings in glans
Vestibule	Space between labia minora that contains vaginal and urethral openings
Vestibular glands	Secrete fluid that moistens and lubricates vestibule

19.5 HORMONAL CONTROL OF FEMALE REPRODUCTIVE FUNCTIONS

The hypothalamus, the anterior pituitary gland, and the ovaries secrete hormones that control maturation of female sex cells, the development and maintenance of female secondary sex characteristics, and changes that occur during the monthly reproductive cycle.

Female Sex Hormones

The female body is reproductively immature until about ten years of age. Then, the hypothalamus begins to secrete increasing amounts of GnRH, which, in turn, stimulates the anterior pituitary to release the gonadotropins FSH and LH. These hormones play primary roles in controlling female sex cell maturation and in producing female sex hormones.

Several tissues, including the ovaries, the adrenal cortices, and the placenta (during pregnancy), secrete female sex hormones belonging to two major groups—**estrogens** (es′tro-jenz) and **progesterone** (pro-jes′tĭ-rōn). *Estradiol* is the most abundant of the estrogens, which also include *estrone* and *estriol*.

The ovaries are the primary source of estrogens (in a nonpregnant female). At puberty, under the influence of the anterior pituitary, the ovaries secrete increasing

amounts of estrogens. Estrogens stimulate enlargement of accessory organs, including the vagina, uterus, uterine tubes, ovaries, and external reproductive structures. Estrogens also develop and maintain the *female secondary sex characteristics,* which include:

1. Development of the breasts and the ductile system of the mammary glands in the breasts.
2. Increased deposition of adipose tissue in the subcutaneous layer generally and in the breasts, thighs, and buttocks particularly.
3. Increased vascularization of the skin.

The ovaries are also the primary source of progesterone (in a nonpregnant female). This hormone promotes changes in the uterus during the female reproductive cycle, affects the mammary glands, and helps regulate the secretion of gonadotropins from the anterior pituitary gland.

Androgen (male sex hormone) concentrations produce certain other changes in females at puberty. For example, increased hair growth in the pubic and axillary regions is due to androgen secreted by the adrenal cortices. Conversely, development of the female skeletal configuration, which includes narrow shoulders and broad hips, is a response to a low androgen concentration.

Practice

36. What stimulates sexual maturation in a female?
37. What is the function of estrogens?
38. What is the function of androgen in a female?

Female Reproductive Cycle

The female reproductive cycle is characterized by regular, recurring changes in the endometrium, which culminate in menstrual bleeding (menses). Such cycles usually begin around age thirteen and continue into the early fifties, then cease.

Elite female athletes and professional dancers may have disturbed reproductive cycles, ranging from diminished menstrual flow (oligomenorrhea) to complete stoppage (amenorrhea). The more active an athlete or dancer, the more likely it is that she will have menstrual irregularities, and this may impair her ability to conceive. Her diminished fat reserves result in decreased secretion of the hormone leptin, an effect which lowers secretion of gonadotropin-releasing hormone from the hypothalamus, which in turn lowers estrogen level. Adipose tissue itself also contains some estrogen, but trim elite athletes and dancers have little fat. Adequate estrogen is necessary for fertility. These girls and women are also at high risk of developing low bone density and cardiovascular impairments.

A female's first reproductive cycle, called **menarche** (mĕ-nar′ke), occurs after the ovaries and other organs of the reproductive control system have matured and begun responding to certain hormones. Then, the hypothalamic secretion of GnRH stimulates the anterior pituitary to release threshold levels of FSH and LH. FSH stimulates maturation of an ovarian follicle. The follicular cells produce increasing amounts of estrogens and some progesterone. LH stimulates certain ovarian cells to secrete precursor molecules (such as testosterone), also used to produce estrogens.

In a young female, estrogens stimulate the development of secondary sex characteristics. Estrogens secreted during subsequent reproductive cycles continue the development and maintenance of these characteristics.

Increasing concentration of estrogens during the first week or so of a reproductive cycle changes the uterine lining, thickening the glandular endometrium (proliferative phase). Meanwhile, the developing follicle fully matures, and by around the fourteenth day of the cycle, the follicle appears on the ovary surface as a blisterlike bulge. Within the follicle, the follicular cells, which surround and connect the secondary oocyte to the inner wall, loosen. Follicular fluid accumulates rapidly.

While the follicle matures, it secretes estrogens that inhibit the release of LH from the anterior pituitary gland but allow LH to be stored in the gland. Estrogens also make anterior pituitary cells more sensitive to the action of GnRH, which is released from the hypothalamus in rhythmic pulses about ninety minutes apart.

Near the fourteenth day of follicular development, the anterior pituitary cells finally respond to the pulses of GnRH and release the stored LH. The resulting surge in LH concentration, which lasts about thirty-six hours, weakens and ruptures the bulging follicular wall, which sends the secondary oocyte and follicular fluid out of the ovary (ovulation).

Following ovulation, the remnants of the follicle within the ovary change rapidly. The space containing the follicular fluid fills with blood, which soon clots. Under the influence of LH, follicular cells expand, forming a temporary glandular structure in the ovary called a **corpus luteum** (kor′pus loot′e-um) ("yellow body").

Follicular cells secrete some progesterone during the first part of the reproductive cycle. However, corpus luteum cells secrete abundant progesterone and estrogens during the second half of the cycle. Consequently, as a corpus luteum forms, the blood progesterone concentration sharply increases.

Progesterone causes the endometrium to become more vascular and glandular. It also stimulates uterine glands to secrete more glycogen and lipids (secretory phase). The endometrial tissues fill with fluids containing nutrients and electrolytes, which provide a favorable environment for an embryo to develop.

High levels of estrogens and progesterone inhibit the anterior pituitary gland's release of LH and FSH. Consequently, no other follicles are stimulated to develop when the corpus luteum is active. However, if the secondary oocyte released at ovulation is not fertilized, the corpus luteum begins to degenerate (regress) (see fig. 19.9) on about the twenty-fourth day of the cycle. Eventually, connective tissue replaces it. The remnant of such a corpus luteum is called a *corpus albicans.*

When the corpus luteum ceases to function, concentrations of estrogens and progesterone rapidly decline, and in response, blood vessels in the endometrium constrict. This reduces the supply of oxygen and nutrients to the thickened endometrium, and these lining tissues soon disintegrate and slough off. At the same time, blood leaves damaged capillaries, creating a flow of blood and cellular debris that passes through the vagina as the *menstrual flow* (menses). This flow usually begins about the twenty-eighth day of the cycle and continues for three to five days, while the concentrations of estrogens are relatively low. The beginning of the menstrual flow marks the end of a reproductive cycle and the beginning of a new cycle. This cycle is summarized in table 19.3 and diagrammed in figure 19.13.

Table 19.3	Major Events in a Reproductive Cycle

1. The anterior pituitary gland secretes follicle-stimulating hormone (FSH) and luteinizing hormone (LH).

2. FSH stimulates maturation of a follicle.

3. Follicular cells produce and secrete estrogens.

   a. Estrogens maintain secondary sex characteristics.

   b. Estrogens cause the endometrium to thicken.

4. The anterior pituitary releases a surge of LH, which stimulates ovulation.

5. Follicular cells become corpus luteum cells, which secrete estrogens and progesterone.

   a. Estrogens continue to stimulate uterine wall development.

   b. Progesterone stimulates the endometrium to become more glandular and vascular.

   c. Estrogens and progesterone inhibit the secretion of FSH and LH from the anterior pituitary gland.

6. If the secondary oocyte is not fertilized, the corpus luteum degenerates and no longer secretes estrogens and progesterone.

7. As the concentrations of estrogens and progesterone decline, blood vessels in the endometrium constrict.

8. The uterine lining disintegrates and sloughs off, producing a menstrual flow.

9. The anterior pituitary gland is no longer inhibited and again secretes FSH and LH.

10. The reproductive cycle repeats.

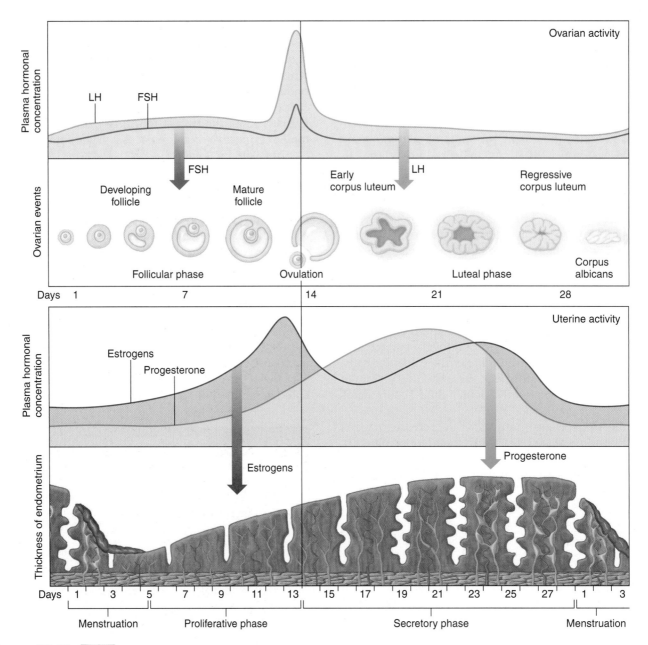

Figure 19.13 AP|R

Major events in the female reproductive cycle.

Low blood concentrations of estrogens and progesterone at the beginning of the reproductive cycle mean that the hypothalamus and anterior pituitary gland are no longer inhibited. Consequently, FSH and LH concentrations soon increase, stimulating a new follicle to mature. As this follicle secretes estrogens, the uterine lining undergoes repair, and the endometrium begins to thicken again.

Menopause

After puberty, reproductive cycles continue at regular intervals into the late forties or early fifties, when they usually become increasingly irregular. Then, within a few months or years, the cycles cease. This period in life is called **menopause** (men′o-pawz), or female climacteric.

Aging of the ovaries causes menopause. After about thirty-five years of cycling, few primary follicles remain to respond to pituitary gonadotropins. Consequently, the follicles no longer mature, ovulation does not occur, and the blood concentration of estrogens plummets.

Reduced concentrations of estrogens and lack of progesterone may change the female secondary sex characteristics. The breasts, vagina, uterus, and uterine tubes may shrink, and the pubic and axillary hair may thin.

Practice

39. Trace the events of the female reproductive cycle.

40. What causes menstrual flow?

41. What are some changes that may occur at menopause?

19.6 MAMMARY GLANDS

The **mammary glands** are accessory organs of the female reproductive system specialized to secrete milk following pregnancy. They are in the subcutaneous tissue of the anterior thorax within elevations called *breasts* (fig. 19.14*a*; reference plate 1, p. 23). The breasts overlie the *pectoralis major* muscles and extend from the second to the sixth ribs and from the sternum to the axillae.

A *nipple* is located near the tip of each breast at about the level of the fourth intercostal space. A circular area of pigmented skin, called the *areola,* surrounds each nipple (fig. 19.14*b*).

A mammary gland is composed of fifteen to twenty lobes. Each lobe contains glands (alveolar glands) and an alveolar duct that leads to a lactiferous duct, which

in turn leads to the nipple and opens to the outside (fig. 19.14). Dense connective and adipose tissues separate the lobes. These tissues also support the glands and attach them to the fascia of the underlying pectoral muscles. Other connective tissue, which forms dense strands called *suspensory ligaments,* extends inward from the dermis of the breast to the fascia, helping support the breast.

The mammary glands of males and females are similar. As children reach puberty, the glands in males do not develop, whereas ovarian hormones stimulate development of the glands in females. The alveolar glands and ducts enlarge, and fat is deposited around and within the breasts. Chapter 20 (pp. 552–553) describes the hormonal mechanism that stimulates mammary glands to produce and secrete milk. Clinical Application 19.3 discusses diagnosis, treatment, and prevention of breast cancer.

Practice

42. Describe the structure of a mammary gland.

43. How does ovarian hormone secretion change the mammary glands?

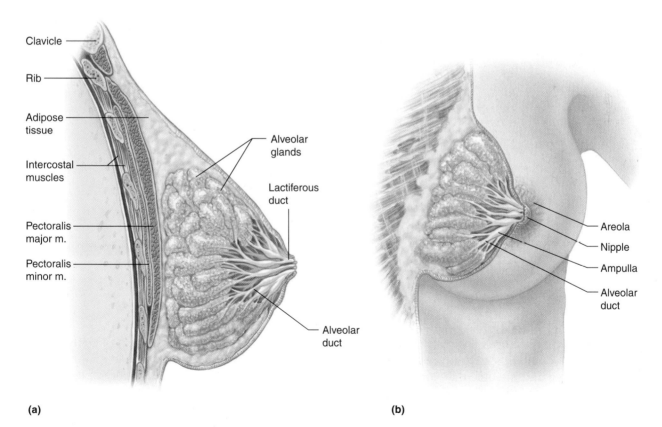

(a) (b)

Figure 19.14 APR

Structure of the female breast and mammary glands. **(a)** Sagittal section. **(b)** Anterior view. (*m.* stands for *muscle.*)

Clinical Application 19.3

Treating Breast Cancer

One in eight women will develop breast cancer at some point in her life (table 19B). About 1% of breast cancer cases are in men. Breast cancer is several illnesses. As research reveals the cellular and molecular characteristics that distinguish subtypes of the disease, treatments old and new are being increasingly tailored to individuals at the time of diagnosis. This "rational" approach may delay progression of the disease and increase the survival rate, while enabling patients to avoid using drugs that are unlikely to work.

Warning Signs

Changes that could signal breast cancer include a small area of thickened tissue, a dimple, a change in contour, or a nipple that is flattened, points in an unusual direction, or produces a discharge. A woman can note these changes by performing a monthly "breast self-exam," in which she lies flat on her back with the arm raised behind her head and systematically feels all parts of each breast. But breast cancer may give no warning at all—early signs of fatigue and feeling ill may occur after the disease has spread beyond the breast.

After finding a lump, the next step is a physical exam, where a health-care provider palpates the breast and does a mammogram, an X-ray scan that can pinpoint the location and approximate extent of abnormal tissue (fig. 19B). An ultrasound scan can distinguish between a cyst (a fluid-filled sac of glandular tissue) and a tumor (a solid mass). If an area

Figure 19B AP|R

Mammogram of a breast with a tumor (arrow).

Table 19B	Breast Cancer Risk		
By Age	**Odds**	**By Age**	**Odds**
25	1 in 19,608	60	1 in 24
30	1 in 2,525	65	1 in 17
35	1 in 622	70	1 in 14
40	1 in 217	75	1 in 11
45	1 in 93	80	1 in 10
50	1 in 50	85	1 in 9
55	1 in 33	95 or older	1 in 8

19.7 BIRTH CONTROL

Birth control is the voluntary regulation of the number of offspring produced and the time they are conceived. This control requires a method of **contraception** (kon″trah-sep′shun) designed to avoid fertilization of an egg cell following sexual intercourse (coitus) or to prevent implantation of a hollow ball of cells (a blastocyst) that will develop into an embryo. The several methods of contraception have varying degrees of effectiveness.

Coitus Interruptus

Coitus interruptus is the practice of withdrawing the penis from the vagina before ejaculation, preventing entry of sperm cells into the female reproductive tract. This method can still result in pregnancy because a male may find it difficult to withdraw just prior to ejaculation. Also, some semen containing sperm cells may reach the vagina before ejaculation occurs.

is suspicious, a thin needle is used to take a biopsy (sample) of the tissue, whose cells are scrutinized for the telltale characteristics of cancer. Further tests can identify estrogen and progesterone receptors on cancer cells, which is information used to guide treatment choices.

Eighty percent of the time, a breast lump is a sign of fibrocystic breast disease, which is benign (noncancerous). The lump may be a cyst or a solid, fibrous mass of connective tissue called a fibroadenoma. Treatment for fibrocystic breast disease includes vitamin E, synthetic androgens, and lowering caffeine intake.

Surgery, Radiation, and Chemotherapies

If biopsied breast cells are cancerous, treatment usually begins with surgery. A lumpectomy removes a small tumor and some surrounding tissue; a simple mastectomy removes a breast; and a modified radical mastectomy removes the breast and surrounding lymph nodes, but preserves the pectoral muscles. A few lymph nodes are typically examined, and if cancer cells are detected, further surgery is performed.

Most breast cancers are then treated with radiation and combinations of chemotherapeutic drugs, plus sometimes newer drugs that are targeted to certain types of breast cancer. Standard chemotherapies kill all rapidly dividing cells, and those used for breast cancer include fluorouracil, doxorubicin, cyclophosphamide, methotrexate, and paclitaxol. Protocols that provide more frequent, lower doses can temper some of the side effects of these powerful drugs.

Newer treatments developed specifically for breast cancer are easier to tolerate and can be extremely effective. Three types of drugs keep signals (estrogen and growth factors) from stimulating cancer cells to divide:

1. Selective estrogen receptor modulators (SERMs), such as tamoxifen and raloxifene, block estrogen receptors. About half of people with breast cancer have receptors for estrogen on their cancer cells.

2. Aromatase inhibitors block an enzyme required for tissues other than those of the ovaries to synthesize estrogens. These drugs are used in women who are past menopause, whose ovaries no longer synthesize estrogen. They are prescribed after a five-year course of a SERM.

3. Trastuzumab can help people whose cancer cells bear too many receptors that bind a particular growth factor. It is a monoclonal antibody, based on an immune system protein. Trastuzumab blocks the growth factor from signaling cell division. Marketed as Herceptin, this drug treats a particularly aggressive form of the disease that strikes younger women.

Prevention Strategies

Public health agencies and physicians are debating exactly at what age women should begin having mammograms, and how frequently thereafter. For this reason, it is important for women to discuss personal mammogram schedules with health-care providers familiar with their family histories. Although a mammogram can detect a tumor up to two years before it can be felt, it can also miss some tumors. Thus, breast self-exam is also important in early detection.

Genetic tests can identify women who have inherited certain variants of genes—such as *BRCA1, BRCA2, p53,* and *HER-2/neu*—that place them at high risk for developing breast cancer. Some of these women have their breasts removed to prevent the disease.

Only 5% to 10% of all breast cancers arise from an inherited tendency. Much research seeks to identify the environmental triggers that contribute to causing the majority of cases. Gene expression profiling is beginning to be used to identify which drugs are most likely to help particular patients. Gene expression profiling of breast cancer cells can also predict recurrence after surgery. This type of information is important in choosing follow-up treatment.

Rhythm Method

The *rhythm method* (also called timed coitus or natural family planning) requires abstinence from sexual intercourse two days before and one day after ovulation. The rhythm method results in a relatively high rate of pregnancy because it is difficult to accurately identify infertile ("safe") times to have intercourse. Another disadvantage of the rhythm method is that it restricts spontaneity in sexual activity.

Mechanical Barriers

Mechanical barrier contraceptives prevent sperm cells from entering the female reproductive tract during sexual intercourse. The *male condom* is a thin latex or natural membrane sheath placed over the erect penis before intercourse to prevent semen from entering the vagina upon ejaculation (fig. 19.15*a*). A *female condom* resembles a small plastic bag. A woman inserts it into her vagina prior to intercourse. The device blocks sperm cells from reaching the cervix (fig. 19.15*a*).

(a)　(b)

(c)　(d)　(e)

Figure 19.15

Devices and substances used for birth control include **(a)** male and female condoms, **(b)** a diaphragm, **(c)** spermicide delivery methods, **(d)** combined hormone contraceptives, and **(e)** an IUD.

Q: *Which of these methods of birth control is statistically most effective?*

Answer may be found in Appendix E on page 568.

Some men feel that a condom decreases the sensitivity of the penis during intercourse, and its use may interrupt spontaneity. However, condoms are inexpensive and may also prevent transmission of some sexually transmitted infections.

Another mechanical barrier is the *diaphragm,* a cup-shaped device with a flexible ring forming the rim. The diaphragm is inserted into the vagina so that it covers the cervix, preventing sperm cells from entering the uterus (fig. 19.15*b*). To be effective, a diaphragm must be fitted for size by a physician, inserted properly, and used with a spermicide applied to the diaphragm surface adjacent to the cervix and to the rim of the diaphragm. It must be left in position for several hours following sexual intercourse. A diaphragm can be inserted up to six hours prior to sexual contact.

Similar to but smaller than the diaphragm is the *cervical cap,* which adheres to the cervix by suction. A woman inserts it with her fingers before intercourse. For centuries, different societies have used cervical caps made of such varied substances as beeswax, lemon halves, paper, and opium poppy fibers.

Chemical Barriers

Chemical barrier contraceptives include creams, foams, and jellies with spermicidal properties (fig. 19.15*c*). These chemicals create an unfavorable environment in the vagina for sperm cells.

Chemical barrier contraceptives are fairly easy to use but have a high failure rate when used alone. They are most effective with a condom or diaphragm.

Combined Hormone Contraceptives

Combined hormone contraceptives deliver estrogen and progestin to prevent pregnancy. Various methods are used to administer the hormones, but all work on the same principle with about the same efficacy, although the amounts of the component hormones may vary. One such method is a small flexible chemical ring inserted deep into the vagina once a month, remaining in place three out of four weeks. A plastic patch impregnated with the hormones may be applied to the skin on the buttocks, stomach, arm, or upper torso once a week for three out of four weeks. The most commonly used method to deliver the hormones is orally, in pill form (fig. 19.15*d*).

Combined hormone contraceptives contain synthetic estrogen-like and progesterone-like chemicals. These drugs disrupt the normal pattern of gonadotropin (FSH and LH) secretion, preventing follicle maturation and the LH surge that triggers ovulation. They also interfere with buildup of the uterine lining necessary for implantation. Most combined hormone contraceptives cause light monthly bleeding. A new type of pill causes only four bleeding periods a year.

If used correctly, combined hormone contraceptives prevent pregnancy nearly 100% of the time. However, they may cause nausea, retention of body fluids, increased skin pigmentation, and breast tenderness. Some women, particularly those over thirty-five years of age who smoke, may develop intravascular blood clots, liver disorders, or high blood pressure when using certain types of these contraceptives.

Injectable Contraception

An intramuscular injection of medroxyprogesterone acetate protects against pregnancy for three months by preventing the maturation and release of a secondary oocyte. It also alters the uterine lining, making it less hospitable for a developing embryo. Medroxyprogesterone acetate is long-acting; it takes ten to eighteen months after the last injection for the effects to wear off. Use of this drug requires a doctor's care because of potential side effects and risks.

Intrauterine Devices

An *intrauterine device (IUD)* is a small, solid object that a physician places in the uterine cavity (fig. 19.15*e*). An IUD interferes with implantation of a blastocyst, perhaps by inflaming the uterine tissues. A levonorgestrel-releasing IUD, combining hormone contraception with an IUD, is now available and 99% effective in preventing pregnancy for up to five years.

An IUD may be spontaneously expelled from the uterus, or produce abdominal pain or excessive menstrual bleeding. It may also harm the uterus or produce other serious health problems and should be checked regularly by a physician.

Surgical Methods

Surgical methods of contraception sterilize the male or female. In the male, a physician removes a small section of each ductus (vas) deferens near the epididymis

and ties (ligates) the cut ends of the ducts. This is a *vasectomy*, an operation that produces few side effects, although it may cause some pain for a week or two.

After a vasectomy, sperm cells cannot leave the epididymis; thus, they are excluded from the semen. However, sperm cells may already be present in the ducts distal to the cuts. Consequently, the sperm count may not reach zero for several weeks.

The corresponding procedure in the female is *tubal ligation.* The uterine tubes are cut and ligated so that sperm cells cannot reach an egg cell.

Neither a vasectomy nor a tubal ligation changes hormonal concentrations or sex drives. These procedures, shown in figure 19.16, are the most reliable forms of contraception. Reversing them requires microsurgery.

(a)

(b)

Figure 19.16

Surgical methods of birth control. **(a)** In a vasectomy, each ductus (vas) deferens is cut and ligated. **(b)** In a tubal ligation, each uterine tube is cut and ligated.

Practice

44. What factors make the rhythm method less reliable than some other methods of contraception?

45. Describe two methods of contraception that use mechanical barriers.

46. How do combined hormone contraceptives prevent pregnancy?

47. How does an IUD prevent pregnancy?

19.8 SEXUALLY TRANSMITTED INFECTIONS

The term **sexually transmitted infection (STI)** is replacing the term *sexually transmitted disease (STD)* because a person can be infected with a pathogen and transmit the pathogen to others but not develop the symptoms of the disease. By the time symptoms appear, it is often too late to prevent complications or spread of the infection to sexual partners. Many STIs have similar symptoms, and some of the symptoms are also seen in diseases or allergies not sexually related, so it is wise to consult a physician if one or a combination of these symptoms appears:

1. Burning sensation during urination.
2. Pain in the lower abdomen.
3. Fever or swollen glands in the neck.
4. Discharge from the vagina or penis.
5. Pain, itching, or inflammation in the genital or anal area.
6. Pain during intercourse.
7. Sores, blisters, bumps, or a rash anywhere on the body, particularly the mouth or genitals.
8. Itchy, runny eyes.

Table 19.4 describes some prevalent STIs. One possible complication of the STIs gonorrhea and chlamydia is **pelvic inflammatory disease,** in which bacteria enter the vagina and spread throughout the reproductive organs. The disease begins with intermittent cramps, followed by sudden fever, chills, weakness, and severe cramps. Hospitalization and intravenous antibiotics can stop the infection. The uterus and uterine tubes are often scarred, resulting in infertility and increased risk of ectopic pregnancy, in which the embryo develops in a uterine tube.

Acquired immune deficiency syndrome (AIDS) is a steady deterioration of the body's immune defenses in which the body is overrun by infection and often cancer. The human immunodeficiency virus (HIV) that causes AIDS is transmitted in body fluids such as semen, blood, and milk. It is most frequently passed during unprotected intercourse or by using a needle containing contaminated blood. Clinical Application 14.1 (p. 389) explores HIV infection further.

Practice

48. Why is the term sexually transmitted infection replacing the term sexually transmitted disease?

49. What are some common symptoms of sexually transmitted infections?

Table 19.4	Some Sexually Transmitted Infections		
Infection	**Cause**	**Symptoms**	**Treatment**
Acquired immune deficiency syndrome	Human immunodeficiency virus	Fever, weakness, infections, cancer	Drugs to treat or delay symptoms; no cure
Chlamydia infection	*Chlamydia trachomatis* bacteria	Painful urination and intercourse, mucous discharge from penis or vagina	Antibiotics
Genital herpes	Herpes simplex 2 virus	Genital sores, fever	Antiviral drug (acyclovir)
Genital warts	Human papilloma virus	Warts on genitals	Chemical or surgical removal
Gonorrhea	*Neisseria gonorrhoeae* bacteria	In women, usually none; in men, painful urination	Antibiotics
Syphilis	*Treponema pallidum* bacteria	Initial chancre usually on genitals or mouth; rash six months later; several years with no symptoms as infection spreads; finally damage to heart, liver, nerves, brain	Antibiotics

Reproductive Systems

Integumentary System

Skin sensory receptors play a role in sexual pleasure.

Skeletal System

Bones can be a temporary source of calcium during lactation.

Muscular System

Skeletal, cardiac, and smooth muscles all play a role in reproductive processes and sexual activity.

Nervous System

The nervous system plays a major role in sexual activity and sexual pleasure.

Endocrine System

Hormones control the production of eggs in the female and sperm in the male.

Cardiovascular System

Blood pressure is necessary for the normal function of erectile tissue in the male and female.

Lymphatic System

Special mechanisms inhibit the female immune system from attacking sperm as foreign invaders.

Digestive System

Proper nutrition is essential for the formation of normal gametes.

Respiratory System

Breathing provides oxygen that assists in the production of ATP needed for egg and sperm development.

Urinary System

Male urinary and reproductive systems share common structures. Kidneys help compensate for fluid loss from the reproductive systems.

Gamete production, fertilization, fetal development, and childbirth are essential for survival of the species.

Summary Outline

19.1 Introduction (p. 506)

Reproductive organs produce sex cells and sex hormones, nurture these cells, or transport them.

19.2 Organs of the Male Reproductive System (p. 506)

The primary male sex organs are the testes, which produce sperm cells and male sex hormones. Accessory sex organs are internal and external.

1. Testes
   a. Structure of the testes
      (1) The testes are separated by connective tissue and filled with seminiferous tubules.
      (2) The seminiferous tubules contain undifferentiated cells that give rise to sperm cells.
      (3) The interstitial cells produce male sex hormones.
   b. Formation of sperm cells
      (1) The epithelium lining the seminiferous tubules includes sustentacular cells and spermatogenic cells.
         (a) Sustentacular cells support and nourish spermatogenic cells.
         (b) Spermatogenic cells give rise to sperm cells.
      (2) Spermatogonia give rise to sperm cells.
      (3) Meiosis reduces the number of chromosomes in sperm cells by one-half (from 46 to 23).
      (4) Spermatogenesis produces four sperm cells from each primary spermatocyte.
   c. Structure of a sperm cell
      A sperm cell consists of a head, midpiece, and tail.
2. Male internal reproductive organs
   a. Epididymides
      (1) Each epididymis is a tightly coiled tube that leads into a ductus deferens.
      (2) They store and nourish immature sperm cells and promote their maturation.
   b. Ductus deferentia
      (1) Each ductus deferens is a muscular tube.
      (2) They pass through the inguinal canal, enter the pelvic cavity, course medially and end behind the urinary bladder.
      (3) They fuse with the ducts from seminal vesicles to form the ejaculatory ducts.
   c. Seminal vesicles
      (1) Each seminal vesicle is a saclike structure attached to a ductus deferens.
      (2) They secrete an alkaline fluid that contains nutrients, such as fructose, and prostaglandins.
   d. Prostate gland
      (1) The prostate gland surrounds the urethra just inferior to the urinary bladder.
      (2) It secretes a thin, milky fluid that neutralizes the pH of semen and the acidic secretions of the vagina.
   e. Bulbourethral glands
      (1) The bulbourethral glands are two small structures inferior to the prostate gland.
      (2) They secrete a fluid that lubricates the penis in preparation for sexual intercourse.
   f. Semen
      (1) Semen consists of sperm cells and secretions of the seminal vesicles, prostate gland, and bulbourethral glands.
      (2) This fluid is slightly alkaline and contains nutrients and prostaglandins.
      (3) Sperm cells in semen begin to swim, but are unable to fertilize egg cells until they are activated in the female reproductive tract.
3. Male external reproductive organs
   a. Scrotum
      The scrotum is a pouch of skin and subcutaneous tissue that encloses the testes for protection and temperature regulation.
   b. Penis
      (1) The penis becomes erect for insertion into the vagina during sexual intercourse.
      (2) Its body is composed of three columns of erectile tissue.
4. Erection, orgasm, and ejaculation
   a. During erection, vascular spaces in the erectile tissue engorge with blood.
   b. Orgasm is the culmination of sexual stimulation and is accompanied by emission and ejaculation.
   c. Semen moves along the reproductive tract as smooth muscle in the walls of the tubular structures contracts by reflex.

19.3 Hormonal Control of Male Reproductive Functions (p. 513)

1. Hypothalamic and pituitary hormones
   a. The male body remains reproductively immature until the hypothalamus releases gonadotropin-releasing hormone (GnRH), which stimulates the anterior pituitary gland to release gonadotropins.
   b. Follicle-stimulating hormone (FSH) stimulates spermatogenesis.
   c. Luteinizing hormone (LH), sometimes known in males as interstitial cell stimulating hormone (ICSH), stimulates interstitial cells to produce male sex hormones.
2. Male sex hormones
   a. Male sex hormones are called androgens, with testosterone the most important.
   b. Androgen production increases rapidly at puberty.
   c. Actions of testosterone
      (1) Testosterone stimulates development of the male reproductive organs.
      (2) It also develops and maintains male secondary sex characteristics.
   d. Regulation of male sex hormones
      (1) A negative feedback mechanism regulates testosterone concentration.
         (a) A rising testosterone concentration inhibits the hypothalamus and reduces the anterior pituitary gland's secretion of gonadotropins.
         (b) As testosterone concentration falls, the hypothalamus signals the anterior pituitary gland to secrete gonadotropins.
      (2) The testosterone concentration remains relatively stable from day to day.

19.4 Organs of the Female Reproductive System (p. 516)

The primary female sex organs are the ovaries, which produce female sex cells and sex hormones. Accessory sex organs are internal and external.

1. Ovaries
   a. Ovary structure
      (1) Each ovary is subdivided into a medulla and a cortex.
      (2) The medulla is composed of connective tissue, blood vessels, lymphatic vessels, and nerves.

(3) The cortex contains ovarian follicles and is covered by cuboidal epithelium.

b. Primordial follicles

(1) During prenatal development, groups of cells in the ovarian cortex form millions of primordial follicles.

(2) Each primordial follicle contains a primary oocyte and a layer of follicular cells.

(3) The primary oocyte begins meiosis, but the process halts until puberty.

(4) The number of primary oocytes steadily declines throughout a female's life.

c. Oogenesis

(1) Beginning at puberty, some primary oocytes are stimulated to continue meiosis.

(2) When a primary oocyte undergoes meiosis, it gives rise to a secondary oocyte in which the original chromosome number is reduced by one-half (from 46 to 23).

(3) Fertilization of a secondary oocyte produces a zygote.

d. Follicle maturation

(1) At puberty, FSH initiates follicle maturation.

(2) During maturation of the follicle, the primary oocyte undergoes meiosis giving rise to a secondary oocyte and first polar body, the follicular cells multiply, and a fluid-filled cavity forms.

(3) Usually, only one follicle at a time fully develops.

e. Ovulation

(1) Ovulation is the release of a secondary oocyte from an ovary.

(2) A rupturing follicle releases the secondary oocyte.

(3) After ovulation, the secondary oocyte is drawn into the opening of the uterine tube.

2. Female internal reproductive organs

a. Uterine tubes

(1) The end of each uterine tube expands, and its margin bears irregular extensions.

(2) Ciliated cells that line the tube and peristaltic contractions in the wall of the tube help transport the secondary oocyte down the uterine tube. Fertilization may occur.

b. Uterus

(1) The uterus receives the embryo and sustains it during development.

(2) The uterine wall includes the endometrium, myometrium, and perimetrium.

c. Vagina

(1) The vagina receives the erect penis, conveys uterine secretions to the outside, and provides an open channel for the fetus during birth.

(2) Its wall consists of mucosal, muscular, and fibrous layers.

3. Female external reproductive organs

a. Labia majora

(1) The labia majora are rounded folds of adipose tissue and skin.

(2) The upper ends form a rounded elevation over the pubic symphysis.

b. Labia minora

(1) The labia minora are flattened, longitudinal folds between the labia majora.

(2) They are well supplied with blood vessels.

c. Clitoris

(1) The clitoris is a small projection at the anterior end of the vulva. It corresponds to the male penis.

(2) It is composed of two columns of erectile tissue.

d. Vestibule

(1) The vestibule is the space between the labia minora.

(2) The vestibular glands secrete mucus into the vestibule during sexual stimulation.

4. Erection, lubrication, and orgasm

a. During periods of sexual stimulation, the erectile tissues of the clitoris and vestibular bulbs engorge with blood and swell.

b. The vestibular glands secrete mucus into the vestibule and vagina.

c. During orgasm, the muscles of the perineum, uterine wall, and uterine tubes contract rhythmically.

19.5 Hormonal Control of Female Reproductive Functions (p. 522)

The hypothalamus, anterior pituitary gland, and ovaries secrete hormones that control sex cell maturation, the development and maintenance of female secondary sex characteristics, and changes that occur during the monthly reproductive cycle.

1. Female sex hormones

a. A female body remains reproductively immature until about ten years of age, when gonadotropin secretion increases.

b. The most important female sex hormones are estrogens and progesterone.

(1) Estrogens develop and maintain most female secondary sex characteristics.

(2) Progesterone prepares the uterus for pregnancy.

2. Female reproductive cycle

a. FSH from the anterior pituitary gland initiates a reproductive cycle by stimulating follicle maturation.

b. Maturing follicular cells secrete estrogens, which maintain the secondary sex characteristics and thicken the uterine lining.

c. Secretion of a relatively large amount of LH by the anterior pituitary gland triggers ovulation.

d. Following ovulation, follicular cells give rise to the corpus luteum.

(1) The corpus luteum secretes estrogens and progesterone, which causes the endometrium to become more vascular and glandular.

(2) If a secondary oocyte is not fertilized, the corpus luteum begins to degenerate.

(3) As concentrations of estrogens and progesterone decline, the uterine lining disintegrates, causing menstrual flow.

e. During this cycle, estrogens and progesterone inhibit the release of LH and FSH. As concentrations of estrogens and progesterone decline, the anterior pituitary gland secretes FSH and LH again, stimulating a new reproductive cycle.

3. Menopause

a. Menopause is termination of reproductive cycles due to aging of the ovaries.

b. Reduced concentrations of estrogens and lack of progesterone may cause regressive changes in female secondary sex characteristics.

19.6 Mammary Glands (p. 525)

1. The mammary glands are in the subcutaneous tissue of the anterior thorax within the breasts.

2. They are composed of lobes that contain glands and a duct.

3. Dense connective and adipose tissues separate the lobes.

4. Ducts connect the mammary glands to the nipple.

5. Ovarian hormones stimulate female breast development.

a. Alveolar glands and ducts enlarge.

b. Fat is deposited around and within the breasts.

19.7 Birth Control (p. 526)

Birth control is voluntary regulation of how many offspring are produced and when they are conceived. It usually involves some method of contraception.

1. Coitus interruptus is withdrawal of the penis from the vagina before ejaculation.
2. The rhythm method is abstinence from sexual intercourse for several days before and after ovulation.
3. Mechanical barriers
   a. Males and females can use condoms.
   b. Females can also use diaphragms and cervical caps.
4. Chemical barriers
   Spermicidal creams, foams, and jellies provide an unfavorable environment in the vagina for sperm survival.
5. Combined hormone contraceptives
   a. A flexible ring inserted deep into the vagina, a plastic patch, or a pill can deliver estrogen and progestin to prevent pregnancy.
   b. They disrupt a female's normal pattern of gonadotropin secretion, preventing follicle maturation, ovulation, and the normal buildup of the uterine lining.
6. Injectable contraception
   Intramuscular injection with medroxyprogesterone every three months acts similarly to oral contraceptives to prevent pregnancy.
7. Intrauterine devices (IUDs)
   An IUD is a solid object inserted in the uterine cavity that prevents pregnancy by interfering with implantation of a blastocyst.
8. Surgical methods
   Vasectomies in males and tubal ligations in females are surgical sterilization procedures.

19.8 Sexually Transmitted Infections (p. 530)

1. Sexually transmitted infections, formerly called sexually transmitted diseases, are passed during sexual contact and may go undetected for years.
2. Many of the sexually transmitted infections share similar symptoms.

Chapter Assessments

19.1 Introduction

1. General functions of the male and female reproductive systems include _____. (p. 506)
   a. producing sex cells
   b. nurturing sex cells
   c. transporting sex cells to sites of fertilization
   d. secreting hormones
   e. all of the above

19.2 Organs of the Male Reproductive System

2. List the organs (both primary and accessory) of the male reproductive system, and explain how each organ's structure affects the organ's function. (pp. 506–512)
3. List the major steps in spermatogenesis. (p. 507)
4. Trace the path of sperm cells from their site of formation to the outside. Indicate composition and when and where secretions are added to produce semen. (pp. 510–511)
5. Distinguish between emission and ejaculation. (p. 512)

19.3 Hormonal Control of Male Reproductive Functions

6. Describe the role of gonadotropin-releasing hormone (GnRH) in the control of male reproductive functions. (p. 513)
7. Discuss the actions of testosterone. (p. 515)

19.4 Organs of the Female Reproductive System

8. List the organs (both primary and accessory) of the female reproductive system, and explain how each organ's structure affects the organ's function. (pp. 516–521)
9. List the major steps in oogenesis. (p. 517)
10. Describe how a follicle matures. (p. 518)
11. Define *ovulation*. (p. 519)

19.5 Hormonal Control of Female Reproductive Functions

12. Describe the role of gonadotropin-releasing hormone (GnRH) in the control of female reproductive functions. (p. 522)

13. Discuss the actions of estrogens. (p. 522)
14. Summarize the major events in the reproductive cycle. (p. 523)

19.6 Mammary Glands

15. Describe the structure of a mammary gland. (p. 525)

19.7 Birth Control

16. Match the birth control method with its description. (pp. 526–529)

    (1) withdrawal
    (2) rhythm method
    (3) condom
    (4) spermicide (foam, gel)
    (5) estrogen/progesterone
    (6) IUD
    (7) vasectomy
    (8) tubal ligation

    A. Kills sperm (not very effective when used alone)
    B. Keeps sperm out of vagina or from entering cervix (additionally, may help prevent disease)
    C. Prevents implantation of blastocyst
    D. No intercourse during fertile times (ineffective)
    E. Penis removed from vagina before ejaculation
    F. Sperm cells never reach penis (very effective)
    G. Prevents follicle maturation and ovulation
    H. Oocytes never reach uterus (very effective)

19.8 Sexually Transmitted Infections

17. Common symptoms of sexually transmitted infections include _____. (p. 530)
    a. a burning sensation during urination
    b. discharge from vagina or penis
    c. pain during intercourse
    d. sores, blisters, or rash on genitals
    e. all of the above

18. If left untreated, a complication of the sexually transmitted infections gonorrhea and chlamydia is _____. (p. 530)

Integrated Assessments/Critical Thinking

OUTCOMES 11.5, 11.8, 19.2, 19.3, 19.4, 19.5, 19.7

1. Understanding the causes of infertility can be valuable in developing new birth control methods. Cite a type of contraceptive based on each of the following causes of infertility:
(a) failure to ovulate due to a hormonal imbalance;
(b) a large fibroid tumor that disturbs the uterine lining;
(c) endometrial tissue blocking uterine tubes;
(d) low sperm count (too few sperm per ejaculate).

OUTCOMES 11.5, 11.8, 19.2, 19.4

2. What changes, if any, would a male who has had one testis removed experience? A female who has had one ovary removed?

OUTCOME 19.2

3. Some men are unable to become fathers because their spermatids do not mature into sperm. Injection of their spermatids into their partners' secondary oocytes sometimes results in conception. A few men have fathered healthy babies this way. Why would this procedure work with spermatids, but not with primary spermatocytes?

OUTCOME 19.4

4. Sometimes a sperm cell fertilizes a polar body rather than a secondary oocyte. An embryo does not develop, and the fertilized polar body degenerates. Why is a polar body unable to support development of an embryo?

WEB CONNECTIONS

Visit the text website at **www.mhhe.com/shieress11** for additional quizzes, interactive learning exercises, and more.

APR

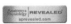

Anatomy & Physiology REVEALED includes cadaver photos that allow you to peel away layers of the human body to reveal structures beneath the surface. This program also includes animations, radiologic imaging, audio pronunciations, and practice quizzing. To learn more visit **www.aprevealed.com**.

20

Pregnancy, Growth, Development, and Genetics

Sperm for the future. Some servicemen left frozen samples of sperm at sperm banks before reporting for active duty, along with written statements expressing their wishes for their wives to use the sperm to conceive children should they not return.

Postmortem sperm retrieval. Bruce and Gaby Vernoff, in their early thirties, had delayed becoming parents, confident that their good health would make pregnancy possible. But Bruce suddenly died of an allergic reaction to a medication. Gaby knew how much he had wanted to be a father, so she requested that physicians take some of Bruce's sperm after his death. Thirty hours after Bruce died, the medical examiner collected a sperm sample and sent it to a sperm bank, where it lay deeply frozen for more than a year. In the summer of 1978, the medical director of the sperm bank used the defrosted sperm to fertilize one of Gaby's eggs. On March 17, their daughter was born. It was the first case of "postmortem sperm retrieval" and use in which the father had not actively participated in the decision. In other cases, the dying men had had time to state, in writing, their wishes to be fathers posthumously.

Postmortem sperm retrieval raises legal and ethical issues. In another case, with her husband's prior consent a woman conceived twins sixteen months after her husband had died of leukemia at age thirty. But

the Social Security Administration refused to provide survivor benefits to their daughters, claiming that the man was not their father, but a sperm donor. The Massachusetts Superior Court reversed this decision.

Postmortem sperm retrieval, like other assisted reproductive technologies, is not regulated at the federal level in the United States. Bioethicists suggest that men document their wishes, identifying situations to avoid:

- Someone other than a spouse wishing to use the sperm
- A too-hasty decision based on grief
- Use of the sperm for monetary gain

Some nations, including Germany, Australia, Canada, and Sweden, ban the procedure.

Learning Outcomes

After studying this chapter, you should be able to do the following:

20.1 Introduction

1. Distinguish between growth and development. (p. 537)

2. Distinguish between prenatal and postnatal periods. (p. 537)

20.2 Pregnancy

3. Describe fertilization. (p. 538)

20.3 Prenatal Period

4. List and provide details of the major events of cleavage. (p. 541)

5. Distinguish between an embryo and a fetus. (p. 541)

6. Discuss the hormonal changes in the maternal body during pregnancy. (p. 542)

7. List the structures produced by each of the primary germ layers. (p. 544)

8. Describe the major events of the embryonic stage of development. (p. 544)

9. Describe the major events of the fetal stage of development. (p. 547)

10. Trace the general path of blood through the fetal cardiovascular system. (p. 549)

11. Explain the role of hormones in the birth process and milk production. (p. 551)

20.4 Postnatal Period

12. Describe the major physiological adjustments required of the newborn. (p. 553)

20.5 Aging

13. Distinguish between passive aging and active aging. (p. 555)

20.6 Genetics

14. Distinguish among the modes of inheritance. (p. 557)

15. Describe the components of multifactorial traits. (p. 560)

Module 14: Reproductive System

 Learn   Practice   Assess

Aids to Understanding Words   *(Appendix A on page 564 has a complete list of Aids to Understanding Words.)*

allant- [sausage] *allant*ois: Tubelike structure extending from the yolk sac into the connecting stalk of the embryo.

chorio- [skin] *chorio*n: Outermost membrane surrounding the fetus and its membranes.

cleav- [to divide] *cleav*age: Period of development when the zygote divides, producing increasingly smaller cells.

hetero- [other, different] *hetero*zygous: Condition in which the members of a gene pair are different.

hom- [same, common] *hom*ozygous: Condition in which the members of a gene pair are the same.

lacun- [pool] *lacun*a: Space between the chorionic villi that fills with maternal blood.

morul- [mulberry] *morul*a: Embryonic structure consisting of a solid ball of about sixteen cells that resembles a mulberry.

nat- [to be born] pre*nat*al: Period of development before birth.

troph- [well fed] *troph*oblast: Cellular layer that surrounds the inner cell mass and helps nourish it.

umbil- [navel] *umbil*ical cord: Structure attached to the fetal navel (umbilicus) that connects the fetus to the placenta.

20.1 INTRODUCTION

A sperm cell and an egg cell (secondary oocyte) unite, forming a zygote, and the journey of prenatal development begins. Following thirty-eight weeks of cell division, growth, and specialization into distinctive tissues and organs, a new human being enters the world.

Humans grow, develop, and age. **Growth** is an increase in size. In humans and other many-celled organisms, growth entails an increase in cell numbers, followed by enlargement of the newly formed cells. **Development,** which includes growth, is the continuous process by which an individual changes from one life phase to another. These life phases include a **prenatal period** (pre-na′tal pe′re-od), which begins with fertilization and ends at birth, and a **postnatal** (pōst-na′tal) **period,** which begins with birth and ends at death.

Practice

1. Describe growth of an individual.
2. When is the beginning and ending of the postnatal period?

20.2 PREGNANCY

The union of a secondary oocyte and a sperm cell is called **fertilization** (fer″tĭ-lĭ-za′shun), or conception, which takes place in a uterine tube. **Pregnancy** (preg′nan-se) is the presence of a developing offspring in the uterus. Pregnancy consists of three periods called trimesters, each about three months long.

Transport of Sex Cells

Each month, a female of reproductive age usually ovulates a secondary oocyte, unless she is anovulatory, as discussed in Clinical Application 20.1. The released egg cell then usually enters a uterine tube. During sexual intercourse, the male deposits semen containing sperm cells in the vagina near the cervix. To reach the secondary oocyte, the sperm cells must move upward through the uterus and uterine tube. Prostaglandins in the semen stimulate lashing of sperm tails, and muscular contractions within the walls of the uterus and uterine tube aid the sperm cells' journey. Also, under the influence of high concentrations of estrogens during the first part of the reproductive cycle, the uterus and cervix secrete a watery fluid that promotes sperm transport and survival. Conversely, during the latter part of the cycle, when the progesterone concentration is high, the female reproductive tract secretes a viscous fluid that hampers sperm transport and survival. Sperm cells reach the upper part of the uterine tube in less than an hour following sexual intercourse. Many sperm cells may reach a secondary oocyte, but usually only one sperm cell fertilizes it (fig. 20.1; see fig. 20.4*a*).

Figure 20.1

Scanning electron micrograph of sperm cells on the surface of a secondary oocyte (1,100×). Only one sperm cell actually fertilizes a secondary oocyte.

Female Infertility

Infertility is the inability to conceive after a year of trying. In 90% of cases, infertility has a physical cause, and 60% of the time the abnormality is in the female's reproductive system.

A common cause of female infertility is insufficient secretion (hyposecretion) of gonadotropic hormones by the anterior pituitary, preventing ovulation (anovulation). Testing the urine for *pregnanediol,* a product of progesterone metabolism, can detect an anovulatory reproductive cycle. Because progesterone concentration normally rises after ovulation, no increase in pregnanediol in the urine during the latter part of the reproductive cycle suggests lack of ovulation.

Fertility specialists can treat anovulation due to hyposecretion of gonadotropic hormones by administering human chorionic gonadotropin (hCG) obtained from human placentas. Another ovulation-stimulating biochemical, human menopausal gonadotropin (hMG), contains luteinizing hormone (LH) and follicle-stimulating hormone (FSH) and is collected from the urine of postmenopausal women. However, either hCG or hMG may overstimulate the ovaries and cause many follicles to release secondary oocytes simultaneously, which may result in multiple births.

Another cause of female infertility is *endometriosis,* in which small pieces of the inner uterine lining (endometrium) move up through the uterine tubes during menstruation and implant in the abdominal cavity. Here, the tissue changes in a similar way to the uterine lining during the reproductive cycle. The misplaced tissue breaks down at the end of the cycle but cannot be expelled. Instead it remains in the abdominal cavity, irritating the lining (peritoneum) and causing considerable pain. This tissue also stimulates formation of fibrous tissue (fibrosis), which may encase the ovary, preventing ovulation or obstructing the uterine tubes.

Sexually transmitted infections (STIs), such as gonorrhea, can cause female infertility. These infections can inflame and obstruct the uterine tubes or stimulate production of viscous mucus that plugs the cervix and prevents sperm entry.

Women become infertile if their ovaries are removed, which may be part of cancer treatment, or are damaged by cancer treatment. To make future pregnancies possible, these women can have strips of ovarian tissue removed before their cancer treatment begins. The strips are frozen and stored, then thawed and reimplanted under the skin of the forearm or abdomen or in the pelvic cavity near the ovaries, when the woman wishes to conceive. When the tissue ovulates, oocytes are collected and fertilized *in vitro.*

Finding the right treatment for a particular patient requires determining the infertility's cause. Table 20A describes diagnostic tests for female infertility.

Table 20A	Tests to Assess Female Infertility
Test	**What It Checks**
Hormone levels	Whether ovulation occurs
Ultrasound	Placement and appearance of reproductive organs and structures
Postcoital test	Cervix examined soon after unprotected intercourse to see if mucus is thin enough to allow sperm through
Endometrial biopsy	Small piece of uterine lining sampled and viewed under microscope to see if it can support an embryo
Hysterosalpingogram	Dye injected into uterine tube and followed with scanner to show if tube is clear or blocked
Laparoscopy	Small, lit optical device inserted near navel to detect scar tissue blocking tubes, which ultrasound may miss

In a procedure called *in vitro fertilization* (IVF), a sperm cell and a secondary oocyte fuse in a laboratory dish. The first "test tube baby," Louise Joy Brown, was born in 1978 in England. Since then IVF has led to the births of 74 million children, and now accounts for 1 in 80 births in the United States.

Fertilization

A sperm cell first invades the follicular cells that adhere to the secondary oocyte's surface (corona radiata), then binds to the *zona pellucida,* a membrane rich in glycoproteins that surrounds the secondary oocyte's cell membrane. The acrosome of the sperm cell releases enzymes (including hyaluronidase) that aid penetration of the sperm head by digesting proteins in the zona pellucida (fig. 20.2). However, at least several hundred sperm cells must be present to produce enough enzymes to enable one to penetrate. This is why males with very low sperm counts are said to be subfertile.

The head portion of one sperm cell enters the oocyte, leaving the mitochondria-rich middle section and tail outside. Sperm entry triggers lysosome-like

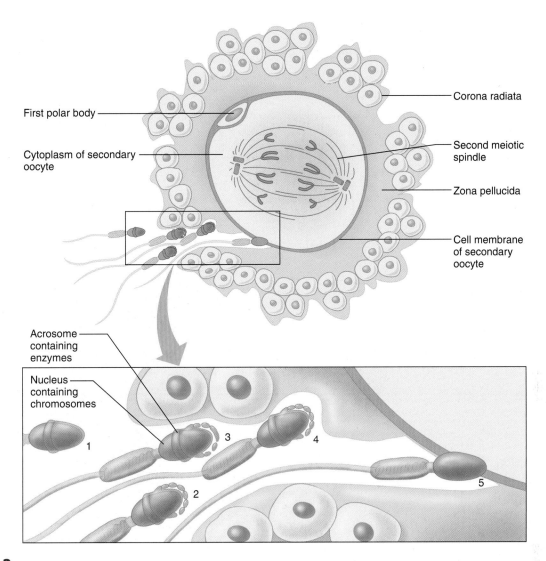

Figure 20.2

Steps in fertilization: **(1)** The sperm cell reaches the corona radiata surrounding the secondary oocyte. **(2)** and **(3)** The acrosome of the sperm cell releases a protein-digesting enzyme. **(4)** The sperm cell penetrates the zona pellucida surrounding the oocyte. **(5)** The sperm cell's membrane fuses with the oocyte's cell membrane.

Q: *How many chromosomes are contained in the secondary oocyte prior to fertilization?*

Answer can be found in Appendix E on page 568.

vesicles just beneath the oocyte cell membrane to release enzymes that harden the zona pellucida. This reduces the chance that other sperm cells will penetrate. The sperm cell nucleus enters the secondary oocyte's cytoplasm and swells. The secondary oocyte then divides unequally to form a large cell, whose nucleus contains the female's genetic contribution, and a tiny second polar body, which is later expelled. Meiosis is completed. The approaching nuclei from the two sex cells are called pronuclei, until they meet and merge (fig. 20.3). Next the pronuclei unite. Their nuclear membranes fall apart, and their chromosomes mingle, completing fertilization. Table 20.1 describes assisted reproductive technologies used to achieve fertilization.

Each sex cell provides 23 chromosomes, so the product of fertilization is a cell with 46 chromosomes—the usual number in a human body cell (somatic cell). This cell, called a **zygote** (zi′gōt), is the first cell of the future offspring.

Practice

3. What factors aid the movements of the secondary oocyte and sperm cells through the female reproductive tract?

4. Where in the female reproductive system does fertilization take place?

5. List the events of fertilization.

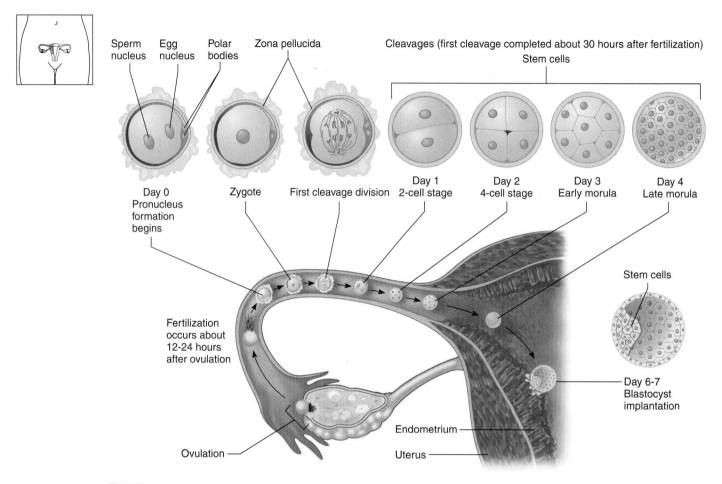

Sperm nucleus Egg nucleus Polar bodies Zona pellucida Cleavages (first cleavage completed about 30 hours after fertilization)

Stem cells

Day 0 Pronucleus formation begins

Zygote

First cleavage division

Day 1 2-cell stage

Day 2 4-cell stage

Day 3 Early morula

Day 4 Late morula

Stem cells

Fertilization occurs about 12-24 hours after ovulation

Day 6-7 Blastocyst implantation

Ovulation

Endometrium

Uterus

Figure 20.3 AP|R

Stages of early human prenatal development.

Table 20.1	Assisted Reproductive Technologies	
Technology	**Condition It Treats**	**Procedure**
Intrauterine insemination	Male infertility—lack of sperm cells or low sperm count	Donated sperm cells are placed into the cervix or uterus.
Surrogate mother	Female infertility—a woman has healthy ovaries, but lacks a uterus	A secondary oocyte fertilized *in vitro* is implanted in a woman other than its donor. The surrogate, or "gestational mother," gives the newborn to the "genetic mother" and her partner, the sperm donor.
Gamete intrafallopian transfer (GIFT)	Female infertility—bypasses blocked uterine tube	Secondary oocytes are removed from a woman's ovary, then placed along with donated sperm cells into a uterine tube past the blockage.
Zygote intrafallopian transfer (ZIFT)	Female infertility—bypasses blocked uterine tube	A secondary oocyte fertilized *in vitro* is placed in a uterine tube past the blockage. It travels to the uterus on its own.
Embryo adoption	Female infertility—a woman has nonfunctional ovaries, but a healthy uterus	A woman is artificially inseminated with sperm cells from a man whose partner cannot ovulate. If the woman conceives, the morula is flushed from her uterus and implanted in the uterus of the sperm donor's partner.
Intracytoplasmic sperm injection	Low sperm count; sperm that cannot mature past spermatid stage; men with paralysis who cannot ejaculate	Sperm or spermatids are injected into secondary oocytes *in vitro*.

20.3 PRENATAL PERIOD

Early Embryonic Development

About thirty hours after the zygote forms, it undergoes *mitosis,* giving rise to two new cells (blastomeres) (fig. 20.4*b*). These cells, in turn, divide into four cells, which divide into eight cells, and so forth. The divisions occur rapidly, with little time for the cells to grow. As a result, each division yields smaller cells. This phase of early rapid cell division is termed **cleavage** (klēv'ij) (see fig. 20.3).

During cleavage, the tiny mass of cells moves through the uterine tube to the uterine cavity. The trip takes about three days, and by then the structure consists of a solid ball, called a *morula* (mor'u-lah), of about sixteen cells (figs. 20.3 and 20.4*c*).

The morula remains free within the uterine cavity for about three days. During this stage, the zona pellucida of the original secondary oocyte degenerates. Then the morula hollows out, forming a *blastocyst,* which begins to attach to the endometrium. By the end of the first week of development, the blastocyst superficially implants in the endometrium (fig. 20.5). Up until this point, the cells that will become the developing offspring are pluripotent stem cells (see fig. 20.3), which means that they can give rise to several specialized types of cells, as well as yield additional stem cells.

An ectopic pregnancy occurs when the very early embryo implants outside the uterus, such as in a uterine tube, on an ovary, on the cervix, or on an organ in the abdominal cavity. A fertilized egg implanting in a uterine tube is a *tubal pregnancy,* a form of ectopic pregnancy. The tube usually ruptures as the embryo enlarges, resulting in severe pain and heavy vaginal bleeding. Treatment is prompt surgical removal of the embryo and repair or removal of the damaged uterine tube.

At about the time of implantation, certain cells on the inner face of the blastocyst organize into a group, called the inner cell mass, that will give rise to the body of the developing offspring. This marks the beginning of the **embryonic stage** of development. The offspring is termed an **embryo** (em'bre-o) until the end of the eighth week, when the basic structural form of the human body is recognizable. After the eighth week and until birth, the offspring is called a **fetus** (fe'tus). Rudiments of all organs are present by the end of embryonic development. These organs and other structures enlarge and specialize during fetal development.

During the embryonic stage, the cells surrounding the embryo, with cells of the endometrium, form a

(a)

(b)

(c)

Figure 20.4

Light micrographs of **(a)** a human secondary oocyte surrounded by follicular cells and sperm cells (250×), **(b)** the two-cell stage (600×), and **(c)** a morula (500×).

(a)

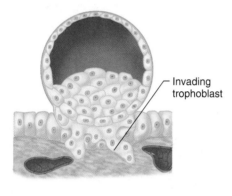

(b)

Figure 20.5

About the sixth day of prenatal development, the blastocyst **(a)** contacts the uterine wall and **(b)** begins to implant. The trophoblast, which will help form the placenta, secretes hCG, a hormone that maintains the pregnancy.

complex vascular structure called the **placenta** (plah-sen'tah). This organ attaches the embryo to the uterine wall and exchanges nutrients, gases, and wastes between maternal blood and the embryo's blood. The placenta also secretes hormones.

Fraternal (dizygotic) twins develop if two ovarian follicles release secondary oocytes simultaneously and both are fertilized. Such twins are no more alike genetically than nontwin siblings.

Identical (monozygotic) twins develop from a single fertilized secondary oocyte if two inner cell masses form within a blastocyst and each produces an embryo. Twins of this type usually share a single placenta, and they are genetically identical. Therefore, they are always the same sex and are similar in appearance.

Practice

6. What is cleavage?

7. What is implantation?

8. How do an embryo and a fetus differ?

Hormonal Changes During Pregnancy

During a typical reproductive cycle, the corpus luteum degenerates about two weeks after ovulation. Consequently, concentrations of estrogens and progesterone decline rapidly, the uterine lining breaks down, and the endometrium sloughs away as menstrual flow. If this occurs following implantation, the embryo is lost in a spontaneous abortion.

The hormone **human chorionic gonadotropin (hCG)** normally prevents spontaneous abortion. It is secreted by cells from the outer blastocyst that form a layer called the trophoblast, which surrounds the developing embryo and later helps form the placenta (fig. 20.5). This hormone, similar in function to luteinizing hormone (LH), maintains the corpus luteum, which continues secreting estrogens and progesterone, stimulating the uterine wall to grow and develop. At the same time, hCG inhibits the anterior pituitary gland's release of follicle-stimulating hormone (FSH) and LH, halting the normal reproductive cycle.

Secretion of hCG continues at a high level for about two months and then declines by the end of four months. Detecting this hormone in urine or blood is the basis of pregnancy tests. The corpus luteum persists throughout pregnancy, but its function as a hormone source becomes less important after the first three months (first trimester), when the placenta secretes sufficient estrogens and progesterone (fig. 20.6).

For the remainder of the pregnancy, *placental estrogens* and *placental progesterone* maintain the uterine wall. The placenta also secretes a hormone called **placental lactogen** that, with placental estrogens and progesterone, stimulates breast development and prepares

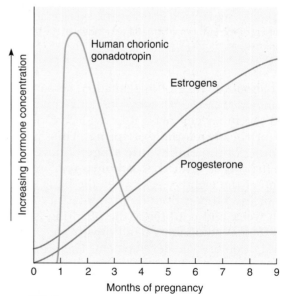

Figure 20.6

The relative concentrations of three hormones in maternal blood change during pregnancy.

the mammary glands for milk secretion. Placental progesterone and a polypeptide hormone called *relaxin* from the corpus luteum inhibit the smooth muscles in the myometrium, suppressing uterine contractions until the birth process begins.

The high concentration of placental estrogens during pregnancy enlarges the vagina and external reproductive organs. Also, relaxin relaxes the ligaments joining the pubic symphysis and sacroiliac joints during the last week of pregnancy, allowing greater movement at these joints and aiding the passage of the fetus through the birth canal.

Other hormonal changes of pregnancy include increased adrenal secretion of aldosterone, which promotes renal reabsorption of sodium and leads to fluid retention. The parathyroid glands secrete parathyroid hormone, which helps maintain a high concentration of maternal blood calcium (see chapter 11, p. 303). Table 20.2 summarizes the hormonal changes of pregnancy.

Practice

9. Which hormone normally prevents spontaneous abortion?

10. What is the source of the hormones that sustain the uterine wall during pregnancy?

11. What other hormonal changes occur during pregnancy?

Very early in pregnancy, while vast hormonal changes sweep a woman's body and the embryo rapidly increases in size and complexity, the woman may not yet realize she is pregnant. Early signs of pregnancy resemble those of approaching menstruation, such as bloating and irritable mood. As the pregnancy continues, the woman's blood volume increases by one-third, and her bones may weaken if she does not receive adequate dietary calcium. Muscle spasms may occur in response to electrolyte imbalances. In the later months, the enlarging uterus pushing against the woman's abdominopelvic organs can produce heartburn, shortness of breath, and frequent urination. Fetal movements become noticeable by the fourth or fifth month, first as slight flutterings, then as jabs, kicks, and squirming movements.

Embryonic Stage

The embryonic stage extends from the beginning of the second week through the eighth week of prenatal development. During this time the placenta forms, the main internal organs develop, and the major external body structures appear.

Early in the embryonic stage, the inner cell mass organizes into a flattened **embryonic disc** with two distinct layers—an outer *ectoderm* and an inner *endoderm*. A short time later the ectoderm and endoderm fold, and a third layer of cells, the *mesoderm,* forms between

Table 20.2	Hormonal Changes During Pregnancy

1. Following implantation, cells of the trophoblast begin to secrete human chorionic gonadotropin (hGC).

2. Human chorionic gonadotropin maintains the corpus luteum, which continues to secrete estrogens and progesterone.

3. The developing placenta secretes abundant estrogens and progesterone.

4. Placental estrogens and progesterone:
   a. stimulate the uterine lining to continue development.
   b. maintain the uterine lining.
   c. inhibit secretion of follicle-stimulating hormone (FSH) and luteinizing hormone (LH) from the anterior pituitary gland.
   d. stimulate development of mammary glands.
   e. inhibit uterine contractions (progesterone).
   f. enlarge the reproductive organs (estrogens).

5. Relaxin from the corpus luteum also inhibits uterine contractions and relaxes the pelvic ligaments.

6. The placenta secretes placental lactogen that stimulates breast development.

7. Aldosterone from the adrenal cortex promotes renal reabsorption of sodium.

8. Parathyroid hormone from the parathyroid glands helps maintain a high concentration of maternal blood calcium.

them. All organs form from these three cell layers, called the **primary germ layers** (pri′mar-e jerm la′erz) (fig. 20.7). A *connecting stalk* attaches the embryonic disc to the developing placenta. The two-week embryo, with its three primary germ layers, is called a **gastrula** (gas′troo-lah). Table 20.3 summarizes the stages of early human prenatal development.

Table 20.3	Stages and Events of Early Human Prenatal Development	
Stage	**Time Period**	**Principal Events**
Zygote	12–24 hours following ovulation	Secondary oocyte is fertilized, meiosis is completed; zygote has 46 chromosomes and is genetically distinct
Cleavage	30 hours to third day	Mitosis increases cell number
Morula	Third to fourth day	Solid ball of cells
Blastocyst	Fifth day through second week	Hollowed ball forms trophoblast (outside) and inner cell mass, which implants and flattens to form embryonic disc
Gastrula	End of second week	Primary germ layers form

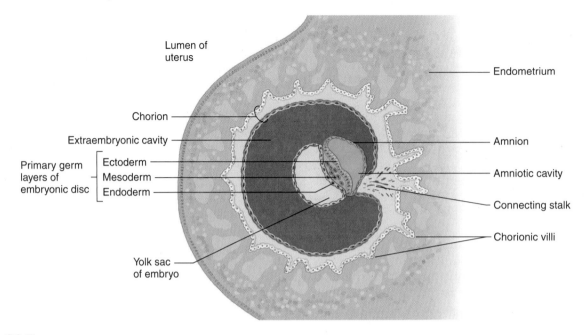

Lumen of
uterus

Chorion

Extraembryonic cavity

Primary germ
layers of
embryonic disc
Ectoderm
Mesoderm
Endoderm

Yolk sac
of embryo

Endometrium

Amnion

Amniotic cavity

Connecting stalk

Chorionic villi

Figure 20.7

Early in the embryonic stage of development, the three primary germ layers form. The ectoderm and the endoderm form first, and then the mesoderm forms between them.

Ectodermal cells give rise to the nervous system, parts of special sensory organs, the epidermis, hair, nails, glands of the skin, and linings of the mouth and anal canal. Mesodermal cells form all types of muscle tissue, bone tissue, bone marrow, blood, blood vessels, lymphatic vessels, internal reproductive organs, kidneys, and the epithelial linings of the body cavities. Endodermal cells produce the epithelial linings of the digestive tract, respiratory tract, urinary bladder, and urethra.

As the embryo implants in the uterus, proteolytic enzymes from the trophoblast break down endometrial tissue, providing nutrients for the developing embryo. A second layer of cells begins to line the trophoblast, and together these two layers form a structure called the **chorion** (ko're-on). Soon slender projections grow out from the trophoblast, including the new cell layer, eroding their way into the surrounding endometrium by continuing to secrete proteolytic enzymes. These projections become increasingly complex and form the highly branched **chorionic villi,** which are well established by the end of the fourth week (fig. 20.7).

As the chorionic villi develop, embryonic blood vessels form within them and are continuous with those passing through the connecting stalk to the body of the embryo. At the same time, irregular spaces called **lacunae** form around and between the villi. These spaces fill with maternal blood that escapes from endometrial blood vessels eroded by enzyme action.

During the fourth week of development, the flat embryonic disc becomes cylindrical. By the end of week four, the head and jaws appear, the heart beats

and forces blood through the blood vessels, and tiny buds form, which will give rise to the upper and lower limbs (fig. 20.8).

During the fifth through seventh weeks, as figure 20.8 shows, the head grows rapidly and becomes rounded and erect. The face develops eyes, nose, and mouth. The upper and lower limbs elongate, and fingers and toes form (fig. 20.9). By the end of the seventh week, all the main internal organs are established. As these structures enlarge, the body takes on a humanlike appearance.

Until about the end of the eighth week, the chorionic villi cover the entire surface of the former trophoblast. However, as the embryo and the surrounding chorion enlarge, only villi that contact the endometrium endure. The others degenerate, and the areas of the chorion where they were attached become smooth. The region of the chorion still in contact with the uterine wall is restricted to a disc-shaped area that becomes the placenta.

A thin **placental membrane** separates embryonic blood in the capillary of a chorionic villus from maternal blood in a lacuna. This membrane is composed of the epithelium of the chorionic villus, some connective tissue, and the epithelial wall of the capillary inside the villus. Through this membrane, substances are exchanged between the maternal blood and the embryo's blood (fig. 20.10). Oxygen and nutrients diffuse from the maternal blood into the embryo's blood, and carbon dioxide and other wastes diffuse from the embryo's blood into the maternal blood. Various substances also cross the placental membrane by active transport and pinocytosis.

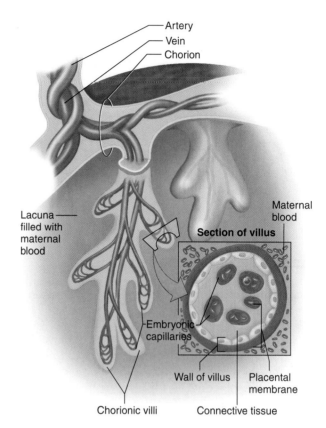

Figure 20.10
The placental membrane consists of the epithelial wall of an embryonic capillary and the epithelial wall of a chorionic villus, as illustrated in the section of the villus (lower part of the figure).

Figure 20.8
In the fifth through the seventh weeks of development, the embryonic body and face develop a humanlike appearance.

Figure 20.9
Human embryo after about six weeks of development (6.5×).

If a pregnant woman repeatedly ingests an addictive substance that crosses the placenta, her newborn may suffer from withdrawal symptoms when amounts of the chemical the fetus was accustomed to receiving suddenly plummet after birth. Newborn addiction occurs with certain drugs of abuse, such as heroin, and with certain prescription drugs used to treat anxiety.

Practice
12. Describe the major events of the embryonic stage of development.
13. Which tissues and structures develop from ectoderm? From mesoderm? From endoderm?
14. What is the function of the placental membrane?
15. How are substances exchanged between the embryo's blood and the maternal blood?

The embryonic portion of the placenta is the chorion and its villi; the maternal portion is the area of the uterine wall (decidua basalis) where the villi attach

Umbilical cord

Umbilical arteries

Umbilical vein

Chorion

Placenta

Endometrium

Embryonic blood vessels

Lacuna

Villi (embryonic portion of placenta)

Decidua basalis (maternal portion of placenta)

Maternal blood vessels

Myometrium

Figure 20.11

The placenta consists of an embryonic portion and a maternal portion.

(fig. 20.11). The fully formed placenta is a reddish-brown disc about 20 centimeters long and 2.5 centimeters thick, and weighs about 0.5 kilogram.

While the placenta is forming from the chorion, a second membrane, called the **amnion** (am′ne-on), develops around the embryo. It appears during the second week. Its margin is attached around the edge of the embryonic disc, and **amniotic fluid** fills the space between the amnion and the embryonic disc. The amniotic fluid allows the embryo to grow freely without compression from surrounding tissues and also

protects the embryo from jarring movements of the woman's body. As the embryo becomes more cylindrical, the margins of the amnion fold, enclosing the embryo in the amnion and amniotic fluid. The amnion envelops the tissues on the underside of the embryo, by which the embryo is attached to the chorion and the developing placenta. In this manner, the **umbilical cord** (um-bil′ĭ-kal kord) forms (fig. 20.12).

The umbilical cord contains three blood vessels—two *umbilical arteries* and one *umbilical vein*—that transport blood between the embryo and the placenta

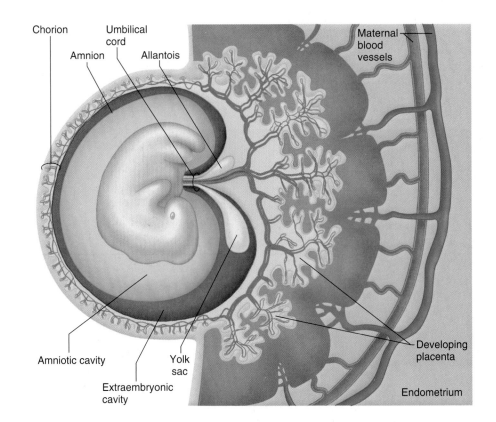

Chorion

Umbilical cord

Amnion

Allantois

Maternal blood vessels

Amniotic cavity

Yolk sac

Extraembryonic cavity

Developing placenta

Endometrium

Figure 20.12

As the amnion develops, it surrounds the embryo, and the umbilical cord begins to form from structures in the connecting stalk.

(see fig. 20.11). The umbilical cord suspends the embryo in the *amniotic cavity.* Two other extra-embryonic membranes form during development—the yolk sac and the allantois (fig. 20.12). ("Extra-embryonic" refers to structures that are distinct from the embryo.) The **yolk sac** forms during the second week and attaches to the underside of the embryonic disc. It forms blood cells in the early stages of development and gives rise to the cells that later become sex cells. The **allantois** (ah-lan'to-is) forms during the third week as a tube extending from the early yolk sac into the connecting stalk of the embryo. It, too, forms blood cells and gives rise to the umbilical arteries and vein.

By the beginning of the eighth week, the embryo is usually 30 millimeters long and weighs less than 5 grams. It is recognizable as human (fig. 20.13). The end of the eighth week concludes the most critical period of development when all the essential external and internal body parts form. Factors that cause congenital malformations by affecting an embryo are called **teratogens.** Such agents include drugs, viruses, radi-

ation, and even large amounts of otherwise healthful substances, such as fat-soluble vitamins.

The specific nature of a birth defect reflects the structures developing when the damage occurs. The time during prenatal development when a genetic mutation or exposure to a teratogen can alter a specific structure is called the critical period. Clinical Application 20.2 discusses some teratogens.

> Some structures, such as fingers and toes, have very short critical periods. In contrast, the brain is sensitive throughout development and even into childhood, so its critical period is very long. This is why many times birth defects are associated with the brain, resulting in intellectual impairment.

Practice

16. What is the function of amniotic fluid?
17. Which blood vessels are in the umbilical cord?
18. What is the function of the yolk sac?

Figure 20.13

By the beginning of the eighth week of development, the embryonic body is recognizable as human (6×).

Fetal Stage

The **fetal stage** begins at the end of the eighth week of development and lasts until birth. During this period, growth is rapid and body proportions change considerably. At the beginning of the fetal stage, the head is disproportionately large and the lower limbs are short. Gradually proportions come to more closely resemble those of a child.

During the third month, body lengthening accelerates but head growth slows. The upper limbs achieve the relative length they will maintain throughout development, and ossification centers appear in most bones. By the twelfth week the external reproductive organs are distinguishable as male or female.

In the fourth month the body grows rapidly and reaches a length of up to 20 centimeters. The lower limbs considerably lengthen, and the skeleton continues to ossify. A four-month-old fetus will startle and turn away from a bright light flashed on a pregnant woman's belly, and may also react to sudden loud noises.

In the fifth month, growth slows. The lower limbs achieve their final relative proportions. Skeletal muscles contract, and the pregnant woman may feel fetal movements. Hair begins to grow on the head. Fine, downy hair and a cheesy mixture of dead epidermal cells and sebum from the sebaceous glands cover the skin.

During the sixth month, the fetus gains substantial weight. Eyebrows and eyelashes grow. The skin is wrinkled and translucent. Blood vessels in the skin cause a reddish appearance.

In the seventh month, the skin becomes smoother as fat is deposited in subcutaneous tissues. The eyelids,

Clinical Application 20.2

Some Causes of Birth Defects

Thalidomide

The idea that the placenta protects the embryo and fetus from harmful substances was tragically disproven between 1957 and 1961, when 10,000 children in Europe were born with flippers in place of limbs due to exposure to a mild tranquilizer drug, *thalidomide,* during the time of limb formation. Although some women in the United States did use thalidomide and had affected children, the country was spared a larger disaster because an astute government physician noted adverse effects of the drug on monkeys in experiments and halted use of the drug. However, thalidomide is used today to treat leprosy, certain blood disorders, and a type of severe nosebleed.

Rubella

The virus that causes rubella (German measles) is a powerful teratogen. Exposure in the first trimester leads to cataracts, deafness, and heart defects, and later exposure causes learning disabilities, speech and hearing problems, and type 1 diabetes mellitus. Successful vaccination programs provide for maternal immunity to the virus, and have since greatly lowered the incidence of "congenital rubella syndrome" in many countries.

Alcohol

A pregnant woman who has just one or two alcoholic drinks a day, or perhaps many drinks at a crucial time in prenatal development, risks *fetal alcohol syndrome (FAS)* in her

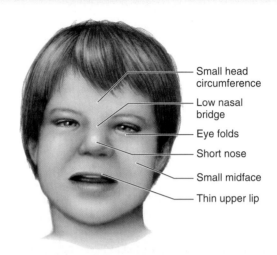

Figure 20A

Fetal alcohol syndrome. Some children whose mothers drank alcohol during pregnancy have characteristic flat faces. Women who drink excessively while pregnant have a 30% to 45% chance of having a child affected to some degree by prenatal exposure to alcohol.

unborn child. The effects of small amounts of alcohol at different stages of pregnancy are not yet well understood, and because each woman metabolizes alcohol slightly differently, women are advised to avoid drinking alcohol entirely when pregnant or when trying to become pregnant.

A child with FAS has a small head, misshapen eyes, and a flat face and nose (fig. 20A). Growth is slow before and after

which fused during the third month, reopen. At the end of this month the fetus is about 40 centimeters long.

In the final trimester, fetal brain cells rapidly form networks, as organs specialize and grow. A layer of fat is laid down beneath the skin. In the male, the testes descend from regions near the developing kidneys, through the inguinal canal, and into the scrotum. The digestive and respiratory systems mature last, which is why premature infants may have difficulty digesting milk and breathing. At the end of the ninth month (on average, 266 days), the fetus is *full-term.* It is about 50 centimeters long and weighs 2.7–3.6 kilograms. The skin has lost its downy hair, but sebum and dead epidermal cells still coat it. Hair usually covers the scalp. The fingers and toes have well-developed nails. The skull bones are largely ossified. As figure 20.14 shows, the fetus is usually positioned upside down, with its head toward the cervix.

Figure 20.14

A full-term fetus is usually positioned with its head near the cervix.

birth. Intellect is impaired, ranging from minor learning disabilities to mental retardation. Many individuals remain at an early grade-school level of intellectual development, and they often lack social and communication skills. People with FAS are also more likely to have seizures.

In the United States today, FAS is the third most common cause of mental retardation in newborns. Each year in the United States, about 5,000 children are born with FAS, and many more are born with milder "alcohol-related effects."

Cigarettes

Chemicals in cigarette smoke stress a fetus. Carbon monoxide crosses the placenta and plugs up the sites on the fetus's hemoglobin molecules that would normally bind oxygen. Other chemicals in smoke prevent nutrients from reaching the fetus. Studies comparing the placentas of smokers and nonsmokers show that smoke-exposed placentas lack important growth factors. The result of these assaults on the fetus is poor growth before and after birth. Cigarette smoking during pregnancy is linked to spontaneous abortion, stillbirth, prematurity, and low birth weight.

Nutrients and Malnutrition

Certain nutrients in large amounts, particularly vitamins, act in the body as drugs. The acne medication *isotretinoin* (Accutane) is a derivative of vitamin A that causes spontaneous abortions and defects of the fetal heart, nervous system, and face. A vitamin A–based drug used to treat psoriasis, as well as excesses of vitamin A itself, also cause birth defects because some forms of the vitamin are stored in body fat for up to three years after ingestion.

Malnutrition during pregnancy causes intrauterine growth retardation (IUGR). Malnutrition before birth also causes shifts in metabolism to make the most of calories from food. This protective action, however, sets the stage for developing obesity and associated disorders, such as type 2 diabetes and cardiovascular disease, in adulthood.

Occupational Hazards

The workplace can be a source of teratogens. Women who work with textile dyes, lead, certain photographic chemicals, semiconductor materials, mercury, and cadmium have increased rates of spontaneous abortion and delivering children with birth defects. Men whose jobs expose them to sustained heat, such as smelter workers, glass manufacturers, and bakers, may produce sperm that can fertilize a secondary oocyte and possibly lead to spontaneous abortion or a birth defect. A virus or a toxic chemical carried in semen may also cause a birth defect.

Practice

19. What major changes occur during the fetal stage of development?
20. How is a fetus usually positioned in the uterus as birth nears?

Fetal Blood and Circulation

Throughout fetal development, the maternal blood supplies oxygen and nutrients and carries away wastes. These substances diffuse between maternal and fetal blood through the placental membrane, and the umbilical blood vessels carry them to and from the fetus.

The fetal blood and cardiovascular system are adapted to intrauterine existence. The concentration of oxygen-carrying hemoglobin in fetal blood is about 50% greater than in maternal blood, and fetal hemoglobin has a greater attraction for oxygen than does adult hemoglobin. At a particular oxygen partial pressure, fetal hemoglobin can carry 20–30% more oxygen than adult hemoglobin. Different genes encode the protein subunits of hemoglobin in embryos, fetuses, and individuals after birth. The different subunits have different affinities for oxygen.

A treatment for severe sickle cell disease is a drug that reactivates silenced fetal hemoglobin genes. In the disease, adult hemoglobin changes shape under low oxygen conditions, causing red blood cells to distort into a sickle-like shape. The abnormal cells block circulation, causing very painful "crises." Hydroxyurea induces production of fetal hemoglobin subunits, which bind some of the adult hemoglobin subunits, preventing sickling and enabling the blood to circulate to reach the lungs and pick up oxygen. Hydroxyurea decreases the frequency of crises and hospitalizations.

Aortic arch

Superior vena cava

Foramen ovale
(becomes fossa ovalis)

Inferior vena cava

Ductus venosus
(becomes ligamentum
venosum)

Hepatic portal vein

Umbilical vein
(becomes ligamentum
teres)

Placenta

Ductus arteriosus
(becomes ligamentum
arteriosum)

Pulmonary artery

Pulmonary veins

Pulmonary trunk

Left atrium

Left ventricle

Abdominal aorta

Left renal artery

Common iliac artery

Umbilical arteries
(become medial
umbilical ligaments)

Internal iliac artery

Umbilical vein

Umbilical arteries

Decreasing
blood
oxygen
level

Figure 20.15

The general pattern of fetal circulation.

Q: *Name the structures that are unique to the fetal circulation.*

Answer can be found in Appendix E on page 568.

Figure 20.15 shows the path of blood in the fetal cardiovascular system. The umbilical vein transports blood rich in oxygen and nutrients from the placenta to the fetus. This vein enters the body and travels along the anterior abdominal wall to the liver. About half the blood it carries passes into the liver, and the rest enters a vessel called the **ductus venosus** (duk′tus ve′no-sus), which bypasses the liver.

The ductus venosus extends a short distance and joins the inferior vena cava. There, oxygenated blood from the placenta mixes with deoxygenated blood from the lower parts of the fetal body. This mixture continues through the inferior vena cava to the right atrium.

In an adult heart, blood from the right atrium enters the right ventricle and is pumped through the pulmo-

nary trunk and arteries to the lungs (see chapter 13, p. 346). The fetal lungs, however, are nonfunctional, and blood largely bypasses them. Much of the blood from the inferior vena cava that enters the fetal right atrium is shunted directly into the left atrium through an opening in the atrial septum called the **foramen ovale** (fo-ra′man ova′le). Blood passes through the foramen ovale because blood pressure is somewhat greater in the right atrium than in the left atrium. Furthermore, a small valve on the left side of the atrial septum overlies the foramen ovale and helps prevent blood from moving in the reverse direction.

The rest of the fetal blood entering the right atrium, including a large proportion of the deoxygenated blood entering from the superior vena cava, passes into the right ventricle and out through the pulmonary trunk. Only a small volume of blood enters the pulmonary circuit, because the lungs are collapsed and their blood vessels have a high resistance to blood flow. However, enough blood does reach lung tissues to sustain them.

Most of the blood in the pulmonary trunk bypasses the lungs by entering a fetal vessel called the **ductus arteriosus** (duk′tus ar-te″re-o′sus), which connects the pulmonary trunk to the descending portion of the aortic arch. As a result of this connection, blood with a relatively low oxygen concentration, returning to the heart through the superior vena cava, bypasses the lungs. At the same time, it is prevented from entering the portion of the aorta that branches to the heart and brain.

The more highly oxygenated blood that enters the left atrium through the foramen ovale mixes with a small amount of deoxygenated blood returning from the pulmonary veins. This mixture moves into the left ventricle and is pumped into the aorta. Some of it reaches the myocardium through the coronary arteries, and some reaches the brain tissues through the carotid arteries.

Blood carried by the descending aorta includes the less oxygenated blood from the ductus arteriosus. Some of the blood is carried into the branches of the aorta that lead to the lower regions of the body. The rest passes into the umbilical arteries, which branch from the internal iliac arteries and lead to the placenta. There the blood is reoxygenated (fig. 20.15).

Table 20.4 summarizes the major features of fetal circulation. At birth, the fetal cardiovascular system must adjust when the placenta ceases to function and the newborn begins to breathe.

Table 20.4	Fetal Cardiovascular Adaptations
Adaptation	**Function**
Fetal blood	Hemoglobin has greater oxygen-carrying capacity than adult hemoglobin
Umbilical vein	Carries nutrient-rich oxygenated blood from placenta to fetus
Ductus venosus	Conducts about half the blood from the umbilical vein directly to the inferior vena cava, bypassing the liver
Foramen ovale	Conveys a large portion of the blood entering the right atrium from inferior vena cava, through the atrial septum, and into the left atrium, bypassing the lungs
Ductus arteriosus	Conducts some blood from the pulmonary trunk to the aorta, bypassing the lungs
Umbilical arteries	Carry blood containing carbon dioxide and wastes from the internal iliac arteries to the placenta

The umbilical cord usually contains two arteries and one vein. On rare occasion a newborn will have only one umbilical artery. Because this condition is often associated with other cardiovascular disorders, the vessels in the severed cord are routinely counted following birth. Some inherited conditions are also associated with an abnormal number of umbilical cord vessels.

Practice

21. Which umbilical vessel carries oxygenated blood to the fetus?

22. What is the function of the ductus venosus?

23. How does fetal circulation allow blood to bypass the lungs?

Birth Process

Pregnancy usually continues for thirty-eight weeks from conception and ends with the *birth process.* During pregnancy, progesterone suppresses uterine contractions. As the placenta ages, the progesterone concentration in the uterus declines, which stimulates synthesis of a prostaglandin that promotes uterine contractions. At the same time, the cervix thins and then opens. Changes in the cervix may begin a week or two before other signs of labor appear.

Stretching of the uterine and vaginal tissues late in pregnancy also stimulates the birth process. This initiates nerve impulses to the hypothalamus, which in turn signals the posterior pituitary gland to release the hormone **oxytocin** (see chapter 11, p. 300). Oxytocin stimulates powerful uterine contractions. Combined with the greater excitability of the myometrium due to the decline in progesterone secretion, stimulation by oxytocin aids *labor* in its later stages.

During labor, rhythmic muscular contractions begin at the top of the uterus and travel down its length. Because the fetus is usually positioned head downward, labor contractions force the head against the cervix (fig. 20.16). This action stretches the cervix, which elicits a reflex that stimulates stronger labor contractions. Therefore, in this *positive feedback system*, uterine contractions stimulate more intense uterine contractions. At the same time, dilation of the cervix reflexly stimulates the posterior pituitary to increase oxytocin release. As labor continues, positive feedback stimulates abdominal wall muscles to contract, helping to propel the fetus through the cervix and vagina to the outside.

An infant passing through the birth canal can stretch and tear the delicate tissues between the vulva and anus (perineum). To avoid a ragged tear, a physician makes an *episiotomy*, which is a clean cut in the perineal tissues.

Figure 20.16

Stages in birth. **(a)** Fetal position before labor, **(b)** dilation of the cervix, **(c)** expulsion of the fetus, **(d)** expulsion of the placenta.

Following birth of the fetus, the placenta separates from the uterine wall and is pushed by uterine contractions through the birth canal. This expelled placenta, called the *afterbirth,* is accompanied by bleeding, because vascular tissues are damaged in the process. However, oxytocin stimulates continued uterine contraction, which compresses the bleeding vessels and minimizes blood loss. Breastfeeding also contributes to returning the uterus to its original, prepregnancy size, because suckling by the newborn stimulates the mother's posterior pituitary gland to release oxytocin.

Practice

24. Describe the role of progesterone in initiating labor.
25. Explain how dilation of the cervix affects labor.

Milk Production and Secretion

During pregnancy, placental estrogens and progesterone stimulate further development of the mammary glands. Estrogens cause the ductile systems to grow and branch and deposit abundant fat around them. Pro-

gesterone stimulates the development of the alveolar glands at the ends of the ducts. Placental lactogen also promotes these changes.

Hormonal activity doubles breast size during pregnancy, and the mammary glands become capable of secreting milk. However, milk secretion does not begin until after birth, because placental progesterone inhibits milk production and placental lactogen blocks the action of *prolactin* (see chapter 11, p. 300).

Following childbirth and the expulsion of the placenta, maternal blood concentrations of placental hormones decline rapidly. In two or three days, prolactin, which is no longer inhibited, stimulates the mammary glands to secrete milk. In the meantime, the glands secrete a thin, watery fluid called *colostrum* that has more protein, but less carbohydrate and fat, than milk. Colostrum contains antibodies from the mother's immune system that protect the newborn from certain infections.

Milk ejection requires contraction of specialized *myoepithelial cells* surrounding the alveolar glands (fig. 20.17). Suckling or mechanical stimulation of sensory receptors in the nipple or areola elicits the reflex action that controls this process. Impulses from these receptors go to the hypothalamus, which signals the posterior

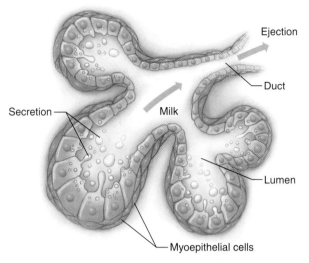

Figure 20.17
Myoepthelial cells contract, releasing milk from an alveolar gland.

pituitary gland to release oxytocin. Oxytocin travels in the bloodstream to the breasts and stimulates myoepithelial cells to contract. In about thirty seconds, milk squirts into a suckling infant's mouth.

As long as milk is removed from the breasts, release of prolactin and oxytocin continues, and the mammary glands produce milk. If milk is not removed regularly, the hypothalamus inhibits prolactin secretion, and within about one week the mammary glands stop producing milk.

Human milk is the best possible food for human babies. The milk of other mammals contains different proportions of nutrients.

Human milk is 4.5% fat and 1.1% protein, which is suitable for supporting synthesis of myelin, the fatty electrical insulation that enables neurons in the developing brain to communicate effectively. In contrast, cow's milk has 3.5% fat and 3.1% protein, because it is more important for survival for the calf to bulk up its muscles than to develop its brain. Milk of the gray seal has a whopping 53% fat, but this animal uses the fat for thermal insulation in its frigid environment.

Practice

26. How does pregnancy affect the mammary glands?
27. What stimulates the mammary glands to produce milk?
28. What causes milk ejection?

20.4 POSTNATAL PERIOD

Following birth, the newborn experiences physiological and structural changes. The postnatal period of development lasts from birth to death.

Neonatal Period

The **neonatal** (ne''o-na'tal) **period** begins abruptly at birth and extends to the end of the first four weeks. At birth, the newborn must make quick physiological adjustments to become self-reliant. It must respire, obtain and digest nutrients, excrete wastes, and regulate body temperature.

A newborn's most immediate need is to obtain oxygen and excrete carbon dioxide. The first breath must be particularly forceful, because the lungs are collapsed and their small airways considerably resistant to air movement. Also, surface tension holds the moist membranes of the lungs together. However, the lungs of a full-term fetus continuously secrete *surfactant* (see chapter 16, p. 451), which reduces surface tension. After the first powerful inhalation inflates the lungs, breathing eases.

Premature infants' survival chances increase directly with age and weight, and parallel the increasing maturity of the lungs. The ability of the alveoli to exchange gases and the presence of surfactant to reduce alveolar surface tension are important. A baby born at twenty-five weeks has a 50% chance of survival; at twenty-four weeks, a 39% chance; and at twenty-three weeks, a 17% chance. The smallest and earliest premature baby to survive in recent times was Amilla Taylor, born at twenty-one weeks, six days, and weighing slightly under 10 ounces.

The newborn has a high metabolic rate. To supplement the liver's supply of glucose to support metabolism, the newborn typically utilizes stored fat for energy.

A newborn's kidneys are usually unable to produce concentrated urine, so they excrete a dilute fluid. For this reason the newborn may become dehydrated and develop a water and electrolyte imbalance. Also, certain homeostatic control mechanisms, such as the temperature-regulating system, may not function adequately.

When the placenta ceases to function and breathing begins, the newborn's cardiovascular system changes. Following birth, the umbilical vessels constrict. The umbilical arteries close first, and if the umbilical cord is not clamped or severed for a minute or so, blood continues to flow from the placenta to the newborn through the umbilical vein, adding to the newborn's blood volume. Similarly, the ductus venosus constricts shortly after birth and appears in the adult as a fibrous cord (ligamentum venosum) superficially embedded in the wall of the liver.

The foramen ovale closes as a result of blood pressure changes in the right and left atria. As blood ceases to flow from the umbilical vein into the inferior vena cava, the blood pressure in the right atrium falls. Also, as the lungs expand with the first breathing movements, resistance to blood flow through the pulmonary circuit decreases, more blood enters the left atrium through the pulmonary veins, and blood pressure in the left atrium increases.

As the blood pressure in the left atrium rises and that in the right atrium falls, the valve on the left side of the atrial septum closes the foramen ovale. In most individuals this valve gradually fuses with the tissues along the margin of the foramen. In an adult, a depression called the *fossa ovalis* marks the site of the past opening.

The ductus arteriosus, like the other fetal vessels, constricts after birth. After this, blood can no longer bypass the lungs by moving from the pulmonary trunk directly into the aorta. In an adult, a cord called the *ligamentum arteriosum* represents the ductus arteriosus.

Changes in the newborn's cardiovascular system are gradual. Constriction of the ductus arteriosus may be functionally complete within fifteen minutes, but the permanent closure of the foramen ovale may take up to a year.

> In *patent ductus arteriosus (PDA),* the ductus arteriosus does not close completely. After birth, the metabolic rate and oxygen consumption in neonatal tissues increase. If the ductus arteriosus remains open, the neonate's blood oxygen concentration may be too low to adequately supply tissues, including the myocardium. If PDA is not corrected surgically, the heart may fail, even though the myocardium is normal.

Fetal hemoglobin production falls after birth, and by the time an infant is four months old, most of the circulating hemoglobin is the adult type. Figure 20.18 illustrates cardiovascular changes in the newborn. Table 20.5 summarizes the major events during the neonatal period as well as those of the later stages of human development.

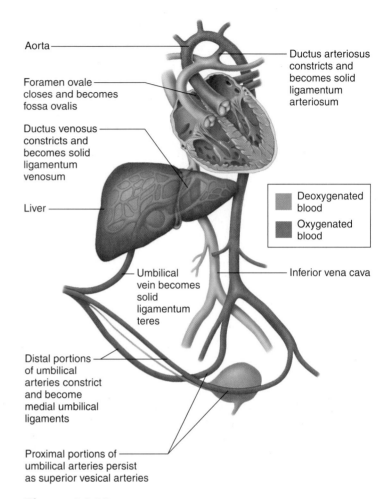

Figure 20.18

Major changes in the newborn's cardiovascular system.

Table 20.5	Stages in Postnatal Development	
Stage	**Time Period**	**Major Events**
Neonatal period	Birth to end of fourth week	Newborn begins to respire, eat, digest nutrients, excrete wastes, regulate body temperature, and make cardiovascular adjustments
Infancy	End of fourth week to one year	Growth rate is high; teeth begin to erupt; muscular and nervous systems mature so that coordinated activities are possible; communication begins
Childhood	One year to puberty	Growth rate is high; primary teeth erupt and are then replaced by secondary teeth; high degree of muscular control is achieved; bladder and bowel controls are established; intellectual abilities mature
Adolescence	Puberty to adulthood	Person becomes reproductively functional and emotionally more mature; growth spurts occur in skeletal and muscular systems; high levels of motor skills are developed; intellectual abilities increase
Adulthood	Adolescence to old age	Person remains relatively unchanged anatomically and physiologically; degenerative changes begin to occur
Senescence	Old age to death	Degenerative changes continue; body becomes less able to cope with demands; death usually results from mechanical disturbances in the cardiovascular system or from diseases that affect vital organs

Practice

29. Why must a newborn's first breath be particularly forceful?

30. What does a newborn use for energy during its first few days?

31. How do the kidneys of a newborn differ from those of an adult?

32. What changes occur in the newborn's cardiovascular system?

20.5 AGING

The aging process is difficult to analyze because of the intricate interactions of the body's organ systems. Breakdown of one structure ultimately affects the functioning of others. Table 20.6 outlines aging-related changes.

Passive Aging

Aging as a passive process is the breakdown of structures and slowing of function. At the molecular level, passive aging is seen as the degeneration of the elastin and collagen proteins of connective tissues, causing skin to sag and muscle to lose firmness.

During a long lifetime, biochemical abnormalities accumulate. Mistakes occur throughout life when DNA replicates in dividing cells. Usually repair enzymes correct this damage immediately. But over many years, exposure to chemicals, viruses, and radiation disrupts DNA repair mechanisms so that the error burden becomes too great to be fixed. The cells may die as a result of faulty genetic instructions.

Another sign of passive aging at the biochemical level is the breakdown of lipids. As aging membranes leak during lipid degeneration, a fatty, brown pigment called lipofuscin accumulates. Mitochondria also begin to break down in older cells, decreasing the supply of chemical energy to power the cell's functions.

The cellular degradation associated with aging may be set into action by highly reactive chemicals called **free radicals.** A molecule that is a free radical has an unpaired electron in its outermost valence shell. This causes the molecule to grab electrons from other

Table 20.6	Aging-Related Changes
Organ System	**Aging-Related Changes**
Integumentary system	Degenerative loss of collagenous and elastic fibers in dermis; decreased production of pigment in hair follicles, hair eventually turns white; reduced activity of sweat and sebaceous glands; skin thins, wrinkles, and becomes drier
Skeletal system	Degenerative loss of bone matrix; bones become thinner, less dense, and more likely to fracture; stature may shorten due to compression of intervertebral discs and vertebrae
Muscular system	Loss of skeletal muscle fibers; degenerative changes in neuromuscular junctions; loss of muscular strength
Nervous system	Degenerative changes in neurons; loss of dendrites and synaptic connections; accumulation of lipofuscin in neurons; decreases in sensation; decreasing efficiency in processing and recalling information; decreasing ability to communicate; diminished sense of smell and taste; loss of elasticity of lenses and consequent loss of ability to accommodate for close vision
Endocrine system	Reduced hormonal secretions; decreased metabolic rate; reduced ability to cope with stress; reduced ability to maintain homeostasis
Cardiovascular system	Degenerative changes in cardiac muscle; decrease in lumen diameters of arteries and arterioles; decreased cardiac output; increased resistance to blood flow; increased blood pressure
Lymphatic system	Decrease in efficiency of immune system; increased incidence of infections and neoplastic diseases; increased incidence of autoimmune diseases
Digestive system	Decreased motility in gastrointestinal tract; reduced secretion of digestive juices; reduced efficiency of digestion
Respiratory system	Degenerative loss of elastic tissue in lungs; fewer alveoli; reduced vital capacity; increase in dead air space; reduced ability to clear airways by coughing
Urinary system	Degenerative changes in kidneys; fewer functional nephrons; reductions in filtration rate, tubular secretion, and tubular reabsorption
Reproductive systems	
Male	Reduced secretion of sex hormones; enlargement of prostate gland; decrease in sexual energy
Female	Degenerative changes in ovaries; decrease in secretion of sex hormones; menopause; regression of secondary sex characteristics

molecules, destabilizing them, and a chain reaction of chemical instability begins that could kill the cell. Free radicals are a by-product of normal metabolism and also form by exposure to radiation or toxic chemicals.

Active Aging

Aging also entails new activities or the appearance of new substances. Lipofuscin granules, for example, may be considered an active sign of aging, but they result from the passive breakdown of lipids. Another example of active aging is autoimmunity, in which the immune system turns against the body, attacking its cells as if they were invading organisms.

Active aging begins before birth, as certain cells die as part of the developmental program encoded in the genes. This process of programmed cell death, called **apoptosis** (ayp-o-toe'sis), occurs regularly in the embryo, degrading certain structures to pave the way for new ones. The number of neurons in the fetal brain, for example, is halved as those that make certain synaptic connections are spared from death. Throughout life, apoptosis enables organs to maintain their characteristic shapes.

Mitosis and apoptosis are opposite, but complementary, processes. As organs grow, the number of cells in some regions increases, but in others, it decreases. Cell death is not a phenomenon only of the aged. It is a normal part of life.

Practice

33. How is aging a passive process?
34. How is aging an active process?

20.6 GENETICS

The newborn enters the world, and the elated parents look for family resemblances. Does she have her father's nose, or her grandmother's curly hair?

As the child grows, a unique mix of traits emerges. Inherited traits are determined by DNA sequences that comprise genes, which instruct cells to synthesize particular proteins, as discussed in chapter 4, section 4.6 (pp. 85–89). When a gene's DNA sequence changes, or *mutates,* illness may result. The Genetics Connection features in chapters 3 (p. 58), 8 (p. 193), 12 (p. 333), and 16 (p. 462) describe illnesses that arise from mutations.

The field of **genetics** (jĕ-net'iks) investigates how genes confer specific characteristics that affect health or contribute to our natural variation, and how genes are passed from generation to generation. As discussed in Genetics Connection 20.1, fetal chromosome checks provide clues to an individual's future health. The envi-

ronment influences how most genes are expressed. For example, inherited gene variants that confer susceptibility to lung cancer might affect health only if a person smokes or is exposed to air pollution for many years.

Chromosomes and Genes Come in Pairs

Charts called karyotypes display by size the 23 chromosome pairs of a human somatic cell (fig. 20.19). Pairs 1 through 22 are **autosomes** (aw'to-somz), which do not carry genes that determine sex. The other two chromosomes, the X and the Y, carry genes that determine sex and are called **sex chromosomes.** Females have two X chromosomes, and males have one X and one Y.

Each chromosome except the tiny Y includes hundreds of genes. Somatic cells have two copies of each autosome, and therefore two copies of each gene. Gene copies can be identical or slightly different in DNA sequence. Such variant forms of a gene are called **alleles** (ah-léels). An individual who has two identical alleles of a gene is **homozygous** (ho"mo-zi'gus) for that gene. A person with two different alleles is **heterozygous** (het'er-o-zi'gus) for it. A heterozygote is also called a *carrier.*

The combination of alleles, for one gene or many, constitutes a person's **genotype** (jē'no-tīp). The appearance, health condition, or other characteristics associated with a particular genotype is the **phenotype** (fe'no-tīp). An allele is *wild type* (indicated with

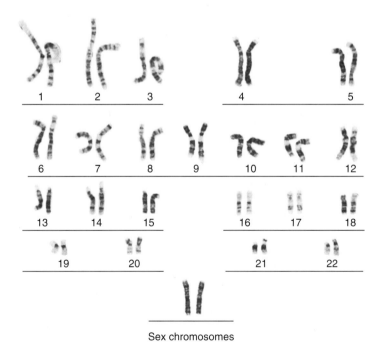

Sex chromosomes

Figure 20.19

The human karyotype. Each somatic cell contains 23 chromosome pairs, or 46 chromosomes. This cell is from a female—it has two X chromosomes.

a plus sign) if its associated phenotype is either normal function or the most common expression in a particular population. An allele that differs from wild type has undergone a *mutation*, which may lead to a *mutant* (abnormal or unusual) phenotype.

Practice

35. Distinguish between autosomes and sex chromosomes.
36. How does a homozygote differ from a heterozygote?
37. Distinguish between genotype and phenotype.

Modes of Inheritance

We can predict the probability that a certain inherited trait will occur in the offspring of two individuals by considering how genes and chromosomes are distributed in meiosis, and the combinations in which they can unite at fertilization. Patterns of inheritance through families are called *modes of inheritance.*

Dominant and Recessive Inheritance

A **dominant** allele masks expression of a **recessive** allele. Dominant alleles are usually indicated with a capital letter. An allele that causes a trait or disease can be recessive or dominant, and inherited in either an autosomal or an X-linked manner. Y-linked conditions are extremely rare because that chromosome has very few genes.

The following generalizations describe modes of inheritance:

1. An autosomal condition affects both sexes. X-linked characteristics affect males much more than females. Y-linked traits are passed only from father to son.
2. A person inherits an autosomal recessive condition from two healthy carrier parents. Recessive conditions can "skip" generations.
3. A person who inherits a dominant condition has at least one affected parent. Therefore, generations are not skipped.

Three major modes of inheritance are autosomal recessive, autosomal dominant, and X-linked recessive. Cystic fibrosis illustrates *autosomal recessive* inheritance, in which two recessive alleles, one from each parent, transmit a trait. Receiving two disease-causing alleles impairs chloride channels in cells lining the pancreas, respiratory tract, intestines, and testes (see Genetics Connection 16.1, p. 449). Half of a heterozygous man's sperm have the disease-causing allele, as do half of the woman's secondary oocytes. Because sperm and oocytes combine at random, each offspring has a 25% chance of inheriting two wild-type alleles, a 50% chance of inheriting a disease-causing allele from either

parent and being a carrier, and a 25% chance of inheriting a disease-causing allele from each parent.

Figure 20.20 illustrates two ways to depict the possible offspring of two carriers of cystic fibrosis. A **Punnett square** symbolizes the logic used to deduce the probabilities of inheriting particular genotypes in offspring. Each box records an allele combination at

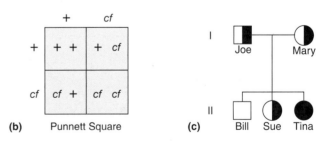

Figure 20.20

Inheritance of cystic fibrosis from carrier parents illustrates autosomal recessive inheritance. **(a)** Each child has a 25% chance of being unaffected and not a carrier, a 50% chance of being an unaffected carrier, and a 25% chance of being affected. Sexes are affected with equal frequency. A Punnett square **(b)** and a pedigree **(c)** are other ways of depicting this information. Symbols in the pedigree with both black and white indicate unaffected carriers (heterozygotes).

Genetics Connection 20.1

Fetal Chromosome Checks

Chromosomes provide clues to health. A chromosome number other than 46 usually signals a serious medical condition, as do chromosomes that have missing or extra material. Sampling cells from a fetus and preparing charts of the chromosomes can detect these conditions.

Ultrasound, in which sound waves bounced off a fetus are converted into an image, can reveal fusion of the eyes, cleft lip and/or palate, malformed nose, and extra fingers and toes that are part of certain chromosomal syndromes, and many other anomalies. Also, blood tests performed on the woman during the fifteenth week of pregnancy detect levels of biomarkers called maternal serum markers (alpha fetoprotein, human chorionic gonadotropin, an estrogen, and certain other biochemicals). (Clinical Application 2.2 on page 47 discusses biomarkers.) Abnormal levels of maternal serum markers can reflect an extra chromosome. Doctors follow up questionable ultrasound or blood test results or a family history of abnormal chromosomes with one of the following procedures, which examine fetal chromosomes.

Chorionic Villus Sampling

Chorionic villus sampling (CVS) (fig. 20B*a*) examines the chromosomes in chorionic villus cells, which are genetically identical to fetal cells because they are derived from the same fertilized egg. On rare occasion the test causes spontaneous abortion. Due to this risk, women are advised to have the procedure only if they have previously had a child with a detectable chromosome abnormality and therefore have a higher risk of having a detectable problem than the risk of the test itself. CVS is performed at the tenth week of gestation.

Amniocentesis

Amniocentesis is performed after the fourteenth week of gestation. A physician uses ultrasound to guide a needle into the amniotic sac and withdraws about 5 milliliters of fluid

(a) Chorionic villus sampling

Chorionic villi — Uterus
Placenta — Amniotic membrane
— Fetus
Catheter —
— Cervix
Vagina —
— Syringe

(b) Amniocentesis

Fetus 15–16 weeks

(c) Fetal cell sorting

Rare fetal cells

Figure 20B

Three ways to check a fetus's chromosomes. **(a)** Chorionic villus sampling (CVS) removes cells of the chorionic villi, whose chromosomes match those of the fetus. **(b)** In amniocentesis, a needle is inserted into the uterus to collect a sample of amniotic fluid, which contains fetal cells. **(c)** Fetal cell sorting separates fetal cells in the woman's circulation, but is still experimental. For all three types of tests, fetal chromosomes are stained and examined, and a genetic counselor interprets the results for patients. Additional tests are required for specific genetic disorders.

(fig. 20B*b*). Fetal fibroblasts in the sample are cultured and their chromosomes are checked. It takes about a week to grow these cells.

Until recently, amniocentesis carried a risk of about 0.5% of being followed by spontaneous abortion. But in the thirty or so years since those statistics were collected, the procedure has become much safer. The current risk of amniocentesis causing a miscarriage is about 1 in 1,600. Previously amniocentesis was offered only to women over age thirty-five, when the risk of conceiving a fetus with abnormal chromosomes is about 0.5% (risk increases with age), and women who had already had a child with detectable chromosomal abnormality. Many more women are likely to be offered the test if the new study results are confirmed.

Fetal Cell Sorting

Fetal cell sorting separates rare fetal cells that normally cross the placenta and enter the woman's bloodstream (fig. 20B*c*). A device called a fluorescence-activated cell sorter can pull out the fetal cells. The technique is safer than CVS or amniocentesis, because the fetus and its membranes are not touched.

Fetal cell sorting traces its roots to 1957, when an autopsy on a pregnant woman revealed cells from a very early embryo lodged in a blood vessel in her lung. Researchers realized that the cells were from an embryo because of the Y chromosomes, which only male cells have. Fetal cells enter the maternal circulation in up to 70% of all pregnancies, and may remain for decades in the woman's body.

Preimplantation Genetic Diagnosis (PGD)

If a couple has a family history of a chromosomal or single gene condition that could affect their offspring, a procedure called preimplantation genetic diagnosis (PGD) can allow them to select early embryos that have not inherited the condition (fig. 20C). After secondary oocytes are fertilized *in vitro* and allowed to divide to the 8-celled stage, one cell from each of several embryos is removed and tested for the disease-causing mutation or chromosome abnormality. If the genes or chromosomes are unaffected, a tested 7-celled embryo is implanted in the woman, and if all goes well, development ensues. PGD has enabled hundreds of children to be born free of inheriting a disease-causing genotype. A famous case was Adam Nash, who was selected as an embryo because his cell surfaces matched those of his sister Molly, who was dying of an inherited anemia, and he did not have the anemia. An umbilical cord stem cell transplant from brother to sister saved Molly's life.

PGD was originally developed to help families with known genetic conditions, but it is increasingly being used in other situations. In the United Kingdom, for example, inherited cancer susceptibility is an approved indication for having PGD. Bioethicists object because these cancers do not begin until adulthood, not everyone who inherits the susceptibility actually develops the cancer, and the cancers may be treatable. Another controversy concerning PGD is its use to select for gender.

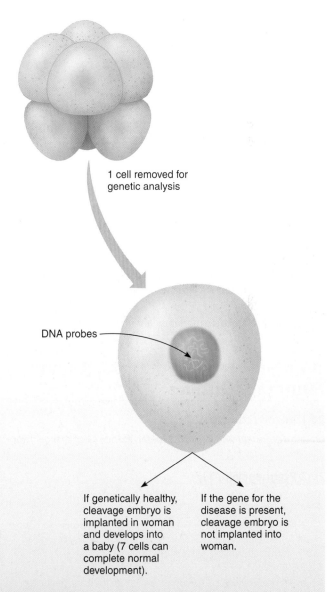

1 cell removed for genetic analysis

DNA probes

If genetically healthy, cleavage embryo is implanted in woman and develops into a baby (7 cells can complete normal development).

If the gene for the disease is present, cleavage embryo is not implanted into woman.

Figure 20C

Preimplantation genetic diagnosis detects disease-causing genotypes in an 8-celled cleavage embryo.

fertilization. A **pedigree** shows family members, how they are related, and their genotypes. Males are squares, females are circles, and the symbols for carriers are half filled in while those for affected individuals are entirely filled in. Geneticists use Punnett squares and pedigrees to predict the outcome in all modes of inheritance.

Only one disease-causing allele is necessary to inherit an *autosomal dominant* condition. Huntington disease, which is inherited in an autosomal dominant manner, is characterized by loss of coordination, uncontrollable dancelike movements, cognitive impairment, and personality changes, typically beginning gradually near age forty. An affected person has an affected parent. Autosomal dominant disorders tend to begin in adulthood; autosomal recessive disorders usually have an early onset.

A third major mode of inheritance is *X-linked recessive*. For a female, X-linked inheritance is like autosomal recessive inheritance, because she has two X chromosomes. That is, she can be a heterozygote or a homozygote. For a male, however, recessive alleles on the lone X chromosome are always expressed. A male with an X-linked condition inherits it from a mother who is either a carrier or affected; he does not inherit an X chromosome from his father (or he wouldn't be male). Colorblindness and the blood-clotting disorder hemophilia A are X-linked recessive conditions.

Multifactorial Traits

Nearly all inherited traits and disorders are influenced by environmental factors, such as nutrition, physical activity, and exposure to toxins and pathogens. Genes also influence each other. Environmental influences are particularly noticeable for traits determined by more than one gene, termed *polygenic*. Usually several genes contribute, in differing degrees, toward molding the overall phenotype of a polygenic trait. Such a trait is said to be "continuously varying," which means that there are many degrees of its expression. Height, skin color, and intelligence are polygenic traits that show great variation.

When individuals with a polygenic trait are categorized into classes and the frequencies of the classes are plotted as a bar graph, a bell-shaped curve emerges. The curves are strikingly similar for different polygenic traits. Figure 20.21 vividly shows the effect of the environment in the bell curve for height—as nutrition improved during the time span between the two photos, the heights of the tallest people increased.

Traits molded by one or more genes plus the environment are termed *multifactorial*, or complex. Height and skin color are multifactorial as well as polygenic, because they are influenced by environmental factors—nutrition and sun exposure, respectively. Most of the more common illnesses, including heart disease, diabetes mellitus, hypertension, and cancers, are multifactorial, as are most polygenic traits. An exception may be eye color, on which the environment has little, if any, impact.

Practice

38. Distinguish between dominant and recessive alleles.
39. How do Punnett squares and pedigrees depict gene transmission?
40. Compare the three modes of inheritance.

Practice

41. What is a polygenic trait?
42. What is a multifactorial trait?

Summary Outline

20.1 Introduction (p. 537)

Growth is an increase in size. Development is the process of changing from one life phase to another.

20.2 Pregnancy (p. 537)

Pregnancy is the presence of a developing offspring in the uterus.

1. Transport of sex cells
   a. A male deposits semen in the vagina during sexual intercourse.
   b. A sperm cell lashes its tail to move and is aided by muscular contractions in the female reproductive tract.
2. Fertilization
   a. An enzyme helps a sperm cell penetrate the zona pellucida.
   b. When a sperm cell head penetrates a secondary oocyte's cell membrane, changes in the membrane and the zona pellucida prevent entry of additional sperm cells.
   c. Completion of meiosis forms the second polar body.
   d. Fusion of the pronuclei from the two sex cells completes fertilization.
   e. The product of fertilization is a zygote with 46 chromosomes.

20.3 Prenatal Period (p. 541)

1. Early embryonic development
   a. Cells undergo mitosis, giving rise to smaller and smaller cells during cleavage.
   b. The developing offspring moves down the uterine tube to the uterus, where it implants in the endometrium.
   c. Once implanted, the offspring is called an embryo through the eighth week of development. Thereafter it is a fetus.
   d. Eventually, embryonic and maternal cells form a placenta.
2. Hormonal changes during pregnancy
   a. Embryonic cells produce human chorionic gonadotropin (hCG), which maintains the corpus luteum.
   b. Placental tissue produces high concentrations of estrogens and progesterone.
      (1) Estrogens and progesterone maintain the uterine wall and inhibit secretion of follicle-stimulating hormone (FSH) and luteinizing hormone (LH).

(a)

(b)

Figure 20.21

Previous editions of this (and other) textbooks have used the photograph in **(a)** to illustrate the continuously varying nature of height. In the photo, taken around 1920, 175 cadets at the Connecticut Agricultural College lined up by height. In 1997, Professor Linda Strausbaugh asked her genetics students at the school, today the University of Connecticut at Storrs, to recreate the scene **(b)**. They did, and confirmed the continuously varying nature of human height. But they also demonstrated how height increased during the twentieth century. Improved nutrition has definitely played a role in expressing genetic potential for height. The tallest people in the old photograph **(a)** are 5'9" tall, whereas the tallest people in the more recent photograph **(b)** are 6'5" tall.

(2) Progesterone and relaxin inhibit contraction of uterine muscles.

(3) Estrogens enlarge the vagina.

(4) Relaxin helps relax the ligaments of the pelvic joints.

c. Placental lactogen stimulates development of the breasts and mammary glands.

d. During pregnancy, increased aldosterone secretion promotes retention of sodium and body fluid. Increased secretion of parathyroid hormone helps maintain a high concentration of maternal blood calcium.

3. Embryonic stage

a. The embryonic stage extends from the beginning of the second week through the eighth week of development.

b. During this stage, the placenta and main internal and external body structures develop.

c. The cells of the inner cell mass organize into primary germ layers.

d. The embryonic disc becomes cylindrical and attaches to the developing placenta.

e. The placental membrane consists of the epithelium of the chorionic villi and the epithelium of the capillaries inside the villi.

(1) Oxygen and nutrients diffuse from maternal blood across the placental membrane and into fetal blood.

(2) Carbon dioxide and other wastes diffuse from fetal blood across the placental membrane and into maternal blood.

f. A fluid-filled amnion develops around the embryo.

g. The umbilical cord forms as the amnion envelops the tissues attached to the underside of the embryo.

h. The yolk sac forms on the underside of the embryonic disc.

i. The allantois extends from the yolk sac into the connecting stalk.

j. By the beginning of the eighth week, the embryo is recognizable as human.

4. Fetal stage

a. The fetal stage extends from the end of the eighth week of development until birth.

b. Existing structures grow and mature. Only a few new parts appear.

c. The fetus is full-term at the end of the ninth month.

5. Fetal blood and circulation
   a. Umbilical vessels carry blood between the placenta and the fetus.
   b. Fetal blood carries a greater concentration of oxygen than does maternal blood because the concentration of oxygen-carrying hemoglobin is greater in fetal blood and fetal hemoglobin has greater affinity for oxygen.
   c. Blood enters the fetus through the umbilical vein and partially bypasses the liver through the ductus venosus.
   d. Blood enters the right atrium and partially bypasses the lungs through the foramen ovale.
   e. Blood entering the pulmonary trunk partially bypasses the lungs through the ductus arteriosus.
   f. Blood enters the umbilical arteries from the internal iliac arteries.
6. Birth process
   a. During pregnancy, placental progesterone inhibits uterine contractions.
   b. A variety of factors promote birth.
      (1) A decreasing progesterone concentration and the release of a prostaglandin initiate the birth process.
      (2) The posterior pituitary gland releases oxytocin.
      (3) Oxytocin stimulates uterine muscles to contract, and labor begins.
   c. Following birth, placental tissues are expelled.
7. Milk production and secretion
   a. Following childbirth, concentrations of placental hormones decline, the action of prolactin is no longer blocked, and the mammary glands begin to secrete milk.
   b. A reflex response to mechanical stimulation of the nipple stimulates the posterior pituitary gland to release oxytocin, which causes the alveolar ducts to eject milk.

20.4 Postnatal Period (p. 553)

1. Neonatal period
   a. The neonatal period extends from birth to the end of the first four weeks.
   b. The newborn must begin to respire, obtain nutrients, excrete wastes, and regulate body temperature.
   c. The first breath must be powerful to expand the lungs.
   d. The liver is immature and unable to supply sufficient glucose, so the newborn depends primarily on stored fat for energy.

   e. A newborn's immature kidneys cannot concentrate urine well.
   f. A newborn's homeostatic mechanisms may function imperfectly, and body temperature may be unstable.
   g. The cardiovascular system changes when placental circulation ceases.
      (1) Umbilical vessels constrict.
      (2) The ductus venosus constricts.
      (3) A valve closes the foramen ovale as blood pressure in the right atrium falls and pressure in the left atrium rises.
      (4) The ductus arteriosus constricts.

20.5 Aging (p. 555)

1. Passive aging entails breakdown of structures and slowing or failure of functions.
2. Active aging involves apoptosis.

20.6 Genetics (p. 556)

1. Chromosomes and genes come in pairs
   a. Karyotypes are charts that display the two copies of each of the 22 autosomes, which do not carry genes that determine sex, and the sex chromosomes (X and Y), which do.
   b. A person with two identical variants, or alleles, for a gene is homozygous. A person with two different alleles is heterozygous.
   c. The combination of alleles is the genotype; their expression as a trait is the phenotype.
2. Modes of inheritance
   a. A dominant allele masks the expression of a recessive allele.
   b. Modes of inheritance include autosomal recessive, in which an affected individual inherits an illness from heterozygous or affected parents; autosomal dominant, in which one affected parent passes on the condition; and X-linked recessive, which affects males more than females because males lack a second X chromosome to mask disease-causing recessive alleles.
   c. Punnett squares and pedigrees depict gene transmission in families.
3. Multifactorial traits
   a. Traits determined by more than one gene are polygenic.
   b. The continuously varying nature of polygenic traits can be depicted in bell-shaped curves.
   c. One or more genes and environmental influences cause a multifactorial trait.

Chapter Assessments

20.1 Introduction

1. _____ is an increase in the size of the individual, whereas _____ is the continuous process by which an individual changes from one life phase to another. (p. 537)

2. _____ is the period of development from fertilization to birth, whereas _____ is the period of development from birth to death. (p. 537)

20.2 Pregnancy

3. Define *pregnancy*. (p. 537)

4. Describe how sperm cells move in the female reproductive tract. (p. 537)

5. Summarize the events occurring after the sperm cell head enters the oocyte's cytoplasm. (p. 538)

20.3 Prenatal Period

6. Describe the process of cleavage. (p. 541)

7. Distinguish between an embryo and a fetus. (p. 541)

8. List hormones associated with pregnancy, and describe the function of each. (p. 542)

9. Ectodermal cells of the developing embryo give rise to _____. (p. 544)
   a. bone tissue   d. the epidermis
   b. the kidneys   e. connective tissues
   c. the lining of the urethra

10. Describe the composition and formation of the placenta. (p. 544)

11. List the function(s) of the placenta. (p. 544)

12. Distinguish between the chorion and the amnion. (p. 544 and 546)

13. Explain the function of the amniotic fluid. (p. 546)

14. Describe the formation of the umbilical cord. (p. 546)

15. List the major changes that occur during fetal development. (p. 547)

16. Describe a full-term fetus. (p. 548)

17. Trace the pathway of blood from the placenta to the fetus and back to the placenta. (p. 549)

18. Explain positive feedback and the role of hormones in the expulsion of the fetus and the afterbirth. (p. 551)

19. Describe milk production and secretion. (p. 552)

20.4 Postnatal Period

20. Explain why a newborn's first breath must be particularly forceful. (p. 553)

21. Relate the difficulties of the fetus in maintaining water/electrolyte and body temperature homeostasis. (p. 553)

22. Describe the changes in the newborn's cardiovascular system. (p. 553)

20.5 Aging

23. Discuss the signs of passive and active aging and the physiological causes of these signs. (p. 555)

20.6 Genetics

24. Distinguish between autosomes and sex chromosomes. (p. 556)

25. An individual with two different alleles is _____. (p. 556)

26. Match the mode of inheritance with its description. (p. 557)

    (1) autosomal recessive   A. Inherited by male from carrier mothers

    (2) autosomal dominant   B. Inherited from one affected parent

    (3) X-linked recessive   C. Inherited from two carrier (unaffected) parents

27. Traits that are polygenic and affected by the environment are termed _____. Name two such traits. (p. 560)

Integrative Assessments/Critical Thinking

OUTCOMES 13.2, 13.6, 20.3, 20.4

1. What symptoms may appear if a newborn's ductus arteriosus fails to close?

OUTCOMES 14.9, 20.2

2. Why can twins resulting from a single fertilized egg exchange blood or receive organ transplants from each other without rejection, while twins resulting from two fertilized eggs sometimes cannot?

OUTCOMES 20.2, 20.3, 20.5

3. What kinds of studies and information are required to determine whether a man's exposure to a potential teratogen can cause birth defects years later? How would such analysis differ if a woman were exposed?

OUTCOMES 20.3, 20.4

4. What technology would enable a fetus born in the fourth month to survive in a laboratory setting? (This is not yet possible.)

OUTCOME 20.5

5. Bob and Joan know from a blood test that they are each heterozygous (carriers) for the autosomal recessive gene that causes sickle cell disease. If their first three children are healthy, what is the probability that their fourth child will have the disease?

WEB CONNECTIONS

Visit the text website at **www.mhhe.com/shieress11** for additional quizzes, interactive learning exercises, and more.

APR

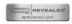

Anatomy & Physiology REVEALED includes cadaver photos that allow you to peel away layers of the human body to reveal structures beneath the surface. This program also includes animations, radiologic imaging, audio pronunciations, and practice quizzing. To learn more visit **www.aprevealed.com**.

AIDS TO UNDERSTANDING WORDS

acetabul-, vinegar cup: *acetabul*um
adip-, fat: *adip*ose tissue
agglutin-, to glue together: *agglutin*ation
aliment-, food: *aliment*ary canal
allant-, sausage: *allant*ois
alveol-, small cavity: *alveol*us
an-, without: *an*aerobic respiration
ana-, up: *ana*bolism
andr-, man: *andr*ogens
append-, to hang something: *append*icular
ax-, axis: *ax*ial skeleton, *ax*on
bil-, bile: *bil*irubin
-blast, bud: osteo*blast*
brady-, slow: *brady*cardia
bronch-, windpipe: *bronch*us
calat-, something inserted: inter*calat*ed disc
calyc-, small cup: major *calyc*es
cardi-, heart: peri*cardi*um
carp-, wrist: *carp*als
cata-, down: *cata*bolism
chondr-, cartilage: *chondr*ocyte
chorio-, skin: *chorio*n
choroid, skinlike: *choroid* plexus
chym-, juice: *chym*e
-clast, break: osteo*clast*
cleav-, to divide: *cleav*age
cochlea, snail: *cochlea*
condyl-, knob: *condyl*e
corac-, a crow's beak: *corac*oid process
cort-, covering: renal *cort*ex
cran-, helmet: *cran*ial
cribr-, sieve: *cribr*iform plate
cric-, ring: *cric*oid cartilage
-crin, to secrete: endo*crin*e
crist-, crest: *crist*a galli
cut-, skin: sub*cut*aneous
cyt-, cell: *cyt*oplasm, osteo*cyt*e
de-, separation from: *de*hydration
decidu-, falling off: *decidu*ous teeth
dendr-, tree: *dendr*ite
derm-, skin: *derm*is
detrus-, to force away: *detrus*or muscle
di-, two: *di*saccharide
diastol-, dilation: *diastol*ic pressure
diuret-, to pass urine: *diuret*ic
dors-, back: *dors*al
ejacul-, to shoot forth: *ejacul*ation
embol-, stopper: *embol*us
endo-, within: *endo*plasmic reticulum, *endo*crine gland
epi-, upon: *epi*thelial tissue, *epi*dermis, *epi*glottis
erg-, work: syn*erg*ist
erythr-, red: *erythr*ocyte
exo-, outside: *exo*crine gland
extra-, outside: *extra*cellular fluid
fimb-, fringe: *fimb*riae

follic-, small bag: hair *follic*le, ovarian *follic*le
fov-, pit: *fov*ea capitis
funi-, small cord or fiber: *funi*culus
gangli-, a swelling: *gangli*on
gastr-, stomach: *gastr*ic gland
-gen, to be produced: aller*gen*
-genesis, origin: spermato*genesis*
glen-, joint socket: *glen*oid cavity
-glia, glue: neuro*glia*
glom-, little ball: *glom*erulus
glyc-, sweet: *glyc*ogen
-gram, something written: electrocardio*gram*
hema-, blood: *hema*tocrit
hemo-, blood: *hemo*globin
hepat-, liver: *hepat*ic duct
hetero-, other, different: *hetero*zygous
hom-, same, common: *hom*ozygous
homeo-, same: *homeo*stasis
humor-, fluid: *humor*al immunity
hyper-, above, more, over: *hyper*tonic, *hyper*trophy, *hyper*thyroidism
hypo-, below: *hypo*tonic, *hypo*thyroidism
im-, not: *im*balance
immun-, free: *immun*ity
inflamm-, set on fire: *inflamm*ation
inter-, among, between: *inter*phase, *inter*calated disc, *inter*vertebral disc
intra-, inside, within: *intra*membranous bone, *intra*cellular fluid
iris, rainbow: *iris*
iso-, equal: *iso*tonic
kerat-, horn: *kerat*in
labi-, lip: *labi*a minora
labyrinth, maze: *labyrinth*
lacri-, tears: *lacri*mal gland
lacun-, pool: *lacun*a
laten-, hidden: *laten*t period
-lemm, rind or peel: neuri*lemm*a
leuko-, white: *leuko*cyte
lingu-, tongue: *lingu*al tonsil
lip-, fat: *lip*ids
-logy, study of: physio*logy*
-lyt, dissolvable: electro*lyt*e
macr-, large: *macr*ophage
macula, spot: *macula* lutea
meat-, passage: auditory *meat*us
melan-, black: *melan*in
mening-, membrane: *mening*es
mens-, month: *mens*es
meta-, change: *meta*bolism
mict-, to pass urine: *mict*urition
mit-, thread: *mit*osis
mono-, one: *mono*saccharide
mons-, mountain: *mons* pubis
morul-, mulberry: *morul*a
moto-, moving: *moto*r neuron

mut-, change: *mut*ation
myo-, muscle: *myo*fibril
nat-, to be born: pre*nat*al
nephr-, pertaining to the kidney: *nephr*on
neutr-, neither one nor the other: *neutr*al
nod-, knot: *nod*ule
nutri-, nourish: *nutri*ent
odont-, tooth: *odont*oid process
olfact-, to smell: *olfact*ory
-osis, abnormal condition: leukocyt*osis*
os-, bone: *os*seous tissue
papill-, nipple: *papill*ary muscle, renal *papill*ae
para-, beside: *para*thyroid glands
pariet-, wall: *pariet*al membrane
path-, disease: *path*ogen
pelv-, basin: *pelv*ic cavity
peri-, around: *peri*cardial membrane, *peri*pheral nervous system, *peri*stalsis
phag-, to eat: *phag*ocytosis
pino-, to drink: *pino*cytosis
pleur-, rib: *pleur*al membrane
plex-, interweaving: choroid *plex*us
-poie, make, produce: hemato*poie*sis, erythro*poie*tin
poly-, many: *poly*unsaturated
pseud-, false: *pseud*ostratified epithelium
puber-, adult: *puber*ty
pyl-, gatekeeper: *pyl*oric sphincter
sacchar-, sugar: mono*sacchar*ide
sarco-, flesh: *sarco*plasm
scler-, hard: *scler*a
seb-, grease: *seb*aceous gland
sens-, feeling: *sens*ory neuron
-som, body: ribo*som*e
squam-, scale: *squam*ous epithelium
-stasis, standing still, halt: homeo*stasis*, hemo*stasis*
strat-, layer: *strat*ified
syn-, together: *syn*thesis, *syn*ergist, *syn*apse, *syn*cytium
systol-, contraction: *systol*ic pressure
tachy-, rapid: *tachy*cardia
tetan-, stiff: *tetan*ic contraction
thromb-, clot: *thromb*ocyte
toc-, birth: oxy*toc*in
-tomy, cutting: ana*tomy*
trigon-, triangle: *trigon*e
-troph, well fed: muscular hyper*troph*y, *troph*oblast
-tropic, influencing: adrenocortico*tropic*
tympan-, drum: *tympan*ic membrane
umbil-, navel: *umbil*ical cord
ventr-, belly or stomach: *ventr*icle
vill-, hairy: *vill*i
vitre-, glass: *vitre*ous humor
-zym, ferment: en*zym*e

METRIC MEASUREMENT SYSTEM AND CONVERSIONS

Measurement	Unit & Abbreviation	Metric Equivalent	Conversion Factor Metric to English (approximate)	Conversion Factor English to Metric (approximate)
Length	1 kilometer (km)	1,000 (10^3) m	1 km = 0.62 mile	1 mile = 1.61 km
	1 meter (m)	100 (10^2) cm 1,000 (10^3) mm	1 m = 1.1 yards = 3.3 feet = 39.4 inches	1 yard = 0.9 m 1 foot = 0.3 m
	1 decimeter (dm)	0.1 (10^{-1}) m	1 dm = 3.94 inches	1 inch = 0.25 dm
	1 centimeter (cm)	0.01 (10^{-2}) m	1 cm = 0.4 inches	1 foot = 30.5 cm 1 inch = 2.54 cm
	1 millimeter (mm)	0.001 (10^{-3}) m 0.1 (10^{-1}) cm	1 mm = 0.04 inches	
	1 micrometer (μm)	0.000001 (10^{-6}) m 0.001 (10^{-3}) mm		
Mass	1 metric ton (t)	1,000 (10^3) kg	1 t = 1.1 ton	1 ton = 0.91 t
	1 kilogram (kg)	1,000 (10^3) g	1 kg = 2.2 pounds	1 pound = 0.45 kg
	1 gram (g)	1,000 (10^3) mg	1 g = 0.04 ounce	1 pound = 454 g 1 ounce = 28.35 g
	1 milligram (mg)	0.001 (10^{-3}) g		
	1 microgram (μg)	0.000001 (10^{-6}) g		
Volume (liquids and gases)	1 liter (L)	1,000 (10^3) mL	1 L = 1.06 quarts	1 gallon = 3.78 L 1 quart = 0.95 L
	1 milliliter (mL)	0.001 (10^{-3}) L 1 cubic centimeter (cc or cm^3)	1 mL = 0.03 fluid ounce 1 mL = 1/5 teaspoon 1 mL = 15–16 drops	1 quart = 946 mL 1 fluid ounce = 29.6 mL 1 teaspoon = 5 mL
Time	1 second (s)	1/60 minute	same	same
	1 millisecond (ms)	0.001 (10^{-3}) s	same	same
Temperature	Degrees Celsius (°C)		°F = 9/5°C + 32	°C = 5/9 (°F − 32)

PERIODIC TABLE OF ELEMENTS

Key

1	Atomic Number
Hydrogen	Name
H	Symbol
1.0079	Atomic Weight

Representative Elements (*s* Series)

Representative Elements (*p* Series)

Transition Metals (*d* Series of Transition Elements)

Period

	IA																	VIIIA
1	1 Hydrogen **H** 1.0079	IIA											IIIA	IVA	VA	VIA	VIIA	2 Helium **He** 4.0026
2	3 Lithium **Li** 6.941	4 Beryllium **Be** 9.0122											5 Boron **B** 10.811	6 Carbon **C** 12.0112	7 Nitrogen **N** 14.0067	8 Oxygen **O** 15.9994	9 Fluorine **F** 18.9984	10 Neon **Ne** 20.179
3	11 Sodium **Na** 22.989	12 Magnesium **Mg** 24.305	IIIB	IVB	VB	VIB	VIIB	VIIIB			IB	IIB	13 Aluminum **Al** 26.9815	14 Silicon **Si** 28.086	15 Phosphorous **P** 30.9738	16 Sulfur **S** 32.064	17 Chlorine **Cl** 35.453	18 Argon **Ar** 39.948
4	19 Potassium **K** 39.098	20 Calcium **Ca** 40.08	21 Scandium **Sc** 44.956	22 Titanium **Ti** 47.90	23 Vanadium **V** 50.942	24 Chromium **Cr** 51.996	25 Manganese **Mn** 54.938	26 Iron **Fe** 55.847	27 Cobalt **Co** 58.933	28 Nickel **Ni** 58.71	29 Copper **Cu** 63.546	30 Zinc **Zn** 65.38	31 Gallium **Ga** 69.723	32 Germanium **Ge** 72.59	33 Arsenic **As** 74.922	34 Selenium **Se** 78.96	35 Bromine **Br** 79.904	36 Krypton **Kr** 83.80
5	37 Rubidium **Rb** 85.468	38 Strontium **Sr** 87.62	39 Yttrium **Y** 88.905	40 Zirconium **Zr** 91.22	41 Niobium **Nb** 92.906	42 Molybdenum **Mo** 95.94	43 Technetium **Tc** (99)	44 Ruthenium **Ru** 101.07	45 Rhodium **Rh** 102.905	46 Palladium **Pd** 106.4	47 Silver **Ag** 107.868	48 Cadmium **Cd** 112.40	49 Indium **In** 114.82	50 Tin **Sn** 118.69	51 Antimony **Sb** 121.75	52 Tellurium **Te** 127.60	53 Iodine **I** 126.904	54 Xenon **Xe** 131.30
6	55 Cesium **Cs** 132.905	56 Barium **Ba** 137.34	*57 Lanthanum **La** 138.91	72 Hafnium **Hf** 178.49	73 Tantalum **Ta** 180.948	74 Tungsten **W** 183.85	75 Rhenium **Re** 186.2	76 Osmium **Os** 190.2	77 Iridium **Ir** 192.2	78 Platinum **Pt** 195.09	79 Gold **Au** 196.967	80 Mercury **Hg** 200.59	81 Thallium **Tl** 204.37	82 Lead **Pb** 207.19	83 Bismuth **Bi** 208.980	84 Polonium **Po** (209)	85 Astatine **At** (210)	86 Radon **Rn** (222)
7	87 Francium **Fr** (223)	88 Radium **Ra** (226)	**89 Actinium **Ac** (227)	104 Rutherfordium **Rf** (261)	105 Hahnium **Ha** (262)	106 Seaborgium **Sg** (263)	107 Neilsbohrium **Ns** (261)	108 Hassium **Hs** (265)	109 Meitnerium **Mt** (266)									

Inner Transition Elements (*f* Series)

*Lanthanides   4f

58 Cerium **Ce** 140.12	59 Praseodymium **Pr** 140.907	60 Neodymium **Nd** 144.24	61 Promethium **Pm** 144.913	62 Samarium **Sm** 150.35	63 Europium **Eu** 151.96	64 Gadolinium **Gd** 157.25	65 Terbium **Tb** 158.925	66 Dysprosium **Dy** 162.50	67 Holmium **Ho** 164.930	68 Erbium **Er** 167.26	69 Thulium **Tm** 168.934	70 Ytterbium **Yb** 173.04	71 Lutetium **Lu** 174.97

**Actinides   5f

90 Thorium **Th** 232.038	91 Protactinium **Pa** (231)	92 Uranium **U** 238.03	93 Neptunium **Np** (237)	94 Plutonium **Pu** 244.064	95 Americium **Am** (243)	96 Curium **Cm** (247)	97 Berkelium **Bk** (247)	98 Californium **Cf** 242.058	99 Einsteinium **Es** (254)	100 Fermium **Fm** 257.095	101 Mendelevium **Md** 258.10	102 Nobelium **No** 259.101	103 Lawrencium **Lr** 260.105

Currently, 92 naturally occurring elements are known and the rest have been synthesized in the laboratory.

CHANGES OCCURRING IN THE HEART DURING A CARDIAC CYCLE

Some changes that occur in the heart during a cardiac cycle, with corresponding ECG pattern and heart sounds.

FIGURE QUESTION ANSWERS

Chapter 1

Figure 1.6: The heat would come on until the temperature reached the new set point.

Figure 1.13: The hand is more lateral than the hip (remember that anatomical position is the reference).

Chapter 2

Figure 2.11: pH 7.4, the same as the pH of human blood

Figure 2.14: Fats are a subgroup of lipids. Other lipids include the steroids and the phospholipids.

Chapter 3

Figure 3.10: chromatin and nucleolus

Figure 3.18: a carrier protein and cellular energy

Chapter 4

Figure 4.5: cytosol

Figure 4.13: messenger RNA

Chapter 5

Figure 5.2: cross section

Figure 5.20: fluid

Chapter 6

Figure 6.2: palms of hands, soles of feet

Figure 6.4: lunula

Chapter 7

Figure 7.8: receptors, control center (set point), effectors

Figure 7.29: Female obturator foramen is more triangular, male is more oval. Female pubic arch is a wider (greater) angle than the male pubic arch. Female ilia are more flared than those on the male hip bones. Female hip bones are lighter, thinner, and have less obvious muscular attachments. The female obturator foramina and acetabular are smaller and further apart than those of a male. The female pelvic cavity is wider in all diameters and is shorter, roomier, and less funnel-shaped. The distance between the female ischial spines and ischial tuberosities are greater than in the male.

Chapter 8

Figure 8.5: by diffusion

Figure 8.8: their length stays the same

Chapter 9

Figure 9.12: ATP

Figure 9.32: inferior and posterior

Chapter 10

Figure 10.6: They are the same.

Figure 10.17: the optic chiasma

Chapter 11

Figure 11.1: specific receptors

Figure 11.15: exocrine

Chapter 12

Figure 12.12: macrophage

Figure 12.16: the conversion of soluble fibrinogen to insoluble fibrin

Chapter 13

Figure 13.14: ventricular depolarization, atrial repolarization

Figure 13.21: impermeant solute, plasma proteins

Figure 13.30: radial artery and ulnar artery

Figure 13.33: axillary vein

Chapter 14

Figure 14.4: thoracic duct

Figure 14.16: secondary immune response

Chapter 15

Figure 15.2: 19.7 inches (based on the measurements in the figure)

Figure 15.19: Pancreas, the pancreas is stimulated to secrete digestive enzymes.

Figure 15.30: The major secretion from these cells in the large intestinal wall is mucus to lubricate and protect against the abrasiveness of material flowing through the tube, to bind fecal matter, and to control pH of the large intestinal contents.

Figure 15.34: BMI will vary from individual to individual. Suggestions for adjusting diet to achieve/maintain healthy weight include: lower than healthy weight—increase calorie intake, especially proteins; healthy weight—continue with your current diet unless your activity level changes; overweight—decrease your calorie intake and increase your activity level; obese—see your physician for advice on diet (lowering calorie intake, particularly carbohydrates and lipids) and exercise (you may have medical conditions that limit your physical activity or require you to slowly build up your activity level).

Chapter 16

Figure 16.8: because it carries deoxygenated blood

Figure 16.15: upward

Chapter 17

Figure 17.9: afferent arteriole

Figure 17.12 : tubular reabsorption

Chapter 18

Figure 18.2: potassium

Figure 18.8: Breathing exhales excess CO_2, so that the right amount of carbon dioxide is in the body combining with water to form carbonic acid to maintain body pH of about 7.4. By avoiding the accumulation of carbonic acid, the body is able to maintain a more constant (around neutral) pH.

Chapter 19

Figure 19.3: It is important for the sperm and egg to each possess 23 chromosomes, so that upon fertilization of the egg by the sperm, the developing zygote will have the full complement of 46 chromosomes (no more, no less) found in human cells.

Figure 19.10: corona radiata

Figure 19.15: combined hormone contraceptives, intrauterine device

Chapter 20

Figure 20.2: 23 chromosomes each with two chromatids

Figure 20.15: umbilical vein, umbilical arteries, ductus venosus, ductus arteriosus, foramen ovale

Glossary

A phonetic guide to pronunciation follows each glossary word. Any unmarked vowel that ends a syllable or stands alone as a syllable has the long sound. Thus, the word *play* is phonetically spelled *plā*. Any unmarked vowel followed by a consonant has the short sound. The word *tough*, for instance, is phonetically spelled *tuf*. If a long vowel appears in the middle of a syllable (followed by a consonant), it is marked with a macron (ˉ), the sign for a long vowel. Thus, the word *plate* is phonetically spelled *plāt*. Similarly, if a vowel stands alone or ends a syllable, but has a short sound, it is marked with a breve (˘).

A

abdominal cavity (ab-dom′ĭ-nal kav′ĭ-te) Space between the diaphragm and the pelvic inlet that contains the abdominal viscera. p. 8

abdominopelvic cavity (ab-dom″ĭ-no-pel′vik kav′ĭ-te) Space between the diaphragm and the pelvic outlet that contains the abdominal and pelvic viscera. p. 8

abduction (ab-duk′shun) Movement of a body part away from the midline. p. 167

accessory organ (ak-ses′o-re or′gan) Organ that supplements the functions of other organs. p. 276

accommodation (ah-kom″o-da′shun) Adjustment of the lens of the eye for close or distant vision. p. 281

acetylcholine (as″ĕ-til-ko′lēn) **(ACH)** Type of neurotransmitter, which is a biochemical secreted into the synaptic cleft at axon ends of neurons. p. 183

acetylcholinesterase (as″ĕ-til-ko″lin-es′ter-ās) Enzyme that catalyzes breakdown of acetylcholine. p. 183

acid (as′id) Substance that ionizes in water to release hydrogen ions. p. 38

acidosis (as″ĭ-do′sis) Decrease in pH of body fluids below pH 7.35. p. 500

ACTH Adrenocorticotropic hormone. p. 300

actin (ak′tin) Protein in a muscle fiber that forms filaments that slide between filaments of the protein myosin, contracting muscle fibers. p. 179

action potential (ak′shun po-ten′shal) Sequence of electrical changes in part of a nerve cell or muscle cell exposed to a stimulus that exceeds threshold. p. 222

active site (ak′tiv sīt) Part of an enzyme molecule that temporarily binds a substrate. p. 79

active transport (ak′tiv trans′port) Process that requires energy and a carrier molecule to move a substance across a cell membrane against the concentration gradient. p. 65

adaptation (ad″ap-tā′shun) Ability of the nervous system to become less responsive to a sustained stimulus. p. 263

adaptive defense (a-dap′tiv dē-fenc) Specific defenses T and B cells carry out. p. 384

adduction (ah-duk′shun) Movement of body part toward the midline. p. 167

adenosine triphosphate (ah-den′o-sēn tri-fos′fāt) **(ATP)** Organic molecule that transfers energy, used in cellular processes. p. 80

ADH Antidiuretic hormone. p. 300

adipose tissue (ad′ĭ-pōs tish′u) Fat-storing tissue. p. 104

adrenal cortex (ah-dre′nal kor′teks) Outer part of the adrenal gland. p. 304

adrenal gland (ah-dre′nal gland) Endocrine gland on the superior portion of each kidney. p. 304

adrenal medulla (ah-dre′nal me-dul′ah) Inner part of the adrenal gland. p. 304

adrenergic fiber (ad″ren-er′jik fi′ber) Axon that secretes norepinephrine at its terminal. p. 254

adrenocorticotropic hormone (ad-re″no-kor″te-ko-trōp′ik hor′mōn) **(ACTH)** Hormone that the anterior pituitary secretes to stimulate activity in the adrenal cortex. p. 300

aerobic respiration (a″er-o′bik res″pĭ-ra′shun) Complete, energy-releasing, breakdown of glucose to carbon dioxide and water, in the presence of oxygen. p. 82

afferent arteriole (af′er-ent ar-te′re-ōl) Vessel that conveys blood to the glomerulus of each kidney nephron. p. 469

agglutination (ah-gloo″tĭ-na′shun) Clumping of blood cells in response to a reaction between an antibody and an antigen. p. 333

agonist (ago′-nist) Prime mover. p. 194

agranulocyte (a-gran′u-lo-sīt) Nongranular leukocyte. p. 324

albumin (al-bu′min) Plasma protein that helps regulate the colloid osmotic pressure of blood. p. 329

aldosterone (al-dos′ter-ōn″) Adrenal cortical hormone that regulates sodium and potassium ion concentrations and fluid volume. p. 306

alimentary canal (al″ĭ-men′tar-e kah-nal′) Tubular part of the digestive tract from the mouth to the anus. p. 401

alkalosis (al″kah-lo′sis) Increase in pH of body fluids above pH 7.45. p. 500

allantois (ah-lan′to-is) Structure in the embryo from which the umbilical cord blood vessels develop. p. 547

allele (ah-le′el) One of two or more forms of a gene. p. 556

allergen (al′er-jen) Chemical that triggers the immune response. p. 393

alveolar duct (al-ve′o-lar dukt) Fine tube that conducts inhaled air to an air sac of the lungs. p. 448

alveolus (al-ve′o-lus) (plural, *alveoli*) Air sac of a lung; a saclike structure. p. 448

amino acid (ah-me′no as′id) Organic compound that includes an amino group (–NH$_2$) and a carboxyl group (–COOH); the structural unit of a protein molecule. p. 44

amnion (am′ne-on) Extraembryonic membrane that encircles a fetus and contains amniotic fluid. p. 546

amniotic fluid (am″ne-ot′ik floo′id) Fluid in the amniotic cavity that surrounds the developing fetus. p. 546

ampulla (am-pul′ah) Expansion at the end of each semicircular canal that houses a crista ampullaris. p. 275

anabolism (ah-nab′o-lizm) Synthesis of larger molecules from smaller ones; anabolic metabolism. p. 77

anaerobic respiration (an″er-o′bik res″pĭ-ra′shun) Energy-releasing reactions that occur in the absence of molecular oxygen. p. 82

anaphase (an′ah-fāz) Stage in mitosis when replicated chromosomes separate and move to opposite poles of the cell. p. 69

anatomy (ah-nat′o-me) Branch of science dealing with the form and structure of body parts. p. 3

androgen (an′dro-jen) Male sex hormone such as testosterone. p. 513

antagonist (an-tag′o-nist) Muscle that opposes a prime mover. p. 194

antebrachial (an″te-bra′ke-al) Pertaining to the forearm. p. 16

antecubital (an″te-ku′bĭ-tal) Region anterior to the elbow joint. p. 16

anterior (an-te′re-or) Pertaining to the front. p. 14

anterior pituitary (an-te′re-or pĭ-tu′ĭ-tār″e) Front lobe of the pituitary gland. p. 297

antibody (an′tĭ-bod″e) Protein that B cells of the immune system produce in response to a nonself antigen; it reacts with the antigen; immunoglobulin. p. 333

anticodon (an″tĭ-ko′don) Three nucleotides of a transfer RNA molecule that are complementary to a specific mRNA codon. p. 87

antidiuretic hormone (an″tĭ-di″u-ret′ik hor′mōn) **(ADH)** Hormone of the posterior pituitary gland that enhances water conservation in the kidneys. p. 500

antigen (an′tĭ-jen) Chemical that triggers an immune response. p. 333

aorta (a-or′tah) Major systemic artery that receives blood directly from the left ventricle. p. 343

aortic sinus (a-or′tik si′nus) Swelling in the aortic wall, behind each cusp of the semilunar valve, that contains baroreceptors. p. 363

aortic valve (a-or′tik valv) Flaplike structures in the wall of the aorta near its origin that prevent blood from returning to the left ventricle of the heart. p. 343

apocrine gland (ap′o-krin gland) Type of gland whose secretions have parts of secretory cells. p. 101

aponeurosis (ap″o-nu-ro′sis) Sheet of connective tissue that attaches muscles to muscles. p. 179

apoptosis (ayp-o-toe′sis) Programmed cell death. p. 71

appendicular (ap″en-dik′u-lar) Pertaining to the upper or lower limbs. p. 8

aqueous humor (a′kwe-us hu′mor) Watery fluid that fills the anterior cavity of the eye. p. 282

arachnoid mater (ah-rak′noid ma′ter) Delicate, weblike middle layer of meninges. p. 233

areolar tissue (ah-re′o-lar tish′u) Connective tissue composed mainly of fibers. p. 104

arrector pili muscle (ah-rek′tor pil′i mus′l) Smooth muscle in the skin associated with a hair follicle. p. 123

arteriole (ar-te′re-ōl) Small branch of an artery that communicates with a capillary network. p. 354

artery (ar′ter-e) Vessel that transports blood from the heart. p. 354

articular cartilage (ar-tik′u-lar kar′tĭ-lij) Hyaline cartilage that covers the ends of bones in synovial joints. p. 133

ascending tract (ah-send′ing trakt) Group of nerve fibers in the spinal cord that carry sensory impulses to the brain. p. 234

association area (ah-so″se-a′shun a′re-ah) Region of the cerebral cortex controlling memory, reasoning, judgment, and emotions. p. 239

astrocyte (as′tro-sīt) Type of neuroglia that connects neurons to blood vessels in the CNS. p. 216

atmospheric pressure (at″mos-fēr′ik presh′ur) Pressure exerted by the weight of air, about 760 millimeters of mercury at sea level. p. 450

atom (at′om) Smallest particle of an element that has the properties of that element. p. 3

atomic number (ah-tom′ik num′ber) Number of protons in an atom of an element. p. 32

atomic weight (ah-tom′ik wāt) The number of protons and neutrons in an atom of an element. p. 32

ATP Adenosine triphosphate. p. 80

ATPase Enzyme that releases energy stored in the terminal phosphate bonds of ATP molecules. p. 183

atrium (a′tre-um) (plural, *atria*) Upper chamber of the heart. p. 343

atrioventricular bundle (a″tre-o-ven-trik′u-lar bun′dl) **(AV bundle)** Group of specialized muscle fibers that conducts impulses from the atrioventricular node to the ventricular muscle of the heart. p. 350

atrioventricular node (a″tre-o-ven-trik′u-lar nōd) **(AV node)** Specialized mass of cardiac muscle fibers in the interatrial septum of the heart that transmits cardiac impulses from the sinoatrial node to the AV bundle. p. 350

auditory ossicle (aw′di-to″re os′i-kl) Bone of the middle ear. p. 271

auditory tube (aw′di-to″re tūb) Tube that connects the middle ear cavity to the pharynx; Eustachian tube. p. 271

auricle (aw′ri-kl) An earlike structure; the part of the heart that forms the wall of an atrium. p. 270

autonomic nervous system (aw″to-nom′ik ner′vus sis′tem) Part of nervous system that controls the viscera. p. 216

autosome (aw′to-sōm) One of the 22 chromosomes that does not include a gene that determines sex. p. 556

axial (ak′se-al) Pertaining to the head, neck, and trunk. p. 8

axillary (ak′sĭ-ler″e) Pertaining to the armpit. p. 16

axon (ak′son) A nerve fiber. It conducts an impulse away from the neuron cell body. p. 111

B

B cell (sel) Lymphocyte that produces and secretes antibodies that bind and destroy foreign antigens; B lymphocyte. p. 386

baroreceptor (bar″o-re-sep′tor) Sensory receptor in a blood vessel wall stimulated by changes in blood pressure. p. 352

basal ganglia (ba′sal gang′lē-ah) Mass of gray matter deep within a cerebral hemisphere of the brain. p. 240

base (bās) Substance that ionizes in water, releasing hydroxide ions (OH⁻) or other ions that combine with hydrogen ions. p. 38

basement membrane (bās′ment mem′brān) Layer of nonliving material that anchors epithelial tissue to underlying connective tissue. p. 95

basophil (ba′so-fil) White blood cell containing cytoplasmic granules that stain with basic dye. p. 325

beta oxidation (ba′tah ok″sĭ-da′shun) Chemical process that breaks fatty acids down to form molecules of acetyl coenzyme A, which enters the citric acid cycle. p. 429

bile (bīl) Fluid secreted by the liver and stored in the gallbladder. p. 418

bile duct (bīl dukt) Tube that transports bile from the cystic duct and common hepatic duct to the duodenum. p. 418

bilirubin (bil″ĭ-roo′bin) A bile pigment produced from hemoglobin breakdown. p. 323

biliverdin (bil″ĭ-ver′din) A bile pigment produced from hemoglobin breakdown. p. 323

blastocyst (blas′to-sist) Early stage of prenatal development when the embryo is a hollow ball of cells. p. 541

blood (blud) A connective tissue consisting of cells in a liquid matrix called plasma that circulate through the heart and vessels carrying substances throughout the body. p. 319

bone (bōn) A hard, mineralized connective tissue that forms the bones of the skeleton. p. 108

brachial (bra′ke-al) Pertaining to the arm. p. 16

brainstem (brān′stem) Part of the brain that includes the midbrain, pons, and medulla oblongata. p. 236

bronchial tree (brong′ke-al tre) The bronchi and their branches that carry air between the trachea and the alveoli of the lungs. p. 446

bronchiole (brong′ke-ōl) Small branches of bronchi in the lung. p. 448

bronchus (brong′kus) (plural, *bronchi*) Branch of the trachea that leads to a lung. p. 446

buccal (buk′al) Pertaining to the mouth and inner lining of the cheeks. p. 16

buffer (buf′er) Substance that can react with a strong acid or base to form a weaker acid or base and thus resist a change in pH. p. 39

bulbourethral gland (bul″bo-u-re′thral gland) Gland that secretes viscous fluid into the male urethra during sexual excitement; Cowper's gland. p. 511

bursa (ber′sah) (plural, *bursae*) Saclike, fluid-filled structure, lined with synovial membrane, near a joint. p. 164

C

calcitonin (kal″sĭ-to′nin) Hormone from the thyroid gland that helps regulate blood calcium concentration. p. 302

calorie (kal′o-re) Unit that measures heat energy and the energy content of foods. p. 428

canaliculus (kan″ah-lik′u-lus) Microscopic canal that connects lacunae of bone tissue. p. 108

capacitation (kah-pas″i-ta′shun) Activation of a sperm cell to fertilize an egg cell. p. 511

capillary (kap′ĭ-lar″e) Small blood vessel that connects an arteriole and a venule. p. 356

carbaminohemoglobin (kar-bam″ĭ-no-he″mo-glo′bin) Bonded carbon dioxide and hemoglobin. p. 461

carbohydrate (kar″bo-hi′drāt) Organic compound consisting of carbon, hydrogen, and oxygen, in a 1:2:1 ratio. p. 41

carbonic anhydrase (kar-bon′ik an-hi′drās) Enzyme that catalyzes the reaction between carbon dioxide and water to form carbonic acid. p. 462

carboxypeptidase (kar-bok″se-pep′tĭ-dās) Protein-splitting enzyme in pancreatic juice. p. 413

cardiac conduction system (kar′de-ak kon-duk′shun sis′tem) System of specialized cardiac muscle fibers that conducts cardiac impulses from the SA node into the myocardium. p. 350

cardiac cycle (kar′de-ak si′kl) Sequence of myocardial contraction and relaxation that constitutes a complete heartbeat. p. 347

cardiac muscle tissue (kar′de-ak mus′l tish′u) Specialized muscle tissue found only in the heart. p. 111

cardiac output (kar′de-ak owt′poot) The volume of blood per minute that the heart pumps (stroke volume in milliliters multiplied by heart rate in beats per minute). p. 360

cardiovascular (kahr″de-o-vas′ku-lur) Pertaining to the heart and blood vessels. p. 12

carpal (kar′pal) Wrist bone. p. 16

cartilage (kar′ti-lij) Type of connective tissue in which cells are in lacunae separated by a semisolid extracellular matrix. p. 107

cartilaginous joint (kar-tĭ-laj′ĭ-nus joint) Two or more bones joined by cartilage. p. 164

catabolism (kă-tab′o-lizm) Breakdown of large molecules; catabolic metabolism. p. 77

catalyst (kat′ah-list) Chemical that increases the rate of a chemical reaction, but is not permanently altered by the reaction. p. 38

celiac (se′le-ak) Pertaining to the abdomen. p. 16

cell (sel) The structural and functional unit of an organism. p. 3

cell body (sel bod′e) Part of a nerve cell that includes a cytoplasmic mass and a nucleus, and from which nerve processes extend. p. 214

cell membrane (sel mem′brān) The selectively permeable outer boundary of a cell consisting of a phospholipid bilayer embedded with proteins. p. 52

cellular immune response (sel′u-lar i-mūn′ ri-spons′) The body's attack of T cells and their secreted products on nonself antigens. p. 386

cellular respiration (sel′u-lar res″pĭ-ra′shun) A biochemical pathway that releases energy from organic compounds. p. 80

cellulose (sel′u-lōs) Polysaccharide abundant in plant tissues that human digestive enzymes cannot break down. p. 429

cementum (se-men′tum) Bonelike material that fastens the root of a tooth into its bony socket. p. 407

central canal (sen′tral kah-nal′) Tiny channel in bone tissue that houses a blood vessel, p. 108; tube in the spinal cord continuous with brain ventricles and contains cerebrospinal fluid. p. 108

central nervous system (sen′tral ner′vus sis′tem) **(CNS)** The brain and spinal cord. p. 214

centriole (sen′tre-ōl) Cellular structure built of microtubules that organizes the mitotic spindle. p. 57

centromere (sen′tro-mēr) Region of chromosome where spindle fibers attach during mitosis. p. 507

centrosome (sen′tro-sōm) Cellular organelle consisting of two centrioles. p. 57

cephalic (sĕ-fal′ik) Pertaining to the head. p. 17

cerebellar cortex (ser″ĕ-bel′ar kor′teks) Outer layer of the cerebellum. p. 244

cerebellum (ser″ĕ-bel′um) Part of the brain that coordinates skeletal muscle movement. p. 236

cerebral cortex (ser′ĕ-bral kor′teks) Outer layer of the cerebrum. p. 238

cerebral hemisphere (ser′ĕ-bral hem′ĭ-sfēr) One of the large, paired structures that constitute the cerebrum. p. 236

cerebrospinal fluid (ser″ĕ-bro-spi′nal floo′id) **(CSF)** Fluid in the ventricles of the brain, subarachnoid space of the meninges, and the central canal of the spinal cord. p. 234

cerebrum (ser′ĕ-brum) Part of the brain in the upper part of the cranial cavity that provides higher mental functions. p. 236

cervical (ser′vĭ-kal) Pertaining to the neck. p. 17

cervix (ser′viks) Narrow, inferior end of the uterus that leads into the vagina. p. 520

chemical bond (kem′ĭkel bond) Attractive force holding atoms together. p. 32

chemistry (kem′is-trē) The study of matter. p. 31

chemoreceptor (ke″mo-re-sep′tor) Receptor stimulated by the binding of certain chemicals. p. 263

chief cell (chēf sel) Cell type in a gastric gland that secretes digestive enzymes. p. 511

cholecystokinin (ko″le-sis″to-ki′nin) Hormone the small intestine secretes that stimulates release of pancreatic juice from the pancreas and bile from the gallbladder. p. 412

cholesterol (ko-les′ter-ol) A lipid that some cells produce and use to synthesize steroid hormones. p. 429

cholinergic fiber (ko″lin-er′jik fi′ber) Axon that secretes acetylcholine at its terminal. p. 251

chondrocyte (kon′dro-sīt) A cartilage cell. p. 107

chordae tendineae (kor′de ten′dĭ-ne) Fibrous strings attached to the cusps of the tricuspid and mitral valves in the heart. p. 344

chorion (ko′re-on) Extraembryonic membrane that forms the outermost covering around a fetus and contributes to formation of the placenta. p. 544

chorionic villus (ko″re-on′ik vil′us) Projection that extends from the outer surface of the chorion and helps attach an embryo to the uterine wall. p. 544

choroid coat (ko′roid kōt) Vascular, pigmented middle layer of the wall of the eye. p. 280

choroid plexus (ko′roid plek′sus) Mass of specialized capillaries that secretes cerebrospinal fluid into a brain ventricle. p. 240

chromatid (kro′mah-tid) One longitudinal half of a replicated chromosome. p. 507

chromatin (kro′mah-tin) DNA and complexed protein that condenses to form chromosomes during mitosis. p. 60

chromatophilic substance (kro″mah-to-fil′ik sub′stans) **(Nissl bodies)** Membranous sacs in the cytoplasm of nerve cells that have ribosomes attached to their surfaces. p. 216

chromosome (kro′mo-sōm) Rodlike structure that condenses from chromatin in a cell's nucleus during mitosis. p. 60

chylomicron (ki″lo-mi′kron) Microscopic droplet of fat in the blood that forms following fat digestion. p. 423

chyme (kīm) Semifluid mass of partially digested food that passes from the stomach to the small intestine. p. 413

chymotrypsin (ki″mo-trip′sin) A protein-splitting enzyme in pancreatic juice. p. 413

cilia (sil′e-ah) Microscopic, hairlike processes on the exposed surfaces of certain epithelial cells. p. 58

ciliary body (sil′e-er″e bod′e) Structure associated with the choroid layer of the eye that secretes aqueous humor and houses the ciliary muscle. p. 280

circadian rhythm (ser″kah-de′an rithm) Pattern of repeated behavior associated with cycles of night and day. p. 309

cisterna (sis-ter′nah) Enlarged portion of the sarcoplasmic reticulum near the actin and myosin filaments of a muscle fiber. p. 182

citric acid cycle (sit′rik as′id si′kl) Series of chemical reactions that oxidizes certain molecules, releasing energy; Krebs cycle. p. 80

cleavage (klēv′ij) Early successive divisions of the zygote into a ball of progressively smaller cells. p. 541

clitoris (klit′o-ris) Small, erectile organ in the anterior vulva; corresponding to the penis. p. 521

clone (klōn) Group of cells that descend from a single cell and are therefore genetically identical to it. p. 388

CNS Central nervous system. p. 214

coagulation (ko-ag″u-la′shun) Blood clotting. p. 331

cochlea (kok′le-ah) Portion of inner ear that has hearing receptors. p. 272

codon (ko′don) Set of three contiguous nucleotides of a messenger RNA molecule that specifies a particular amino acid. p. 87

coenzyme (ko-en′zīm) Nonprotein organic molecule required for the activity of a particular enzyme. p. 80

cofactor (ko′fak-tor) Small molecule or ion that must combine with an enzyme for activity. p. 80

collagenous fiber (kol′ah-jen-us fi′ber) White fiber consisting of the protein collagen, common in connective tissues, including bone matrix. p. 103

common hepatic duct (kom′mon hepat′ik dukt) Tube that transports bile from the liver to the bile duct. p. 415

compact bone (kom′-pakt bōn) Dense tissue in which cells are organized in osteons without apparent spaces. p. 133

complement (kom′ple-ment) Group of proteins activated when an antibody binds an antigen; enhances reaction against nonself substances. p. 385

complete protein (kom-plēt′ pro′te-in) Protein that contains adequate amounts of the essential amino acids to maintain body tissues and to promote normal growth and development. p. 431

compound (kom′pownd) Substance composed of two or more chemically bonded elements. p. 36

cone (kōn) Color receptor in the retina of the eye. p. 285

conformation (kon″for-ma″shen) Three-dimensional form of a protein. p. 44

conjunctiva (kon″junk-ti′vah) Membranous covering on the anterior surface of the eye and lining the eyelids. p. 277

connective tissue (kŏ-nek′tiv tish′u) Basic type of tissue that consists of cells within an extracellular matrix. Connective tissues include bone, cartilage, and blood. p. 102

convergence (kon-ver′jens) Two or more neurons forming synapses with the same neuron. p. 229

cornea (kor′ne-ah) Transparent anterior portion of the outer layer of the eye wall. p. 279

coronary artery (kor′o-na″re ar′ter-e) An artery that supplies blood to the wall of the heart. p. 346

coronary sinus (kor′o-na″re si′nus) Large vessel on the posterior surface of the heart into which cardiac veins drain. p. 347

corpus callosum (kor′pus kah-lo′sum) Mass of white matter in the brain composed of nerve fibers connecting the right and left cerebral hemispheres. p. 236

corpus luteum (kor′pus loot′e-um) Structure that forms from the tissues of a ruptured ovarian follicle and secretes female hormones. p. 523

cortisol (kor′tĭ-sol) Glucocorticoid secreted by the adrenal cortex. p. 306

costal (kos′tal) Pertaining to the ribs. p. 16

covalent bond (ko′va-lent bond) Chemical bond formed by electron sharing between atoms. p. 35

cranial nerve (kra′ne-al nerv) Nerve that arises from the brain or brainstem. p. 246

creatine phosphate (kre′ah-tin fos′fāt) Muscle biochemical that stores energy. p. 185

crista ampullaris (kris′tah am-pul′ar-is) Sensory organ in a semicircular canal that functions in the sense of dynamic equilibrium. p. 275

cubital (ku′bi-tal) Pertaining to the elbow. p. 17

cutaneous (ku-ta′ne-us) Pertaining to the skin. p. 110

cyclic AMP (cAMP) (si′klik a-em-pē) A second messenger molecule in a signal transduction pathway. p. 295

cystic duct (sis′tik dukt) Tube that connects the gallbladder to the bile duct. p. 418

cytocrine secretion (si′to-krin se-kre′shun) Transfer of melanin granules from melanocytes into epithelial cells. p. 119

cytokinesis (si″to-kĭ-ne′sis) Division of the cytoplasm during cell division. p. 69

cytoplasm (si′to-plazm) The contents of a cell including the gel like cytosol and organelles, excluding the nucleus, enclosed by the cell membrane. p. 52

cytoskeleton (si′to-skel′e-ten) A cell's framework of protein filaments and tubules. p. 55

D

deamination (de-am″ĭ-na′shun) Removing amino groups (–NH$_2$) from amino acids. p. 431

decomposition (de″-kom-po-zish′un) Breakdown of molecules. p. 38

dehydration synthesis (de″hi-dra′shun sin′thĕ-sis) Anabolic process that joins small molecules by releasing the equivalent of a water molecule; synthesis. p. 77

dendrite (den′drīt) Process of a neuron that receives input from other neurons. p. 214

dense connective tissue (dens kō-nek′tiv tish′u) A connective tissue with many collagenous fibers, a fine network of elastin fibers, and sparse fibroblasts. p. 103

dentin (den′tin) Bonelike substance that forms the bulk of a tooth. p. 407

deoxyhemoglobin (de-ok″se-he″mo-glo′bin) Hemoglobin that has not bound oxygen. p. 319

deoxyribonucleic acid (de-ok′si-ri″bo-nu-kle″ik as′id) **(DNA)** The genetic material; a double-stranded polymer of nucleotides, each containing a phosphate group, a nitrogenous base (adenine, thymine, guanine, or cytosine), and the sugar deoxyribose. p. 46

depolarization (de-po″lar-ĭ-za′shun) A change in membrane potential from negative to more positive. p. 225

dermis (der′mis) The thick layer of the skin beneath the epidermis. p. 117

descending tract (de-send′ing trakt) Group of nerve fibers that carries nerve impulses from the brain down through the spinal cord. p. 235

detrusor muscle (de-trūz′or mus′l) Muscular wall of the urinary bladder. p. 483

diapedesis (di″ah-pĕ-de′sis) Squeezing of leukocytes between the cells of blood vessel walls. p. 325

diaphragm (di′ah-fram) A sheetlike structure largely composed of skeletal muscle and connective tissue that separates the thoracic and abdominal cavities, p. 8; a contraceptive device inserted in the vagina. p. 451

diaphysis (di-af′ĭ-sis) Shaft of a long bone. p. 133

diastole (di-as′to-le) Phase of the cardiac cycle when a heart chamber wall relaxes. p. 347

diastolic pressure (di-a-stol′ik presh′ur) Lowest arterial blood pressure reached during the diastolic phase of the cardiac cycle. p. 359

diencephalon (di″en-sef′ah-lon) Part of the brain in the region of the third ventricle that includes the thalamus and hypothalamus. p. 236

differentiation (dif″er-en″she-a′shun) Cell specialization. p. 71

diffusion (dĭ-fu′zhun) Random movement of molecules from a region of higher concentration toward one of lower concentration. p. 60

digestion (di-jest′yun) Breaking down of large nutrient molecules into molecules small enough to be absorbed; hydrolysis. p. 401

dipeptide (di-pep′tīd) Molecule composed of two joined amino acids. p. 78

disaccharide (di-sak′ah-rīd) Sugar produced by the union of two monosaccharides. p. 41

distal (dis′tal) Further from a point of attachment; opposite of *proximal*. p. 15

divergence (di-ver′jens) Spreading apart. A single neuron forming synapses with two or more post synaptic cells. p. 229

DNA Deoxyribonucleic acid; the genetic material. p. 46

dominant (dom′eh-nant) A type of allele that is expressed over a recessive allele. p. 557

dorsal (dors′al) Pertaining to the back surface of a body part. p. 17

dorsal root (dor′sal root) Sensory branch of a spinal nerve by which it joins the spinal cord. p. 249

ductus arteriosus (duk′tus ar-te″re-o′sus) Blood vessel that connects the pulmonary artery and the aorta in a fetus. p. 551

ductus deferens (duk′tus def″er-ens) (plural, *ductus deferentia*) Tube that leads from the epididymis to the urethra of the male reproductive tract. p. 506

ductus venosus (duk′tus ven-o′sus) Blood vessel that connects the umbilical vein and the inferior vena cava in a fetus. p. 550

dura mater (du′rah ma′ter) Tough outer layer of the meninges. p. 233

dynamic equilibrium (di-nam′ik e″kwĭ-lib′re-um) Maintenance of balance when the head and body are suddenly moved or rotated. p. 275

E

eardrum (er′drum) Tympanic membrane; a thin membrane that separates the external ear from the middle ear. p. 270

eccrine gland (ek′rin gland) Sweat gland that maintains body temperature. p. 124

ECG Electrocardiogram. p. 351

ectoderm (ek′to-derm) Outermost primary germ layer of the embryo. p. 543

edema (ĕ-de′mah) Fluid accumulation in tissue spaces. p. 64

effector (e-fek′tor) A muscle or gland that effects change in the body. p. 6

efferent arteriole (ef′er-ent ar-te′re-ōl) The vessel that conducts blood away from the glomerulus of each kidney nephron. p. 469

ejaculation (e-jak″u-la′shun) Discharge of sperm-containing semen from the male urethra. p. 512

elastic cartilage (ē-las-tik kar′-ti-lij) Opaque, flexible connective tissue with many elastic fibers. p. 107

elastic fibers (e-las′tic fiberz) Stretchy yellow connective tissue fibers consisting of the protein elastin. p. 103

electrocardiogram (e-lek″tro-kar′de-o-gram″) **(ECG)** Recording of the electrical activity associated with the cardiac cycle. p. 351

electrolyte (e-lek′tro-līt) Substance that ionizes in a water solution. p. 38

electrolyte balance (e-lek′tro-līt bal′ans) Condition when electrolytes entering the body equal those leaving it. p. 493

electron (e-lek′tron) Small, negatively charged particle that encircles the nucleus of an atom. p. 32

electron transport chain (e-lektron tranz′port chān) Series of metabolic reactions that takes high-energy electrons from glycolysis and the citric acid cycle to form ATP, water, CO$_2$, and heat. p. 80

element (el′ĕ-ment) Chemical substance with only one type of atom. p. 31

embolus (em′bo-lus) Blood clot or gas bubble carried in circulation that may obstruct a blood vessel. p. 331

embryo (em′bre-o) A prenatal stage of development after primary germ layers form and rudiments of all organs are present. p. 541

emission (e-mish′un) Movement of sperm cells from the ductus deferentia into the ejaculatory ducts and urethra. p. 512

emulsification (e-mul″sĭ-fĭ-ka′shun) Breaking up of fat globules into smaller droplets by the action of bile salts. p. 419

enamel (e-nam′el) Hard covering on the exposed surface of a tooth. p. 407

endocardium (en″do-kar′de-um) Inner lining of the heart chambers. p. 343

endochondral bone (en″do-kon′dral bōn) Bone that begins as hyaline cartilage that is subsequently replaced by bone tissue. p. 136

endocrine gland (en′do-krin gland) Gland that secretes hormones directly into the blood. p. 101

endocytosis (en″do-si-to′sis) Process by which a cell membrane envelops a substance and draws it into the cell in a vesicle. p. 65

endoderm (en′do-derm) The innermost primary germ layer in the embryo. p. 543

endolymph (en′do-limf) Fluid in the membranous labyrinth of the inner ear. p. 272

endometrium (en″do-me′tre-um) Inner lining of the uterus. p. 520

endomysium (en″do-mis′e-um) Sheath of connective tissue surrounding each skeletal muscle fiber. p. 179

endoplasmic reticulum (en′do-plaz mik rĕ-tik′u-lum) Organelle composed of a network of connected membranous tubules and vesicles. p. 55

endosteum (en-dos′te-um) Tissue lining the medullary cavity in a bone. p. 134

endothelium (en″do-the′le-um) Layer of epithelial cells that forms the inner lining of blood vessels and heart chambers. p. 354

energy (en′er-je) An ability to move something and thus do work. p. 80

enzyme (en′zīm) Protein that catalyzes a specific biochemical reaction. p. 77

eosinophil (e″o-sin′o-fil) White blood cell containing cytoplasmic granules that stain with acidic dye. p. 325

ependyma (ĕ-pen′dĭ-mah) Neuroglia that line the ventricles of the brain. p. 216

epicardium (ep″ĭ-kar′de-um) Visceral part of the pericardium on the surface of the heart. p. 343

epidermis (ep″i-der′mis) Outer epithelial layer of the skin. p. 98

epididymis (ep″ĭ-did′ĭ-mis) (plural, *epididymides*) Highly coiled tube that leads from the seminiferous tubules of the testis to the ductus deferens. p. 506

epidural space (ep″ĭ-du′ral spās) Space between the dural sheath of the spinal cord and the bone of the vertebral canal. p. 233

epigastric region (ep″ĭ-gas′trik re′jun) Upper middle part of the abdomen. p. 16

epiglottis (ep″ĭ-glot′is) Flaplike, cartilaginous structure at the back of the tongue near the entrance to the trachea. p. 446

epimysium (ep″i-mis′e-um) Outer sheath of connective tissue surrounding a skeletal muscle. p. 179

epinephrine (ep″ĭ-nef′rin) Hormone the adrenal medulla secretes during times of stress. p. 304

epiphyseal plate (ep″ĭ-fiz′e-al plāt) Cartilaginous layer between the epiphysis and diaphysis of a long bone that grows, lengthening the bone. p. 136

epiphysis (ĕ-pif′ĭ-sis) End of a long bone. p. 133

epithelial tissue (ep″ĭ-the′le-al tish′u) Tissue that covers all external and lines all internal body surfaces. Varieties are classified by cell shape (squamous, cuboidal, or columnar) and number of layers (simple, stratified, or pseudostratified). p. 95

erythrocyte (ĕ-rith′ro-sīt) Red blood cell. p. 319

erythropoietin (ĕ-rith″ro-poi′ĕ-tin) Kidney hormone that promotes red blood cell formation. p. 310

esophagus (ĕ-sof′ah-gus) Tubular part of the digestive tract connecting the pharynx to the stomach. p. 410

essential amino acid (ĕ-sen′shal ah-me′no as′id) Amino acid required for health that body cells cannot synthesize in adequate amounts. p. 431

essential fatty acid (ĕ-sen′shal fat′e as′id) Fatty acid required for health that body cells cannot synthesize in adequate amounts. p. 430

estrogens (es′tro-jenz) Group of hormones that stimulates the development of female secondary sex characteristics. p. 522

eumelanin (u-mel′ah-nin) Brownish-black pigment. p. 123

eversion (e-ver′zhun) Turning the sole of the foot outward away from the midline. p. 167

exchange reaction (eks-chănj re-ak′shun) Chemical reaction in which parts of two kinds of molecules trade positions. p. 38

exocrine gland (ek′so-krin gland) Gland that secretes its products into ducts or onto an outside body surface. p. 101

exocytosis (ek″so-si-tōsis) Transport of substances out of a cell in membrane-bounded vesicles. p. 65

expiration (ek″spī-ra′shun) Breathing out; exhalation. p. 450

extension (ek-sten′shun) Movement increasing the angle between parts at a joint. p. 167

extracellular fluid (eks″trah-sel′u-lar floo′id) Body fluid outside cells. p. 490

extracellular matrix (eks″trah-sel′u-lar ma′triks) Molecules that fill spaces between cells, consisting mostly of protein fiber networks. p. 102

extrapyramidal tract (ek″strah-pĭ-ram′i-dal trakt) Nerve tracts, other than the corticospinal tracts, that transmit impulses from the cerebral cortex to the spinal cord. p. 235

F

facilitated diffusion (fah-sil″ĭ-tāt′ed dĭ-fu′zhun) Diffusion in which a carrier molecule transports a substance across a cell membrane from a region of higher concentration to a region of lower concentration. p. 63

facilitation (fah-sil″ĭ-ta′shun) Subthreshold stimulation of a neuron that increases responsiveness to further stimulation. p. 229

fascia (fash′e-ah) Sheet of fibrous connective tissue that encloses a muscle. p. 177

fat (fat) Adipose tissue; an organic molecule that includes glycerol and fatty acids. p. 42

fatty acid (fat′e as′id) Building block of a fat molecule. p. 42

feces (fe′sēz) Material expelled from the digestive tract during defecation. p. 427

fertilization (fer″tĭ-lĭ-za′shun) Union of an egg cell and a sperm cell. p. 537

fetus (fe′tus) Prenatal human after eight weeks of development. p. 541

fibrin (fi′brin) Insoluble, fibrous protein formed from fibrinogen during blood coagulation. p. 331

fibrinogen (fi-brin′o-jen) Plasma protein converted into fibrin during blood coagulation. p. 329

fibroblast (fi′bro-blast) Cell that produces fibers in connective tissues. p. 102

fibrocartilage (fi′bro kar′tĭ-lij) Strongest and most durable cartilage; made up of cartilage cells and many collagenous fibers. p. 108

fibrous joint (fi′brus joint) Two or more bones joined by dense connective tissue. p. 164

filtration (fil-tra′shun) Movement of material through a membrane as a result of hydrostatic pressure. p. 64

fissure (fish'ur) Narrow cleft separating parts, such as the lobes of the cerebrum. p. 237

flagellum (fla-jel'um) Relatively long, motile process that extends from the surface of a sperm cell. p. 58

flexion (flek'shun) Bending at a joint that decreases the angle between bones. p. 167

follicle-stimulating hormone (fol'ĭ-kl stim'u-la"ting hor'mōn) **(FSH)** Hormone secreted by the anterior pituitary to stimulate development of an ovarian follicle in a female or sperm cell production in a male. p. 300

follicular cell (fŏ-lik'u-lar sel) Ovarian cell that surrounds a developing oocyte and secretes female sex hormones. p. 516

fontanel (fon"tah-nel') Membranous region between certain developing cranial bones in the skull of a fetus or infant. p. 149

foramen magnum (fo-ra'men mag'num) Opening in the occipital bone of the skull through which the spinal cord passes. p. 144

foramen ovale (fo-ra'men o-val'e) Opening in the interatrial septum of the fetal heart. p. 551

fovea centralis (fo've-ah sen-tral'is) Depressed region of the retina, consisting of densely packed cones, that provides the greatest visual acuity. p. 282

free nerve endings (frē nerv end-ingz) Receptors abundant in epithelium, associated with sensing temperature and pain. p. 264

frontal (frun'tal) Plane that divides a structure into anterior and posterior portions, pertaining to the forehead. p. 15

FSH Follicle-stimulating hormone. p. 300

functional syncytium (funk'shun-al sin-sish'e-um) Merging cells that function as a unit; those of the heart are joined electrically. p. 349

G

gallbladder (gawl'blad-er) Saclike organ associated with the liver that stores and concentrates bile. p. 416

ganglion (gang'glē-on) (plural, *ganglia*) Mass of neuron cell bodies, usually outside the central nervous system. p. 220

gastric gland (gas'trik gland) Gland in the stomach wall that secretes gastric juice. p. 411

gastric juice (gas'trik jōōs) Secretion of the gastric glands in the stomach. p. 411

gastrin (gas'trin) Hormone secreted by the stomach lining that stimulates gastric juice secretion. p. 412

gene (jēn) Part of a DNA molecule that encodes information to synthesize a protein, a control sequence, or tRNA or rRNA; the unit of inheritance. p. 84

genetic code (jě-net'ik kōd) Information for synthesizing proteins encoded in the nucleotide sequence of DNA molecules. p. 83

genome (je'nōm) Complete set of genetic instructions for an organism. p. 84

genotype (jē-no-tīp) The alleles (gene variants) of a particular gene in an individual. p. 556

GH Growth hormone. p. 298

globulin (glob'u-lin) Type of protein in blood plasma. p. 329

glomerular capsule (glo-mer'u-lar kap'sūl) Double-walled enclosure of the glomerulus of a nephron; Bowman's capsule. p. 470

glomerulus (glo-mer'u-lus) Filtering capillary tuft in the glomerular capsule of a nephron. p. 470

glottis (glot'is) Slitlike opening between the true vocal folds or vocal cords. p. 446

glucagon (gloo'kah-gon) Hormone secreted by the pancreatic islets that releases glucose from glycogen. p. 308

glucocorticoid (gloo"ko-kor'tĭ-koid) Any of several hormones that the adrenal cortex secretes that affects carbohydrate, fat, and protein metabolism. p. 306

glucose (gloo'kōs) Monosaccharide in blood that is the primary source of cellular energy. p. 429

gluteal (gloo'te-al) Pertaining to the buttocks. p. 17

glycerol (glis'er-ol) Organic compound that is a building block for fat molecules. p. 42

glycogen (gli'ko-jen) Polysaccharide that stores glucose in the liver and muscles. p. 429

glycolysis (gli-kol'ĭ-sis) The energy-releasing breakdown of glucose to produce 2 pyruvic acid molecules and a net of 2 ATP. p. 80

goblet cell (gob'let sel) Epithelial cell specialized to secrete mucus in the respiratory tract and intestines. p. 97

Golgi apparatus (gol'je ap"ah-ra'tus) Organelle that prepares and modifies cellular products for secretion. p. 55

gonadotropin (go-nad"o-trōp'in) Hormone that stimulates activity in the gonads (testes and ovaries). p. 513

granulocyte (gran'u-lo-sīt) Leukocyte with granules in the cytoplasm. p. 324

growth hormone (grōth hor'mōn) **(GH)** Hormone released by the anterior pituitary that promotes growth of the organism; somatotropin. p. 298

gyrus (ji'rus) Elevation on the brain's surface; convolution. p. 237

H

hair follicle (hār fol'ĭ-kl) Tubelike depression in the epidermis in which a hair develops. p. 122

haploid (hap'loyd) Sex cell with half of the normal number of chromosomes, in humans 23. p. 509

hapten (hap'ten) Small molecule that combines with a larger one, forming an antigen. p. 386

hematocrit (he-mat'o-krit) The percentage by volume of red blood cells in a sample of whole blood. p. 319

hematopoiesis (he"-mă-to-poi-e'sis) Production of blood cells from dividing stem and progenitor cells. p. 140

heme (hēm) Iron-containing part of a hemoglobin molecule. p. 323

hemoglobin (he"mo-glo'bin) Oxygen-carrying pigment in red blood cells. p. 140

hemostasis (he"mo-sta'sis) Stoppage of bleeding. p. 330

hepatic lobule (hě-pat'ik lob'ul) Functional unit of the liver. p. 415

heterozygous (het'er-o-zi'gus) Different alleles in a gene pair. p. 556

holocrine gland (ho'lo-krin gland) Gland whose secretion contains entire secretory cells. p. 101

homeostasis (ho"me-o-sta'sis) Dynamic state in which the body's internal environment is maintained in the normal range. p. 5

homozygous (ho'mo-zi'gus) Identical alleles in a gene pair. p. 556

hormone (hor'mōn) Substance secreted by an endocrine gland and transported in the blood which has actions on target cells. p. 292

humoral immune response (hu'mor-al i-mūn' ri-spons') Circulating antibodies' destruction of cells bearing nonself antigens. p. 391

hyaline cartilage (hī-ah-lin kar'ti-lij) Semitransparent, flexible connective tissue with an ultra-fine fiber matrix. p. 107

hydrogen bond (hi'dro-jen bond) Weak bond between a hydrogen atom and an atom of oxygen or nitrogen between molecules or different regions of a very large molecule. p. 36

hydrolysis (hi-drol′ĭ-sis) Enzymatically adding a water molecule to split a molecule. p. 78

hydrostatic pressure (hy″dro-stat′ik presh′ur) Pressure exerted by fluids, such as blood pressure. p. 64

hymen (hi′men) Membranous fold of tissue that partially covers the vaginal opening. p. 520

hypertonic (hi″per-ton′ik) Solution with a greater osmotic pressure than the solution (usually body fluids) to which it is compared. p. 64

hyperventilation (hi″per-ven″tĭ-la′shun) Deep and rapid breathing that lowers blood CO_2 levels. p. 458

hypochondriac regions (hi″po-kon′dre-ak re′jun) Portion of the abdomen on either side of the upper middle or epigastric region. p. 16

hypogastric region (hi″po-gas′trik re′jun) Lower middle portion of the abdomen. p. 16

hypothalamus (hi″po-thal′ah-mus) Part of the brain below the thalamus and forming the floor of the third ventricle. p. 242

hypotonic (hi″po-ton′ik) Solution with a lower osmotic pressure than the solution (usually body fluids) to which it is compared. p. 64

I

iliac region (il′e-ak re′jun) Portion of the abdomen on either side of the lower middle or hypogastric region. p. 16

ilium (il′e-um) One of the bones making up the hip bone. p. 158

immunity (ĭ-mu′nĭ-te) Resistance to the effects of specific disease-causing agents. p. 384

immunoglobulin (im″u-no-glob′u-lin) Globular plasma protein that functions as an antibody. p. 390

incomplete protein (in″kom-plēt′ pro′te-in) Protein with inadequate amounts of essential amino acids. p. 431

inferior (in-fer′e-or) Situated below something else; pertaining to the lower surface of a part. p. 14

inflammation (in″flah-ma′shun) Tissue response to stress that dilates blood vessels and accumulates fluid in the affected region. p. 125

inguinal (ing′gwĭ-nal) Pertaining to the groin region. p. 17

innate defense (in′ate dē-fens) Inborn, nonspecific defense that blocks entry of or destroys pathogens. p. 384

inorganic (in″or-gan′ik) Chemical substances that do not include carbon and hydrogen atoms. p. 37

insertion (in-ser′shun) End of a muscle attached to a movable part. p. 192

inspiration (in″spĭ-ra′shun) Breathing in; inhalation. p. 450

insula (in′su-lah) Cerebral lobe deep within the lateral sulcus. p. 238

insulin (in′su-lin) Hormone the pancreatic islets secrete that stimulates cells to take up glucose. p. 308

integumentary system (in-teg-u-men′tar-e sis′tem) The skin and its accessory structures. p. 12

intercalated disc (in-ter″kah-lāt′ed disk) Membranous boundary between cardiac muscle cells. p. 111

internal environment (in-ter′nĕl en-vi-ruhment) Conditions and elements that make up the inside of the body, surrounding the cells. p. 5

interneuron (in″ter-nu′ron) Neuron between a sensory neuron and a motor neuron; internuncial or association neuron. p. 220

interphase (in′ter-fāz) Period between cell divisions when a cell metabolizes and prepares to divide. p. 69

interstitial cell (in″ter-stish′al sel) Hormone-secreting cell between seminiferous tubules of the testis. p. 506

intervertebral disc (in″ter-ver′tĕ-bral disk) Layer of fibrocartilage between bodies of adjacent vertebrae. p. 142

intestinal gland (in-tes′tĭ-nal gland) Tubular gland at the base of a villus in the intestinal wall. p. 421

intestinal villus (in-tes′tĭ-nal vil′us) (plural, *intestinal villi*) Fingerlike extension of the small intestinal lining. p. 420

intracellular fluid (in″trah-sel′u-lar floo′id) The liquid portion of the cell; cytosol. p. 490

intramembranous bone (in″trah-mem′brah-nus bōn) Bone that forms from membranelike layers of primitive connective tissue. p. 135

intrinsic factor (in-trin′sik fak′tor) Substance produced by the gastric glands necessary for absorption of vitamin B12. p. 411

inversion (in-ver′zhun) Turning the sole of the foot inward toward the midline. p. 167

ion (i′on) Electrically charged atom or molecule. p. 33

ionic bond (i-on′ik bond) Chemical bond that results when two ions form by transfer of electrons. p. 35

iris (i′ris) Colored, muscular part of the eye around the pupil that regulates its size. p. 282

isotonic (i″so-ton′ik) Solution with the same osmotic pressure as the solution (usually body fluids) to which it is compared. p. 64

isotope (i′so-tōp) Atom that has the same number of protons as other atoms of the same element but a different number of neutrons in its nucleus. p. 33

J

joint (joy′-nt) Union of two or more bones; articulation. p. 164

juxtaglomerular apparatus (juks″tah-glo-mer′u-lar ap″ah-ra′tus) A group of cells in the wall of the afferent arteriole in the kidney that plays a role in the control of renin secretion. p. 471

K

keratin (ker′ah-tin) Intracellular protein in epidermis, hair, and nails. p. 98

keratinization (ker″ah-tin′i-za′shun) Process by which cells form fibrils of keratin and harden. p. 119

ketone body (ke′tōn bod′e) Compound produced during fat catabolism. p. 429

Kupffer cell (koop′fer sel) Large, fixed phagocyte in the liver that removes bacterial cells from the blood. p. 370

L

labyrinth (lab′ĭ-rinth) System of connecting tubes in the inner ear, including the cochlea, vestibule, and semicircular canals. p. 272

lacrimal gland (lak′rĭ-mal gland) Tear-secreting gland. p. 277

lactase (lak′tās) Enzyme that catalyzes the breakdown of lactose into glucose and galactose. p. 422

lacteal (lak′te-al) Lymphatic capillary associated with a villus of the small intestine. p. 378

lactic acid (lak′tik as′id) Organic compound formed from pyruvic acid in the anaerobic pathway of cellular respiration. p. 185

lacuna (lah-ku′nah) Hollow cavity. p. 107

lamella (lah-mel′ah) Layer of matrix surrounding the central canal of an osteon. p. 108

laryngopharynx (lah-ring″go-far′ingks) Lower part of the pharynx posterior to the larynx. p. 409

larynx (lar′inks) Structure superior to the trachea that houses the vocal cords. p. 445

latent period (la′tent pe′re-od) Time between application of a stimulus and the beginning of a response in a muscle fiber. p. 189

lateral (lat′er-al) Pertaining to the side. p. 14

leukocyte (lu′ko-sīt) White blood cell. p. 324

lever (lev′er) Simple mechanical device consisting of a rod, fulcrum, weight, and a source of energy that is applied to some point on the rod. p. 140

ligament (lig′ah-ment) Cord or sheet of connective tissue binding two or more bones at a joint. p. 103

limbic system (lim′bik sis′tem) Connected structures in the brain that produce emotional feelings. p. 243

lingual frenulum (ling′gwahl fren′u-lum) Fold of tissue that anchors the tongue to the floor of the mouth. p. 403

lipase (li′pās) Fat-digesting enzyme. p. 423

lipid (lip′id) Chemical group that includes fats (triglycerides), steroids, and phospholipids. p. 42

lumbar (lum′bar) Pertaining to the region of the lower back. p. 16

lumen (lu′men) Space in a tubular structure such as a blood vessel or intestine. p. 96

luteinizing hormone (lu′te-in-īz″ing hor′mōn) **(LH)** A hormone that the anterior pituitary secretes that controls formation of the corpus luteum in females and testosterone secretion in males. p. 300

lymph (limf) Fluid carried in lymphatic vessels. p. 379

lymphatic pathway (lim-fat′ik path′wā) Connected vessels that transport lymph. p. 378

lymph node (limf nōd) Mass of lymphoid tissue located along the course of a lymphatic vessel. p. 379

lymphocyte (lim′fo-sīt) Type of white blood cell that provides immunity; B cell or T cell. p. 325

lysosome (li′so-sōm) Organelle that contains digestive enzymes. p. 56

M

macromolecule (mak-rō-mol′ĕ-kūl) Very large molecule, such as protein or nucleic acid. p. 3

macronutrient (mak-rō-nu′tree-ent) Nutrient (carbohydrate, lipid, and protein) required in large amount. p. 428

macrophage (mak′ro-fāj) Large phagocytic cell. p. 102

macula lutea (mak′u-lah lu′te-ah) Yellowish depression in the retina associated with acute vision. p. 282

malnutrition (mal″nu-trish′un) Symptoms resulting from lack of specific nutrients. p. 436

mammary (mam′er-e) Pertaining to the breast. p. 17

marrow (mar′o) Connective tissue in spaces in bones that includes blood-forming stem and progenitor cells. p. 134

mast cell (mast sel) Cell to which antibodies, formed in response to allergens attach, bursting the cell and releasing allergy mediators, which cause symptoms. p. 103

matter (mat′er) Anything that has weight and occupies space. p. 31

mechanoreceptor (mek″ah-no-re-sep′tor) Sensory receptor sensitive to mechanical stimulation, such as changes in pressure or tension. p. 263

medial (me′de-al) Toward or near the midline. p. 14

mediastinum (me″de-as-ti′num) Tissues and organs of the thoracic cavity that form a septum between the lungs. p. 8

medulla oblongata (mĕ-dul′ah ob″long-gah′tah) Part of the brainstem between the pons and the spinal cord. p. 243

medullary cavity (med′u-lār″e kav′ĭ-te) Cavity containing red or yellow marrow within the diaphysis of a long bone. p. 134

megakaryocyte (meg″ah-kar′e-o-sīt) Large cell in red bone marrow that fragments to yield blood platelets. p. 327

meiosis (mi-o′sis) Cell division that halves the genetic material, resulting in egg or sperm. p. 507

melanin (mel′ah-nin) Dark pigment normally found in skin and hair. p. 119

melanocyte (mel′ah-no-sīt) Melanin-producing cell. p. 119

melatonin (mel″ah-to′nin) Hormone that the pineal gland secretes. p. 309

memory cell (mem′o-re sel) B cell or T cell produced in the primary immune response that can be activated rapidly if the same nonself antigen is encountered in the future. p. 388

menarche (mĕ-nar′ke) A female's first reproductive cycle. p. 523

meninx (me′ninks) (plural, *meninges*) Any of the 3 layers of membrane that cover the brain and spinal cord. p. 232

meniscus (mĕ-nis′kus) (plural, *menisci*) Fibrocartilage that separates the articulating surfaces of bones in the knee. p. 164

menopause (men′o-pawz) Cessation of the female reproductive cycles. p. 524

merocrine gland (mer′o-krin gland) A structure whose cells remain intact while secreting. p. 101

mesentery (mes′en-ter″e) Fold of peritoneal membrane that attaches an abdominal organ to the abdominal wall. p. 420

mesoderm (mez′o-derm) Middle primary germ layer of the embryo. p. 543

messenger RNA (mes′in-jer) **(mRNA)** RNA that transmits information for a protein's amino acid sequence from the nucleus to the cytoplasm. p. 86

metabolic pathway (mĕ-tab′o-lik path-wa) Series of linked, enzymatically controlled chemical reactions. p. 82

metabolism (mĕ-tab′o-lizm) The combined chemical reactions in cells that use or release energy. p. 4

metacarpal (met″ah-kar′pal) Bone of the hand between the wrist bones and finger bones. p. 143

metaphase (met′ah-fāz) Stage in mitosis when chromosomes align in the middle of the cell. p. 69

metatarsal (met″ah-tar′sal) Foot bone between the ankle bones and the toe bones. p. 143

microfilament (mi″kro-fil′ah-ment) Rod of actin protein in the cytoplasm that provides structural support or movement. Part of the cytoskeleton. p. 57

microglia (mi-krog′le-ah) Neuroglia of the CNS that support neurons and phagocytize. p. 216

micronutrient (mi-kro-nu′tree-ent) Nutrient (vitamins and minerals) required in small amount. p. 428

microtubule (mi″kro-tu′būl) Hollow rod constructed of many molecules of the protein tubulin. Part of the cytoskeleton. p. 57

microvillus (mi″kro-vil′us) Tiny, cylindrical process that extends from some epithelial cell membranes, increasing membrane surface area. p. 97

micturition (mik″tu-rish′un) Urination. p. 483

midbrain (mid′brān) Small region of the brainstem between the diencephalon and the pons. p. 243

mineral (min′er-al) Inorganic element essential in human metabolism. p. 433

mineralocorticoid (min″er-al-o-kor′tĭ-koid) Hormone the adrenal cortex secretes that affects electrolyte concentrations in body fluids. p. 306

mitochondrion (mi″to-kon′dre-on) (plural, *mitochondria*) Organelle housing enzymes that catalyze the aerobic reactions of cellular respiration. p. 56

mitosis (mi-to′sis) Division of a somatic cell, forming two genetically identical somatic cells. p. 69

mitral valve (mi′trul valv) Heart valve between the left atrium and the left ventricle; bicuspid valve. p. 344

mixed nerve (mikst nerv) Nerve that includes both sensory and motor nerve fibers. p. 230

molecular formula (mo-lek′u-lar fŏr′mu-lah) Abbreviation for the number of atoms of each element in a compound. p. 37

molecule (mol′ĕ-kūl) Particle composed of two or more joined atoms. p. 3

monocyte (mon′o-sīt) Type of white blood cell that is a phagocyte. p. 325

monosaccharide (mon″o-sak′ah-rīd) Simple sugar, such as glucose or fructose. p. 41

motor area (mo′tor a′re-ah) Region of the brain that sends impulses to muscles or glands. p. 239

motor end plate (mo′tor end plāt) Specialized part of a muscle fiber membrane at a neuromuscular junction. p. 182

motor nerve (mo′tor nerv) Nerve that consists of motor nerve fibers. p. 230

motor neuron (mo′tor nu′ron) Neuron that transmits impulses from the central nervous system to an effector. p. 182

motor speech area (mo′tor spēch ār′e-ah) Region of the frontal lobe that coordinates complex muscular actions of mouth, tongue, and larynx, making speech possible; Broca's area. p. 239

motor unit (mo′tor u′nit) A motor neuron and its associated muscle fibers. p. 190

mucosa (mu-ko′sah) Innermost layer of the alimentary canal. p. 401

mucous cell (mu′kus sel) Glandular cell that secretes mucus. p. 408

mucous membrane (mu′kus mem-brān) Membrane that lines cavities and tubes that open to the outside of the body. p. 110

mucus (mu′kus) Fluid secretion of the mucous cells. p. 97

muscle tissue (mus′el tish′u) Contractile tissue consisting of filaments of actin and myosin, which slide past each other, shortening cells. p. 110

myelin (mi′ĕ-lin) Lipid material that forms a sheathlike covering around some axons. p. 216

myelin sheath (mi′ĕ-lin shēth) Lipid layer formed from certain neuroglia that surrounds an axon, providing insulation. p. 216

myocardium (mi″o-kar′de-um) Muscle tissue of the heart. p. 343

myofibril (mi″o-fi′bril) Contractile fiber in muscle cells. p. 179

myoglobin (mi″o-glo′bin) Pigmented compound in muscle tissue that stores oxygen. p. 185

myometrium (mi″o-me′tre-um) Layer of smooth muscle tissue in the uterine wall. p. 520

myosin (mi′o-sin) Protein that, with actin, forms the filaments that interact to contract muscle fibers. p. 179

N

nasal cavity (na′zal kav′ĭ-te) Space in the nose. p. 8

nasal concha (na′zal kong′kah) Shell-like bone extending outward from the wall of the nasal cavity; turbinate bone. p. 443

nasal septum (na′zal sep′tum) Wall of bone and cartilage that separates the nasal cavity into two parts. p. 443

nasopharynx (na″zo-far′inks) Part of the pharynx posterior to the nasal cavity. p. 409

negative feedback (neg′ah-tiv fēd′bak) A mechanism that restores the level of a biochemical or other balance in the internal environment p. 6

neonatal (ne″o-na′tal) The first four weeks after conception. p. 553

nephron (nef′ron) Functional unit of a kidney, consisting of renal corpuscle and renal tubule. p. 469

nervous tissue (ner′vus tish′u) Neurons and neuroglia. p. 111

neurilemma (nu″rĭ-lem′ah) Sheath formed from Schwann cells on the exterior of some axons. p. 217

neurofibril (nu″ro-fi′bril) Fine, cytoplasmic thread that extends from the cell body into the processes of a neuron. p. 216

neuroglia (nu-rog′le-ah) Specialized cells of the nervous system that produce myelin, communicate between cells, maintain the ionic environment, and nurture the differentiation of neurons. p. 112

neuromuscular junction (nu″ro-mus′ku-lar jungk′shun) Synapse between a

motor neuron and a skeletal muscle fiber. p. 182

neuron (nu′ron) Nerve cell. p. 111

neurotransmitter (nu″ro-trans′mit-er) Chemical that an axon secretes that stimulates or inhibits an effector (muscle or gland) or other neuron. p. 222

neutral (nu′tral) Neither acidic nor alkaline; pH7. p. 39

neutron (nu′tron) Electrically neutral subatomic particle. p. 32

neutrophil (nu′tro-fil) A type of phagocytic leukocyte. p. 325

nonelectrolyte (non″e-lek′tro-līt) Substance that does not dissociate into ions when dissolved in water. p. 39

nonprotein nitrogenous substance (non-pro′te-in ni-troj′ĕ-nus sub′stans) A substance, such as urea or uric acid, that contains nitrogen but is not a protein. p. 330

norepinephrine (nor″ep-ĭ-nef′rin) Neurotransmitter released from the axons of some nerve cells. p. 254

nuclear envelope (nu′kle-er ahn-veh-lop) Double membrane surrounding the cell nucleus and separating it from the cytoplasm. p. 60

nucleic acid (nu-kle′ik as′id) A molecule that is a polymer of nucleotides; RNA or DNA. p. 44

nucleolus (nu-kle′o-lus) (plural, *nucleoli*) Small structure in the cell nucleus that contains RNA and proteins. p. 60

nucleotide (nu′kle-o-tīd″) Building block of a nucleic acid molecule, consisting of a sugar, a nitrogenous base, and a phosphate group. p. 44

nucleus (nu′kle-us) (plural, *nuclei*) The dense core of an atom, composed of protons and neutrons, p. 32. Cellular organelle enclosed by double-layered, porous membrane and containing DNA, p. 52. Masses of interneuron cell bodies in the central nervous system. p. 220.

nutrient (nu′tre-ent) Chemical that the body requires from the environment. p. 428

O

occipital (ok-sip′ĭ-tal) Pertaining to the lower, back part of the head. p. 17

olfactory (ol-fak′to-re) Pertaining to the sense of smell. p. 267

olfactory nerves (ol-fak′to-re nervz) The first pair of cranial nerves, which conduct impulses associated with the sense of smell. p. 246

oligodendrocyte (ol″ĭ-go-den′dro-sīt) Type of neuroglia that produces myelin in the CNS. p. 216

oocyte (o′o-sīt) Cell formed by oogenesis. Egg cell. p. 506

oogenesis (o″o-jen′ĕ-sis) Formation of an oocyte (egg cell). p. 517

optic chiasma (op′tik ki-az′mah) X-shaped structure on the underside of the brain formed by optic nerve fibers (axons) that partially cross over. p. 242

optic disc (op′tik disk) Region in the retina where nerve fibers (axons) exit, becoming part of the optic nerve. p. 284

oral (o′ral) Pertaining to the mouth. p. 8

organ (or′gan) Structure consisting of a group of tissues with a specialized function. p. 3

organelle (or″gah-nel′) A structure in a cell that has a specialized function. p. 3

organic (or-gan′ik) Chemicals that contain both carbon and hydrogen. p. 39

organism (or′gah-nizm) An individual living thing. p. 3

organ system (or′gan sis′tem) Group of organs coordinated to carry on a specialized function. p. 3

orgasm (or′gazm) An intense sensation that is the peak of sexual excitement. p. 512

origin (or′ĭ-jin) End of a muscle that attaches to a relatively immovable part. p. 192

oropharynx (o″ro-far′inks) Part of the pharynx posterior to the oral cavity. p. 409

osmosis (oz-mo′sis) Movement of water through a semipermeable membrane toward a concentration of an impermeant solute. p. 63

osmotic pressure (oz-mot′ik presh′ur) Pressure needed to stop osmosis; a solution's potential pressure caused by impermeant solute particles in the solution. p. 64

ossification (os″ĭ-fĭ-ka′shun) Formation of bone tissue. p. 136

osteoblast (os′te-o-blast″) Bone-forming cell. p. 135

osteoclast (os′te-o-klast″) Cell that breaks down bone. p. 136

osteocyte (os′te-o-sīt) Mature bone cell. p. 108

osteon (os′te-on) Cylinder-shaped unit containing bone cells that surround a central canal; Haversian system. p. 108

oval window (o′val win′do) Connection between the stapes and the inner ear. p. 271

ovary (o′var-e) Primary female reproductive organ that produces oocytes. p. 300

ovulation (o″vu-la′shun) Release of an egg cell from a mature ovarian follicle. p. 319

oxidation (ok″sĭ-da′shun) Process by which oxygen combines with another chemical; removal of hydrogen or the loss of electrons; opposite of reduction. p. 80

oxygen debt (ok′sĭ-jen det) Amount of oxygen required following physical exercise to react accumulated lactic acid to form glucose. p. 186

oxyhemoglobin (ok″sĭ-he″mo-glo′bin) Compound formed when oxygen binds to hemoglobin. p. 319

oxytocin (ok″sĭ-to′sin) Hormone released by the posterior pituitary that contracts smooth muscles in the uterus and mammary glands. p. 300

P

pacemaker (pās′māk-er) Mass of specialized cardiac muscle tissue that controls the rhythm of the heartbeat; sinoatrial node. p. 350

pain receptor (pān re″sep′tor) Sensory nerve ending that transmits impulses interpreted as pain. p. 263

palate (pal′at) Roof of the mouth. p. 404

palatine (pal′ah-tīn) Pertaining to the palate. p. 404

palmar (pahl′mar) Pertaining to the palm of the hand. p. 17

pancreas (pan′kre-as) Glandular organ in the abdominal cavity that secretes hormones and digestive enzymes. p. 306

pancreatic juice (pan″kre-at′ik joōs) Digestive secretions of the pancreas. p. 413

papilla (pah-pil′ah) Tiny, nipplelike projection. p. 404

papillary muscle (pap′ĭ-ler″e mus′l) Muscle that extends inward from the ventricular walls of the heart and to which the chordae tendineae attach. p. 344

paranasal sinus (par″ah-na′zal si-nus) Air-filled cavity in a cranial or facial bone lined with mucous membrane and connected to the nasal cavity. p. 444

parasympathetic division (par″ah-sim″pah-thet′ik de-vijh′in) Part of the autonomic nervous system that arises from the brain and sacral region of the spinal cord. p. 250

parathyroid gland (par″ah-thi′roid gland) One of four small endocrine glands embedded in the posterior part of the thyroid gland. p. 303

parathyroid hormone (par″ah-thi′roid hor′mōn) **(PTH)** Hormone secreted by the parathyroid glands that helps regulate the levels of blood calcium and phosphate ions. p. 303

parietal (pah-ri′ĕ-tal) Pertaining to the wall of a cavity. p. 10

parietal cell (pah-ri′ĕ-tal sel) Cell of a gastric gland that secretes hydrochloric acid and intrinsic factor. p. 411

parietal pericardium (pah-ri′ĕ-tal per″ĭ-kar′de-um) Membrane that forms the outer wall of the pericardial cavity. p. 342

parietal peritoneum (pah-ri′ĕ-tal per″ĭ-to-ne′-um) Membrane that forms the outer wall of the peritoneal cavity. p. 10

parietal pleura (pah-ri′ĕ-tal ploo′rah) Membrane that forms the outer wall of the thoracic cavity. p. 450

parotid gland (pah-rot′id gland) Large salivary gland on the side of the face just in front and below the ear. p. 408

partial pressure (par′shal presh′ur) Pressure one gas produces in a mixture of gases. p. 459

pathogen (path′o-jen) Disease-causing agent. p. 384

pectoral (pek′tor-al) Pertaining to the chest. p. 17

pectoral girdle (pek′tor-al ger′dl) Part of the skeleton that supports and attaches the upper limbs. p. 143

pedigree (ped′eh-gree) Chart that displays relationships among family members and their inherited traits and disorders. p. 560

pelvic cavity (pel′vik kav′ĭ-te) Hollow place within the ring formed by the sacrum and hip bones that encloses the terminal part of the large intestine, the urinary bladder, and the internal reproductive organs. p. 8

pelvic girdle (pel′vik ger′dl) Part of the skeleton to which the lower limbs attach. p. 143

pelvis (pel′vis) Bony ring formed by the sacrum and hip bones (bony pelvis); the pelvic cavity. p. 158

penis (pe′nis) Male external reproductive organ through which the urethra passes. p. 512

pepsin (pep′sin) Protein-splitting enzyme that the gastric glands secrete. p. 411

pepsinogen (pep-sin′o-jen) Inactive form of pepsin. p. 411

pericardial cavity (per″ĭ-kar′de-al kav′ĭ-te) Potential space between the visceral and parietal pericardial membranes. p. 342

pericardium (per″ĭ-kar′de-um) Serous membrane that surrounds the heart. p. 342

perichondrium (per″ĭ-kon′dre-um) Layer of dense connective tissue that encloses cartilaginous structures. p. 107

perilymph (per′ĭ-limf) Fluid in the space between the membranous and osseous labyrinths of the inner ear. p. 272

perimetrium (per-ĭ-me′tre-um) Outer serosal layer of the uterine wall. p. 520

perimysium (per″ĭ-mis′e-um) Sheath of connective tissue that encloses a bundle of skeletal muscle fibers. p. 179

periodontal ligament (per″e-o-don′tal lig′ah-ment) Fibrous membrane around a tooth that attaches it to the jawbone. p. 407

periosteum (per″e-os′te-um) Dense connective tissue covering the surface of a bone. p. 133

peripheral nervous system (pĕ-rif′er-al ner′vus sis′tem) **(PNS)** Parts of the nervous system outside the central nervous system. p. 214

peripheral resistance (pĕ-rif′er-al re-zis′ tans) Combined peripheral resistance to blood flow due to friction between the blood and blood vessel walls. p. 360

peristalsis (per″ĭ-stal′sis) Rhythmic waves of muscular contraction in the walls of certain tubular organs. p. 191

peritoneal cavity (per″ĭ-to-ne′al kav′ĭ-te) Potential space between the visceral and parietal peritoneal membranes. p. 10

peritubular capillary (per″ĭ-tu′bu-lar kap′ĭ-ler″e) Capillary that surrounds a renal tubule and functions in tubular reabsorption and tubular secretion during urine formation. p. 471

peroxisome (pĕ-roks′ĭ-sōm) Membranous sac abundant in kidney and liver cells that contains enzymes that catalyze reactions that decompose hydrogen peroxide. p. 56

pH (pH) A shorthand system that indicates the acidic or alkaline condition of a solution; values range from 0 to 14. Technically the negative log of the hydrogen ion concentration. p. 39

phagocytosis (fag″o-si-to′sis) Process by which a cell engulfs and digests solids. p. 66

phalanx (fa′langks) (plural, *phalanges*) Bone of a finger or toe. p. 143

pharynx (far′inks) Part of the digestive tube posterior to the nasal and oral cavities, as well as the larynx. p. 409

phenotype (fe′no-tīp) The expression of a gene or genes. p. 557

pheomelanin (fe″o-mel′ah-nīn) A reddish-yellow pigment. p. 123

phospholipid (fos″fo-lip′id) Molecule consisting of two fatty acid molecules and a phosphate group bound to a glycerol molecule. p. 42

photoreceptor (fo″to-re-sep′tor) Sensory receptor sensitive to light energy; rods and cones of the eyes. p. 263

physiology (fiz″e-ol′o-je) The study of body functions. p. 3

pia mater (pi′ah ma′ter) Inner layer of meninges that encloses the brain and spinal cord. p. 234

pineal gland (pin′e-al gland) A small structure in the central brain that secretes the hormone melatonin, which controls certain biological rhythms. p. 242

pinocytosis (pi″no-si-to′sis) Process by which a cell engulfs droplets of fluid from its surroundings. p. 66

pituitary gland (pĭ-tu′ĭ-tār″e gland) Endocrine gland attached to the base of the brain consisting of anterior and posterior lobes. p. 297

placenta (plah-sen′tah) Structure that attaches the fetus to the uterine wall, delivering nutrients to and removing wastes from the fetus. p. 310

plantar (plan′tar) Pertaining to the sole of the foot. p. 17

plasma (plaz′mah) Fluid portion of the blood. p. 319

plasma cell (plaz′mah sel) Antibody-producing cell that forms when activated B cells proliferate. p. 390

plasma protein (plaz′mah pro′te-in) Protein dissolved in blood plasma. p. 329

platelet (plāt′let) Cellular fragment formed in red bone marrow that helps blood clot. p. 108

pleural cavity (ploo′ral kav′ĭ-te) Potential space between pleural membranes. p. 450

pleural membrane (ploo′ral mem′brān) Serous membrane that encloses the lungs and lines the chest wall. p. 10

plexus (plek′sus) Network of interlaced nerves or blood vessels. p. 249

polar body (pō′lar bod′e) Small, nonfunctional cell that is a product of meiosis in the female. p. 35

polarization (po″lar-ĭ-za′shun) Voltage difference across a cell membrane due to an unequal distribution of

positive and negative ions on either side of the membrane. p. 222

polysaccharide (pol″e-sak′ah-rīd) Carbohydrate composed of many joined monosaccharides. p. 41

polyunsaturated fatty acid (pol″e-un-sach′ĕ-ra-ted fat′e as′id) Fatty acid with more than one double carbon bond. p. 42

pons (ponz) Part of the brainstem above the medulla oblongata and below the midbrain. p. 243

popliteal (pop″lĭ-te′al) Pertaining to the region behind the knee. p. 17

positive feedback system (poz′ĭ-tiv fēd′bak sis′tem) Process by which changes cause additional similar changes, producing unstable conditions. p. 331

posterior (pos-tēr′e-or) Toward the back; the opposite of *anterior*. p. 14

posterior pituitary (pos-tēr′e-or pĭ-tu′ĭ-tār″e) The lobe of the pituitary gland that secretes oxytocin and antidiuretic hormone (vasopressin). p. 297

postganglionic fiber (pōst″gang-gle-on′ik fi′ber) Axon of an autonomic neuron on the distal side of a ganglion. p. 251

postnatal (pōst-na′tal) After birth. p. 537

preganglionic fiber (pre″gang-gle-on′ik fi′ber) Axon of an autonomic neuron on the proximal side of a ganglion. p. 251

pregnancy (preg′nan-se) Condition in which a female has a developing offspring in her uterus. p. 537

prenatal (pre-na′tal) Before birth. p. 537

primary follicle (pri′ma-re fol′ĭ-kl) Primordial follicle that begins to mature in response to hormonal changes in a female. p. 518

primary germ layers (pri′mar-e jerm la′erz) Three layers of cells in the embryo that divide and differentiate into specific tissues and organs; ectoderm, mesoderm, and endoderm. p. 543

primary sex organs (pri′ma-re seks or′ganz) Sex-cell-producing parts; testes in males and ovaries in females. p. 506

prime mover (prīm moov′er) Muscle that provides a particular body movement. Also called an agonist. p. 194

progenitor cell (pro-jen′ĭ-tor sel) Daughter cell of a stem cell whose own daughter cells are restricted to follow specific lineages. p. 71

progesterone (pro-jes′tĭ-rōn) Female hormone secreted by the corpus luteum and placenta. p. 522

projection (pro-jek′shun) Process by which the brain causes a sensation to seem to come from the region of the body being stimulated. p. 263

prolactin (pro-lak′tin) **(PRL)** Hormone secreted by the anterior pituitary that stimulates milk production in the mammary glands. p. 300

pronation (pro-na′shun) Downward or inward rotation of the palm. p. 167

prophase (pro′fāz) Stage of mitosis when chromosomes become visible when stained and viewed under a microscope. p. 69

prostaglandins (pros″tah-glan′dins) A group of compounds with powerful, hormonelike effects. p. 296

prostate gland (pros′tāt gland) Gland surrounding the male urethra below the urinary bladder that secretes into semen prior to ejaculation. p. 510

protein (pro′tēn) Nitrogen-containing organic compound consisting of bonded amino acid molecules. p. 43

prothrombin (pro-throm′bin) Plasma protein that functions in blood clotting. p. 331

proton (pro′ton) Positively charged particle in an atomic nucleus. p. 32

protraction (pro-trak′shun) Forward movement of a body part. p. 167

proximal (prok′sĭ-mal) Closer to the point of attachment; opposite of *distal*. p. 15

pseudostratified columnar epithelium (soo″do-strat′ĭ-fīd co-lum′nar ep″ĭ-the′lē-um) Single layer of cells appearing as more than one layer because the nuclei occupy different positions in the cells. p. 98

PTH Parathyroid hormone. p. 303

puberty (pu′ber-te) Stage of development in which the reproductive organs become functional. p. 513

pulmonary circuit (pul′mo-ner″e ser′kit) System of blood vessels that carries blood between the heart and the lungs. p. 341

pulse (puls) Surge of blood felt through the walls of arteries due to the contraction of the heart ventricles. p. 359

Punnett square (pun-it sqware) A grid diagram that displays possible progeny that can be predicted based on parental gametes. p. 557

pupil (pu′pil) Opening in iris through which light enters the eye. p. 282

Purkinje fibers (pur-kin′je fi′berz) Specialized cardiac muscle fibers that conduct cardiac impulses from the AV bundle into the ventricular walls. p. 350

pyruvic acid (pi-roo′vik as′id) Intermediate product of carbohydrate oxidation. p. 185

R

radioactive (ra″de-o-ak′tiv) Property of some atoms that releases energy at a constant rate. p. 33

rate-limiting enzyme (rāt lim′i-ting en′zīm) Enzyme, usually present in small amounts, that controls the rate of a metabolic pathway by regulating one step. p. 82

receptor (re-sep′tor) Specialized cell that provides information about the environment. Also, cell membrane structure that binds specific molecules, called ligands, thereby sending a signal inside the cell. p. 6

recessive (re-sess′iv) A gene whose expression is masked by another (dominant). p. 557

recruitment (re-krōōt′ment) Increase in the number of motor units activated as stimulation intensity increases. p. 190

red marrow (red mar′o) Blood-cell-forming tissue in spaces within bones. p. 140

referred pain (re-ferd′ pān) Pain that feels as if it is originating from a part other than the site being stimulated. p. 265

reflex (re′fleks) A rapid, automatic response to a stimulus. p. 231

reflex arc (re′fleks ark) Nerve pathway, consisting of a sensory receptor, a sensory neuron, interneuron, motor neuron, and an effector that forms the structural and functional bases for a reflex. p. 231

refraction (re-frak′shun) Bending of light as it passes between media of different densities. p. 284

relaxin (re-lak′sin) Hormone from the corpus luteum that inhibits uterine contractions during pregnancy. p. 543

renal corpuscle (re′nal kor′pusl) Part of a nephron that consists of a glomerulus and a glomerular capsule. p. 470

renal cortex (re′nal kor′teks) Outer part of a kidney. p. 469

renal medulla (re′nal mĕ-dul′ah) Inner part of a kidney. p. 469

renal pelvis (re′nal pel′vis) Cavity in a kidney that channels urine to the ureter. p. 468

renal tubule (re′nal tu′būl) Part of a nephron that extends from the renal corpuscle to the collecting duct. p. 470

renin (re′nin) Enzyme that kidneys release that maintains blood pressure and blood volume. p. 476

replication (rep″lĭ-ka′shun) Copying of a DNA molecule. p. 84

respiration (res″pĭ-ra′shun) Breathing. p. 443

respiratory capacity (re-spi′rah-to″re kah-pas′ĭ-te) The sum of any two or more respiratory volumes. p. 455

respiratory membrane (re-spi′rah-to″re mem′brān) Membrane composed of a capillary wall, an alveolar wall, and their basement membranes through which blood and inspired air exchange gases. p. 459

resting potential (res′ting po-ten′shal) Difference in electrical charge between the inside and the outside of an undisturbed nerve cell membrane. p. 223

reticular fiber (rĕ-tik′u-lar fi′ber) Thin collagenous fiber. p. 103

reticular formation (rĕ-tik′u-lar for-ma′shun) Complex network of nerve fibers in the brainstem that arouses the cerebrum. p. 244

retina (ret′ĭ-nah) Inner layer of the eye wall that includes the visual receptors. p. 282

retraction (rĕ-trak′shun) Movement of a part toward the back. p. 167

retroperitoneal (ret″ro-per″i-to-ne′al) Behind the peritoneum. p. 468

reversible reaction (re-ver′sĭ-bl re-ak′shun) Chemical reaction in which the products react, reforming the reactants. p. 38

rhodopsin (ro-dop′sin) Light-sensitive pigment in the rods of the retina; visual purple. p. 286

ribonucleic acid (ri″bo-nu-kle′ik as′id) **(RNA)** Single stranded polymer of nucleotides in which each nucleotide includes the sugar ribose, a phosphate group, and a nitrogenous base (adenine, uracil, guanine, or cytosine). p. 46

ribosome (ri′bo-sōm) Organelle composed of RNA and protein that is a structural support for protein synthesis. p. 55

RNA Ribonucleic acid. p. 46

rod (rod) Type of light receptor that provides colorless (black and white) vision. p. 279

rotation (ro-ta′shun) Movement turning a body part on its longitudinal axis. p. 167

round window (rownd win′do) Membrane-covered opening between the inner ear and the middle ear. p. 273

S

saccule (sak′ūl) An enlarged region of the membranous labyrinth of the inner ear. p. 275

sagittal (saj′ĭ-tal) Plane or section that divides a structure into right and left portions. p. 15

salivary amylase (sal′i-ver-e am′i-lās) Enzyme that hydrolyzes (digests) starch in the mouth. p. 408

salt (salt) Compound produced by the reaction of an acid and a base. p. 40

SA node (nōd) Sinoatrial node. p. 340

sarcomere (sar′ko-mēr) Structural and functional unit of a myofibril. p. 181

sarcoplasmic reticulum (sar″ko-plaz′mik rĕ-tik′u-lum) Membranous network of channels and tubules of a muscle fiber, corresponding to the endoplasmic reticulum of other cells. p. 181

saturated fatty acid (sat′u-rāt″ed fat′e as′id) Fatty acid molecule that includes maximal hydrogens and therefore has no double carbon bonds. p. 42

Schwann cell (shwahn sel) Type of neuroglia that surrounds an axon of a peripheral neuron, forming the neurilemmal sheath and myelin. p. 216

sclera (skle′rah) White, fibrous outer layer of the eyeball. p. 280

scrotum (skro′tum) Pouch of skin that encloses the testes. p. 512

sebaceous gland (se-ba′shus gland) Skin gland that secretes sebum. p. 124

sebum (se′bum) Oily secretion of sebaceous glands. p. 124

secretin (se-kre′tin) Hormone from the small intestine that stimulates the pancreas to release pancreatic juice. p. 413

selectively permeable (se-lek′tiv-le per′me-ah-bl) Membrane that allows some types of molecules through but not others; semipermeable. p. 53

semen (se′men) Fluid containing sperm cells and secretions discharged from the male reproductive tract at ejaculation. p. 511

semicircular canal (sem″ĭ-ser′ku-lar kah-nal′) Tubular structure in the inner ear that houses receptors providing the sense of dynamic equilibrium. p. 272

seminiferous tubule (sem″ĭ-nif′er-us tu′būl) Tubule in the testes where sperm cells form. p. 506

sensation (sen-sa′shun) A feeling resulting from the brain's interpretation of sensory nerve impulses. p. 263

sensory area (sen′so-re a′re-ah) Part of the cerebral cortex that receives and interprets sensory nerve impulses. p. 238

sensory nerve (sen′so-re nerv) A nerve composed of sensory nerve fibers. p. 230

sensory neuron (sen′so-re nu′ron) Neuron that transmits impulses from sensory receptors to the central nervous system. p. 220

sensory receptor (sen′so-re re″sep′tor) Specialized structure associated with the peripheral end of a sensory neuron specific to detecting a particular sensation and triggering a nerve impulse in response. p. 214

sensory speech area (sen′so-re spēch ār′e-ah) Region of the parietal lobe and the temporal lobe just posterior to the lateral sulcus, that is necessary for understanding written and spoken language; Wernicke's area. p. 239

serosa (sĕ′ro-sah) Outer covering of the alimentary canal. p. 401

serotonin (se″ro-to′nin) Vasoconstrictor released from blood platelets when blood vessels break, controlling bleeding. Also a neurotransmitter. p. 267

serous cell (ser′us sel) Glandular cell that secretes a watery fluid (serous fluid) with high enzyme content. p. 408

serous membrane (ser′us mem′brān) Membrane that lines a cavity that does not open to the outside of the body. p. 110

set point (set′ point) Target value of a physiological measure maintained in the body by homeostasis. p. 6

sex chromosome (seks crō-mo-some) Chromosome that carries genes responsible for the development of characteristics associated with maleness or femaleness; an X or Y chromosome. p. 356

simple sugar (sim′pl shoog′ar) Monosaccharide. p. 41

sinoatrial node (si″no-a′tre-al nōd) **(SA node)** Specialized tissue in the wall of the right atrium that initiates cardiac cycles; pacemaker. p. 340

skeletal muscle tissue (skel′ĭ-tal mus′l tish′u) Type of voluntary muscle tissue in muscles attached to bones. p. 110

smooth muscle tissue (smooth mus′l tish′u) Type of involuntary muscle tissue. p. 110

solute (sol′ūt) Chemical dissolved in a solution. p. 41

solvent (sol′vent) Liquid portion of a solution in which a solute is dissolved. p. 41

somatic nervous system (so-mat′ik ner′vus sis′tem) Motor pathways of the peripheral nervous system that lead to skeletal muscles. p. 216

special sense (spesh′al sens) Sense that stems from receptors associated with specialized sensory organs, such as the eyes and ears. p. 263

spermatid (sper′mah-tid) Intermediate stage in sperm cell formation. p. 507

spermatogenesis (sper″mah-to-jen′ĕ-sis) Sperm cell production. p. 507

spermatogonium (sper″mah-to-go′ne-um) Undifferentiated spermatogenic cell in the outer part of a seminiferous tubule. p. 507

spinal cord (spi′nal kord) Part of the central nervous system extending from the brainstem below the foramen magnum through the vertebral canal. p. 234

spinal nerve (spi′nal nerv) Nerve that arises from the spinal cord. p. 234

spleen (splēn) Large organ in the upper left region of the abdomen that is a blood reservoir. p. 383

spongy bone (spun′jē bōn) Bone that consists of bars and plates separated by irregular spaces; cancellous bone. p. 133

static equilibrium (stat′ik e″kwĭ-lib′re-um) Maintenance of balance when the head and body are motionless. p. 275

stem cell (stem sel) Undifferentiated cell that can divide to yield two daughter stem cells, or a stem cell and a progenitor cell. p. 71

steroid (ste′roid) Type of lipid including complex rings of carbon and hydrogen atoms. p. 42

stomach (stum′ak) Digestive organ between the esophagus and small intestine. p. 410

stratum basale (strat′tum ba′sal-e) Deepest layer of the epidermis, where cells divide; stratum germinativum. p. 119

stratum corneum (stra′tum kor′ne-um) Outer, horny layer of the epidermis. p. 119

stressor (stres′or) Factor capable of stimulating a stress response. p. 311

stroke volume (strōk vol′ūm) Volume of blood the ventricle discharges with each heartbeat. p. 360

structural formula (struk′cher-al for′mu-lah) Representation of the way atoms bond to form a molecule, using symbols for each element and lines to indicate chemical bonds. p. 37

subarachnoid space (sub″ah-rak′noid spās) Space in the meninges between the arachnoid mater and the pia mater. p. 234

subcutaneous layer (sub″ku-ta′ne-us la′yer) Loose connective tissue layer that is mostly fat and beneath the skin. p. 117

sublingual (sub-ling′gwal) Beneath the tongue. p. 408

submucosa (sub″mu-ko′sah) Layer of the alimentary canal underneath the mucosa. p. 401

substrate (sub′strāt) Target of enzyme action. p. 79

sucrase (su′krās) Digestive enzyme that catalyzes the breakdown of sucrose. p. 422

sulcus (sul′kus) (plural, *sulci*) Shallow groove, such as that between gyri on the brain surface. p. 237

summation (sum-ma′shun) Increased force of contraction by a skeletal muscle fiber when a twitch occurs before the previous twitch relaxes. p. 190

superior (soo-pe′re-or) Structure above another structure. p. 14

surface tension (sur′fis ten′shun) Force that adheres moist membranes due to the attraction of water molecules. p. 452

surfactant (ser-fak′tant) Substance produced by the lungs that reduces the surface tension in alveoli. p. 452

sweat gland (swet gland) Exocrine gland in skin that secretes a mixture of water, salt, urea, and other bodily substances. p. 124

sympathetic nervous system (sim″pah-thet′ik ner′vus sis′tem) Part of the autonomic nervous system that arises from the thoracic and lumbar regions of the spinal cord. p. 250

synapse (sin′aps) Functional connection between the axon of a neuron and the dendrite or cell body of another neuron or the membrane of another cell type. p. 182

synaptic cleft (sĭ-nap′tik kleft) Space between two cells forming a synapse. p. 221

synaptic knob (sĭ-nap′tik nob) Tiny enlargement at the end of an axon that secretes a neurotransmitter. p. 222

synergist (sin′er-jist) Muscle that assists the action of a prime mover. p. 194

synovial joint (sĭ-no′ve-al joint) Freely movable joint. p. 164

synovial membrane (sĭ-no′ve-al mem′brān) Membrane that forms the inner lining of the capsule of a freely movable joint. p. 110

synthesis (sin′thĕ-sis) Building large molecules from smaller ones. p. 37

systemic circuit (sis-tem′ik ser′kit) Vessels that conduct blood between the heart and all body tissues except the lungs. p. 341

systole (sis′to-le) Phase of the cardiac cycle when a heart chamber wall contracts. p. 347

systolic pressure (sis-tol′ik presh′ur) Arterial blood pressure reached during the systolic phase of the cardiac cycle. p. 359

T

target cell (tar′get sel) Cell with specific receptors on which a hormone exerts its effect. p. 292

tarsal (tahr′sul) Ankle bone. p. 17

taste bud (tāst bud) Organ containing receptors associated with the sense of taste. p. 269

T cell (sel) A type of lymphocyte that interacts directly with antigens, producing the cellular immune response. p. 382

telophase (tel′o-fāz) Stage in mitosis when newly formed cells separate. p. 69

tendon (ten′don) Cordlike or bandlike mass of white fibrous connective tissue that connects a muscle to a bone. p. 103

testis (tes′tis) (plural, *testes*) Primary male reproductive organ; sperm-cell-producing organ. p. 300

testosterone (tes-tos′tĕ-rōn) Male sex hormone secreted by the interstitial cells of the testes. p. 513

tetanic contraction (tĕ-tan′ik kon-trak′shun) Continuous, forceful muscular contraction without relaxation. p. 190

thalamus (thal′ah-mus) Mass of gray matter at the base of the cerebrum in the wall of the third ventricle. p. 242

thermoreceptor (ther″mo-re-sep′tor) Sensory receptor sensitive to temperature changes; heat and cold receptors. p. 263

thoracic cavity (tho-ras′ik kav′ĭ-te) Hollow space inside the chest containing the thoracic organs. p. 8

threshold potential (thresh′old po-ten′shul) Stimulation level that must be exceeded to elicit an action potential or a muscle contraction. p. 225

thrombus (throm′bus) Blood clot that remains where it forms in a blood vessel. p. 331

thymosins (thi′mo-sinz) Group of peptides the thymus secretes that increases production of certain types of white blood cells. p. 310

thymus (thi′mus) Glandular organ in the mediastinum behind the sternum and between the lungs. p. 309

thyroid gland (thi′roid gland) Endocrine gland located just below the larynx and in front of the trachea that secretes thyroid hormones. p. 301

thyroid-stimulating hormone (thi-roid stim-ū-lay-ting hor-mone) **(TSH)** Hormone secreted from the anterior pituitary that controls secretion from the thyroid gland. p. 300

thyroxine (thi-rok′sin) Hormone secreted by the thyroid gland; T4. p. 301

tissue (tish′u) Group of similar cells that performs a specialized function. p. 3

trachea (tra′ke-ah) Tubular organ that leads from the larynx to the bronchi. p. 446

transcellular fluid (trans″sel′u-lar floo′id) Part of the extracellular fluid; includes the fluid within special body cavities. p. 490

transcription (trans-krip′shun) Manufacturing a complementary RNA from DNA. p. 86

transfer RNA (trans′fer) **(tRNA)** RNA molecule that carries an amino acid to a ribosome in protein synthesis. p. 87

translation (trans-la′shun) Assembly of an amino acid chain according to the sequence of base triplets in an mRNA molecule. p. 87

transverse (tranz-vers′) Plane that divides a structure into superior and inferior portions. p. 15

transverse tubule (trans-vers′ tu′būl) Membranous channel that extends inward from a muscle fiber membrane. p. 181

tricuspid valve (tri-kus′pid valv) Heart valve between the right atrium and the right ventricle. p. 344

triglyceride (tri-glis′er-īd) Lipid composed of three fatty acids and a glycerol molecule; a fat or oil. p. 429

triiodothyronine (tri″i-o″do-thi′ro-nēn) Type of thyroid hormone; T3. p. 301

trypsin (trip′sin) Enzyme in pancreatic juice that breaks down protein molecules. p. 413

twitch (twich) Single contraction of a muscle fiber followed by relaxation. p. 188

U

umbilical cord (um-bil′ĭ-kal kord) Cordlike structure, containing one vein and two arteries, that connects the fetus to the placenta. p. 546

umbilical region (um-bil′ĭ-kal re′jun) Central portion of the abdomen. p. 17

unsaturated fatty acid (un-sat′u-rāt″ed fat′e as′id) Fatty acid molecule with one or more double carbon bonds. p. 42

urea (u-re′ah) Nonprotein nitrogenous substance resulting from protein metabolism. p. 431

ureter (u-re′ter) Muscular tube that carries urine from the kidney to the urinary bladder. p. 481

urethra (u-re′thrah) Tube leading from the urinary bladder to the outside of body. p. 484

urine (u′rin) Wastes and excess water and electrolytes removed from the blood and excreted by the kidneys into the ureters, to the urinary bladder, and out of the body through the urethra. p. 472

uterine tube (u′ter-in tūb) Tube that extends from the uterus on each side toward an ovary and transports sex cells; oviduct or Fallopian tube. p. 519

uterus (u′ter-us) Hollow, muscular organ in the female pelvis where a fetus develops. p. 520

utricle (u′trĭ-kl) Enlarged part of the membranous labyrinth of the inner ear. p. 275

uvula (u′vu-lah) Fleshy part of the soft palate that extends down above the root of the tongue. p. 404

V

vaccine (vak′sēn) Preparation that includes antigens that stimulate an immune response to prevent an infectious disease. p. 393

vagina (vah-ji′nah) Tubular organ that leads from the uterus to the vestibule of the female reproductive tract. p. 520

vasoconstriction (vas″o-kon-strik′shun) Decrease in the diameter of a blood vessel. p. 354

vasodilation (vas″o-di-la′shun) Increase in the diameter of a blood vessel. p. 354

vein (vān) Vessel that carries blood toward the heart. p. 358

vena cava (ve′nah kav′ah) One of two large veins (superior and inferior) that convey deoxygenated blood to the right atrium of the heart. p. 369

ventral root (ven′tral root) Motor branch of a spinal nerve by which it connects with the spinal cord. p. 249

ventricle (ven′trĭ-kl) Cavity, such as a brain ventricle filled with cerebrospinal fluid, or heart ventricle that contains blood. pp. 240, 343

venule (ven′ūl) Vessel that carries blood from capillaries to a vein. p. 358

vertebral (ver′te-bral) Pertaining to the bones of the spinal column. p. 8

vesicle (ves′ĭ-kal) Membranous cytoplasmic sac formed by an infolding of the cell membrane. p. 59

viscera (vis′er-ah) Organs in a body cavity. p. 8

visceral pericardium (vis′er-al per″ĭ-kar′de-um) Membrane that covers the surface of the heart. p. 342

visceral peritoneum (vis′er-al per″ĭ-to-ne′-um) Membrane that covers organ surfaces in the abdominal cavity. p. 10

visceral pleura (vis′er-al ploo′rah) Membrane that covers the surfaces of the lungs. p. 450

viscosity (vis-kos′ĭ-te) Tendency for a fluid to resist flowing due to the internal friction of its molecules. p. 360

vitamin (vi′tah-min) Organic compound other than carbohydrate, lipid, or protein needed for normal metabolism that the body cannot synthesize in adequate amounts and must therefore be obtained in the diet. p. 432

vitreous humor (vit′re-us hu′mor) Fluid between the lens and the retina of the eye. p. 284

vocal cord (vo′kal kord) Fold of tissue of the larynx that produces sound when it vibrates. p. 445

vulva (vul′vah) External female reproductive parts that surround the vaginal opening. p. 521

W

water balance (wot′er bal′ans) When the volume of water entering the body is equal to the volume leaving it. p. 492

water of metabolism (wot′er uv mĕ-tab′o-lizm) Water produced as a by-product of aerobic metabolism. p. 492

Y

yellow marrow (yel′o mar′o) Fat storage tissue in certain bone cavities. p. 140

Z

zygote (zi′gōt) Cell produced when an egg and a sperm fuse; fertilized ovum. p. 517

zymogen granule (zi-mo′jen gran′ūl) Cellular structure that stores inactive forms of protein-splitting enzymes in a pancreatic cell. p. 413

Photo Credits

Contents

p. iii(1): ©Ed Reschke; p. iii(2): ©Michael Klein/Peter Arnold/Photolibrary; p. iii(3): UK Medical Research Council (2006) used by kind permission. From Owen, A.M., et.al. 8 September 2006. "Detecting Awareness in the Vegetative State." Science. 313(5792) p. 1402. ©2006 American Association for the Advancement of Science; p. iii(4): ©Bill Longcore/Photo Researchers; p. iii(5): ©Biophoto Associates/Photo Researchers; p. iii(6): ©Profs. P.M. Motta & J. Van Blerkom/Photo Researchers; p. xv(top): ©Ed Reschke; p. xv(bottom): ©Michael Klein/Peter Arnold/Photolibrary; p. xvi(top): UK Medical Research Council (2006) used by kind permission. From Owen, A.M., et.al. 8 September 2006. "Detecting Awareness in the Vegetative State." Science. 313(5792) p. 1402. ©2006 American Association for the Advancement of Science; p. xvi(bottom): ©Bill Longcore/Photo Researchers; p. xvii(top): ©Biophoto Associates/Photo Researchers; p. xvii(bottom): ©Profs. P.M. Motta & J. Van Blerkom/Photo Researchers.

Chapter Preview

p. xxviii: ©Royalty-Free/Corbis; p. xxxi: ©McGraw-Hill Higher Education, Inc./Jim Womack.

Chapter 1

Opener(all): ©Andreas G. Nerlich; figure 1.1: From Vesalius, Andreas. 1543. De Humani Corporis Fabrica, second book; figure 1.14(top, left): ©McGraw-Hill Higher Education, Inc.; figure 1.14(center): ©The McGraw-Hill Companies, Inc./Joe DeGrandis; figure 1.14(right): ©McGraw-Hill Higher Education, Inc.; figure 1.14(bottom, left): ©Paul Reimann; photographer; figure Fig. 1A(both): ©SPL/Photo Researchers.

Chapter 2

Opener: ©Bruno Vincent/Getty Images; figure 2A: ©Eric Hildebrandt/Photo Researchers; figure 2B(a): ©CNRI/Photo Researchers; figures 2.9a-b, 2.19: Courtesy, John W. Hole, Jr.

Chaper 3

Opener: ©Ed Reschke; figure 3.4a: ©Don Fawcett/Photo Researchers; figure 3.5a: ©Gordon Leedale/Biophoto Associates; figure 3.6a: ©Bill Longcore/Photo Researchers; figure 3.8a: ©Don W. Fawcett/Visuals Unlimited; figure 3.9a: ©Oliver Meckes/Photo Researchers; figure 3.9b: ©Manfred Kage/Peter Arnold/Photolibrary; figure 3.10b: ©Stephen L. Wolfe; figure 3.16a-c: ©David M. Phillips/Visuals Unlimited; figure 3A: ©Dr. Tony Brain/SPL/Photo Researchers; figure 3.22a-e: ©Ed Reschke.

Chapter 4

Opener: ©Sucheta Das/Reuters/Corbis; figure 4.9: ©Royalty-Free/Corbis.

Chapter 5

Opener: ©OBSTETRICS & GYNAECOLOGY/Photo Researchers; Opener: ©Keith R. Porter/Photo Researchers; figures 5.1b, 5.1d, 5.2b, 5.3b: ©The McGraw-Hill Companies, Inc./Al Telser, photographer; figure 5.4b: ©The McGraw-Hill Companies, Inc./Dennis Strete, photographer; figures 5.5b, 5.6b, 5.7b: ©The McGraw-Hill Companies, Inc./Al Telser, photographer; figures 5.8b, d: ©Ed Reschke; figure 5.10: ©David M. Phillips/Visuals Unlimited; figure 5.12: ©Microworks Color/Phototake; figure 5.11: ©Manfred Kage/Peter Arnold/Photolibrary; figure 5.13b: ©The McGraw-Hill Companies, Inc./Dennis Strete, photographer; figure 5.14b: ©The McGraw-Hill Companies, Inc./Al Telser, photographer; figure 5.15b: ©The McGraw-Hill Companies, Inc./Dennis Strete, photographer; figures 5.16b, 5.17b, 5.18b: ©The McGraw-Hill Companies, Inc./Al Telser, photographer; figure 5.19b: ©The McGraw-Hill Companies, Inc./Dennis Strete, photographer; figure 5.19c: ©Prof. P. Motta/Univ."La Sapienza"/Photo Researchers; figure 5.20b: ©The McGraw-Hill Companies, Inc./Dennis Strete, photographer; figure 5.21b:

©The McGraw-Hill Companies, Inc./Al Telser, photographer; figure 5.22b: ©The McGraw-Hill Companies, Inc./Dennis Strete, photographer; figure 5.24b: ©Royalty-Free/Corbis; figure 5.23b: ©The McGraw-Hill Companies, Inc./Al Telser, photographer.

Chapter 6

Opener: ©Imagesource/PictureQuest; figure 6.1b: ©Image Source/Getty Images; figure 6.2b: ©The McGraw-Hill Companies, Inc./Al Telser, photographer; figure 6.3a: ©M. Schliwa/Visuals Unlimited; figure 6A(a-c): ©SPL/Photo Researchers; figures 6.5b, 6.7: ©The McGraw-Hill Companies, Inc./Al Telser, photographer; figure 6.6: ©CNRI/SPL/Photo Researchers.

Chapter 7

Opener: ©Michael Klein/Peter Arnold/Photolibrary; figure 7.2: ©Ed Reschke; figure 7.4: ©Biophoto Associates/Photo Researchers; figure 7.6: ©James Shaffer; figure 7.22b: Courtesy Eastman Kodak; figure 7.27b: ©Martin Rotker; figures 7.38(all), 7.39(all), 7.40(all): ©The McGraw-Hill Companies, Inc./Womack Photography, Ltd.; Plates 8-11: ©McGraw-Hill Higher Education, Inc./Jim Womack.

Chapter 8

Opener: ©Royalty-Free/CORBIS; figures 8.3a-b: ©H.E. Huxley; figure 8A: ©Gianni Giansanti/Corbis; figure 8B: ©Muscular Dystrophy Association; p. 211(both): ©Royalty-Free/Corbis.

Chapter 9

Opener: ©UK Medical Research Council (2006) used by kind permission. From Owen, A.M., et.al. 8 September 2006. "Detecting Awareness in the Vegetative State." Vol. 313., No. 5792, p. 1402. ©2006 American Association for the Advancement of Science; figure 9.1: ©Ed Reschke; figure 9.10: ©Don Fawcett/Photo Researchers; figure 9.24b: ©Carolina Biological Supply Company/PhotoTake; figure 9.27b: ©Martin Rotker/Photo Researchers.

Chapter 10

Opener: ©David Young-Wolff/PhotoEdit; figures 10.1b-c: ©Ed Reschke; figure 10.7: ©The McGraw-Hill Companies, Inc./ Jim Womack, photographer; figure 10.10: ©Fred Hossler/Visuals Unlimited; figure 10.22: ©Per H. Kjeldsen; figure 10.23b: ©Mediscan/Corbis; figure 10.25c: ©Frank S. Werblin, Ph.D.

Chapter 11

Opener: ©Superstock; figure 11.16: ©Biodisc/Visuals Unlimited; figure 11A(both): Courtesy, Eli Lilly & Company Archives.

Chapter 12

Opener: ©Photodisc/Getty Images; figure 12.1(1, 3): ©Keith Brofsky/Getty Images; figure 12.1(2): ©Comstock/ Punchstock; figure 12.3b: ©Bill Longcore/Photo Researchers; figure 12.4b: ©The McGraw-Hill Companies, Inc./Al Telser, photographer; figure 12.7-12.11: ©Ed Reschke; figure 12A(a): ©Joaquin Carrillo-Farga/Photo Researchers; figure 12A(b): ©Andrew Syred/SPL/Photo Researchers; figure 12.14: ©SPL/Photo Researchers; figure 12.15a-b: ©The McGraw-Hill Companies, Inc./Al Telser, photographer; figure 12.18c: ©G.W. Willis/Visuals Unlimited; figure 12.18d: ©George W. Wilder/ Visuals Unlimited.

Chapter 13

Opener: Courtesy of Medtronic, Inc.; figure 13.3: ©The McGraw-Hill Companies, Inc./Photo and dissection by Christine Eckel; figure 13.6a: ©The McGraw-Hill Companies, Inc./ University of Michigan Biomedical Communications; figure 13A(a): ©Ed Reschke/Peter Arnold/Photolibrary; figure 13A(b): ©J. & L. Weber/Peter Arnold/Photolibrary; figure 13A©:

©Alfred Pasieka/Peter Arnold/ Photolibrary; figure 13.20: ©Don. W. Fawcett/Visuals Unlimited; figure 13.22: ©The McGraw-Hill Companies, Inc./Al Telser, photographer.

Chapter 14

Opener: USDA Photo by: Ken Hammond; figure 14.3: ©The McGraw-Hill Companies, Inc./Dennis Strete, photographer; figure 14.7: ©Dr. Kent M. Van De Graaff; figure 14.9b: ©The McGraw-Hill Companies, Inc./Dennis Strete, photographer; figure 14.10b: ©The McGraw-Hill Companies, Inc./Al Telser, photographer; figure 14.11: Courtesy of Nina Lampen, Sloan-Kettering.

Chapter 15

Opener: ©Centers for Disease Control and Prevention; figure 15.7: ©The McGraw-Hill Companies, Inc./Rebecca Gray, photographer; figure 15.12b: ©The McGraw-Hill Companies, Inc./Al Telser, photographer; figure 15.17c: ©Victor B. Eichler, Ph.D; figure 15.20: ©Carroll Weiss/Camera M.D. Studios; figure 15.22: ©Armed Forces Institute of Pathology; figure 15.25: ©The McGraw-Hill Companies, Inc./Al Telser, photographer; figure 15.28: ©Jim Wehtje/Getty Images; figure 15.30: ©Ed Reschke.

Chapter 16

Opener: ©Brand X Pictures; figure 16.3b: ©Biophoto Associates/Photo Researchers; figure 16.5c: ©CNRI/ PhotoTake; figure 16.9: ©The McGraw-Hill Companies, Inc./Bob Coyle, photographer; figure 16A(a-b): ©Victor B. Eichler, Ph.D.

Chapter 17

Opener: ©Burke Triolo Productions / Getty Images; figure 17.4: ©R.G. Kessel and R.H. Kardon, Tissues and Organs:

A Text-Atlas of Scanning Electron Microscopy, 1979 W. H. Freeman and Company, all rights reserved; figure 17.5a: ©Biophoto Associates/Photo Researchers; figure 17.5b: ©Manfred Kage/Peter Arnold/Photolibrary; figure 17A: ©Stephen J. Krasemann/Photo Researchers; figure 17.15: ©Per H. Kjeldsen; figure 17.17: ©Ed Reschke.

Chapter 18

Opener: ©Tetra Images/Getty Images.

Chapter 19

Opener: ©Profs. P.M. Motta & J. Van Blerkom/Photo Researchers; figures 19.2c, 19.10: ©The McGraw-Hill Companies, Inc./Al Telser, photographer; figure 19.12: ©McGraw-Hill Higher Education, Inc./Carol D. Jacobson, Ph.D., Department of Veterinary Anatomy, Iowa State University; figure 19B: ©Courtesy, Southern Illinois University School of Medicine; figures 19.15a-b: ©The McGraw-Hill Companies, Inc./ Jill Braaten, photographer; figure 19.15c: ©PhotoLink/Getty Images; figure 19.15d: ©Don Farrall/Getty Images; figure 19.15e: ©The McGraw-Hill Companies, Inc./Jill Braaten, photographer.

Chapter 20

Opener: ©J. Meyer/Custom Medical Stock Photo; figure 20.1: Tegner, M., Epel, D. (1973) "Sea Urchin Sperm." Science, 16(179), 685-688. ©1973 American Association for the Advancement of Science; figure 20.4a: ©A. Tsiara/Photo Researchers; figure 20.4b: ©Omikron/Photo Researchers; figure 20.4c: ©Petit Format/Nestle/Photo Researchers; figures 20.9, 20.13: ©Donald Yaeger/Camera M.D. Studios; figure 20.19: Courtesy Genzyme Corporation; figure 20.21a: Library of Congress; figure 20.21b: Peter Morenus/University of Connecticut at Storrs.

Application Index

Boxed Readings

Atherosclerosis, 355
Beyond the Human Genome
    Project: Personalizing
    Medicine, 90–91
Biological Rhythms, 312
Biomarkers, 47
Bone Fractures, 138–139
Burns, 126–127
Cancer, 68
Coagulation Disorders, 332
Computerized Tomography, 16
Cystic Fibrosis, 449
Dental Carries, 407
Diabetes Mellitus, 310–311
Drug Abuse, 245
Eating Extremes: Undereating
    and Overeating, 436
Emphysema and Lung
    Cancer, 454
Exercise and Breathing, 458
Exercise and the
    Cardiovascular System, 362
Factors Affecting Synaptic
    Transmission, 223
Female Infertility, 538
Fetal Chromosome Checks,
    558–559
Headache, 281
Hepatitis, 417
Immunity Breakdown:
    HIV/AIDS, 389
Inflammatory Bowel
    Disease, 428
Inherited Diseases of
    Muscle, 193
Joint Disorders, 168
Kidney Stones, 482
Leukemia, 328
Lysosomal Storage Diseases, 58
Male Infertility, 514
Mutations, 86
Persisting Fetal Cells and
    Autoimmunity, 395
Prostate Cancer, 511
Radioactive Isotopes: Helpful
    or Harmful, 34
Skin Cancer, 121
Sodium and Potassium
    Imbalances, 496
Some Causes of Birth Defects,
    548–549
Steroids and Athletes—An
    Unhealthy Combination, 188
Synesthesia: Connected
    Senses, 268

The Body's Glue: The
    Extracellular Matrix,
    104–105
Too Little or Too Much Pain, 54
Treating Breast Cancer, 526–527
Use and Disuse of Skeletal
    Muscles, 189
Water Balance Disorders,
    494–495

Clinical Applications

ABO blood types, 334
Acromegaly, 298
Active immunity, 393
Acute pancreatitis, 413
Addiction, newborn, 545
Addison disease, 306
Adrenal medulla, tumors in, 305
Agglutination, 333
All-or-none response, 189
Anaphylactic shock, 394
Aneurysms, 354
Angina pectoris, 347
Angiotensin-converting
    enzyme inhibitor, 476
Anosmia, 269
Atelectasis, 453
Autologous stem cell
    transplants, 141
Autonomic neuropathy, 267
Baldness, 123
Bili lights, 334
Bone marrow transplants, 141
Botox, 117
Bradycardia, 352
Brain tumors, 216
Carbon dioxide poisoning, 462
Carcinoma *in situ*, 520
Carcinomas, 101
Cataracts, 280
Celiac disease, 424
Cholecystectomy, 418
Cleft palates, 148
Clot-busting biochemicals, 331
Collapsed lung, 453
Colorectal cancer, 426
Concussions, 246
Conductive hearing loss, 275
Corneal transplants, 280
Coughing, 455
Crystallins, 280
Cushing syndrome, 306
Cyanosis, 319
Cystic fibrosis, 54
Cystitis, 483
Deviated septum, 444

Decubitus ulcers, 122
Dental implants, 408
Diabetes insipidus, 300
Dialysis, 62–63
Diseasome, 3
Dissecting aneurysms, 354
Ectopic pregnancy, 541
Edema, 329
Endocarditis, 345, 349
Endocrine paraneoplastic
    syndrome, 306
Endocytosis, 65
Episiotomy, 551
Erythroblastosis fetalis, 336
Escherichia coli, 484
Exocytosis, 65
Fainting, 458
Fiberoptic colonoscope, 426
Floaters in the vitreous
    humor, 284
Gallstones, 418
Gigantism, 298
Glaucoma, 282
Glomerulonephritis, 475
Glycyrrhizic, 360
Goiters, 303
Gout, 480
Hearing loss, 275
Heart transplant, 347
Hemodialysis, 62–63
Hemolytic jaundice, 418
Hemorrhoids, 426
Hepatocellular jaundice, 418
Hereditary neuropathy, 267
Hiatal hernia, 410
Hiccuping, 455
Hip fractures, 162
Hydroxyurea, 323, 549
Hyperparathyroidism, 303
Hypertension, 362, 476
Hyperthermia, 125
Hyperthyroidism, 303
Hyperventilation, 458
Hypoparathyroidism, 303
Hypothermia, 125
Hypothyroidism, 303
Hypoxemia, 221
Immediate-reaction
    allergies, 394
In vitro fertilization, 538
Injections, 119
Intradermal injections, 119
Ischemia, 221, 347
Jaundice, 334, 418
Kidney transplant, 470
Lactose intolerance, 422

Laryngitis, 446
Leber congenital amaurosis, 286
Left ventricular assist
    device, 347
Lumbar punctures, 242
Lymphadenitis, 382
Lymphadenopathy, 382
Lymphangitis, 382
Malabsorption, 424
Methionine, 431
Mitral valve prolapse, 345
Motion sickness, 276
Mucous membrane
    infections, 272
Muscle pulls, 191
Muscle strains, 182
Myocardial infarction, 347
Neuroglia, 216
Newborn addiction, 545
Night blindness, 286
Obstructive jaundice, 418
Pap smear, 520
Parathyroid disorders, 303
Patent ductus arteriosus, 554
Pericarditis, 343
Peripheral neuropathy, 267
Physiologic jaundice, 334
Pituitary dwarfism, 298
Pneumothorax, 453
Pressure ulcer, 122
Prothrombin time, 331
Respiratory distress
    syndrome, 452
RhoGAM, 336
Scurvy, 433
Sensorineural hearing loss, 275
Shock, 475
Sickle cell disease, 323, 549
Sinus headaches, 444
Smell, loss of, 269
Sneezing, 455
Spina bifida, 149
Spontaneous ejaculation, 513
Spontaneous emission, 513
Streptokinase, 331
Subcutaneous injections, 119
Subdural hematoma, 234
Tachycardia, 352
Tattoos, removal of, 122
Telomere shortening, 69
Temporomandibular joint
    syndrome, 197
Tendinitis, 179
Tenosynovitis, 179
Testicular cancer, 506
Thyroid disorders, 303

Tonsillitis, 405
Transcytosis, 65
Transdermal patch, 119
Ulcers, 122
Umbilical cord blood banks, 335
Urinary tract infection, 484
Varicose veins, 358
Venipuncture, 362
Vital signs, 5
Vomiting, 413
Whiplash, 250
Wrinkles, treatments for facial, 117
Yawning, 455

Tables

A Comparison Between the Nervous System and the Endocrine System, 292
A Comparison of DNA and RNA Molecules, 86
A Comparison of T Cells and B Cells, 387
Aging-Related Changes, 555
Amino Acids in Foods, 431
Antigens and Antibodies of the ABO Blood Group, 334
Assisted Reproductive Technologies, 540
Atomic Structures of Elements 1 Through 12, 32
Bones of the Adult Skeleton, 143
Cellular Components of Blood, 327
Characteristics of Blood Vessels, 359
Characteristics of Life, 5
Chemical Buffer System, 498
Codons (mRNA Three-Base Sequences), 87
Comparative Effects of Epinephrine and Norepinephrine, 305
Components of Connective Tissue, 103
Connective Tissues, 109
Differences Between Female and Male Skeletons, 160

Effects of Neurotransmitter Substances on Visceral Effectors or Actions, 256
Elements in the Human Body, 31
Epithelial Tissues, 102
Events Leading to the Conduction of a Nerve Impulse, 227
Factors Associated with Edema, 495
Fat-Soluble Vitamins, 432
Fetal Cardiovascular Adaptations, 551
Functions of Cranial Nerves, 248
Functions of Female Reproductive System, 522
Functions of Nephron Components, 481
Functions of the Male Reproductive Organs, 513
Gases Transported in Blood, 462
Heart Valves, 345
HIV Transmission, 389
Hormonal Changes During Pregnancy, 543
Hormones of the Adrenal Cortex, 307
Hormones of the Pituitary Gland, 301
Hormones of the Thyroid Gland, 302
Important Groups of Lipids, 43
Inflammation, 125
Inorganic Substances Common in the Body, 40
Intestinal Absorption of Nutrients, 424
Major Branches of the Aorta, 365
Major Branches of the External and Internal Carotid Arteries, 366
Major Components of Gastric Juice, 412

Major Events in a Reproductive Cycle, 523
Major Events of Muscle Contraction and Relaxation, 185
Major Functions of the Liver, 418
Major Minerals, 434
Movements Through Cell Membranes, 67
Muscle and Nervous Tissues, 113
Muscle Metabolism, 187
Muscles Associated with the Eyelids an Eyes, 279
Muscles of Facial Expression, 195
Muscles of Mastication, 197
Muscles of the Abdominal Wall, 203
Muscles of the Pelvic Outlet, 203
Muscles That Move the Arm, 199
Muscles That Move the Foot, 207
Muscles That Move the Forearm, 201
Muscles That Move the Hand, 202
Muscles That Move the Head, 197
Muscles That Move the Leg, 205
Muscles That Move the Pectoral Girdle, 198
Muscles That Move the Thigh, 204
Organic Compounds in Cells, 46
Parts of a Reflex Arc, 232
Parts of the Respiratory System, 451
Plasma Proteins, 329
Practical Classification of Immunity, 393
Preferred and Permissible Blood Types for Transfusions, 335

Primary and Secondary Teeth, 407
Relative Concentrations of Substances in the Plasma, Glomerular Filtrate, and Urine, 475
Respiratory Air Volumes and Capacities, 456
Role of ADH in Regulating Urine Concentration and Volume, 480
Sodium and Potassium Imbalances, 496
Some Neurotransmitters and Representative Actions, 229
Some Particles of Matter, 37
Some Sexually Transmitted Infections, 530
Stages and Events of Early Human Prenatal Development, 543
Stages in Postnatal Development, 554
Steps in Antibody Production, 391
Steps in the Generation of Sensory Impulses from the Ear, 274
Structures and Functions of Cell Parts, 61
Subdivisions of the Nervous System, 246
Summary of the Major Digestive Enzymes, 422
Terms Used to Describe Skeletal Structures, 144
Tests to Assess Female Infertility, 538
Trace Elements, 435
Types of Exocrine Glandular Secretions, 101
Types of Hormones, 293
Types of Joints, 167
Types of Muscle Tissue, 192
Types of Tissue, 95
Water-Soluble Vitamins, 433

Subject Index

Page numbers followed by an *f* refer to figures, page numbers followed by a *t* refer to tables, and page numbers followed by a *p* refer to plates.

A

A band, 180*f*, 181
ABCDE rule, 121
abdominal aorta, 29*p*, 363, 364*f*, 365*t*, 468*f*, 550*f*
abdominal cavity, 8, 9*f*, 29*p*
abdominal lymph node, 382*f*
abdominal region, defined, 16, 17*f*, 18*f*
abdominal viscera, veins from, 370, 371*f*
abdominal wall
  arteries to, 367
  muscles of the, 201, 203*t*, 453, 453*f*
  pectoral girdle, muscles that move the, 197, 198*f*, 198*t*, 199*f*
  veins from, 370
abdominopelvic cavity, 8, 9*f*, 10
abdominopelvic membrane, 10, 11*f*
abducens nerve (VI), 247, 247*f*, 248*t*
abduction, 167, 169*f*
ABO blood group, 333, 334–336, 334*t*, 335*f*
abortion, spontaneous, 542
absorption, 5*t*
  of bone tissue, 137
  gastric, 412
  in large intestine, 427
  in small intestine, 422–424, 423*f*, 424*t*
accessory nerve (XI), 247, 247*f*, 248*t*
accessory organ, visual, 276, 277–279, 278*f*
accommodation, 281, 282*f*
acetabulum, 158, 160*f*
acetoacetic acid, 501
acetone, 501
acetylcholine (Ach), 412
  beta oxidation of, 429–430, 430*f*
  in cardiac cycle, 352
  as neurotransmitter, 182*f*, 228, 229*t*, 251, 254*f*
  in skeletal muscle contraction, 183, 184*f*, 185*f*
acetylcholinesterase, 228
acetyl CoA (acetyl coenzyme A)
  in cellular respiration, 81*f*, 82
  in citric acid cycle, 81*f*
  conversion to ketone bodies, 429–430, 430*f*
  in deamination of proteins, 431, 431*f*
  formation of, 431
acid, 38–39, 39*f*, 497
  acid-base balance (*see* acid-base balance)
  acid-base imbalance (*see* acid-base imbalance)

  in destruction of tooth enamel, 407
  strengths of, 497
acid-base balance
  acids and bases, strengths of, 497
  buffer systems, 498–499, 498*t*, 499*f*
  hydrogen ion concentration, regulation of, 498–499
  hydrogen ions, sources of, 497, 497*f*
  renal excretion of hydrogen ions, 499, 499*f*
  respiratory center regulation of carbon dioxide, 499, 499*f*
  time course of hydrogen ion regulation, 499, 499*f*
acid-base imbalance, 500–502, 500*f*, 501*f*, 502*f*
  acidosis, 500–501, 500*f*, 501*f*
  alkalosis, 500, 500*f*, 501–502, 501*f*, 502*f*
  buffer systems, 500–502
acidic ketone body, 497, 497*f*
acidosis, 39, 310–311, 496, 498, 500–501, 500*f*, 501*f*
A1c measurements, 311
acne, 124, 127
acquired immune deficiency syndrome (AIDS). *See* AIDS (acquired immune deficiency syndrome)
acromegaly, 298
acromial region, defined, 16, 18*f*
acromion process, 155, 156*f*
acrosome, 510, 510*f*, 539*f*
actin, 57, 58*f*, 179, 180*f*, 431
  in cardiac muscle, 191–192
  fast twitch muscle fibers and, 189
  in rigor mortis, 187
  in skeletal muscle contraction, 183, 183*f*, 184*f*
  in smooth muscle contraction, 191
action potential, 222, 225–227, 225*f*, 226*f*, 227, 227*f*
activation energy, 79
active immunity, 393, 393*t*
active protein, 295*f*
active sites, 79, 79*f*
active surveillance of prostate cancer, 511
active transport, 65, 67*t*
  crossing of placental membrane by, 544
  through cell membranes, 65, 65*f*
  in tubular reabsorption, 477*f*, 478–479
  in tubular secretion, 479
activity sites, 294–295, 295*f*
acute pain fiber, 265
adaptation, sensory, 263
adaptive (specific) defense, 384, 386
  allergic reactions, 393–394
  antibody (*see* antibody)
  antigen (*see* antigen)
  autoimmunity, 394–395
  cellular immune response, 386–388, 387*f*, 387*t*, 388*f*

  humoral immune response, 390–391, 390*f*
  immune response, 392–393, 392*f*
  lymphocyte (*see* lymphocyte)
  transplantation and tissue rejection, 394
addiction
  drug abuse, 245
  newborn, 545*f*, 548
Addison disease, 306, 496
adduction, 167, 169*f*
adductor brevis muscle, 28*p*
adductor longus muscle, 26*p*, 27*p*, 28*p*, 29*p*, 195*f*, 204*f*, 204*t*, 205*f*
adductor magnus muscle, 29*p*, 195*f*, 204*f*, 204*t*
adenine, 84, 84*f*, 85, 85*f*, 86
adenoid, 404, 405, 406*f*
adenosine triphosphate (ATP), 56
adenylate cyclase, 295, 295*f*
adipose tissue, 104, 106, 106*f*, 109*t*, 117, 118*f*, 123*f*
  in female breast, 525*f*
  in labia majora, 521
  lack of, on elite female athletes, 522
  in mammary gland, 525
  storage of fat molecules in, 430
adolescence, stages in development, 554*t*
ADP (adenosine diphosphate)
  in cellular respiration, 80, 81*f*
  in skeletal muscle contraction, 183, 184*f*, 185–186, 186*f*
  in smooth muscle contraction, 191
adrenal androgen, 306, 307*t*
adrenal cortex, 304, 305*f*, 313*f*
  female sex hormone, secretion of, 522
  hormones of the, 306, 307*f*, 307*t*
  secretion rates, difficulties with, 306
  structure of, 304, 305*f*
adrenal gland, 28*p*, 293, 293*f*, 300, 305*f*
  hormones of the adrenal cortex, 306, 307*f*, 307*t*
  hormones of the adrenal medulla, 304–305, 305*t*
  structure of, 304, 305*f*
  with urinary system, 468*f*, 480
adrenalin, 394
adrenal medulla, 297, 304, 305*f*, 313*f*
  hormones of the, 304–305, 305*t*
  structure of, 304, 305*f*
  tumors in, 305
adrenal sex hormone, 306, 307*t*
adrenal tumor, 306
adrenergic fiber, 251, 254, 256*t*
adrenocorticotropic hormone, 297, 300, 301*t*, 307*f*, 309, 312, 313*f*
adulthood, stages in development, 554*t*
advanced sleep phase syndrome, 312
aerobic cellular respiration, 81*f*, 82
aerobic respiration of glucose, 186–187, 186*f*, 497, 497*f*

afferent arteriole, 469, 470*f*, 472*f*, 473, 473*f*, 474*f*, 475, 477*f*
afferent fiber, 230
afferent lymphatic vessel, 380*f*, 381*f*
afferent neuron. *See* sensory neuron
afterbirth, 552
agglutination, 333, 335*f*, 392*f*
agglutinin. *See* antibody
agglutinogen, 333
aging, 555–556, 555*t*
  active, 556
  passive, 555–556
agranulocyte, 322*f*, 324–325, 327*t*
AIDS (acquired immune deficiency syndrome), 389, 530, 530*t*
  leukopenia from, 327
  treatment, 389
  universal precautions for, 318
air pressure
  in breathing process, 450–451, 451*f*
  in pleural cavity, 452
airway, effects of epinephrine and norepinephrine on, 305*t*
alanine, 431*t*
alarm stage, 311, 313*f*
albinism, 123
albumin, 320*f*, 329, 329*t*
alcohol
  birth defects from, 548–549, 548*f*
  promotion of urine production by, 493
aldosterone, 306, 307*f*, 307*t*
  changes during pregnancy, 543
  as regulator of electrolyte imbalances, 493, 496
  as regulator of urine, 307*f*, 476*f*, 480
alimentary canal, 401, 402*f*
  functions of, 403
  infections of, 420
  tube, movements of the, 403, 405*f*
  wall, structure of the, 401, 404*f*
alkaline taste sensation, 270
alkalosis, 39, 458, 498, 500, 500*f*, 501–502, 501*f*, 502*f*
allantois, 547
alleles, 556
allergen, 393–394
allergic reactions, 393–394, 536
all-or-none response
  in muscles, 189
  in nerve impulse conduction, 227–228
alopecia, 127
alpha cell, 307–308, 309*f*
alpha fetoprotein, 558
alpha globulin, 329, 329*t*
alpha radiation, 33
alveolar, 451, 451*f*, 454
  alveolar gas exchange by, 459–460, 459*f*
  in bronchial tree, 447*f*, 448, 448*f*, 449*f*
  as member of cardiovascular system, 346, 346*f*
alveolar arch, 148, 149
alveolar capillary, 346*f*

alveolar duct, 447f, 448, 448f, 525f
alveolar gas exchange, 459–460, 459f, 460f
alveolar gland, 525, 525f, 552–553, 553f
alveolar process, 148, 408f
alveolar sac, 448, 448f
alveoli, 453
amenorrhea, 522
amine, 293f, 293t
amino acid, 44, 44f, 45f
    absorption of, in small intestine, 424t
    active transport across cell membranes of, 65
    deamination of, 431, 431f
    in dehydration synthesis, 77–78, 77f, 78f
    diffusion through cell membranes, 63
    essential, 431
    in foods, 430–431
    as neurotransmitters, 228, 229t
    as nonprotein nitrogenous substance, 330
    oxidation of, 497
    as plasma nutrients, 329
    in protein synthesis, 85–88, 87t, 88f, 89f
    reabsorption of, 478
    as source of hydrogen ions, 497, 497f
amino group, 44
ammonia, 442
amniocentesis, 558–559, 558f
amnion, 544f, 546–547, 546f
amniotic cavity, 544f, 546f
amniotic fluid, 546, 548f
amniotic sac, 552f, 558
amoeboid motion, 325
ampulla, 272f, 275, 277f, 525f
amyotrophic lateral sclerosis (ALS), 50, 182
anabolic steroid, 188, 245
anabolism, 77–78
anaerobic cellular respiration, 82
anaerobic respiration of glucose, 186–187, 186f, 497, 497f
anal canal, 424, 425f, 426, 426f
anal column, 426, 426f
anal sphincter muscle, 202f
    external, 426, 426f
    internal, 426, 426f
anaphase, 68f, 69, 70f
anaphylactic shock, 394
anastomosis, 347
anatomical neck, of humerus, 155
anatomical position, defined, 14
anatomic dead space, 456
anatomy, 3
androgen, 513, 515f, 522
androstadienone, 291
anemia, 323
anemic hypoxia, 460
aneurysm, 354
angina pectoris, 347
angiogenesis, 68
angiotensin I, 476, 476f
angiotensin II, 476, 476f, 480
angiotensin-converting enzyme, 476, 476f
angiotensin-converting enzyme inhibitor, 476
angiotensinogen, 476, 476f
anorexia, 436
anosmia, 269

anovulation, 538
antagonist, 194
antebrachial region, defined, 16, 18f
antecubital region, defined, 16, 18f
anterior, defined, 14, 14f, 18f
anterior articular facet, 153f
anterior cardiac vein, 348f
anterior cavity of the eye, 280f
anterior cerebral artery, 366f
anterior chamber of the eye, 280f, 282
anterior choroid artery, 365f
anterior circumflex artery, 367f
anterior crest, 162f
anterior facial vein, 369f, 372f
anterior fontanel, 150f
anterior funiculus, 234, 235f
anterior horn, 234, 235f
anterior inferior iliac spine, 160f
anterior intercostal artery, 367
anterior interventricular artery, 348f
anterior median fissure, 234, 235f
anterior sacral foramen, 152, 153, 154f
anterior superior iliac spine, 23p, 25p, 29p, 158, 160f
anterior tibial artery, 367, 368f
anterior tibial vein, 370–371, 372f
anterolateral fontanel, 150f
antibody, 43, 325, 333, 334f, 390–391, 390f
    actions of, 391–392
    antigen-antibodies reaction, 391–392, 392f
    antigen-antibody reaction, 333, 334
    in colostrum, 552
    passage of antibodies from mother to placenta, 391, 393
    production, 389, 390–391, 390f, 391t
    types of, 391
anticodon, 87–88
antidiuretic, 300
antidiuretic hormone, 300, 301t, 306, 312
    as regulator of urine, 480, 480t
    as regulator of water output, 493
antigen, 333, 334f, 386–388, 388f
    antibody-antigen reaction, 333, 334, 391–392, 392f
    antigen-presenting cell, 386–388, 388f
antigen A, 334–336, 334f, 334t, 335f, 335t
antigen-antibody reaction, 333, 334, 391–392, 392f
antigen B, 334–336, 334f, 334t, 335f, 335t
antigen D, 336
antigenetically similar tissues, 470
antigen-presenting cell, 386–388, 388f
antigen receptors, 390, 390f
antihistamine, 377
anti-Rh antibody, 336
antrum, 518
anus, 202f, 401, 402f, 403f, 424, 425f, 426, 426f, 507f
aorta, 11f, 27p, 341f, 342f, 344–345, 344f, 346, 353f, 368f
    abdominal, 29p, 363, 364f, 365t, 468f, 550f
    ascending, 363, 365t
    branches of, 363–364, 364f, 365t

descending, 28p, 29p, 363, 365t, 551
    newborn circulation, 553–554, 554f
    thoracic, 28p, 365t
aortic arch, 352, 364f, 550f, 551
aortic baroreceptor, 353f
aortic body, 457
aortic sinus, 363
aortic valve, 344f, 345t, 346, 347f, 349f
apex of the heart, 342, 348f
apical papilla, 408
apical surface, 95
apocrine gland, 101–102, 101f, 101t, 124
aponeurosis, 179, 199f
apoptosis, 71, 556
appendicitis, 426
appendicular portion of body, 8
appendicular skeleton, 142f, 143, 143t
appendix, 26p, 27p, 403f, 420f, 425f, 426
    referred pain, 266f
    vermiform, 26p, 27p
aqueous humor, 280f, 282, 490
arachidonic acid, 296
arachnoid mater, 233–234, 233f, 241f
arch of aorta, 26p, 28p, 29p, 365t
arcuate artery, 469, 470f
arcuate vein, 470f
areola, 23p, 525, 525f
areolar tissue, 104, 105f, 109t
arginine, 431t
arm. See upper limb
arrector pili muscle, 118f, 123, 123f
arrhythmia, 362
arsenicosis, 76
arterial blood pressure, 359–362, 359f, 360f, 361f
arterial system, 363–364, 364f, 365f, 365t, 366–367, 366f, 366t, 367f, 368f
    aorta, branches of, 363–364, 364f, 365t
    major vessels of, 368f
    neck, head and brain, arteries to, 363–364, 364f, 365t
    pelvis and lower limb, arteries to, 367, 368f
    shoulder and upper limb, arteries to, 366–367, 366f
    thoracic and abdominal walls, arteries to, 367
arteriole, 341, 354, 355, 355f, 356, 356f, 358t, 379f, 421f
artery, 341, 354, 354f, 355, 356, 358f, 358t
    characteristics of, 354, 354f, 356
    system of (see arterial system)
    weakening of, 354
artery wall, 333f
arthritis, 168
arthroscope, 170f
arthroscopy, 170f
articular cartilage, 133, 134f, 136f, 137, 165f
artificially acquired active immunity, 393, 393f
artificially acquired passive immunity, 393, 393f, 393t
ascending aorta, 363, 365t
ascending colon, 26p, 27p, 371f, 425f, 426
ascending limb of nephron loop, 471, 472f, 473f, 476, 481t

ascending lumbar vein, 370, 372f
ascending tract, 234–235, 236f
ascorbic acid, 432–433, 433t, 478
    tubular reabsorption by, 478
-ase, 79
asparagine, 431t
aspartic acid, 431t
assimilation, 5t
association areas of cerebrum, 238f, 239
association neuron, 220
asthma, 442
astrocyte, 216, 217f
atherosclerosis, 43, 104–105, 333f, 354, 355, 355f
athletes, disturbed reproductive cycles by elite female, 522
athlete's foot, 127
atlas, 150, 153f, 166f
atmospheric pressure
    in breathing process, 450–451, 451f
    partial pressure, 459–460, 460f
    in pleural cavity, 452
atom, 3, 4f, 31–32, 37t
    atomic structure, 32–33, 32f, 32t, 33f
    bonding of, 33, 35–36, 35f, 36f, 37f
atomic number, 32
atomic structure, 32–33, 32f, 32t, 331f
atomic weight, 32–33, 33f
ATP (adenosine triphosphate), 46, 65, 80
    in cellular respiration, 80, 81f, 82
    in citric acid cycle, 431, 431f
    in cystic fibrosis, 449
    in digestion of proteins, 431f
    in glycolysis, 80, 81f
    in nonsteroid hormone, 295, 295f
    in rigor mortis, 187
    in skeletal muscle contraction, 183, 184f, 185–186, 186f
ATPase, 183
atrial diastole, 347, 349f
atrial natriuretic peptide, 310, 476
atrial septum, 554
atrial syncytium, 349, 350, 350f
atrial systole, 347, 349f
atrioventricular node (AV node), 350, 350f, 352, 353f
atrioventricular valve, 344, 345
atrium, 343–345, 344f, 345f, 348f, 350
    in cardiac cycle, 347, 349–350, 349f, 350f, 351f
    left, 27p, 341f, 344f, 346, 551
    right, 27p, 341f, 344f, 346, 550–551
atrophy, 189
auditory area of cerebral cortex, 238f
auditory cortex, 262
auditory nerve pathway, 274–275
auditory ossicle, 270, 271, 271f
auditory tube, 271–272, 271f, 406f, 444f
aura, 281
auricle, 270, 271f, 342f, 343, 348f
auricular surface, 154f
autoantibody, 394
autocrine secretion, 292
autoimmune disorder, 311, 394, 395
autoimmunity, 394–395
autologous stem cell transplant, 141
automatic bladder, 484
autonomic ganglion, 251f

autonomic nervous system, 216, 246, 250, 251*f*, 252*f*, 253*f*, 254
  autonomic neurons, 250–251, 251*f*
  characteristics, 250
  control of, 254
  divisions, 250
  nerve fibers, 361
  neurotransmitters, 251, 254, 254*f*
  parasympathetic division, 251, 253*f*
  sympathetic division, 251, 252*f*
autonomic neuron, 250–251, 251*f*
autonomic neuropathy, 267
autosomal dominant condition, 560
autosomal recessive inheritance, 557
autosome, 556
AV bundle, 350, 350*f*
AV node (atrioventricular node), 350, 350*f*, 353*f*
axial portion of body, 8
axial skeleton, 142–143, 142*f*, 143*t*
axillary artery, 26*p*, 366, 367*f*, 368*f*, 370*f*
axillary lymph node, 382*f*
axillary nerve, 249*f*, 250
axillary region, defined, 16, 18*f*
axillary vein, 26*p*, 369, 372*f*
axis, 150, 153*f*, 166*f*
axon, 111, 214, 215*f*, 217, 218*f*, 219, 219*f*
  of motor neurons, 182, 182*f*
  in muscle fibers, 227*f*
  in synapses, 221–222, 221*f*, 222*f*
axon hillock, 217, 218*f*, 225, 231, 231*f*
azygos vein, 370, 372*f*

B

bacteria, 382, 384. *See also specific bacterium*
  as bioweapon, 444
  as cause of ulcers, 412
  in destruction of tooth enamel, 407
  hemolytic uremic syndrome, 467
  intestinal, 422
ball-and-socket joint, 165, 166*f*, 167*t*
bands, 325
Bangladesh, arsenic poisoning in, 76
barbiturate drug, 244, 245
bariatric surgery, 436
baroreceptor, 352, 353*f*, 361–362, 361*f*
baroreceptor reflex, 352
Barrett, Syd, 268
Barrett's esophagus, 410
basal cell carcinoma, 121, 121*f*
basal ganglia, 240
basal metabolic rate, 301
basal nuclei, 240, 240*f*
base, 38–39, 39*f*, 497
  acid-base balance (*see* acid-base balance)
  acid-base imbalance (*see* acid-base imbalance)
  strengths of, 497
basement membrane, 95, 96*f*, 97*f*, 508*f*, 509*f*
  of hair follicle, 123*f*
  of skin, 117, 118*f*, 119*f*
base of the heart, 342
basilar artery, 364, 365*f*, 366*f*
basilar membrane, 273–274, 273*f*
basilic vein, 369, 370*f*, 372*f*
basophil, 320*f*, 322*f*, 325, 326*f*, 327*t*

basophilic band cell, 322*f*
basophilic myelocyte, 322*f*
B cell, 322*f*, 381, 381*f*, 386–388, 387*f*, 387*t*, 388*f*
  activation, 386–388, 387*f*, 387*t*, 388*f*
  in antibody production, 391*t*
  compared to T cells, 387*t*
  and humoral immune response, 390–391, 390*f*
Becker muscular dystrophy, 193
benzene, 442
benzodiazepine, 245
beta-carotene, 122
beta cell, 307–308, 309*f*, 311
beta globulin, 329, 329*t*
beta-hydroxybutyric acid, 501
beta oxidation, 429, 430*f*
beta radiation, 33
bicarbonate
  chemical buffer system, 498–499, 498*t*, 499*f*, 501
  in extracellular fluid, 491*f*
  in glomerular filtrate, 475*t*
  in plasma, 475*t*
  in urine, 475*t*, 501
bicarbonate ion, 461, 461*f*, 462
  as constituent of cells, 40, 40*t*
  in electrolyte intake, 493
  in extracellular fluid, 491*f*
  in intracellular fluid, 491*f*
  in pancreatic juice, 413, 414*f*
  reabsorption of, 478, 478*f*
biceps brachii muscle, 24*p*, 192–193, 194, 194*f*, 195*f*
  explanation of joint, 140, 140*f*
  as moving the forearm, 200, 200*f*, 201*t*
biceps femoris muscle, 195*f*, 205, 205*f*, 205*t*, 207*f*
biceps muscle, 192
bicuspid (teeth), 406*f*, 407, 407*t*
bicuspid valve. *See* mitral valve
bifid spinous process, 152*f*
bilateral, defined, 15
bilayer structure of cell membrane, 53, 53*f*
bile, 323–324, 324*f*, 418–419, 419*f*
  composition of, 418
  release, regulation of, 418
  salts, functions of, 418–419
bile caniculus, 415, 416*f*
bile duct, 308*f*, 414*f*, 415*f*, 416*f*
  common, 419*f*
bile ductule, 415, 416*f*
bile pigment, 418
bile salt, 418–419, 430, 432
bili light, 324
bilirubin, 122, 323–324, 324*f*, 418
biliverdin, 323–324, 324*f*, 418
binding site, 294–295, 295*f*
binge-and-purge cycle, 436
biogenic amine, 228
biological rhythm, 312
biomarker, 47
biome, 400
biotin, 433
bipolar neuron, 220, 220*f*, 283*f*
birth control, 526–530, 528*f*, 529*f*
  chemical barriers, 528
  coitus interruptus, 526
  contraception, 526
  contraceptives, combined hormone, 528–529, 528*f*
  injectable contraception, 529
  intrauterine devices, 529

mechanical barriers, 527–528, 528*f*
  rhythm method, 527
birth defect, 547, 548–549
birth process, 551–552, 552*f*
bisphosphonate, 132
bitter, taste sensation, 270
blackhead, 124
bladder. *See* urinary bladder
blastocyst, 540*f*, 541–542, 542*f*, 543*t*
blastomere, 541
blindness
  causes of, 280, 282
  from glaucoma, 282
  legal blindness, 281
  monocular blindness, 278
  night blindness, 286
blind spot, 284
blood, 108, 109*t*, 110*f*, 318–326
  centrifuged blood sample, 320*f*
  cholesterol in, 355
  coagulation, 331–332, 331*f*, 333, 333*f*
  coagulation disorders, 332
  composition, 108, 110*f*, 319, 320*f*, 327*t*
  distribution, effects of neurotransmitters on, 256*t*
  in embryonic development, 544, 546–547, 546*f*
  fetus, circulation in, 549–551, 550*f*, 551*t*
  filtering by liver, 415, 418*f*
  formation of angiotensin in bloodstream, 476
  functions of, 319, 320*f*, 321, 325
  path of, through the heart, 346, 346*f*
  pH of, 39
  production and control, 140
  red blood cell (*see* red blood cell)
  smear, 320*f*
  supply to the heart, 346–347, 347*f*, 348*f*
  volume, 319
  white cell (*see* white blood cell)
blood clot, 130, 331–332, 331*f*, 333, 333*f*, 347, 385
blood gas, 329–330
blood glucose, 63, 256*t*, 297, 306, 308–311, 309*f*, 329, 415
blood groups, 333–336, 334*f*, 334*t*, 335*f*, 335*t*
blood plasma. *See* plasma
blood platelet. *See* platelet
blood pressure, 243, 359–363, 359*f*, 360*f*, 361*f*
  arterial pressure, 359–362, 359*f*, 360*f*, 361*f*
  control of, 7, 296, 305, 305*t*, 361–362, 361*f*
  diastolic, 359, 359*f*
  elevation of, 361, 476
  in filtration, 64
  formula for, 361
  influences on, 357, 357*f*, 360, 360*f*
  in newborn, 553–554
  shock from low blood pressure, 475
  systolic, 359, 359*f*
  venus blood flow, 362–363
blood serum cholesterol, 42
blood-testis barrier, 509*f*

blood transfusion, 333–336, 334*f*, 334*t*, 335*f*, 335*t*
blood transport, 387*f*
blood urea, 63
blood vessel, 353, 354*f*, 355, 356*f*, 357*f*, 358*f*, 359*t*
  effects of epinephrine and norepinephrine on, 305*t*
  miles of blood vessels in humans, 353
  spasm, 330, 333*f*
  weakening of, 354
blood viscosity, 360, 360*f*
blood volume, 360, 360*f*
B lymphocyte. *See* B cell
body cavity, 8, 9*f*, 10
body fluid, 490–492, 490*f*, 491*f*
  composition, 491
  distribution of, 490–492, 490*f*, 491*f*
  fluid compartments, 490, 490*f*
  movement of fluid between compartments, 491–492, 492*f*
body mass index, 436, 437*f*
body region
  of the penis, 512
  of the stomach, 410
  terminology, 16–17, 17*f*, 18*f*
  of the uterus, 520
body section, terminology, 14*f*, 15–16, 15*f*
boils, 127
bolus, 409
bonding of atoms, 33, 35–36, 35*f*, 36*f*, 37*f*
bone, 132–141, 324*f*
  body movement, 140, 140*f*
  classification, 133
  development and growth, 108, 109*f*, 132, 135–137, 135*f*, 136*f*
  endochondral bone, 135*f*, 136–137, 137*f*
  fractures, 132, 138–139, 138*f*, 139*f*, 162
  functions of, 137, 140–141, 140*f*, 141*f*
  healthy, 132
  homeostasis of bone tissue, 137
  hyaline cartilage in growth of, 107
  inorganic salts, storage of, 141, 141*f*
  intramembranous bone, 135–136, 135*f*
  loss of, 137
  structure of, 131–132, 132*f*, 133–134, 134*f*, 135*f*
  support and protection, 137
  tissue, 108, 109*f*, 109*t*
bone marrow, 134, 137, 140
  blood cell formation, 140–141
  hematopoietic stem cells in, 322*f*
  red marrow, 108, 134*f*, 140, 141, 323, 324, 324*f*, 328, 386, 387, 387*f*
  transplants, 141, 323
  yellow marrow, 134*f*, 140
bone matrix, 135*f*
bone scan, 132
bony callus, 139, 139*f*
brachial artery, 26*p*, 366, 368*f*
  deep, 367*f*, 368*f*

brachialis muscle, 195*f*, 200, 200*f*, 201*t*
brachial plexus, 26*p*, 249*f*, 250
brachial region, defined, 16, 18*f*
brachial vein, 369, 370*f*, 372*f*
brachiocephalic artery, 27*p*, 29*p*, 363, 364*f*, 365*f*, 368*f*
brachiocephalic vein, 26*p*, 28*p*, 369, 369*f*, 370, 370*f*, 372*f*
brachioradialis muscle, 195*f*, 200, 200*f*, 201*f*, 201*t*
bradycardia, 352
brain, 215*f*, 232, 233*f*, 236–246
   arteries to, 355, 364, 365*f*, 366, 366*t*
   blood clot formation in, 331
   divisions, 236, 237*f*
   effects of epinephrine and norepinephrine on, 305*t*
   effects of hyperventilation on, 458
   embryonic development, 547, 548
   injury to, 213, 234, 246
   tumors, 216
   veins from, 369, 369*f*
brainstem, 236, 236*f*, 237*f*, 238*f*, 240*f*, 243–244, 244*f*
   preganglionic fibers arising from, 253*f*
breast, 23*p*, 525, 525*f*
   development during pregnancy, 542–543
   lactating, 300
   self exam, 526
breastbone, 155
breast cancer, 47, 380*f*, 526–527, 526*t*
breastfeeding, 300, 552
breath analysis, 460
breathing, 443, 450
   air volumes and capacities, 455–456, 456*t*
   control of, 456–458, 457*f*, 458*f*
   and exercise, 458
   factors affecting, 456–458, 458*f*
   first breath, 453, 553
breathing mechanism, 450–458
   expiration, 453–455, 453*f*, 454*f*, 455–456, 455*f*, 456*t*
   inspiration, 450–453, 451*f*, 452*f*, 455–456, 455*f*, 456*t*
Broca's area, 238*f*, 239
bronchi, 28*p*, 446, 447*f*, 448, 448*f*
bronchial drainage, 449
bronchial tree, 443, 446, 447*f*, 448–450, 448*f*, 449*f*, 451*t*
bronchiole, 256*t*, 448, 448*f*
bronchitis, 442
bronchogenic carcinoma, 454
buccal region, defined, 16, 18*f*
buccinator muscle, 194, 195*t*, 196*f*
buffer, 39
buffer systems
   with acid-base balance, 498–499, 498*t*, 499*f*
   with acid-base imbalance, 500–502
buffy coat of blood, 320*f*
bulbospongiosus muscle, 202*f*, 203, 520
bulbourethral gland, 507*f*, 511–512, 513*t*
bulimia, 436
bundle branch, 350*f*
bundle of His, 350
burns, 126, 127*f*
bursae, 164–165, 165*f*

bursitis, 168
B vitamins, 432
bypass graft surgery, 355

C

cadaver graft, 354
cadmium, birth defects from, 549
caffeine, 223, 493
calcaneal tendon, 195*f*, 206, 207*f*
calcaneus, 163, 163*f*, 207*f*
calcitonin, 141, 141*f*, 302, 302*t*, 303
calcium, 31*t*
   active transport across cell membranes of, 65
   in blood clot formation, 331
   in bone development, 132, 136, 137
   in cardiac cycle, 352
   dietary calcium in pregnant women, importance of, 543
   in electrolyte intake, 493
   in electrolyte output, 493, 496
   in extracellular fluid, 491*f*
   in glomerular filtrate, 475*t*
   in hair cells, 274
   in hemostasis, 333*f*
   in intracellular fluid, 491*f*
   ion channels, 54
   ions in heatstroke, research on, 489
   in kidney stones, 482
   as major mineral in human body, 434, 434*t*
   in parathyroid hormone, 303, 304*f*
   in plasma, 475*t*
   in rigor mortis, 187
   in skeletal muscle contraction, 183, 184*f*
   storage of, by bones, 141, 141*f*
   in urine, 475*t*
calcium carbonate, 275
calcium ion, 40, 40*t*
calcium oxalate, 482
calcium phosphate, 303, 304*f*
   in bone, 141
   in kidney stones, 482
callus, 119
calorie, 400, 428
canaliculus, 108, 109*f*, 134, 135*f*
   inferior, 278*f*
   superior, 278*f*
canal of Schlemm, 282
cancer, 68, 454
   breast, 47, 380*f*, 526–527, 526*t*
   carcinoma *in situ*, 520
   carcinomas, 101
   as cause of peripheral neuropathy, 267
   cervical, 94
   colorectal, 426
   endocrine paraneoplastic syndrome in, 309
   extracellular matrix in, 104
   in female infertility, 538
   leukemia, 94, 327
   lung, 454, 456
   from nonendocrine tissue, 309
   primary pulmonary cancers, 454
   prostate, 94, 511
   radiation as treatment of, 34
   skin, 121, 121*f*
   testicular, 506
   treatment of, 328

canine (teeth), 406*f*, 407, 407*t*
cannabinoid, 245
capacitation, 511
capillary, 97, 341, 346, 356–358, 356*f*, 357*f*, 358*t*
   in alveolar gas exchange, 459, 459*f*, 460*f*
   exchanges in, 356*f*, 357–358, 357*f*
   in gas transport, 462, 462*f*
   glomerular, 473, 474*f*, 475
   in kidney, 472*f*, 473, 474*f*, 475, 477, 477*f*
   lymphatic, 378*f*, 379, 379*f*, 380*f*
   myocardial, 347
   network on alveoli, 448*f*
   in spleen, 384
   systemic, 346*f*, 378*f*
capillary bed, 299*f*, 379*f*
capitate, 159*f*
capitulum, 155, 157*f*
capsaicin, 270
capsule
   of lymph node, 381*f*, 382
   of spleen, 384
carbaminohemoglobin, 461, 462
carbohydrate, 41–42, 41*f*, 46*t*, 47, 428–429
   breakdown of, 82, 83*f*
   chemical digestion of, 408
   hydrolysis of, 78, 78*f*
   in liver, metabolism of, 415, 417, 418*t*
   as macronutrient, 428
   requirements, 429
   in small intestine, digestion of, 422
   sources of, 428–429
   synthesis hampered by diabetes mellitus, 310
   use of, 429
carbon, 31*t*
   atomic structure, 32*t*
   bonding in, 35
   in carbohydrates, 41, 41*f*
   in cellular respiration, 81*f*, 82
   in dehydration synthesis, 77–78, 77*f*, 78*f*
   in nucleic acids, 44
   in proteins, 43–44, 44*f*
   in steroids, 294
carbonate ion, 40, 40*t*
carbon dioxide, 320*f*
   affecting blood pressure, 361–362
   in alkalosis, 501–502, 501*f*
   atmospheric, 457, 457*f*
   in blood, 341, 341*f*, 346, 347
   as blood gas, 329
   blood levels of, 457–458, 458*f*
   in capillaries, 357
   in cerebrospinal fluid, 457
   in citric acid cycle, 81*f*, 430*f*
   as constituent of cells, 40, 40*t*
   diffusion from blood, 448, 449*f*
   diffusion through cell membranes, 61, 62*f*
   in digestion of proteins, 431*f*
   formula for, 37*f*
   gas transport of, 460–462, 461*f*, 462*f*, 462*t*
   partial pressure of, 459–460, 460*f*
   production during exercise, 458
   in respiratory acidosis, 500–501, 500*f*
   respiratory center regulation of, 499, 499*f*

carbonic acid, 462
   respiratory acidosis, 500
   as source of hydrogen ions, 497, 497*f*
carbonic anhydrase, 462
carbon monoxide, 462, 549
carboxyl group, 44, 498
carboxypeptidase, 413, 422*t*
carbuncle, 129
carcinogen, 442
carcinoma, 101. *See also* cancer
carcinoma *in situ*, 520
cardia, 410, 410*f*
cardiac arrest, sudden, 340
cardiac center, 243, 353*f*, 361, 361*f*
cardiac control center, 352
cardiac cycle, 347, 349–350, 350*f*, 351*f*, 352, 353*f*
cardiac muscle, 111, 112*f*, 113, 113*t*, 179, 192*t*
   conduction system, 349–351, 350*f*, 351*f*
   contraction and relaxation, 191, 192
   fibers, 191–192, 349, 351*f*
cardiac output, 360, 360*f*
cardiac plexus, 253*f*
cardiac sphincter, 410, 410*f*
cardiac vein, 347, 348*f*
   anterior, 348*f*
   great, 348*f*
   small, 348*f*
cardioaccelerator reflex, 361
cardioinhibitor reflex, 361, 361*f*
cardiology, 17
cardiovascular disease
   genetic study of, 88–89
   heart attack (*see* myocardial infarction)
cardiovascular efficiency, 312
cardiovascular system, 5*t*, 6*t*, 12, 13*f*, 340–373
   aging-related changes, 556*t*
   circulation, paths of, 363
   conduction system, 349–351, 350*f*, 351*f*
   and exercise, 362, 458
   fetus, circulation in, 549–551, 550*f*, 551*t*
   neonates, cardiovascular system of, 553–554
   pulmonary circuit, 363
   systemic circuit, 363
carina, 447*f*
carotid artery, 364
   baroreceptors in, 361, 361*f*
   common, 24*p*, 25*p*, 27*p*, 28*p*, 29*p*, 353*f*, 363, 364*f*, 365*f*, 366, 367*f*, 368*f*
   external, 365*f*, 366, 366*t*, 368*f*
   internal, 365*f*, 366, 366*f*, 366*t*, 368*f*
carotid baroreceptor, 353*f*
carotid body, 457
carotid sinus, 352, 353*f*, 361, 361*f*, 365*f*, 366
carpal, 159*f*, 166*f*
carpal bone, 142*f*, 143, 157, 159*f*
carpal region, defined, 16, 18*f*
carpus, 157, 159*f*
carrier gene, 556
cartilage, 107–108, 107*f*, 109*t*
   elastic, 107, 107*f*
   fibrocartilage, 108, 108*f*
   hyaline cartilage, 107, 107*f*

cartilaginous callous, 138–139
cartilaginous joint, 164, 167*t*
cartilaginous ring, 446, 447*f*
cat, sense of taste in, 270
catabolism, 77, 78, 78*f*
catalase, 79
catalyst, 38, 79
cataract, 280
cauda equina, 249, 249*f*
caudate nucleus, 240, 240*f*
CCR5, 389
CD4, 389
cecum, 26*p*, 27*p*, 403*f*, 424, 425*f*
celiac artery, 28*p*, 363, 364*f*, 368*f*
celiac disease, 424
celiac ganglion, 252*f*
celiac plexus, 252*f*, 253*f*
celiac region, defined, 16
cell, 3, 4*f*, 50–71, 72*f*
    chemical constituents of,
        39–47
    death of, 71
    structure of, 51–60, 60*f*, 61*t*
cell body of neuron, 214, 215,
        218*f*, 221*f*
cell cycle, 67–72
cell differentiation, 71, 72*f*
cell division, 69, 70*f*
cell-mediated immunity, 386
cell membrane, 52, 52*f*, 53–54, 53*f*,
        57*f*, 58*f*, 61*t*
    characteristics, 53
    ion channels, 54
    movement through, 60–67, 67*t*
    permeability, 53
    as separating fluid compartments,
        490, 490*f*
    structure of, 53–54, 53*f*
cell membrane potential, 222–223,
        223*f*, 224*f*, 225–227,
        225*f*, 226*f*
    action potential, 225–227,
        225*f*, 226*f*
    ions, distribution of, 222, 223*f*
    potential changes, 225, 225*f*
    resting potential, 223, 224*f*, 225
cells of Leydig, 506, 508*f*
cellular adhesion molecule (CAM), 54
cellular degradation, 555–556
cellular immune response, 386–388,
        387*f*, 387*t*, 388*f*
cellular metabolism. *See* metabolism
cellular reprogramming, 50
cellular respiration, 80, 81*f*, 82, 186,
        186*f*, 443
cellular turnover, 421
cellulose, 427, 429
cementum, 408*f*
central canal, 108, 109*f*, 134, 135*f*,
        234, 235*f*
central chemoreceptor, 457
central nervous system, 214,
        215*f*, 221*f*
    depression by drugs, 245
    subdivisions, 246*t*
central sulcus, 238–239, 238*f*
central vein, 415, 416*f*
centrifugation, 320*f*
centriole, 52*f*, 57–58, 59*f*, 70*f*
centromere, 69, 70*f*, 507
centrosome, 57–58, 59*f*, 61*t*
cephalic region, defined, 17, 18*f*
cephalic vein, 370*f*, 372*f*
cerebellar peduncle, 242*f*,
        244–245, 244*f*

cerebellum, 233*f*, 236, 237*f*, 238*f*,
        240*f*, 244–246, 244*f*, 275
cerebral aqueduct, 240, 241*f*
cerebral arterial circle, 364
cerebral artery
    anterior, 366*f*
    posterior, 364, 366*f*
cerebral cortex, 238–239, 238*f*, 244–
        245, 265, 267, 298*f*, 458
    functional regions of,
        238–239, 240*f*
    injuries to, 239
cerebral hemispheres, 236–237, 240*f*
cerebral thrombosis, 331
cerebral vascular accident, 355
cerebrospinal fluid, 234, 240, 241*f*,
        242, 457, 490
cerebrum, 233*f*, 236–240, 236*f*, 237*f*,
        238*f*, 240*f*, 241*f*, 242
    in cardiac cycle, 352, 353*f*
    functions of, 238–239, 240*f*
    hemisphere dominance,
        239–240, 240*f*
    structure of, 236–238, 237*f*, 238*f*
    ventricles and cerebrospinal
        fluid, 240, 241*f*, 242
ceruminous gland, 124
cervical cancer, 94, 520
cervical cap, 528
cervical enlargement, 234, 234*f*
cervical lymph node, 382*f*
cervical nerve, 248–249, 249*f*
cervical orifice, 519*f*, 520
cervical plexus, 249*f*, 250
cervical region, defined, 17, 18*f*
cervical vertebra, 152*f*
cervical vertebrae, 150–151, 151*f*, 152*f*
cervix, 514, 516*f*, 519*f*, 520
    birth control, 529*f*
    in birth process, 548*f*,
        551–552, 552*f*
chambers of the heart, 343–345,
        344*f*, 345*f*
Charcot-Marie-Tooth disease, 193
cheek, 403
chemical barrier
    as birth control, 528
    against pathogens, 385
chemical bond, 32, 35–36, 35*f*, 36*f*
chemical buffer system
    with acid-base balances,
        498–499, 498*t*, 499*f*
    with acid-base imbalances,
        500–502
chemical digestion, 401
chemical energy, release of, 80, 80*f*
chemical peel, 117
chemical reaction, 37–38
chemistry, defined, 31
chemoreceptor, 263, 458
chemosensitive areas, 457
chemotaxis, 385, 392*f*
chest
    abdominal wall, muscles of the,
        201, 203*t*
    pectoral girdle, muscles that
        move the, 197, 198*f*,
        198*t*, 199*f*
chewing, 409
chief cell, 411*f*
childhood, stages in
        development, 554*t*
chili pepper, as stimulating pain
        receptors, 270
chlamydia, 530, 530*t*

chloride
    bonding in, 33, 35, 35*f*
    in cystic fibrosis, 449
    in electrolyte intake, 493
    in extracellular fluid, 491*f*
    in glomerular filtrate, 475*t*
    in intracellular fluid, 491*f*
    ion channels, 54
    in plasma, 330, 475*t*
    tubular reabsorption by, 478, 478*f*
    in urine, 475*t*, 476
chloride ions as constituent of
        cells, 40
chlorine
    bonding in, 33, 35*f*
    as element in human body, 31*t*
    as major mineral in human body,
        434, 434*t*
chlorolabe, 286
cholecystectomy, 418
cholecystokinin, 412, 415, 418,
        418*t*, 419*f*
cholesterol
    in atherosclerosis, 355
    in bile, 418
    in cell membranes, 54
    HDL, 355
    intestinal absorption of, 423
    LDL, 355
    in lipoproteins, 430
cholinergic fiber, 251, 254, 254*f*, 256*f*
chondrocyte, 107, 107*f*
chordae tendineae, 344, 344*f*
chorion, 544, 544*f*, 545, 545*f*
chorionic villus, 544, 545, 545*f*
chorionic villus sampling, 558, 558*f*
choroid artery, 365*f*
choroid coat, 280, 280*f*, 281*f*, 283*f*
choroid plexus, 216, 240, 241*f*
chromatid, 70*f*, 507, 509
chromatin, 52*f*, 60, 60*f*, 61*t*, 70*f*, 510
chromatophilic substance, 216, 218*f*
chromium, as trace element in
        human body, 31*t*,
        434, 435*t*
chromosome, 59*f*, 60
    in fertilization, 539, 539*f*, 540*f*
    fetal chromosome checks,
        558–559, 558*f*, 559*f*
    in mitosis, 69, 70*f*
    in oogenesis, 517, 517*f*
    paired with genes, 556–557, 556*f*
    sex chromosome, 556
    in spermatogenesis, 507, 508*f*,
        509, 509*f*
chronic hepatitis, 417
chronic myeloid leukemia, 328
chronic obstructive pulmonary
        disease, 454
chronic pain fiber, 165
chronobiology, 312
chylomicron, 423, 423*f*
chyme, 413, 414*f*, 415, 419*f*, 424, 427
chymotrypsin, 413
cigarette
    birth defects from, 549
    secondhand smoke, dangers
        of, 442

in simple columnar epithelial
        tissue, 97
    in uterine tube, 519–520
ciliary body, 280, 280*f*
ciliary ganglion, 253*f*
ciliary muscle, 280*f*, 281, 282, 282*f*
ciliary process, 280, 281*f*
circadian rhythm, 309, 312
circle of Willis, 354, 364, 366
circular fiber, 401
circular fold, 404*f*
circular muscle, 404*f*
circular set, 282
circulation. *See* cardiovascular
        system
circumcision, 512
circumduction, 167, 169*f*
circumflex artery, 348*f*
    anterior, 367*f*
    posterior, 367*f*
cisterna chyli, 380*f*
cisternae, 181*f*, 182
citric acid, tubular reabsorption
        by, 478
citric acid cycle, 80, 81*f*, 83*f*,
        429–430, 430*f*
clavicle, 23*p*, 24*p*, 142*f*, 143, 155, 156*f*,
        200*f*, 525*f*
clavicular notch, 154*f*
cleavage, 540*f*, 541, 543*t*
cleavage furrow, 69, 70*f*
cleft palate, 148
clitoris, 202*f*, 516*f*, 521, 522*t*
clock gene, 312
cloning cells, 388, 390*f*, 392
Clostridium botulinum, 117
clot busting, 331, 332
clotting factor, 331, 332
clotting mechanism, 333*f*
cobalt
    as major mineral in human body,
        434, 434*t*
    as trace element in human body,
        31*t*, 434, 435*t*
cocaine, 245
coccygeal nerve, 248–249, 249*f*
coccygeus, 202*f*
coccyx, 142, 142*f*, 154*f*, 159*f*,
        160*t*, 202*f*
cochlea, 271*f*, 272–273, 272*f*, 273*f*
cochlear branch, 247
cochlear duct, 272*f*, 273, 273*f*
cochlear nerve, 272*f*, 273*f*
codon, 87–88, 87*f*, 88*f*, 89*f*
coenzyme, 80
cofactor, 80
coitus, 526
coitus interruptus, 526
cold receptor, 264–265
collagen, 46*f*, 103, 104–105, 105*f*, 117,
        134, 432–433
collagen fiber, 330*f*
collagenous fiber, 103, 103*t*
collarbone. *See* clavicle
collateral, 218*f*
collecting duct, 379, 380*f*, 469*f*, 471,
        471*f*, 472*f*, 481*t*, 493
colloid, 301, 302*f*
colloid osmotic pressure, 329, 357,
        357*f*, 491
colon, 424, 425*f*, 426
    ascending, 26*p*, 27*p*, 371*f*,
        425*f*, 426
    descending, 26*p*, 27*p*, 28*p*, 371*f*,
        425*f*, 426

colon (continued)
referred pain, 266*f*
sigmoid, 27*p*, 28*p*, 425*f*, 426
transverse, 26*p*, 27*p*, 421*f*, 425*f*, 426
colony-stimulating factor, 387
colorblindness, 285, 296
colorectal cancer, 426
colostrum, 391, 552
columnar epithelial tissue, 95–96, 102*t*, 421*f*
pseudostratified, 98, 98*f*
simple, 97, 97*f*, 519
stratified, 98, 100*f*
supporting olfactory receptor cells, 268*f*
in uterine tube, 519
in uterus, 520
comatose state, 244
combined hormone contraceptive, 528–529, 528*f*
comedone, 124
comminuted fracture, 138, 138*f*
common bile duct, 419*f*
common carotid artery, 24*p*, 25*p*, 27*p*, 28*p*, 29*p*, 353*f*, 363, 364*f*, 365*f*, 366, 367*f*, 368*f*
common hepatic duct, 414*f*, 415, 418, 419*f*
common iliac artery, 27*p*, 28*p*, 364, 364*f*, 367, 368*f*, 550*f*
common iliac vein, 371, 372*f*
communicating artery, posterior, 366*f*
community, ecological, 400
compact bone, 133–134, 134*f*, 135*f*, 136*f*, 137, 139*f*
compartment, fluid, 490, 490*f*
compensated acidosis, 501
complement, 385
complementary base pairing, 84, 84*f*
complete protein, 431
compound, 36, 37*f*
compound fracture, 138
computed tomographic colonography, 426
computer-aided sperm analysis, 514, 514*f*
computerized tomography (CT), 1, 16
concentration gradient, 61, 62*f*, 357*f*
concussion, 246
condom, 527–528, 528*f*
conductive hearing loss, 275
condylar joint, 165, 166*f*, 167*t*
condyle, 144*t*, 155
cone, 283*f*, 284, 285–286, 285*f*
conformation of a protein, 44, 45*f*
conjunctiva, 277, 278*f*, 280
connecting stalk, 543–544, 544*f*, 546*f*
connective tissue, 95, 95*t*, 102–108, 109*f*, 109*t*, 110*f*, 354*f*
bone (*see* bone)
cartilage (*see* cartilage)
categories of, 104–109, 110*f*
cell types, 102–103, 102*f*, 103*f*
characteristics, 102–105, 102*f*, 103*f*, 103*t*
dense, 106, 106*f*, 109*t*
extracellular matrix, 102, 104–105, 105*f*
fibers, 103, 103*t*
in labium minora, 521
loose, 104–106, 105*f*, 106*f*, 109*t*
in mammary gland, 525
sun, exposure to, 103

contraception, 526
contralateral, defined, 15
convergence, 229, 230*f*
copper, as trace element in human body, 31*t*, 434, 435*t*
copy number variants, 91, 91*f*
coracobrachialis muscle, 24*p*, 25*p*, 26*p*, 197, 199*t*
coracoid process, 155, 156*f*, 194
corn, 119
cornea, 278*f*, 279–280, 280*f*
corneal transplant, 280
coronal, defined, 15, 15*f*
coronal suture, 144, 145*f*, 146*f*, 148*f*, 150*f*, 175*p*, 177*p*
corona radiata, 538, 539*f*
coronary artery, 345, 346–347, 347*f*, 348*f*, 364*f*, 368*f*
coronary embolism, 355
coronary ligament, 415*f*
coronary sinus, 344, 344*f*, 347, 348*f*
coronary thrombosis, 331, 347, 355
coronoid fossa, 155, 157, 157*f*
coronoid process, 146*f*, 149, 157, 158*f*
corpora cavernosa, 512, 521
corpora spongiosum, 512
corpus albicans, 518*f*, 523, 524*f*
corpus callosum, 236–237, 237*f*, 238, 244*f*
corpus cavernosum, 507*f*
corpus luteum, 518*f*, 523, 523*t*, 524*f*, 542, 543
corpus spongiosum, 507*f*
cortex of adrenal gland, 300
cortex of ovary, 516
cortical radiate artery, 469, 470*f*, 472*f*
cortical radiate vein, 470*f*, 472*f*
corticospinal tract, 235
corticotropin-releasing hormone, 300, 307*f*, 313*f*
cortisol, 306, 307*f*, 307*t*, 312, 313*f*
costal cartilage, 154*f*, 156*f*
costal region, defined, 17
coughing, 455
covalent bond, 35, 36*f*
cow, milk of, 553
Cowper's gland, 511
coxal region, defined, 17, 18*f*
cramp, muscle, 187
cranial branch, 247
cranial cavity, 8, 9*f*, 10, 10*f*, 176*p*, 242
cranial nerve, 215*f*, 246–248, 247*f*, 248*t*
cranial nerve (III), 253*f*
cranial nerve (IX), 253*f*
cranial nerve (VII), 253*f*
cranial parasympathetic neuron, 254*f*
cranium, 142–143, 142*f*, 144, 145*f*, 146*f*, 147, 148*f*, 233*f*
C-reactive protein, 47
creatine
as nonprotein nitrogenous substance, 330
tubular reabsorption by, 478
creatine phosphate, 185–186, 186*f*, 330
creatinine, as nonprotein nitrogenous substances, 330
crest, 144*t*
cretinism, 303
cribriform plate, 147, 147*f*, 148*f*, 176*p*, 268*f*
cricoid cartilage, 445, 445*f*
crista ampullaris, 275, 277*f*
cristae, 56, 57*f*

crista galli, 147, 147*f*, 148*f*, 176*p*
critical periods of development, 547
Crohn disease, 428
cross bridge, myosin, 183, 183*f*, 184*f*
cross section, defined, 15, 16*f*
crown (teeth), 407, 408*f*
crural region, defined, 17, 18*f*
cubital region, defined, 17, 18*f*
cubital vein, median, 369, 370*f*, 372*f*
cuboid, 163*f*
cuboidal epithelial tissue, 95–96, 102*t*
simple, 96, 97*f*
stratified, 98, 99*f*
cupola, 277*f*
curvature
greater, 410*f*
lesser, 410*f*
Cushing syndrome, 306, 496
cusp, 344, 347*f*
cuspid (teeth), 406*f*, 407, 407*t*
cutaneous carcinoma, 121
cutaneous membrane, 110
cyanide, 80, 442, 460
cyanocobalamin, 433
cyanosis, 122, 319
cyclic adenosine monophosphate (cAMP), 295, 295*f*
cyclic AMP (cAMP) (cyclic adenosine monophosphate), 295, 295*f*
cyclophosphamide, 527
cyst, 526, 527
cysteine, 44, 44*f*, 431*t*
cystic duct, 27*p*, 414*f*, 415*f*, 418, 418*f*, 419*f*
cystic fibrosis, 54, 449, 557, 557*f*
cystic fibrosis transmembrane regulator, 449
cystitis, 483
cysts, 129
cytocrine secretion, 119
cytokine, 387, 388, 388*f*, 390*f*, 428
cytokinesis (cytoplasmic division), 67, 68*f*, 69, 70*f*, 71, 72*f*
cytology, 17
cytoplasm, 52, 52*f*, 55–59, 60*f*, 112*f*, 295*f*, 517, 539, 539*f*
cytoplasmic division (cytokinesis), 67, 68*f*, 69, 70*f*, 71, 72*f*
cytosine, 84, 84*f*, 85, 85*f*, 86
cytoskeleton, 55, 58
cytosol, 53, 57*f*, 82
cytotoxic T cell, 388, 388*f*, 389

D

daughter cell, 66, 67*f*, 509*f*
deafness, 274
deamination, 431, 431*f*
death
signs of, 5
skeletal muscles after death, change in, 187
decibel, 274
decidua basalis, 546*f*
decomposition, 38
decubitus ulcer, 122
dedifferentiation, in cancer, 68
deep, defined, 15
deep brachial artery, 367*f*, 368*f*
deep femoral artery, 368*f*
deep volar artery, 367*f*
defecation reflex, 427
defibrillator, used for treatment of cardiac arrest, 340

De Humani Corporis Fabrica (Vesalius), 2*f*
dehydration, 300, 310, 494
dehydration synthesis, 77–78, 77*f*, 78*f*
dehydrocholesterol, 117
delayed-reaction allergy, 393
deltoid muscle, 23*p*, 24*p*, 194, 195*f*, 198, 198*f*, 199*f*, 199*t*, 200*f*
deltoid tuberosity, 155, 157*f*
den, 151, 153*f*, 166*f*
denatured shape, 44
dendrite, 214, 215*f*, 217, 218, 218*f*, 219*f*, 221*f*
dense connective tissue, 106, 106*f*, 109*t*
dental alveoli, 148
dental arch, 148
dental caries, 407
dentin, 407, 408*f*
deoxygenated blood, 341, 341*f*
deoxyhemoglobin, 319
deoxyribonucleic acid. *see* DNA (deoxyribonucleic acid)
deoxyribose, 429
depolarization, 225, 225*f*, 226*f*
deposition of matrix, 137
depressant, CNS, 245
depression, 170*f*
dermabrasion, 117
dermal blood vessel, 123*f*
dermal papillae, 118*f*, 119*f*, 122
dermatitis, 129, 394
dermatology, 17
dermis, 117, 118*f*, 119*f*, 120*f*, 122, 264*f*
descending aorta, 28*p*, 29*p*, 363, 365*t*, 551
descending colon, 26*p*, 27*p*, 28*p*, 371*f*, 425*f*, 426
descending limb of nephron loop, 471, 472*f*, 481*t*
descending thoracic aorta, 28*p*
descending tract, 234–235, 236*f*
detoxification by liver, 417, 418*t*
detrusor muscle, 483, 483*f*
development, 537
of bone, 108, 109*f*, 135–137, 135*f*, 136*f*
neonatal period, 553–554, 554*f*, 554*t*
prenatal period (*see* prenatal period)
deviated septum, 444
diabetes insipidus, 300, 496
diabetes mellitus, 310–311, 460, 480, 501, 501*f*
as cause of peripheral neuropathy, 267
gene chip for, 91
tests for, 311
type 1, 310, 394
type 2, 310
diacylglycerol, 296
dialysis, 62–63
diapedesis, 325, 326*f*
diaphoresis, 347
diaphragm, 8, 9*f*, 25*p*, 26*p*, 27*p*, 28*p*, 29*p*, 342, 342*f*
in expiration, 453, 453*f*, 456, 457*f*
hiatal hernia protruding through, 410
in inspiration, 451, 452, 452*f*, 456, 457*f*
referred pain, 266*f*
veins that drain, 383*f*

diaphragm (contraception), 528, 528*f*
diaphysis, 133, 134*f*
diarrhea, 424, 501, 501*f*
diastole, 347, 349, 349*f*
diastolic blood pressure, 359, 359*f*
diencephalon, 236, 237*f*, 242–243, 242*f*
differentially permeable membrane, 53
differential white blood cell count, 327
differentiation, cell, 71, 72*f*
diffusion, 60–61, 67*t*
   in capillaries, 357
   through cell membranes, 61, 62*f*, 63
diffusional equilibrium, 61, 62*f*
digestion, 5*t*
   microflora in, 400
   smooth muscle tissue in, 110–111, 111*f*
   types of, 401
digestive system, 12, 13*f*, 400–437
   aging-related changes, 556*t*
   embryonic development, 548
   functions of, 401
   hormones of the, 418*t*
   microorganisms in, 400
digestive tract, smooth muscle tissue along, 111*f*
digital artery, 367*f*
digital region, defined, 17, 18*f*
dihydrogen phosphate, 498
dioxin, 30
dipeptide, 78, 78*f*
diphenhydramine, 276
diphenylhydantoin, 223
diplopia, 278
diuretic, 493
disaccharide, 41, 408, 413
   in dehydration synthesis, 77*f*
   hydrolysis of, 78
diseasome, 3
dissecting aneurysm, 354
distal, defined, 14*f*, 15
distal convoluted tubule, 470*f*, 471, 473*f*, 481*t*, 493
distal epiphysis, 134*f*
distal gut microbial community, 400
distal phalanx, 159*f*, 163*f*
diuretic, 300
divergence, 229–230, 230*f*
dizygotic twins, 542
DNA (deoxyribonucleic acid), 46, 83, 556
   in AIDS, 389
   analysis of human gut microflora using, 400
   in cell nucleus, 60, 60*f*
   genetic code in, 85
   in mitosis, 69, 70*f*
   molecular structure, 84, 84*f*
   mutating of, 86
   protein synthesis, DNA provided instructions for (*see* protein synthesis)
   replication of, 69, 70*f*, 71, 84–85, 85*f*
   RNA compared with, 86*t*
   of skin cells, 119, 120*f*
   in steroid hormone, 294, 294*f*
DNA chip, 90–91, 90*f*
DNA chip technology, 328
DNA microarray, 90–91, 90*f*, 328
DNA polymerase, 84

dog
   sense of smell in, 267
   sense of taste in, 270
dominant hemisphere, 239–240, 240*f*
dominant inheritance, 557, 557*f*, 560
dopamine, 228, 229*t*, 240
dorsal, defined, 14, 14*f*
dorsal arch, 370*f*
dorsalis pedis artery, 368*f*
dorsal region, defined, 17
dorsal respiratory group, 456, 457*f*
dorsal root, 233*f*, 235*f*
dorsal root ganglion, 235*f*, 249, 249*f*, 251*f*, 266*f*
dorsiflexion, 167, 169*f*
dorsum region, defined, 18*f*
doubled muscle, 178
double vision, 278
drug abuse, 245
drug tolerance, 245
Duchenne muscular dystrophy, 193
ductus arteriosus, 550*f*, 551, 551*t*, 554
ductus deferens, 26*p*, 483*f*, 506, 507*f*, 510, 513*t*, 529*f*
ductus venosus, 550, 550*f*, 551*t*, 554*f*
duodenal ulcer, 412
duodenum, 27*p*, 28*p*, 410*f*, 420, 420*f*, 421*f*, 422
   in alimentary canal, 403*f*
   chyme entering, 413, 414*f*, 415, 419*f*
   mucous membrane in, 413, 415
dural sinus, 233*f*
dural space, 241*f*
dura mater, 233, 233*f*, 241*f*
dynamic equilibrium, 275–276
dyspnea, 347
dystrophin, 193

E
ear
   auditory tube, 271–272
   elastic cartilage in external, 107*f*
   generation of sensory impulses from the, 274*f*
   growth of, 108
   hearing (*see* hearing, sense of)
   inner ear, 272–274, 272*f*, 273*f*
   middle, 271, 272
   outer, 270, 271*f*
eardrum, 270, 271*f*, 275
eccrine sweat gland, 123*f*, 124, 125
ecosystem, 400
ectoderm, 543, 544, 544*f*
ectopic pregnancy, 541
eczema, 129
edema, 30, 64, 329, 381, 475, 495, 495*t*
effector, 6–8, 7*f*, 8*f*, 8*f*, 216, 221*f*
   in cardiac cycle, 353*f*
   in patellar reflex, 231, 231*f*
   in reflex arc, 231, 232*t*
   in regulation of blood glucose concentration, 309*f*
   visceral, 254*f*
   in withdrawal reflex, 232, 232*f*
efferent arteriole, 470*f*, 471, 472*f*, 473, 473*f*, 474*f*, 475, 477*f*
efferent fiber, 230
efferent lymphatic vessel, 380*f*, 381*f*
efferent neuron. *See* motor neuron
egg, female. *See* oocyte
ejaculation, 510, 511, 512–513
ejaculatory duct, 507*f*, 510
elastic cartilage, 107, 107*f*, 109*t*
elastic fibers, 103, 103*t*

elbow, 140, 140*f*, 369
electrocardiogram, 351–352, 351*f*, 352*f*, 362
electrolyte, 38–39, 490
   absorption of, in large intestine, 427
   absorption of, in small intestine, 423, 423*f*, 424*t*
   in bile, 418
   as intracellular fluid, 490
   movement of, between fluid compartments, 491–492, 492*f*
   plasma, 330
   tubular reabsorption of, 478, 478*f*
electrolyte balance, 493, 496–497, 496*f*
   intake, 493
   output, 493, 496–497, 496*f*
   in pregnancy, importance of, 543
electromyography, 193
electron, 32, 32*f*, 37*t*
electron transport chains, 80, 81*f*, 83*f*
element, 31–32
   in human body, 31–32, 31*t*
   trace, 31*t*
   trace elements, 434, 435*t*
elevation, 170*f*
ellipsoidal joint, 165
embolus, 331, 347
embryo, 541, 542, 544–547, 544*f*, 545*f*, 546*f*, 547*f*
embryo adoption, 540*t*
embryonic development, 548
embryonic disc, 543, 544, 544*f*, 546
embryonic stage of development, 543–547, 544*f*, 545*f*, 546*f*, 547*f*
emission, 512, 513
emphysema, 454, 456, 460
emulsification, 418–419
enamel, teeth, 407, 408*f*
endocarditis, 345, 349
endocardium, 343, 343*f*
endochondral bone, 135*f*, 136–137, 137*f*
endocrine glands, 101, 292, 293*f*, 296*f*
endocrine paraneoplastic syndrome, 309
endocrine system, 12, 13*f*, 291–313. *See also* hormone
   aging-related changes, 556*t*
   characteristics, 292–293, 293*f*
   compared to nervous system, 292–293, 292*t*, 293*f*
   prostaglandins, 296
   stress (*see* stress)
endocrinology, 17
endocytosis, 65–66, 66*f*, 67*t*
endoderm, 543, 544, 544*f*
endogenous factor, 312
endolymph, 272, 272*f*, 273*f*, 274, 277*f*
endometrial biopsy, 538*t*
endometriosis, 538
endometrium, 519*f*, 520, 520*f*
   in embryonic development, 540*f*, 541, 542*f*, 544*f*, 546*f*
   in reproductive cycle, 523–524, 524*f*
endomysium, 179, 180*f*
endoneurium, 230, 230*f*
endoplasmic reticulum, 55, 56*f*, 57*f*, 61*t*, 390*f*, 423, 423*f*
   rough, 52*f*, 55, 56*f*, 57*f*, 58*f*, 59
   smooth, 52*f*, 55, 56*f*

endorphin, 228, 229*t*, 267
endosteum, 134, 134*f*, 135*f*
endothelial cell, 356*f*, 474*f*
endothelial lining, 330*f*
endothelium, 354, 354*f*, 356, 356*f*
energy, 80
   from carbohydrates, 428, 429
   from fats, 429, 430, 430*f*
   forms, 80
   from proteins, 431, 431*f*
enkephalin, 229*t*, 267
enterokinase, 413, 422*t*
entry inhibitor, 389
environmental tobacco smoke, 442
enzyme, 43, 44*f*, 77
   action of, 79–80, 79*f*
   angiotensin-converting, 476, 476*f*
   as chemical barriers from pathogens, 385
   enzyme-catalyzed reaction, 79, 79*f*
   factors that alter, 80
   fat-splitting, 419
   in pancreatic juice, 413, 415
   protein-digesting enzyme by sperm cell, release of, 538–539, 539*f*
   in small intestine, 422, 422*t*, 423
   urine formation by, 476
enzyme catalyzed reaction, 79, 79*f*
eosinophil, 320*f*, 322*f*, 325, 326*f*, 327*t*
eosinophilic band cell, 322*f*
eosinophilic myelocyte, 322*f*
ependymal cell, 216, 217*f*
epicardium, 10, 342, 342*f*, 343, 343*f*
epicondyle, 144*t*, 155
epicranial aponeurosis muscle, 196*f*
epicranius muscle, 194, 195*t*, 196*f*
epidemiology, 17
epidermis, 117, 118*f*, 119–122, 119*f*, 120*f*, 121*f*, 264*f*
epididymis, 26*p*, 506, 507*f*, 508*f*, 510, 513*t*
epidural space, 233, 233*f*
epigastric region, defined, 16, 17*f*
epiglottic cartilage, 445, 445*f*, 446, 446*f*
epiglottis, 406*f*, 409, 443*f*, 446, 446*f*
epilepsy, 223
epimysium, 179, 180*f*, 230
epinephrine, 304–305, 305*f*, 305*t*
   in control of blood pressure, 362
   fight-or-flight response, 313*f*
   as neurotransmitter, 228
   as treatment for anaphylactic shock, 394
   as treatment for peanut allergy reactions, 377
epineurium, 230, 230*f*
epiphyseal plate, 136, 137, 137*f*
epiphysis, 133, 134*f*, 136, 137
episiotomy, 551
epithelial cell, 51, 51*f*, 264*f*
   environmental tobacco smoke, damage from breathing, 442
   in kidney, 471, 471*f*, 475, 480
   in lining of gallbladder, 418
   in respiratory tract, 443–444, 445*f*, 454, 454*f*
   in small intestine, 421, 422, 423, 423*f*
epithelial membrane, 110

epithelial tissue, 95–102, 95*t*, 102*t*
 in alimentary canal, 401, 404*f*
 cancers in, 101
 characteristics, 95–96
 columnar epithelial (*see* columnar epithelial tissue)
 cuboidal epithelial (*see* cuboidal epithelial tissue)
 glandular, 101–102, 101*f*, 101*t*
 in respiratory tract, 454*f*
 of seminiferous tubule, 507
 as separating extracellular fluid from extracellular fluid, 490
 simple, 95–97, 96*f*, 97*f*, 102*t*
 squamous epithelial (*see* squamous epithelial tissue)
 stratified epithelial tissue, 98, 99*f*, 100*f*, 102*t*
 transitional, 99, 101*f*
equilibrium, sense of, 275–276, 276*f*
 dynamic equilibrium, 275–276
 static equilibrium, 275
erection
 female, 521
 male, 512
erythema, 126, 129
erythroblastosis fetalis, 336
erythrocyte. *See* red blood cell
erythrolabe, 286
erythropoiesis, 321
erythropoietin, 310, 321, 323*f*, 469
Escherichia coli, 400, 467, 484
esophageal hiatus, 410
esophageal mucosa, 410
esophageal sphincter, lower, 410*f*
esophagus, 11*f*, 28*p*, 29*p*, 303*f*, 406*f*, 410, 410*f*
 in alimentary canal, 401, 402*f*, 403*f*
 cancer of the, 410
 hiatal hernia in, 410
 as member of digestive system, 401, 402*f*, 410
 as member of respiratory system, 443*f*
 in respiratory tract, 444*f*
 stratified squamous epithelial tissue in, 98, 99*f*
essential amino acid, 431
essential fatty acid, 430
essential hypertension, 362
essential nutrient, 428
estradiol, 522
estriol, 522
estrogen, 306, 522, 523, 523*t*, 524, 524*f*, 537
 changes during pregnancy, 542, 542*f*, 543*t*
 in combined hormone contraceptives, 528–529, 528*f*
 as maternal serum marker, 558
 in milk production and secretion, 552
 placental estrogen, 542–543
estrogen receptor, 527
estrone, 522
ethmoidal sinus, 145*f*, 147
ethmoid bone, 145*f*, 146*f*, 147, 147*f*, 148*f*, 175*p*, 176*p*, 177*p*, 444
eumelanin, 123
Eustachian tube, 271–272, 271*f*, 406*f*, 444*f*

eversion, 167, 170*f*
exchange reaction, 38
excitatory neurotransmitter, 222, 222*f*, 228
excretion, 5*t*, 12, 14
exercise, 189
 adapting of cardiovascular system to, 362
 benefits of, 178, 189
 and breathing, 458
 lack of, 109, 189
 resistance training, 178
exocrine gland, 101–102, 292, 490
exocytosis, 55, 65–66, 66*f*, 67*t*
exogenous factor, 312
exons, 86
exophthalmia, 303
expiration, 453–456, 453*f*, 454*f*, 455*f*, 456*t*
expiratory intercostal muscle, 453, 453*f*, 457*f*
expiratory reserve volume, 456*t*
extension, 167, 169*f*, 192–193
extensor carpi radialis brevis muscle, 200, 201*f*, 202*t*
extensor carpi radialis longus muscle, 200, 200*f*, 202*t*
extensor carpi ulnaris muscle, 201, 201*f*, 202*t*
extensor digitorum longus muscle, 195*f*, 205, 206, 207*t*
extensor digitorum muscle, 194, 201, 201*f*, 202*t*
external acoustic meatus, 146*f*, 147, 177*p*, 270, 271*f*
external anal sphincter muscle, 202*f*, 426, 426*f*
external carotid artery, 365*f*, 366, 366*t*, 368*f*
external iliac artery, 367, 368*f*
external iliac vein, 371, 372*f*
external intercostal muscle, 24*p*, 25*p*, 29*p*, 199*f*, 457*f*
external intercostal muscles, 452, 452*f*
external jugular vein, 25*p*, 369, 369*f*, 370*f*, 372*f*
external oblique muscle, 23*p*, 24*p*, 25*p*, 194, 195*f*, 199*f*, 201, 203*t*
external respiration, 443
external urethral sphincter, 483, 483*f*, 484
external vertebral orifice, 484
extracellular fluid, 490–492, 490*f*, 491, 491*f*
extracellular fluid compartment, 490, 490*f*, 494, 494*f*
extracellular matrix (ECM), 102, 104–105, 105*f*
 in bone, 108
 in cartilage, 107, 107*f*
extracorporeal shock-wave lithotripsy, 482
extraembryonic cavity, 544*f*, 546*f*
extrafollicular cell, 302*f*
extrapyramidal tract, 235
extrinsic muscle, 278, 279
eye
 aging of the, 284
 blindness (*see* blindness)
 blind spots, 284
 disorders, 286
 inner layer, 282, 283*f*, 284, 284*f*
 light refraction, 284–285, 284*f*

middle layer, 280–282, 281*f*, 282*f*
 muscles associated with, 279*t*
 nerve pathways, visual, 286–287, 286*f*
 outer layer, 279–280
 photopigments, 286
 photoreceptors, 285–286, 285*f*
 sight, sense of (*see* sight, sense of)
 structure of, 278*f*, 279–282, 279*f*, 280*f*, 281*f*, 282*f*, 283*f*, 284, 284*f*
eyelash, 278*f*
eyelid, 277, 278*f*, 279*t*

F

face, 142*f*
 bones of the, 142–143, 142*f*, 175*p*, 177*p*
 embryonic development, 544, 545*f*
 facial expression, muscles of, 194–195, 194*f*, 194*t*, 196*f*, 197
 wrinkles, treatments for facial, 117
facet, 144*t*, 150, 152*f*, 153*f*
facial artery, 365*f*
facial nerve (VII), 247, 247*f*, 248*t*
facial skeleton, 145*f*, 146*f*, 148–149
facial vein, anterior, 369*f*, 372*f*
facilitated diffusion, 63, 63*f*, 67*t*
facilitation, 229
factor V Leiden, 332
fainting, 458
falciform ligament, 25*p*
false rib, 153, 154*f*, 155
false vocal cord, 446, 446*f*
fascia, 179, 180*f*
fascicle
 of muscle, 179, 180*f*
 of nerves, 230
fasting, 106
fast-twitch muscle fiber, 189
fat, 42, 42*f*
 absorption of, in small intestine, 423, 423*f*
 breakdown of, 83*f*
 deposits on arteries, 355
 fat globules, action of bile salts on, 418–419
 in human milk, proportion of, 553
 sources of, 429
 synthesis hampered by diabetes mellitus, 310–311
 use of, 429–430, 430*f*
fat cells, in adults, 106
fatigue, muscle, 187
fat-soluble vitamin, 432, 432*t*
fatty acid, 42, 42*f*, 429–430, 430*f*
 in cell membranes, 53–54, 53*f*
 in dehydration synthesis, 77–78, 77*f*, 78*f*
 essential, 430
 incomplete oxidation of, 497
 intestinal absorption of, 423*f*, 424*t*
 oxidation of, 497
 as source of hydrogen ions, 497, 497*f*
feces, 427, 492*f*, 493
female
 blood counts, 515
 blood volume, 319
 pelvic outlet, muscles of the, 201, 202*f*, 203, 203*t*

pelvis, 160, 160*t*, 161*f*
 percentage of water by weight, 490
 sex chromosomes, 556, 557
female condom, 527–528, 528*f*
female reproductive system, 516–526, 556*t*
 accessory sex organs, 516
 eggs, selling of, 505
 erection, lubrication, and orgasm, 521
 external reproductive organs, 521
 hormonal control of, 522–525, 523*t*, 524*t*
 internal reproductive organs, 519–521, 519*f*, 520*f*
 menopause, 524–525
 milk production and secretion, 552–553, 553*f*
 primary sex organs, 516
 reproductive cycle, 522–524, 523*t*, 524*f*, 537, 537*f*
 sex hormone, 522
femoral artery, 24*p*, 25*p*, 27*p*, 367, 368*f*
 deep, 368*f*
femoral nerve, 24*p*, 25*p*, 26*p*, 249*f*, 250
femoral region, defined, 17, 18*f*
femoral vein, 23*p*, 24*p*, 25*p*, 27*p*, 371, 372*f*
femur, 29*p*, 159*f*, 161–162, 162*f*, 165*f*
 in appendicular skeleton, 142*f*, 143
 structure of, 133–134, 134*f*
fenestrae, 474*f*
fertilization, 538–539, 539*f*, 540*f*, 540*t*
 avoiding, with birth control, 526
 occurring in uterine tube, 519–520, 537, 537*f*
 in vitro, 559
fetal alcohol syndrome, 548–549, 548*f*
fetal cell sorting, 558*f*, 559
fetus, 541. *see also* prenatal period
 fetal chromosome checks, 558–559, 558*f*, 559*f*
 fetal stage of development, 547–551, 548*f*, 550*f*, 551*t*
 full-term fetus, 548
fever, as defense against pathogens, 385–386
fiber, 429
fiberoptic colonoscope, 426
fibrin, 331, 331*f*, 333*f*
fibrinogen, 320*f*, 329, 329*t*, 333*f*, 385
fibroadenoma, 527
fibroblast, 102, 102*f*, 103*t*, 113, 331, 385
fibroblast growth factor, 355
fibrocartilage, 108, 108*f*, 109*t*, 138–139, 139*f*, 153
fibrocystic breast disease, 527
fibrosis, 538
fibrous coat, of ureter, 481
fibrous joint, 164, 164*f*, 167*t*
fibrous pericardium, 342, 342*f*, 343*f*
fibula, 142*f*, 143, 162–163, 162*f*, 163*f*, 206
fibular artery, 368*f*
fibularis longus muscle, 195*f*, 206, 207*f*, 207*t*
fibularis tertius muscle, 205, 206, 207*t*
fight-or-flight response, 250, 305, 311, 313*f*

filtration, 64, 67*t*
  in capillaries, 357
  through cell membranes, 64, 65*f*
filtration rate, 475–476, 476*f*
fimbriae, 516*f*, 519, 519*f*
finger. *See also* metacarpal bone
  embryonic development, 544,
    545*f*, 547, 548
fingerprint, 122
first line of defense against
    pathogens, 385
first line of defense against shifts in
    pH, 499, 499*f*
first meiotic division, 507, 508*f*
first messenger, 295
first polar body, 517, 518*f*, 539*f*
fissure, 237
fissured fracture, 138, 138*f*
flagella, 52*f*, 58, 59*f*, 61*t*, 510
flared ilium, 161*f*
flat bone, 133
flatus, 427
flexion, 167, 169*f*, 192–193
flexor carpi radialis muscle, 200,
    200*f*, 201*f*, 202*t*
flexor carpi ulnaris muscle, 200,
    200*f*, 202*t*
flexor digitorum longus muscle, 206,
    207*f*, 207*t*
flexor digitorum profundus muscle,
    200, 202*t*
floater, 284
floating rib, 154*f*, 155
flourescence-activated cell sorter, 559
fluid compartments, 490, 490*f*
fluorine
  atomic structure, 32*t*
  as trace element in human body,
    31*t*, 434, 435*t*
fluorouracil, 527
folacin. *See* folic acid
folic acid, 149, 323, 324*f*, 433
folinic acid, 432–433
follicle, 301, 519
follicle-stimulating hormone, 297,
    300, 301*t*, 538
  disruption of, by
    contraceptives, 529
  in female reproductive cycle, 505,
    522–524, 523*t*, 524*f*
  in male reproduction, 513, 515*f*
  in oogenesis, 518
  pregnancy, changes during, 542
follicular cell, 301, 302*f*, 516–517, 518,
    518*f*, 519, 523*t*
follicular fluid, 518, 518*f*
follicular phase, 524*f*
fontanel, 144*t*, 149, 150*f*
food
  amino acids in, 430–431,
    431*f*, 431*t*
  intake of water through, 492*f*
  as requirement for life, 5
food poisoning, 467
food pyramid, 434, 435*f*
foot, 163, 163*f*
  muscles that move the, 205–206,
    206*f*, 207*f*, 207*t*
foramen, 144*t*, 152*f*
foramen magnum, 144, 146*f*, 147,
    147*f*, 148*f*, 176*p*, 234, 234*f*
foramen ovale, 550*f*, 551*t*,
    553–554, 554*f*
forearm. *See* upper limb
foreskin, 512

formaldehyde, 442
formed element, 108
formulas
  molecular, 37, 38*f*
  structural, 37, 37*f*, 38*f*
fossa, 144*t*
fossa ovalis, 550*f*, 554
fourth ventricle, 240, 241*f*, 242*f*, 457*f*
fovea, 144*t*
fovea capitis, 161, 162*f*
fovea centralis, 282, 284, 284*f*, 286
fracture, 132, 138–139, 138*f*, 139*f*, 162
fragility fracture, 132
Frank-Starling law of the heart, 361
fraternal twins, 542
free radical, 489, 555–556
free sensory nerve ending, 264, 264*f*
free surface, 95
friction, in capillaries, 357
frontal, defined, 15, 15*f*
frontal bone, 144, 145*f*, 146*f*, 148*f*,
    150*f*, 175*p*, 176*p*, 177*p*, 444
frontal eye field, 238*f*, 239
frontalis muscle, 195*f*, 196*f*
frontal lobe, 238, 238*f*, 239
frontal region, defined, 17, 18*f*
frontal sinus, 10*f*, 144, 145*f*, 148*f*, 406*f*
  as member of respiratory
    system, 443*f*
  in respiratory tract, 444*f*
frontal suture, 150*f*
fructose, 429, 510
full-term fetus, 548
full-thickness burn, 126, 127*f*
fuminant hepatitis, 417
functional MRI, 213
functional residual capacity,
    455–456, 456*f*
functional syncytium, 349
fundus, 410, 410*f*, 520
fungus, 384
funiculus, 234, 235*f*

G

GABA (γ-aminobutyric acid), 245
galactose, 429
gallbladder, 11*f*, 25*p*, 26*p*, 27*p*, 308*f*,
    415*f*, 418, 419, 419*f*
  as member of digestive system,
    401, 402*f*, 413, 414*f*, 418,
    419, 419*f*
  muscles, effects of
    neurotransmitters on, 256*t*
  referred pain, 266*f*
  secretions from, 413, 414*f*
  veins that drain, 371*f*
gallium-67, 34
gallstone, 418, 419
gamete intrafallopian transfer, 540*t*
gametogenesis, 69
gamma-amino butyric acid, 228, 229*t*
gamma globulin, 329, 329*t*, 391, 393
gamma radiation, 33
ganglia, 220, 233*f*
ganglion cell, 283*f*
gas
  in blood, 329
  as neurotransmitters, 228, 229*t*
gas transport, 460–462, 461*f*,
    462*f*, 462*t*
  of carbon dioxide, 460–462, 461*f*,
    462*f*, 462*t*
  of oxygen, 460, 461*f*
gastric absorption, 412–413
gastric artery, 363, 364*f*

gastric fold, 410*f*
gastric gland, 411, 411*f*, 412
gastric juice, 385, 411–412, 411*f*,
    412*f*, 412*t*
gastric pepsin, 413
gastric pit, 411, 411*f*
gastric secretion, 411–412, 411*f*, 412*t*
gastric ulcer, 412
gastric vein, 370, 371*f*
gastrin, 412, 412*f*, 418*t*
gastritis, 412
gastrocnemius muscle, 195*f*, 205*f*,
    206, 207*f*, 207*t*
gastroenterology, 18
gastrula, 543–544, 543*t*, 544*f*
gated membrane channel, 223*f*, 225*f*
gelatinous material, 276*f*
gene, 44, 84
  copy number variants, 91, 91*f*
  as determining fingerprint
    patterns, 122
  as determining hair color, 123
  dominant and recessive
    inheritance, 557, 557*f*, 560
  fetal chromosome checks,
    558–559, 558*f*, 559*f*
  gene expression, 88–89, 88*f*
  human genome, 3, 83–84, 90–91
  identifying differences, 91
  modes of inheritance, 557,
    557*f*, 560
  multifactorial traits, 560
  mutations with cystic
    fibrosis, 449
  paired with chromosomes,
    556–557, 556*f*
  Punnett square, 557, 557*f*, 560
gene expression, 88–89, 88*f*, 511
gene expression profiling, 527
general adaptation syndrome, 311
genetic code, 83, 85
genetics, 556–560, 556*f*, 557*f*,
    558*f*, 559*f*
genetic testing
  for breast cancer, 527
  for Huntington disease, 86
genital herpes, 530*t*
genital region, defined, 17, 18*f*
genital wart, 530*t*
genome, human, 3, 84, 90–91
genotype, 556–557, 560
geriatrics, 18
germ cell, 506, 509
gerontology, 18
GH release-inhibiting hormone, 298
GH-releasing hormone, 298
gigantism, 298
ginger
  root, as remedy for motion
    sickness, 276
  as stimulating pain receptors, 270
gingiva, 408*f*
girdle, 155
glabrous skin, 119
gland, 101
glandular cell, 292–293, 293*f*
glandular epithelial tissue, 101–102,
    101*f*, 101*t*, 102*t*
glans penis, 507*f*, 512
glaucoma, 282
glenoid cavity, 155, 156*f*
gliadin, 431
gliding joint, 165
glioma, 216
globulin, 320*f*, 329, 329*t*

globus pallidus, 240, 240*f*
glomerular capillary, 473, 474*f*, 475
glomerular capsule, 470–471, 471*f*,
    473, 473*f*, 477*f*, 481*t*
glomerular filtrate, 473, 474*f*, 475*t*,
    477, 477*f*, 478, 478*f*
glomerular filtration, 472–473, 474*f*
glomerular filtration rate,
    475–476, 476*f*
glomerular hydrostatic pressure,
    475–476, 475*f*
glomeruli, 470–471, 471*f*, 472*f*, 473*f*
glomerulonephritis, 475
glomerulus, 477*f*, 478*f*, 481*t*
glossopharyngeal nerve (IX), 247,
    247*f*, 248*t*
glottis, 413, 446, 446*f*
glucagon, 297, 307, 308, 309*f*,
    312, 313*f*
glucocorticoid, 306
glucose, 37, 38*f*
  absorption of, from digestive
    tract, 429
  aerobic respiration of, 186–187,
    186*f*, 497, 497*f*
  anaerobic respiration of, 186–187,
    186*f*, 497, 497*f*
  blood, 63, 256*t*, 297, 306,
    308–311, 309*f*, 329, 415
  in cellular respiration, 82
  diffusion through cell
    membranes, 63
  formation, 306, 307*f*
  in glomerular filtrate, 475*t*
  in glycolysis, 81*f*
  hydrolysis of, 430, 430*f*
  molecule of, 41*f*
  oxidation of, 80
  in plasma, 475*t*
  requirement for, 429
  in skeletal muscle contraction,
    186–187, 186*f*
  as source of hydrogen ions, 497*f*
  transport by plasma, 329
  in urine, 475*t*, 478, 480
glutamic acid, 228, 229*t*, 431*t*
glutamine, 431*t*
gluteal region, defined, 17, 18*f*
gluteal tuberosity, 162*f*
gluten, reactions to, 424
gluteus maximus muscle, 195*f*, 197,
    202*f*, 205*f*
gluteus medius muscle, 29*p*, 195*f*,
    204*f*, 204*t*, 205*f*
gluteus minimus muscle, 204*t*
glycerol, 42, 42*f*, 43*f*
  in citric acid cycle, 429–430, 430*f*
  in dehydration synthesis, 77–78,
    77*f*, 78*f*
  intestinal absorption of, 423, 424*t*
glycine, 431*t*
glycogen, 42, 310, 329
  conversion of, in liver, 415
  fast-twitch muscle fibers and, 189
  formation of, 429
glycolipid, 386
glycolysis, 80, 81*f*, 82, 83*f*,
    186–187, 186*f*
glycoprotein, 293*t*, 518
  altered in Golgi apparatus, 55
  antigen as, 386
  as cause of muscular
    dystrophy, 193
  in extracellular matrix,
    104–105, 105*f*

glycoprotein (continued)
   formation, 54
   functions of, 43
   in peanuts, 377
glycosuria, 310
glycyrrhizic acid, 460
goblet cell, 97, 97f
   in large intestine, 427
   in respiratory tract, 443–444, 454f
   in small intestine, 404f, 421f, 422
goiter, 303
Golgi apparatus, 52f, 55, 57f, 61t
gonad, 300, 506
gonadal artery, 363–364, 364f, 368f
gonadal vein, 372f
gonadotropin, 300, 513
gonadotropin-releasing hormone
   (GnRH), 505, 513, 515f,
   522–523, 538
gonorrhea, 530, 530t, 538
gout, 480
gp120, 389
G$_1$ phase, 68f
G$_2$ phase, 68f
G protein, 295, 295f, 296
gracilis muscle, 26p, 27p, 28p, 29p,
   195f, 204p, 204t, 205f, 207f
grafting vein, 354
graft-versus-host disease, 141, 395
granulation, 126
granulocyte, 322f, 324, 327t
granulocytic leukemia, 328
gray commissure, 234, 235f
gray matter, 217, 233f, 234, 235f
great cardiac vein, 348f
greater curvature, 410f
greater omentum, 25p, 420, 421f
greater sciatic notch, 160f
greater trochanter, 161, 162f
greater tubercle, 155, 157f
great saphenous vein, 23p, 24p, 27p,
   371, 372
greenstick fracture, 138, 138f
ground substance, 102
growth, 5t, 537
growth hormone, 297, 298, 301t, 312
guanine, 84, 84f, 85, 85f, 86
gums (gingiva), 408f
gynecology, 18
gyri, 237, 237f

H
hair, embryonic development, 548
hair cell, 273f, 274, 274f, 275,
   276f, 277f
hair follicle, 118f, 122–123,
   123f, 124f
hair root, 123, 123f
hair shaft, 118f, 123, 123f
half-life, 33
hallucination, 245
hallucinogen, 245
hamate, 159f
hamstring group, 205t
hand
   bones of the, 157–158, 159f
   functions of, 3
   muscles that move the, 200–201,
      200f, 201f, 202
haploid cell, 508
hapten, 386
hard palate, 148, 404, 405f, 406f
   as member of respiratory
      system, 443f
   in respiratory tract, 444f

haustra, 425f
Haversian canal, 108, 134
HDL cholesterol, 355
head, 144t
   arteries to, 355, 364, 365f,
      366, 366t
   cavities, 8, 9f, 10, 10f, 176p
   embryonic development, 544,
      545f, 547
   facial expression, muscles of,
      194–195, 194f, 194t,
      196f, 197
   mastication, muscles of, 196f,
      197, 197t
   muscles that move the, 196f,
      197, 197t
   veins from, 369, 369f
headache, 281, 444
head malleolus, 162
head region
   of muscles, 192
   of sperm cell, 510, 510f
hearing, sense of, 270–275, 271f,
      272f, 273f, 274f, 274t
   auditory nerve pathways,
      274–275
   auditory tube, 271–272
   damage to, 274, 275
   inner ear, 272–274, 272f, 273f
   middle ear, 271, 272
   outer ear, 270, 271f
   pitch sensitivity in hearing
      sounds, 260
   sensory impulses from the ear,
      generation of, 274f
heart, 11f, 26p, 341–359, 450f
   actions, 347, 349–352, 353f
   blood clot formation in, 331
   blood supply to the, 346–347,
      347f, 348f
   chambers, 343–345, 344f, 345f
   conduction system, 349–351,
      350f, 351f
   coverings of, 342–343, 342f, 343f
   effects of epinephrine and
      norepinephrine on, 305t
   embryonic development, 544,
      545f, 550–551, 551f
   exercise, changes due to, 362
   fibers, 349, 351f
   membranes of, 10
   pain impulses from, 266f
   path of blood through the,
      346, 346f
   referred pain, 265, 266f
   size and location, 342, 343f
   skeleton of the, 345, 345f
   sounds, 349
   structure of, 342–347, 342f, 343f,
      344f, 345f, 345t, 346f,
      347f, 348f
   transplantation of, 347
   valves, 343–345, 344f, 345f, 345t,
      349, 354f
   veins that drain, 383f
   wall of the, 342f, 343, 343f
heart attack. See myocardial
      infarction
heartbeat, 349, 352
heartburn, 410
heart failure, 104–105
heart murmur, 349
heart rate, 256t, 362
heat, as requirement for life, 5
heat index, 489

heat production by muscle
      contraction, 187
heatstroke, 489
HeLa cell, 94
Helicobacter pylori, 412
helium, 32t, 33f
helper T cell, 388, 388f, 389
hematocrit, 319
hematology, 18
hematoma, 138, 139f
hematopoiesis, 140
hematopoietic stem cell, 321, 322f
hemiazygos vein
   inferior, 370
   superior, 370
hemochromatosis, 323
hemocytoblast, 321
hemodialysis, 62–63, 116
hemoglobin, 44, 140, 186, 311
   in acidosis, 500–501
   in bile production, 418
   in blood, 319, 323, 324f
   broken down by bile
      pigments, 418
   in fetal blood, 549
   in gas transport, 460, 461f, 462
   in mismatched blood
      transfusions, 333
   in newborn, 554
   in sickle cell anemia, 323, 549
hemolysin, 336
hemolytic disease in, 336
hemolytic jaundice, 418
hemolytic uremic syndrome, 467
hemophilia, 332
hemorrhage, 358, 360
hemorrhagic fever, 318
hemorrhoid, 426
hemostasis, 330–332, 330f, 333f
   blood coagulation, 331–332, 331f,
      333, 333f
   blood vessel spasm, 330, 333f
   coagulation disorders, 332
   platelet plug formation, 330,
      330f, 333f
heparin, 103, 103f, 325
hepatic artery, 363, 364f, 415f, 416f
hepatic cell, 416f
hepatic duct, 414f, 415, 415f
   common, 418, 419f
hepatic lobule, 415, 416f
hepatic portal system, 370
hepatic portal vein, 370, 371f, 415,
      415f, 416f, 550f
hepatic sinusoid, 415, 416f
hepatic vein, 370
hepatitis, 417
hepatitis A, 417
hepatitis B, 318, 417
hepatitis C, 417
hepatitis E, 417
hepatocellular jaundice, 418
hepatopancreatic ampulla, 414f
hepatopancreatic sphincter, 413,
      414f, 418
hereditary idiopathic dilated
      cardiomyopathy, 193
hereditary sensory neuropathy, 267
hernias, hiatal, 410
herniated disc, 153
heroin, 245
herpes, 129, 530t
heterozygous gene, 556
hiatal hernia, 410
hiccup, 455

high-density lipoprotein, 47,
      423, 430
hilum, 381f, 468, 468f
hinge joint, 165, 166f, 167t
hip bone, 142f, 159f, 166f
hip fracture, 162
histamine, 103, 103t, 229t, 325, 358,
      394, 495
histidine, 431t
histocompatibility, 470
histology, 18
HIV (human immunodeficiency
      virus), 389
   as cause of peripheral
      neuropathy, 267
   immunity from, 86
   method of entering body, 65
   transmission, 389t
hives, 394
holocrine glands, 101–102,
      101f, 101t
homeostasis, 5–8, 6f, 7f, 8f
   of bone tissue, 137
   with electrolyte balance, 493
   with stress, 311
   with water balance, 492
homeostatic mechanism, 6–8, 6f, 7f
homozygous gene, 556
horizontal, defined, 15, 15f
hormone, 12, 14, 291, 292
   action of, 293–296, 294f
   of the adrenal medulla,
      304–305, 305t
   of the anterior pituitary lobe,
      297–298, 298f, 299f,
      300, 301t
   in bone development, growth,
      and repair, 137
   breast size during pregnancy,
      controlled by, 552
   contraceptives, combined
      hormone, 528–529, 528f
   of the digestive system, 418,
      418t, 419f
   in female infertility, 538, 538t
   female reproductive functions,
      control of, 522–525, 523t,
      524t, 537
   female sex hormone, 522
   male reproductive functions,
      hormonal control of, 513,
      514, 514f, 515, 515f
   male sex hormone, 513, 515, 515f
   nonsteroid hormone,
      294–296, 295f
   of the pancreatic islets, 308–309,
      309f, 310–311
   of the parathyroid gland,
      303, 304f
   of the posterior pituitary lobe,
      300, 301t
   pregnancy, hormonal changes
      during, 542–543,
      542f, 543t
   secretions, control of hormonal,
      296–297, 296f
   steroid hormone, 294, 294f
   of the thyroid gland,
      301–303, 302t
   types of, 293t
   in urine formation, 476
hormone-receptor complex, 294,
      294f, 295f
hormone-secreting islet cell, 308f
horn, 234

human body
  internal environment, 5, 6*f*
  organization of, 8–14, 9*f*, 10*f*,
    11*f*, 13*f*
human chorionic gonadotropin
  (hCG), 538, 542, 542*f*,
    543*t*, 558
human genome, 3, 84, 90–91
human immunodeficiency virus
  (HIV). *See* HIV (human
  immunodeficiency virus)
human menopausal gonadotropin
  (hMG), 538
humerus, 26*p*, 142*f*, 143, 155, 156*f*,
    157, 157*f*, 166*f*
humoral immune response,
  390–391, 390*f*
Huntington disease, 86, 240, 560
hyaline cartilage, 107, 107*f*, 109*t*,
    136, 446
hyaluronidase, 538
hybrid fixator, 139
hydrochloric acid, 38–39, 411,
    412*f*, 412*t*
  as chemical barrier to
    pathogens, 385
hydrogen
  active transport across cell
    membranes of, 65
  affecting blood pressure, 361–362
  atomic structure, 32*t*, 33*f*
  blood levels of, 457–458, 458*f*
  bonding in, 35, 36, 36*f*, 37*f*
  in carbohydrates, 41, 41*f*
  in cerebrospinal fluid, 457
  in dehydration synthesis, 77–78,
    77*f*, 78*f*
  in DNA, 84
  as element in human body, 31*t*
  formula for, 37*f*
  in nucleic acids, 44
  in pH scale, 39, 39*f*
  in proteins, 43–44, 44*f*
  in steroids, 294
  tubular secretion of, 479, 479*f*
  in water molecules, 36, 37*f*
hydrogen ion, 38–39
  acid-base balance (*see* acid-base
    balance)
  acid-base imbalance (*see*
    acid-base imbalance)
  concentration, regulation of,
    498–499
  in electrolyte intake, 493
  in metabolic alkalosis, 502
  renal excretion of, 499, 499*f*
  sources of, 497, 497*f*
  time course of hydrogen ion
    regulation, 499, 499*f*
  tubular secretion, 479, 479*f*
hydrogen peroxide, 79
hydrolysis, 78, 78*f*, 429–430, 430*f*, 431*f*
hydrostatic pressure
  of blood, 475
  body fluid movement by,
    491–492, 491*f*
  as driving lymph movement, 381
  in filtration, 64
hydroxide ion, 39, 497
hydroxyurea, 323, 549
hymen, 520
hyoid, 143*t*
hyoid bone, 142, 142*f*, 406*f*, 409,
    444*f*, 445*f*
hyperextension, 167, 169*f*

hyperglycemia, 310
hyperkalemia, 496
hypernatremia, 306
hyperparathyroidism, 303
hyperplasia, 68
hyperpolarization, 226, 226*f*
hypertension, 362, 476
hyperthermia, 125, 489, 494
hyperthyroidism, 303
hypertonic solutions, 64, 64*f*
hyperventilation, 458, 501–502, 501*f*
hypochondriac region, defined,
    16, 17*f*
hypodermic injection, 119
hypodermis, 117
hypogastric region, defined, 16, 17*f*
hypoglossal nerve (XII), 247–248,
    247*f*, 248*t*
hypoglycemia, 306
hypokalemia, 496
hyponatremia, 495, 496
hypoparathyroidism, 303
hypophyseal portal vein, 297, 299*f*
hypoproteinemia, 475, 495
hyposecretion, 538
hypothalamus, 240*f*, 242, 243, 293*f*,
    298, 298*f*, 299*f*, 300
  antidiuretic hormone, production
    of, 480
  body temperature, control of, 125
  in cardiac cycle, 352, 353*f*
  corticotropin-releasing hormone,
    secretion of, 307*f*
  endocrine system, control of,
    296–297, 296*f*
  female reproductive functions,
    control of, 522–524
  fight-or-flight response, 311, 313*f*
  male reproductive functions,
    control of, 513, 515, 515*f*
  micturition reflex, facilitation
    of, 484
  in milk production and
    secretion, 552
  in stress response, 312, 313*f*
  thirst center located in, 492
hypothermia, 125
hypothyroidism, 303
hypotonic extracellular solution, 306
hypotonic solution, 64, 64*f*
hypoxemia, 221, 460
hypoxia, 319, 460
hypoxyurea, 323
hysterosalpingogram, 538*t*
H zone, 180*f*, 181

I

I band, 180*f*, 181
icterus. *See* jaundice
identical twins, 542
idiopathic hypertension, 362
ileocecal sphincter, 424, 425*f*
ileum, 27*p*, 403*f*, 420, 420*f*, 424
iliac artery, 470
  common, 27*p*, 364, 364*f*, 367,
    368*f*, 550*f*
  external, 367, 368*f*
  internal, 367, 368*f*, 550*f*
iliac crest, 29*p*, 158, 160*f*
iliac fossa, 160*f*, 470
iliac region, defined, 16, 17*f*
iliac spine, anterior superior, 23*p*,
    25*p*, 29*p*, 158
Iliacus muscle, 204*f*, 204*t*
iliacus muscle, 29*p*

iliac vein, 470
  common, 371, 372*f*
  external, 371, 372*f*
  internal, 371, 372*f*
iliotibial tract, 204*f*
ilium, 158, 160, 470
illusions, 245
immunity, 325, 384–395
  adaptive (specific) defense (*see*
    adaptive (specific) defense)
  classification of, 393, 393*t*
  innate (nonspecific) defense
    (*see* innate (nonspecific)
    defense)
immunity privileged, 280
immunoglobulin A, 391
immunoglobulin D, 391
immunoglobulin E, 391, 394
immunoglobulin G, 391, 392, 393
immunoglobulin M, 391, 392
immunoglobulins. *See* antibody
immunology, 18
immunosuppressive drugs, 394
impetigo, 129
implantable cardioverter
  defibrillator, 340
implanted insulin pump, 311
inactive protein, 295*f*
in cardiac cycle, 353*f*
incisor (teeth), 406*f*, 407, 407*t*
incomplete protein, 431
incus, 271, 271*f*
India, arsenic poisoning in, 76
in electrolyte output, 493, 496
inert electron, 33
infarction, 331
infancy, stages in development, 554*t*
infant
  neonates (*see* neonate)
  premature infants, 553
  skull of, 149, 150*f*
infarction, 331
infection, 384
  of alimentary canal, 420
  body defenses against, 384
  inflammatory bowel disease
    from, 428
  of lymphatic vessels, 382
  of lymph nodes, 382
  middle ear, 272
inferior, defined, 14, 14*f*
inferior articular process, 149, 153*f*
inferior canaliculus, 278*f*
inferior hemiazygos vein, 370
inferior hypogastric plexus, 253*f*
inferior lobe, 447*f*
inferior mesenteric artery, 28*p*, 363,
    364*f*, 368*f*
inferior mesenteric ganglion, 252*f*
inferior mesenteric vein, 370, 371*f*
inferior nasal concha, 145*f*, 148*f*, 149,
    175*p*, 444*f*
inferior oblique muscle, 279*t*
inferior peduncle, 244, 244*f*
inferior rectus muscle, 279*t*
inferior vena cava, 28*p*, 29*p*, 344,
    344*f*, 346*f*, 348*f*, 415*f*
  carrying deoxygenated blood
    from lower half of body,
    369, 372*f*
  embryonic development, 550, 550*f*
  in newborn, 554*f*
  used by urinary system, 468*f*, 469
infertility, 505
  female, 538
  male, 514, 514*f*, 514*t*

inflammation, 125, 125*t*
  as defense against pathogens, 385
  with edema, 495, 495*t*
  from excess uric acid, 480
  of urinary bladder, 483, 483*f*
inflammatory bowel disease, 428
inflation reflex, 457
infradian rhythm, 312
infraorbital foramen, 145*f*, 175*p*
infrapatellar bursa, 165*f*
infraspinatus muscle, 195*f*, 198, 198*f*,
    199*t*, 200*f*
infraspinous fossa, 156*f*
infundibulum, 242, 519, 519*f*
inguinal canal, 25*p*, 510
inguinal lymph node, 382*f*
inguinal region, defined, 17, 18*f*
inhibin, 513, 515*f*
inhibitory neurotransmitter, 222,
    222*f*, 228
injectable contraception, 529
injection, 119
innate (nonspecific) defense,
    384–386
  chemical barriers, 385
  fever, 385–386
  inflammation, 385
  mechanical barriers, 385
  natural killer cells, 385
  phagocytosis, 385, 386
  species resistance, 384–385
inner ear, 272–274, 272*f*, 273*f*
inorganic constituents of cells,
    39–40, 40*t*
inositol triphosphate, 296
insertion, 167
insertion, muscle, 192–193
  in skeletal muscle contraction, 184*f*
insomnia, 312
inspiration, 450–453, 451*f*, 452*f*,
    455–456, 455*f*, 456*t*
inspiratory capacity, 455–456
inspiratory intercostal muscle, 452,
    452*f*, 457*f*
inspiratory reserve volume, 455, 456*t*
insula, 238, 238*f*
insulin, 63, 297, 308–309, 309*f*,
    310–311
integrative function of nervous
  system, 215
integrin, 104–105
integumentary system, 12, 13*f*,
    116–128, 117–128, 556
interatrial septum, 350, 350*f*
intercalated disc, 111, 112*f*, 192
intercondylar eminence, 162*f*
intercondylar fossa, 162*f*
intercostal artery, 368*f*
  anterior, 367
  posterior, 367
intercostal muscle, 525*f*
  external, 24*p*, 25*p*, 29*p*, 452,
    452*f*, 457*f*
  internal, 25*p*, 29*p*, 453, 453*f*, 457*f*
intercostal nerve, 249*f*, 250
intercostal vein, 370
  posterior, 370
interferon, 385
interior lobar bronchus, 447*f*
interleukin, 324
interleukin-1, 387
interleukin-2, 387, 388*f*
interleukin-10, 428
interlobular artery, 469, 470*f*
interlobular vein, 470*f*

intermediate cuneiform, 163f
intermediate lobe in pituitary
    gland, 297
internal acoustic meatus, 148f
internal anal sphincter muscle,
    426, 426f
internal carotid artery, 365f, 366,
    366f, 366t, 368f
internal iliac artery, 367, 368f, 550f
internal iliac vein, 371, 372f
internal intercostal muscle, 25p, 29p,
    199f, 453, 453f, 457f
internal jugular vein, 24p, 25p, 28p,
    369, 369f, 370f, 372f, 380f
internal oblique muscle, 24p, 25p,
    199f, 201, 203t
internal respiration, 443
internal thoracic artery, 367
internal thoracic vein, 370
internal urethral sphincter, 483f
internal vertebral sphincter, 483, 483f
interneuron, 220, 221f, 231, 232,
    232f, 251f
    in reflex arc, 231, 232t
    in withdrawal reflex, 232, 232f
internuncial neuron, 220
interphase, 67, 68f, 69, 70f, 509f
interstitial cell, 506, 508f, 513t, 515
interstitial cell stimulating hormone,
    297, 513
interstitial fluid, 490
    as extracellular fluid, 490, 490f
    movement of, between fluid
        compartments, 491, 491f
intertubercular groove, 155, 157f
interventricular foramen, 241f
interventricular septum, 344f, 346,
    350, 350f
intervertebral disc, 29p, 142, 149,
    151f, 153, 164, 234f
intervertebral foramen, 150, 249, 249f
intestinal flora, 427
intestinal gland, 404f, 421, 421f, 422
intestinal lipase, 422, 422t
intestinal villus, 404f, 420–421,
    421f, 422
intestine
    calcium, absorption of, 304f
    effects of neurotransmitters
        on, 256t
intracellular fluid, 490–492, 490f, 491f
intracellular fluid compartment, 490,
    490f, 494, 494f
intracellular protein receptor, 294f
intracytoplasmic sperm
    injection, 540t
intradermal injection, 119
intramembranous bone, 135–136, 135f
intramuscular injection, 119
intraspinous fossa, 156f
intrauterine device, 529
intrauterine growth retardation, 549
intrauterine insemination, 540t
intrinsic factor, 411, 412f
intron, 86
inversion, 167, 170f
in vitro fertilization, 505, 536,
    538, 559
involuntary muscle tissue, 111, 111f
iodine, 31t, 301, 434, 435t
iodine-131, 34, 34f
ion, 33, 37t
    bonds, ionic, 35, 35f
    cell membrane transport of,
        222, 223f

ion channel, 54
ipsilateral, defined, 15
iris, 280f, 282, 283f
iron
    in blood cell production, 323
    as trace element in human body,
        31t, 434, 435t
irregular bone, 133
ischemia, 221, 265, 347
ischial spine, 160f
ischial tuberosity, 160, 160f
ischiocavernosus muscle, 202f, 203
ischium, 158, 160, 160f
islet of Langerhans, 307, 308–309,
    308f, 309f, 310–311
isoleucine, 431t
isotonic solution, 64, 478
isotope
    defined, 33
    radioactive, 33, 34, 34f
isthmus, 301, 302f
itching, 116

J
jaundice, 122, 324, 333, 418
jejunum, 27p, 403f, 420, 420f
jet lag, 309, 312
joint, 164–170
    disorders, 168
    movements, types of, 165, 166f,
        167, 167t, 169f, 170, 170f
    tissue types, 164–165, 164f, 165f
joint capsule, 164, 165f
jugular vein
    external, 25p, 369, 369f,
        370f, 372f
    internal, 24p, 25p, 28p, 369, 369f,
        370f, 372f, 380f
juxtaglomerular apparatus,
    471–473, 473f
juxtaglomerular cell, 471, 473f

K
keloid, 129
keratin, 119
keratinization, 119, 119f, 122, 123, 123f
ketone body, 429–430, 430f, 497,
    497f, 501, 501f
ketones, in urine, 480
ketonuria, 501
kidney, 11f, 28p, 293, 293f, 468–470,
    468f, 469f, 470f
    acidosis, 500, 501
    adrenal glands, association of,
        304, 305f
    capillaries, 472, 472f, 473,
        474f, 475
    conservation of calcium in,
        303, 304f
    damage or failure, 62–63, 467
    embryonic development, 548
    formation of angiotensin in,
        476, 476f
    functions of, 469
    location of, 468, 468f
    low blood oxygen, response to,
        321, 323f
    as member of urinary system, 468f
    in newborn, 553
    referred pain, 266f
    regulation of hydrogen ion
        concentration in, 499, 499f
    renal blood vessels, 469–470, 470f
    simple cuboidal epithelial tissue
        in, 97f

structure of, 468–469, 468f, 469f
    transplantation of, 470
kidney disease, 501, 501f
kidney stone, 482
knee, 164–165, 165f
kneecap. See patella
Krabbe disease, 58
Kupffer cell, 370, 415, 416f

L
labium majus, 516f, 521, 522t
labium minus, 516f, 521, 522t
labor, 552
labor contraction, 552
labyrinth, 272–273, 272f
Lacks, Henrietta, 94
lacrimal apparatus, 277, 278f
lacrimal bone, 145f, 146f, 149,
    175p, 177p
lacrimal gland, 277, 278f
lacrimal sac, 278, 278f
lactase, 422, 422t
lacteal, 378, 404f, 421, 421f, 423, 423f
lactic acid, 186–187, 186f, 478,
    497, 497f
Lactobacillus, 400, 407
lactogen, placental, 542–543
lactose, 429
lactose intolerance, 422
lacunae, 107, 107f, 108, 109f, 135,
    135f, 544, 545f, 546f
lambdoid suture, 144, 146f, 148f, 177p
lamellae, 108, 109f
lamellated corpuscle, 264, 264f
lamina, 149, 152f
laparascopic adjustable gastric
    band, 436
laparoscopy, 538t
large intestine, 11f, 424, 425f,
    426–428, 426f, 427f, 507f
    in alimentary canal, 401,
        402f, 403f
    effects of neurotransmitters
        on, 256t
    feces, 427
    functions of, 426–527, 527f
    as member of digestive system,
        401, 402f, 424, 425f,
        426–428, 426f, 427f
    microorganisms in, 400
    movements of, 427
    wall, structure of the, 426
laryngeal muscle, 446
laryngitis, 446
laryngopharynx, 406f, 409, 444f, 445
larynx, 24p, 27p, 302f, 383f, 406f,
    409, 445–446, 445f, 446f
    elastic cartilage in external, 107
    infection or irritation of, 446
    as member of respiratory system,
        443f, 445–446, 445f, 446f,
        447f, 451t
    in respiratory tract, 443, 444f
laser angioplasty, 355
latent period of skeletal muscle
    contraction, 188f, 189
lateral, defined, 14, 14f, 15
lateral condyle, 162, 162f
lateral cuneiform, 163f
lateral epicondyle, 157f, 162f
lateral funiculus, 234, 235f
lateral horn, 234, 235f
lateral malleolus, 162, 162f
lateral rectus muscle, 279t, 280f

lateral sulcus, 238f, 239
lateral ventricle, 240, 241f
latissimus dorsi muscle, 24p, 25p,
    195f, 197, 198f, 199t
laughing, 455
LDL cholesterol, 355
Leber congenital amaurosis, 286
left hypochondriac region, defined,
    16, 17f
left iliac region, defined, 16, 17f
left lower quadrant of abdominal
    area, 17f
left lumbar region, defined, 16, 17f
left optic tract, 287
left upper quadrant of abdominal
    area, 17f
left ventricular assist device, 347
leg. See lower limb
lens, eye, 280–281, 280f, 281f, 282f
lens crystallins, 280
lens fiber, 280
leptin, 522
lesser curvature, 410f
lesser sciatic notch, 160f
lesser trochanter, 161, 162f
lesser tubercle, 155, 157f
leukemia, 94, 327
leukocyte. See white blood cell
leukocytosis, 325, 327
leukopenia, 327
leukotriene, 394
levator ani muscle, 201, 202f, 426f
levator palpebrae superioris muscle,
    277, 278f, 279t
levator scapulae muscle, 197, 198f,
    198t, 200f
lever, 140, 140f
life
    characteristics of, 4, 5t
    maintenance of, 5–8, 6f, 7f, 8f
ligament, 103, 106f
ligamentum arteriosum, 550f, 554
ligamentum teres, 550f
ligand, 66
light-dark cycle, 312
ligmentum capitus, 161
limbic system, 243
limb of nephron loop
    ascending, 471
    descending, 471
limbus, 280
limited transport capacity, 478
linea alba, 24p, 199f, 201
linea aspera, 162f
lingual artery, 365f
lingual frenulum, 403–404, 405f
lingual tonsils, 406f
linolineic acid, 430
lip, 405f
lipase, 79
lipid, 42–43, 42f, 43f, 43t, 46f, 47,
    429–430, 430f
    in cell membranes, 53–54, 53f
    in cytoplasm, 55
    effects of liver on, 415,
        417, 418t
    in endoplasmic reticulum,
        synthesis of, 55, 56f
    hydrolysis of, 78, 78f
    as macronutrient, 428
    as plasma nutrients, 329
    requirements, 430
    sources of, 429
    in steroid hormone diffusion,
        294, 294f

lipofuscin, 556
lipophilic molecule, 46
lipoprotein, 430
lipoprotein lipase, 423
lips (mouth), 403, 405f
lissencephaly, 238
lithium, 32t, 32t, 33f
liver, 11f, 25p, 26p, 27p, 415, 416f,
    417–418, 418t
    destruction of red blood cells in,
        323, 324f
    effects of epinephrine and
        norepinephrine on, 305t
    formation of angiotensin in,
        476, 476f
    functions of, 415, 417, 418t,
        429, 430
    low blood oxygen, response to,
        321, 323f
    as member of digestive system,
        401, 402f, 415, 416f,
        417–418, 418t, 429
    in newborn, 554f
    production of prothrombin, 331
    referred pain, 266f
    regeneration of, 417
    response to fevers, 385
    secretions from, 413
    structure of, 415, 415f, 416f
    veins that drain, 371f, 383f, 421f
    weight of, 415
liver failure, 417
liver fibrosis, extracellular matrix
    in, 104
lobar bronchus, 446, 447f, 448
lobes of the liver, 415, 415f
lobule
    of the liver, 415, 415f
    of the spleen, 383
    of testes, 506
    of testis, 506
    of thymus, 383
    of thymus gland, 383, 383f
local current, 227
local hormone, 292
long bone, 133–134, 134f,
    136–137, 136f
long head biceps brachii muscle, 24p
longitudinal fiber, 401
longitudinal fissure, 237, 240f, 244f
longitudinal muscle, 404f
longitudinal section, defined, 15, 16f
long QT syndrome, 362
loose connective tissue, 104–106,
    105f, 106f, 109t
Lou Gehrig's disease, 50, 182
low-density lipoprotein, 47, 423, 430
lower esophageal sphincter, 410, 410f
lower limb, 143, 143t
    arteries to, 367, 368f
    bones of the, 161–163, 162f, 163f
    foot, muscles that move the,
        205–206, 206f, 207f, 207t
    thigh, muscles that move the,
        203, 204f, 204t
    veins from, 370–371, 372f
lower lobe, 447f
lower motor neuron, 239
lower respiratory infection, 442
lower respiratory tract, 443
LSD (lysergic acid diethylamide), 245
lumbar artery, 364f, 367, 368f
lumbar enlargement, 234, 234f
lumbar nerve, 248–249, 249f
lumbar puncture, 242

lumbar region, defined, 16, 17,
    17f, 18f
lumbar vein, ascending, 370, 372f
lumbar vertebrae, 151f, 152, 152f
lumbar vertebrae, fifth, 29p
lumbosacral plexus, 249f, 250
lumen, 96
    of alimentary canal, 401, 421, 421f
    of arteries and veins, 333f, 354f,
        355, 355f
    of intestine, 414f, 423
    of kidneys, 97f
    of male urethra, 100f
    of salivary glands, 98, 99f
    of seminiferous tubule, 508f, 509f
    of urethra, 484f
    of uterus, 520f, 544f
    of vagina, 520
lumpectomy, 527
lunate, 159f
lung, 11f, 25p, 26p, 27p, 383f
    blood clot formation in, 331
    bronchial tree, 443, 446, 447f,
        448–450, 448f, 449f, 451t
    expiration, 453–455, 453f, 454f,
        455f, 456, 456t
    fetal, 550–551f
    formation of angiotensin in,
        476, 476f
    inspiration, 450–453, 451f, 452f,
        455–456, 455f, 456t
    lobes, 447f, 450, 450f
    as member of respiratory system,
        443f, 450, 450f, 451t
    membranes of, 10, 450
    in newborn, 553, 554
    oxygenation of blood in, 341f
    in respiratory tract, 443
    simple squamous epithelial tissue
        in, 96, 96f
    water lost through, 492f, 493
lung cancer, 454, 456
lunula, 122, 123f
luteal phase, 524f
luteinizing hormone, 297, 300,
    301t, 538
    disruption of, by
        contraceptives, 529
    in female reproductive cycle, 505,
        519, 522–524, 523t, 524f
    in male reproduction, 513,
        515, 515f
    pregnancy, changes during, 542
Lyme arthritis, 168
lymph, 379, 380f, 423f
    as extracellular fluid, 490, 490f
    formation and function, 380
    movement, 381, 381f
    movement of, between fluid
        compartments, 491, 491f
    obstruction of, 495f
lymphadenopathy, 382
lymphangitis, 382
lymphatic capillary, 357f, 358, 378f,
    379, 379f, 380f
lymphatic pathway, 378–379,
    379f, 380f
lymphatic system, 12, 13f,
    378–384, 556t
lymphatic trunk, 379, 380f
lymphatic vessel, 378, 378f, 379, 379f,
    380f, 382
    afferent, 380f, 381f
    efferent, 380f, 381f
lymph gland, 381

lymph node, 328, 378, 379f, 380f,
    381–382, 381f, 382f, 383,
    387, 404f
lymph nodule, 381f, 382
lymphoblast B cell precursor, 322f
lymphoblastic leukemia, 328
lymphoblast T cell precursors, 322f
lymphocyte, 320f, 325, 326f, 327t,
    381, 382
    functions of, 386–388, 387f, 388f
    interferon production by, 385
    natural killer cells, 385
    origins, 386, 386f, 387f
    precursors, 386, 387f
    in the spleen, 383
    T lymphocyte (see T cell)
lymphoid leukemia, 328
lymphoid stem cell, 322f
lymph sinus, 382
lymph vessel, 421f
lysergic acid diethylamide (LSD), 245
lysine, 431, 431t
lysis, 392f
lysosomal storage disease, 58
lysosome, 52f, 56, 61t, 325
lysozyme, 278

M

macromolecule, 3, 4f
macronutrient, 82, 428
macrophage, 102–103, 103f, 103t,
    322f, 325, 459f
    in allergic reactions, 393–394
    destruction of red blood cells by,
        323, 324f, 333
    in lymph nodes, 381, 381f
    in phagocytosis, 385
    in spleen, 383, 384
    in T cell activation, 386–388,
        387f, 387t, 388f
macula densa, 471, 471f, 473f, 476
maculae, 272f, 275, 276f
macula lutea, 284f
magnesium, 31t
    atomic structure, 32t
    in electrolyte intake, 493
    in extracellular fluid, 491f
    in glomerular filtrate, 475t
    in intracellular fluid, 491f
    in plasma, 475t
    in urine, 475t
magnesium ion, as constituent of
    cells, 40, 40t
magnesium phosphate, in kidney
    stones, 482
main bronchus, 446, 447f, 448, 448f
mainstream smoke, 442
major calyx, 468, 469f
major duodenal papilla, 414f
major histocompatibility complex, 386
major mineral, 433
malabsorption, 424
malaria, 418
male
    blood counts, 515
    blood volume, 319
    pelvic outlet, muscles of the, 201,
        202f, 203, 203t
    pelvis, 160, 160t, 161f
    percentage of water by
        weight, 490
    sex chromosomes, 556
male climacteric, 515
male condom, 527–528, 528f
male infertility, 514, 514f, 514t

male reproductive system,
    506–515, 556t
    accessory sex organs, 506
    erection, orgasm, and ejaculation,
        512–513
    external reproductive organs,
        512–513, 513t
    hormonal control of, 513, 514,
        514f, 515, 515f
    internal reproductive organs,
        510–512
    organs of, 506–513
    primary sex organs, 506
    sex hormone, 513, 515, 515f
    sexual stimulation, 512–513
male secondary sex characteristic,
    515, 515f
male sex hormone, 513, 515, 515f
malignant hyperthermia, 489
malignant melanoma, 121, 121f
malleolus, 162
malleus, 271, 271f
malnutrition, birth defects from, 549
maltase, 422, 422t
maltose, 422, 422t
mamillary body, 42, 242, 242f
mammary gland, 23p, 525, 525f
    development during pregnancy,
        542–543
    lymphatics of, 380f
    milk production and secretion,
        552–553, 553f
    as modified sweat glands, 124
    similarity of female and
        male, 525
    stratified cuboidal epithelial
        tissue in, 98
mammary region, defined, 17, 18f
mammogram, 526
mandible, 144, 145f, 146f, 148f,
    149, 150f, 175p, 177p,
    406f, 409f
mandibular condyle, 146f, 149, 177p
mandibular division, 247
mandibular fossa, 146f, 147, 176p
manganese, 31t, 434, 435t
manometer, 242
manubrium, 154f, 155
marrow, bone. See bone marrow
mass, 31
masseter gland, 409f
masseter muscle, 196f, 197, 197t
mass movements in large
    intestine, 427
mast cell, 103, 103f, 103t, 394
mastectomy, 527
mastication, 403
    muscles of, 196f, 197, 197t
mastoid fontanel, 150f
mastoid process, 146f, 147,
    148f, 177p
maternal serum marker, 558
mating behavior, 291
matrix, 137
matter, structure of, 31–33
maxillae, 145f, 146f, 148, 148f, 150f,
    175p, 176p, 177p, 406f
maxillary artery, 365f
maxillary bone, 444
maxillary division, 247
maxillary sinus, 145f, 148
measles, German, 548
meatus, 144t
mechanical barrier
    as birth control, 527–528, 528f
    against pathogens, 385

mechanical digestion, 401
mechanoreceptor, 263
medial, defined, 14, 14f
medial condyle, 162, 162f
medial cuneiform, 163f
medial epicondyle, 157f, 162f
medial malleolus, 162, 162f
medial rectus muscle, 279t, 280f
medial septum of scrotum, 512
medial umbilical ligament, 550f, 554f
median, defined, 15, 15f
median cubital vein, 369, 370f, 372f
median nerve, 249f, 250
median sacral crest, 154f
mediastinum, 8, 9f, 11f, 342, 450
medication, allergic reactions to, 536
medulla
    adrenal (see adrenal medulla)
    of brainstem, 236f
    of lymph node, 381f
    oblongata (see medulla
        oblongata)
    of ovary, 516
    in respiration, 456, 457f
medulla oblongata, 236f, 237f, 242f,
        243–444
    in breathing, 456, 457, 457f
    in cardiac cycle, 352
    in controlling blood pressure, 361
medullary cavity, 134, 134f, 136f,
        137, 387f
medullary respiratory center,
        456, 457f
megakaryoblast, 322f
megakaryocyte, 327
meiosis, 69, 421
    in fertilization, 539, 539f, 540f
    in primordial follicle,
        516–517, 517f
    in spermatogenesis, 507, 508f,
        509, 509f
meiosis I, 507, 509
meiosis II, 507, 509
Meissner's corpuscle, 264, 264f
melanin, 119, 120f, 297
melanocyte, 119, 120, 120f, 121
melatonin, 309
membrane-bound receptor
        molecule, 295f
membranes, types of, 110
membranous labyrinth, 272, 272f
memory cell, 388, 388f, 390f
menarche, 523
menisci, 164, 165f
menix, 232–234, 233f, 237f
menopause, 132, 524–525
menses, 523
menstrual flow, 523, 523t
menstrual irregularity, 522
menstruation, 524t
mental foramen, 145f, 146f, 175p
mental region, defined, 17, 18f
mercury, birth defects from, 549
merocrine gland, 101–102, 101f, 101t
mesenteric artery
    inferior, 28p, 363, 364f, 368f
    superior, 28p, 363, 364f, 368f
mesenteric vein
    inferior, 370, 371f
    superior, 28p, 370, 371f
mesentery, 27p, 404f, 420, 420f, 421f
mesoderm, 543, 544, 544f
messenger RNA (mRNA), 86, 88, 88f,
        89f, 294, 294f
metabolic acidosis, 500–501, 501f

metabolic alkalosis, 501–502, 502f
metabolic pathway, 82, 83f
metabolic reaction
    cellular respiration, 80, 81f, 82
    chemical energy, release of,
        80, 80f
    control of, 79–80, 79f, 80f
    energy for, 80, 80f, 81f, 82
    types of, 77–78, 77f, 78f
metabolism, 4, 76–91, 77
    cellular, 77–91
    effects of epinephrine and
        norepinephrine on, 305t
    muscle, 187, 187t
    in newborn, 553
    water of metabolism, 492, 492f
metacarpal bone, 142f, 143, 157–158,
        159f, 166f
metacarpus, 157–158
metallic taste sensation, 270
metaphase, 68f, 69, 70f
metastasis, in cancer, 68
metatarsal bone, 142f, 143, 163, 163f
metatarsus, 163, 163f
methionine, 431, 431t
metopic suture, 150f
microbial biome, 400
microfilament, 57, 58f, 61t
microflora, 400
microglial cell, 216, 217f
microgravity, 137
micronutrient, 428
microtubule, 52f, 57, 58f, 61t, 70f
microvillus, 52f, 97, 97f, 404f, 477
micturition reflex center, 483
midbrain, 236f, 237f, 243, 457f
middle ear, 271, 272
middle ear bone, 143t
middle ear cavity, 10, 10f
middle lobar bronchus, 447f
middle lobe, 447f
middle nasal concha, 145f, 147,
        175p, 444f
middle peduncle, 244f
middle phalanx, 159f, 163f
middle sacral artery, 364f
midpiece of sperm cell, 510, 510f
midsagittal, defined, 15, 15f
migraine headache, 281
milk
    production and secretion,
        552–553, 553f
    proportions of nutrients in
        human, 553
mineralocorticoid, 306
minerals (dietary), 433–434, 434f
    characteristics of, 433–434
    major, 434, 434t
    as micronutrient, 428
    trace elements, 434, 435t
minor calyx, 468, 469f
minor nerve, 230
mitochondria, 52f, 56, 57f, 58f, 61t,
        538–539
    in cardiac muscle, 192
    in neurons, 222, 223f
    in sperm cells, 510
mitosis, 67, 68f, 71, 72f, 421, 509f, 518
    during early embryonic
        development, 541
    during maturation of primary
        oocyte, 518
mitral valve, 344, 344f, 345, 345t,
        346, 349f
mitral valve prolapse, 345
mixed nerve, 230

mixing movements of the alimentary
        canal, 403, 424
M line, 180f, 181
model organism, 385
modes of inheritance, 557, 557f, 560
molar (tooth), 406f, 407, 407t
mole, 121, 129
molecular formula, 37, 38f
molecule, 3, 4f, 36, 37f, 37t
monoamine, as neurotransmitters, 229t
monoblast, 322f
monocular blindness, 278
monocyte, 320f, 322f, 325, 326f,
        327t, 385
monohydrogen phosphate, 498
mononuclear phagocytic system, 385
monosaccharide, 41, 428–429
    absorption in small intestine,
        422, 424t
    in dehydration synthesis, 77, 77f
monozygotic twins, 542
mons pubis, 23p, 521
morula, 541, 541f, 543t
motion sickness, 276f
motor areas of cerebral cortex, 239
motor cortex, 236f
motor end plate, 182, 182f
motor fiber, 230, 236f
motor function of nervous
        system, 215
motor nerve, 230
motor neuron, 182, 182f,
        220–221, 221f
    in cardiac cycle, 352, 353f
    reflex arc, 231f, 232t
motor speech area of cerebral
        cortex, 238f
motor speech areas, 238f, 239
motor unit, recruitment of, 190, 190f
mouth, 403–408
    in alimentary canal, 401,
        402f, 403f
    embryonic development, 544, 545f
    functions of, 3, 403
    as member of digestive system,
        401, 402f, 403–408
movement, 5t
mucin, 101
mucosa. See mucous membrane
mucous cell, 408, 410, 411, 411f
mucous coat
    of ureter, 481
    of urinary bladder, 482–483, 483f
mucous gland, 404f, 410
mucous membrane, 110
    in alimentary canal, 401, 404f
    infections, 272
    in large intestine, 425f, 426, 427,
        427f, 428
    in rectum, 428
    in respiratory system, 443–444,
        446, 448
    in respiratory tract, 443–444, 446
    in salivary glands, 408
    in small intestine, 413, 414f, 415,
        421–422
    in stomach, 410, 411, 411f
    ulcers in, 412
    in urethra, 484, 484f
    in uterus, 520
    in vagina, 520
mucous neck cell, 411
mucus, 97, 97f, 101, 412f
    in cervix, 514, 537
    in digestion, 408

esophageal, 410
mucus-secreting gland, 421–422
    in respiratory tract, 444,
        445f, 454f
    thick, in cystic fibrosis, 449
    in urethral gland, 484, 484f
    in uterine tube, 519
    in uterus, 537
    in vagina, 520
multifactorial trait, 560
multipolar neuron, 219, 220f
multiunit smooth muscle, 191
murmur, heart, 349
muscle
    cardiac (see cardiac muscle)
    double muscles, 178
    metabolism, 187, 187t
    skeletal (see skeletal muscle)
    smooth (see smooth muscle)
muscle cell, 51, 51f
muscle fatigue, 187
muscle fiber, 110, 230
muscle pull, 191
muscle strain, 182
muscle tissue, 95, 95t, 110–111,
        111f, 112f
    cardiac, 111, 112f, 113, 113t, 192t
    skeletal, 110, 111f, 113, 113t, 192t
    smooth, 110–111, 111f, 113,
        113t, 192t
    types, 110
muscle tone, 190–191
muscular coat
    of ureter, 481
    of urinary bladder, 483, 483f
muscular dystrophy, 193
muscular hypertrophy, 189
muscular layer of alimentary canal,
        401, 403f, 404f
muscular system, 12, 13f,
        178–208, 556t
musculocutaneous nerve, 26p,
        249f, 250
mutagen, 86
mutation, 86
    as causes of cancers, 68
    of DNA sequences, 556
myelin, 58, 216, 218f, 219, 227, 553
myelin sheath, 216, 217, 217f, 219f
myeloblast, 322f
myeloid leukemia, 328
myeloid stem cell, 322f
myocardial capillary, 347
myocardial fiber, 361
myocardial infarction, 347
    defibrillator as treatment of, 340
    referred pain with, 265
myocardial muscle fiber, 349, 351f
myocardium, 343, 343f, 347, 551
myoepithelial cell, 300, 552–553, 553f
myofibril, 57, 179, 180f, 181,
        181f, 182f
myoglobin, 186
myogram, 188–189, 188f, 190f
myometrium, 519f, 520, 520f
myosin, 179, 180f, 431
    in cardiac muscle, 191–192
    fast-twitch muscle fibers and, 189
    in rigor mortis, 187
    in skeletal muscle contraction,
        183, 183f, 184f
    in smooth muscle
        contraction, 191
myosin filament, 183, 183f
myostatin, 178

N

nail beds, 122, 123f
nail plate, 122, 123f
nails, fingers and toes, 122, 123f
nasal bone, 145f, 146f, 148f, 149, 150f, 175p, 177p
nasal cavity, 8, 10f, 268f, 291, 406f, 443–444, 443f, 445f
    as member of respiratory system, 443–444, 443f, 451t
    in respiratory tract, 443, 445f
nasal concha, 443–444, 443f
    inferior, 145f, 148f, 149, 175p, 444f
    middle, 145f, 147, 175p, 444f
    in respiratory tract, 444f
    superior, 147, 268f, 444f
nasal region, defined, 17, 18f
nasal septum, 148f
nasolacrimal duct, 278, 278f
nasopharynx, 271f, 406f, 409, 444f, 445
natural family planning, 527
natural killer cell, 385
naturally acquired immunity, 393, 393t
navicular, 163f
neck
    arteries to, 355, 364, 365f, 366, 366t
    head, muscles that move the, 196f, 197, 197t
    injury to, 250
    muscles of the, 195f, 196f
    veins from, 369, 369f
necrosis, 122
negative feedback, 6–7
    in control of hormonal secretions, 296, 297, 297f
    in control of testosterone concentrations, 515f
    in cortical release, 306, 307f
    in red blood cell formation, 321
    in secretion of glucagon, 308
neonatal period, 553–554, 554f, 554t
neonate
    addiction, newborn, 548
    cardiovascular system of, 553–554
    first breath, 453, 553
    newborn addiction, 548
    premature infants, 553
neonatology, 18
nephrology, 18
nephron, 469, 469f, 470–472, 471f, 472f, 473f, 493
    blood supply of, 471
    functions of, 480, 481t
    juxtaglomerular apparatus, 471–473, 473f
    regulation of hydrogen ion concentration in, 499, 499f
    structure, 470–471, 470f, 471f, 472f
nephron loop, 472f, 473f
    ascending limb, 472f, 473f, 476, 481t
    descending limb, 472f, 481t
nerve, 230, 230f
nerve cell. See neuron
nerve cell process, 118f
nerve conduction velocity, 193
nerve fiber, 230, 273f
    sensing pain with, 265
    with taste impulse, 270
    with vision, 283f, 284, 285

nerve impulse, 214, 227–228, 227f
    all-or-none response, 227–228
    events leading to, 227t
    impulse conduction, 227, 227t
    processing, 228–230, 230f
nerve pathway, 231–232, 231f, 232f, 232t
    reflex arcs, 231, 231f, 232t
    reflex behavior, 231–232, 232f
    for smell, 267, 268f
    for taste, 269–270
    visual, 286–287, 286f
nervous system, 12, 13f, 213–256
    aging-related changes, 556t
    compared to endocrine system, 292–293, 292t, 293f
    divisions, 214, 215f
    endocrine system, control of, 296–297, 296f
    general functions, 215–216
    injury to, 232
    prenatal development of, 219
    subdivisions, 246t
    synapse, 221–222, 221f, 223
nervous tissue, 95, 95t, 111–112, 112f, 113, 113t
net filtration pressure, 475
neural stem cell, 219
neurilemma, 217, 219f
neurilemmal sheath, 217, 219f
neurofibril, 216, 217, 218f, 219f
neuroglia, 112, 112f, 214, 215f, 216, 217f
neuroimaging, 213
neurology, 18
neuromuscular junction, 182, 182f
neuron, 51, 51f, 214, 215f, 293f
    autonomic, 250–251, 251f
    cells, 214, 215f
    cellular reprogramming of, 50
    characteristics, 214, 215f
    classification, 219–221, 220f, 221f
    in the dermis, 122
    exocytosis in, 65
    in fetal brain, 556
    functions of, 111–112, 112f
    neuronal pools, 228–229, 230
    retinal, 286–287
    as sensitive to changes in pH, 498
    structure of, 216–217, 218f, 219, 219f
    synapse, 221–222, 221f, 223
neuronal pool, 228–229, 230f
neuropathic pain, 267
neuropeptide, 228, 229f, 229t, 267
neurosecretory cell, 299f
neurotransmitter, 182, 222, 222f, 228, 229t, 292–293, 293f
    autonomic, 251, 254, 254f
    excitatory and inhibitory, 222, 222f, 228
    in hair cells, 274
    release of, events leading to, 228, 229t
neutralization, 392f
neutral solution, 39, 39f
neutron, 32, 32f, 37t
neutrophil, 320f, 322f, 325, 326f, 327t, 385
neutrophilic band cell, 322f
neutrophilic myelocyte, 322f
nevus, 121
newborn addiction, 545f

niacin, 433
nicotinic acid, 433
night blindness, 286
nipple, 23p, 525, 525f
Nissl body, 216
nitric oxide, 228, 229t, 354, 512
nitrogen, 31t, 320f
    atomic structure, 32t
    as blood gas, 329
    in DNA, 84
    in nucleic acids, 44
    partial pressure of, 459–460, 460f
nitrogenous base, in nucleic acids, 44
nitrogenous waste, removal by urinary system, 468
nodes of Ranvier, 217, 218f
nonelectrolyte, 39
nonessential amino acid, 431
nonprotein nitrogenous substance, 330
nonrespiratory movement, 455
nonsteroidal anti-inflammatory drug, 168
nonsteroid hormone, 294–296, 295f
norepinephrine, 254, 304–305, 305f, 305t, 313f
    in cardiac cycle, 352
    in control of blood pressure, 362
    as neurotransmitter, 228, 229t, 254f
normoblast, 322f
nose
    growth of, 108
    hyaline cartilage in, 107
    as member of respiratory system, 443, 443f, 451t
nostril
    as member of respiratory system, 443, 443f
    in respiratory tract, 444f
nuclear envelope, 52f, 57f, 60, 60f, 61t, 70f
nuclear pore, 60, 60f
nuclease, 413, 422t
nucleic acid, 44, 45f, 46, 46t, 47, 47f
    in pancreatic juice, 413
    as source of hydrogen ions, 497, 497f
    in viruses, 384
nucleolus, 52f, 60, 60f, 61t, 218f
nucleoplasm, 60
nucleotide, 44, 45f, 46, 46f
    DNA, 84, 85f
    as plasma nutrients, 329
    in protein synthesis, 85
nucleus, cell, 52, 52f, 57f, 58f
    of atomic structure, 32, 32f
    of cardiac muscle tissue, 111, 112f
    division of, 69, 70f
    of interneurons, 220
    of neurons, 215, 215f, 219f
    of skeletal muscle fiber, 180f, 181f
    of smooth muscle tissue, 110, 111f
    of sperm cell, 510, 510f
nutrient, 428–437
    birth defects from, 549
    in blood, 329
    essential, 428
    intestinal absorption of, 423, 424t
    macronutrients, 428
    micronutrients, 428

nutrition, 428–437
    body mass index, 436, 437f
    malabsorption, 424
    malnutrition, 436
    in nervous system development, 219

O

obesity, 436
oblique fracture, 138, 138f
oblique muscle
    external, 23p, 24p, 25p, 194, 195f
    inferior, 279t
    internal, 24p, 25p
    superior, 279t
oblique section, defined, 15, 16f
obstetrics, 18
obstructive jaundice, 418
obturator foramen, 29p, 159f, 160, 160f
obturator nerve, 249f, 250
occipital artery, 365f
occipital bone, 144, 146f, 147f, 148f, 150f, 176p, 177p
occipital condyle, 146f, 147, 176p
occipitalis muscle, 195f, 196f
occipital lobe, 238, 238f, 239, 268, 286f
occipital region, defined, 17, 18f
occupational hazards, birth defects from, 549
oculomotor nerve (III), 246, 247f, 248t
odontoid process, 153f
olecranon fossa, 155, 157, 157f, 158f
olfactory bulb, 246, 247f, 267, 268f
olfactory nerve (I), 246, 247f, 248t
olfactory nerve pathway, 267
olfactory organ, 267, 268f
olfactory receptor, 243, 267–269, 268f
olfactory receptor cell, 246, 267, 268f
olfactory stimulation, 268–269
olfactory tract, 246, 247f, 267, 268f
oligodendrocyte, 216, 217f
oligomenorrhea, 522
omentum, 420
    greater, 420, 421f
oncology, 18
oocyte, 506, 516–519, 517f, 518f
    follicle maturation, 518, 518f
    as largest cell, 517
    in oogenesis, 517, 517f, 519
    ovulation, 505, 518f, 519, 519f, 523, 524f, 537, 540f
    primary, 516–517, 517f, 518, 518f, 519
    secondary (see secondary oocyte)
    selling of, 505
oogenesis, 519
open fracture, 138
ophthalmic division, 247
ophthalmic vein, superior, 369f
ophthalmology, 18
opiate, 245
opsin, 286
opsonization, 392, 392f
optic chiasma, 242, 242f, 286, 287f, 298f, 299f
optic disc, 280f, 284, 284f
optic foramen, 246
optic nerve, 242f, 280, 280f, 281, 286–287, 286f, 298f
optic nerve (II), 246, 247f, 248t
optic radiation, 286f, 287

optic tract, 242, 242*f*, 247*f*, 286*f*
  left, 287
  right, 287
oral cavity, 8, 10*f*, 403, 406*f*, 443*f*
oral region, defined, 17, 18*f*
orbicularis oculi muscle, 194, 195*f*, 195*t*, 196*f*, 277, 278*f*, 279*t*
orbicularis oris muscle, 194, 195*f*, 195*t*, 196*f*
orbital cavity, 10
orbital region, defined, 17, 18*f*
orchiectomy, 506
organ, 4*f*, 117
organelle, 3, 4*f*, 52–53, 52*f*, 55
organic constituents of cells, 39, 41–47
organism
  defined, 3, 4*f*
  requirements of, 5
organ of Corti, 274
organ system, 3, 4*f*, 12, 13*f*, 14
  absorption and excretion, 12, 13*f*, 14
  body covering, 12, 13*f*
  integration and coordination, 12, 13*f*
  reproduction, 13*f*, 14
  support and movement, 12, 13*f*
  transport, 12, 13*f*
orgasm
  female, 521
  male, 512–513
origin, 167
origin, muscle, 192–193
oropharynx, 406*f*, 409, 445
orthopedic, 18
orthopedic surgeon, 139
osmoreceptor, 300, 492
osmosis, 63, 63*f*, 67*t*
  in capillaries, 357
  dehydration by, 494, 494*f*
  through cell membranes, 63–64, 63*f*
  urine formation by, 472, 474*f*
osmotic pressure, 329
  body fluid movement by, 491–492, 491*f*
  in plasma, 329
  of tubular fluid, 493
  tubular reabsorption by, 478, 478*f*
osseous labyrinth, 272*f*
ossification, 136
osteoarthritis, 168
osteoblast, 135–136, 137, 138
osteoclast, 136, 137, 138, 139*f*, 303
osteocyte, 108, 109*f*, 134, 135*f*, 136
osteon, 134, 135*f*
osteopenia, 132
osteoporosis, 132, 132*f*
ostreon, 108
otic ganglion, 253*f*
otic region, defined, 18*f*
otolaryngology, 19
otolith, 276*f*
outer ear, 270, 271*f*
oval window, 271, 271*f*, 272*f*
ovarian follicle, 516, 523
ovary, 27*p*, 28*p*, 505, 516*f*, 522*t*, 529*f*
  in female infertility, 538
  female sex hormone, secretion of, 522
  follicle maturation, 518, 518*f*
  as gonad, 300
  hormones secreted by, 522

oogenesis, 517, 517*f*
ovulation, 518*f*, 519, 519*f*
  primordial follicle, 516–517, 517*f*
  referred pain, 266*f*
  as reproductive gland, 293, 293*f*
  simple cuboidal epithelial tissue in, 97*f*
  structure of, 516, 516*f*
overeating, 436
overnutrition, 436
oviduct, 519
ovulation, 505, 518*f*, 519, 519*f*, 523, 524*f*, 537, 540*f*
oxaloacetic acid, 81*f*
oxidation, 80
oxygen, 31, 31*t*, 320*f*
  affecting blood pressure, 361–362
  atomic structure, 32*t*
  in blood, 321, 341, 341*f*, 346, 347
  as blood gas, 329
  blood levels of, 457–458, 458*f*
  in capillaries, 357
  in carbohydrates, 41, 41*f*
  in cellular respiration, 80, 81*f*, 82
  as constituent of cells, 40, 40*t*
  deficiency, cyanosis from, 319
  deficiency in neurons, 221
  in dehydration synthesis, 77–78, 77*f*, 78*f*
  diffusion from blood, 448, 449*f*
  diffusion through cell membranes, 61, 62*f*
  formula for, 37*f*
  gas transport of, 460, 461*f*
  low blood oxygen, 321
  in nucleic acids, 44
  partial pressure of, 459–460, 460*f*
  in proteins, 43–44, 44*f*
  as requirement for life, 5
  in skeletal muscle contraction, 186–187, 186*f*
  slow-twitch muscle fibers and, 189
  utilization during exercise, 458
  in water molecules, 36, 37*f*
oxygenated blood, 341, 341*f*
oxygen-carrying capacity, 321
oxygen debt, 186–187, 186*f*
oxyhemoglobin, 319, 460, 461*f*
oxytocin, 300, 301*t*, 512, 551–552

P

pacemaker, SA node as, 350
Pacinian corpuscles, 264, 264*f*
pain, 265, 266*f*, 267
  impulses, regulation of, 265, 267
  ion channels in, 54
  nerve fibers, 265
  neuropathic, 267
  referred, 265, 266*f*
  suppression of, 245
  transmission of impulse, 232, 232*f*
  visceral, 265, 266*f*
pain receptors, 263
palate, 404–405, 405*f*, 406*f*
palatine bone, 146*f*, 148, 148*f*, 176*p*
palatine process, 146*f*, 148, 148*f*, 176*p*
palatine tonsil, 404, 405, 405*f*, 406*f*, 444*f*
palmaris longus muscle, 200, 200*f*, 202*t*
palmar region, defined, 17, 18*f*

pancreas, 11*f*, 28*p*, 293, 293*f*, 297, 421*f*
  hormones of the pancreatic islet, 308–309, 308*f*, 309*f*, 310–311
  as member of digestive system, 401, 402*f*, 413, 414*f*, 415
  pancreatic juice, 413
  referred pain, 266*f*
  regulation of secretions, 413, 414*f*, 415
  structure of, 307–308, 308*f*, 413, 414*f*
  veins that drain, 371*f*
pancreatic acinar cell, 413
pancreatic amylase, 413, 422*t*
pancreatic duct, 308*f*, 413, 414*f*, 419
pancreatic islet, 307, 308–309, 308*f*, 309*f*, 310–311
pancreatic juice, 413, 415
pancreatic lipase, 413, 422*t*
pancreatitis, 30, 413
pantothenic acid, 433
papilla
  renal, 469, 469*f*
  taste buds, 269, 269*f*
papillary muscle, 344, 344*f*
Pap (Papanicolaou) smear, 520
paracrine secretion, 292
paranasal sinus, 144, 145*f*, 443, 444
  as member of respiratory system, 451*t*
parasagittal, defined, 15, 15*f*
parasympathetic division of autonomic nervous system, 250
  in cardioinhibitor reflex, 361, 361*f*
  control of heart rate, 352, 353, 353*f*
  in the pancreas, 413, 415
  regulation of gastric juices by, 412, 412*f*
  release of salivary secretions, 408
  in release of stomach secretions, 412
parasympathetic ganglion, 283*f*
parasympathetic motor nerve fiber, 283*f*
parasympathetic nerve, 408
parathyroid gland, 293, 293*f*, 303, 303*f*, 304*f*
  disorders of, 303*f*
  parathyroid hormone, 303, 304*f*
  structure of, 303, 303*f*
parathyroid hormone, 141, 543*t*
paravertebral ganglion, 266*f*
parietal, 10
parietal bone, 144, 145*f*, 146*f*, 147*f*, 148*f*, 150*f*, 175*p*, 176*p*, 177*p*
parietal cell, 411, 411*f*
parietal lobe, 238–239, 238*f*, 268
parietal pericardium, 342, 342*f*, 343*f*
parietal peritoneum, 421*f*
parietal pleura, 10, 11*f*, 450, 450*f*
parietal pleural membrane, 10, 11*f*
Parkinson disease, 240
parotid gland, 408, 409*f*
paroxysmal extreme pain disorder, 54
partially complete proteins, 431
partial pressure, 459–460, 460*f*, 461, 461*f*
partial prothrombin time, 331
passive aging, 555–556

passive immunity, 393, 393*t*
passive transport, 478
patella, 142*f*, 143, 161–162, 162*f*, 204*f*, 206, 231*f*
patellar ligament, 162, 204*f*, 206, 231, 231*f*
patellar reflex, 231
patellar region, defined, 17, 18*f*
patent ductus arteriosus, 554
pathogen, 384
  adaptive (specific) defense (*see* adaptive (specific) defense)
  first line of defense against, 385
  innate (nonspecific) defense (*see* innate (nonspecific) defense)
pathology, 19
peanut allergy, 377
pectoral girdle, 143, 155, 156*f*
  muscles that move the, 197, 198*f*, 198*t*, 199*f*
pectoralis major muscle, 23*p*, 24*p*, 194, 195*f*, 197, 199*f*, 199*t*, 525, 525*f*
pectoralis minor muscle, 24*p*, 197, 198*t*, 199*f*, 452, 452*f*, 525*f*
pectoral region, defined, 17, 18*f*
pedal region, defined, 17, 18*f*
pediatrics, 19
pedicle, 149, 152*f*
pediculosis, 129
pedigree, 560
pelvic brim, 160, 161*f*
pelvic cavity, 8, 9*f*, 160*t*
pelvic diaphragm, 201
pelvic girdle, 143, 143*t*, 158, 159*f*, 160–161, 160*f*, 160*t*, 161*f*
pelvic inflammatory disease, 530, 530*t*
pelvic lymph node, 382*f*
pelvic nerve, 253*f*
pelvic outlet, muscles of the, 201, 202*f*, 203, 203*t*
pelvic region, defined, 17
pelvic sacral foramen, 29*p*
pelvis, 143, 158, 159*f*
  arteries to, 367, 368*f*
  female, 160, 161*t*
  male, 160, 161*t*
  veins from, 370–371, 372*f*
penis, 25*p*, 26*p*, 202*f*, 253*f*, 507*f*, 512, 513*t*, 520, 521
pepsin, 385, 411, 412*f*, 422*t*
pepsinogen, 411, 412*f*
peptidase, 422, 422*t*
peptide bonds, 78, 78*f*
peptides, 293*t*
perception, 263, 268
percutaneous transluminal angioplasty, 355
perforating canal, 134, 135*f*
perforin, 385, 388
pericardial cavity, 9*f*, 10, 11*f*, 342–343, 342*f*, 343*f*, 450*f*
pericardial membrane, 10, 11*f*
pericardial sac, 25*p*
pericarditis, 343
pericardium, 342–343, 342*f*, 343*f*
  fibrous, 342, 342*f*, 343*f*
  parietal, 10, 11*f*, 342, 342*f*, 343*f*
  visceral, 10, 11*f*, 342, 342*f*, 343*f*
perichondrium, 107
perilymph, 272, 272*f*, 273*f*
perimetrium, 519*f*, 520, 520*f*

perimysium, 179, 180*f*, 230
perineal region, defined, 17, 18*f*
perineal tissue, 551
perineurium, 230, 230*f*
period mutation gene, 312
periodontal ligament, 407, 408*f*
periosteum, 133, 134*f*, 135*f*, 136*f*, 138
peripheral, defined, 15
peripheral blood smear, 320*f*
peripheral chemoreceptor, 457–458, 458*f*
peripheral nervous system, 214, 215*f*, 216, 221*f*, 246–250, 246*f*, 247*f*, 249*f*
   cranial nerve (*see* cranial nerve)
   subdivisions, 246*t*
peripheral neuropathy, 267
peripheral resistance
   in arterial blood pressure, 360, 360*f*
   in capillaries, 357
peristalsis, 191, 403, 405*f*
   movement of sperm by, 510, 512
   in release of bile, 418
   in small intestine, 424
   in stomach, 413
   in swallowing, 409, 410
   in ureter, 481
   in uterine tube, 519–520
peristaltic rush, 424
peritoneal cavity, 10, 11*f*, 421*f*
peritoneal membrane, 10, 11*f*, 420, 421*f*
peritoneum, 538
peritonitis, 426
peritubular capillary, 470*f*, 472*f*, 474*f*, 477, 477*f*, 478, 478*f*
peritubular capillary system, 471
permeability, in cells, 53–54
peroxisome, 56, 61*t*
perpendicular plate, 145*f*, 147, 148*f*
persistent vegetative state, 213
perspiration, water lost from, 492*f*, 493
Peyer's patches, 382
pH, 39, 39*f*
   in acid-base imbalance, 500, 500*f*, 501, 502
   first line of defense against shifts in, 499, 499*f*
   hydrogen ion concentrations, regulation of, 498, 499, 499*f*
   neurons as sensitive to changes in, 498
   scale, 39*f*, 500*f*
   second line of defense against shifts in, 499, 499*f*
   of semen, 511
phagocytes, 66
phagocytosis, 66, 66*f*, 67*t*, 112, 385, 386
phalange, 159, 166*f*
   of fingers, 142*f*, 143, 158, 159*f*
   joint, 166*f*
   of toes, 142*f*, 143, 163, 163*f*
pharmacology, 19
pharyngeal tonsil, 404, 406*f*, 444*f*
pharyngeal wall, 409
pharynx, 303*f*, 408–409, 445
   in alimentary canal, 401, 402*f*, 403*f*
   as member of digestive system, 401, 402*f*, 408–409, 445, 451*t*

in respiratory tract, 443
   structure of, 409
   swallowing mechanism, 409
phenotype, 556–557
phenylalanine, 44, 44*f*, 431*t*
pheomelanin, 123
pheromone, 291
phosphate
   in ATP, 80, 81*f*, 82
   buffer system, 498–499, 498*t*, 499*f*
   in DNA, 84
   in electrolyte intake, 493
   in extracellular fluid, 491*f*
   in glomerular filtrate, 475*t*
   in intracellular fluid, 491*f*
   in parathyroid gland, 303, 304*f*
   in plasma, 475*t*
   in skeletal muscle contraction, 183, 185–186, 186*f*
   tubular reabsorption by, 478
   in urine, 475*t*
phosphate group, in nucleic acids, 44
phosphate ion, as constituent of cells, 40, 40*t*
phosphodiesterase, 295
phospholipid, 42, 43*f*, 43*t*, 429
   in blood clot formation, 331
   in cell membranes, 53–54, 53*f*
phosphoprotein, as source of hydrogen ions, 497, 497*f*
phosphoric acid, 497, 497*f*
phosphorus, 31*t*
   as major mineral in human body, 434, 434*t*
   in nucleic acids, 44
   storage of, by bones, 141
photophobia, 281
photopigment, 286
photoreceptor, 263, 282, 283*f*, 285–286, 285*f*
phrenic artery, 367
phrenic nerve, 249*f*, 250, 452
physical activity. *See* exercise
physiological buffer system, 499, 499*f*, 500
physiologic jaundice, 324
physiology, 3
pia mater, 233, 233*f*, 234, 241*f*
pigmented epithelium, 283*f*
pimple, 124
pineal gland, 42, 242, 242*f*, 293, 293*f*, 309
pinocytosis, 66, 66*f*, 67*t*, 544
pisiform, 159*f*
pitch, 262
pituitary dwarfism, 298
pituitary gland, 242, 242*f*, 293, 293*f*, 301*t*, 307*f*, 313*f*, 366*f*
   anterior pituitary hormone, 297–298, 298*f*, 299*f*, 300, 301*t*, 519
   anterior pituitary lobe, 297, 298*f*, 299*f*
   endocrine system, control of, 296–297, 296*f*
   in female infertility, 538
   female reproductive functions, control of, 519, 522–524, 523*t*
   intermediate lobe, 297
   male reproductive functions, control of, 513, 515, 515*f*
   in milk production and secretion, 552–553

posterior pituitary hormone, 300, 301*t*
posterior pituitary lobe, 240, 297, 298*f*
pituitary stalk, 298*f*
pivot joint, 165, 166*f*, 167*t*
placenta, 541–542, 543*t*, 544–547, 544*f*, 545*f*, 546*f*, 547*f*, 550*f*. *See also* prenatal period
   in birth process, 551–552, 552*f*
   female sex hormone, secretion of, 522
   hormone production by, 310
   passage of antibodies from mother, 391, 393
placental estrogen, 542–543
placental lactogen, 542–543, 552
placental membrane, 544, 545*f*, 549
placental progesterone, 542–543
plane joint, 165, 166*f*, 167*t*
plantar flexion, 167, 169*f*
plantar region, defined, 17, 18*f*
plant protein, 431
plaque, 333*f*, 355, 355*f*
plasma, 108, 110*f*, 319, 320*f*
   electrolytes, 330
   endocrine system, control of, 296–297, 296*f*
   as extracellular fluid, 490, 490*f*, 491
   gases and nutrients, 329–330
   in gas transport, 460, 461*f*, 462, 462*f*
   as intracellular fluid, 491
   lymph originating in, 380
   movement of, between fluid compartments, 491, 491*f*
   nonprotein nitrogenous substances, 330
   protein concentrations in, 329, 329*f*, 329*t*, 475, 495*t*, 498
   tubular secretion of, 479
   in urine formation, 472, 473, 475*t*
plasma cell, 322*f*, 390–391, 390*f*, 391
plasma colloid osmotic pressure, 380, 475–476, 475*f*
plasma cortisol, 312
plasma membrane, 53
plasma nutrient, 329–330
plasmin, 331
plasminogen, 331
platelet, 108, 322*f*, 327, 327*t*
   in buffy coat of blood, 320*f*
   development of, 327
   function of, 319, 320*f*
   plug formation, 330, 330*f*
   production of, 322*f*
platelet count, 327
platelet plug, 330, 333*f*
platysma muscle, 194, 195*t*, 196*f*
pleural cavity, 9*f*, 10, 11*f*, 28*p*, 450*f*
pleural membrane, 10, 11*f*, 452
plexus, 249–250
pneumonia, 442, 456, 460
pneumotaxic center, 456
podiatry, 19
podocyte, 473*f*, 474*f*
polar body, 540*f*
polarized cell membrane, 222, 223*f*
polar molecule, 35–36, 36*f*, 46
polio, 94
polycythemia, 321
polydipsia, 310
polygenic trait, 560
polymorphism, 91
polypeptide, 44, 45*f*, 78, 78*f*, 411

polysaccharide, 41–42, 386
polysome, 55
polyunsaturated fat, 42, 43*f*
pons, 236*f*, 237*f*, 242*f*, 243
   in breathing, 456, 457*f*
   micturition reflex, facilitation of, 484
pontine respiratory group, 456, 457, 457*f*
popliteal artery, 367, 368*f*
popliteal region, defined, 17, 18*f*
popliteal vein, 371, 372*f*
positive feedback system
   in birth process, 551
   with blood coagulation, 331
   defined, 7–8
postcoital testing for female infertility, 538*t*
posterior, defined, 14, 14*f*, 18*f*
posterior cavity of the eye, 280*f*, 284
posterior cerebral artery, 364, 366*f*
posterior chamber of the eye, 280*f*, 282
posterior circumflex artery, 367*f*
posterior communicating artery, 366*f*
posterior fontanel, 150*f*
posterior funiculus, 234, 235*f*
posterior horn, 234, 235*f*
posterior inferior iliac spine, 160*f*
posterior intercostal artery, 367
posterior intercostal vein, 370
posterior interventricular artery, 348*f*
posterior median sulcus, 234, 235*f*
posterior pituitary gland, 242
posterior sacral foramen, 152, 154*f*
posterior superior iliac spine, 160*f*
posterior tibial artery, 367, 368*f*
posterior tibial vein, 370–371, 372*f*
posterolateral fontanel, 150*f*
postganglionic fiber, 251, 251*f*, 254*f*
postganglionic neuron, 252*f*, 253*f*
postmortem sperm retrieval, 536
postnatal period, 537
postsynaptic membrane, 222*f*, 223*f*
postsynaptic neuron, 221, 221*f*, 225*f*
postsynaptic neurotransmitter receptor, 222*f*
potassium
   active transport across cell membranes of, 65
   in cardiac cycle, 352
   cell membrane transport of, 222, 223, 224*f*, 225, 226*f*
   concentrations of, in glomerular filtrate, 475*t*
   effect on aldosterone, 306
   in electrolyte intake, 493
   in electrolyte output, 493, 496
   as element in human body, 31*t*
   in extracellular fluid, 491*f*
   high potassium concentration, 306
   imbalance, 496, 496*f*
   in intracellular fluid, 491*f*, 496
   ion channels, 54
   low potassium concentration, 306
   as major mineral in human body, 434, 434*t*
   in nerve impulse conduction, 227–228, 227*t*
   in plasma, 475*t*
   tubular secretion of, 479, 479*f*, 496, 496*f*
   in urine, 475*t*
   in urine formation, 476

potassium ion
 as constituent of cells, 40, 40*t*
 in extracellular fluids, 493,
  496, 496*f*
 tubular secretion, 479, 479*f*
potential difference, 223, 224*f,* 225
PQ interval, 351
precapillary sphincter, 356, 356*f,* 357
precocious puberty, 296
preganglionic fiber, 251*f,* 252, 252*f,*
  253*f,* 254*f*
preganglionic neuron, 252*f,* 253*f*
pregnancy, 522, 537–539, 537*f,*
  538*f,* 538*t,* 539*f,* 540*f,*
  540*t. See also* prenatal
  period
 addictive substances, use of, 545
 birth process, 551–552, 552*f*
 calcium, importance of
  dietary, 543
 contractions, stimulating, 300
 early embryonic development,
  541–543, 541*f,* 542*f,* 543*t*
 early signs of, 543
 ectopic, 541
 embryonic stage, 543–547, 544*f,*
  545*f,* 546*f,* 547*f*
 expansion of uterus during, 520
 fetal cells persisting in
  women, 395
 fetal chromosome checks,
  558–559, 558*f,* 559*f*
 fetal movements, 543
 fetal stage, 547–551, 548*f,*
  550*f,* 551*t*
 hormonal changes during,
  542–543, 542*f,* 543*t*
 malnutrition, effects of, 549
 milk production and secretion,
  552–553, 553*f*
 passage of antibodies from
  mother to placenta,
  391, 393
 protein intake, requirements
  for, 432
 Rh compatibility, 338
 transport of sex cells,
  537–538, 537*f*
pregnanediol, 538
preimplantation genetic
  diagnosis, 559
premature infants, survival chances
  of, 553
premolar (teeth), 406*f,* 407, 407*t*
prenatal period, 537, 541–553
 birth process, 551–552, 552*f*
 bone development in, 135,
  135*f,* 136*f*
 critical periods, 547
 early embryonic development,
  541–543, 541*f,* 542*f,* 543*t*
 embryonic stage, 541–542,
  543–547, 544*f,* 545*f,*
  546*f,* 547*f*
 fetal cells persisting in
  women, 395
 fetal movements, 543
 fetal stage, 547–551, 548*f,*
  550*f,* 551*t*
 hemolytic disease of, 336
 intermediate lobe in, 297
 of nervous system, 219
 release of precursor lymphocytes
  by red marrow, 387*f*
 Rh incompatibility, 336, 336*f*

stages and events of early, 543*t*
 twins, development of, 542
prepatellar bursa, 165*f*
prepuce, 507*f,* 512
pressoreceptor, 352
pressure
 as requirement for life, 5
 sense of, 264, 264*f*
pressure ulcer, 122
presynaptic neuron, 221, 221*f,* 222,
  222*f,* 225
primary bronchus, 446, 447*f,*
  448, 448*f*
primary follicle, 518*f,* 524
primary germ layer, 543–544, 544*f*
primary hypertension, 362
primary immune response,
  392–393, 392*f*
primary oocyte, 516–517, 517*f,* 518,
  518*f,* 519
primary ossification center, 136, 136*f*
primary pulmonary cancer, 454
primary spermatocyte, 507, 508*f,*
  509*f,* 515*f*
primary teeth, 406*f*
prime mover, muscle, 194
primordial follicle, 516–517, 518*f*
probiotics, 400
process (bone), 133, 144*t*
proerythroblast, 322*f*
progenitor cell, 71, 72*f*
progesterone, 522, 523, 523*t,* 524,
  524*f,* 537, 538
 changes during pregnancy, 542,
  542*f,* 543*t*
 in combined hormone
  contraceptives,
  528–529, 528*f*
 as is initiating birth process, 551
 in milk production and
  secretion, 552
 placental progesterone, 542–543
 uterine contractions, suppressing
  of, 551
progesterone receptor, 527
progranulocyte, 322*f*
projection, 263
prolactin, 297, 300, 301*t,* 552, 553
proliferative phase of female
  reproductive cycle, 524*f*
proline, 431*t*
prolymphocyte, 322*f*
promonocyte, 322*f*
pronation, 167, 169*f*
pronator quadratus muscle, 200,
  200*f,* 201*t*
pronator teres muscle, 200, 200*f,* 201*t*
pronucleus, 539, 540*f*
propelling movements of the
  alimentary canal, 403
prophase, 68*f,* 69, 70*f*
proprioceptor, 458
prostaglandin, 296, 510, 511, 537
prostaglandin D₂, 394
prostate cancer, 94, 511
prostate gland, 483*f,* 507*f,*
  510–511, 513*t*
prostate-specific antigen, 511
protease, 104
protease inhibitor, 389
proteasome, 88
protein, 43–44, 44*f,* 45*f,* 46*t,* 47,
  430–432, 431*f,* 431*t,* 432*t*
 abnormal, in muscular
  dystrophies, 193

actin (*see* actin)
 antigens as, 386
 in blood, 357
 in blood clot formation, 331
 breakdown of, 83*f*
 buffer system, 498–499,
  498*t,* 499*f*
 in cell membranes, 53–54, 53*f*
 complete, 431
 concentrations in plasma, 475
 in cytoplasm, 55
 in diet, deficiencies of, 329
 digestion in small intestine, 423
 effects of liver on, 415, 417, 418*t*
 in endoplasmic reticulum,
  synthesis of, 55, 56*f*
 in extracellular matrix,
  104–105, 105*f*
 in human milk, proportion
  of, 553
 hydrolysis of, 78, 78*f*
 incomplete, 431
 as macronutrient, 428
 in muscle function, 179, 183, 183*f*
 as nonsteroid hormone,
  294–296, 295*f*
 in nuclear envelope, 60
 partially complete, 431
 in plants, 431
 in plasma, 329, 329*f,* 475, 495
 in plasma, protein concentrations
  in, 498
 requirements, 432
 sources, 430–431
 synthesis hampered by diabetes
  mellitus, 310–311
 as type of hormone, 293*t*
 in urine, 480
 use of, 431, 431*f*
 in viruses, 384
protein fiber, 102
protein synthesis, 85–88, 86*t,* 87*f,*
  87*t,* 88*f,* 89*f*
 genetic code in, 85
 transcription, 85–86, 86*t,* 87*f*
 translation, 87–88, 87*f,* 88*f,* 89*f*
 in viruses, 384
proteolytic, 413
proteolytic enzyme, 422*t,* 544
prothrombin, 331, 333*f*
prothrombin activator, 331, 333*f*
prothrombin time, 331
proton, 32, 32*f,* 37*t*
protozoan, 384
protraction, 167, 170*f*
proximal, defined, 14*f,* 15
proximal convoluted tubule, 470*f,*
  471, 472*f,* 473*f,* 474*f,*
  478*f,* 481*f*
proximal epiphysis, 134*f*
proximal phalanx, 159*f,* 163*f*
pruritus, 129
Pseudomonas bacteria, 449
psoas major muscle, 29*p,* 204*f,* 204*t*
psoriasis, 129
psychiatry, 19
puberty, 507, 509, 513, 515, 522
pubic arch, 160, 161*f*
pubic symphysis, 28*p,* 159*f,* 160, 161*f,*
  202*f,* 516*f,* 521, 552*f*
pubis, 158, 160, 160*f*
pulmonary artery, 27*p,* 344, 344*f,*
  346, 348*f,* 448*f*
pulmonary circuit, 341, 341*f,* 363
pulmonary embolism, 331

pulmonary plexus, 252*f,* 253*f*
pulmonary trunk, 26*p,* 27*p,* 342*f,*
  344, 346, 348*f*
pulmonary valve, 344, 345*t,* 346, 349*f*
pulmonary vein, 27*p,* 341*f,* 344, 344*f,*
  346, 348*f,* 448*f*
pulp
 red, 383, 384*f*
 white, 383, 384*f*
pulp (teeth), 407
pulp cavity, 408*f*
pulse, 359
pump, in active transport, 65
Punnett square, 557, 557*f,* 560
pupil, 256*t,* 280*f,* 282, 283*f*
Purkinje fiber, 343, 350, 350*f*
pus, 385
pustule, 124, 129
putamen, 240, 240*f*
P wave, 351, 351*f*
pyloric antrum, 410*f*
pyloric canal, 410–411, 410*f*
pyloric sphincter, 410–411, 410*f,* 414*f*
pylorus, 410, 410*f*
pyramidal cell, 239
pyramidal tract, 235, 242*f*
pyruvic acid, 186–187, 186*f*
 in cellular respiration, 81*f,* 82
 in citric acid cycle, 81*f*

Q

QRS complex, 351, 351*f,* 352*f*
quadrate lobe of liver, 415*f*
quadratus lumborum muscle, 29*p*
quadriceps femoris group, 205, 205*t,*
  231, 231*f*
quaternary structure, 44, 45*f*
Q wave, 351, 351*f*

R

radial artery, 359, 367, 367*f,* 368*f*
radial nerve, 249*f,* 250
radial notch, 158*f*
radial recurrent artery, 367*f*
radial set, 282
radial tuberosity, 157, 158*f*
radial vein, 370*f,* 372*f*
radiation, forms of, 34
radical mastectomy, 527
radioactive isotope, 33, 34, 34*f*
radiology, 19
radiopaque, 425*f*
radius, 140, 140*f,* 142*f,* 143, 156*f,* 157,
  158*f,* 159*f,* 166*f,* 194
rate-limiting enzymes, 82
receptor, 6–8, 7*f,* 8*f,* 220, 221*f*
 in cardiac cycle, 353*f*
 in patellar reflex, 231, 231*f*
 in reflex arc, 232*t*
 in regulation of blood glucose
  concentration, 308, 309*f*
 in withdrawal reflex, 232, 232*f*
receptor ends of sensory neurons,
  220, 221*f*
receptor-mediated endocytosis, 66,
  66*f,* 67*t,* 423
recessive inheritance, 557, 557*f,* 560
recommended daily allowance of
  vitamins, 432, 433
recruitment of motor unit, 190, 190*f*
rectal vein, 426
rectum, 27*p,* 29*p,* 202*f,* 371*f,* 421*f,*
  424, 425*f,* 426, 426*f*
 in alimentary canal, 401,
  402*f,* 403*f*

in birth process, 552*f*
in female reproductive
system, 516*f*
as member of digestive system,
401, 402*f*
rectus abdominis muscle, 23*p*, 24*p*,
195*f*, 199*f*, 201, 203*t*
rectus femoris muscle, 24*p*, 26*p*,
27*p*, 28*p*, 195*f*, 204*f*,
205, 205*t*
rectus muscle
inferior, 279*t*
lateral, 279*t*
medial, 279*t*
superior, 279*t*
recurrent artery
radial, 367*f*
ulnar, 367*f*
recurrent laryngeal nerve, 446
red blood cell, 108, 110*f*, 140, 320*f*,
321*f*, 323*f*
agglutination in, 334–335, 335*f*
in bile production, 418
cellular components, 327*t*
clumping of, 330–332, 330*f*,
331*f*, 333*f*
counts, 321, 515
destruction of, 323–324, 324*f*
dietary factors affecting
production of, 323
functions of, 319, 320*f*, 321
in hypertonic solutions, 64*f*
in hypotonic solutions, 64*f*
life cycle of, 324
in mismatched blood
transfusions, 333
production, 321, 322*f*, 469
Rh incompatibility of, 336, 336*f*
in urine, 480
red blood cell count, 321, 515
red bone marrow, 386, 387, 387*f*
red pulp, 383, 384*f*
referred pain, 265, 266*f*
reflex, 231–232, 231*f*, 232*f*, 235, 455
reflex arc, 231, 231*f*, 232*t*
reflex center, 231
refraction, 284–285, 284*f*
refractory period, 228
regulatory enzyme, 82
relaxin, 543, 543*t*
releasing hormone, 297, 299
renal artery, 364*f*, 368*f*, 468*f*, 469,
470, 470*f*, 472*f*, 550*f*
renal blood vessel, 469–470, 470*f*
renal capsule, 469*f*
renal column, 469*f*
renal corpuscle, 469*f*, 470–471,
471*f*, 481*t*
renal cortex, 469*f*, 470*f*, 471*f*, 472*f*
renal medulla, 469*f*, 470*f*, 471*f*, 472*f*
renal papilla, 469, 469*f*
renal pelvis, 468, 469*f*, 470*f*
renal plasma threshold, 478
renal pyramid, 469*f*
renal sinus, 468, 469, 469*f*
renal tubule, 469*f*, 470–471, 471*f*, 477,
477*f*, 481*t*
reabsorption by, 472–473, 474*f*,
478–479, 478*f*, 479*f*
secretion by, 472–473, 474*f*
renal vein, 372*f*, 468*f*, 469, 470*f*
renin, 476, 476*f*
repolarization, 226, 226*f*
reproduction, 5*t*
reproductive gland, 293, 293*f*

reproductive system, 13*f*, 14
aging-related changes, 556*t*
female (*see* female reproductive
system)
male (*see* male reproductive
system)
reproductive technology, 539, 540*t*
residual volume, 455, 456, 456*t*
resistance stage, 311, 313*f*
resistance to disease, 384–385
resistance training, benefits of, 178
resorption of matrix, 137
respiration, 5*t*, 443
air volumes and capacities,
455–456, 455*f*, 456*t*
breathing, factors affecting,
456–458, 458*f*
cellular, 80, 81*f*, 82, 443
events of, 443
external, 443
internal, 443
respiratory areas, 456, 457*f*
venous blood moved in, 362–363
respiratory acidosis, 500, 500*f*
respiratory alkalosis, 501–502, 501*f*
respiratory area, 456, 457*f*, 458
respiratory bronchiole, 447*f*, 448*f*
respiratory capacity, 455–456,
455*f*, 456*t*
respiratory center, 243–244, 499, 499*f*
respiratory cycle, 455
respiratory distress syndrome, 452
respiratory group, 456, 457*f*
respiratory membrane, 460
alveolar gas exchange in,
459, 459*f*
diffusion, factors affecting,
459–460, 460*f*
respiratory muscle, 457*f*
respiratory rate, 457–458
respiratory system, 12, 13*f*, 442–463
aging-related changes, 556*t*
divisions of, 443
embryonic development, 548
and exercise, 458
organs of, 443–450, 451*t*
pseudostratified columnar
epithelial tissue in, 98*f*
respiratory tube, cilia lining of, 58, 59*f*
respiratory volume, 455–456,
455*f*, 456*t*
responsiveness, 5*t*
resting neuron, 222
resting potential, 223, 224*f*, 225, 226*f*
resting tidal volume, 455
reticular connective tissue, 106, 106*f*
reticular fiber, 103, 103*t*
reticular formation, 244
reticulocyte, 322*f*
reticuloendothelial system, 385
retina, 280*f*, 281*f*, 282, 283*f*, 285–286
retinal, 286
retinal neuron, 286–287
retinitis pigmentosa, 296
retraction, 167, 170*f*
retroperitoneal positioning, 468
reversible reaction, 38
R group, 44
Rh blood group, 333, 336, 336*f*
rhesus monkey, 336
rheumatoid arthritis, 168, 394
Rh incompatibility, 336, 336*f*
Rh-negative, 336, 336*f*
rhodopsin, 286
RhoGAM, 336

rhomboid major muscle, 197,
198*f*, 198*t*
rhomboid muscle, 195*f*
Rh-positive, 336, 336*f*
rhythmicity, 191
rhythm method, 527
rib, 11*f*, 29*p*, 142*f*, 143, 153, 154*f*, 155,
156*f*, 525*f*
riboflavin, 433, 433*t*
ribonucleic acid. *See* RNA
(ribonucleic acid)
ribosome, 55, 56*f*, 58*f*, 61*t*
artificial, creation of, 55
in protein synthesis, 87–88,
88*f*, 89*f*
in steroid hormone synthesis,
294, 294*f*
right hypochondriac region, defined,
16, 17*f*
right iliac region, defined, 16, 17*f*
right lower quadrant of abdominal
area, 17*f*
right lumbar region, defined, 16, 17*f*
right lymphatic duct, 379, 380*f*
right optic tract, 287
right upper quadrant of abdominal
area, 17*f*
rigor mortis, 187
RNA (ribonucleic acid), 46, 60
in AIDS, 389
DNA compared with, 86*t*
messenger RNA, 86, 88, 88*f*, 89*f*
in protein synthesis, 86–88, 87*f*,
88*f*, 89*f*
ribosomes in, 55
transfer RNA, 87–88, 89*f*
RNA polymerase, 86
rod, 283*f*, 285–286, 285*f*
root (teeth), 407, 408*f*
root canal, 407, 408*f*
rotation, 167, 169*f*
roughage, 429
round bone, 133
round ligament
of liver, 415*f*
of uterus, 27*p*
round window, 271*f*, 272*f*, 273
rubella, 548
rugae, 410
"rule of nines," 126
ruptured disc, 153
R wave, 351, 351*f*

S

saccule, 272*f*, 275
sacral artery, middle, 364*f*
sacral canal, 153, 154*f*
sacral curvature, 161*f*
sacral foramen, anterior, 153
sacral hiatus, 153, 154*f*
sacral nerve, 248–249, 249*f*
sacral parasympathetic neuron, 254*f*
sacral promontory, 154*f*, 161*f*
sacral region, defined, 17, 18*f*
sacroiliac, 158
sacrum, 29*p*, 142, 142*f*, 152–153, 154*f*,
159*f*, 160*t*
saddle joint, 165, 166*f*, 167*t*
sagittal, defined, 15
sagittal suture, 144, 150*f*
saliva, 408, 409*f*
salivary amylase, 408, 422
salivary gland, 408, 409*f*
effects of neurotransmitters
on, 256*t*

as member of digestive system,
401, 402*f*, 408, 409*f*
secretions of, 408
stratified cuboidal epithelial
tissue in, 98, 99*f*
Salmonella, 400
salt, 35*f*
as constituent of cells, 40, 40*t*
removal by urinary system, 468
salty, taste sensation, 270
SA node (sinoatrial node), 350, 350*f*,
352, 353*f*, 361, 361*f*
saphenous vein
great, 23*p*, 24*p*, 27*p*, 371, 372
small, 371, 372*f*
sarcolemma, 179, 182*f*
sarcomere, in sliding filament model,
180*f*, 181, 181*f*, 183,
183*f*, 185*f*
sarcoplasm, 179, 181, 181*f*
sarcoplasmic reticulum, 180*f*, 181,
181*f*, 183, 184*f*, 192
sartorius muscle, 23*p*, 24*p*, 25*p*, 27*p*,
28*p*, 195*f*, 197, 204*f*, 205,
205*f*, 205*t*, 207*f*
saturated fat, 42, 43, 43*f*
scab, 126
scabies, 129
scala, 272*f*
scala tympani, 272–273, 273*f*
scala vestibuli, 272–273, 273*f*
scalp, 233*f*
scaphoid, 159*f*
scapula, 142*f*, 143, 155, 156*f*, 200*f*
scar, 126
scar tissue, 113
Schwann cell, 216, 217, 218*f*, 219*f*
sciatic nerve, 249*f*, 250
sciatic spine, 160*f*
scintillation counter, 34, 34*f*
sclera, 280, 280*f*, 281*f*, 283*f*
scleral venus sinus, 282
scleroderma, 395
scopolamine, 276
scratching, 116
scrotum, 26*p*, 202*f*, 253*f*, 506, 507*f*,
512, 513*t*, 529*f*
scurvy, 433
sebaceous gland, 118*f*, 123*f*,
124–125, 124*f*
sebum, 124, 548
secondary immune response,
392–393, 392*f*
secondary oocyte, 519*f*, 523*t*,
537, 537*f*
after zygote forms, 541, 541*f*
in fertilization, 537*f*, 538–539,
539*f*, 540*f*
in oogenesis, 517, 517*f*, 518,
518*f*, 519
in vitro fertilization, 538
secondary ossification center,
136, 136*f*
secondary sex characteristic
female, 522, 524
male, 515, 515*f*
secondary spermatocyte, 509*f*
secondary teeth, 406*f*
second line of defense against
pathogens, 385
second line of defense against shifts
in pH, 499, 499*f*
second meiotic division, 507,
508*f*, 509
second messenger, 295, 296

second polar body, 517*f*
secretin, 414*f*, 418*t*
secretory cell of anterior pituitary gland, 299*f*
secretory phase of female reproductive cycle, 524*f*
segmental bronchus, 447*f*
segmentation in alimentary canal, 403, 405*f*
selective estrogen receptor moderator, 527
selective serotonin reuptake inhibitor, 223
selenium, 434, 435*t*
self-monitoring of blood glucose, 311
sellar joint, 165
sella turcica, 147*f*, 147*f*, 148*f*, 176*p*, 298*f*, 299*f*
semen, 511, 514*t*, 537
semicircular canal, 271*f*, 272–273, 272*f*, 277*f*
semilunar valve, 345
semimembranosus muscle, 195*f*, 205, 205*t*, 207*f*
seminal vesicle, 483*f*, 507*f*, 510, 513*t*
seminiferous tubule, 506, 508*f*, 513*t*
semipermeable membrane, 53–54
semispinalis capitis muscle, 196*f*, 197, 197*t*
semitendinosus muscle, 195*f*, 205, 205*f*, 205*t*, 207*f*
senescence, stages in development, 554*t*
sensation, 263, 268, 270
senses, 262–287
  somatic (*see* somatic senses)
  special (*see* special senses)
sensorineural hearing loss, 275
sensory adaptation, 263
sensory area of cerebral cortex, 238*f*
sensory area of the cerebrum, 238–239, 238*f*
sensory fibers, 230, 236*f*
sensory nerve, 230
sensory nerve fiber, 264*f*, 266*f*, 269*f*, 277*f*, 521
sensory neuron, 220, 221*f*, 231*f*, 232*t*
  in autonomic nervous system, 250–251, 251*f*
  in cardiac cycle, 352, 353
sensory receptor, 215, 215*f*, 263
sensory speech area, 238*f*, 239
septum, 343–344
serine, 431*t*
serosa. *See* serous membrane
serotonin, 228, 229*t*, 267, 330
serous cell, 408
serous coat, of urinary bladder, 483, 483*f*
serous fluid, 101, 110, 490
serous membrane, 110
  in alimentary canal, 401, 404*f*
  in large intestine, 425*f*
  in respiratory system, 450
  of scrotum, 512
  in stomach, 411, 411*f*
  in uterus, 520
serratus anterior muscle, 23*p*, 24*p*, 195*f*, 197, 198*t*, 199*f*
Sertoli cell, 507
serum, 331
sesamoid bone, 133
set point, 6–8, 7*f*, 8*f*
sex chromosome, 556

sexual intercourse, 526, 537
sexually transmitted infection, 530, 530*t*
sexual stimulation
  females, 521
  male, 512–513
shells, of electrons, 33, 33*f*
shigatoxin, 467
shivering, 7
shock, 475
short bone, 133
short head biceps brachii muscle, 24*p*
shoulder. *See also* upper limb
  arm, muscles that move the, 197–198, 198*f*, 199*f*, 199*t*, 200*f*
  arteries to, 366–367, 366*f*
  veins from, 369, 370*f*
shoulder blade. *See* scapula
shoulder girdle, 155
sickle cell disease, 323, 549
side chain group, 44
sidestream smoke, 442
sight, sense of, 276–287
  accessory organs, visual, 276, 277–279, 278*f*
  disorders, 286
  double vision, 278
  eye, structure of the, 278*f*, 279–282, 279*f*, 280*f*, 281*f*, 282*f*, 283*f*, 284, 284*f*
  light refraction, 284–285, 284*f*
  loss of (*see* blindness)
  photoreceptors, 285–286, 285*f*
sigmoid colon, 27*p*, 28*p*, 425*f*, 426
signal transduction, 295, 295*f*
simple diffusion, 61, 62*f*, 63
simple epithelial tissue, 95–97, 96*f*, 97*f*, 102*t*
simple squamous epithelial cell, in bronchial tree, 448*f*
simple squamous epithelial tissue, 459
single nucleotide polymorphism (SNP), 91
sinoatrial node (SA node), 350, 350*f*, 352, 353*f*, 361, 361*f*
sinus, 10, 10*f*, 144*t*, 381*f*
sinus headache, 444
skeletal muscle, 179–208, 192*t*
  abdominal wall, muscles of the, 201, 203*t*
  actions, 192–194, 194*f*
  arm, muscles that move the, 197–198, 198*f*, 199*f*, 199*t*, 200*f*
  connective tissue coverings, 179, 180*f*
  contraction (*see* skeletal muscle contraction)
  embryonic development, 547
  facial expression, muscles of, 194, 194*f*, 196*f*, 197
  fibers, 179, 180*f*, 181–182, 181*f*
  foot, muscles that move the, 205–206, 206*f*, 207*f*, 207*t*
  forearm, muscles that move the, 198, 200*f*, 201*f*
  hand, muscles that move the, 200–201, 200*f*, 201*f*, 202
  head, muscles that move the, 196*f*, 197, 197*t*
  interaction of, 193–194
  major, 194–207

mastication, muscles of, 196*f*, 197, 197*t*
neuromuscular junction, 182, 182*f*
  number of, 197
  origin and insertion, 192–193
  pectoral girdle, muscles that move the, 197, 198*f*, 198*t*, 199*f*
  pelvic outlet, muscles of the, 201, 202*f*, 203, 203*t*
  resistance training, benefits of, 178
  structure of, 179, 180*f*, 181–182, 181*f*, 182*f*
  superficial, 195*f*
  thigh, muscles that move the, 203, 204*f*, 204*t*
  use and disuse of, 189
skeletal muscle contraction, 182–191
  actin and myosin, role of, 183, 183*f*
  all-or-none response, 189
  cellular respiration, 186, 186*f*
  energy sources for, 184–186, 186*f*
  heat production, 187
  major events, 185*t*
  motor units, recruitment of, 190, 190*f*
  muscle fatigue, 187
  oxygen debt, 186–187, 186*f*
  oxygen supply, 186, 186*f*
  recording, 188–189, 188*f*
  resistance training, benefits of, 178
  responses, 187–191, 188*f*, 190*f*
  stimulus for, 183–184, 184*f*, 185*f*
  summation, 190
  sustained contractions, 190–191
  threshold stimulus, 187–188
skeletal muscle relaxation, 183, 184*f*, 185*t*, 188–189, 188*f*
skeletal muscle tissue, 110, 111*f*, 113, 113*t*, 192*t*
skeletal system, 12, 13*f*, 132–171
  aging-related changes, 555*t*
  female and male skeletons, differences between, 160, 160*t*, 161*f*
  organization, 142–144, 142*f*, 143*t*, 144*t*
  terms used to describe, 144*t*
skeleton of the heart, 345, 345*f*
skin, 110, 233*f*
  accessory structures of, 122–125, 123*f*, 124*f*
  body temperature, regulation of, 125
  burns, 126, 127*f*
  color, 120–122, 120*f*
  disorders, 127, 129
  embryonic development, 547, 548
  lymphatic capillaries in, 379
  sunlight, exposure to, 103, 117, 121, 126
  tissues of, 117–122
  water lost through, 492*f*, 493
  wounds, healing of, 125–127, 125*t*, 127*f*
  wrinkles, treatments for, 117
skin cancer, 121, 121*f*
skull, 142–149, 142*f*, 145*f*, 146*f*, 147*f*, 148*f*, 237*f*
  differences between female and male, 160*t*
  facial skeleton, 145*f*, 146*f*, 148–149

infantile, 149, 150*f*
  male and female, differences between, 161*t*
sliding filament model, 183, 183*f*, 184*f*
slit in capillary, 356*f*
slow-twitch muscle fiber, 189
small cardiac vein, 348*f*
small intestine, 11*f*, 25*p*, 26*p*, 308*f*, 420–424, 420*f*, 421*f*, 422*t*, 423*f*, 424*t*, 425*f*
  absorption in, 422–424, 423*f*, 424*t*
  in alimentary canal, 401, 402*f*, 403*f*
  breakdown of food in, 422
  effects of neurotransmitters on, 256*t*
  intestinal wall, structure of the, 420–421, 421*f*
  lacteals in, 378
  as member of digestive system, 401, 402*f*, 413, 420–424, 420*f*, 421*f*, 422*t*, 423*f*, 424*t*, 425*f*
  microorganisms in, 400
  movements of, 424, 425*f*
  mucous membrane in, 413, 414*f*, 415
  parts of, 420, 420*f*, 421*f*
  in red blood cell production, 323, 324*f*
  referred pain, 266*f*
  replacement of epithelial cells, 421
  simple columnar epithelial tissue in, 97, 97*f*
  ulcers in, 412
  veins that drain, 371*f*
small saphenous vein, 371, 372*f*
smell, sense of, 243, 267–269, 268*f*
  loss of, 269
  olfactory nerve pathways, 267
  olfactory organs, 267, 268*f*
  olfactory receptors, 267–269, 268*f*
  olfactory stimulation, 268–269
smooth brain, 238
smooth muscle, 110–111, 111*f*, 113, 113*t*, 179, 191, 192*t*, 356, 356*f*
  in alimentary canal, 401
  contraction, 191
  in esophagus, 410
  fibers, 191
  in labia majora, 521
  in respiratory tubes, 448*f*
  in scrotum, 512
  of ureter, 481
  in urethra, 484
  in uterus, 520
  in vagina, 520
sneeze, 455
sodium
  active transport across cell membranes of, 65
  atomic structure, 32*t*
  bonding in, 35, 35*f*
  cell membrane transport of, 222, 223, 224*f*, 225, 225*f*, 226*f*
  in cystic fibrosis, 449
  effect on aldosterone, 306
  in electrolyte intake, 493
  in electrolyte output, 493, 496
  as element in human body, 31*t*
  in extracellular fluid, 491*f*, 492, 496

in glomerular filtrate, 475*t*

high sodium concentration, 306

imbalance, 496, 496*f*

in intracellular fluid, 491*f*, 496

ion channels, 54

low blood, 495

low sodium concentration, 496

as major mineral in human body, 434, 434*t*

in nerve impulse conduction, 227–228, 227*t*

in plasma, 330, 475*t*

reabsorption, 478–479, 478*f*

tubular reabsorption by, 478

tubular secretion of, 479, 479*f*

in urine, 475*t*, 476

sodium bicarbonate, 497, 502

sodium chloride, 38

sodium hydroxide, 39, 497

sodium ion

as constituent of cells, 40, 40*t*

in extracellular fluids, 493, 496, 496*f*

soft palate, 404, 405*f*, 406*f*, 409, 443*f*

soleus muscle, 195*f*, 206, 207*f*, 207*t*

solubility, 38–39

solute, 40

solvent, 40

somatic motor neuron, 250–251, 251*f*

somatic nervous system, 216, 246

somatic senses

pain (*see* pain)

temperature sensation, 264–265

touch and pressure senses, 264, 264*f*

sour, taste sensation, 270

special senses, 263, 267–287

equilibrium (*see* equilibrium, sense of)

hearing (*see* hearing, sense of)

sight (*see* sight, sense of)

smell (*see* smell, sense of)

taste (*see* taste, sense of)

species resistance, 384–385

sperm, 58, 59*f*, 508*f*, 511–513, 529*f*

analysis, 514*t*

fertilization by, 538–539, 539*f*, 540*f*

formation of, 507, 508*f*, 509

infertility, male, 514, 514*t*

postmortem retrieval, 536

in semen, 511

as smallest cell, 517

in spermatogenesis, 507, 508*f*, 509

structure of, 510, 510*f*

subfertility, 538

in vitro fertilization, 505, 538

spermatic cord, 25*p*

spermatid, 508, 509

spermatocyte, 509

primary, 507, 508*f*, 509*f*, 515*f*

secondary spermatocyte, 509*f*

spermatogenesis, 507, 508*f*, 509, 513

spermatogenic cell, 506, 507, 508*f*

spermatogonium, 508*f*

spermicide, 228*f*, 528

spermiogram, 514

sphenoidal sinus, 10*f*, 145*f*, 147, 148*f*, 298*f*, 406*f*, 444*f*

sphenoid bone, 146*f*, 147, 147*f*, 148*f*, 150*f*, 175*p*, 176*p*, 177*p*

paranasal sinuses in, 145*f*, 444

pituitary gland in sella turcica of, 298*f*

sphenoid fontanel, 150*f*

sphenopalatine ganglion, 253*f*

spheroidal joint, 165

sphygmomanometer, 359

spina bifida, 149

spinal branch, 247

spinal column, fibrocartilage in, 108, 108*f*

spinal cord, 11*f*, 215*f*, 232, 233*f*, 234–236, 234*f*, 237*f*

in cardiac cycle, 353*f*

in cerebral arterial circle, 366*f*

damage to, 484

function of, 234–235, 235*f*, 236*f*

intervertebral discs, pain from rupturing of, 153

motor fibers passing through, 251*f*

preganglionic fibers arising from, 253*f*

structure of, 234, 235*f*

spinal nerve, 215*f*, 233*f*, 234, 235*f*, 248–250, 249*f*

brachial plexuses, 250

cervical plexuses, 250

functions of, 249–250

injury to, 153, 250

lumbosacral plexuses, 250

names of, 248–249, 249*f*

spinal reflex, 235

spindle fiber, 70*f*

spine, 144*t*, 155, 156*f*

spinothalamic tract, 235

spinous process, 149, 152*f*, 153*f*

spiral fracture, 138, 138*f*

spiral organ, 274

spirometer, 456

spleen, 11*f*, 26*p*, 27*p*, 28*p*, 383–384, 383*f*, 384*f*

destruction of red blood cells in, 323, 324*f*

response to fevers, 385

veins that drain, 371*f*, 383*f*

splenic artery, 363, 364*f*, 384*f*

splenic sinus, 384

splenic vein, 370, 384*f*

splenius capitis muscle, 196*f*, 197, 197*t*

spongy bone, 133–134, 134*f*, 135*f*, 136*f*, 139*f*, 165*f*

spontaneous abortion, 542

spontaneous ejaculation, 513

spontaneous emission, 513

sprain, 168

squamous cell carcinoma, 121, 121*f*

squamous epithelial cell, 448, 448*f*

squamous epithelial tissue, 95–96, 102*t*

simple, 96, 96*f*, 459

stratified, 98, 99*f*

squamous suture, 145*f*, 146*f*, 147, 148*f*, 177*p*

stapedius muscle, 197

stapes, 271, 271*f*

static equilibrium, 275

statins, 355

stem cell, 71, 72*f*, 540*f*

autologous transplants of, 141

in bone marrow transplants, 141

for cornea transplants, 280

in hair follicles, 123

hematopoietic, 321, 322*f*

neural, 219

in red bone marrow, 388*f*

in umbilical cord, 335

umbilical cord stem cells, transplant of, 323

from wisdom teeth, 408

sternal region, defined, 17, 18*f*

sternocleidomastoid, 452, 452*f*

sternocleidomastoid muscle, 23*p*, 24*p*, 195*f*, 196*f*, 197, 197*t*, 199*f*

sternum, 11*f*, 25*p*, 142*f*, 143, 154*f*, 155, 156*f*, 452, 452*f*

steroid hormone, 42, 43*f*, 43*t*, 46, 293, 293*t*, 294, 294*f*

abuse, dangers of, 188

anabolic, 188, 245

stimulant, CNS, 245

stomach, 11*f*, 25*p*, 26*p*, 27*p*, 410–413, 411*f*, 412*f*, 412*t*, 421*f*

in alimentary canal, 401, 402*f*, 403*f*

gastric absorption, 412–413

gastric secretions, 411–412, 411*f*, 412*t*

as member of digestive system, 401, 402*f*, 410–413, 411*f*, 412*f*, 412*t*

microorganisms in, 400

mixing and emptying actions, 413

referred pain, 266*f*

regions of the, 409, 410*f*

simple columnar epithelial tissue in, 97

veins that drain, 371*f*, 383*f*

stratified epithelial tissue, 96, 98, 99*f*, 100*f*, 102*t*, 506, 520

stratum basale, 118*f*, 119, 119*f*

stratum corneum, 119, 119*f*

stratum granulosum, 119, 119*f*

stratum lucidum, 119, 119*f*

stratum spinosum, 119, 119*f*

Streptococcus, 345, 394

Streptococcus mutans, 407

streptokinase, 331

stress, 311–312, 313*f*

as cause of headaches, 281

responses to, 311–312, 313*f*

types of, 311

stressor, 311

stress response, 311–312, 313*f*

stress test, 362

striation, 110, 111*f*, 112*f*, 179

stroke volume, 360, 360*f*, 361, 362

strong acid, 497, 498

structural formula, 37, 37*f*, 38*f*

styloid process, 146*f*, 147, 148*f*, 157, 158*f*

subarachnoid space, 233*f*, 234, 241*f*

subcapsule, 381*f*

subclavian artery, 26*p*, 27*p*, 28*p*, 29*p*, 363, 364, 364*f*, 367*f*, 368*f*

subclavian vein, 25*p*, 26*p*, 28*p*, 369, 369*f*, 372*f*, 380*f*

subcutaneous injection, 119

subcutaneous layer, 117, 118*f*

subcutaneous tissue, 233*f*

subdural hematoma, 234

subfertility, 538

sublingual gland, 408, 409*f*

submandibular ganglion, 253*f*

submandibular gland, 408, 409*f*

submucosa, 401, 404*f*

submucous coat, of urinary bladder, 483, 483*f*

subpatellar fat, 165*f*

subscapularis muscle, 25*p*, 198, 199*t*

substance P, 279*t*

substrate chemical, 79, 79*f*

sucrase, 422*t*

sucrose, 36, 422, 422*t*

sudden cardiac arrest, 340

sugar, 41

absorption in small intestine, 422

active transport across cell membranes of, 65

diffusion through cell membranes, 61, 62*f*

in nucleic acids, 44

as plasma nutrients, 329

sulcus, 237, 237*f*

sulfate

in electrolyte intake, 493

in glomerular filtrate, 475*t*

in plasma, 475*t*

in urine, 475*t*

sulfate ion

as constituent of cells, 40, 40*t*

in extracellular fluid, 491*f*

in intracellular fluid, 491*f*

sulfur

as element in human body, 31*t*

as major mineral in human body, 434, 434*t*

sulfuric acid, as source of hydrogen ions, 497, 497*f*

summation, 190

sunlight, exposure of skin to, 103, 117, 121, 126

superficial, defined, 15

superficial partial-thickness burn, 126, 127*f*

superficial temporal artery, 365*f*, 368*f*

superficial temporal vein, 372*f*

superficial transversus perinei muscle, 202*f*, 203

superficial volar arch artery, 367*f*

superior, defined, 14, 14*f*

superior articular facet, 152*f*, 153*f*

superior articular process, 149, 152*f*, 154*f*

superior border, 156*f*

superior canaliculus, 278*f*

superior hemiazygos vein, 370

superior hypogastric plexus, 253*f*

superior lobar bronchus, 447*f*

superior lobe, 447*f*

superior mesenteric artery, 28*p*, 363, 364*f*, 368*f*

superior mesenteric ganglion, 252*f*

superior mesenteric vein, 28*p*, 370, 371*f*

superior nasal concha, 147, 268*f*, 444*f*

superior oblique muscle, 279*t*

superior ophthalmic vein, 369*f*

superior orbital fissure, 175*p*

superior peduncle, 244*f*

superior pubic ramus, 507*f*

superior rectus muscle, 279*t*

superior sagittal sinus, 233*f*

superior thyroid artery, 365*f*

superior vena cava, 27*p*, 28*p*, 342*f*, 344, 344*f*, 346*f*, 348*f*

carrying deoxygenated blood from upper half of body, 369, 370*f*, 372*f*

in embryonic development, 550–551, 550*f*

supination, 167, 169*f*

supinator muscle, 200, 200*f*, 201*t*

supporting cell, 273*f*, 276*f*, 277*f*

suppression amblyopia, 278

suprachiasmatic nuclei, 312
supraorbital foramen, 144, 145*f*
supraorbital notch, 144, 175*p*
suprapatellar bursa, 165*f*
suprarenal artery, 363–364, 364*f*, 368*f*
suprascapular notch, 156*f*
supraspinatus muscle, 198, 198*f*, 199*t*, 200*f*
supraspinous fossa, 156*f*
supratrochlear lymph node, 382*f*
sural region, defined, 17, 18*f*
surface tension, in breathing process, 452, 453
surfactant, 452, 459*f*, 553
surgery
    as birth control method, 529, 529*f*
    as breast cancer treatment, 527
surgical neck, of humerus, 155, 157*f*
surrogate mother, 540*t*
suspensory ligament, 280, 280*f*, 281, 281*f*, 282*f*, 525
sustained contraction, 190–191
sustentacular cell, 507, 509*f*
suture, 144, 144*t*, 145*f*, 164, 164*f*
swallowing mechanism, 409
S wave, 351, 351*f*
sweat, water lost from, 492*f*, 493
sweat gland, 118*f*, 124
sweat gland duct, 118*f*
sweat gland pore, 118*f*
sweet, taste sensation, 270
swelling. *See* edema
sympathetic chain ganglion, 252*f*
sympathetic division of autonomic nervous system, 250, 251, 252*f*
sympathetic ganglia, 251
sympathetic motor nerve fiber, 283*f*
sympathetic nerve, 353*f*
sympathetic neuron, 254*f*
sympathetic trunk, 353*f*
symphysis pubis, 28*p*, 29*p*
synapse, 182, 221–222, 221*f*, 222*f*, 223
synaptic cleft, 182*f*, 221, 221*f*, 222*f*, 223*f*
synaptic knob, 218*f*, 222, 222*f*, 223, 223*f*
synaptic transmission, 221
synaptic vesicle, 182*f*, 222, 222*f*, 223*f*
synergist, 194
synesthesia, 245, 268
synovial fluid, 164, 165*f*, 490
synovial joint, 164–165, 165*f*, 167*t*
synovial membrane, 110, 164, 165*f*
synthesis, 37
synthetic graft, 354
syphilis, 530*t*
systemic capillary, 346*f*, 378*f*
systemic circuit, 341, 341*f*, 363
systemic lupus erythematosus, 394
systole, 347, 349, 349*f*
systolic blood pressure, 359, 359*f*

T

tachycardia, 352
tactile corpuscle, 264, 264*f*
tail of sperm cell. *See* flagella
talus, 163, 163*f*
tamoxifen, 527
target cell, 12, 292, 293, 293*f*, 294, 294*f*, 296*f*, 299*f*
tarsal, 163*f*
tarsal bone, 142*f*, 143, 163, 163*f*

tarsal region, defined, 17, 18*f*
tarsus, 142*f*, 143, 163, 163*f*
taste, sense of, 269–270
    nerve pathways, 270
    sensations, taste, 270
    taste receptors, 269–270
taste bud, 269–270, 269*f*, 270
taste cell, 269, 269*f*
taste hair, 269–270, 269*f*
taste nerve pathway, 270
taste pore, 269–270, 269*f*
taste receptor, 269–270
taste sensation, 270
tattoo, 122
T cell, 322*f*, 381, 381*f*, 386–388, 387*f*, 387*t*, 388*f*
    activation, 386–388, 387*f*, 387*t*, 388*f*
    in AIDS, 389
    in allergic reactions, 393–394
    in antibody production, 391*t*
    compared to B cells, 387*t*
    cytotoxic, 388, 388*f*, 389
    helper, 388, 388*f*, 389
    and humoral immune response, 390
    produced in thymus gland, 382
tear gland, 256*t*
tectorial membrane, 273*f*, 274
telomere, 69
telophase, 68*f*, 69, 70*f*
temperature
    as affecting heart rate, 352
    altering of enzymes by, 80
    body, regulation of, 7–8, 8*f*, 125, 312
    sensation, 264–265
temporal artery, superficial, 365*f*, 368*f*
temporal bone, 145*f*, 146*f*, 147, 147*f*, 148*f*, 150*f*, 175*p*, 176*p*, 177*p*
temporalis muscle, 195*f*, 196*f*, 197, 197*t*
temporal lobe, 238, 238*f*, 239, 268
temporal process, 146*f*, 149, 177*p*
temporal vein, superficial, 372*f*
temporomandibular joint, 197
temporomandibular joint syndrome, 197
tendon, 103, 106*f*
tendonitis, 179
teniae coli, 426
tenosynovitis, 179
tension headache, 281
tensor fasciae latae muscle, 24*p*, 27*p*, 28*p*, 195*f*, 204*f*, 204*t*
teratogen, 547, 549
teres major muscle, 25*p*, 195*f*, 197, 198*f*, 199*t*, 200*f*
teres minor muscle, 195*f*, 198, 198*f*, 199*t*, 200*f*
terminal bronchiole, 447*f*, 448*f*
terminology, anatomical, 2, 14–19
    body regions, 16–17, 17*f*, 18*f*
    body sections, 14*f*, 15–16, 15*f*
    relative positions, 14–15
ternia coli, 425*f*
tertiary bronchus, 447*f*
testicular cancer, 506
testis, 26*p*, 506–507, 507*f*, 508*f*, 509–510, 509*f*, 510*f*
    embryonic development, 548
    enlargement of, by testosterone, 515

functions of, 513*t*
    as gonad, 300
    in male infertility, 514, 514*t*
    as reproductive gland, 293, 293*f*
testosterone, 188, 513, 515, 515*f*
test tube baby, 538
tetanic contraction, 190, 190*f*
tetanus, 190
thalamus, 236*f*, 240*f*, 242–243, 244*f*, 265, 286*f*, 287
thalidomide, 548
thallium-201, 34
thermoreceptor, 263
thiamine, 433*t*
third line of defense, 386
third ventricle, 240, 241*f*, 299*f*
thirst, as regulator of water intake, 492, 492*f*
thirst center, 492
thoracic aorta, 363, 365*t*
thoracic artery, internal, 367
thoracic cage, 142–143, 143*t*, 153, 154*f*, 155
thoracic cavity, 8, 9*f*, 29*p*
thoracic duct, 379, 380*f*
thoracic lymph node, 382*f*
thoracic membrane, 10
thoracic nerve, 248–249, 249*f*
thoracic vein, internal, 370
thoracic vertebrae, 151–152, 151*f*, 152*f*, 154*f*, 233*f*
thoracic wall
    arteries to, 367
    veins from, 370
threonine, 431*t*
threshold potential, 225, 226*f*
threshold stimulus, 187–188
throat. *See* pharynx
thrombin, 331, 332
thrombocyte. *See* platelet
thrombocytopenia, 328
thrombopoietin, 327
thrombus, 331, 347
thymine, 84, 84*f*, 85, 85*f*, 86
thymosin, 310
thymus gland, 293, 293*f*, 309–310, 382, 383*f*, 387*f*
thyrocervical artery, 366
thyrocervical axis, 364
thyroid artery, superior, 365*f*
thyroid cartilage, 25*p*, 26*p*, 445, 445*f*
thyroid gland, 24*p*, 25*p*, 26*p*, 293, 293*f*, 300, 301–303, 302*f*, 383*f*
    disorders of the, 296, 303
    radioactive isotopes used to study, 34, 34*f*
    structure of, 301, 302*f*
    thyroid hormone, 301–303, 302*t*
thyroid-stimulating hormone, 297, 299, 299*f*, 300, 301*t*
thyrotropin-releasing hormone, 299*f*, 300
thyroxine, 301–302, 302*t*
tibia, 142*f*, 143, 162, 162*f*, 163*f*, 165*f*
tibial artery
    anterior, 367, 368*f*
    posterior, 367, 368*f*
tibialis anterior muscle, 195*f*, 205, 206, 207*t*
tibialis posterior muscle, 206, 207*t*
tibial tuberosity, 162, 162*f*
tibial vein
    anterior, 370–371, 372*f*
    posterior, 370–371, 372*f*
tidal volume, 455, 456*t*, 457

timed coitus, 527
tissue, 4*f*, 94–113, 113*t*
    donating for research, 94
    repair, 113
    types, 95, 95*t*
tissue cells, 357*f*, 379*f*
tissue engineer, 482
tissue fluid, 356*f*, 380
tissue plasminogen activator, 331
tissue rejection, 394
tissue thromboplastin, 331
titin, 179
T lymphocyte. *See* T cell
toe. *See also* metatarsal bone
    embryonic development, 544, 545*f*, 547, 548
tongue, 269, 269*f*, 403–404, 403*f*, 405*f*, 406*f*, 409, 409*f*, 444*f*
tonometer, 282
tonsil, 382
    lingual tonsil, 406*f*
    palatine, 404, 405, 405*f*, 406*f*, 444*f*
    pharyngeal, 404, 406*f*, 444*f*
tonsillectomy, 405
tonsillitis, 405
tooth, 405, 406*f*, 407–408, 407*t*, 408*f*
    enamel destruction, 407
    extraction and replacement of, 408
    grinding, 197
total lung capacity, 456, 456*t*
touch, sense of, 264, 264*f*
toxicology, 19
toxin, 30
trabeculae, 134, 135*f*, 381*f*
trace-amine-associated receptor, 291
trace element, 31, 434, 435*t*
trachea, 25*p*, 27*p*, 28*p*, 303*f*, 383*f*, 445*f*, 446
    in digestion, 406*f*, 413
    as member of respiratory system, 443*f*, 446, 447*f*, 451*t*
    in respiratory tract, 443, 444*f*
tracheotomy, 394
tract, 234–235, 236*f*
transcellular fluid, 490, 490*f*, 491, 491*f*
transcription, 85–86, 86*t*, 87*f*
transcytosis, 65
transdermal patch, 119
trans fat, 43
transfer RNA (tRNA), 87–88, 89*f*
transfusion, blood, 333–336, 334*f*, 334*t*, 335*f*, 335*t*, 360
transitional epithelial tissue, 99, 101*f*
translation, 87–88, 87*f*, 88*f*, 89*f*
transplantation, 394
    of bone marrow, 141, 323
    of corneas, 280
    heart, 347
    kidney, 470
    tissue rejection, 394
    of umbilical cord stem cells, 141, 323
transverse, defined, 15, 15*f*
transverse colon, 26*p*, 27*p*, 421*f*, 425*f*, 426
transverse fissure, 237, 237*f*
transverse foramen, 150, 152*f*, 153*f*
transverse fracture, 138, 138*f*
transverse ligament, 164*f*
transverse process, 149, 152*f*, 153*f*
transverse tubule, 181, 181*f*, 182, 192
transversus abdominis muscle, 24*p*, 25*p*, 29*p*, 199*f*, 201, 203*t*

trapezium, 159f, 166f
trapezius muscle, 23p, 195f, 197, 198, 198f, 199f, 199t, 200f
trapezoid, 159f
triceps brachii muscle, 140, 140f, 195f, 200, 200f, 201f, 201t
tricuspid valve, 344, 344f, 345, 345t, 346, 349f
trigeminal nerve (V), 246, 247f, 248t
trigger zone, 225, 225f
triglyceride, 42, 42f, 43t, 46, 329, 423, 429
trigone, 482, 483f
triiodothyronine, 301–302, 302t
triquetrum, 157, 159f
trochanter, 144t
trochlea, 155, 157f
trochlear nerve (IV), 246, 247f, 248t
trochlear notch, 157, 158f
trochoid joint, 165
trophoblast, 542, 542f, 544
tropomyosin, 183, 183f, 184f
troponin, 183, 183f, 184f
true rib, 153, 154f, 155
true vocal cord, 446, 446f
trypsin, 413, 422t
tryptophan, 431, 431t
tubal ligation, 529, 529f
tubal pregnancy, 541
tubercle, 144t, 152, 154f, 155
tuberosity, 144t
tubular, 479
tubular fluid, 474f
tubular necrosis, 475
tubular reabsorption, 472–473, 474f, 477–478, 477f
tubular secretion, 472–473, 474f, 477f, 479, 479f
tubulin, 57, 58f
tunica externa, 354, 354f
tunica interna, 354, 354f
tunica media, 354, 354f
T waves, 351, 351f
twins, development of, 542
twitch, muscle, 188–189, 188f, 190f
tympanic cavity, 271, 271f
tympanic membrane, 270, 271f
tyosine, 431t
type AB blood group, 334–336, 334f, 334t, 335f, 335t
type A blood group, 334–336, 334f, 334t, 335f, 335t
type B blood group, 334–336, 334f, 334t, 335f, 335t
type O blood group, 334–336, 334f, 334t, 335f, 335t
tyrosine kinase, 328

U

ulcerative colitis, 428
ulcers, 129, 412
    duodenal, 412
    gastric, 412
ulna, 140f, 142f, 143, 156f, 157, 158f, 159f, 166f
ulnar artery, 367, 367f, 368f
ulnar nerve, 249f, 250
ulnar notch, 157, 158f
ulnar recurrent artery, 367f
ulnar vein, 370f
ultradian rhythms, 312
ultrasound, 538t
umbilical, 550–551, 551f
umbilical artery, 546–547, 546f, 550f, 551, 551t, 554f

umbilical cord, 141, 335, 546–547, 546f, 548f, 551
umbilical region, defined, 16, 17, 17f, 18f
umbilical vein, 546–547, 546f, 550, 550f, 551t, 554f
umbilicus, 23p
unami, taste sensation, 270
undereating, 436
undernutrition, 436
unipolar neuron, 220, 220f, 221f
universal donor, 335
universal precautions, 318
universal recipient, 335
unmyelinated sheath, 217, 219f
unsaturated fat, 42, 42f
upper limb, 143, 143t. See also shoulder
    arm, muscles that move the, 197–198, 198f, 199f, 199t, 200f
    arteries to, 366–367, 366f
    bones, 155, 156f, 157–158, 157f, 158f
    forearm, muscles that move the, 198, 200f, 201t
    movement, 140, 140f
    veins from, 369, 370f
upper lobar bronchus, 447f
upper motor neuron, 239
upper respiratory tract, 443
uracil nucleotide, 86, 87f
urea, 431, 431f, 480
    in glomerular filtrate, 475t
    in plasma, 475t
    in urine, 475t
uremic acidosis, 501
ureter, 27p, 28p, 468, 468f, 469f, 470, 483f, 507f
    as member of urinary system, 468f
    referred pain, 266f
    transitional epithelial tissue in, 98, 100f
    urine elimination by, 481, 482f
ureteritis, 483
urethra, 29p, 202f, 483f, 484, 484f, 552f
    in female reproductive system, 512, 516f
    in male reproductive system, 507f, 511
    as member of urinary system, 468f
    prostate gland pushing on, 511
    stratified columnar epithelial tissue in, 100f
    transitional epithelial tissue in, 98
urethral gland, 484, 484f
urethral orifice, 202f
urethral sphincter
    external, 483, 483f, 484
    internal, 483f
uric acid, 480
    in glomerular filtrate, 475t
    as nonprotein nitrogenous substance, 330
    tubular reabsorption by, 478
    in urine, 475t
urinary bladder, 25p, 26p, 27p, 28p, 421f, 470, 507f, 552f
    in females, 516f
    in males, 483
    as member of urinary system, 468f

muscles, effects of neurotransmitters on, 256t
    referred pain, 266f
    replacements, 482
    transitional epithelial tissue in, 98, 100f
    urine elimination by, 481–483
urinary meatus, 484
urinary system, 13f, 14, 468–485, 556t
urinary tract infection, 484
urination, 483–484
urine, 472
    alkaline, 502
    altering of electrolyte output by production of, 493, 496–497, 496f
    composition, 480–481, 481t
    concentrations and volume, regulation of, 480
    concentrations of substances, 473, 475t
    elimination, 481–484, 482f, 483f, 484f
    filtration pressure, 473, 475, 475f
    filtration rate, 475–476, 476f
    formation of, 472–481
    glucose in, 310
    micturition, 483–484
    sodium and water reabsorption, 478–479, 478f
    tubular reabsorption, 472–473, 474f, 477–478, 477f
    tubular secretion, 472–473, 474f, 477f, 479, 479f
    water lost in, 492f, 493
urogenital diaphragm, 201, 202f, 203
urology, 19
urticaria, 129
uterine tube, 27p, 516f, 518f, 519–520, 519f
    cleavage, cell movement during, 541
    fertilization occurring in, 519–520, 529f
    functions of, 522t
    in pregnancy, 541
    transport of sex cells through, 537
uterine wall, 548f
uterus, 27p, 28p, 516f, 519f, 522–524, 522t
    in birth process, 551–552, 552f
    contractions, stimulation of, 300
    development of embryo in, 520, 520f
    endometriosis, 538
    implantation of blastocyst in wall of, 541–542, 542f
    simple columnar epithelial tissue in, 97
    transport of sex cells through, 537
utricle, 272f, 275, 276f
uvula, 404, 405f, 406f, 444f

V

vaccine, 393
vagina, 29p, 202f, 519f, 520–521, 522t, 529f
    in birth process, 551–552, 552f
    enlargement of during pregnancy, 543
    layers, 520
    during sexual intercourse, 512, 537

vaginal orifice, 202f, 516f, 520
vagus nerve (X), 247, 247f, 248t, 253f, 353f, 446, 457, 458f
valine, 431t
valve
    heart, 343–345, 344f, 345f, 345t, 354, 354f, 358
    in lymphatic vessels, 379, 379f, 381
valve heart, 358
varicose vein, 358
vascular headache, 281
vas deferens. See ductus deferens
vasectomy, 529, 529f
vasoconstriction, 330, 354, 357, 362, 363
vasodilation, 125, 354, 357
vasomotor center, 243, 361
vasomotor fiber, 354
vasospasm, 330
vastus intermedius muscle, 28p, 205, 205f
vastus lateralis muscle, 26p, 27p, 28p, 195f, 204f, 205, 205f, 205t, 206
vastus medialis muscle, 26p, 27p, 195f, 204f, 205, 205t
vegetative state, 213
vein, 341, 353, 354f, 355, 358, 358f, 358t, 359t
    characteristics of, 358, 358f
    system of (see venous system)
    weakening of, 354f
vein grafting, 354
vena cava
    inferior (see inferior vena cava)
    superior (see superior vena cava)
venipuncture, 369
venoconstriction, 362, 363
venous sinus, 33, 369f
venous system
    abdominal and thoracic walls, veins from, 370
    abdominal viscera, veins from, 370, 371f
    brain, head and neck, veins from, 369, 369f
    characteristics of, 369
    lower limb and pelvis, veins from, 370–371, 372f
    upper limb and shoulder, veins from, 369, 370f
ventilation. See breathing
ventral, defined, 14, 14f
ventral respiratory group, 456, 457f
ventral root, 233f, 235f, 249, 249f
ventricle, 240, 241, 241f, 242f, 343–345, 344f, 345f
    in cardiac cycle, 347, 349–350, 349f, 350f, 351f
    left, 27p, 341f, 342f, 344f, 346, 348f, 359, 359f, 550f
    right, 27p, 341f, 342f, 344f, 346, 348f, 550–551
ventricular diastole, 347, 349f, 359
ventricular syncytium, 349, 350, 350f
ventricular systole, 347, 349, 349f
ventricular tachycardia, 340
venule, 341, 356f, 358, 358f, 358t, 359t, 379f, 421f
venus blood flow, 362–363
venus sinus, 369f
vermiform appendix, 26p
vermis, 244
vertebra, 11f, 132, 149, 151f, 233f
vertebral arch, 149

vertebral artery, 364, 365*f*, 366*f*, 368*f*
vertebral canal, 8, 9*f*, 149, 234, 234*f*
vertebral column, 142, 142*f*, 143*t*, 149–153, 150*f*, 151*f*, 152*f*, 153*f*
vertebral foramen, 149, 152*f*, 153*f*
vertebral orifice, external, 484
vertebral region, defined, 17, 18*f*
vertebral rib, 155
vertebral vein, 369*f*
vertebrochondral rib, 154*f*, 155
vertebrosternal rib, 153, 154*f*
vertigo, 129
very low-density lipoprotein, 423, 430
vesicle, 55, 57*f*, 58*f*, 59, 61*t*
    in exocytosis, 65
    in receptor-mediated endocytosis, 66, 66*f*
vestibular branch, 247
vestibular bulb, 521
vestibular gland, 521, 522*t*
vestibular membrane, 273, 273*f*
vestibular nerve, 271*f*, 272*f*
vestibule, 275, 403*f*, 405*f*, 406*f*, 521, 522*t*
vestibulocochlear nerve (VIII), 247, 247*f*, 248*t*
villus
    intestinal, 404*f*, 420–421, 421*f*, 422
    placental, 544, 545, 545*f*, 546*f*
vinyl chloride, 442
viral load, 389
virus. *See also specific viruses*
    as being alive, 384
    universal precautions against, 318
viscera, 8, 9*f*, 251*f*
visceral pain, 265, 266*f*
visceral pericardium, 10, 342, 342*f*, 343, 343*f*
visceral peritoneum, 11*f*, 401, 421*f*
visceral pleural membrane, 10, 11*f*, 450, 450*f*
visceral smooth muscle, 191
viscosity, 360
vision. *See* sight, sense of

visual accessory organ, 276, 277–279, 278*f*
visual area of cerebral cortex, 238*f*
visual cortex, 286*f*
visual purple, 286
vital capacity, 455, 456*t*
vital sign, 5
vitamin, 432–433, 432*t*, 433*t*
    fat-soluble, 432, 432*t*
    as micronutrient, 428
    recommended daily allowances, 432, 433
    water-soluble, 432–433, 433*t*
vitamin A, 286, 432, 432*t*, 549
vitamin B$_1$, 433*t*
vitamin B$_2$, 433, 433*t*
vitamin B$_3$, 433
vitamin B$_5$, 433
vitamin B$_6$, 433
vitamin B$_{12}$, 323, 324*f*, 433
vitamin B complex, 432
vitamin C, 432–433, 478
vitamin D, 432, 432*t*
    activation of, 469
    in bone deformation, 132, 137
    production by skin cells, 117
vitamin deficiency, as cause of peripheral neuropathy, 267
vitamin E, 432, 432*t*
vitamin K, 432, 432*t*
vitreous body, 284, 284*f*
vitreous humor, 280*f*, 283*f*, 284*f*
    as transcellular fluid, 490
vocal cord, 445, 446, 446*f*
    false, 446, 446*f*
    true, 446, 446*f*
vocal fold, 446, 446*f*
volar arch artery, superficial, 367*f*
volar artery, deep, 367*f*
Volkmann's canals, 134
voluntary muscle tissue, 110, 111*f*
vomer bone, 145*f*, 146*f*, 148*f*, 149, 175*p*, 176*p*
vomiting, 413, 501, 501*f*, 502
vomiting center, 413
von Willebrand disease, 332
vulva, 521

W

warm receptor, 264–265
warts, 129
wasting, 188
water
    absorption by large intestine, 427
    absorption by small intestine, 423, 424*t*
    in the body, 490
    in citric acid cycle, 430*f*
    as constituent of cells, 40, 40*t*
    in digestion of proteins, 431*f*
    filtration, 64
    formula for, 37, 38*f*
    in humans, 490
    intake, 492–493, 492*f*
    as intracellular fluid, 490
    molecules, bonding in, 36, 36*f*, 37*f*
    movement of, between fluid compartments, 491–492, 492*f*
    osmosis, movement of water by, 63–64, 63*f*
    output, 492*f*, 493
    reabsorption of, 478–479, 478*f*
    as requirement for life, 5
    solubility, 38–39, 46
water balance, 492–493, 492*f*
    disorders, 494–496, 494*f*, 495*t*
    water intake, 492–493, 492*f*
    water output, 492*f*, 493
water intoxication, 495
water of metabolism, 492, 492*f*
wave, in electrocardiograms, 351, 351*f*
weak acid, 497, 498
weight, 31
weight loss, 436
Wernicke's area, 238*f*, 239
whiplash, 250
white blood cell, 108, 110*f*, 140, 319, 320*f*
    cellular components, 327*t*
    counts, 325, 327
    functions of, 325

    in leukemia, 328
    as phagocytes, 66
    production of, 117, 322*f*, 324
    types of, 324–325
    in urine, 480
    white blood cell counts, 325, 327
white blood cell count, 325, 327
white matter, 217, 233*f*, 234, 235*f*
white pulp, 383, 384*f*
wild type allele, 556–557
windpipe. *See* trachea
wisdom teeth, 408
withdrawal reflex, 232, 232*f*
Wright, Frank Lloyd, 268
wrinkles, treatments for facial, 117

X

xiphoid process, 154*f*, 155
X-linked recessive, 560

Y

yawning, 455
yolk sac, 140, 544*f*, 546*f*, 547

Z

zinc
    as major mineral in human body, 434, 434*t*
    as trace element in human body, 31*t*, 434, 435*t*
Z line, 180*f*, 181, 193
zona fasciculata, 305*f*
zona glomerulosa, 305*f*
zona pellucida, 518, 518*f*, 538, 539, 539*f*, 541
zona reticularis, 305*f*
zygomatic arch, 146*f*, 149
zygomatic bone, 145*f*, 146*f*, 147, 148–149, 150*f*, 175*p*, 176*p*, 177*p*
zygomatic process, 146*f*, 147, 177*p*
zygomaticus muscle, 194, 195*f*, 195*t*, 196*f*
zygote, 517, 517*f*, 539, 540*f*, 543*t*
zygote intrafallopian transfer, 540
zymogen granule, 413